T0178463

FUZZY AUTOMATA and LANGUAGES

Theory and Applications

COMPUTATIONAL MATHEMATICS SERIES

Series Editor Mike J. Atallah

Published Titles

Inside the FFT Black Box: Serial and Parallel Fast Fourier Transform Algorithms
Eleanor Chu and Alan George

Mathematics of Quantum Computation
Ranee K. Brylinski and Goong Chen

Fuzzy Automata and Languages: Theory and Applications
John N. Mordeson and Davender S. Malik

Forthcoming Titles

Cryptanalysis of Number Theoretic Ciphers
Samuel S. Wagstaff

FUZZY AUTOMATA and LANGUAGES

Theory and Applications

John N. Mordeson
Davender S. Malik

CRC Press
Taylor & Francis Group
Boca Raton London New York

CRC Press is an imprint of the
Taylor & Francis Group, an **informa** business

A CHAPMAN & HALL BOOK

CRC Press
Taylor & Francis Group
6000 Broken Sound Parkway NW, Suite 300
Boca Raton, FL 33487-2742

First issued in paperback 2019

© 2002 by Taylor & Francis Group, LLC
CRC Press is an imprint of Taylor & Francis Group, an Informa business

No claim to original U.S. Government works

ISBN-13: 978-1-58488-225-1 (hbk)
ISBN-13: 978-0-367-39627-5 (pbk)

Library of Congress Cataloging-in-Publication Data

Mordeson, John N.
   Fuzzy automata and languages: theory and applications / John N. Mordeson and Davender
S. Malik.
        p.   cm. — (Computational mathematics series)
   Includes bibliographical references and index.
   ISBN 1-58488-225-5 (alk. paper)
   1. Fuzzy automata. 2. Fuzzy languages. I. Malik, D. S. II. Title. III. Series.

QA267.5.F89 M67 2002
511.3—dc21   2002017475

Visit the CRC Press Web site at www.crcpress.com

Library of Congress Card Number 2002017475

Visit the Taylor & Francis Web site at
http://www.taylorandfrancis.com

and the CRC Press Web site at
http://www.crcpress.com

Preface

In 1965, L. A. Zadeh introduced the notion of a fuzzy subset of a set as a method for representing uncertainty. His ideas have been applied to a wide range of scientific areas. One such area is automata theory and language theory first introduced by W. G. Wee in [249]. This is the area that is dealt with in this book. Our purpose is to give an up-to-date treatment of fuzzy automata theory and fuzzy language theory when the set of truth values is the closed interval [0, 1]. When the interval is replaced by a lattice or a semiring or some other type of structure, the reader is referred to [247].

There are many ways that ordinary automata and languages have been fuzzified. Consequently, a wide variety of terminology is needed. Since no industry standards have been set as to the terminology used to describe these ways, we decided in most cases to use the terminology of the authors on whose papers the book depends. Some modifications were made when different terminology was used to describe the same concept or when the same terminology was used to describe different concepts. A considerable number of symbols are needed due to the large number of concepts involved. Consequently, we have decided not to include symbols in the symbols list if their use is localized.

At the end of each chapter, we provide a few exercises. Some of the exercises present material that is not covered in detail in the book, while others test the readers' understanding of the material. The exercises should be useful if the book is used in a course on fuzzy automata and fuzzy languages. Examples are provided in each chapter to illustrate the concepts developed. The book should be of interest to research mathematicians, engineers, and computer scientists.

Chapter 1 provides some basic material needed for an understanding of the book. This material includes results from set theory, fuzzy set theory, abstract algebra, automata, and language theory. Nevertheless, the reader would find a background in these areas useful.

Chapter 2 begins the study of max-min machines and examines their behavior. We consider equivalences and homomorphisms of max-min automata in order to determine their reductions. A max-min algebra is developed in order to have a structure in which the study of max-min machines can be placed.

Chapter 3 introduces max-product machines and considers their irreducibility and minimality. Max-product grammars and languages are studied with special attention to weak regular max-product grammars and languages. A max-product algebra is developed in order to have a structure in which to carry out the study. This algebra resembles the algebra developed in the previous chapter, but it has major differences.

Natural languages lack the precision of formal languages. It is hoped that the introduction of fuzziness into the structure of formal languages will help close the gap between the two. Chapter 4 studies fuzzy context-free grammars and languages. Special attention is given to trees, fuzzy dendrolanguage generating systems, and normal forms. Concerning trees, we are particularly interested in fuzzy tree automata and fuzzy tree transducers.

Chapter 5 deals with probabilistic automata and their approximation by nonprobabilistic automata. Various types of probabilistic grammars and automata are studied. These types include programmed and time-variant grammars, weak regular probabilistic grammars, context-free probabilistic grammars, asynchronous probabilistic automata, and probabilistic pushdown automata. Realization of fuzzy languages by probabilistic automata, max-product automata, and max-min automata is examined.

Chapter 6 is concerned with algebraic fuzzy automata theory. In this chapter, we are particularly interested in semigroups of fuzzy finite state machines including fuzzy transformation semigroups. We examine the structure of fuzzy finite state machines through products and covers and stress the concepts of submachines, retrievability, separability, and connectedness of fuzzy finite state machines.

Chapter 7 presents additional results on fuzzy languages. The concept of a partial fuzzy automaton is introduced in order to study certain types of fuzzy regular languages. We consider fuzzy recognizers including the construction of recognizers and recognizable sets. The chapter concludes with a presentation of the algebraic properties of fuzzy regular languages, adjunctive languages, and dense languages.

In Chapter 8, we consider the minimization of fuzzy automata. We examine first the equivalence, reduction, and minimization of finite fuzzy automata by an algebraic approach. Some ideas presented in Chapters 2 and 3 are completed here. We then consider the minimization of a fuzzy finite automaton. The automata here have distinct transition and output functions. The chapter concludes by considering an approach to the study of finite fuzzy automata by using the ability to solve systems of linear equations over a bounded chain. A polynomial time algorithm for solving such systems is given.

In Chapter 9, we continue our study of the recognition of fuzzy languages. We study cutpoint languages and recursive languages. We replace the interval [0,1] by a lattice in this chapter. The purpose of this chapter is to give the reader only an introduction to some of the ideas used in fuzzy language theory when the interval [0,1] is replaced by a lattice.

Chapter 10 is devoted to applications. We consider a formulation of fuzzy automata and their application as a model of learning systems. We also apply the concept of fractionally fuzzy grammars to pattern recognition. Stability and fault tolerance of fuzzy state automata is also examined. A fuzzy automaton as a clinical monitor is presented. An application to data base theory is also given.

The authors are grateful to the editorial and production staff of Chapman Hall/CRC Press, especially Robert Stern. We are indebted to Paul Wang, Azriel Rosenfeld, and Hu Cheng-ming for their support of fuzzy mathematics. We are also appreciative of the support of Dr. Timothy Austin, Dean of Creighton College of Arts and Sciences, Dr. and Mrs. George Haddix, benefactors of our research center, and Lynn Schneiderman of the Creighton Alumni Library.

John N. Mordeson
Davender S. Malik

Authors

John N. Mordeson, Ph.D., is Professor of Mathematics at Creighton University. He received his B.S., M.S., and Ph.D. degrees from Iowa State University. At Creighton he has received the Distinguished Faculty Award and was a recipient of the Burlington Northern Scholar of the Year Award and the College Reasearch Award of the year. He is on the editorial board of several journals including *Information Sciences, Fuzzy Sets and Systems,* and the *Journal of Fuzzy Mathematics.* Professor Mordeson has published more than 150 papers, chapters, and lecture notes series. He has authored six books. He is a member of the American Mathematical Society, the board of directors of the Berkley Initiative in Soft Computing, and the board of directors of the Association for Intelligent Machinery.

Davender S. Malik, Ph.D., is Professor of Mathematics and Computer Science at Creighton University. He received his B.A. and M.A. from University of Delhi, and Ph.D. from Ohio University specializing in ring theory. At Creighton University, he teaches both mathematics and computer science courses. Professor Malik has authored more than 45 papers and 5 books.

CONTENTS

Preface   v

Authors   ix

List of Symbols   xvii

1 Introduction   **1**
- 1.1 Sets 1
- 1.2 Relations 2
- 1.3 Functions 6
- 1.4 Fuzzy Subsets 8
- 1.5 Semigroups 9
- 1.6 Finite-State Machines 13
- 1.7 Finite-State Automata 15
- 1.8 Languages and Grammars 20
- 1.9 Nondeterministic Finite-State Automata 24
- 1.10 Relationships Between Languages and Automata 30
- 1.11 Pushdown Automata 36
- 1.12 Exercises 42

2 Max-Min Automata   **45**
- 2.1 Max-Min Automata 45
- 2.2 General Formulation of Automata 45
- 2.3 Classes of Automata 46
- 2.4 Behavior of Max-Min Automata 48
- 2.5 Equivalences and Homomorphisms of Max-Min Automata . 51
- 2.6 Reduction of Max-Min Automata 55
- 2.7 Definite Max-Min Automata 56
- 2.8 Reduction of Max-Min Machines 58
- 2.9 Equivalences 67
- 2.10 Irreducibility and Minimality 75
- 2.11 Nondeterministic and Deterministic Case 86
- 2.12 Exercises 88

3 Fuzzy Machines, Languages, and Grammars   **91**
   3.1 Max-Product Machines   91
   3.2 Equivalences   95
   3.3 Irreducibility and Minimality   98
   3.4 Max-Product Grammars and Languages 102
   3.5 Weak Regular Max-Product Grammars 108
   3.6 Weak Regular Max-Product Languages 116
   3.7 Properties of £ 120
   3.8 Exercises 125

4 Fuzzy Languages and Grammars   **127**
   4.1 Fuzzy Languages 127
   4.2 Types of Grammars 130
   4.3 Fuzzy Context-Free Grammars 131
   4.4 Context-Free Max-Product Grammars 137
   4.5 Context-Free Fuzzy Languages 141
   4.6 On the Description of the Fuzzy Meaning of Context-Free
       Languages 146
   4.7 Trees and Pseudoterms 147
   4.8 Fuzzy Dendrolanguage Generating Systems 148
   4.9 Normal Form of F-CFDS 150
   4.10 Sets of Derivation Trees of Fuzzy Context-Free Grammars . 154
   4.11 Fuzzy Tree Automaton 158
   4.12 Fuzzy Tree Transducer 162
   4.13 Fuzzy Meaning of Context-Free Languages 166
   4.14 Exercises 170

5 Probabilistic Automata and Grammars   **173**
   5.1 Probabilistic Automata and Their Approximation 173
   5.2 ϵ-Approximating by Nonprobability Devices 177
   5.3 ϵ-Approximating by Finite Automata 180
   5.4 Applications 181
   5.5 The $\overline{P}_\epsilon$ Relation 183
   5.6 Fuzzy Stars Acceptors and Probabilistic Acceptors 186
   5.7 Characterizations and the $\overline{R}_\epsilon$-Relation 187
   5.8 Probabilistic and Weighted Grammars 191
   5.9 Probabilistic and Weighted Grammars of Type 3 197
   5.10 Interrelations with Programmed and Time-Variant Grammars 202
   5.11 Probabilistic Grammars and Automata 204
   5.12 Probabilistic Grammars 205
   5.13 Weakly Regular Grammars and Asynchronous Automata . 209
   5.14 Type-0 Probabilistic Grammars and Probabilistic Turing Ma-
       chines 213
   5.15 Context-Free Probabilistic Grammars and Pushdown Auto-
       mata 217

5.16 Realization of Fuzzy Languages by Various Automata . . . 222
5.17 Properties of L_k, $k = 1, 2, 3$ 228
5.18 Further Properties of L_3 231
5.19 Exercises 234

6 Algebraic Fuzzy Automata Theory   237
6.1 Fuzzy Finite State Machines 237
6.2 Semigroups of Fuzzy Finite State Machines 237
6.3 Homomorphisms 242
6.4 Admissible Relations 245
6.5 Fuzzy Transformation Semigroups 247
6.6 Products of Fuzzy Finite State Machines 253
6.7 Submachines of a Fuzzy Finite State Machine 268
6.8 Retrievability, Separability, and Connectivity 272
6.9 Decomposition of Fuzzy Finite State Machines 274
6.10 Subsystems of Fuzzy Finite State Machines 277
6.11 Strong Subsystems 283
6.12 Cartesian Composition of Fuzzy Finite State Machines . . . 287
6.13 Cartesian Composition 289
6.14 Admissible Partitions 296
6.15 Coverings of Products of Fuzzy Finite State Machines . . . 304
6.16 Associative Properties of Products 306
6.17 Covering Properties of Products 308
6.18 Fuzzy Semiautomaton over a Finite Group 312
6.19 Exercises 322

7 More on Fuzzy Languages   325
7.1 Fuzzy Regular Languages 325
7.2 On Fuzzy Recognizers 334
7.3 Minimal Fuzzy Recognizers 347
7.4 Fuzzy Recognizers and Recognizable Sets 353
7.5 Operations on (Fuzzy) Subsets 354
7.6 Construction of Recognizers and Recognizable Sets 358
7.7 Accessible and Coaccessible Recognizers 362
7.8 Complete Fuzzy Machines 364
7.9 Fuzzy Languages on a Free Monoid 366
7.10 Algebraic Character and Properties of Fuzzy Regular Languages 370
7.11 Deterministic Acceptors of Regular Fuzzy Languages 383
7.12 Exercises 388

8 Minimization of Fuzzy Automata   391
8.1 Equivalence, Reduction, and Minimization of Finite Fuzzy Automata 391
8.2 Equivalence of Fuzzy Automata: An Algebraic Approach . 396

8.3   Reduction and Minimization of Fuzzy Automata 400
8.4   Minimal Fuzzy Finite State Automata 404
8.5   Behavior, Reduction, and Minimization of Finite L-Automata 410
8.6   Matrices over a Bounded Chain 410
8.7   Systems of Linear Equivalences over a Bounded Chain . . . 412
8.8   Finite L-Automata-Behavior Matrix 414
8.9 ε-Equivalence 416
8.10 ε-Irreducibility 418
8.11 Minimization 419
8.12 Exercises 421

9 *L*-Fuzzy Automata, Grammars, and Languages   423
9.1   Fuzzy Recognition of Fuzzy Languages 423
9.2   Fuzzy Languages 424
9.3   Fuzzy Recognition by Machines 426
9.4   Cutpoint Languages 430
9.5   Fuzzy Languages not Fuzzy Recognized by Machines in DT_2 435
9.6   Rational Probabilistic Events 436
9.7   Recursive Fuzzy Languages 438
9.8   Closure Properties 439
9.9   Fuzzy Grammars and Recursively Enumerable Fuzzy Languages 441
9.10 Recursively Enumerable *L*-Subsets 442
9.11 Various Kinds of Automata with Weights 446
9.12 Exercises 461

10 Applications   463
10.1 A Formulation of Fuzzy Automata and Its Application as a
      Model of Learning Systems 463
10.2 Formulation of Fuzzy Automata 463
10.3 Special Cases of Fuzzy Automata 466
10.4 Fuzzy Automata as Models of Learning Systems 468
10.5 Applications and Simulation Results 471
10.6 Properties of Fuzzy Automata 479
10.7 Fractionally Fuzzy Grammars and Pattern Recognition . . . 481
10.8 Fractionally Fuzzy Grammars 484
10.9 A Pattern Recognition Experiment 490
10.10 General Fuzzy Acceptors for Syntactic Pattern Recognition 494
10.11 ε-Equivalence by Inputs 496
10.12 Fuzzy-State Automata: Their Stability and Fault Tolerance 499
10.13 Relational Description of Automata 500
10.14 Fuzzy-State Automata 504
10.15 Stable and Almost Stable Behavior of Fuzzy-State Automata 508
10.16 Fault Tolerance of Fuzzy-State Automata 511
10.17 Clinical Monitoring with Fuzzy Automata 516

10.18 Fuzzy Systems 522

10.19 Exercises 526

References **529**

INDEX **547**

List of Symbols

$\mathbb{N}$	the set of positive integers, p. 1
$\mathbb{Z}$	the set of integers, p. 1
$\mathbb{Q}$	the set of rational numbers, p. 1
$\mathbb{R}$	the set of real numbers, p. 1
$\mathbb{R}^+$	the set of positive real numbers, p. 1
$|S|$	the cardinality of a set S, p. 1
$\in$	belongs to, p. 1
$\notin$	does not belong to, p. 1
$\subseteq$	subset, p. 1
$\subset$	proper subset, p. 1
$\supseteq$	contains
$\supset$	properly contains
$\emptyset$	the empty set, p. 1
$\cup$	unions, p. 2, 8, 127
$\cap$	intersection, p. 2, 8, 127
$A \backslash B$	relative complement of B in A, p. 2
$\overline{A}$	relative complement of A in its universe, p. 2
$A \times B$	Cartesian cross-product of A and B, p. 2
$\mathcal{P}(X)$	power set of X, p. 2
$\mathrm{Dom}(R)$	domain of a relation R, p. 3
$\mathrm{Im}(R)$	image of a relation R, p. 3
$[x]$	equivalence class of an element x, p. 3
$\circ$	composition of crisp and fuzzy relations, p. 5, 9
μ_c	c-cut of the fuzzy subset μ, p. 8
$\mathrm{Supp}(\mu)$	support of the fuzzy subset μ, p.8
$\mathcal{FP}(X)$	the fuzzy power set of X, p. 8
$\overline{\mu}$	complement of the fuzzy subset μ, p. 8
$\wedge$	infimum, p. 9
$\vee$	supremum, p. 9
Λ	empty string, empty tape, or identity of the semigroup, p. 14
S^*	free monoid generated by S with operation, concatenation, p. 14
S^+	$S^* \backslash \{\Lambda\}$, p. 14
$|x|$	length of a string, p. 20
PDA	p. 37
$r\mu$	response function, p. 48

Ω^*	max-min table, p. 50
θ^*	max-min pre-table, p. 50
$\Omega_1^* + \Omega_2^*$	direct sum of max-min tables, p. 50
$A_1^* \times A_2^*$	direct product of max-min automata, p. 50
$\sim$	p. 52, 53, 73, 74, 75, 97, 98, 245, 341, 394, 415
$\approx$	p. 55, 74, 97, 239, 394, 407
$\otimes$	max-min binary operation, p. 59
$C(X)$	convex span of X, p. 60
MSLM	max-min sequential-like machine, p. 67
$(X \times Y)^*$	p. 68, 95, 415
$\rho(A)$	collection of all rows of a matrix A, p. 69, 97
sd	state distribution, p. 73
IMSLM	initialized max-min sequential-like machine, p. 73
r^I	response function, p.73, 97
$\simeq$	p. 74, 97, 240
$\rho(M)$	p. 83, 101
$\overline{\rho}(M)$	p. 83, 101
NSLM	nondeterministic sequential-like machine, p. 86
DSLM	deterministic sequential-like machine, p. 86
$\odot$	max-product binary operation, p. 91
$C(X)$	the convex MP-span of X, p. 92
MPSM	max-product sequential-like machine, p. 95
i.d.	initial distribution, p. 96
IMPSM	initialized max-product sequential-like machine, p. 96
$\equiv$	p. 100, 238, 369
$\mathbb{R}^{\geq 0}$	the set of all nonnegative real numbers, p. 103
$\mathbb{R}_\infty^{\geq 0}$	$\mathbb{R}_\infty^{\geq 0} = \mathbb{R}^{\geq 0} \cup \{\infty\}$, p. 103
AMA	p. 110
MA	p. 114
$L(\lambda, r, >)$	$L(\lambda, r, >) = \{x \in T^* \mid f(x) > r\}$, p. 116, 141
$L(\lambda, r, \geq)$	$L(\lambda, r, \geq) = \{x \in T^* \mid f(x) \geq r\}$, p. 116, 142
$L(\lambda, r, =)$	$L(\lambda, r, =) = \{x \in T^* \mid f(x) = r\}$, p. 116, 142
$\mathcal{R}$	the family of all regular languages, p. 120, 144
$M^{-1}(L)$	p. 123
λ^∞	Kleene closure, p. 128
CMG	context-free max-product grammar, p. 137
CFFL	context-free fuzzy language, p. 139
$\mathcal{C}$	the family of all CFL, p. 144
$\mathcal{L}$	p. 144
F-CFDS	p. 148
F-CFDL	p. 149
F-CFG	p. 154
fa	p. 175
fac	p. 175
pa	p. 176
pac	p. 177
APA	p. 209
PTM	p. 213
PPA	p. 218

ffsm	p. 237
$E(M)$	p. 238
$[[x]]$	p. 239
$\underbrace{E(M)}$	p. 239
x^M	p. 240
S_M	p. 240
$\prec x \succ$	p. 240
$M_1 \leq M_2$	covering of M_1 by M_2, p. 253
$M_1 \omega M_2$	cascade product, p. 259, 294
$M_1 \circ M_2$	wreath product, p. 261
$M_1 \cdot M_2$	Cartesian composition, p. 289
M/π	quotient fuzzy finite state machine, p. 298
$\cong$	p. 304
$M_1 \preceq M_2$	strong covering of M_1 by M_2, p. 305
$\oplus$	direct sum, p. 305
$+$	sum, p. 305
$M_1 \preceq_w M_2$	weak covering of M_1 by M_2, p. 310
$\sim_A$	p. 341
$\#$	p. 356
$\circ$	p. 367, 393, 494
$\mathbf{FR}(X^*)$	family of F-regular languages, p. 368
fqfa	finite quasi-fuzzy automata, p. 384
L(G)	regular fuzzy language, p. 384
L(c)	$L(c) = \{x \mid x \in X^*, \ \mu(x) \geq c\}$, p. 385
$\underset{\sim}{\prec}$	p. 394
$\underset{\approx}{\prec}$	p. 394
ffa	p. 404
$\approx_{Q_1 Q_2}$	p. 406
$\approx_{Q_1}$	p. 406
FG	p. 482
$\mathbb{N}^0$	p. 500
$\blacksquare$	end of proof

Chapter 1

Introduction

1.1 Sets

In this section, we review some results from set theory. We use the notation $\mathbb{N}$ for the set of positive integers, $\mathbb{Z}$ for the set of integers, $\mathbb{Q}$ for the set of rational numbers, $\mathbb{R}$ for the set of real numbers, and $\mathbb{R}^+$ for the set of positive real numbers.

We assume that the reader is familiar with the basics of set theory. Nevertheless, we give a brief review of some basics of set theory. We think of a **set** as a collection of objects. We let $|S|$ denote the cardinality of S, i.e., the number of elements of S. If S is a set such that $|S| < \infty$, then S is called a **finite** set; otherwise S is called an **infinite** set.

Given a set S, we use the notation $x \in S$ and $x \notin S$ to mean x is a member of S and x is not a member of S, respectively.

A set A is said to be a **subset** of a set S if every element of A is an element of S. In this case, we write $A \subseteq S$ and say that A is contained in S. If $A \subseteq S$, but $A \neq S$, then we write $A \subset S$ and say that A is properly contained in S or that A is a **proper subset** of S.

The **empty set** or **null set** is the set with no elements. We denote the empty set by $\emptyset$. The empty set is a subset of every set.

We also describe sets in the following manner. Given a set S, the notation

$$A = \{x \mid x \in S, P(x)\}$$

or

$$A = \{x \in S \mid P(x)\}$$

means that A is the set of all elements x of S such that x satisfies the property P.

Sets can be combined in several ways.

1

Definition 1.1.1 *The **union** of two sets A and B, written $A \cup B$, is defined to be the set*

$$A \cup B = \{x \mid x \in A \text{ or } x \in B\}.$$

Definition 1.1.2 *The **intersection** of two sets A and B, written $A \cap B$, is defined to be the set*

$$A \cap B = \{x \mid x \in A \text{ and } x \in B\}.$$

The union and intersection for any finite number of sets can be defined in a similar manner. That is, suppose that $A_1, A_2, \ldots, A_n$ are n sets. The union of $A_1, A_2, \ldots, A_n$, denoted by $\cup_{i=1}^{n} A_i$ or $A_1 \cup A_2 \cup \ldots \cup A_n$, is the set of all elements x such that x is an element of some A_i, where $1 \leq i \leq n$. The intersection of $A_1, A_2, \ldots, A_n$, denoted by $\cap_{i=1}^{n} A_i$ or $A_1 \cap A_2 \cap \ldots \cap A_n$, is the set of all elements x such that $x \in A_i$ for all i, $1 \leq i \leq n$.

We say that a set I is an **index set** for a collection of sets $\mathcal{A}$ if for any $i \in I$, there exists a set $A_i \in \mathcal{A}$ and $\mathcal{A} = \{A_i \mid i \in I\}$.

The **union** of the sets A_i, $i \in I$, is defined to be the set $\{x \mid x \in A_i$ for at least one $i \in I\}$ and is denoted by $\cup_{i \in I} A_i$. The **intersection** of the sets A_i, $i \in I$, is defined to be the set $\{x \mid x \in A_i$ for all $i \in I\}$ and is denoted by $\cap_{i \in I} A_i$.

Definition 1.1.3 *Given two sets A and B, the **relative complement** (or **set difference**) of B in A, written $A \backslash B$, is the set*

$$A \backslash B = \{x \mid x \in A, \text{ but } x \notin B\}.$$

We sometimes write $\overline{A}$ for the relative complement of a set A in the universal set.

If A and B are sets, we let $A \times B$ denote the **Cartesian cross-product** of A and B, i.e., $A \times B$ is the set of all ordered pairs (a, b), where $a \in A$ and $b \in B$. We write A^n for the set of all ordered n-tuples of elements from A, where $n \in \mathbb{N}$. We let $\mathcal{P}(X)$ denote the **power set** of a set X, i.e., $\mathcal{P}(X)$ is the set of all subsets of X.

1.2 Relations

An important and fundamental concept in mathematics is the notion of a relation.

Definition 1.2.1 *A **binary relation** or simply a **relation** R from a set A into a set B is a subset of $A \times B$.*

Let R be a relation from a set A into a set B. If $(x, y) \in R$, we sometimes write xRy or $R(x) = y$. If $A = B$, then R is called a **binary relation** on A.

Definition 1.2.2 *Let R be a relation from a set A into a set B. Then the* **domain** *of R, written $Dom(R)$, is defined to be the set*

$$\{x \mid x \in A \text{ and there exists } y \in B \text{ such that } (x, y) \in R\}.$$

The **image** *of R, written $Im(R)$, is defined to be the set*

$$\{y \mid y \in B \text{ and there exists } x \in A \text{ such that } (x, y) \in R\}.$$

Definition 1.2.3 *Let R be a binary relation on a set A. Then R is called*
(1) **reflexive** *if for all $x \in A$, xRx,*
(2) **symmetric** *if for all $x, y \in A$, xRy implies yRx,*
(3) **transitive** *if for all $x, y, z \in A$, xRy and yRz imply xRz.*

Definition 1.2.4 *A binary relation E on a set A is called an* **equivalence relation** *on A if E is reflexive, symmetric, and transitive.*

Definition 1.2.5 *Let E be an equivalence relation on a set A. For all $x \in A$, let $[x]$ denote the set*

$$[x] = \{y \in A \mid yEx\}.$$

The set $[x]$ is called the **equivalence class** *(with respect to E) determined by x.*

We prove some basic properties of equivalence classes in the following theorem.

Theorem 1.2.6 *Let E be an equivalence relation on the set A. Then the following properties hold:*
(1) for all $x \in A$, $[x] \neq \emptyset$,
(2) for all $x, y \in A$, if $y \in [x]$, then $[x] = [y]$,
(3) for all $x, y \in A$, either $[x] = [y]$ or $[x] \cap [y] = \emptyset$,
(4) $A = \cup_{x \in A}[x]$, i.e., A is the union of all equivalence classes with respect to E.

Proof. (1) Let $x \in A$. Since E is reflexive, xEx. Hence $x \in [x]$ and so $[x] \neq \emptyset$.

(2) Let $y \in [x]$. Then yEx and by the symmetric property of E, xEy. Let $u \in [y]$. Then uEy. Since uEy and yEx, the transitivity of E implies that uEx. Hence $u \in [x]$. Thus $[y] \subseteq [x]$. Now let $u \in [x]$. Then uEx. Since uEx and xEy, uEy by transitivity and so $u \in [y]$. Hence $[x] \subseteq [y]$. Consequently, $[x] = [y]$.

(3) Let $x, y \in A$. Suppose $[x] \cap [y] \neq \emptyset$. Then there exists $u \in [x] \cap [y]$. Thus $u \in [x]$ and $u \in [y]$. By (2), $[x] = [u] = [y]$.

(4) Let $x \in A$. Then $x \in [x] \subseteq \cup_{x \in A}[x]$. Thus $A \subseteq \cup_{x \in A}[x]$. Also, $\cup_{x \in A}[x] \subseteq A$. Hence $A = \cup_{x \in A}[x]$. ∎

One of the objectives of this section is to study the relationship between an equivalence relation and a partition of a set. We now turn our attention to partitions.

Definition 1.2.7 *Let A be a set and $\mathcal{P}$ be a collection of nonempty subsets of A. Then $\mathcal{P}$ is called a **partition** of A if the following properties hold:*
(1) for all $B, C \in \mathcal{P}$, either $B = C$ or $B \cap C = \emptyset$,
(2) $A = \cup_{B \in \mathcal{P}} B$. ∎

The next theorem follows easily from Theorem 1.2.6.

Theorem 1.2.8 *Let E be an equivalence relation on the set A. Then*

$$\mathcal{P} = \{[x] \mid x \in A\}$$

is a partition of A. ∎

Given an equivalence relation E on a set A, the set of all equivalence classes forms a partition of A by Theorem 1.2.8. We now prove that corresponding to any partition, we can associate an equivalence relation.

Theorem 1.2.9 *Let $\mathcal{P}$ be a partition of the set A. Define a relation E on A by for all $x, y \in A$, xEy if there exists $B \in \mathcal{P}$ such that $x, y \in B$. Then E is an equivalence relation on A and the equivalence classes are precisely the elements of $\mathcal{P}$.*

Proof. Since $\mathcal{P}$ is a partition of A, $A = \cup_{B \in \mathcal{P}} B$. First we show that E is reflexive. Let x be any element of A. Then there exists $B \in \mathcal{P}$ such that $x \in B$. Since $x, x \in B$, we have xEx. Hence E is reflexive. We now show that E is symmetric. Let xEy. Then $x, y \in B$ for some $B \in \mathcal{P}$. Thus $y, x \in B$ and so yEx. Hence E is symmetric. We now establish the transitivity of E. Let $x, y, z \in A$. Suppose xEy and yEz. Then $x, y \in B$ and $y, z \in C$ for some $B, C \in \mathcal{P}$. Since $y \in B \cap C$, $B \cap C \neq \emptyset$. Also, since $\mathcal{P}$ is a partition and $B \cap C \neq \emptyset$, we have $B = C$ so that $x, z \in B$. Hence xEz. This shows that E is transitive. Therefore, E is an equivalence relation.
We now show that the equivalence classes determined by E are precisely the elements of $\mathcal{P}$. Let $x \in A$. Consider the equivalence class $[x]$. Since $A = \cup_{B \in \mathcal{P}} B$, there exists $B \in \mathcal{P}$ such that $x \in B$. We show that $[x] = B$. Let $u \in [x]$. Then uEx and so $u \in B$ since $x \in B$. Thus $[x] \subseteq B$. Also, since $x \in B$, we have yEx for all $y \in B$ and so $y \in [x]$ for all $y \in B$. This implies that $B \subseteq [x]$. Hence $[x] = B$. Finally, note that if $C \in \mathcal{P}$, then $C = [u]$ for all $u \in C$. Thus the equivalence classes are precisely the elements of $\mathcal{P}$. ∎

The relation E in Theorem 1.2.9 is called the **equivalence relation** on A **induced by the partition** $\mathcal{P}$.

Given a relation R from a set A into a set B and a relation S from B into a set C, there is a relation T from A into C that arises in a natural way as follows: $(a, c) \in T$ for $a \in A$ and $c \in C$ if and only if there exists $b \in B$ such that $(a, b) \in R$ and $(b, c) \in S$. This relation T is called the **composition** of R and S and is denoted by $S \circ R$. More formally we have the following definition.

Definition 1.2.10 *Let R be a relation from a set A into a set B and S be a relation from B into a set C. The **composition** of R and S, denoted by $S \circ R$, is the relation from A into C defined by*

$$x(S \circ R)z \text{ if there exists } y \in B \text{ such that } xRy \text{ and } ySz$$

for all $x \in A$, $z \in C$.

It is important to note that composition is associative.

Let R be a relation on a set A. We define recursively relations R^n on A, $n \in \mathbb{N}$, as follows: Let

$$R^1 = R.$$

Suppose R^n has been defined for $n \in \mathbb{N}$. Define

$$R^{n+1} = R \circ R^n.$$

Definition 1.2.11 *Let R be a relation from a set A into a set B. The **inverse** of R, denoted by R^{-1}, is the relation from B into A defined by*

$$xR^{-1}y \text{ if } yRx$$

for all $x \in B$, $y \in A$.

We are now interested in defining a lattice.

Definition 1.2.12 *A relation R on a set S is called a **partial order** on S if it satisfies the following conditions:*

*(1) $(x, x) \in R$ for all $x \in S$ (i.e., R is **reflexive**).*

*(2) $\forall x, y \in S$, if $(x, y) \in R$ and $(y, x) \in R$, then $x = y$ (i.e., R is **antisymmetric**).*

*(3) $\forall x, y, z \in S$, if $(x, y) \in R$ and $(y, z) \in R$, then $(x, z) \in R$ (i.e., R is **transitive**).*

A partial on a set S is usually denoted by $\leq$. We often write $x \leq y$ rather than $(x, y) \in \leq$.

Let R be a partial order on S. Then R is called a **total order** or **linear order** if $\forall x, y \in S$, either $(x, y) \in R$ or $(y, x) \in R$.

Definition 1.2.13 *A set S together with a partial order is called a **par-tially ordered set (poset)**. A set S together with a total order is called a **totally ordered set** or a **chain**.*

Definition 1.2.14 *Let $(S, \leq)$ be a poset and $\{x, y\} \subseteq S$. Then $z \in S$ is called an **upper bound** of $\{x, y\}$ if $x \leq z$ and $y \leq z$. An element $w \in S$ is called a **lower bound** of $\{x, y\}$ if $w \leq x$ and $w \leq y$.*

Definition 1.2.15 *Let $(S, \leq)$ be a poset and $\{x, y\} \subseteq S$. Then $z \in S$ is called a **least upper bound** of $\{x, y\}$ if z is an upper bound of $\{x, y\}$ and if $z' \in S$ is an upper bound of $\{x, y\}$, then $z \leq z'$. An element $w \in S$ is called a **greatest lower bound** of $\{x, y\}$ if w is a lower bound of $\{x, y\}$ and if $w' \in S$ is a lower bound of $\{x, y\}$, then $w' \leq w$.*

Definition 1.2.16 *Let $(L, \leq)$ be a poset. Then $(L, \leq)$ (or simply L) is called a **lattice** if for every $x, y \in L$, x and y have a least upper bound, denoted $x \vee y$, and greatest lower bound, denoted $x \wedge y$.*

A lattice $(L, \leq)$ is called **distributive** if $x \wedge (y \vee z) = (x \wedge y) \vee (x \wedge z)$ for all $x, y, z \in L$.

1.3   Functions

In this section, we review some of the basic properties of functions.

Definition 1.3.1 *Let A and B be nonempty sets. A relation f from A into B is called a **function** (or **mapping**) from A into B if*
    (1) $Dom(f) = A$ and
    (2) for all $(x, y), (x', y') \in f$, $x = x'$ implies $y = y'$.
    *When (2) is satisfied by a relation f, we say that f is **well defined** or **single-valued**.*

We use the notation $f : A \rightarrow B$ to denote a function f from a set A into a set B. For $(x, y) \in f$, we usually write $f(x) = y$ and say that y is the **image** of x under f and x is a **preimage** of y under f.

If f is a relation from A into B such that (2) of Definition 1.3.1 holds, but (1) is not required, then f is called a **partial function** from A into B.

Definition 1.3.2 *Let f be a function from a set A into a set B. Then*
    *(1) f is called **one-one** if for all $x, x' \in A$, $f(x) = f(x')$ implies $x = x'$. (In this case, f is sometimes called **injective**.)*
    *(2) f is called **onto** B (or f **maps** A **onto** B) if $Im(f) = B$. (In this case, f is sometimes called **surjective**.)*
    *(3) If f is both injective and surjective, it is called **bijective**.*

Let A be a nonempty set. The function $i : A \to A$ defined by $i(x) = x$ for all $x \in A$ is a one-one function of A onto A. i is called the **identity map** on A.

Let A, B, and C be nonempty sets and $f : A \to B$ and $g : B \to C$. By Definition 1.2.10, the **composition** $\circ$ of f and g, written $g \circ f$, is the relation from A into C defined as follows:

$$g \circ f = \{(x, z) \mid x \in A, \; z \in C, \text{ there exists } y \in B$$
$$\text{such that } f(x) = y \text{ and } g(y) = z\}.$$

Let $f : A \to B$ and $g : B \to C$ and $(x, z) \in g \circ f$, i.e., $(g \circ f)(x) = z$. Then by the definition of composition of functions, there exists $y \in B$ such that $f(x) = y$ and $g(y) = z$. Now

$$z = g(y) = g(f(x)).$$

Hence $(g \circ f)(x) = g(f(x))$.

Theorem 1.3.3 *Suppose that* $f : A \to B$ *and* $g : B \to C$. *Then the following properties hold:*
(1) $g \circ f : A \to C$, *i.e.,* $g \circ f$ *is a function from* A *into* C;
(2) If f *and* g *are one-one, then* $g \circ f$ *is one-one;*
(3) If f *is onto* B *and* g *is onto* C, *then* $g \circ f$ *is onto* C.

Proof. (1) Let $x \in A$. Since f is a function and $x \in A$, there exists $y \in B$ such that $f(x) = y$. Since g is a function and $y \in B$, there exists $z \in C$ such that $g(y) = z$. Hence $(g \circ f)(x) = g(f(x)) = g(y) = z$, i.e., $(x, z) \in g \circ f$. Thus $x \in \text{Dom}(g \circ f)$. Consequently, $A \subseteq \text{Dom}(g \circ f)$. However, $\text{Dom}(g \circ f) \subseteq A$ and so $\text{Dom}(g \circ f) = A$. We now show that $g \circ f$ is well defined.

Suppose that $(x, z) \in g \circ f$, $(x_1, z_1) \in g \circ f$, and $x = x_1$, where $x, x_1 \in A$ and $z, z_1 \in C$. By the definition of composition of functions, there exist $y, y_1 \in B$ such that $f(x) = y$, $g(y) = z$, $f(x_1) = y_1$, and $g(y_1) = z_1$. Since f is a function and $x = x_1$, we have $y = y_1$. Similarly, since g is a function and $y = y_1$, we have $z = z_1$. Hence $g \circ f$ is well defined. Thus $g \circ f$ is a function from A into C.

(2) Let $x, x' \in A$. Suppose $(g \circ f)(x) = (g \circ f)(x')$. Then $g(f(x)) = g(f(x'))$. Since g is one-one, $f(x) = f(x')$. Since f is one-one, $x = x'$. Hence $g \circ f$ is one-one.

(3) Let $z \in C$. Then there exists $y \in B$ such that $g(y) = z$ since g is onto C. Since f is onto B, there exists $x \in A$ such that $f(x) = y$. Thus $(g \circ f)(x) = g(f(x)) = g(y) = z$. Hence $g \circ f$ is onto C. $\blacksquare$

Theorem 1.3.4 *Let* $f : A \to B$, $g : B \to C$, *and* $h : C \to D$. *Then*

$$h \circ (g \circ f) = (h \circ g) \circ f.$$

That is, composition of functions is associative.

Proof. Clearly, $h \circ (g \circ f) : A \to D$ and $(h \circ g) \circ f : A \to D$. Let $x \in A$. Then $[h \circ (g \circ f)](x) = h((g \circ f)(x)) = h(g(f(x))) = (h \circ g)(f(x)) = [(h \circ g) \circ f](x)$. Thus $h \circ (g \circ f) = (h \circ g) \circ f$. ∎

1.4   Fuzzy Subsets

The notion of a fuzzy subset of a set was introduced by Zadeh in 1965, [255]. This introduction was an important step in the evolution of the modern concept of uncertainty.

Let X be a set and A be a subset of X. The **characteristic function** of A is the function χ_A of X into $\{0,1\}$ defined by $\chi_A(x) = 1$ if $x \in A$ and $\chi_A(x) = 0$ if $x \notin A$. We sometimes write χ if A is understood. The characteristic function can be used to indicate either members or nonmembers of a subset of X. This notion can be generalized in a way that introduces the notion of a fuzzy subset of X.

Definition 1.4.1 *A **fuzzy subset** μ of X is a function of X into the closed interval $[0,1]$, [255].*

Let μ be a fuzzy subset of a set X. For all $x \in X$, $\mu(x)$ can be thought of as the degree of membership of x in μ. We sometimes use the notation μ_A for a fuzzy subset of X, where A is thought of as a fuzzy set and μ_A gives the grade of membership of elements of X in A. At times A may be merely a description of a fuzzy subset μ of X.

Definition 1.4.2 *Let μ be a fuzzy subset of X.*
*(1) Let $c \in [0,1]$. Define $\mu_c = \{x \in X \mid \mu(x) \geq c\}$. We call μ_c a **c-cut**.*
*(2) The **support** of μ is defined to be the set,*

$$Supp(\mu) = \{x \in X \mid \mu(x) > 0\}.$$

μ_c in Definition 1.4.2 is also called a **level set**.

We let $\mathcal{FP}(X)$ denote the **fuzzy power set** of X, i.e., the set of all fuzzy subsets of X.

Definition 1.4.3 *Let μ and ν be fuzzy subsets of a set X. Define $\overline{\mu}$, $\mu \cap \nu$, $\mu \cup \nu$ as follows:*

$$
\begin{aligned}
\overline{\mu}(x) &= 1 - \mu(x), \\
(\mu \cap \nu)(x) &= \min\{\mu(x), \nu(x)\}, \\
(\mu \cup \nu)(x) &= \max\{\mu(x), \nu(x)\},
\end{aligned}
$$

*for all $x \in X$. $\overline{\mu}$ is called the **complement** of μ, $\mu \cap \nu$ is called the **intersection** of μ and v, and $\mu \cup \nu$ is called the **union** of μ and v.*

We some times use $\wedge$ to denote min and infimum and $\vee$ to denote max and supremum. Using these symbols, $(\mu \cap \nu)(x) = \mu(x) \wedge \nu(x)$ and $(\mu \cup \nu)(x) = \mu(x) \vee \nu(x)$ for all $x \in X$.

We can extend the notion of union and intersection to a family of fuzzy subsets of X. Let $\{\mu_i\}_{i \in I}$ be a family of fuzzy subsets of X, where I is an index set. Define $\cap_{i \in I}\mu_i$ and $\cup_{i \in I}\mu_i$ as follows:

$$
\begin{aligned}
(\cap_{i \in I}\mu_i)(x) &= \wedge_{i \in I}\mu_i(x), \\
(\cup_{i \in I}\mu_i)(x) &= \vee_{i \in I}\mu_i(x).
\end{aligned}
$$

Thus if I is a finite set, say $I = \{1, 2, \ldots, n\}$, then $\cap_{i \in I}\mu_i = \mu_1 \cap \mu_2 \cap \ldots \cap \mu_n$ and $\cup_{i \in I}\mu_i = \mu_1 \cup \mu_2 \cup \ldots \cup \mu_n$. In this case, we sometimes write

$$(\cap_{i \in I}\mu_i)(x) = \mu_1(x) \wedge \mu_2(x) \wedge \ldots \wedge \mu_n(x)$$

and

$$(\cup_{i \in I}\mu_i)(x) = \mu_1(x) \vee \mu_2(x) \vee \ldots \vee \mu_n(x).$$

Definition 1.4.4 *Let X, Y, and Z be nonempty sets and let μ be a fuzzy subset of $X \times Y$ and ν be a fuzzy subset of $Y \times Z$. Define the fuzzy subset $\mu \circ \nu$ of $X \times Z$ by*

$$(\mu \circ \nu)(x, z) = \vee\{\mu(x, y) \wedge \nu(y, z) \mid y \in Y\}$$

$\forall x \in X$ *and* $\forall z \in Z$.

If μ is a fuzzy subset of $X \times X$, we define $\mu^1 = \mu$ and $\mu^{n+1} = \mu \circ \mu^n$ for all $n \in \mathbb{N}$.

A detailed study of fuzzy subsets can be found in [51, 108, 110, 114, 115, 266].

1.5 Semigroups

In this section, we provide some basic results concerning semigroups that are needed later for our presentation of fuzzy automata and fuzzy languages.

Let X be a nonempty set. Then a function from $X \times X$ into X is called a **binary operation** on X. If $*$ is a binary operation on X, then the pair $(X, *)$ is called a **mathematical system**. If $(X, *)$ is a mathematical system such that $\forall a, b, c \in X$, $(a*b)*c = a*(b*c)$, then $*$ is called **associative** and $(X, *)$ is called a **semigroup**. Let $(X, *)$ be a mathematical system. If there exists $e \in X$ such that $\forall a \in X$, $a * e = a = e * a$, then e is called an **identity** of $(X, *)$ and $(X, *)$ is said to have an identity. If $(X, *)$ has an identity, then it is easily seen that the identity is unique. A semigroup $(X, *)$ is called a **monoid** if it has an identity.

Let $(X, *)$ be a semigroup and $\equiv$ be an equivalence relation on X. Then $\equiv$ is called a **right** (**left**) **congruence relation** on X if $\forall a, b, c \in X$, $a \equiv b$ implies $a * c \equiv b * c$ ($c * a \equiv c * b$). A right and left congruence relation $\equiv$ on X is called a **congruence relation**. The number of congruence classes of $\equiv$, i.e., equivalence classes of $\equiv$, is called the **index** of $\equiv$.

Let $(X, *)$ be a semigroup and let S be a nonempty subset of X. Then $(S, *)$ is said to be a **subsemigroup** of $(X, *)$ if $(S, *)$ is a mathematical system. (Here for $(S, *)$, we mean $*$ restricted to $S \times S$.) Let $(X, *)$ be a monoid and let $(S, *)$ be a subsemigroup of $(X, *)$. If e is the identity of $(X, *)$ and $e \in S$, then $(S, *)$ is called a **submonoid** of $(X, *)$. We often write X for $(X, *)$ when the operation $*$ is understood. Let $a \in X$. We define $a^1 = a$ and if a^n is defined for $n \in \mathbb{N}$, we define $a^{n+1} = a^n * a$.

We next state some well-known theorems and definitions.

Theorem 1.5.1 *Let X be a semigroup (monoid). The intersection of any collection of subsemigroups (submonoids) of X is a subsemigroup (submonoid) of X.* ∎

Definition 1.5.2 *Let X be a semigroup (monoid) and let S be a subset of X. Let $\langle S \rangle$ denote the intersection of all subsemigroups (submonoids) of X that contain S. Then $\langle S \rangle$ is called the subsemigroup (submonoid) of X* **generated** *by S. If $\langle S \rangle = X$, then S is called a set of* **generators** *for X.*

Let $(X, *)$ be a semigroup and let e be an element not in X. Let X' denote $X \cup \{e\}$. Extend $*$ from $X \times X$ into X to $*'$ from $X' \times X'$ into X' as follows: $a *' b = a * b$ if $a, b \in X$; $a *' e = a = e *' a$ if $a \in X$; $e *' e = e$. Then $(X', *')$ is a monoid and X is a subsemigroup of X'. Even if X were a monoid, it is not a submonoid of X' since its identity is not e.

Definition 1.5.3 *Let $(X, *)$ and $(Y, \cdot)$ be semigroups. A function f from X into Y is called a* **homomorphism** *if $\forall a, b \in X$, $f(a * b) = f(a) \cdot f(b)$. Let f be a homomorphism of X into Y. If f is one-one, then f is called a* **monomorphism**. *If f is onto Y, then f is called an* **epimorphism**. *If f is both a monomorphism and epimorphism, then f is called an* **isomorphism** *and X and Y are said to be* **isomorphic**.

Theorem 1.5.4 *Let X, Y, and Z be semigroups. If f is a homomorphism of X into Y and g is a homomorphism of Y into Z, then $g \circ f$ is a homomorphism of X into Z.* ∎

Let S be a set. A **free semigroup** on the set S is a semigroup F together with a function $f : S \to F$ such that for any semigroup X and every function $g : S \to X$, there exists a unique homomorphism $h : F \to X$ such that $h \circ f = g$.

Theorem 1.5.5 *Let F be a free semigroup on a set S together with a function $f : S \to F$. Then f is one-one and $\langle f(S) \rangle = F$.*

Proof. Let $a, b \in S$ be such that $a \neq b$. Let X be a semigroup containing more than one element and consider a function $g : S \to X$ such that $g(a) \neq g(b)$. Now $h \circ f = g$. Since $h(f(a)) = g(a) \neq g(b) = h(f(b))$, it follows that $f(a) \neq f(b)$. Thus f is one-one.

We now show that $\langle f(S) \rangle = F$. Let $\langle f(S) \rangle = A$. Then the function f defines a function $g : S \to A$ such that $i \circ g = f$, where i is the identity homomorphism $i : A \to F$. By the definition of a free semigroup, there exists a homomorphism $h : F \to A$ such that $h \circ f = g$. Consider the diagram:

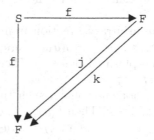

where j is the identity isomorphism and $k = i \circ h$. Since $j \circ f = f$, $k \circ f = i \circ h \circ f = i \circ g = f$, it follows that $i \circ h = k = j$ from the uniqueness property defining free semigroups. The inclusion homomorphism i must be an epimorphism. Hence $A = X$ and so $f(S)$ generates F. ∎

Theorem 1.5.6 *Let (F, f) and (F', f') be free semigroups on the same set S. Then there exists a unique isomorphism $j : F \to F'$ such that $j \circ f = f'$.*

Proof. Since (F, f) is a free semigroup on the set S, there exists a homomorphism $j : F \to F'$ such that $j \circ f = f'$. Similarly, there exists a homomorphism $k : F' \to F$ such that $k \circ f' = f$. Let $h = k \circ j$ and i be the identity isomorphism of F. In the diagram,

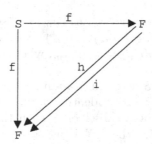

it follows that $h \circ f = k \circ j \circ f = k \circ f' = f$, $i \circ f = f$. From the uniqueness, it follows that $k \circ j = h = i$. Since i is an isomorphism, it follows that j is a monomorphism. Similarly, we can show that $j \circ k$ is the identity endomorphism on F'. Thus j is also an epimorphism. Hence j is an isomorphism. ∎

Theorem 1.5.7 *Let S be a set. Then there exists a free semigroup on S.*

Proof. Let F denote the set of all finite sequences of elements (repetitions allowed) of S. Let $x = (a_1, \dots, a_m)$, $y = (b_1, \dots, b_n) \in F$. Define

$$xy = (a_1, \dots, a_m, b_1, \dots, b_n).$$

Clearly, for any $x, y, z \in F$, $x(yz) = (xy)z$. Under this operation, F is a semigroup.

Define $f : S \to F$ as follows: For all $a \in S$, define $f(a) = (a)$, the sequence consisting of the element a. We now show that (F, f) is a free semigroup on S.

Let $g : S \to X$ be an arbitrary function from S into a semigroup X. Define $h : F \to X$ by $h(a_1, \dots, a_m) = g(a_1) \dots g(a_m)$ for all $(a_1, \dots, a_m) \in F$. Clearly, h is a homomorphism. Let $a \in S$. Then $(h \circ f)(a) = h(f(a)) = h((a)) = g(a)$. Hence $h \circ f = g$.

Let $k : F \to X$ be a homomorphism such that $k \circ f = g$. We show that $h = k$. Let $(a_1, \dots, a_m) \in F$. Then $k(a_1, \dots, a_m) = k((a_1) \dots (a_m)) = k((a_1)) \dots k((a_m)) = k(f(a_1)) \dots k(f(a_m)) = g(a_1) \dots g(a_m) = h(a_1, \dots, a_m)$. Thus $h = k$. Therefore, (F, f) is a free semigroup on S. ∎

Theorem 1.5.8 *Let X be a semigroup and let S be a set of generators of X. Then every element of X can be written as the product of a finite sequence of elements in S.*

Proof. Let (F, f) be the free semigroup on S as constructed in the proof of Theorem 1.5.7. Then by the definition of a free semigroup, there exists a homomorphism $h : F \to X$ such that $h \circ f = g$, where g is the inclusion map $g : S \to X$.

We now prove that h is an epimorphism. Consider the image $h(F)$. Clearly, $h(F)$ is a subsemigroup of X. Since $S = g(S) = (h \circ f)(S) = h(f(S)) \subseteq h(F)$ and S generates X, it follows that $h(F) = X$. Thus h is an epimorphism.

Let $x \in X$. Then there exists $(a_1, \dots, a_m) \in F$ such that $h(a_1, \dots, a_m) = x$. Now $x = h(a_1, \dots, a_m) = h((a_1) \dots (a_m)) = h((a_1)) \dots h((a_m)) = h(f(a_1)) \dots h(f(a_m)) = g(a_1) \dots g(a_m) = a_1 \dots a_m$, where $a_1, \dots, a_m \in S$. ∎

Let S be a set. Then S determines a unique free semigroup (F, f). Since $f : S \to F$ is injective, we may identify S with its image $f(S)$ in F. We can then consider S to be a subset of F such that S generates F. Also, any function $g : S \to X$, where X is a semigroup, extends to a unique homomorphism $h : F \to X$. We call F the **free semigroup generated** by S.

Consider the monoid $F^* = F \cup \{e\}$, where F is the free semigroup generated by S. Then every function $g : S \to X$, where X is a monoid, extends to a unique proper homomorphism $h : F^* \to X$. This monoid F^* is called the **free monoid generated** by the set S.

1.6 Finite-State Machines

The theory of machines has had a major impact on the development of computer systems and their associated languages and software. It has also found applications in such areas of science as biology, biochemistry, pyschology, and others.

In Sections 1.6−1.10, we review some basic results of automata and language theory. We are indebted to [103].

Definition 1.6.1 *A six-tuple $M = (Q, X, Y, f, g, s)$ is called a **finite-state machine** if Q, X, and Y are finite nonempty sets, $f : Q \times X \rightarrow Q$, $g : Q \times X \rightarrow Y$, and $s \in Q$.*

The elements of Q are called states. The elements of X and Y are called **input** and **output symbols**, respectively. The functions f and g are called the **state transition** and **output functions**, respectively. The state s is called the **initial state**.

Example 1.6.2 *Let $Q = \{q_0, q_1\}$, $X = \{a, b\}$, and $Y = \{0, 1\}$. Define the functions $f : Q \times X \rightarrow Q$ and $g : Q \times X \rightarrow Y$ as described in Table 1.1.*

	f		g	
$Q \setminus X$	a	b	a	b
q_0	q_0	q_1	0	1
q_1	q_1	q_1	1	0

Table 1.1

Then $M = (Q, X, Y, f, g, q_0)$ is a finite-state machine. The interpretation of Table 1.1 is as follows:

$$
\begin{aligned}
f(q_0, a) &= q_0 & g(q_0, a) &= 0 \\
f(q_0, b) &= q_1 & g(q_0, b) &= 1 \\
f(q_1, a) &= q_1 & g(q_1, a) &= 1 \\
f(q_1, b) &= q_1 & g(q_1, b) &= 0
\end{aligned}
$$

The next-state and output functions can also be defined by a transition diagram.

Example 1.6.3 *We draw the transition diagram for the finite-state machine of Example 1.6.2. See Figure 1.1.*

*The transition diagram is known to be what is called a **digraph** in graph theory. The vertices are the states. The initial state is indicated by an arrow as shown. If the finite-state machine is in state q and inputting x causes output y and a move to state q', a directed edge is drawn from the vertex q to the vertex q' and labelled x/y. The transition diagram of Figure 1.1 is* *obtained.*

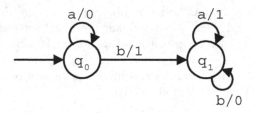

Figure 1.1

If S is a set such that $1 \leq |S| < \infty$, let S^* denote the free monoid generated by S with operation, concatenation. The elements of S^* are called **strings** or **words**. The empty string, Λ, is identity of S^*. Let $S^+ = S^* \backslash \{\Lambda\}$.

Definition 1.6.4 *Let $M = (Q, X, Y, f, g, q)$ be a finite-state machine. An*

input string *for M is a string over X. The string*

$$y_1 \ldots y_n$$

*is the **output string** for M corresponding to the input string*

$$x = x_1 \ldots x_n$$

if there exists states $q_0 \ldots q_n \in Q$ such that

$$\begin{aligned}
q_0 &= q & \\
q_i &= f(q_{i-1}, x_i) & \text{for } i = 1, \ldots, n_i \\
y_i &= g(q_{i-1}, x_i) & \text{for } i = 1, \ldots, n.
\end{aligned}$$

Example 1.6.5 *We now determine the output string corresponding to the input string (read from left to right)*

$$aababba \qquad\qquad (1.1)$$

for the finite-state machine of Example 1.6.2. We start in the state q_0. The symbol a is inputted and a 0 is outputted. We stay at q_0. The symbol a is again inputted and once again we remain in state q_0. Next the symbol b is inputted, 1 is outputted, and the machine goes to state q_1. Continuing this process until the last a is inputted, we obtain 0011001 as the corresponding output sequence.

1.7 Finite-State Automata

A **finite-state automaton** is a special kind of finite-state machine. It is of special interest due to its relationship to languages as shown in Section 1.10.

Definition 1.7.1 *A **finite-state automaton** $A = (Q, X, Y, f, g, s)$ is a finite-state machine such that the set of output symbols Y is $\{0, 1\}$ and where the current state determines the last output. Those states for which the last output was 1 are called **accepting states**.*

Example 1.7.2 *The transition diagram of the finite-state machine A is defined by the table. The initial state is q_0. A is a finite-state automation and the set of accepting states is $\{q_1, q_2\}$.*

	f		g	
$Q \backslash X$	a	b	a	b
q_0	q_0	q_1	0	1
q_1	q_0	q_2	0	1
q_2	q_0	q_2	0	1

The transition diagram is shown in Figure 1.2. If the finite-state machine is in state q_0, the last output was 0. If the machine is in either state q_1 or q_2, the last output was 1. Consequently, A is a finite-state automaton. The accepting states are q_1 and q_2.

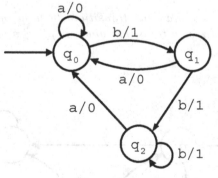

Figure 1.2

We see from Example 1.7.2 that the finite-state machine defined by a transition diagram is a finite-state automaton if the set of output symbols is $\{0, 1\}$ and if, for each state q, all incoming edges to q have the same output label.

The transition diagram of a finite-state automaton is often drawn with the accepting states in double circles and with the output symbols omitted.

If we draw the transition diagram of Figure 1.2 in this way, we obtain the transition diagram of Figure 1.3.

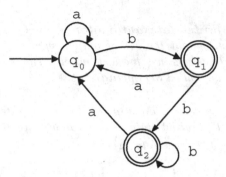

Figure 1.3

In Figure 1.3, accepting states are in double circles and output symbols are omitted.

Example 1.7.3 *We now draw the transition diagram of the finite-state automaton of Figure 1.4 as a transition diagram of a finite-state machine.*

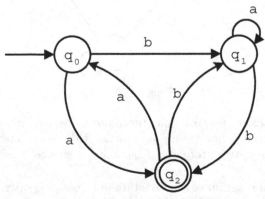

Figure 1.4

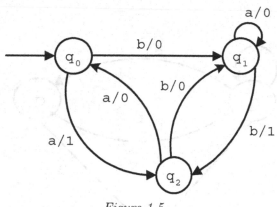

Figure 1.5

Since q_2 is an accepting state, we label all its incoming edges with output 1 as in Figure 1.5. Since the states q_0 and q_1 are not accepting, all their incoming edges are labelled with output 0. The transition diagram of Figure 1.5 is obtained.

An alternative to Definition 1.7.1 can be obtained by regarding a finite-state automaton A as consisting of

(1) a finite set Q of *states*,

(2) a finite set X of *input symbols*,

(3) a *next-state function* f from $Q \times X$ into Q,

(4) a subset A of Q of *accepting states*,

(5) an *initial state* $q \in Q$.

In this case, we write $A = (I, S, f, A, q)$.

Example 1.7.4 *The transition diagram of the finite-state automaton $A = (Q, X, f, A, q)$, where $Q = \{q_0, q_1, q_2\}$, $X = \{a, b\}$, $A = \{q_2\}$, $q = q_0$, and f is given by the following table,*

	f	
$Q \backslash X$	a	b
q_0	q_1	q_0
q_1	q_2	q_0
q_2	q_2	q_0

is shown in Figure 1.6.

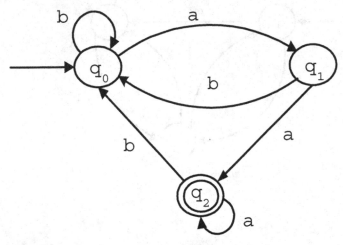

Figure 1.6

If a string is input to a finite-state automaton, we end at either an accepting or a nonaccepting state. The status of this final state determines whether the string is accepted by the finite-state automaton.

Definition 1.7.5 *Let $A = (Q, X, f, \mathcal{A}, q)$ be a finite-state automaton. Let $x = x_1 \cdots x_n$ be a string over X. If there exists states $q_0, \dots, q_n$ satisfying*
 (1) $q_0 = q$;
 (2) $f(q_{i-1}, x_i) = q_i$ for $i = 1, ..., n$;
 (3) $q_n \in \mathcal{A}$;
then x is said to be accepted by A. The empty string is accepted if and only if $q \in \mathcal{A}$. Let $Ac(A)$ denote the set of strings accepted by A. Then A is said to accept $Ac(A)$.

Let $x = x_1 \cdots x_n$ be a string over X. Define states $q_0, \dots, q_n$ by conditions (1) and (2) above. Then the (directed) path $(q_0, \dots, q_n)$ is called the **path representing** x in A.

It follows from Definition 1.7.5 that if the path P represents the string x in a finite-state automaton A, then A accepts x if and only if P ends at an accepting state.

The next two examples illustrate design problems.

Example 1.7.6 *We design a finite-state automaton that accepts precisely those strings over $\{a, b\}$ that contain no a's.*
 We use two states:
 A : There exists an a.
 B : There does not exist an a.

The state B is the initial state and the only accepting state. The finite-state automaton in Figure 1.7 correctly accepts the empty string.

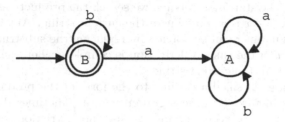

Figure 1.7

Example 1.7.7 *We design a finite-state automaton that accepts precisely those strings over* $\{a, b\}$ *that contain an odd number of a's.*

We use two states:

E : An even number of a's are in the string.

O : An odd number of a's are in the string.

The initial state is E and the accepting state is O. We obtain the transition diagram shown in Figure 1.8.

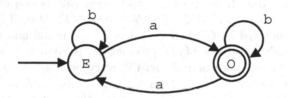

Figure 1.8

A finite-state automaton is essentially an algorithm to decide whether or not a given string is accepted.

Definition 1.7.8 *Finite-state automata A and A' are called* **equivalent** *if* $Ac(A) = Ac(A')$.

Example 1.7.9 *It can be shown that the finite-state automata of Figures 1.7 and 1.9 are equivalent.*

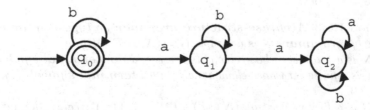

Figure 1.9

1.8   Languages and Grammars

A derivation within a grammar proceeds as follows: Starting with the current string with the starting symbol s, we search the productions for a rule whose left-hand side x is a substring of the current string. Any rule that is matched in this manner can be applied by replacing the substring with the right-hand side y of the rule. A derivation is complete when there are only terminal symbols in the current string.

Grammars are classified according to the form of the production rules used (Chomsky hierarchy). These grammars are sometimes described as follows: Unrestricted or type 0 grammars have no restrictions on the form of the rules. Context-sensitive or type 1 grammars require that if $x \rightarrow y$ is a production, then the length of the string x, denoted by $|x|$, must be not greater than the length $|y|$ of the string y. A grammar is said to be context-free or type 2 if every production is of the form $A \rightarrow y$, where $A \in N$. A grammar is said to be regular (left linear) if all its productions are of the form $A \rightarrow B$, $A \rightarrow b$, or $A \rightarrow bB$, where $A, B \in N$ and $b \in T$. It is sometimes useful to write a grammar in a particular form. Two forms for context-free grammars have been commonly used: Chomsky normal form and Greibach normal form. A context-free grammar is said to be in **Chomsky normal form** if every production rule is one of the form $A \rightarrow BC$ or $A \rightarrow a$, where $A, B, C \in N$ and $a \in T$. Also, if $\Lambda \in L(G)$, the language generated by G, then $s \rightarrow \Lambda$ is a production and s does not appear on the right-hand side of any production. A context-free grammar is said to be in **Greibach normal form** if every production rule is of the form $A \rightarrow az$, where $A \in N$, $a \in T$, and $z \in N^*$. Also, if $\Lambda \in L(G)$, then $s \rightarrow \Lambda$ is a production and s does not appear on the right-hand side of any production.

Definition 1.8.1 *Let A be a finite set. A (**formal**) **language** L over A is a subset of A^*, the set of all strings over A.*

Example 1.8.2 *Let $A = \{a, b\}$. The set L of all strings over A containing an odd number of a's is a language over A. As we saw in Example 1.7.7, L is precisely the set of strings over A accepted by the finite-state automaton of Figure 1.8.*

One method to define a language is to give a list of rules that the language must obey.

Definition 1.8.3 *A **phrase-structure grammar** or **type 0 grammar** (or, simply, **grammar**) G is a 4-tuple $G = (N, T, P, s)$, where*
    *(1) N is a finite set whose elements are called **nonterminal symbols**,*
    *(2) T is a finite set whose elements are called **terminal symbols**, where $N \cap T = \emptyset$,*
    *(3) P is a finite subset of $[(N \cup T)^* \backslash T^*] \times (N \cup T)^*$, called the **set of productions**,*

*(4) $s \in N$ and is called the **starting symbol**.*

Example 1.8.4 *Let $N = \{s, S\}$, $T = \{a, b\}$, and $P = \{s \to bs, s \longrightarrow aS, S \longrightarrow bS, S \longrightarrow b\}$. Then $G = (N, T, P, s)$ is a grammar.*

Given a grammar G, a language $L(G)$ can be constructed from G by using the productions to derive the strings that make up $L(G)$. We begin with the starting symbol and then repeatedly use productions until a string of terminal symbols is obtained. The language $L(G)$ is the set of all such strings of terminals.

Definition 1.8.5 *Let $G = (N, T, P, s)$ be a grammar. If $z \longrightarrow w$ is a production and $xzy \in (N \cup T)^*$, then xwy is said to be **directly derivable** from xzy. In this case, we write*

$$xzy \Longrightarrow xwy.$$

*If $z_i \in (N \cup T)^*$ for $i = 1, \dots, n$, and z_{i+1} is directly derivable from z_i for $= 1, \dots, n-1$, we say that z_n is **derivable** from z_1 and write*

$$z_1 \Longrightarrow z_n.$$

We call

$$z_1 \Longrightarrow z_2 \Longrightarrow \cdots \Longrightarrow z_n$$

*the **derivation** of z_n (from z_1). By convention, any element of $(N \cup T)^*$ is derivable from itself.*

The **language generated** by G, written $L(G)$, consists of all strings over T derivable from s.

Example 1.8.6 *Let G be the grammar of Example 1.8.4. The string $abSbb$ is directly derivable from $aSbb$, written*

$$aSbb \Longrightarrow abSbb,$$

where only the production $S \longrightarrow bS$ is used.
     The string $bbab$ is derivable from s, written

$$s \Longrightarrow bbab.$$

The derivation is

$$s \Longrightarrow bs \Longrightarrow bbs \Longrightarrow bbaS \Longrightarrow bbab.$$

The only derivations from s are given as follows:

$$s \quad \Longrightarrow \quad bs$$
$$\vdots$$
$$\Longrightarrow \quad b^n s \qquad\qquad n \geq 0$$
$$\Longrightarrow \quad b^n a S$$
$$\vdots$$
$$\Longrightarrow \quad b^n a b^{m-1} S$$
$$\Longrightarrow \quad b^n a b^m \qquad n \geq 0, \quad m \geq 1.$$

Hence $L(G)$ consists of the strings over $\{a, b\}$ containing precisely one a that end with at least one b.

Grammars are classified according to the types of productions that define the grammars.

Definition 1.8.7 *Let G be a grammar.*
(1) If every production is of the form

$$zAw \longrightarrow zvw, \qquad where\ z, w \in (N \cup T)^*, \quad A \in N, \qquad (1.2)$$
$$v \in (N \cup T)^* \backslash \{\Lambda\},$$

*then G is called a **context-sensitive** (or **type 1**) **grammar**.*
(2) If every production is of the form

$$A \to v,\ where\ A \in N,\ v \in (N \cup T)^*, \qquad (1.3)$$

*then G is called a **context-free** (or **type 2**) **grammar**.*
(3) If every production is of the form

$$A \to x\ or\ A \to xB\ or\ A \to \Lambda\ \ where\ A, B, \in N, \quad x \in T,$$

*then G is called a **regular** (or **type 3**) **grammar**.*

The definition of a context-sensitive grammar comes from the normal form of a more general definition of such grammars.

According to (1.2), in a context-sensitive grammar, A may be replaced by v if A is in the context of x and y. In a context-free grammar, (1.2) states that A may be replaced by v anytime. A regular grammar has especially simple substitution rules: A nonterminal symbol is replaced by a terminal symbol, by a terminal symbol followed by a nonterminal symbol, or by the null string.

It is important to note that a regular grammar is a context-free grammar and that a context-free grammar with no productions of the form $A \to \Lambda$ is a context-sensitive grammar.

Some definitions allow x to be replaced by a string of terminals in Definition 1.8.7(3). However, it can be shown that the two definitions produce the same languages.

Let G be a context-free grammar. A derivation $w_0 \Rightarrow w_1 \Rightarrow \ldots \Rightarrow w_n$ is called a **leftmost derivation** if for $i = 0, 1, \ldots, n$, $w_i = xAy$, $w_{i+1} = xzy$, and $A \to z$ is a production, where $x \in T^*$, $A \in N$, $z \in L(G)$, and $y \in (N \cup T)^*$. G is called **ambiguous** if it generates an element of $L(G)$ by two or more distinct left derivations.

Example 1.8.8 *Consider the grammar G defined as follows:*

$$T = \{a, b, c\}, \quad N = \{s, A, B, C, D, E\}$$

with productions

$$
\begin{array}{llll}
s \to aAB, & s \to aB, & A \to aAC, & A \to aC, \\
B \to Dc, & D \to b, & CD \to CE, & CE \to DE, \\
DE \to DC, & Cc \to Dcc, & &
\end{array}
$$

and starting symbol s. Then G is context-sensitive. For example, the production $CE \to DE$ ($\Lambda CE \to \Lambda DE$) allows C to be replaced by D if C is followed by E and the production $Cc \to Dcc$ ($\Lambda Cc \to \Lambda Dcc$) allows C to be replaced by Dc if C is followed by c.

DC can be derived from CD since

$$CD \Longrightarrow CE \Longrightarrow DE \Longrightarrow DC.$$

The string $a^3 b^3 c^3$ is in $L(G)$, since we have

$$
\begin{array}{llllll}
s & \Longrightarrow & aAB & \Longrightarrow & aaACB & \Longrightarrow & aaaCCDc \\
  & \Longrightarrow & aaaDCCc & \Longrightarrow & aaaDCDcc & \Longrightarrow & aaaDDCcc \\
  & \Longrightarrow & aaaDDDccc & \Longrightarrow & aaabbbccc.
\end{array}
$$

It follows that

$$L(G) = \{a^n b^n c^n \mid n = 1, 2, \ldots\}.$$

Definition 1.8.9 *A language L is **context sensitive** (respectively, **context-free, regular**) if there is a context-sensitive (respectively, context-free, regular) grammar G with $L = L(G)$.*

Example 1.8.10 *By Example 1.8.8, the language*

$$L(G) = \{a^n b^n c^n \mid n = 1, 2, \ldots\}$$

is context-sensitive. In [96, p.127], it is shown that there is no context-free grammar G with $L = L(G)$. Thus L is not context-free.

Example 1.8.11 *Consider the grammar G defined as follows*

$$T = \{a, b\}, \qquad N = \{s\},$$

with productions

$$s \rightarrow asb, \qquad s \rightarrow ab$$

and starting symbol s. Then G is context-free. The only derivations of s are

$$s \implies asb$$
$$\vdots$$
$$\implies a^{n-1}sb^{n-1}$$
$$\implies a^{n-1}abb^{n-1} = a^n b^n.$$

Hence $L(G)$ consists of the strings over $\{a, b\}$ of the form $a^n b^n$, $n = 1, 2 \ldots$. This language is context-free. In Section 1.10, we show that $L(G)$ is not regular.

It follows from Examples 1.8.10 and 1.8.11 that the set of context-free languages that do not contain the null string is a proper subset of the set of context-sensitive languages and that the set of regular languages is a proper subset of the set of context-free languages. It also follows that there are languages that are not context-sensitive.

Example 1.8.12 *The grammar G defined in Example 1.8.4 is regular. Hence the language*

$$L(G) = \{b^n ab^m \mid n = 0, 1, \ldots ; m = 1, 2, \ldots \}$$

it generates is regular.

Definition 1.8.13 *Grammars G and G′ are called **equivalent** if $L(G) = L(G′)$.*

1.9   Nondeterministic Finite-State Automata

In this section and Section 1.10, we show that regular grammars and finite-state automata are essentially the same in that either is a specification of a regular language. We next illustrate how a finite-state automaton can be converted to a regular grammar.

Example 1.9.1 *We show how to write the regular grammar given by the finite-state automaton of Figure 1.8.*

Let the terminal symbols be the input symbols a, b. Let the nonterminal symbols be the states E and O. The initial state E becomes the starting symbol. *The productions correspond to the directed edges. If there is an edge labeled x from S to R, we create the production*

$$S \rightarrow xR.$$

Hence we obtain the productions

$$E \rightarrow bE, \quad E \rightarrow aO, \quad O \rightarrow aE, \quad O \rightarrow bO. \qquad (1.4)$$

If S is an accepting state, we include the production

$$S \rightarrow \Lambda.$$

In this example, we obtain the additional production

$$O \rightarrow \Lambda. \qquad (1.5)$$

Then the grammar $G = (N, T, P, E)$, where $N = \{O, E\}$, $T = \{a, b\}$, and P consists of the productions (1.4) and (1.5), generates the language $L(G)$. $L(G)$ is the same as the set of strings accepted by the finite-state automaton of Figure 1.8.

Theorem 1.9.2 *Let A be a finite-state automaton given by a transition diagram. Let s be the initial state. Let T denote the set of input symbols and let N denote the set of states. Define productions*

$$S \rightarrow xR$$

if there is an edge labeled x from S to R, and

$$S \rightarrow \Lambda$$

if S is an accepting state. Let G be the grammar

$$G = (N, T, P, s).$$

Then G is regular and the set of strings accepted by A is equal to $L(G)$.

Proof. Clearly, G is regular. We first show that $Ac(A) \subseteq L(G)$. Let $x \in Ac(A)$. If x is the empty string, then s is an accepting state. In this case, G contains the production

$$s \rightarrow \Lambda.$$

The derivation

$$s \Longrightarrow \Lambda \qquad (1.6)$$

yields $x \in L(G)$.

Suppose $x \in Ac(A)$ and x is not the empty string. Then $x = x_1 \cdots x_n$, where $x_i \in T$, $i = 1, 2, \ldots, n$. Since x is accepted by A, there is a path $(s, S_1, \ldots, S_n)$, where S_n is an accepting state, with edges successively labeled $x_1, \ldots, x_n$. It follows that G contains the productions

$$s \rightarrow x_1 S_1$$
$$S_{i-1} \rightarrow x_i S_i \text{ for } i = 2, \ldots, n.$$

Since S_n is an accepting state, G also contains the production

$$S_n \rightarrow \Lambda.$$

From the derivation

$$
\begin{aligned}
s &\Longrightarrow x_1 S_1 \\
&\Longrightarrow x_1 x_2 S_2 \\
&\vdots \\
&\Longrightarrow x_1 \cdots x_n S_n \\
&\Longrightarrow x_1 \cdots x_n,
\end{aligned}
\tag{1.7}
$$

we have that $x \in L(G)$. Thus $Ac(A) \subseteq L(G)$.

Suppose that $x \in L(G)$. If x is the empty string, x must result from the derivation (1.6) since a derivation that starts with any other production would yield a nonempty string. Thus the production $s \rightarrow \Lambda$ is in the grammar. Therefore, s is an accepting state in A. It follows that $x \in Ac(A)$.

Now suppose $x \in L(G)$ and x is not the empty string. Then $x = x_1 \ldots x_n$, where $x_i \in T$, $i = 1, 2, \ldots, n$. It follows that there is a derivation of the form (1.7). If, in the transition diagram, we start at s and have the path $(s, S_i, \ldots, S_n)$, we can generate the string x. The last production used in (10.4.4) is $S_n \rightarrow \Lambda$. Thus the last state reached is an accepting state. Therefore, $x \in Ac(A)$ and so $L(G) \subseteq Ac(A)$. Hence $Ac(A) = L(G)$. ∎

We now consider the reverse situation, i.e., given a regular grammar G, we want to construct a finite-state automaton A such that $L(G)$ is precisely the set of strings accepted by A. The procedure of Theorem 1.9.2 cannot simply be reversed as the next example shows.

Example 1.9.3 *Consider the regular grammar defined as follows:*

$$T = \{a, b\}, \quad N = \{s, q\}$$

with productions

$$s \rightarrow bs, \; s \rightarrow aq, \;\; q \rightarrow bq, \;\; q \rightarrow b$$

and starting symbol s.

Let the nonterminal symbols be the states with s as the initial state. For each production of the form

$$S \rightarrow xR,$$

we draw an edge from state S to state R and label it x. The productions

$$s \to bs, \quad s \to aq, \quad q \to bq$$

give the graph shown in Figure 1.10.

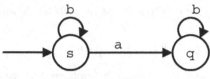

Figure 1.10

The production $q \to b$ is equivalent to the two productions

$$q \to bF, \quad F \to \Lambda,$$

where F is an additional nonterminal symbol. The productions

$$s \to bs, \quad s \to aq, \quad q \to bq, \quad q \to bF$$

give the graph shown in Figure 1.11. From the production

$$F \to \Lambda,$$

it follows that F should be an accepting state.

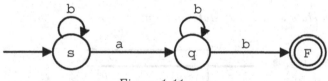

Figure 1.11

However, the graph of Figure 1.11 is not a finite-state automaton for several reasons. Vertex C has no outgoing edge labeled a and vertex F has no outgoing edges at all. Also, vertex C has two outgoing edges labeled b. A diagram like that of Figure 1.11 yields a different kind of automaton called a **nondeterministic finite-state automaton.** The word "nondeterministic" is used since if in Figure 1.11 the automaton is in state C and b is input, a choice of next states exists, i.e., the automaton either remains in state C or goes to state F.

Definition 1.9.4 *A **nondeterministic finite-state automaton** A is a 5-tuple $A = (Q, X, f, A, s)$, where*
    (1) Q is a finite set of states,
    (2) X is a finite set of input symbols,
    (3) f is a next-state function from $Q \times X$ into $\mathcal{P}(Q)$,
    (4) A is a subset of Q, the accepting states,
    (5) $s \in Q$ is the initial state.

The main difference between a nondeterministic finite-state automaton and a finite-state automaton is that in a finite-state automaton the next-state function maps a state, input pair to a uniquely defined state, while in a nondeterministic finite-state automaton the next-state function maps a state, input pair to a set of states.

Example 1.9.5 *For the nondeterministic finite-state automaton of Figure 1.11, we have*

$$Q = \{s, q, F\}, \quad X = \{a, b\}, \quad \mathcal{A} = \{F\}.$$

The initial state is s and the next-state function f is given by

$Q \backslash X$	a	b
s	$\{q\}$	$\{s\}$
q	$\emptyset$	$\{q, F\}$
F	$\emptyset$	$\emptyset$

The transition diagram of a nondeterminate finite-state automaton is drawn similarly to that of a finite-state automaton. An edge from state q to each state in the set $f(q, x)$ is drawn and each is labeled x.

Example 1.9.6 *Consider the nondeterministic finite-state automaton given as follows:*

$$Q = \{s, q, p\}, \quad X = \{a, b\}, \quad \mathcal{A} = \{q, p\}$$

with initial state s and next-state function

$Q \backslash X$	a	b
s	$\{s, q\}$	$\{p\}$
q	$\emptyset$	$\{q\}$
p	$\{q, p\}$	$\emptyset$

Its transition diagram is shown in Figure 1.12. It is the transition diagram of the nondeterministic automaton of this example.

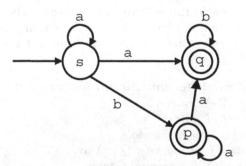

Figure 1.12

A string x is accepted by a nondeterministic finite-state automaton A if there is some path representing x in the transition diagram of A beginning at the initial state and ending in an accepting state.

Definition 1.9.7 *Let $A = (Q, X, f, \mathcal{A}, q)$ be a nondeterministic finite-state automaton. The empty string is **accepted** by A if and only if $s \in \mathcal{A}$. If $x = x_1 \cdots x_n$ is a nonempty string over X and there exist states $q_0, \ldots, q_n$ such that*
  (1) $q_0 = q$;
  (2) $q_i \in f(q_{i-1}, x_i)$ for $i = 1, \ldots, n$;
  (3) $q_n \in \mathcal{A}$;
*then x is said to be **accepted** by A. Let $Ac(A)$ denote the set of strings accepted by A.*

*If A and A' are nondeterministic finite-state automata and $Ac(A) = Ac(A')$, then A and A' are said to be **equivalent**.*

*If $x = x_1 \cdots x_n$ is a string over X and there exists states $q_0, \ldots, q_n$ satisfying conditions (1) and (2), then the path $(q_0, \ldots, q_n)$ is called a **path representing** x in A.*

Example 1.9.8 *The string*

$$x = bbabb$$

is accepted by the nondeterministic finite-state automaton of Figure 1.11. This follows since the path (s, s, s, q, q, F), which ends at an accepting state, represents x. The path $P = (s, s, s, q, q, q)$ also represents x, but P does not end at an accepting state. However, the string x is still accepted since there is at least one path representing x that ends at an accepting state. A string y is not accepted if no path represents y or every path representing y ends at a nonaccepting state.

Example 1.9.9 *The string $x = aabaabbb$ is accepted by the nondeterministic finite-state automaton of Figure 1.12. The string $x = abba$ is not accepted.*

Theorem 1.9.10 *Let $G = (N, T, P, s)$ be a regular grammar. Let $X = T$, $Q = N \cup \{F\}$, where $F \notin N \cup T$,*

$$f(q, x) = \{q' \mid q \to xq' \in P\} \cup \{F \mid q \to x \in P\},$$

and $\mathcal{A} = \{F\} \cup \{q \mid q \to \Lambda \in P\}$.

Then the nondeterministic finite-state automaton $A = (Q, X, f, \mathcal{A}, s)$ accepts precisely the strings $L(G)$. ∎

In the next section we show that given a nondeterministic finite-state automaton A, there is a finite-state automaton that is equivalent to A.

1.10   Relationships Between Languages and Automata

In Section 1.9, we showed that if A is a finite-state automaton, then there is a regular grammar G such that $L(G) = Ac(A)$. A partial converse is given by Theorem 1.9.10 where it is shown that if G is a regular grammar, then there is a nondeterministic finite-state automaton A such that $L(G) = Ac(A)$. In this section, we show that if G is a regular grammar, then there is a finite-state automaton A such that $L(G) = Ac(A)$. This result follows from Theorem 1.9.10 once it has been established that any nondeterministic finite-state automaton can be converted to an equivalent finite-state automaton. The method is illustrated by the following example.

Example 1.10.1 *We construct a finite-state automaton equivalent to the nondeterministic finite-state automaton of Figure 1.11.*

The set of input symbols is the same. The states consist of all subsets

$$\emptyset, \ \{s\}, \{q\}, \{F\}, \{s,q\}, \{s,F\}, \{q,F\}, \{s,q,F\}$$

of the original set $Q = \{s, q, F\}$ of states. The initial state is $\{s\}$. The accepting states are all subsets of Q that contain an accepting state of the original nondeterministic finite-state automaton, namely

$$\{F\}, \ \{s,F\}, \ \{q,F\}, \ \{s,q,F\}.$$

Let $X, Y \subseteq Q$. An edge is drawn from X to Y and labeled x if $X = \emptyset = Y$ or if

$$\cup_{q \in x} f(q,x) = Y.$$

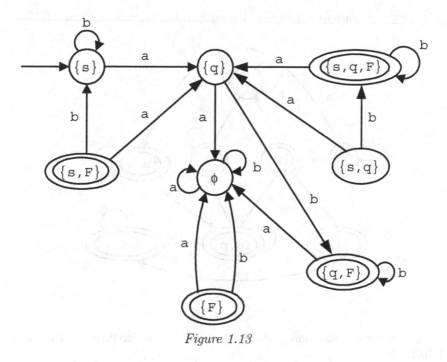

Figure 1.13

The finite-state automaton of Figure 1.13 *is obtained. The states*

$$\{s, F\}, \ \{s, q\}, \quad \{s, q, F\}, \ \{F\}$$

can never be reached and are thus deleted. This yields the simplified, equivalent finite-state automaton of Figure 1.14.

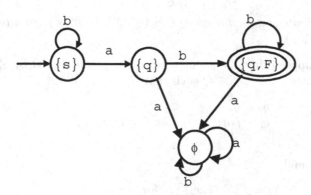

Figure 1.14

Example 1.10.2 *The finite-state automaton equivalent to the nondeter-*

ministic finite-state automaton of Example 1.9.6 is given in Figure 1.15.

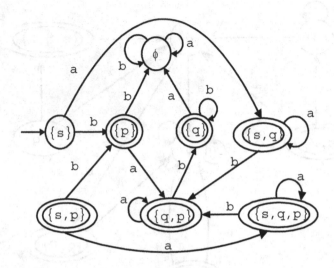

Figure 1.15

The following theorem justifies the method of Examples 1.10.1 and 1.10.2.

Theorem 1.10.3 *Let $A = (Q, X, f, \mathcal{A}, s)$ be a nondeterministic finite-state automaton. Let*

 (1) $Q' = \mathcal{P}(Q)$,

 (2) $X' = X$,

 (3) $s' = \{s\}$,

 (4) $\mathcal{A}' = \{S \subseteq Q \mid S \cap \mathcal{A} \neq \emptyset\}$,

 (5) $f'(S, x) = \begin{cases} \emptyset & \text{if } S = \emptyset \\ \cup_{q \in S} f(q, x) & \text{if } S \neq \emptyset. \end{cases}$

Then the finite-state automaton $A' = (Q', X', f', \mathcal{A}', s')$ is equivalent to A.

Proof. Suppose that the string $x = x_1 \cdots x_n$ is accepted by A. Then there exist states $q_0, \ldots, q_n \in Q$ such that

$$q_0 = q;$$
$$q_i \in f(q_{i-1}, x_i) \quad \text{for } i = 1, \ldots, n;$$
$$q_n \in \mathcal{A}.$$

Set $Y_0 = \{q_0\}$ and

$$Y_i = f'(Y_{i-1}, x_i) \quad \text{for } i = 1, \ldots, n.$$

Since

$$Y_1 = f'(Y_0, x_1) = f'(\{q_0\}, x_1) = f(q_0, x_1),$$

we have that $q_1 \in Y_1$. Now

$$q_2 \in f(q_1, x_2) \subseteq \cup_{S \in Y_1} f(S, x_2) = f'(Y_1, x_2) = Y_2.$$

Also,

$$q_3 \in f(q_2, x_3) \subseteq \cup_{S \in Y_2} f(S, x_3) = f'(Y_2, x_3) = Y_3.$$

We see that the argument may be formalized, using induction, to show that $q_n \in Y_n$. Since q_n is an accepting state in A, Y_n is an accepting state in A'. Thus in A', we have

$$
\begin{aligned}
f'(q', x_1) &= f'(Y_0, x_1) &= Y_1 \\
f'(Y_1, x_2) &= Y_2 \\
&\vdots \\
f'(Y_{n-1}, x_n) &= Y_n.
\end{aligned}
$$

Consequently, x is accepted by A'.

Now suppose that the string $x = x_1 \cdots x_n$ is accepted by A'. Then there exist subsets $Y_0, \ldots, Y_n$ of Q such that

$$
\begin{aligned}
Y_0 &= s' &= \{s\}, \\
f'(Y_{i-1}, x_i) &= Y_i &\text{for } i = 1, \ldots, n,
\end{aligned}
$$

and there exists a state $q_n \in Y_n \cap A$.

Since

$$q_n \in Y_n = f'(Y_{n-1}, x_n) = \cup_{S \in Y_{n-1}} f(S, x_n),$$

there exists $q_{n-1} \in Y_{n-1}$ such that $q_n \in f(q_{n-1}, x_n)$. Similarly, since

$$q_{n-1} \in Y_{n-1} = f'(Y_{n-2}, x_{n-1}) = \cup_{S \in Y_{n-2}} f(S, x_{n-1}),$$

there exists $q_{n-2} \in Y_{n-2}$ such that $q_{n-1} \in f(q_{n-2}, x_{n-1})$. Continuing in this manner, we obtain

$$q_i \in Y_i \quad \text{for } i = 0, \ldots, n,$$

such that

$$q_i \in f(q_{i-1}, x_i) \quad \text{for } i = 1, \ldots, n.$$

In particular,

$$q_0 \in Y_0 = \{s\}.$$

Thus $q_0 = s$, the initial state in A. Since q_n is an accepting state in A, the string x is accepted by A. ∎

We can now state the following result.

Theorem 1.10.4 *A language L is regular if and only if there exists a finite-state automaton that accepts precisely the strings in L.*

Proof. This theorem follows from Theorems 1.9.2, 1.9.10, and 1.10.3.
∎

Example 1.10.5 *In the example, we determine a finite-state automaton A that accepts precisely the strings generated by the regular grammar G having productions*

$$s \to bs, s \to aq, q \to bq, q \to b.$$

The starting symbol is s, the set of terminal symbols is $\{a, b\}$, and the set of nonterminal symbols is $\{s, q\}$.

The nondeterministic finite-state automaton A' that accepts $L(G)$ is shown in Figure 1.11. A finite-state automaton equivalent to A' is shown in Figure 1.13 and an equivalent simplified finite-state automaton A is shown in Figure 1.14. The finite-state automaton A accepts precisely the strings generated by G.

We close this section by giving some applications of the methods and theory we have developed.

Example 1.10.6 *In this example, we show that the language*

$$L = \{a^n b^n \mid n = 1, 2, \dots\}$$

is not regular.

Suppose that L is regular. Then there exists a finite-state automaton A such that $Ac(A) = L$. Assume that A has k states. The string $x = a^k b^k$ is accepted by A. Let P be the path that represents x. Since there are k states, some state q is revisited on the part of the path representing a^k. Thus there is a cycle C, all of whose edges are labeled a, that contains q. We change the path P to obtain a path P' as follows. When we arrive at q in P, we traverse C. After returning to q on C, we continue on P to the end. If the length of C is j, the path P' represents the string $x' = a^{j+k} b^k$. Since P and P' end at the same state q' and q' is an accepting state, x' is accepted by A. This is impossible since x' is not of the form $a^n b^n$. Hence L is not regular.

Of course L in Example 1.10.6 can be shown not to be regular by the pumping lemma for regular languages. The statement of this lemma appears in the Exercises. We also consider the pumping lemma for context-free languages in the Exercises. For more details, the reader is referred to [96].

Example 1.10.7 *Let L be the set of strings accepted by the finite-state automaton A of Figure* 1.16.

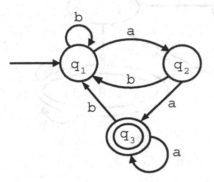

Figure 1.16

We construct a finite-state automaton that accepts the set of strings

$$L^R = \{x_n \ldots x_1 \mid x_1 \ldots x_n \in L\}.$$

We convert A to a finite-state automaton that accepts L^R. The string $x = x_1 \ldots x_n$ is accepted by A if there is a path P in A representing x that starts at q_1 and ends at q_3. If we start at q_3 and trace P in reverse, we end at q_1 and process the edges in order $x_n, \ldots, x_1$. Thus it suffices to reverse all arrows in Figure 1.16 and make q_3 the starting state and q_1 the accepting state (see Figure 1.17). The result is a nondeterministic finite-state automaton that accepts L^R.

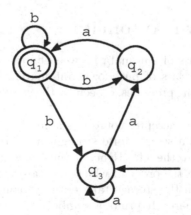

Figure 1.17

After finding an equivalent finite-state automaton and eliminating the unreachable states, the equivalent finite-state automaton of Figure 1.18 is obtained.

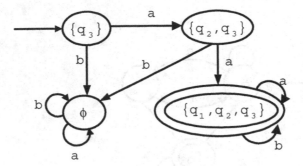

Figure 1.18

Let L be the set of strings accepted by a finite-state automaton A with more than one final state. A procedure to construct L^R is as follows:

First construct a nondeterministic finite-state automaton equivalent to A with one accepting state. Then use the method of Example 1.10.7. Examples can be found in [103].

A finite state machine remembers which state it is in. In this sense, we say that it has internal memory. If external memory is allowed on which the machine can read and write, more powerful machines can be defined. By allowing the machine to scan the input string in either direction and by allowing the machine to alter the input string, other enhancements can be obtained. One can then characterize the classes of machines that accept context-free languages, context-sensitive languages, and languages generated by phase-structure grammars.

1.11   Pushdown Automata

The most important class of automata between finite-state machines and Turing machines is the class of pushdown automata. Their operation is related to many computing processes, especially the analysis and translation of artificial languages.

A pushdown automata acceptor comprises a finite-state control, a semi-infinite input tape, and a semi-infinite storage tape. It is not permitted to move its input head to the left. Hence it must examine the symbols of its input tape strictly in the order in which they are written on the tape. The machine starts with the storage tape entirely blank. The symbol # is inscribed in every square. It prints a symbol on the tape each time it moves the storage head to the right. The machine reads a symbol from the storage tape each time it moves the storage head to the left. Information written to the right of the head cannot be retrieved since it is overwritten when the head again moves right.

The storage tape string may be arbitrarily long. Since it can affect the behavior of the automaton, a pushdown acceptor has unbounded memory. However, the information most recently written into the memory must be the first to be retrieved. This form of limited-access storage mechanism is called a **pushdown stack** since it implements a "last-in, first-out" retrieval rule. This restricted form of unbounded memory results in a language-defining ability intermediate between that of finite-state machines and Turing machines.

Definition 1.11.1 *A pushdown acceptor (PDA) is a six-tuple*

$$M = (Q, X, U, P, I, F),$$

where

    Q is a finite set of control states,
    X is a finite input alphabet,
    U is a finite stack alphabet,
    P is the program of M,
    $I \subseteq Q$ is a set of initial states,
    $F \subseteq Q$ is a set of final or accepting states.
    The program P is a finite sequence of instructions, each taking one of the following forms:

$$q] \ scan \ (a, q'),$$
$$q] \ write \ (u, q'),$$
$$q] \ read \ (u, q'),$$

where $q, q' \in Q$, $a \in X$, and $u \in U$. In each case, q is the label of the instruction and q' is the successor state. Each state in Q labels at most one type of instruction, read, write, or scan.

    *If $q \in Q$, $x \in X^*$, and $y \in U^*$, then (q, x, y) is called a **configuration** of M. The string y is called the **stack** and the symbol under the stack head the **top stack symbol**.*

A configuration (q, x, y) completely describes the total state of a pushdown acceptor at some point in its analysis of an input tape. The control unit is in state q; the prefix x of the entire input string w has been scanned, and the input head is positioned at the last symbol of x; the stack y is the content of the storage tape, and the stack head is positioned at the last symbol of y.

The next definition specifies how the execution of instructions by a pushdown acceptor takes it from one configuration to another. The scan instructions read successive symbols from the input tape, write instructions load symbols into the stack, and read instructions retrieve symbols from the stack.

Definition 1.11.2 *Let M be a pushdown acceptor with program P. Suppose that the string $w \in X^*$ is written on the input tape. Instructions of M's program are* **applicable** *in configurations according to the following rules:*

*(1) An instruction q] **scan** (a, q') applies in any configuration (q, x, y) if xa is a prefix of w. In executing this instruction, M moves its input head one square to the right, observes the symbol a inscribed therein, and enters state q'. A scan move is represented by the notation*

$$(q, x, y) \xrightarrow{S} (q, xa, y).$$

*(2) An instruction q] **write** (u, q') applies in any configuration (q, x, y). In executing this instruction, M moves the stack head one square to the right, prints the symbol u therein, and goes to state q'. A write move is represented by the notation*

$$(q, x, y) \xrightarrow{W} (q, x, yu).$$

*(3) An instruction q] **read** (u, q') applies in any configuration (q, x, y) in which $y = y'u$. In executing this instruction, M observes the symbol u under the stack head, moves the stack head one square to the left, and goes to state q'. A read move is represented by the notation*

$$(q, x, y'u) \xrightarrow{R} (q', x, y').$$

A move sequence,

$$(q_0, x_0, y_0) \to (q_1, x_1, y_1) \to \ldots \to (q_k, x_k, y_k),$$

where each move is a scan, read, or write move, is sometimes shortened to

$$(q_0, x_0, y_0) \Rightarrow (q_k, x_k, y_k).$$

The operation of a pushdown acceptor begins with the control in an initial state and the heads positioned at the initial sharps of their tapes. The machine passes through a sequence of configurations. Each configuration results from the execution of an instruction applicable in the preceding configuration. Operation continues until a configuration is reached in which there is no applicable instruction. If at any point all symbols of some input string x have been scanned, the stack is empty, and the control unit is in a final state, then the string x is accepted by M.

Definition 1.11.3 *An **initial configuration** of a pushdown acceptor M is any configuration of the form (q, Λ, Λ), where q is an initial state of M. A **final configuration** of M is any configuration of the form (q', x, Λ), where q' is a final state of M and x is a prefix of the string written on M's input tape. The string x is **accepted** by M if M has a move sequence*

$$(q, \Lambda, \Lambda) \Rightarrow (q', x, \Lambda), \quad q \in I, \ q' \in F.$$

*The **language recognized** by M is the set of accepted strings.*

Example 1.11.4 *Figure* 1.19 *shows the program and the state diagram of a pushdown acceptor* M_{cm} *with alphabets* $X = \{a, b, c\}$ *and* $U = \{a, b\}$.

Program of M_{cm}

```
1] scan   (a,2) (b,3) (c,4)
2] write  (a,1)
3] write  (b,1)
4] scan   (a,5) (b,6)
5] read   (a,4)
6] read   (b,4)
```

State Diagram of M_{cm}

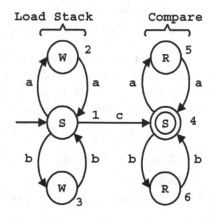

Figure 1.19 Example of a pushdown acceptor.

Instructions having the same label are sometimes combined. For example, the notion

$$1] \ scan \ (a, 2)(b, 3)(c, 4)$$

is used in place of

$$1] \ scan \ (a, 2)$$
$$1] \ scan \ (b, 3)$$
$$1] \ scan \ (c, 4).$$

Unless we specify otherwise, state 1 is the initial state.

    The nodes of the state diagram represent states of the control unit, and each node is inscribed with $S, R,$ or W according to the type of instruction labeled by the state. The initial states and the accepting states are identified in the same manner as for finite-state automata.

    At most one instruction is applicable to any configuration of M_{cm}. Hence the behavior of the machine is uniquely determined by its input string w.

The machine begins operation in state 1 and passes through the following two stages: in the load stack stage it copies into the stack the portion of w up to the symbol c; in the compare stage it matches the stack symbols against the remaining symbols of the input.

If the portion of w following the letter c is exactly the reverse of the string loaded into the stack, then M will empty the storage tape immediately after scanning the final letter of w, leaving M_{cm} in the accepting configuration $(4, w, \Lambda)$. Hence M_{cm} accepts each string $w = xcx^R$, where $x \in \{a, b\}^*$. For example, the move sequence by which M_{cm} accepts $w = abcba$ is as follows:

$$
\begin{aligned}
(1, \Lambda, \Lambda) &\xrightarrow{S} (2, a, \Lambda) \xrightarrow{W} (1, a, a) \\
&\xrightarrow{S} (3, ab, a) \xrightarrow{W} (1, ab, ab) \\
&\xrightarrow{S} (4, abc, ab) \\
&\xrightarrow{S} (6, abcb, ab) \xrightarrow{R} (4, abcb, a) \\
&\xrightarrow{S} (5, abcba, a) \xrightarrow{R} (4, abcba, \Lambda) \quad [accept].
\end{aligned}
$$

If a letter scanned by M_{cm} in state 4 does not match the last letter of the stack, or if symbols remain in the stack after all of w has been scanned, M_{cm} will stop with a nonempty stack and reject the input. The rejection of $w = abcb$ is illustrated by the following move sequence:

$$
\begin{aligned}
(1, \Lambda, \Lambda) &\xrightarrow{S} (2, a, \Lambda) \xrightarrow{W} (1, a, a) \\
&\xrightarrow{S} (3, ab, a) \xrightarrow{W} (1, ab, ab) \\
&\xrightarrow{S} (4, abc, ab) \\
&\xrightarrow{S} (5, abca, ab) \quad [stop\ and\ reject].
\end{aligned}
$$

Thus

$$
L_{cm} = \{xcx^R \mid x \in \{a, b\}^*\}
$$

is the language recognized by M_{cm}. The language is known as the mirror-image language with center marker, or the center-marked palindrome language. It is generated by the following context-free grammar:

$$
\begin{aligned}
G_{cm} : \quad & S \to A \\
& A \to aAa \\
& A \to bAb \\
& A \to c.
\end{aligned}
$$

It is known that a context-free grammar exists for any language recognized by a pushdown acceptor. In fact, the equivalence of the class of context-free languages and the class of languages recognized by these acceptors can be established.

The pushdown acceptor M_{cm} is deterministic since there is never a choice of move for any configuration. A slight modification of the language

L_{cm} requires a nondeterministic pushdown acceptor. Consider, for example, the language

$$L_{mi} = \{xx^R \mid x \in \{a,b\}^*\}.$$

This language is known simply as the **mirror-image language**. There is no special symbol in sentences of L_{mi} to indicate when a pushdown acceptor should switch from a load-stack mode to a compare mode as there is in the language L_{cm}. We note that there exists a pushdown acceptor for L_{mi} by modifying M_{cm} to obtain an automaton with an accepting move sequence for every string of L_{mi}. Rather than waiting for a symbol c to be scanned, the machine M_{mi} described in Figure 1.20 is allowed to switch to its compare mode whenever it writes a scanned symbol into its stack. This type of behavior must be allowed since there is no way for the machine to determine when it has scanned the first half of a mirror-image string. The machine must be permitted to "guess" after each symbol scanned whether it should or should not switch to compare mode.

```
1] scan   (a,2) (b,3)
2] write  (a,1) (a,4)
3] write  (b,1) (b,4)
4] scan   (a,5) (b,6)
5] read   (a,4)
6] read   (b,4)
```

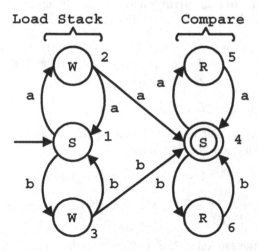

Figure 1.20 Example of a nondeterministic pushdown acceptor.

1.12   Exercises

1. Prove Theorem 1.5.1.

2. Prove Theorem 1.5.4.

3. Design a finite state machine that outputs 1 if an even number of 1's have been input; otherwise outputs 0.

4. Design a finite state machine that outputs 1 when it sees 101 and thereafter; otherwise outputs 0.

5. Design a finite state machine that outputs 1 when it sees the first 0 and until it sees another 0; thereafter, outputs 0; in all other cases outputs 0.

6. Show that there is no finite state machine that receives a bit string and outputs 1 whenever the number of 1's input equals the number of 0's input and outputs 0 otherwise.

7. Let L be a finite set of strings over $\{0,1\}$. Show that there is a finite-state automaton that accepts L.

8. Let L_i be a finite set of strings accepted by the finite-state automaton $A_i = (Q_i, X, f_i, \mathcal{A}_i, s_i)$, $i = 1, 2$. Let $A = (Q_1 \times Q_2, X, f, \mathcal{A}, s)$, where

$$
\begin{aligned}
f((q_1, q_2), x) &= (f_1(q_1, x), f_2(q_2, x)), \\
\mathcal{A} &= \{(q_1, q_2) \mid q_1 \in \mathcal{A}_1 \text{ and } q_2 \in \mathcal{A}_2\}, \\
s &= (s_1, s_2).
\end{aligned}
$$

Show that $A_c(A) = L_1 \cap L_2$.

9. Let L_i be a finite set of strings accepted by the finite-state automaton $A_i = (Q_i, X, f_i, \mathcal{A}_i, s_i)$, $i = 1, 2$. Let $A = (Q_1 \times Q_2, X, f, \mathcal{A}, s)$, where

$$
\begin{aligned}
f((q_1, q_2), x) &= (f_1(q_1, x), f_2(q_2, x)), \\
\mathcal{A} &= \{(q_1, q_2) \mid q_1 \in \mathcal{A}_1 \text{ or } q_2 \in \mathcal{A}_2\}, \\
s &= (s_1, s_2).
\end{aligned}
$$

Show that $A_c(A) = L_1 \cup L_2$.

10. Let G be a grammar and let Λ denote the empty string. Show that if every production is of the form

$$A \rightarrow x \text{ or } A \rightarrow xB \text{ or } A \rightarrow \Lambda,$$

where $A, B \in N$, $x \in T^* \backslash \{\Lambda\}$, then there is a regular grammar G' with $L(G) = L(G')$.

11. Show that the language $L = \{a^n b^n c^k \mid n, k = 1, 2, \ldots\}$ is a context-free language.

12. Design a nondeterministic finite-state automaton that accepts strings over $\{a, b\}$ having the property of having each b preceded and followed by an a.

13. Write a regular grammar that generates the strings of Exercise 12.

14. Locate the path in representing the string $x = aabaabbb$ in Example 1.9.9.

15. Show that the string $x = abba$ is not accepted by the nondeterministic finite state automaton of Figure 1.12.

16. Prove Theorem 1.9.10.

17. Show that the set $L = \{x_1 \ldots x_n \mid x_1 \ldots x_n = x_n \ldots x_1\}$ of strings over $\{a, b\}$ is not a regular language.

18. Show that if L_1 and L_2 are regular languages over X and S is the set of all strings over X, then each of $S \backslash L_1$, $L_1 \cup L_2$, L_1^+, and $L_1 L_2$ is a regular language.

19. Show, by example, that there are context-free languages L_1 and L_2 such that $L_1 \cap L_2$ is not context-free.

20. Determine the validity of the following statement: If L is a regular language, then so is $L' = \{u^n \mid u \in L, n = 1, 2, \ldots\}$.

21. Determine the possible move sequence of M_{mi} of Example 1.11.4 for the input string $aabbaa$.

22. (Pumping lemma for regular languages.) Let L be a regular language. Prove that $\exists n \in \mathbb{N}$ such that $\forall z \in L$, $|z| \geq n$ implies that $\exists u, v, w$ such that $z = uvw$, $|uv| \leq n$, $|v| \geq 1$, and $\forall i \in \mathbb{N} \cup \{0\}$, $uv^i w \in L$. Moreover, n is not larger than the number of states of the smallest finite-state automaton accepting L.

23. (Pumping lemma for context-free languages.) Let L be a context-free language. Prove that $\exists n \in \mathbb{N}$ such that $\forall z \in L$, $|z| \geq n$ implies that $\exists u, v, w, x, y$ such that $z = uvwxy$, $|vx| \geq 1$, $|vwx| \leq n$, and $\forall i \in \mathbb{N} \cup \{0\}$, $uv^i wx^i y \in L$.

Chapter 2

Max-Min Automata

2.1 Max-Min Automata

This chapter is based on the results of [193] and [203]. A general formulation of automata is given which is similar to that of sequential machines introduced in [212]. We show how deterministic, nondeterministic, and probabilistic automata fit into this formulation.

The max-min automata include both deterministic and nondeterministic automata as special cases. Moreover, its state transition function as well as initial distribution may be interpreted as grade of membership functions of fuzzy subsets.

Despite its generality, most of the basic concepts of existing automata, e.g., equivalences, reduction, behaviors, etc. may be carried over to max-min automata with appropriate modifications. Moreover, results related to these basic concepts can be generalized to max-min automata.

There is a strong resemblance between max-min automata and probabilistic automata [165]. Hence one can introduce similar concepts definable for probabilistic automata, but not present in deterministic and nondeterministic automata.

Recall that effective procedure means algorithm. It is within this context that we use the terminology "effectively constructed."

2.2 General Formulation of Automata

Definition 2.2.1 *A **pseudoautomaton** A is a 5-tuple* (Q, X, μ, F, σ) *such that the following properties hold:*

(1) Q is a finite nonempty set (states),

(2) X is a finite nonempty set (inputs),

(3) μ is a function from $Q \times X \times Q \times T$ into $[0, 1]$ where T is a subset of $\mathbb{R}$ (state transition function with members of T representing time),

(4) F is a subset of Q (final states),
(5) σ is a function from Q × T into [0, 1] (initial distribution).

If $T = \mathbb{N}$, the set of all natural numbers, then A is said to be **discrete**. If μ and σ are independent of T, then A is said to be **stationary**.

Elements of X are called input symbols. Finite sequences of input symbols are called input strings, words, or tapes. Recall that the empty tape Λ has the property that $x\Lambda = x = \Lambda x$ for every tape x. The collection of all input tapes is denoted by X^*. Let x be a tape. Then $|x|$ denotes the length of x. The tape y is a k-suffix of the tape x if $x = zy$ for some tape z and $|y| = k$.

Definition 2.2.2 *An **automaton** A^* is a 5-tuple (Q, X, μ^*, F, σ) where Q, X, F, and σ are the same as that given above. μ^* is the same as μ given in Definition 2.2.1 with X^* replacing X.*

To each automaton A^* there is associated in a natural manner a pseudoautomaton A, i.e., μ is the restriction of μ^* to $Q \times X \times Q \times T$.

A pseudo automaton is said to be of class $\mathcal{C}(C)$ if μ and σ satisfy a set C of constraints.

Definition 2.2.3 *If the pseudo automaton A associated with the automaton A^* is of class $\mathcal{C}(C)$ and if A^* can be obtained from A by a rule of extension R, consistent with C, which extends μ uniquely into μ^*, then A^* is said to belong to class $\mathcal{C}(C, R)$.*

2.3   Classes of Automata

We now present several classes of automata.

1. Probabilistic automata $\mathcal{C}(C_P, R_P)$:

The constraints in C_P are as follows: for every $q \in Q$, $u \in X$, and $n \in \mathbb{N}$,

$$\sum_{q' \in Q} \mu(q, u, q', n) = 1 \quad \text{and} \quad \sum_{q' \in Q} \sigma(q', n) = 1.$$

The rule of extension R_P is as follows: μ^* is defined recursively on $|x|$, $x \in X^*$,

$$\mu^*(q, \Lambda, q', n) = \begin{cases} 1 & \text{if} \quad q = q' \\ 0 & \text{if} \quad q \neq q' \end{cases}$$

$$\mu^*(q, xu, q', n) = \sum_{q'' \in Q} \mu^*(q, x, q'', n)\mu(q'', u, q', n + |x|).$$

2. Deterministic automata $\mathcal{C}(C_D, R_D)$: C_D is the same as C_P with the additional constraints that $\text{Im}(\mu) \subseteq \{0, 1\}$ and $\text{Im}(\sigma) \subseteq \{0, 1\}$. R_D is the same as R_P.

3. Max-min automata $\mathcal{C}(C_A, R_A) : C_A$ is an empty set, i.e., there are no constraints.

$$\mu^*(q, \Lambda, q', n) = \begin{cases} 1 & \text{if } q = q' \\ 0 & \text{if } q \neq q' \end{cases}$$

$$\mu^*(q, xu, q', n) = \bigvee_{q' \in Q} \wedge \{\mu^*(q, x, q'', n), \mu(q'', u, q', n + |x|)\}.$$

An interesting subclass of $\mathcal{C}(C_A, R_A)$ is the class of all max-min automata satisfying the constraints

$$\bigvee_{q' \in Q} \mu(q, u, q', n) = 1 \text{ and } \bigvee_{q' \in Q} \sigma(q', n) = 1$$

for all $q \in Q$, $u \in X$, and $n \in \mathbb{N}$. Automata belonging to this subclass are called **restricted max-min automata**.

4. Nondeterministic automata $\mathcal{C}(C_N, R_N)$: The only constraint in C_N is that $\text{Im}(\mu) \subseteq \{0, 1\}$ and $\text{Im}(\sigma) \subseteq \{0, 1\}$. R_N is the same as R_A.

5. Min-max automata $\mathcal{C}(C_I, R_I) : C_I$ is empty.

$$\mu^*(q, \Lambda, q', n) = \begin{cases} 0 & \text{if } q = q' \\ 1 & \text{if } q \neq q' \end{cases}$$

$$\mu^*(q, xu, q', n) = \bigwedge_{q'' \in Q} \vee \{\mu^*(q, x, q'', n), \mu(q'', u, q', n + |x|)\}.$$

6. Composite automata $\mathcal{C}(C_C, R_C) : C_C$ is empty. $\mu_C^* = a\mu_A^* + b\mu_I^*$, where a and b are nonnegative real numbers such that $a + b = 1$, and the subscripts C, A, and I refer to $\mathcal{C}(C_C, R_C)$, $\mathcal{C}(C_A, R_A)$, and $\mathcal{C}(C_I, R_I)$, respectively.

7. Max-product automata $\mathcal{C}(C_T, R_T) : C_T$ is empty.

$$\mu^*(q, \Lambda, q', n) = \begin{cases} 1 & \text{if } q = q' \\ 0 & \text{if } q \neq q' \end{cases}$$

$$\mu^*(q, xu, q', n) = \bigvee_{q'' \in Q} \mu(q, x, q'', n)\mu(q'', u, q', n + |x|).$$

One may consider, in a similar manner, restricted nondeterministic, min-max, composite, and max-product automata.

8. Fixed automata $\mathcal{C}(C_X, R_X) : C_X$ is empty.

$$\mu^*(q, ux, q', t) = \mu(q, u, q', t).$$

2.4   Behavior of Max-Min Automata

In this and the following sections, we confine our discussion to stationary max-min automata. The symbol A^*, with or without subscripts, always denotes a max-min automaton. All automata are assumed to have the same input set X. Thus an automaton is represented by (Q, μ^*, F, σ) with X being suppressed. We note that max-min automata include both deterministic and nondeterministic automata as special cases.

The **response function** of A^* is defined to be

$$r\mu(A^*, x, q) = \vee\{\sigma(q') \wedge \mu^*(q', x, q) \mid q' \in Q\}.$$

We simply write $r\mu(x, q)$ rather than $r\mu(A^*, x, q)$ when A^* is understood. If $\sigma(q)$ is concentrated at q_k, i.e., $\sigma(q_k) = 1$ and $\sigma(q) = 0$ for $q \neq q_k$, then

$$r\mu(x, q) = \mu^*(q_k, x, q).$$

Let $q \in Q$ and $0 \leq c < 1$. Then q is called an **accessible state** of A^* with threshold c if there exists $x \in X^*$ such that $r\mu(x, q) > c$. Let $\overline{Q}_c$ denote the set of all such states. The state q is an accessible state of A^* if and only if for all $0 \leq c < 1$, q is an accessible state of A^* with threshold c. Let $\overline{Q}$ denote the set of all such states. Clearly, $\overline{Q} = \cap\{\overline{Q}_c \mid 0 \leq c < 1\}$.

Theorem 2.4.1 *If q is an accessible state of A^* with threshold c, then there exists $x \in X^*$ such that $r\mu(x, q) > c$ and $|x| < |Q|$.*

Proof. By hypothesis, there exists $x_0 \in X^*$ such that $r\mu(x_0, q) > c$. Let $x_0 = u_{i_1} u_{i_2} \ldots u_{i_n}$. If $n < |Q|$, the result follows trivially. Thus assume $n \geq |Q|$. From the definition of $r\mu$, there exist $q_{i_0}, q_{i_1}, q_{i_2}, \ldots, q_{i_n}$ such that $\sigma(q_{i_0}) > c$ and $\mu^*(q_{i_{k+1}}, u_{i_k}, q_{i_k}) > c$ for $k = 1, 2, \ldots, n$, where $q_{i_n} = q$. Since $n \geq |Q|$, at least two of $q_{i_0}, q_{i_1}, \ldots, q_{i_n}$ are equal, say $q_{i_j} = q_{i_l}$, where $j < l$. Let $x_1 = u_{i_1} \ldots u_{i_j} u_{i_{l+1}} \ldots u_{i_n}$. Then $r\mu(x_1, q) > c$. If $|x_1| \geq |Q|$, repeat the same argument. In this manner, the desired x is obtained. ∎

Theorem 2.4.2 *If q is an accessible state of A^*, then there exists $x \in X^*$ such that $r\mu(x, q) = 1$ and $|x| < |Q|$.*

Proof. By Theorem 2.4.1, there exists $x_n \in X^*$ such that $r\mu(x_n, q) > 1 - (1/n)$ and $|x_n| < |Q|$ for $n = 1, 2, \ldots$. Since the set of all x, where $|x| < |Q|$, is finite, there exists a j such that $r\mu(x_j, q) > 1 - (1/n)$ for infinitely many n. Thus $r\mu(x_j, q) = 1$. ∎

The above theorems provide a practical method for determining $\overline{Q}_c$ and $\overline{Q}$.

A generalization of the concept of accessible states is the concept of accessible initial distributions. An initial distribution σ is said to be an

accessible initial distribution of A^* **with threshold** c if there exists $x \in X^*$ such that for every $q \in Q$, $\sigma(q) > c \Leftrightarrow r\mu(x, q) > c$. σ is an **accessible initial distribution** of A^* if for every $0 \leq c < 1$, σ is an accessible initial distribution of A^* with threshold c. σ is called a **strictly accessible initial distribution** of A^* if there exists $x \in X^*$ such that $\sigma(q) = r\mu(x, q)$ for all $q \in Q$. Clearly, if σ is a strictly accessible initial distribution of A^*, then σ is an accessible initial distribution of A^*.

Example 2.4.3 Let $Q = \{q_1, q_2\}$ and $X = \{u\}$. Define $\mu : Q \times X \times Q \to [0, 1]$ and $\sigma : Q \to [0, 1]$ as follows:

$$
\begin{aligned}
\mu(q_1, u, q_1) &= 0 \\
\mu(q_1, u, q_2) &= 1 \\
\mu(q_2, u, q_1) &= \tfrac{1}{2} \\
\mu(q_2, u, q_2) &= 0 \\
\sigma(q_1) &= \tfrac{1}{2} \\
\sigma(q_2) &= 1.
\end{aligned}
$$

Now

$$
r\mu(A^*, u^n, q_2) = \vee \{\sigma(q') \wedge \mu(q', u^n, q_2) \mid q' \in Q\} = \frac{1}{2}
$$

and $r\mu(A^*, u^n, q_1) = \tfrac{1}{2}$ $\forall n \in \mathbb{N}$. *It follows that* $\overline{Q_c} = Q$ *for* $0 \leq c < \tfrac{1}{2}$ *and* $\overline{Q_c} = \emptyset$ *for* $\tfrac{1}{2} \leq c < 1$. *Hence* q_1 *and* q_2 *are accessible with threshold* c *for* $0 \leq c < \tfrac{1}{2}$.

Example 2.4.4 Let Q, X, μ, and σ be defined as in Example 2.4.3. Then

$$
\begin{aligned}
\sigma(q_1) &= \tfrac{1}{2}, & r\mu(A^*, u^n, q_1) &= \tfrac{1}{2}, \\
\sigma(q_2) &= 1, & r\mu(A^*, u^n, q_2) &= \tfrac{1}{2}.
\end{aligned}
$$

Hence it follows that σ *is not accessible with threshold* $\tfrac{1}{2}$. *In fact,* σ *is accessible with threshold* c *for* $0 \leq c < \tfrac{1}{2}$, *but not accessible with threshold* c *for* $\tfrac{1}{2} \leq c < 1$. *Clearly,* σ *is not strictly accessible.*

If instead of defining $\sigma(q_2) = 1$, *we define* $\sigma(q_2) = \tfrac{1}{2}$, *then* σ *is strictly accessible.*

Theorem 2.4.5 *If* σ *is an accessible initial distribution of* A^* *with threshold* c, *then there exists* $x \in X^*$ *such that* $|x| < 2^{|Q|}$ *and for every* $q \in Q$, $\sigma(q) > c$ *if and only if* $r\mu(x, q) > c$.

Proof. The proof is similar to that of Theorem 2.4.1. ∎

It is sometimes convenient to consider what are called in [193] max-min tables and max-min pre-tables. **Max-min tables** are max-min automata without an initial distribution. **Max-min pre-tables** are max-min tables without the final states. (We use different terminology in subsequent chapters.) They are well defined since the rule of extension R_A does not depend

on the initial distribution nor the final states. The symbols Ω^* and θ^*, with
or without subscripts, are used to denote max-min tables and pre-tables,
respectively. A max-min automaton having Ω^* as its max-min table and σ
its initial distribution is denoted by (Ω^*, σ). If σ is concentrated at q, we
also write (Ω^*, q) for (Ω^*, σ). A max-min table having θ^* as its pre-table
and F its final states is denoted by (θ^*, F).

Let $\Omega_1^* = (Q_1, \mu_1^*, F_1)$ and $\Omega_2^* = (Q_2, \mu_2^*, F_2)$, where $Q_1 \cap Q_2 = \emptyset$. The
direct sum of Ω_1^* and Ω_2^* is a max-min table $\Omega_1^* + \Omega_2^*$ defined by (Q, μ^*, F),
where $Q = Q_1 \cup Q_2$, $F = F_1 \cup F_2$, and for every $u \in X$,

$$\mu(q, u, q') = \begin{cases} \mu_1(q, u, q') & \text{if } q, q' \in Q_1, \\ \mu_2(q, u, q') & \text{if } q, q' \in Q_2, \\ 0 & \text{otherwise.} \end{cases}$$

Let $A_1^* = (Q_1, \mu_1^*, F_1, \sigma_1)$ and $A_2^* = (Q_2, \mu_2^*, F_2, \sigma_2)$. The **direct product** of A_1^* and A_2^* is the max-min automaton $A_1^* \times A_2^*$ defined by (Q, μ^*, F, σ)
where $Q = Q_1 \times Q_2$, $F = F_1 \times F_2$, and for every $q_1, q_1' \in Q_1, q_2, q_2' \in Q_2, x \in X$,

$$\begin{aligned} \mu((q_1, q_2), x, (q_1', q_2')) &= \mu_1(q_1, x, q_1') \wedge \mu_2(q_2, x, q_2') \\ \sigma((q_1, q_2)) &= \sigma_1(q_1) \wedge \sigma_2(q_2). \end{aligned}$$

The behavior of A^* with threshold c is defined to be the set

$$B(A^*, c) = \{x \in X^* \mid \bigvee_{q \in F} r\mu(x, q) > c\}.$$

Theorem 2.4.6 $B(A^*, c) \neq \emptyset$ *if and only if there exists* $x \in B(A^*, c)$ *such
that* $|x| < |Q|$.

Proof. The proof follows from Theorem 2.4.1 ∎

Theorem 2.4.7 $B(A_1^*, c) \cap B(A_2^*, c) \neq \emptyset$ *if and only if there exists* $x \in
B(A_1^*, c) \cap B(A_2^*, c)$ *such that* $|x| < |Q_1| \cdot |Q_2|$.

Proof. Let $A^* = A_1^* \times A_2^*$. The desired result follows from Theorem
2.4.6 and the fact that $x \in B(A^*, c)$ if and only if $x \in B(A_1^*, c) \cap B(A_2^*, c)$.
∎

Theorem 2.4.8 *For every* A^* *and* $0 \le c < 1$*, there exists a nondeterministic automaton* $\widetilde{A}_c^*$ *such that* $B(A^*, c) = B(\widetilde{A}_c^*, 0)$.

Proof. Let $A^* = (Q, \mu^*, F, \sigma)$ and define $\widetilde{A}_c^* = (Q, \widetilde{\mu}_c^*, F, \widetilde{\sigma}_c)$ by $\forall q, q' \in
Q$ and $\forall u \in X$,

$$\widetilde{\mu}_c(q, u, q') = \begin{cases} 1 & \text{if } \mu(q, u, q') > c \\ 0 & \text{if } \mu(q, u, q') \le c \end{cases}$$

$$\tilde{\sigma}_c(q) = \begin{cases} 1 & \text{if} \quad \sigma(q) > c \\ 0 & \text{if} \quad \sigma(q) \le c. \end{cases}$$

Clearly, $\tilde{A}_c^*$ is nondeterministic. Moreover, $B(A^*, c) = B(\tilde{A}_c^*, 0)$. ■

Let $Y \subseteq X^*$. Then Y is said to be **regular** if $Y = B(D^*, 0)$ for some deterministic automaton D^*.

Theorem 2.4.9 *For every* A^* *and* $0 \le c < 1$, $B(A^*, c)$ *is regular.*

Proof. Let $A^* = (Q, \mu^*, F, \sigma)$ and define $D^* = (Q_D, \mu_D^*, F_D, \sigma_D)$ where $Q_D = \mathcal{P}(Q)$, the power sets of Q,

$$F_D = \{Q' \in Q_D \mid Q' \cap F \neq \emptyset\}$$

and for every $u \in X$, $Q' \subseteq Q$, $Q'' \subseteq Q$,

$$\mu_D(Q', u, Q'') = \begin{cases} 1 & \text{if } Q'' = \{q \in Q \mid \mu(q', u, q) > c \text{ for some } q' \in Q'\} \\ 0 & \text{otherwise} \end{cases}$$

$$\sigma_D(Q') = \begin{cases} 1 & \text{if } Q' = \{q \in Q \mid \sigma(q) > c\} \\ 0 & \text{otherwise.} \end{cases}$$

Clearly D^* is deterministic. Moreover, $B(A^*, c) = B(D^*, 0)$. ■

The interested reader should note the "equivalence" between max-min sequential machines [194] and restricted max-min automata. The proof is similar to that given in [197], where the equivalence between stochastic sequential machines and probabilistic automata was shown. One merely replaces summation and product by maximum and minimum, respectively. Due to this equivalence, max-min sequential machines and max-min automata behave similarly in many respects.

2.5 Equivalences and Homomorphisms of Max-Min Automata

Let $A^* = (Q, \mu^*, F, \sigma)$ be a max-min automaton, k a nonnegative integer, and $0 \le c < 1$. The k-**behavior** of A^* with threshold c is the set

$$B_k(A^*, c) = \{x \in X^* \mid \vee_{q \in F} r\mu(x, q) > c \text{ and } |x| \le k\}.$$

Clearly, $B(A^*, c) = \cup_{k=0}^{\infty} B_k(A^*, c)$.

We now define various types of equivalences.

1. $A_1^* \overset{c,k}{\sim} A_2^* \Leftrightarrow B_k(A_1^*, c) = B_k(A_2^*, c)$. Otherwise, we write $A_1^* \overset{c,k}{\not\sim} A_2^*$.

2. $A_1^* \overset{c}{\sim} A_2^* \Leftrightarrow A_1^* \overset{c,k}{\sim} A_2^*$ for $k = 0, 1, 2 \ldots$. Otherwise, we write $A_1^* \overset{c}{\not\sim} A_2^*$.

3. $A_1^* \overset{k}{\sim} A_2^* \Leftrightarrow A_1^* \overset{c,k}{\sim} A_2^*$ for all $0 \leq c < 1$. Otherwise, we write $A_1^* \overset{k}{\approx} A_2^*$.

4. $A_1^* \sim A_2^* \Leftrightarrow A_1^* \overset{c}{\sim} A_2^*$ for all $0 \leq c < 1$. Otherwise, we write $A_1^* \approx A_2^*$.

The following lemmas are immediate consequences of the above definitions.

Lemma 2.5.1 $A_1^* \overset{c}{\sim} A_2^*$ *if and only if* $B(A_1^*, c) = B(A_2^*, c)$. ■

Lemma 2.5.2 $A_1^* \overset{k}{\sim} A_2^*$ *if and only if* $\underset{q \in F}{\vee} r\mu(A_1^*, x, q) = \underset{q \in F}{\vee} r\mu(A_2^*, x, q)$ *for all* $x \in X^*$ *where* $|x| \leq k$. ■

Lemma 2.5.3 $A_1^* \sim A_2^*$ *if and only if* $\underset{q \in F}{\vee} r\mu(A_1^*, x, q) = \underset{q \in F}{\vee} r\mu(A_2^*, x, q)$ *for all* $x \in X^*$. ■

A^* is said to be **connected with threshold** c if $Q = \overline{Q}_c$. A^* is **connected** if for every $0 \leq c < 1$, A^* is connected with threshold c. Clearly, A^* is connected if and only if $Q = \overline{Q}$. A^* is **totally connected with threshold** c if every initial distribution of A^* is accessible with threshold c. A^* is **totally connected** if for every c, A^* is totally connected with threshold c. A^* is **strictly connected** if every initial distribution of A^* is accessible. Clearly, A^* is totally connected if and only if every initial distribution of A^* is accessible. Moreover, if A^* is strictly connected, then A^* is totally connected and connected.

Example 2.5.4 *Let* $A = (Q, X, \mu, \sigma)$, *where* $Q = \{q_1, q_2\}$, $X = \{u\}$, *and* $\mu : Q \times X \times Q \to [0, 1]$ *is defined as follows:* $\mu(q_1, u, q_2) = 1 = \mu(q_2, u, q_1)$ *and* $\mu(q_1, u, q_1) = 0 = \mu(q_2, u, q_2)$. *Then* $r\mu(A^*, uu, q_1) = (\mu^*(q_1, uu, q_1) \wedge \sigma(q_1)) \vee (\mu^*(q_2, uu, q_1) \wedge \sigma(q_2)) = (1 \wedge \sigma(q_1)) \vee (0 \wedge \sigma(q_2)) = \sigma(q_1)$ *and* $r\mu(A^*, uu, q_2) = (\mu^*(q_1, uu, q_2) \wedge \sigma(q_1)) \vee (\mu^*(q_2, uu, q_2) \wedge \sigma(q_2)) = (0 \wedge \sigma(q_1)) \vee (1 \wedge \sigma(q_2)) = \sigma(q_2)$. *Hence* σ *is accessible. Since* σ *is arbitrary,* A^* *is strictly connected.*

Theorem 2.5.5 *For every* A^* *and* $0 \leq c < 1$, *there exists a max-min automaton* $\overline{A}_c^*$ *that is connected with threshold* c *and is such that* $A^* \overset{c}{\sim} \overline{A}_c^*$.

Proof. Let $A^* = (Q, \mu^*, F, \sigma)$. Define $\overline{A}_c^* = (\overline{Q}_c, \overline{\mu}_c^*, \overline{F}_c, \overline{\sigma}_c)$ as follows: $\overline{F}_c = \overline{Q}_c \cap F$ and $\forall q, q' \in Q$ and $\forall u \in X$,

$$\overline{\mu}_c(q, u, q') = \begin{cases} \mu(q, u, q') & \text{if} \quad \mu(q, u, q') > c \\ 0 & \text{if} \quad \mu(q, u, q') \leq c \end{cases}$$

$$\overline{\sigma}_c(q) = \begin{cases} \sigma(q) & \text{if} \quad \sigma(q) > c \\ 0 & \text{if} \quad \sigma(q) \leq c. \end{cases}$$

Then $\overline{A}_c^*$ has the desired properties. ■

Theorem 2.5.6 *If* $(\Omega^*, \sigma) \overset{c,k-1}{\sim} (\Omega^*, \sigma')$ *and* $(\Omega^*, \sigma) \overset{c,k}{\not\sim} (\Omega^*, \sigma')$, *then for every* $j = 1, 2, \ldots, k-1$, *there exist* σ_j *and* σ'_j *such that* $(\Omega^*, \sigma_j) \overset{c,j-1}{\sim}$ (Ω^*, σ'_j) *and* $(\Omega^*, \sigma_j) \overset{c,j}{\not\sim} (\Omega^*, \sigma'_j)$.

Proof. By hypothesis, there exists $x_0 \in X^*$ such that $|x_0| = k$ and say $\vee_{q \in Q} r\mu(A'^* x_0, q) > c$, but $\vee_{q \in Q} r\mu(A'^*, x_0, q) \leq c$, where $A^* = (\Omega^*, \sigma)$ and $A'^* = (\Omega^*, \sigma')$. Let $x_0 = u_{i_1} u_{i_2} \ldots u_{i_k}$ and for $1 \leq j \leq k-1$, let $x_j = u_{i_1} u_{i_2} \ldots u_{i_r}$, where $r = k - j$. Define $\sigma_j(q) = r\mu(A^*, x_j, q)$ and $\sigma'_j(q) = r\mu(A'^*, x_j, q)$ for all $q \in Q$. Then the initial distributions σ_j and σ'_j have the desired properties. ∎

Two initial distributions σ_1 and σ_2 are said to be **indistinguishable with threshold** c if for every $q \in Q$, $\sigma_1(q) > c$ if and only if $\sigma_2(q) > c$. In symbols, $\sigma_1 \overset{c}{=} \sigma_2$. Clearly, $\sigma_1 = \sigma_2$ if and only if for every $0 \leq c < 1, \sigma_1 \overset{c}{=} \sigma_2$.

Given Ω^* and initial distributions σ_1 and σ_2 define:

1. $\sigma_1 \overset{c,k}{\sim} \sigma_2$ if and only if $(\Omega^*, \sigma_1) \overset{c,k}{\sim} (\Omega^*, \sigma_2)$. Otherwise, we write $\sigma_1 \overset{c,k}{\not\sim} \sigma_2$.

2. $\sigma_1 \overset{c}{\sim} \sigma_2$ if and only if $\sigma_1 \overset{c,k}{\sim} \sigma_2$ for $k = 0, 1, 2, \ldots$. Otherwise, we write $\sigma_1 \overset{c}{\not\sim} \sigma_2$.

3. $\sigma_1 \sim \sigma_2$ if and only if $\sigma_1 \overset{c}{\sim} \sigma_2$ for every $0 \leq c < 1$. Otherwise, we write $\sigma_1 \not\sim \sigma_2$.

If σ_1 and σ_2 are concentrated at q_1 and q_2, respectively, then we also write $q_1 \overset{c,k}{\sim} q_2$, $q_1 \overset{c}{\sim} q_2$, and $q_1 \sim q_2$ for $\sigma_1 \overset{c,k}{\sim} \sigma_2$, $\sigma_1 \overset{c}{\sim} \sigma_2$, and $\sigma_1 \sim \sigma_2$, respectively. Clearly, $\sigma_1 \overset{c}{=} \sigma_2$ implies $\sigma_1 \overset{c}{\sim} \sigma_2$.

Theorem 2.5.7 *If* $(\Omega^*, \sigma_1) \overset{c,m}{\sim} (\Omega^*, \sigma_2)$, *where* $m = 2^{|Q|} - 1$, *then* $(\Omega^*, \sigma_1) \overset{c}{\sim}$ (Ω^*, σ_2).

Proof. Let H denote the set of all initial distributions. Let P_k and P denote, respectively, the equivalence classes of H under $\overset{c,k}{\sim}$ and $\overset{c}{\sim}$. Clearly $|P_k| \leq |P_{k+1}| \leq P$ for every k. Since $\sigma_1 \overset{c}{=} \sigma_2$ implies $\sigma_1 \overset{c}{\sim} \sigma_2$, P is finite. In fact, $|P| \leq 2^{|Q|} = m+1$. By the Theorem 2.5.6, if $P_k = P_{k+1}$ for some k, then $P_k = P_{k+j} = P$ for all nonnegative integers j. Thus $P_m = P_{m+1} = P$ and the desired result holds. ∎

Corollary 2.5.8 *If* $(\Omega^*, \sigma_1) \overset{m}{\sim} (\Omega^*, \sigma_2)$ *where* $m = 2^{|Q|} - 1$, *then* $(\Omega^*, \sigma_1) \sim (\Omega^*, \sigma_2)$. ∎

Theorem 2.5.9 $A_1^* \overset{c}{\sim} A_2^*$ *if and only if* $A_1^* \overset{c,r}{\sim} A_2^*$, *where* $r = 2^{|Q_1| + |Q_2|} - 1$.

Proof. We assume without loss of generality that $Q_1 \cap Q_2 = \emptyset$. Let $A_1^* = (\Omega_1^*, \sigma)$, $A_2^* = (\Omega_2^*, \sigma_2)$, $\Omega^* = \Omega_1^* + \Omega_2^*$, and

$$\sigma'_1(q) = \begin{cases} \sigma_1(q) & \text{if } q \in Q_1 \\ 0 & \text{if } q \in Q_2 \end{cases}$$

$$\sigma_2'\,(q) = \begin{cases} \sigma_2(q) & \text{if } q \in Q_2 \\ 0 & \text{if } q \in Q_1. \end{cases}$$

Since $(\Omega^*, \sigma_1') \sim (\Omega_1^*, \sigma_1)$ and $(\Omega^*, \sigma_2') \sim (\Omega_2^*, \sigma_2)$, the desired result follows from Theorem 2.5.7. ■

Corollary 2.5.10 $A_1^* \sim A_2^*$ if and only if $A_1^* \overset{r}{\sim} A_2^*$, where $r = 2^{|Q_1|+|Q_2|} - 1$. ■

Let $A_1^* = (Q_2, \mu_2^*, F_2, \sigma_2)$. A **homomorphism** from A_1^* onto A_2^* with **threshold** c is a function f from Q_1 onto Q_2 such that for every $q', q'' \in Q_1$ and $u \in X$, the following conditions hold:

(1) $\mu_1(q', u, q'') > c$ if and only if $\mu_2(f(q'), u, f(q'')) > c$;

(2) $\sigma_1(q') > c$ if and only if $\sigma_2(f(q')) > c$;

(3) $q \in F_1$ implies $f(q) \in F_2$.

f is called a **strong homomorphism with threshold** c if f is a homomorphism with threshold c and $q \in F_1$ if and only if $f(q) \in F_2$. f is called an **isomorphism with threshold** c if and only if f is a strong homomorphism with threshold c that is one-one. f is a homomorphism (strong homomorphism, isomorphism) if and only if for every $0 \leq c < 1$, f is a homomorphism (strong homomorphism, isomorphism) with threshold c. The symbols $\overset{c}{\longrightarrow}, \overset{c}{\Longrightarrow}$, and $\overset{c}{\longleftrightarrow}$ will denote homomorphic, strong homomorphic, and isomorphic with threshold c, respectively. The symbols $\longrightarrow, \Longrightarrow$, and $\longleftrightarrow$ are similarly defined.

Theorem 2.5.11 $A_1^* \overset{c}{\Longrightarrow} A_2^*$ implies $A_1^* \overset{c}{\sim} A_2^*$.

Proof. Clearly, $B(A_1^*, c) \subseteq B(A_2^*, c)$. Let $x \in B(A_2^*, c)$. Then there exists $q \in F$ such that $r\mu(A_2^*, x, \varphi(q)) > c$. Since $A_1^* \overset{c}{\Longrightarrow} A_2^*$, $q \in F_1$. Thus $x \in B(A_1^*, c)$. ■

Corollary 2.5.12 $A_1^* \Longrightarrow A_2^*$ implies $A_1^* \sim A_2^*$. ■

Theorem 2.5.13 If A_1^* is totally connected with threshold c and $A_1^* \overset{c}{\longrightarrow} A_2^*$, then $A_1^* \overset{c}{\sim} A_2^*$ implies $A_1^* \overset{c}{\Longrightarrow} A_2^*$.

Proof. Let $q_0 \in F_1$ and σ the initial distribution concentrated at q_0. Since A_1^* is totally connected with threshold c, there exists $x \in X^*$ such that $r\mu(A_1^*, x, q_0) > c$ and $r\mu(A_2^*, x, f(q)) \leq c$ for $q \neq q_0$. Moreover, $x \in B(A_1^*, c) = B(A_2^*, c)$. Hence $f(q_0) \in F_2$. ■

Corollary 2.5.14 If A_1^* is totally connected and $A_1^* \longrightarrow A_2^*$, then $A_1^* \sim A_2^*$ implies $A_1^* \Longrightarrow A_2^*$. ■

2.6   Reduction of Max-Min Automata

Two max-min tables Ω_1^* and Ω_2^* are called **statewise equivalent with threshold** $c \Leftrightarrow$ for every $q \in Q_1$, there exists $q' \in Q_2$ such that $(\Omega_1^*, q) \overset{c}{\sim} (\Omega_2^*, q')$ and vice versa. In symbols, $\Omega_1^* \overset{c}{\sim} \Omega_2^*$. Otherwise, $\Omega_1^* \overset{c}{\not\sim} \Omega_2^*$. $\Omega_1^* \sim \Omega_2^* \Leftrightarrow$ for every $0 \le c < 1$, $\Omega_1^* \overset{c}{\sim} \Omega_2^*$. Otherwise, $\Omega_1^* \not\sim \Omega_2^*$. Ω_1^* and Ω_2^* are **distributionwise equivalent with threshold** $c \Leftrightarrow$ for every σ_1, there exists σ_2 such that $(\Omega_1^*, \sigma_1) \overset{c}{\sim} (\Omega_2^*, \sigma_2)$ and vice versa. In symbols, $\Omega_1^* \overset{c}{\approx} \Omega_2^*$. $\Omega_1^* \approx \Omega_2^* \Leftrightarrow$ for every $0 \le c < 1$, $\Omega_1^* \overset{c}{\approx} \Omega_2^*$.

A max-min table Ω^* is **statewise (distributionwise) minimal** with threshold c if it is not statewise (distributionwise) equivalent with threshold c to any max-min table with a fewer number of states. Ω^* is **statewise irreducible with threshold** c if $q', q'' \in Q$ such that $q' \overset{c}{\sim} q''$ implies $q' = q''$. Ω^* is **distributionwise irreducible with threshold** c if $\sigma_1 \overset{c}{\sim} \sigma_2$ implies $\sigma_1 \overset{c}{=} \sigma_2$. Ω^* is statewise (distributionwise) minimal (irreducible) if and only if for every $0 \le c < 1$, Ω^* is statewise (distributionwise) minimal (irreducible) with threshold c. Clearly, distributionwise irreducible (with threshold c) implies statewise irreducible (with threshold c).

Theorem 2.6.1 Ω^* *is statewise minimal with threshold* c *if and only if* Ω^* *is statewise irreducible with threshold* c.

Proof. Let $\Omega^* = (Q, \mu^*, F)$. Suppose Ω^* is not statewise irreducible with threshold c. Then there exist $q_0, q_1 \in Q$ such that $q_0 \ne q_1$ and $q_0 \overset{c}{\sim} q_1$. Define $\Omega_1^* = (Q_1, \mu_1^*, F_1)$ where $Q_1 = Q \backslash \{q_0\}$, $F_1 = F \backslash \{q_0\}$ and for every $q, q' \in Q_1$, $u \in X$,

$$\mu_1(q, u, q') = \mu(q, u, q') \text{ if } q' \ne q_1$$

$$\mu_1(q, u, q_1) = \vee \{\mu(q, u, q_0), \mu(q, u, q_1)\}.$$

Clearly, $\Omega^* \overset{c}{\sim} \Omega_1^*$, where $|Q_1| < |Q|$. Thus Ω^* is not statewise minimal with threshold c. Conversely, suppose Ω^* is not statewise minimal with threshold c. Then $\Omega^* \overset{c}{\sim} \Omega_1^*$ for some Ω_1^* with $|Q_1| < |Q|$. Thus there exist $q \in Q_1$ and $q', q'' \in Q$ with $q' \ne q''$ such that $(\Omega^*, q') \overset{c}{\sim} (\Omega_1^*, q)$ and $(\Omega^*, q'') \overset{c}{\sim} (\Omega_1^*, q)$. Hence $(\Omega^*, q') \overset{c}{\sim} (\Omega^*, q'')$ or $q' \overset{c}{\sim} q''$. Consequently, Ω^* is not statewise irreducible with threshold c. ∎

Corollary 2.6.2 Ω^* *is statewise minimal if and only if* Ω^* *is statewise irreducible.* ∎

Theorem 2.6.3 *If* Ω^* *is distributionwise irreducible with threshold* c, *then* Ω^* *is distributionwise minimal with threshold* c.

Proof. Let $\Omega^* = (Q, \mu^*, F)$. Suppose $\Omega^* \overset{c}{\approx} \Omega_1^*$ for some Ω_1^* such that $|Q_1| < |Q|$. Let H and H_1 denote, respectively, the collections of all initial distributions of Ω^* and Ω_1^*. Also, let P and P_1 denote, respectively, the equivalence classes of H and H_1 under $\overset{c}{=}$. Clearly, $|P_1| < |P|$. Thus there exist $\sigma_1 \in H_1$ and $\sigma', \sigma'' \in H$ with $\sigma' \neq \sigma''$ such that $(\Omega^*, \sigma') \overset{c}{\sim} (\Omega_1^*, \sigma_1)$ and $(\Omega^*, \sigma'') \overset{c}{\sim} (\Omega_1^*, \sigma_1)$. Therefore, $(\Omega^*, \sigma') \overset{c}{\sim} (\Omega^*, \sigma'')$ or $\sigma' \overset{c}{\sim} \sigma''$, a contradiction. ∎

Corollary 2.6.4 *If Ω^* is distributionwise irreducible, then Ω^* is distributionwise minimal.* ∎

Theorem 2.6.5 $\Omega_1^* \overset{c}{\sim} \Omega_2^*$ *implies* $\Omega_1^* \overset{c}{\approx} \Omega_2^*$.

Proof. For every $q \in Q_1$, define $T(q) = \{q' \in Q_2 \mid (\Omega_1^*, q) \overset{c}{\sim} (\Omega_2^*, q')\}$. Then $T(q) \neq \emptyset$ by hypothesis. Given σ_1, define σ_2 as follows:

$$\sigma_2(q) = \begin{cases} 1 & \text{if for some } q' \in Q_1, \sigma_1(q') > c \text{ and } q \in T(q') \\ 0 & \text{otherwise.} \end{cases}$$

Clearly, $(\Omega_1^*, \sigma_1) \overset{c}{\sim} (\Omega_2^*, \sigma_2)$. In a similar manner, it follows that for every σ_2, there exists σ_1 such that $(\Omega_1^*, \sigma_1) \overset{c}{\sim} (\Omega_2^*, \sigma_2)$. Thus $\Omega_1^* \overset{c}{\approx} \Omega_2^*$. ∎

Corollary 2.6.6 $\Omega_1^* \sim \Omega_2^*$ *implies* $\Omega_1^* \approx \Omega_2^*$. ∎

Corollary 2.6.7 *If Ω^* is distributionwise minimal (with threshold c), then Ω^* is statewise minimal (with threshold c).* ∎

2.7 Definite Max-Min Automata

Let $p \geq 0$ and $Y \subseteq X^*$. Then Y is said to be **weakly p definite** if for all $x \in X^*$ such that $|x| \geq p$, $x \in Y$ if and only if the p-suffix of x belongs to Y. For $p \geq 1$, Y is p **definite** if Y is weakly p definite but not weakly $p - 1$ definite. Y is 0 definite if and only if Y is weakly 0 definite. Y is **definite** if Y is p definite for some p.

A max-min automata A^* is said to be **definite** (p **definite, weakly p definite**) **with threshold** c if $B(A^*, c)$ is definite (p definite, weakly p definite). A^* is said to be **definite** (p **definite, weakly p definite**) if for all $0 \leq c < 1$, A^* is definite (p definite, weakly p definite) with threshold c.

A max-min table Ω^* is said to be **weakly p definite** with threshold c if for every initial distributions σ_1 and σ_2 and $x \in X^*$ with $|x| \geq p$,

$$\vee_{q \in F} r\mu(A_1^*, x, q) > c \iff \vee_{q \in F} r\mu(A_2^*, x, q) > c,$$

where $A_1^* = (\Omega^*, \sigma_1)$ and $A_2^* = (\Omega^*, \sigma_2)$. A max-min pre-table θ^* is said to be **weakly p definite with threshold** c if for all initial distributions σ_1 and σ_2 and $x \in X^*$, with $|x| \geq p, q \in Q$,

$$\vee \wedge_{q' \in Q} \{\sigma_1(q'), \mu^*(q', x, q)\} > c \iff \vee_{q' \in Q} \wedge \{\sigma_2(q'), \mu^*(q', x, q)\} > c.$$

For $p \geq 1, \Omega^*(\theta^*)$ is p definite with threshold c if and only if $\Omega^*(\theta^*)$ is weakly p definite with threshold c, but not weakly $p - 1$ definite with threshold c. $\Omega^*(\theta^*)$ is 0 definite with threshold c if and only if $\Omega^*(\theta^*)$ is weakly 0 definite with threshold c. $\Omega^*(\theta^*)$ is definite with threshold c if and only if $\Omega^*(\theta^*)$ is p definite with threshold c for some p. $\Omega^*(\theta^*)$ is definite (p definite, weakly p definite) if and only if for $0 \leq c < 1$, $\Omega^*(\theta^*)$ is definite (p definite, weakly p definite) with threshold c.

Clearly if θ^* is weakly p definite (with threshold c), then $\Omega^* = (\theta^*, F)$ is weakly p definite (with threshold c).

Theorem 2.7.1 *If $\Omega^* = (\theta^*, F)$ is distributionwise irreducible with threshold c and weakly p definite with threshold c, then θ^* is weakly p definite with threshold c.*

Proof. Suppose Ω^* is weakly p definite with threshold c. Then for every σ_1, σ_2, and x_0, $x \in X^*$ such that $|x| \geq p$,

$$\vee_{q \in F} r\mu(A_1^*, x_0 x, q) > c \iff \vee_{q \in F} r\mu(A_2^*, x_0 x, q) > c,$$

where $A_1^* = (\Omega^*, \sigma_1)$ and $A_2^* = (\Omega^*, \sigma_2)$. Let $\sigma_3(q) = r\mu(A_1^*, x_0, q)$, $\sigma_4(q) = r\mu(A_2^*, x_0, q)$, and $A_3^* = (\Omega^*, \sigma_3)$, $A_4^* = (\Omega^*, \sigma_4)$. Then

$$\vee_{q \in F} r\mu(A_3^*, x, q) > c \iff \vee_{q \in F} r\mu(A_4^*, x, q) > c$$

or $\sigma_3 \overset{c}{\sim} \sigma_4$. Since Ω^* is distributionwise irreducible with threshold c, $\sigma_3 \overset{c}{=} \sigma_4$. Hence θ^* is weakly p definite with threshold c. ∎

Theorem 2.7.2 *If $A^* = (\Omega^*, \sigma)$ is totally connected with threshold c and weakly p definite with threshold c, then Ω^* is weakly p definite with threshold c.*

Proof. For every σ_1 and σ_2, there exist $x_1, x_2 \in X^*$ such that $\sigma_1 \overset{c}{\sim} r\mu(A^*, x_1, q)$ and $\sigma_2 \overset{c}{\sim} r\mu(A^*, x_2, q)$. Since A^* is weakly p definite with threshold c, for every $x \in X^*$ with $|x| \geq p$,

$$\vee_{q \in F} r\mu(A^*, x_1 x, q) > c \iff \vee_{q \in F} r\mu(A^*, x_2 x, q) > c.$$

Let $A_1^* = (\Omega^*, \sigma_1)$ and $A_2^* = (\Omega^*, \sigma_2)$, then

$$\vee_{q \in F} r\mu(A_1^*, x, q) > c \iff \vee_{q \in F} r\mu(A_2^*, x, q) > c.$$

Thus Ω^* is weakly p definite with threshold c. ∎

Theorem 2.7.3 *If Ω^* is weakly p definite with threshold c, then $A^* = (\Omega^*, \sigma)$ is weakly p definite with threshold c.*

Proof. Let $x_1, x_2, x \in X^*$ be such that $|x| \geq p$, $\sigma_1(q) = r\mu(A^*, x_1, q)$, and $\sigma_2(q) = r\mu(A^*, x_2, q)$. Let $A_1^* = (\Omega^*, \sigma_1)$ and $A_2^* = (\Omega^*, \sigma_2)$. Then

$$\vee_{q \in F} r\mu(A^*, x_1 x, q) = \vee_{q \in F} r\mu(A_1^*, x, q)$$

and

$$\vee_{q \in F} r\mu(A_1^*, x_2 x, q) = \vee_{q \in F} r\mu(A_2^*, x, q).$$

Since $\vee_{q \in F} r\mu(A_1^*, x, q) > c \Leftrightarrow \vee_{q \in F} r\mu(A_2^*, x, q) > c$, it follows that

$$\vee_{q \in F} r\mu(A^*, x_1 x, q) > c \Leftrightarrow \vee_{q \in F} r\mu(A^*, x_2 x, q) > c.$$

Thus A is weakly p definite with threshold c. ∎

Theorem 2.7.4 *If $A^* = (\Omega^*, \sigma) = (\theta^*, F, \sigma)$, where A^* is totally connected (with threshold c) and Ω^* is distributionwise irreducible (with threshold c), then the following properties are equivalent:*
(1) A is p definite (with threshold c);
(2) Ω^* is p definite (with threshold c);
(3) θ^* is p definite (with threshold c).

Proof. The proof follows from Theorems 2.7.1, 2.7.2, and 2.7.3. ∎

Theorem 2.7.5 *If Ω^* is 0 definite with threshold c, then $F = Q$ or $F = \emptyset$.*

Proof. For every $q', q'' \in Q$ and $x \in X^*$, $\vee_{q \in F} r\mu(A_1^*, x, q) > c \Leftrightarrow \vee_{q \in F} r\mu(A_2^*, x, q) > c$, where $A_1^* = (\Omega^*, q')$ and $A_2^* = (\Omega^*, q'')$. Take $x = \Lambda$. Then $q' \in F$ if and only if $q'' \in F$. Thus $F = Q$ or $F = \emptyset$. ∎

Theorem 2.7.6 *If Ω^* is p definite with threshold c, where $p \geq 1$, then there exist σ_1 and σ_2 such that $\sigma_1 \overset{c}{\neq} \sigma_2$, but $\sigma_1 \overset{c,1}{\sim} \sigma_2$.*

Proof. Since Ω^* is not $p - 1$ definite with threshold c there exist $x_0 \in X^*$ with $|x_0| \geq p - 1$ and σ_1', σ_2' such that $\vee_{q \in F} r\mu(A_2^*, x_0, q) > c$, but $\vee_{q \in F} r\mu(A_2^*, x_0, q) \leq c$, where $A_1'^* = (\Omega^*, \sigma_1)$ and $A_2'^* = (\Omega^*, \sigma_2')$. However, Ω^* is weakly p definite. Hence for every $u \in X$, $\vee_{q \in F} r\mu(A_1'^*, x_0 u, q) > c \Leftrightarrow \vee_{q \in F} r\mu(A_2'^*, x_0, q) > c$. Let $\sigma_1(q) = r\mu(A_1'^*, x_0, q)$ and $\sigma_2(q) = r\mu(A_2'^*, x_0, q)$. Then σ_1 and σ_2 have the desired properties. ∎

2.8 Reduction of Max-Min Machines

The material in this section is from [203].

The reduction problem of max-min sequential machines and max-min automata was studied in the previous section. In this section, we continue the study.

Three types of equivalence relations are considered, namely, statewise, compositewise, and distributionwise equivalence. We show that the last two are equivalent. From the first two equivalence relations, two minimal forms are defined. The counterparts of these two minimal forms in the theory of stochastic sequential machines are the reduced [29] and minimal state forms [10].

For stochastic sequential machines, (convex) linear algebra is a useful theory. The next section is devoted to the development of a type of algebra, called the **max-min algebra**. The max-min algebra is very useful for dealing with max-min machines. Although max-min algebra resembles linear algebra and the max-product algebra of Section 3.1 in certain respects, they are almost completely unrelated.

Most of the results have counterparts in the theory of stochastic sequential machines and max-product machines. These counterparts are either known to be true [10,29,167,202] or can be shown to be true. See also Chapter 3. The section concludes with a look at nondeterministic and deterministic machines which are special cases of max-min machines. Many of the results for max-min machines are strengthened for these particular cases.

Definition 2.8.1 *Let* $A = [a_{ij}]$ *be an* $n \times p$ *matrix and* $B = [b_{ij}]$ *be a* $p \times m$ *matrix of nonnegative real numbers. Let* $A \otimes B$ *be the* $n \times m$ *matrix* $[c_{ij}]$, *where*

$$c_{ij} = \vee\{a_{ik} \wedge b_{kj} \mid k = 1, \ldots, p\}.$$

Definition 2.8.1 is applicable even if n or m is infinite.

Clearly, the operation $\otimes$ is associative.

In the remainder of the chapter, a, b, and c (with or without subscripts) denote real numbers, and x and y (with or without subscripts) denote (finite or infinite) sequences of real numbers. A superscript is used to denote the particular term of a sequence, e.g., x^k denotes the k-th term of the sequence x. X and Y (with or without subscripts) denote collections of (finite or infinite) sequences of real numbers. Here $a \wedge x$ denotes the sequence whose k-th term is $a \wedge x^k$. When taking max-min combinations of sequences, we assume the sequences are of the same length.

Definition 2.8.2 *Let* $X = \{x_1, x_2, \ldots, x_n\}$. *A* ***max-min combination*** *of* X *is an expression of the form*

$$\vee_{i=1}^{n}(a_i \wedge x_i), \qquad (2.1)$$

where a_i *is a nonnegative real number,* $i = 1, 2, \ldots, n$.

If $0 \leq a_i \leq 1$ *for* $i = 1, 2, \ldots, n$, *then (2.1) is called a* ***convex max-min combination*** *of* X.

Definition 2.8.3 *Let T_x be the set of all distinct terms of x. Then x is called **admissible** if T_x is finite and T_x can be effectively constructed from x. X is called **admissible** if every x in X is admissible.*

Proposition 2.8.4 *Let $x = \vee_{i=1}^{n} a_i \wedge x_i$. If T_x is finite, then*

$$x = \vee_{i=1}^{n} b_i \wedge x_i,$$

where $b_i \in T_x$, $i = 1, 2, \dots, n$.

Proof. Choose b_i to be the largest number in T_x such that $b_i \leq a_i$. ∎

Example 2.8.5 *Let $x_1 = (1, \frac{1}{4}, 1)$, $x_2 = (\frac{1}{8}, 1, 1)$, $x_3 = (1, 1, \frac{1}{4})$, $a_1 = 1$, $a_2 = \frac{1}{2}$, and $a_3 = \frac{1}{3}$. Then $a_1 \wedge x_1 = (1, \frac{1}{4}, 1)$, $a_2 \wedge x_2 = (\frac{1}{8}, \frac{1}{2}, \frac{1}{2})$, and $a_3 \wedge x_3 = (\frac{1}{3}, \frac{1}{3}, \frac{1}{4})$. Thus $x = (a_1 \wedge x_1) \vee (a_2 \wedge x_2) \vee (a_3 \wedge x_3) = (1, \frac{1}{2}, 1)$. Hence $T_x = \{1, \frac{1}{2}\}$. Let $b_i = \vee \{b \in T_x \mid b \leq a_i\}$, $i = 1, 2, 3$. Then $b_1 = 1$ and $b_2 = \frac{1}{2}$. However, there does not exist $b \in T_x$ such that $b \leq \frac{1}{3} \leq a_3$. Since $\vee \emptyset = 0$, we take $b_3 = 0$. Hence $x = (b_1 \wedge x_1) \vee (b_2 \wedge x_2) \vee (b_3 \wedge x_3)$.*

Proposition 2.8.6 *Suppose that T_x is admissible and X is finite. Then it is decidable whether or not x is a (convex) max-min combination of X.*

Proof. The proof follows from Proposition 2.8.4. ∎

Definition 2.8.7 *The **(convex) max-min span** of X is the collection of all (convex) max-min combinations of finite subsets of X.*

Let $C(X)$ denote the convex max-min span of X. Note that if $X = \{(a, b)\}$ with $a > 1$, then $X \not\subseteq C(X)$ since $1 \wedge (a, b) \neq (a, b)$ and so $(a, b) \notin C(X)$.

Proposition 2.8.8 *(1) $X \subseteq C(X)$ if $x \in X$ implies $0 \leq x^k \leq 1$ for each k;*

(2) If $X_1 \subseteq X_2$, then $C(X_1) \subseteq C(X_2)$;
(3) $C(C(X)) = C(X)$. ∎

We can consider C as a function from $\mathcal{P}(V)$ into $\mathcal{P}(V)$, where $V = [0, 1]^n$ for some $n \in \mathbb{N}$. Then C is said to satisfy the Exchange Property if $\forall X \subseteq V$ and $\forall x, y \in V$; if $y \in C(X \cup \{x\})$ and $y \notin C(X)$, then $x \in C(X \cup \{y\})$. This is a key property that holds in some situations. It is mainly due to the Exchange Property that one obtains the existence of bases of unique cardinality for certain algebraic structures. It is important to note that it does not hold here.

Example 2.8.9 *Let $X = \{(1, \frac{1}{8})\}$, $x = (\frac{1}{8}, 1)$, and $y = (\frac{1}{2}, \frac{1}{4})$. Then $y \notin C(X)$. Now $y \in C(X \cup \{x\})$ since $y = (\frac{1}{2} \wedge (1, \frac{1}{8})) \vee (\frac{1}{4} \wedge x)$. However, $x \notin C(X \cup \{y\})$. Thus the Exchange Property does not hold.*

Definition 2.8.10 Y *is called a* **convex max-min set** *if for every* y_1, $y_2 \in Y$, *all convex max-min combinations of* $\{y_1, y_2\}$ *are also in* Y.

Proposition 2.8.11 Y *is a convex max-min set if and only if* $Y = C(Y)$. ∎

Proposition 2.8.12 *For every* X, $C(X)$ *is a convex max-min set.*

Proof. The proof follows from Propositions 2.8.11 and 2.8.8(2). ∎

In the rest of this section, Y always denotes a convex max-min set.

Definition 2.8.13 *Let* $X \subseteq Y$.
(1) X *is called a* **set of generators** *of* Y *if* $Y = C(X)$.
(2) If X *does not contain any proper subset which is itself a set of generators of* Y, *then* X *is called a* **set of vertices** *of* Y.

Clearly, X is a set of generators of $C(X)$.
Note that if $x \in C(X \backslash \{x\})$, then $X \subseteq C(X \backslash \{x\})$ and so $C(X) \subseteq C(C(X \backslash \{x\})) = C(X \backslash \{x\})$.

Proposition 2.8.14 *Let* $X \subseteq Y$. *Then* X *is a set of vertices of* Y *if and only if*
(1) $Y = C(X)$ *and*
(2) if $x \in X$, *then* $x \notin C(X \backslash \{x\})$. ∎

Proposition 2.8.15 *Suppose that* X *is a finite set of generators of* Y. *Then there exists* $X' \subseteq X$ *such that* X' *is a set of vertices of* Y. *Moreover,* X' *can be effectively constructed provided* X *is admissible.*

Proof. The proof follows from Propositions 2.8.14, 2.8.6, and the fact that X is finite. ∎

Definition 2.8.16 Y *is called* **finitary** *if it contains a set of generators that is finite.*

Proposition 2.8.17 *Every finitary convex max-min set has at least one set of vertices.*

Proof. The proof follows from Proposition 2.8.15. ∎

Proposition 2.8.17 is not true in general for arbitrary convex max-min sets.

Proposition 2.8.18 *Suppose that* Y *is finitary. Then every set of generators of* Y *contains a finite subset that is a set of vertices of* Y.

Proof. Let X' be a set of generators of Y. By Proposition 2.8.17, there exists a set X of vertices of Y which is finite. Now $\forall x \in X$, x is a convex max-min combination of a finite subset of X', say X'_x. Let $X'' = \cup_{x \in X} X'_x$. Clearly, X'' is finite and $Y = C(X) \subseteq C(X'') \subseteq Y$. Thus X'' generates Y. The result now follows from Proposition 2.8.15. ∎

Proposition 2.8.19 *Suppose that Y is finitary. Then every set of vertices of Y is finite.*

Proof. Let X be a set of vertices of Y. By Proposition 2.8.18, there is a finite subset X' of X such that X' generates Y. By the minimality of X, $X' = X$ and so X is finite. ∎

Definition 2.8.20 *Let X be a set of vertices of Y. X is called **fundamental** if for every $x \in X$, there does not exist $y \in Y$ such that $x \neq y = a \wedge x$ for some $0 \leq a \leq 1$ and $(X \setminus \{x\}) \cup \{y\}$ is a set of vertices of Y.*

We write $x \leq a$ if $x^k \leq a$ for all k, $x_1 \leq x_2$ means $x_1^k \leq x_2^k$ for all k, and $x = a$ means $x^k = a$ for all k.

Lemma 2.8.21 *Let X be a set of vertices of Y. Then X is fundamental if and only if for every $x \in X$,*

$$x = (a \wedge x) \vee y,$$

where $y \in C(X \setminus \{x\})$ and $0 \leq a \leq 1$ implies $x = a \wedge x$.

Proof. Let $x \in X$ be such that $x = (a \wedge x) \vee y$, where $y \in C(X \setminus \{x\})$ and $0 \leq a \leq 1$, but $x \neq a \wedge x$. Let $y' = a \wedge x$ and $X' = (X \setminus \{x\}) \cup \{y'\}$. Then $x \in C(X \setminus \{x\}, y')$ and so $X \subseteq C(X \setminus \{x\}, y')$. Thus $Y = C(X) = C(C(X')) = C(X')$ and so it follows that X' is a set of vertices of Y. Thus X is not fundamental. Conversely, suppose that X is not fundamental. Then there exists $x \in X$, $y' \in Y$ such that $x \neq y' = a \wedge x$ for some $0 \leq a \leq 1$ and $(X \setminus \{x\}) \cup \{y'\}$ is a set of vertices of Y. Thus x is a convex max-min combination of a finite subset X' of $(X \setminus \{x\}) \cup \{y'\}$. By Proposition 2.8.14, $x \in C(X')$, i.e., $x = (b \wedge y') \vee y$, where $y \in C(X' \setminus \{y'\})$ and $0 \leq b \leq 1$. Let $c = a \wedge b$. Then $x = (b \wedge (a \wedge x)) \vee y = (c \wedge x) \vee y$. Since $c \wedge x = b \wedge y' \leq y' = a \wedge x \leq x$ and $a \wedge x \neq x$, $x \neq c \wedge x$. ∎

Example 2.8.22 *Let $X = \{(1, \frac{1}{8}), (\frac{1}{8}, 1)\}$. Let $x = (1, \frac{1}{8})$ and $y = a \wedge (1, \frac{1}{8})$ be such that $x \neq y$. Then $0 \leq a < 1$. Now $(X \setminus \{x\}) \cup \{y\}$ is not a set of vertices of $Y = C(X)$ since there does not exist $a_1, a_2 \in [0, 1]$ such that $(1, \frac{1}{8}) = (a_1 \wedge y) \vee (a_2 \wedge (\frac{1}{8}, 1)) = (a_1 \wedge (a \wedge (1, \frac{1}{8}))) \vee (a_2 \wedge (\frac{1}{8}, 1))$. A similar argument holds for $x = (\frac{1}{8}, 1)$. Hence X is fundamental for Y.*

Example 2.8.23 *Let $X = \{(1, 0), (\frac{1}{2}, \frac{1}{4})\}$. Then X is a set of vertices of $Y = C(X)$. Now $(\frac{1}{2}, \frac{1}{4}) \in C(\{a \wedge x, (1, 0)\})$, where $a = \frac{1}{4}$ and $x = (\frac{1}{2}, \frac{1}{4})$ since $(\frac{1}{2}, \frac{1}{4}) = (\frac{1}{4} \wedge (\frac{1}{2}, \frac{1}{4})) \vee (\frac{1}{2} \wedge (1, 0))$. Thus letting $y = a \wedge x$, we see by Definition 2.8.20 that X is not fundamental. If we let $y = \frac{1}{2} \wedge (1, 0)$, then we see that X is not fundamental by Lemma 2.8.21.*

Proposition 2.8.24 *Suppose that X is a fundamental set of vertices of $C(X)$ and $x \in X$. Then $X \backslash \{x\}$ is a fundamental set of vertices of $C(X \backslash \{x\})$.*

Proof. The proof follows from Proposition 2.8.14 and Lemma 2.8.21. ∎

Proposition 2.8.25 *Let X' and X'' be fundamental sets of vertices of Y and $\bar{x} \in X' \cap X''$. Then $C(X' \backslash \{\bar{x}\}) = C(X'' \backslash \{\bar{x}\})$.*

Proof. Let $x \in X' \backslash \{\bar{x}\}$. Now

$$x = \vee_{i=0}^{n} a_i \wedge x_i'', \tag{2.2}$$

where for $i = 0, 1, 2, \ldots, n$, $x_i'' \in X''$ and $0 \le a_i \le 1$. If $\bar{x}$ does not occur among the x_i'', then $x \in C(X'' \backslash \{\bar{x}\})$. Suppose that $\bar{x}$ appears among the x_i'', say $x_0'' = \bar{x}$. Now for $i = 1, 2, \ldots, n$,

$$x_i'' = \vee_{j=0}^{m} b_{ij} \wedge x_j', \tag{2.3}$$

where for $j = 0, 1, 2, \ldots m$, $x_j' \in X'$ and $0 \le b_{ij} \le 1$. Thus

$$x = (a_0 \wedge \bar{x}) \vee (\vee_{j=0}^{m} c_j \wedge x_j'),$$

where

$$c_j = \vee_{i=1}^{n} (a_i \wedge b_{ij}).$$

Since X' is a set of vertices of Y and $x \in X' \backslash \{\bar{x}\}$, x must appear among the x_j', say x_0'. Also, since X' is fundamental, by Lemma 2.8.21,

$$x = c_0 \wedge x.$$

Therefore,

$$x \le c_0 = \vee \{a_i \wedge b_{i0} \mid i = 1, 2, \ldots, n\} = a_k \wedge b_{k0}$$

for some $1 \le k \le n$. Thus $x \le a_k$ and $x \le b_{k0}$. By (2.2) and (2.3),

$$x \ge a_k \wedge x_k'' \ge a_k \wedge b_{k0} \wedge x_0' = c_0 \wedge x = x.$$

Hence $x = a_k \wedge x_k''$ or $x \in C(X'' \backslash \{\bar{x}\})$. Thus

$$C(X' \backslash \{\bar{x}\}) \subseteq C(X'' \backslash \{\bar{x}\}).$$

Similarly,

$$C(X''\backslash\{\bar{x}\}) \subseteq C(X'\backslash\{\bar{x}\}).$$

Consequently,

$$C(X'\backslash\{\bar{x}\}) = C(X''\backslash\{\bar{x}\}). \blacksquare$$

Let $\{x\}$ and X' be sets of vertices for Y. Then $\forall x' \in X'$, there exists $a' \in [0,1]$ such that $x' = a' \wedge x$. Thus $\forall x', x'' \in X'$, either $x' \leq x''$ or $x'' \leq x'$. Hence it follows that $|X'| = 1$, say $X' = \{x'\}$. Then there exists $a, b \in [0,1]$ such that $x' = a \wedge x$ and $x = b \wedge x'$. Thus $x' \leq x$ and $x \leq x'$. Hence $x = x'$.

Theorem 2.8.26 *If Y is finitary, then the fundamental set of vertices of Y is unique.*

Proof. Let X' and X'' be fundamental sets of vertices of Y. By Proposition 2.8.19, X' and X'' are finite. Let $X' = \{x'_1, x'_2, \ldots, x'_n\}$ and $X'' = \{x''_1, x''_2, \ldots, x''_m\}$, where $n \leq m$. We prove the result by induction on n. The $n = 1$ case follows from the comments preceding the theorem. Suppose that the theorem is true for all $n < p$. We show that it is also true for $n = p$. Let

$$x'_i = \vee_{j=1}^{m} a_{ij} \wedge x''_j$$

and

$$x''_j = \vee_{k=1}^{n} b_{jk} \wedge x'_k.$$

Then

$$x'_i = \vee_{k=1}^{n} c_{ik} \wedge x'_k,$$

where

$$c_{ik} = \vee_{j=1}^{m} a_{ij} \wedge b_{jk}.$$

Since X' is fundamental, by Lemma 2.8.21, $x'_1 = c_{11} \wedge x'_1$. Hence

$$x'_1 \leq c_{11} = \vee_{j=1}^{m}(a_{1j} \wedge b_{j1}) = a_{1j_1} \wedge b_{j_1 1}$$

for some j_1. This implies that $x'_1 \leq a_{1j_1}$ and $x'_1 \leq b_{j_1 1}$. Thus

$$x''_{j_1} \geq b_{j_1 1} \wedge x'_1 = x'_1.$$

Similarly, there exists j_2 such that $x'_{j_2} \geq x''_{j_1}$. By repeating this process, we obtain a sequence $j_1, j_2, \ldots, j_n, \ldots$, such that

$$x''_{j_{2k-1}} \leq x'_{j_{2k}} \leq x''_{j_{2k+1}} \leq x'_{j_{2k+2}}.$$

Since X'' is finite, we must have $x''_{j_{2k-1}} = x''_{j_{2l-1}}$, for some k and l, where $l > k$. Thus $x''_{j_{2k-1}} = x'_{j_{2k}}$. Let $\bar{x} = x''_{j_{2k-1}} = x'_{j_{2k}}$. Now consider $X' \backslash \{\bar{x}\}$ and $X'' \backslash \{\bar{x}\}$. By Propositions 2.8.24 and 2.8.25 and the induction hypothesis, $X' \backslash \{\bar{x}\} = X'' \backslash \{\bar{x}\}$. Hence $X' = X''$. ∎

Proposition 2.8.27 *Let Y be admissible and $X = \{x_1, x_2, \ldots, x_n\}$ be a set of vertices of Y. Then there exists a fundamental set X' of vertices of Y such that $|X'| = |X|$. Furthermore, X' can be effectively constructed from X.*

Proof. Define x'_i, $i = 1, 2, \ldots n$, recursively as follows: Let

$$\begin{aligned} A_1 &= \{a \mid x_1 = (a \wedge x_1) \vee (\vee_{i>1} a_i \wedge x_i)\\ &\quad \text{for some } 0 \leq a_i \leq 1,\ i \neq 1\}, \end{aligned}$$

and $a'_1 = \wedge A_1$. Define $x'_1 = a'_1 \vee x_1$. Suppose that x'_{k-1} has been defined. Let

$$\begin{aligned} A_k &= \{a \mid x_k = (a \wedge x_k) \vee (\vee_{i<k} a_i \wedge x'_i) \vee (\vee_{i>k} a_i \wedge x_i)\\ &\quad \text{for some } 0 \leq a_i \leq 1,\ i \neq k\}, \end{aligned}$$

$\alpha'_k = \wedge A_k$ and $x'_k = a'_k \wedge x_k$. Let $X' = \{x'_1, x'_2, \ldots, x'_n\}$. Clearly $|X'| = |X|$. By Proposition 2.8.6, X' can be effectively constructed from X. Moreover, $a'_k \in A_k$ for all k. Thus X' is a set of vertices of Y. We now show that X' is fundamental. Let

$$x'_k = (a_k \wedge x'_k) \vee y,$$

where $y \in C(X' \backslash \{\bar{x}'_k\})$. Let $X_k = \{x'_1, x'_2, \ldots, x'_{k-1}, x_{k+1}, \ldots, x_n\}$. Since $x_k = x'_k \vee y'$, where $y' \in C(X_k)$, it follows that $x_k = (a_k \wedge a'_k \wedge x_k) \vee y''$, where $y'' \in C(X_k)$. By the definition of a'_k, $a'_k \leq a_k \wedge a'_k$ or $a_k \geq a'_k$. Thus

$$x'_k \geq a_k \wedge x'_k = a_k \wedge a'_k \wedge x_k = a'_k \wedge x_k = x'_k.$$

Hence $x'_k = a_k \wedge x'_k$. By Lemma 2.8.21, X' is fundamental. ∎

We now illustrate Proposition 2.8.27.

Example 2.8.28 Let $X = \{(1,0), (\frac{1}{2}, \frac{1}{4})\}$. Let $x_1 = (1,0)$ and $x_2 = (\frac{1}{2}, \frac{1}{4})$. Then $A_1 = \{a \mid (1,0) = (a \wedge (1,0)) \vee (a_2 \wedge (\frac{1}{2}, \frac{1}{4}))$ for some $a_2, 0 \le a_2 \le 1\} = \{1\}$. Thus $a_1' = \wedge\{1\} = 1$ and $x_1' = 1 \wedge (1,0) = (1,0)$. Now $A_2 = \{a \mid (\frac{1}{2}, \frac{1}{4}) = (a \wedge (\frac{1}{2}, \frac{1}{4})) \vee (a_1 \wedge (1,0))$ for some $a_1, 0 \le a_1 \le 1\} = [\frac{1}{4}, 1]$. Hence $a_2' = \wedge [\frac{1}{4}, 1] = \frac{1}{4}$ and $x_2' = \frac{1}{4} \wedge (\frac{1}{2}, \frac{1}{4}) = (\frac{1}{4}, \frac{1}{4})$. Thus $\{(\frac{1}{4}, \frac{1}{4}), (1,0)\}$ is fundamental.

Now let $x_1 = (\frac{1}{2}, \frac{1}{4})$ and $x_2 = (1,0)$. Then $A_1 = \{a \mid (\frac{1}{2}, \frac{1}{4}) = (a \wedge (\frac{1}{2}, \frac{1}{4})) \vee (a_2 \wedge (1,0))$ for some $a_2, 0 \le a_2 \le 1\} = [\frac{1}{4}, 1]$. Thus $a_1' = \wedge [\frac{1}{4}, 1] = \frac{1}{4}$ and $x_1' = \frac{1}{4} \wedge (\frac{1}{2}, \frac{1}{4}) = (\frac{1}{4}, \frac{1}{4})$. Now $A_2 = \{a \mid (1,0) = (a \wedge (1,0)) \vee (a_1 \wedge (\frac{1}{2}, \frac{1}{4}))$ for some $a_1, 0 \le a_1 \le 1\} = \{1\}$. Hence $a_2' = \wedge\{1\} = 1$ and $x_2' = 1 \wedge (1,0) = (1,0)$.

Theorem 2.8.29 Suppose that Y is finitary and admissible. Then every set of vertices of Y has the same number of elements.

Proof. The proof follows from Propositions 2.8.19, 2.8.27, and Theorem 2.8.26. ∎

Proposition 2.8.30 Let Y be finitary and admissible and let X_1, X_2 be sets of generators of Y. If $|X_1| > |X_2|$, then there exists $x \in X_1$ such that $x \in C(X_1 \backslash \{x\})$.

Proof. By Proposition 2.8.18, there exist $X_1' \subseteq X_1$ and $X_2' \subseteq X_2$, such that X_1' and X_2' are sets of vertices of Y. By Theorem 2.8.29, $|X_1'| = |X_2'|$. Thus $|X_1'| = |X_2'| \le |X_2| < |X_1|$. This implies that X_1' is a proper subset of X_1. The result now follows from Proposition 2.8.14. ∎

Proposition 2.8.31 Let X be a collection of sequences consisting of 0 and 1. If X is a set of vertices of $C(X)$, then X is fundamental.

Proof. Let $x \in X$ and $x = (a \wedge x) \vee y$, where $0 \le a \le 1$ and $y \in C(X \backslash \{x\})$. Suppose that $a < 1$. Let $x_0 = a \wedge x$. If $x^k = 0$, then $y^k = 0$. If $x^k = 1$, then since $x_0^k = a < 1$, it follows that $y^k = 1$. Thus $x = y$, which is a contradiction. Hence $a = 1$ and $x = x_0$. By Lemma 2.8.21, X is fundamental. ∎

Definition 2.8.32 Y is called **fundamental** if every set of vertices of Y is fundamental.

It follows that not every convex max-min set is fundamental.

Theorem 2.8.33 Let Y be finitary. Then Y is fundamental if and only if Y contains only one set of vertices.

Proof. The proof follows from Proposition 2.8.17 and Theorem 2.8.26. ∎

Definition 2.8.34 *Let X_1, $X_2 \subseteq Y$. Then X_1 is called a **basis** of X_2 if every $x \in X_2$ can be expressed uniquely as a convex max-min combination of a unique finite subset of X_1.*

Let $Y = [0,1] \times [0,1]$. Then clearly $\{(1,0), (0,1)\}$ is a basis of Y.

Proposition 2.8.35 *Let X be a basis of Y. Then X is a fundamental set of vertices of Y.*

Proof. The proof follows from Proposition 2.8.14 and Lemma 2.8.21. ∎

Proposition 2.8.36 *Let Y be finitary and have a basis. Then Y is fundamental.*

Proof. Let $X = \{x_1, x_2, \ldots, x_n\}$ be a basis of Y. By Proposition 2.8.35, X is a fundamental set of vertices of Y. Suppose that Y is not fundamental. Then by Theorem 2.8.33, Y has a set X' of vertices such that $X' \neq X$. By Proposition 2.8.27, we may assume, without loss of generality, that

$$X' = \{x_1, x_2, \ldots, x_{n-1}, x'_n\},$$

where $x'_n \neq x_n$. Furthermore,

$$x_n = a \wedge x'_n,$$

for some $0 \leq a \leq 1$ and

$$x'_n = x_n \vee y,$$

where $y \in C(X \backslash \{x_n\})$. Thus

$$x_n = (a \wedge x_n) \vee (a \wedge y).$$

Since X is a basis of Y, $a = 1$. Hence $x'_n = x_n$, which is a contradiction. Therefore, Y is fundamental. ∎

2.9 Equivalences

Definition 2.9.1 *A **max-min sequential-like machine** (MSLM) is a quadruple (Q, X, Y, μ), where Q, X, Y are finite nonempty sets and μ is a function from $Q \times X \times Y \times Q$ into $[0,1]$.*

The set X is the input alphabet. The set Y is the output alphabet. Q is the set of internal states and μ is the transition function. We may interpret $\mu(q, u, v, q')$ as the grade of membership that the MSLM will enter state q' and produce output v given that the present state is q and the input is u.

As before, finite sequences of elements of $X(Y)$ are called input (output) tapes. The collection of all input (output) tapes is denoted by $X^*(Y^*)$. We denote the set

$$\{(x, y) \mid x \in X^*,\ y \in Y^*,\ |x| = |y|\}$$

by $(X \times Y)^*$.

In the following, the symbol M, with or without subscripts, always denotes an MSLM.

We assume that all MSLMs have the same input set X and output set Y. Therefore, by suppressing X and Y, an MSLM may be represented by (Q, μ). Also, $(X \times Y)^*$ is ordered in such a way that $|x_1| < |x_2|$ implies $(x_1, y_1) < (x_2, y_2)$. The order of $(X \times Y)^*$ is the same for all MSLMs and will be kept fixed throughout the remainder of the chapter.

Definition 2.9.2 *Let $M = (Q, \mu)$ be a MSLM. The **extended transition function** μ^* of M is a function from $Q \times (X \times Y)^* \times Q$ into $[0, 1]$ defined recursively on $|x|$, $x \in X^*$, as follows:*

$$\mu^*(q', \Lambda, \Lambda, q'') = \begin{cases} 1 & if \quad q' = q'' \\ 0 & if \quad q' \neq q'', \end{cases}$$

$$\mu^*(q', ux, vy, q'') = \vee_{q \in Q}\{\mu(q', u, v, q) \wedge \mu^*(q, x, y, q'')\},$$

where $u \in X$, $v \in Y$, and $y \in Y^*$ with $|x| = |y|$. Furthermore, the overall transition function q^M of M is a function from $Q \times (X \times Y)^*$ into $[0, 1]$ defined as follows:

$$q^M(q, x, y) = \vee_{q' \in Q}\mu^*(q, x, y, q').$$

If no confusion arises, we write q for q^M.

For ease of notation, we assume that $Q = \{q_1, q_2, \ldots, q_n\}$. Furthermore,

(1) $P^M(x, y) = [b_{ij}]$ is a matrix, where $b_{ij} = \mu^*(q_i, x, y, q_j)$;

(2) $Q^M(x, y) = [a_i]$ is the column matrix, where the i-th row $a_i = q^M(q_i, x, y)$;

(3) E is the column matrix with all entries 1.

Proposition 2.9.3 *Let $M = (Q, \mu)$ be an MSLM. Then for every (x_1, y_1), $(x_2, y_2) \in (X \times Y)^*$,*

(1) $P^M(x_1x_2,\ y_1y_2) = P^M(x_1, y_1) \otimes P^M(x_2, y_2)$, and

(2) $Q^M(x_1x_2,\ y_1y_2) = P^M(x_1, y_1) \otimes Q^M(x_2, y_2) = P^M(x_1x_2,\ y_1y_2) \otimes E$.

Proof. The proof follows immediately since $\otimes$ is associative. ∎

(1) Let A^M denote the matrix whose columns are $Q^M(x, y)$ arranged in the order of $(X \times Y)^*$.

(2) For any nonnegative integer n, let A_n^M denote the submatrix of A^M consisting of only those columns corresponding to $Q^M(x, y)$ with $|x| \leq n$.

(3) Let B^M and B_n^M denote, respectively, the matrix obtained from A^M and A_n^M by omitting all columns that are convex max-min combinations of previous columns.

(4) Let A be a (finite or infinite) matrix. Then $|A|$ denotes the number of columns of A and $\rho(A)$ denotes the set of distinct rows of A.

Example 2.9.4 *Let* $M = (Q, X, Y, \mu)$, *where* $Q = \{q_1, q_2, q_3\}$, $X = \{u, v\}$, $Y = \{0, 1\}$, *and*

$$\mu : Q \times X \times Y \times Q \to [0, 1]$$

is defined as follows:

$$
\begin{aligned}
\mu(q_1, u, 1, q_2) &= \tfrac{3}{4}, & \mu(q_1, v, 0, q_2) &= \tfrac{3}{8}, \\
\mu(q_1, u, 1, q_3) &= \tfrac{1}{2}, & \mu(q_1, v, 1, q_3) &= \tfrac{5}{8}, \\
\mu(q_2, u, 1, q_1) &= \tfrac{3}{4}, & \mu(q_2, v, 0, q_1) &= \tfrac{3}{8}, \\
\mu(q_2, u, 1, q_3) &= \tfrac{1}{2}, & \mu(q_2, v, 1, q_3) &= \tfrac{5}{8}, \\
\mu(q_3, u, 0, q_1) &= \tfrac{1}{4}, & \mu(q_3, v, 0, q_1) &= \tfrac{1}{8}, \\
\mu(q_3, u, 0, q_2) &= \tfrac{1}{4}, & \mu(q_3, v, 0, q_2) &= \tfrac{1}{8}.
\end{aligned}
$$

Define $\mu(q, w, z, q') = 0$ *for all other* (q, w, z, q'). *Then*

$$
P^M(u, 1) = \begin{bmatrix} 0 & \tfrac{3}{4} & \tfrac{1}{2} \\ \tfrac{3}{4} & 0 & \tfrac{1}{2} \\ 0 & 0 & 0 \end{bmatrix}, \qquad
P^M(u, 0) = \begin{bmatrix} 0 & 0 & 0 \\ 0 & 0 & 0 \\ \tfrac{1}{4} & \tfrac{1}{4} & 0 \end{bmatrix},
$$

$$
P^M(v, 1) = \begin{bmatrix} 0 & 0 & \tfrac{5}{8} \\ 0 & 0 & \tfrac{5}{8} \\ 0 & 0 & 0 \end{bmatrix}, \qquad
P^M(v, 0) = \begin{bmatrix} 0 & \tfrac{3}{8} & 0 \\ \tfrac{3}{8} & 0 & 0 \\ \tfrac{1}{8} & \tfrac{1}{8} & 0 \end{bmatrix}.
$$

$$
\begin{aligned}
q^M(q_1, u, 1) &= \mu(q_1, u, 1, q_1) \vee \mu(q_1, u, 1, q_2) \vee \mu(q_1, u, 1, q_3) &= \tfrac{3}{4}, \\
q^M(q_2, u, 1) &= \mu(q_2, u, 1, q_1) \vee \mu(q_2, u, 1, q_2) \vee \mu(q_2, u, 1, q_3) &= \tfrac{3}{4}, \\
q^M(q_3, u, 1) &= \mu(q_3, u, 1, q_1) \vee \mu(q_3, u, 1, q_2) \vee \mu(q_3, u, 1, q_3) &= 0, \\
q^M(q_1, u, 0) &= \mu(q_1, u, 0, q_1) \vee \mu(q_1, u, 0, q_2) \vee \mu(q_1, u, 0, q_3) &= 0, \\
q^M(q_2, u, 0) &= \mu(q_2, u, 0, q_1) \vee \mu(q_2, u, 0, q_2) \vee \mu(q_2, u, 0, q_3) &= 0, \\
q^M(q_3, u, 0) &= \mu(q_3, u, 0, q_1) \vee \mu(q_3, u, 0, q_2) \vee \mu(q_3, u, 0, q_3) &= \tfrac{1}{4}, \\
q^M(q_1, v, 1) &= \mu(q_1, v, 1, q_1) \vee \mu(q_1, v, 1, q_2) \vee \mu(q_1, v, 1, q_3) &= \tfrac{5}{8}, \\
q^M(q_2, v, 1) &= \mu(q_2, v, 1, q_1) \vee \mu(q_2, v, 1, q_2) \vee \mu(q_2, v, 1, q_3) &= \tfrac{5}{8}, \\
q^M(q_3, v, 1) &= \mu(q_3, v, 1, q_1) \vee \mu(q_3, v, 1, q_2) \vee \mu(q_3, v, 1, q_3) &= 0, \\
q^M(q_1, v, 0) &= \mu(q_1, v, 0, q_1) \vee \mu(q_1, v, 0, q_2) \vee \mu(q_1, v, 0, q_3) &= \tfrac{3}{8}, \\
q^M(q_2, v, 0) &= \mu(q_2, v, 0, q_1) \vee \mu(q_2, v, 0, q_2) \vee \mu(q_2, v, 0, q_3) &= \tfrac{3}{8}, \\
q^M(q_3, v, 0) &= \mu(q_3, v, 0, q_1) \vee \mu(q_3, v, 0, q_2) \vee \mu(q_3, v, 0, q_3) &= \tfrac{1}{8}.
\end{aligned}
$$

Thus

$$
Q^M(u, 1) = \begin{bmatrix} \tfrac{3}{4} \\ \tfrac{3}{4} \\ 0 \end{bmatrix}, \qquad
Q^M(v, 1) = \begin{bmatrix} \tfrac{5}{8} \\ \tfrac{5}{8} \\ 0 \end{bmatrix},
$$

$$
Q^M(u, 0) = \begin{bmatrix} 0 \\ 0 \\ \tfrac{1}{4} \end{bmatrix}, \qquad
Q^M(v, 0) = \begin{bmatrix} \tfrac{3}{8} \\ \tfrac{3}{8} \\ \tfrac{1}{8} \end{bmatrix}.
$$

Now

$$
\begin{aligned}
\mu^*(q_1, uv, 10, q_1) &= \vee\{0 \wedge 0, \tfrac{3}{4} \wedge \tfrac{3}{8}, \tfrac{1}{2} \wedge \tfrac{1}{8}\} &= \tfrac{3}{8}, \\
\mu^*(q_1, uv, 10, q_2) &= \vee\{0 \wedge \tfrac{3}{8}, \tfrac{3}{4} \wedge 0, \tfrac{1}{2} \wedge \tfrac{1}{8}\} &= \tfrac{1}{8}, \\
\mu^*(q_1, uv, 10, q_3) &= \vee\{0 \wedge 0, \tfrac{3}{4} \wedge 0, \tfrac{1}{2} \wedge 0\} &= 0, \\[4pt]
\mu^*(q_2, uv, 10, q_1) &= \vee\{\tfrac{3}{4} \wedge 0, 0 \wedge \tfrac{3}{8}, \tfrac{1}{2} \wedge \tfrac{1}{8}\} &= \tfrac{1}{8}, \\
\mu^*(q_2, uv, 10, q_2) &= \vee\{\tfrac{3}{4} \wedge \tfrac{3}{8}, 0 \wedge 0, \tfrac{1}{2} \wedge \tfrac{1}{8}\} &= \tfrac{3}{8}, \\
\mu^*(q_2, uv, 10, q_3) &= \vee\{\tfrac{3}{4} \wedge 0, 0 \wedge 0, \tfrac{1}{2} \wedge 0\} &= 0, \\[4pt]
\mu^*(q_3, uv, 10, q_1) &= \vee\{0 \wedge 0, 0 \wedge \tfrac{3}{8}, 0 \wedge \tfrac{1}{8}\} &= 0, \\
\mu^*(q_3, uv, 10, q_2) &= \vee\{0 \wedge \tfrac{3}{8}, 0 \wedge 0, 0 \wedge \tfrac{1}{8}\} &= 0, \\
\mu^*(q_3, uv, 10, q_3) &= \vee\{0 \wedge \tfrac{1}{8}, 0 \wedge 0, 0 \wedge 0\} &= 0.
\end{aligned}
$$

$$q^M(q_1, uv, 10) = \mu^*(q_1, uv, 10, q_1) \vee \mu^*(q_1, uv, 10, q_2) \vee \mu^*(q_1, uv, 10, q_3)$$
$$= (\vee\{\mu(q_1, u, 1, q) \wedge \mu(q, v, 0, q_1) \mid q \in Q\}) \vee$$
$$(\vee\{\mu(q_1, u, 1, q) \wedge \mu(q, v, 0, q_2) \mid q \in Q\}) \vee$$
$$(\vee\{\mu(q_1, u, 1, q) \wedge \mu(q, v, 0, q_3) \mid q \in Q\})$$
$$= (\vee\{0 \wedge 0, \tfrac{3}{4} \wedge \tfrac{3}{8}, \tfrac{1}{2} \wedge \tfrac{1}{8}\}) \vee (\vee\{0 \wedge \tfrac{3}{8}, \tfrac{3}{4} \wedge 0, \tfrac{1}{2} \wedge \tfrac{1}{8}\}) \vee$$
$$(\vee\{0 \wedge 0, \tfrac{3}{4} \wedge 0, \tfrac{1}{2} \wedge 0\})$$
$$= \tfrac{3}{8}.$$

$$q^M(q_2, uv, 10) = \mu^*(q_2, uv, 10, q_1) \vee \mu^*(q_2, uv, 10, q_2) \vee \mu^*(q_2, uv, 10, q_3)$$
$$= (\vee\{\mu(q_2, u, 1, q) \wedge \mu(q, v, 0, q_1) \mid q \in Q\}) \vee$$
$$(\vee\{\mu(q_2, u, 1, q) \wedge \mu(q, v, 0, q_2) \mid q \in Q\}) \vee$$
$$(\vee\{\mu(q_2, u, 1, q) \wedge \mu(q, v, 0, q_3) \mid q \in Q\})$$
$$= (\vee\{\tfrac{3}{4} \wedge 0, 0 \wedge \tfrac{3}{8}, \tfrac{1}{2} \wedge \tfrac{1}{8}\}) \vee (\vee\{\tfrac{3}{4} \wedge \tfrac{3}{8}, 0 \wedge 0, \tfrac{1}{2} \wedge \tfrac{1}{8}\}) \vee$$
$$(\vee\{\tfrac{3}{4} \wedge 0, 0 \wedge 0, \tfrac{1}{2} \wedge 0\})$$
$$= \tfrac{3}{8}.$$

$$q^M(q_3, uv, 10) = \mu^*(q_3, uv, 10, q_1) \vee \mu^*(q_3, uv, 10, q_2) \vee \mu^*(q_3, uv, 10, q_3)$$
$$= (\vee\{\mu(q_3, u, 1, q) \wedge \mu(q, v, 0, q_1) \mid q \in Q\}) \vee$$
$$(\vee\{\mu(q_3, u, 1, q) \wedge \mu(q, v, 0, q_2) \mid q \in Q\}) \vee$$
$$(\vee\{\mu(q_3, u, 1, q) \wedge \mu(q, v, 0, q_3) \mid q \in Q\})$$
$$= (\vee\{0 \wedge 0, 0 \wedge \tfrac{3}{8}, 0 \wedge \tfrac{1}{8}\}) \vee (\vee\{0 \wedge \tfrac{3}{8}, 0 \wedge 0, 0 \wedge \tfrac{1}{8}\}) \vee$$
$$(\vee\{0 \wedge \tfrac{1}{8}, 0 \wedge 0, 0 \wedge 0\})$$
$$= 0.$$

$$P^M(uv, 10) = \begin{bmatrix} \tfrac{3}{8} & \tfrac{1}{8} & 0 \\ \tfrac{1}{8} & \tfrac{3}{8} & 0 \\ 0 & 0 & 0 \end{bmatrix}.$$

Since

$$q^M(q_1, uv, 10) = \tfrac{3}{8},$$
$$q^M(q_2, uv, 10) = \tfrac{3}{8},$$
$$q^M(q_3, uv, 10) = 0,$$

$$Q^M(uv, 10) = \begin{bmatrix} \tfrac{3}{8} \\ \tfrac{3}{8} \\ 0 \end{bmatrix}.$$

If $(X \times Y)^*$ is ordered so that the first four members are $(u, 0) < (u, 1) < (v, 0) < (v, 1)$, then

$$A_1^M = \begin{bmatrix} 0 & \tfrac{3}{4} & \tfrac{3}{8} & \tfrac{5}{8} \\ 0 & \tfrac{3}{4} & \tfrac{3}{8} & \tfrac{5}{8} \\ \tfrac{1}{4} & 0 & \tfrac{1}{8} & 0 \end{bmatrix}.$$

Since

$$\begin{bmatrix} \frac{3}{8} \\ \frac{3}{8} \\ \frac{1}{8} \end{bmatrix} = (\frac{3}{8} \wedge \begin{bmatrix} \frac{3}{4} \\ \frac{3}{4} \\ 0 \end{bmatrix}) \vee (\frac{1}{8} \wedge \begin{bmatrix} 0 \\ 0 \\ \frac{1}{4} \end{bmatrix})$$

and

$$\begin{bmatrix} \frac{5}{8} \\ \frac{5}{8} \\ \frac{5}{8} \\ 0 \end{bmatrix} = \frac{5}{8} \wedge \begin{bmatrix} \frac{3}{4} \\ \frac{3}{4} \\ 0 \end{bmatrix},$$

we have

$$B_1^M = \begin{bmatrix} 0 & \frac{3}{4} \\ 0 & \frac{3}{4} \\ \frac{1}{4} & 0 \end{bmatrix}.$$

If the first four elements of $(X \times Y)^*$ are such that $(v,0) < (v,1) < (u,0) < (u,1)$, then

$$A_1^M = \begin{bmatrix} \frac{3}{8} & \frac{5}{8} & 0 & \frac{3}{4} \\ \frac{3}{8} & \frac{5}{8} & 0 & \frac{3}{4} \\ \frac{1}{8} & 0 & \frac{1}{4} & 0 \end{bmatrix}.$$

In this case,

$$B_1^M = \begin{bmatrix} \frac{3}{8} & \frac{5}{8} & 0 & \frac{3}{4} \\ \frac{3}{8} & \frac{5}{8} & 0 & \frac{3}{4} \\ \frac{1}{8} & 0 & \frac{1}{4} & 0 \end{bmatrix}.$$

Proposition 2.9.5 *If $B_{n-1}^M = B_n^M$ for some n, then $B_m^M = B_{n-1}^M$ for all $m \geq n - 1$.*

Proof. It suffices to show that $B_{n+1}^M = B_n^M$. Let $(x,y) \in (X \times Y)^*$, where $|x| = n$. Then $Q^M(x,y)$ is a convex max-min combination of columns of B_{n-1}^M. Since

$$Q^M(ux, vy) = P^M(u,v) \otimes Q^M(x,y),$$

the result now follows from Proposition 2.9.3 and the fact that $\otimes$ is associative. ∎

Theorem 2.9.6 *Let $M = (Q, \mu)$ be an MSLM. Then $B^M = B_m^M$, where $m \leq d^{|Q|} - 1$ and d is the number of distinct entries in A^M.*

Proof. Clearly, $\mid B_n^M \mid \, \leq \, \mid B_{n+1}^M \mid \, \leq \, \mid B^M \mid \, \leq d^{|Q|}$ for every n. Thus $B_{m-1}^M = B_m^M$, where $m = d^{|Q|} - 1$. The result now follows from Proposition 2.9.5. ∎

Proposition 2.9.7 *The matrix B^M can be constructed effectively from M.* ∎

Proposition 2.9.8 *$C[\rho(A^M)]$ is admissible and finitary.* ∎

Definition 2.9.9 *Let $M = (Q, \mu)$ be a MSLM. A **state distribution** (sd) of M is a function η from Q into $[0, 1]$. η is said to be **concentrated** at $q \in Q$ if $\eta(q) = 1$ and $\eta(q') = 0 \; \forall q' \in Q \backslash \{q\}$.*

We will also use the symbol η to denote the row matrix whose i-th row is $\eta(q_i)$.

Definition 2.9.10 *An **initialized max-min sequential-like machine** (**IMSLM**) is an ordered pair (M, η), where M is an MSLM and η is an sd of M.*

If η is concentrated at q, we write (M, q) for (M, η).

Definition 2.9.11 *Let $I = (M, \eta)$ be an IMSLM. The **response function** r^I of I is a function from $(X \times Y)^*$ into $[0, 1]$ such that*

$$r^I(x, y) = \vee_{q \in Q} \{\eta(q) \wedge q^M(q, x, y)\}.$$

Let $R^I = \eta \otimes A^M$. Then R^I is a row matrix whose entries are $r^I(x, y)$. Furthermore, R^I is a convex max-min combination of $\rho(A^M)$.

Definition 2.9.12 *Let I_1 and I_2 be IMSLMs. Then I_1 and I_2 are called **equivalent** ($\sim$) if $r^{I_1} = r^{I_2}$.*

Definition 2.9.13 *Let η_1 and η_2 be state distributions of M. Then η_1 and η_2 are called M-**equivalent** ($\overset{M}{\sim}$) if $(M, \eta_1) \sim (M, \eta_2)$.*

If no confusion arises, we use the symbol $\sim$ for $\overset{M}{\sim}$.
If η_1 is concentrated at q, then we write $q \sim \eta_2$ for $\eta_1 \sim \eta_2$.

Proposition 2.9.14 *Let η_1 and η_2 be state distributions of M. Then $\eta_1 \sim \eta_2$ if and only if*

$$\eta_1 \otimes B^M = \eta_2 \otimes B^M. \; ∎$$

Definition 2.9.15 *Let $M_1 = (Q_1, \mu_1)$ and $M_2 = (Q_2, \mu_2)$ be MSLMs.*

*(1) M_1 and M_2 are called **statewise equivalent** ($\sim$) if for every $q' \in Q_1$, there exists $q'' \in Q_2$ such that $(M_1, q') \sim (M_2, q'')$ and vice versa.*

*(2) M_1 and M_2 are called **compositewise equivalent** ($\simeq$) if for every $q \in Q_1$, there exists an sd η of M_2 such that $(M_1, q) \sim (M_2, \eta)$ and vice versa.*

*(3) M_1 and M_2 are called **distributionwise equivalent** ($\approx$) if for every sd η_1 of M_1, there exists an sd η_2 of M_2 such that $(M_1, \eta_1) \sim (M_2, \eta_2)$ and vice versa.*

Theorem 2.9.16 *Let M_1 and M_2 be MSLMs.*

(1) $M_1 \sim M_2$ if and only if $\rho(A^{M_1}) = \rho(A^{M_2})$.

(2) $M_1 \simeq M_2$ if and only if $\rho(A^{M_1}) \subseteq C[\rho(A^{M_2})]$ and $\rho(A^{M_2}) \subseteq C[\rho(A^{M_1})]$.

(3) $M_1 \approx M_2$ if and only if $C[\rho(A^{M_1})] = C[\rho(A^{M_2})]$. ∎

Proposition 2.9.17 *Let M_1 and M_2 be MSLMs such that $M_1 \sim M_2$. Then $M_1 \simeq M_2$.*

Proof. The proof follows from Propositions 2.8.8(1) and Theorem 2.9.16. ∎

Proposition 2.9.18 *Let M_1 and M_2 be MSLMs. Then $M_1 \simeq M_2$ if and only if $M_1 \approx M_2$.*

Proof. It follows from Theorem 2.9.16(2) and (3) that $M_1 \approx M_2$ implies $M_1 \simeq M_2$. That $M_1 \simeq M_2$ implies $M_1 \approx M_2$ follows from Proposition 2.8.8(2) and (3) and Theorem 2.9.16. ∎

Example 2.9.19 *Let $M_2 = (Q, X, Y, \nu)$, where $Q = \{s_0, s_3\}$, $X = \{u, v\}$, $Y = \{0, 1\}$, and*

$$\nu : Q \times X \times Y \times Q \to [0, 1]$$

is defined as follows:

$$
\begin{aligned}
\nu(s_0, u, 1, s_0) &= \tfrac{3}{4}, & \nu(s_0, v, 0, s_0) &= \tfrac{3}{8}, \\
\nu(s_0, u, 1, s_3) &= \tfrac{1}{2}, & \nu(s_0, v, 1, s_3) &= \tfrac{5}{8}, \\
\nu(s_3, u, 0, s_0) &= \tfrac{1}{4}, & \nu(s_3, v, 0, s_0) &= \tfrac{1}{8}, \\
\nu(s, w, z, s') &= 0,
\end{aligned}
$$

for all other (s, w, z, s').

Let $\eta_2 : S \to [0,1]$. Then

$$
\begin{aligned}
r^{I_2}(u,1) &= [\eta_2(s_0) \wedge (\nu(s_0,u,1,s_0) \vee \nu(s_0,u,1,s_3))] \vee \\
&\quad [\eta_2(s_3) \wedge (\nu(s_3,u,1,s_0) \vee \nu(s_3,u,1,s_3))] \\
&= [\eta_2(s_0) \wedge (\tfrac{3}{4} \vee \tfrac{1}{2})] \vee [\eta_2(s_3) \wedge (0 \vee 0)] \\
&= [\eta_2(s_0) \wedge \tfrac{3}{4}] \vee [\eta_2(s_3) \wedge 0] \\
&= \eta_2(s_0) \wedge \tfrac{3}{4}.
\end{aligned}
$$

Consider $M = (Q, X, Y, \mu)$ of Example 2.9.4. Let $\eta : Q \to [0,1]$. Then

$$
\begin{aligned}
r^{I}(u,1) &= [\eta(q_1) \wedge (\mu(q_1,u,1,q_1) \vee \mu(q_1,u,1,q_2) \vee \mu(q_1,u,1,q_3))] \vee \\
&\quad [\eta(q_2) \wedge (\mu(q_2,u,1,q_1) \vee \mu(q_2,u,1,q_2) \vee \mu(q_2,u,1,q_3))] \vee \\
&\quad [\eta(q_3) \wedge (\mu(q_3,u,1,q_1) \vee \mu(q_3,u,1,q_2) \vee \mu(q_3,u,1,q_3))] \\
&= [\eta(q_1) \wedge (0 \vee \tfrac{3}{4} \vee \tfrac{1}{2})] \vee [\eta(q_2) \wedge (\tfrac{3}{4} \vee 0 \vee \tfrac{1}{2})] \vee \\
&\quad [\eta(q_3) \wedge (0 \vee 0 \vee 0)] \\
&= [\eta(q_1) \wedge \tfrac{3}{4}] \vee [\eta(q_2) \wedge \tfrac{3}{4}].
\end{aligned}
$$

2.10 Irreducibility and Minimality

Definition 2.10.1 *Let $M = (Q, \mu)$ be an MSLM.*

*(1) M is called **statewise irreducible** if for every q', $q'' \in Q$, $q' \sim q''$ implies $q' = q''$.*

*(2) M is called **compositewise irreducible** if for every $q \in Q$ and sd η of M, $q \sim \eta$ implies $\eta(q) > 0$.*

*(3) M is called **distributionwise irreducible** if for every sd η_1 and η_2 of M, $\eta_1 \sim \eta_2$ implies $\eta_1 = \eta_2$.*

Example 2.10.2 *Let $X = \{u\}$ and $Y = \{1\}$. Let $Q_1 = \{q_1, q_2\}$ and $Q_2 = \{s_1, s_2\}$. Define $\mu_1 : Q_1 \times X \times Y \times Q_1 \to [0,1]$ and $\mu_2 : Q_2 \times X \times Y \times Q_2 \to [0,1]$ as follows:*

$$
\begin{aligned}
\mu_1(q_1,u,1,q_2) &= \tfrac{1}{3} & \mu_1(q_2,u,1,q_1) &= \tfrac{2}{3} \\
\mu_1(q_1,u,1,q_1) &= \tfrac{1}{2} & \mu_1(q_2,u,1,q_2) &= 0 \\
\mu_2(s_1,u,1,s_2) &= \tfrac{1}{2} & \mu_2(s_2,u,1,s_1) &= \tfrac{2}{3} \\
\mu_2(s_1,u,1,s_1) &= 0 & \mu_2(s_2,u,1,s_2) &= 0.
\end{aligned}
$$

Let $M_1 = (Q_1, \mu_1)$, $M_2 = (Q_2, \mu_2)$, $I_1 = (M_1, \eta_1)$, and $I_2 = (M_2, \eta_2)$. Then

$$
\begin{aligned}
r^{I_1}(u,1) &= [\eta_1(q_1) \wedge ((\mu_1(q_1,u,1,q_1) \vee \mu_1(q_1,u,1,q_2))] \vee \\
&\quad [\eta_1(q_2) \wedge ((\mu_1(q_2,u,1,q_1) \vee \mu_1(q_2,u,1,q_2))] \\
&= (\eta_1(q_1) \wedge (\tfrac{1}{2} \vee \tfrac{1}{3})) \vee (\eta_1(q_2) \wedge (\tfrac{2}{3} \vee 0)) \\
&= (\eta_1(q_1) \wedge \tfrac{1}{2}) \vee (\eta_1(q_2) \wedge \tfrac{2}{3})
\end{aligned}
$$

and

$$r^{I_2}(u, 1) = (\eta_2(s_1) \wedge \frac{1}{2}) \vee (\eta_2(s_2) \wedge \frac{2}{3}).$$

In fact, it follows that

$$r^{I_1}(u^n, 1^n) = (\eta_1(q_1) \wedge \frac{1}{2}) \vee (\eta_1(q_2) \wedge \frac{1}{2})$$

and

$$r^{I_2}(u^n, 1^n) = (\eta_2(s_1) \wedge \frac{1}{2}) \vee (\eta_2(s_2) \wedge \frac{1}{2})$$

for $n = 2, 3, \ldots$.
   We also have that

$$P^{M_1}(u, 1) = \begin{bmatrix} \frac{1}{2} & \frac{1}{3} \\ \frac{2}{3} & 0 \end{bmatrix},$$

$$q^{M_1}(q_1, u, 1) = \mu(q_1, u, 1, q_1) \vee \mu(q_1, u, 1, q_2) = \frac{1}{2} \vee \frac{1}{3} = \frac{1}{2},$$

$$q^{M_1}(q_2, u, 1) = \mu(q_2, u, 1, q_1) \vee \mu(q_2, u, 1, q_2) = \frac{2}{3} \vee 0 = \frac{2}{3},$$

$$Q^{M_1}(u, 1) = \begin{bmatrix} \frac{1}{2} \\ \frac{2}{3} \end{bmatrix},$$

$$A_1^{M_1} = \begin{bmatrix} \frac{1}{2} \\ \frac{2}{3} \end{bmatrix}.$$

Now

$$\begin{array}{llll}
\mu_1^*(q_1, uu, 11, q_1) & = & \frac{1}{2}, & \mu_1^*(q_1, uu, 11, q_2) & = & \frac{1}{3} \\
\mu_1^*(q_2, uu, 11, q_1) & = & \frac{1}{2}, & \mu_1^*(q_2, uu, 11, q_2) & = & \frac{1}{3}.
\end{array}$$

Thus

$$P^{M_1}(q_1, uu, 11) = \begin{bmatrix} \frac{1}{2} & \frac{1}{3} \\ \frac{1}{2} & \frac{1}{3} \end{bmatrix},$$

$$q^{M_1}(q_1, uu, 11) = \frac{1}{2} \vee \frac{1}{3} = \frac{1}{2},$$

$$q^{M_1}(q_2, uu, 11) = \frac{1}{2} \vee \frac{1}{3} = \frac{1}{2},$$

$$Q^{M_1}(uu, 11) = \begin{bmatrix} \frac{1}{2} \\ \frac{1}{2} \end{bmatrix},$$

$$A_2^{M_1}(uu, 11) = \begin{bmatrix} \frac{1}{2} & \frac{1}{2} \\ \frac{2}{3} & \frac{1}{2} \end{bmatrix}.$$

It is easy to see that

$$P^{M_1}(u^n, 1^n) = \begin{bmatrix} \frac{1}{2} & \frac{1}{3} \\ \frac{1}{2} & \frac{1}{3} \end{bmatrix}$$

for $n = 2, 3, \ldots$. Hence

$$A^{M_1} = \begin{bmatrix} \frac{1}{2} & \frac{1}{2} & \frac{1}{2} & \cdots \\ \frac{2}{3} & \frac{1}{2} & \frac{1}{2} & \cdots \end{bmatrix}$$

and

$$B^{M_1} = \begin{bmatrix} \frac{1}{2} \\ \frac{2}{3} \end{bmatrix}.$$

Now

$$P^{M_2}(u, 1) = \begin{bmatrix} 0 & \frac{1}{2} \\ \frac{2}{3} & 0 \end{bmatrix},$$

$$q^{M_2}(s_1, u, 1) = 0 \vee \frac{1}{2} = \frac{1}{2},$$

$$q^{M_2}(s_2, u, 1) = \frac{2}{3} \vee 0 = \frac{2}{3},$$

$$Q^{M_2}(u, 1) = \begin{bmatrix} \frac{1}{2} \\ \frac{2}{3} \end{bmatrix},$$

$$A_1^{M_2} = \begin{bmatrix} \frac{1}{2} \\ \frac{2}{3} \end{bmatrix}.$$

Now

$$\mu_2^*(s_1, uu, 11, s_1) = \tfrac{1}{2}, \qquad \mu_2^*(s_1, uu, 11, s_2) = 0$$
$$\mu_2^*(s_2, uu, 11, s_1) = 0, \qquad \mu_2^*(s_2, uu, 11, s_2) = \tfrac{1}{2}.$$

Thus

$$P^{M_2}(uu, 11) = \begin{bmatrix} \tfrac{1}{2} & 0 \\ 0 & \tfrac{1}{2} \end{bmatrix},$$

$$q^{M_2}(s_1, uu, 11) = \frac{1}{2} \vee 0 = \frac{1}{2},$$

$$q^{M_2}(s_2, uu, 11) = 0 \vee \frac{1}{2} = \frac{1}{2},$$

$$Q^{M_2}(uu, 11) = \begin{bmatrix} \tfrac{1}{2} \\ \tfrac{1}{2} \end{bmatrix},$$

$$A_2^{M_2}(uu, 11) = \begin{bmatrix} \tfrac{1}{2} & \tfrac{1}{2} \\ \tfrac{2}{3} & \tfrac{1}{2} \end{bmatrix}.$$

It follows that

$$P^{M_1}(u^n, 1^n) = \begin{bmatrix} \tfrac{1}{2} & 0 \\ 0 & \tfrac{1}{2} \end{bmatrix}$$

for $n = 2, 4, 6, \ldots$ and

$$P^{M_2}(u^n, 1^n) = \begin{bmatrix} 0 & \tfrac{1}{2} \\ \tfrac{1}{2} & 0 \end{bmatrix}$$

for $n = 3, 5, 7, \ldots$. Hence

$$A^{M_2} = \begin{bmatrix} \tfrac{1}{2} & \tfrac{1}{2} & \tfrac{1}{2} & \cdots \\ \tfrac{2}{3} & \tfrac{1}{2} & \tfrac{1}{2} & \cdots \end{bmatrix}$$

and

$$B^{M_2} = \begin{bmatrix} \tfrac{1}{2} \\ \tfrac{2}{3} \end{bmatrix}.$$

Now $(1, 0) \otimes B^{M_1} = (1, 0) \otimes B^{M_2}$ and $(0, 1) \otimes B^{M_1} = (0, 1) \otimes B^{M_2}$. Thus $q_1 \sim s_1$ and $q_2 \sim s_2$.

Suppose that $\widehat{\eta} : Q_2 \to [0, 1]$ is such that $\widehat{\eta}(s_1) = 1$ and $\widehat{\eta}(s_2) = 0$. Let η be any sd of M_2. If $\eta(s_1) = 0$ and $\eta(s_2) = \tfrac{1}{2}$, then $\widehat{\eta} \sim \eta$ since $(1, 0) \otimes B^{M_2} = (0, \tfrac{1}{2}) \otimes B^{M_2}$. However, $\eta(s_1) \not> 0$. Thus M_2 is not compositewise irreducible.

Example 2.10.3 *Let $X = \{u\}$ and $Y = \{0,1\}$. Let $Q = \{q_1, q_2\}$. Define $\mu : Q \times X \times Y \times Q \to [0,1]$ as follows:*

$$\mu(q_1, u, 1, q_2) = \frac{1}{2} \text{ and } \mu(q_2, u, 0, q_1) = \frac{2}{3},$$

with μ of any other element equal to 0. Let $M = (Q, \mu)$ and $I = (M, \eta)$. Then

$$r^I(u, 1) = (\eta(q_1) \wedge (0 \vee \frac{1}{2})) \vee (\eta(q_2) \wedge (0 \vee 0)) = \eta(q_1) \wedge \frac{1}{2},$$

$$r^I(u, 0) = (\eta(q_1) \wedge (0 \vee 0)) \vee (\eta(q_2) \wedge (\frac{2}{3} \vee 0)) = \eta(q_2) \wedge \frac{2}{3},$$

$$
\begin{aligned}
r^I(uu, 11) &= 0, \\
r^I(uu, 10) &= \eta(q_1) \wedge \tfrac{1}{2}, \\
r^I(uu, 01) &= \eta(q_2) \wedge \tfrac{1}{2}, \\
r^I(uu, 00) &= 0,
\end{aligned}
$$

$$
\begin{aligned}
r^I(uuu, 111) &= 0, \\
r^I(uuu, 110) &= 0, \\
r^I(uuu, 101) &= \eta(q_1) \wedge \tfrac{1}{2}, \\
r^I(uuu, 100) &= 0, \\
r^I(uuu, 011) &= 0, \\
r^I(uuu, 010) &= \eta(q_2) \wedge \tfrac{1}{2}, \\
r^I(uuu, 001) &= 0, \\
r^I(uuu, 000) &= 0.
\end{aligned}
$$

We see that $r^I(u \ldots u, 1010 \ldots) = \eta(q_1) \wedge \frac{1}{2}$ and $r^I(u \ldots u, 0101 \ldots) = \eta(q_2) \wedge \frac{1}{2}$. Hence if $\eta(q_1) > \frac{1}{2} < \eta(q_2)$,

$$A^M = \begin{bmatrix} \frac{1}{2} & 0 & 0 & \frac{1}{2} & 0 & 0 & 0 & 0 & \frac{1}{2} & 0 & 0 & 0 & 0 & 0 & \cdots \\ 0 & \frac{2}{3} & 0 & 0 & \frac{1}{2} & 0 & 0 & 0 & 0 & 0 & 0 & \frac{1}{2} & 0 & 0 & \cdots \end{bmatrix},$$

and

$$B^M = \begin{bmatrix} \frac{1}{2} & 0 \\ 0 & \frac{2}{3} \end{bmatrix},$$

where $(u, 1) < (u, 0) < (uu, 11) < (uu, 10) < (uu, 01) < (uu, 00) < (uuu, 111) < \ldots < (uuu, 000) < \ldots$. Thus it follows that M is compositewise irreducible.

Theorem 2.10.4 *Let M be an MSLM. Then the following assertions hold.*

(1) M is statewise irreducible if and only if no two rows of B^M are identical.

(2) M is compositewise irreducible if and only if $\rho(B^M)$ is a set of vertices of $C[\rho(B^M)]$, i.e., no row of B^M is a convex max-min combination of the other rows of B^M.

(3) M is distributionwise irreducible if and only if $\rho(B^M)$ is a basis of $C[\rho(B^M)]$.

Proof. (1) The proof follows from Proposition 2.9.14.

(2) The proof follows from Propositions 2.9.14 and 2.8.14.

(3) The proof follows from Proposition 2.9.14 and the definition of basis.

∎

All assertions of Theorem 2.10.4 are also valid if B^M is replaced by A^M.

Proposition 2.10.5 *Let M be an MSLM. If M is distributionwise irreducible, then M is compositewise irreducible.* ∎

Proposition 2.10.6 *Let M be an MSLM. If M is compositewise irreducible, then M is statewise irreducible.* ∎

Definition 2.10.7 *Let M be an MSLM. M is called **statewise (compositewise) minimal** if M is not statewise (compositewise) equivalent to an MSLM with a fewer number of states.*

Let $M = (Q, p)$. The cardinality of M is defined by $|M| = |Q|$.

Theorem 2.10.8 *Let M be an MSLM. Then M is statewise minimal if and only if M is statewise irreducible.*

Proof. Let $M = (Q, \mu)$. Suppose that M is not statewise minimal. Then there exists an MSLM M' with $|M'| < |M|$, $M' \sim M$. By Theorems 2.9.16(1) and 2.10.4(1), M is not statewise irreducible. Conversely, suppose that M is not statewise irreducible. Then there exist $q', q'' \in Q$ such that $q' \sim q''$, but $q' \neq q''$. By renumbering the elements of Q, if necessary, we may assume that $q' = q_{n-1}$ and $q'' = q_n$, where $n = |Q|$. Let $Q' = Q \setminus \{q_n\}$ and $M' = (Q', \mu')$, where for $i = 1, 2, \ldots, n - 1$,

$$\mu'(q_i, u, v, q_j) = \begin{cases} \mu(q_i, u, v, q_j) & \text{if } j = 1, 2, \ldots, n-2 \\ \mu(q_i, u, v, q_{n-1}) \vee \mu(q_i, u, v, q_n) & \text{if } j = n - 1. \end{cases}$$

Now for every $(x, y) \in (X \times Y)^*$,

$$q^M(q_i, x, y) = \begin{cases} q^{M'}(q_i, x, y) & \text{if } i = 1, 2, \ldots, n-1 \\ q^{M'}(q_{n-1}, x, y) & \text{if } i = n. \end{cases}$$

Therefore, $\rho(A^M) = \rho(A^{M'})$. By Theorem 2.9.16(1), $M \sim M'$. Thus M is not statewise minimal. ∎

In the next example, we illustrate the proof of Theorem 2.10.8.

Example 2.10.9 *Let* $M_1 = (Q, X, Y, \mu)$ *be defined as in Example 2.9.4. Rather than* $q_{n-1} \sim q_n$ *as in the proof of Theorem 2.10.8, we have* $q_1 \sim q_2$. *Thus for* $x \in X$ *and* $y \in Y$,

$$\mu'(q_i, x, y, q_j) = \begin{cases} \mu(q_i, x, y, q_j) & \text{if } j = 3 \\ \mu(q_i, x, y, q_2) \vee \mu(q_3, x, y, q_1) & \text{if } j = 2. \end{cases}$$

Thus

$$\begin{aligned} \mu'(q_2, x, y, q_3) &= \mu(q_2, x, y, q_3) \\ \mu'(q_3, x, y, q_3) &= \mu(q_3, x, y, q_3) \\ \mu'(q_2, x, y, q_2) &= \mu(q_2, x, y, q_2) \vee \mu(q_2, x, y, q_1) \\ \mu'(q_3, x, y, q_2) &= \mu(q_3, x, y, q_2) \vee \mu(q_3, x, y, q_1). \end{aligned}$$

Hence

$$\begin{aligned} \mu'(q_2, u, 1, q_3) &= \tfrac{1}{2}, \\ \mu'(q_2, v, 1, q_3) &= \tfrac{5}{8}, \\ \mu'(q_2, u, 1, q_2) &= 0 \vee \tfrac{3}{4} = \tfrac{3}{4}, \\ \mu'(q_2, v, 0, q_2) &= 0 \vee \tfrac{3}{8} = \tfrac{3}{8}, \\ \mu'(q_3, u, 0, q_2) &= \tfrac{1}{4} \vee \tfrac{1}{4} = \tfrac{1}{4}, \\ \mu'(q_3, v, 0, q_2) &= \tfrac{1}{8} \vee \tfrac{1}{8} = \tfrac{1}{8}. \end{aligned}$$

Theorem 2.10.10 *Let M be an MSLM. Then there exists an effective procedure for constructing a statewise minimal MSLM that is statewise equivalent to M.*

Proof. Consider the following procedure:

1. Construct B^M.

2. If there exist two rows of B^M that are identical, then proceed to Step 3; otherwise stop.

3. Construct M' as given in the proof of Theorem 2.10.8. Return to Step 1 with M' replacing M.

Since $|M|$ is finite, the procedure must terminate in a finite number of steps. By Proposition 2.9.7, B^M can be effectively constructed. Thus the procedure is effective. By Theorem 2.10.8, the resulting MSLM is the desired MSLM. ∎

Theorem 2.10.11 *Let M be an MSLM. Then M is compositewise minimal if and only if M is compositewise irreducible.*

Proof. Let $M = (Q, \mu)$. Suppose that M is not compositewise minimal. Then there exists an MSLM M' with $|M'| < |M|$ such that $M' \simeq M$. By Theorems 2.9.16(3), 2.10.4(2), and Propositions 2.9.18, 2.8.30, and 2.9.8, M is not compositewise irreducible. Conversely, suppose that M is not compositewise irreducible. Then there exist $q' \in Q$ and an sd η of M such that $q' \sim \eta$, but $\eta(q') = 0$. By renumbering the elements of Q, if necessary, we may assume that $q' = q_n$, where $n = |Q|$. Since $q_n \sim \eta$ and $\eta(q_n) = 0$,

$$q^M(q_n, x, y) = \vee \{\eta(q_i) \wedge q^M(q_i, x, y) \mid i = 1, 2, \ldots, n\},$$

for all $(x, y) \in (X \times Y)^*$. Let $M' = (Q', \mu')$, where $Q' = Q \backslash \{q_n\}$ and for $i, j = 1, 2, \ldots, n - 1$,

$$\mu'(q_i, u, v, q_j) = \mu(q_i, u, v, q_j) \vee (\eta(q_j) \wedge \mu(q_i, u, v, q_n)).$$

Now for every $(x, y) \in (X \times Y)^*$,

$$q^M(q_i, x, y) = \begin{cases} q^{M'}(q_i, x, y) & \text{if } i = 1, \ldots, n-1 \\ \vee \{\eta(q_j) \wedge q^{M'}(q_j, x, y) \mid j = 1, 2, \ldots, n\} & \text{if } i = n. \end{cases}$$

Hence $C[\rho(A^M)] = C[\rho(A^{M'})]$. By Theorem 2.9.16(3) and Proposition 2.9.18, $M' \simeq M$. Thus M is not compositewise minimal. ∎

Theorem 2.10.12 *Let M be an MSLM. Then there exists an effective procedure for constructing a compositewise minimal MSLM that is compositewise equivalent to M.*

Proof. Consider the following procedure:

1. Construct B^M.

2. If there exists a row of B^M that is a convex max-min combination of the other rows of B^M, then proceed to Step 3; otherwise stop.

3. Construct M' as given in the proof of Theorem 2.10.11. Return to Step 1 with M' replacing M.

Since $|M|$ is finite, the procedure must terminate in a finite number of steps. By Proposition 2.9.7, B^M can be constructed effectively. By Propositions 2.8.6 and 2.9.8 and the fact that $|M|$ is finite, Step 2 can be carried out effectively. Thus the procedure is effective. By Theorem 2.10.11, the resulting MSLM is the desired MSLM. ∎

Proposition 2.10.13 *Let M_1 and M_2 be MSLMs. Let $\rho(B^{M_1})$ and $\rho(B^{M_2})$ be fundamental sets of vertices of $C[\rho(B^{M_1})]$ and $C[\rho(B^{M_2})]$, respectively. Then $M_1 \simeq M_2$ if and only if $M_1 \sim M_2$.*

Proof. The proof follows from Theorems 2.9.16(3), 2.8.26, Proposition 2.9.18, and Theorem 2.9.16(1) and (2). ∎

Definition 2.10.14 *Let M be an MSLM. Then M is called **fundamental** if $C[\rho(B^M)]$ is fundamental.*

Proposition 2.10.15 *Let M_1 and M_2 be MSLMs. If M_1, M_2 are compositewise minimal, M_1 is fundamental, and $M_1 \simeq M_2$, then M_2 is fundamental and $M_1 \sim M_2$.*

Proof. The proof follows from Theorems 2.8.26, 2.10.4(2), 2.10.11, and Proposition 2.10.13. ∎

Proposition 2.10.16 *Let M_1 and M_2 be MSLMs. If M_1 is distribution-wise irreducible, M_2 is compositewise minimal, and $M_1 \simeq M_2$, then M_2 is fundamental and $M_1 \sim M_2$.*

Proof. The proof follows from Propositions 2.8.36 and 2.10.15. ∎

Proposition 2.10.17 *Let M_1 and M_2 be MSLMs. If M_1, M_2 are statewise minimal, and $M_1 \sim M_2$, then for every $(x,y) \in (X \times Y)^*$,*

$$P^{M_1}(x,y) \otimes B^{M_1} = P^{M_2}(x,y) \otimes B^{M_1},$$

after an appropriate rearrangement of states.

Proof. From Theorems 2.9.16(1), 2.10.4(1), and 2.10.8, $A^{M_1} = A^{M_2}$ after an appropriate rearrangement of states. ∎

Definition 2.10.18 *Let M_1 and M_2 be MSLMs. Then M_1 and M_2 are called **isomorphic** ($\equiv$) if they are equal up to a permutation of states.*

Definition 2.10.19 *Let M be statewise (compositewise) minimal. Then M is called **statewise (compositewise) simple** if there does not exist a statewise (compositewise) minimal MSLM that is statewise (compositewise) equivalent to M, but not isomorphic to M.*

Let M be an MSLM. We introduce the following notation.
(1) $\rho(M) = \cup_{u \in X} \cup_{v \in Y} \rho[P^M(u,v)]$,
(2) $\bar{\rho}(M) = \{j \otimes B^M \mid j \in \rho(M)\}$.

Theorem 2.10.20 *Let M be statewise minimal. Then M is statewise simple if and only if $\rho(B^M)$ is a basis of $\bar{\rho}(M)$.*

Proof. Suppose that $\rho(B^M)$ is a basis of $\bar{\rho}(M)$ and M' is a statewise minimal MSLM such that $M' \sim M$. By Proposition 2.10.17, for every $u \in X$, $v \in Y$,

$$P^M(u,v) = P^M(u,v),$$

after an appropriate rearrangement of states. Hence $M' \equiv M$. Thus M
is statewise simple. Conversely, suppose $\rho(B^M)$ is not a basis of $\bar{\rho}(M)$.
There exist $j \in \rho(M)$ such that

$$j \otimes B^M = j' \otimes B^M, \qquad (2.4)$$

for some $j' \neq j$. Let j be the i-th row of the matrix $P^M(u, v)$. Construct
M' from M by replacing the i-th row of the matrix $P^M(u, v)$ by j' and
leaving the rest unchanged. By (2.4), we have $A^M = A^{M'}$. By Theorems
2.9.16(1), 2.10.4(1), and 2.10.8, M' is statewise minimal and $M' \sim M$.
However, $M' \not\equiv M$. Thus M is not statewise simple. ∎

Proposition 2.10.21 *Suppose that M is statewise minimal. If M is distributionwise irreducible, then M is statewise simple.*

Proof. The proof follows from Theorem 2.10.4(3) and 2.10.20. ∎

Theorem 2.10.22 *Suppose that M is compositewise minimal and fundamental. Then M is compositewise simple if and only if $\rho(B^M)$ is a basis of $\bar{\rho}(M)$.*

Proof. Let $\rho(B^M)$ be a basis of $\bar{\rho}(M)$ and M' be a compositewise
minimal MSLM such that $M' \simeq M$. By Proposition 2.10.6 and Theorems
2.10.8 and 2.10.11, both M and M' are statewise minimal. By Proposition
2.10.15, $M' \sim M$. The remainder of the proof is similar to Theorem 2.10.20.
∎

Proposition 2.10.23 *Suppose that M is compositewise minimal. If M is distributionwise irreducible, then M is compositewise simple.*

Proof. The proof follows from Theorems 2.10.4(3), 2.10.22, and Proposition 2.8.36. ∎

In the next example, we illustrate Theorems 2.10.20 and 2.10.22.

Example 2.10.24 *Let $M = (Q, X, Y, \mu)$, where $Q = \{q_1, q_2\}$, $X = \{u\}$,
$Y = \{0, 1\}$, and*

$$\mu : Q \times X \times Y \times Q \to [0, 1]$$

is defined as follows:

$$\begin{array}{llll}
\mu(q_1, u, 1, q_2) & = & \frac{1}{2}, & \mu(q_2, u, 0, q_1) & = & \frac{3}{4}, \\
\mu(q, w, z, q') & = & 0, &
\end{array}$$

for all other (q, w, z, q'). Let $\eta : Q \to [0, 1]$ be arbitrary.

We see that $q^M(q_1, u^n, y) > 0$ if and only if y is an alternating sequence of 1's and 0's beginning with 1 such that $|y| = n$ and $q^M(q_2, u^n, y) > 0$ if and only if y is an alternating sequence of 0's and 1's beginning with 0 such that $|y| = n$. Also

$$
\begin{array}{llll}
q^M(q_1, u, 1) & = \tfrac{1}{2}, & q^M(q_2, u, 0) & = \tfrac{3}{4}, \\
q^M(q_1, uu, 10) & = \tfrac{1}{2}, & q^M(q_2, uu, 01) & = \tfrac{1}{2}, \\
q^M(q_1, uuu, 101) & = \tfrac{1}{2}, & q^M(q_2, uuu, 010) & = \tfrac{1}{2},
\end{array}
$$

$$\vdots \qquad\qquad\qquad \vdots$$

With the order

$$(u, 1) < (u, 0) < (uu, 11) < (uu, 10) < (uu, 01) < (uu, 00) < (uuu, 111) < \ldots$$

on $(X \times Y)^*$,

$$
A^M = \begin{bmatrix} \tfrac{1}{2} & 0 & 0 & \tfrac{1}{2} & 0 & 0 & 0 & \cdots \\ 0 & \tfrac{3}{4} & 0 & 0 & \tfrac{1}{2} & 0 & 0 & \cdots \end{bmatrix}.
$$

Thus

$$
B^M = \begin{bmatrix} \tfrac{1}{2} & 0 \\ 0 & \tfrac{3}{4} \end{bmatrix}.
$$

With the order

$$(u, 0) < (u, 1) < (uu, 00) < (uu, 01) < (uu, 10) < (uu, 11) < (uuu, 000) < \ldots,$$

$$
B^M = \begin{bmatrix} 0 & \tfrac{1}{2} \\ \tfrac{3}{4} & 0 \end{bmatrix}.
$$

Let $\eta : Q \to [0, 1]$. Then

$$
\begin{array}{lll}
r^I(u, 1) & = (\eta(q_1) \wedge (0 \vee \tfrac{1}{2})) \vee (\eta(q_2) \wedge (0 \vee 0)) & = \eta(q_1) \wedge \tfrac{1}{2}, \\
r^I(u, 0) & = (\eta(q_1) \wedge (0 \vee 0)) \vee (\eta(q_2) \wedge (\tfrac{3}{4} \vee 0)) & = \eta(q_2) \wedge \tfrac{3}{4}, \\
r^I(uu, 11) & = (\eta(q_1) \wedge (0 \vee 0)) \vee (\eta(q_2) \wedge (0 \vee 0)) & = 0, \\
r^I(uu, 10) & = (\eta(q_1) \wedge (\tfrac{1}{2} \vee 0)) \vee (\eta(q_2) \wedge (0 \vee 0)) & = \eta(q_1) \wedge \tfrac{1}{2}, \\
r^I(uu, 01) & = (\eta(q_1) \wedge (0 \vee 0)) \vee (\eta(q_2) \wedge (\tfrac{1}{2} \vee 0)) & = \eta(q_2) \wedge \tfrac{1}{2}, \\
r^I(uu, 00) & = (\eta(q_1) \wedge (0 \vee 0)) \vee (\eta(q_2) \wedge (0 \vee 0)) & = 0, \\
r^I(uuu, 111) & = 0, \\
r^I(uuu, 110) & = 0, \\
r^I(uuu, 101) & = \eta(q_1) \wedge \tfrac{1}{2}, \\
r^I(uuu, 100) & = 0, \\
r^I(uuu, 011) & = 0, \\
r^I(uuu, 010) & = \eta(q_2) \wedge \tfrac{1}{2}, \\
r^I(uuu, 111) & = 0, \\
r^I(uuu, 00) & = 0.
\end{array}
$$

Also,

$$P^M(u,1) \quad = \quad \begin{bmatrix} 0 & \frac{1}{2} \\ 0 & 0 \end{bmatrix},$$

$$P^M(u,0) \quad = \quad \begin{bmatrix} 0 & 0 \\ \frac{3}{4} & 0 \end{bmatrix},$$

$$P^M(uu,11) \quad = \quad \begin{bmatrix} 0 & 0 \\ 0 & 0 \end{bmatrix},$$

$$P^M(uu,10) \quad = \quad \begin{bmatrix} \frac{1}{2} & 0 \\ 0 & 0 \end{bmatrix},$$

$$P^M(uu,01) \quad = \quad \begin{bmatrix} 0 & 0 \\ 0 & \frac{1}{2} \end{bmatrix},$$

$$P^M(uu,00) \quad = \quad \begin{bmatrix} 0 & 0 \\ 0 & 0 \end{bmatrix}.$$

It follows that

$$\rho(M) = \{(0,\frac{1}{2}),(0,0),(\frac{3}{4},0)\}.$$

Now

$$\overline{\rho}(M) \quad = \quad \{(0,\tfrac{1}{2}) \otimes \begin{bmatrix} \frac{1}{2} & 0 \\ 0 & \frac{3}{4} \end{bmatrix}, (0,0) \otimes \begin{bmatrix} \frac{1}{2} & 0 \\ 0 & \frac{3}{4} \end{bmatrix},$$
$$(\tfrac{3}{4},0) \begin{bmatrix} \frac{1}{2} & 0 \\ 0 & \frac{3}{4} \end{bmatrix}\}$$
$$= \quad \{(0,\tfrac{1}{2}),(0,0),(\tfrac{1}{2},0)\}$$

and

$$\rho(B^M) = \{(\frac{1}{2},0),(0,\frac{3}{4})\}.$$

Clearly, $\rho(B^M)$ is not a basis of $\overline{\rho}(M)$. (Note $(0,\frac{1}{2})$ is a unique max-min convex combination of $(0,\frac{3}{4})$, but $(\frac{1}{2},0) = a \wedge (\frac{1}{2},0)$ for $\frac{1}{2} \leq a \leq 1$.)

2.11 Nondeterministic and Deterministic Case

Definition 2.11.1 *Let (Q,μ) be an MSML such that $Im(\mu) \in \{0,1\}$. Then (Q,μ) is called a **nondeterministic sequential-like machine (NSLM)**. If for every $q \in Q$, $u \in X$, there exist unique $v \in Y$ and $q' \in Q$ such that $\mu(q,u,v,q') = 1$, then (Q,μ) is called a **deterministic sequential-like machine (DSLM)**.*

Proposition 2.11.2 *Let M be a DSLM. Then for every $q \in Q$, $x \in X^*$, there exists unique $y \in Y^*$ with $|x| = |y|$ such that $q^M(q,x,y) = 1$.* ∎

Due to the special characters of NSLMs and DSLMs, several of the above results can be strengthened.

Proposition 2.11.3 *Let M be a NSLM. Then M is statewise minimal if and only if M is not statewise equivalent to any NSLM with a fewer number of states.*

Proof. The proof follows from the proof of Theorem 2.10.8. ■

Proposition 2.11.4 *Let M be a NSLM. Then there exists an effective procedure for constructing a statewise minimal NSLM that is statewise equivalent to M.* ■

Proposition 2.11.5 *Let M be a NSLM. Then M is compositewise minimal if and only if M is not compositewise equivalent to any NSLM with a fewer number of states.*

Proof. The proof follows from Proposition 2.8.4 and the proof of Theorem 2.10.11. ■

Proposition 2.11.6 *Let M be a NSLM. Then there exists an effective procedure for constructing a compositewise minimal NSLM that is compositewise equivalent to M.* ■

Proposition 2.11.7 *Let M be a DSLM. Then M is statewise minimal if and only if M is not statewise equivalent to a DSLM with a fewer number of states.*

Proof. The proof follows from the proof of Theorem 2.10.8. ■

Proposition 2.11.8 *Let M be a DSLM. Then there exists an effective procedure for constructing a statewise minimal DSLM that is statewise equivalent to M.* ■

Theorem 2.11.9 *Let M be a DSLM. Then M is compositewise minimal if and only if M is statewise minimal.*

Proof. Suppose that M is not compositewise minimal. Then there exists a row of B^M, say the n-th row, such that $n = |M|$ and which is a convex max-min combination of the other rows of B^M. By Proposition 2.8.4, we may assume, without loss of generality, that

$$x_n = \vee\{x_i \mid i = 1, 2, \ldots, m\},$$

where $m < n$, after an appropriate rearrangement of states, and x_i denotes the i-th row of B^M. Since $x_n^k = 0$ implies $x_i^k = 0$ for $i = 1, 2, \ldots, m$, we have by Proposition 2.11.2 that $x_n = x_i$ for $i = 1, 2, \ldots, m$. Thus M is not statewise minimal. The converse is straightforward. ■

Proposition 2.11.10 *Let M_1 and M_2 be compositewise minimal NSLMs. Then $M_1 \simeq M_2$ if and only if $M_1 \sim M_2$.*

Proof. The proof follows from Propositions 2.8.31 and 2.10.13. ∎

Definition 2.11.11 *Let M be statewise (compositewise) minimal NSLM. Then M is called **statewise (compositewise) ND-simple** if there exists no statewise (compositewise) minimal NSLM that is statewise (compositewise) equivalent to M, but not isomorphic to M.*

Definition 2.11.12 *Let M be a statewise minimal DSLM. Then M is called **statewise D-simple** if there exists no statewise minimal DSLM that is statewise equivalent to M, but not isomorphic to M.*

Theorem 2.11.13 *Let M be a statewise (compositewise) minimal NSLM. Then M is statewise (compositewise) ND-simple if and only if for every $x_0 \in \bar{\rho}(M)$, there exists a unique subset L of $\rho(B^M)$ such that*

$$x_0 = \vee\{x \mid x \in L\}.$$

Proof. The proof is similar to the proofs of Theorems 2.10.20 and 2.10.22. For the compositewise case, Proposition 2.11.10 is needed. ∎

Theorem 2.11.14 *Let M be statewise (compositewise) minimal DSLM. Then M is statewise (compositewise) ND-simple.*

Proof. The proof follows from Theorem 2.11.13 and the fact that every row of any matrix $P^M(u, v)$ has at most one nonzero entry and this nonzero entry is a 1. ∎

Corollary 2.11.15 *Let M be a statewise minimal DSLM. Then M is statewise D-simple.* ∎

2.12   Exercises

1. Let $X = \{(x_1, 0),\ (0, x_2)\}$. Show that X is not a basis of $C(X)$ if either $x_1 < 1$ or $x_2 < 1$.

2. Let $X = \{(1, 0),\ (1, 1)\}$. Show that X is not a basis of $C(X)$.

3. Prove Theorem 2.4.5.

4. Show that $\overline{A}_c^*$ defined in Theorem 2.5.5 has the desired properties.

5. Let V be a vector space over a field F. Prove that the intersection of any collection of subspaces of V is a subspace of V.

6. Let V be a vector space over a field F. Define $s : \mathcal{P}(V) \to \mathcal{P}(V)$ by $\forall X \in \mathcal{P}(V)$, $s(X) = $ the intersection of all subspaces of V that contain X. Show that s satisfies the Exchange Property.

7. Show that I and I_2 of Example 2.9.19 are equivalent if $\eta(q_1) = \eta(q_2) = \eta_2(s_0)$ and $\eta(q_3) = \eta_2(s_3)$.

8. Prove that all the assertions of Theorem 2.10.4 are valid if B^M is replaced by A^M.

Chapter 3

Fuzzy Machines, Languages, and Grammars

3.1 Max-Product Machines

In this and the next section, we present the work given in [202].

Max-product machines may be considered as models of fuzzy systems [256]. In this section, we consider the minimization problem of max-product machines. We present various types of equivalence relations and minimal forms. Among the minimal forms considered are those that are similar to the reduced [29] and minimal state [10] forms of stochastic machines.

We present a type of algebra, called the **max-product algebra**, in order to examine the properties of max-product machines. The role played by max-product algebra in the theory of max-product machines is the same as that played by (convex) linear algebra in the theory of stochastic machines.

We present complete solutions for the minimization problem of max-product machines for both the equivalence relations and minimal forms considered.

Definition 3.1.1 *Let* $A = [a_{ij}]$ *be* $n \times p$ *and* $B = [b_{ij}]$ *be* $p \times m$ *matrices of nonnegative real numbers. Let* $A \odot B$ *be the* $n \times m$ *matrix* $[c_{ij}]$, *where*

$$c_{ij} = \vee\{a_{ik}b_{kj} \mid k = 1, 2, \dots, p\}.$$

Definition 3.1.1 is applicable even if n or m is infinite.

Proposition 3.1.2 *Let* A_1 *be* $n \times p$, A_2 *be* $p \times m$, *and* A_3 *be* $m \times t$ *matrices of nonnegative real numbers. Then*

$$(A_1 \odot A_2) \odot A_3 = A_1 \odot (A_2 \odot A_3),$$

that is, the operation $\odot$ is associative.

We introduce the following notation: a, b, and c (with or without subscripts) will denote nonnegative real numbers; x and y (with or without subscripts) will denote (finite or infinite) sequences of nonnegative real numbers. Superscripts will be used to denote the particular term of the sequence, e.g., x^k will denote the k-th term of the sequence x. X and Y (with or without subscripts) will denote collections of (finite or infinite) sequences of nonnegative real numbers. That is, if S denotes the set of all sequences of nonnegative real numbers, then $X, Y \in \mathcal{P}(S)$.

In the remainder of the chapter, we assume that S denotes a set of sequences of nonnegative real numbers of a fixed length. We also assume that S is our universal set.

(1) Let ax denote the sequence whose k-th term is ax^k.

(2) Let $x_1 \vee x_2 \vee \ldots \vee x_n$ or $\vee\{x_i \mid i = 1, \ldots, n\}$ denote the sequence whose k-th term is $x_1^k \vee x_2^k \vee \ldots \vee x_n^k$.

(3) MP stands for max-product.

Definition 3.1.3 *Let $X = \{x_1, x_2, \ldots, x_n\}$. An **MP-combination** of X is an expression of the form*

$$\vee_{i=1}^n a_i x_i, \tag{3.1}$$

where a_i is a nonnegative real number, $i = 1, 2, \ldots, n$.

*If $0 \leq a_i \leq 1$ for $i = 1, 2, \ldots, n$, then (3.1) is called a **convex MP-combination** of X.*

Definition 3.1.4 *The (**convex**) **MP-span** of X is the collection of all (convex) MP-combinations of finite subsets of X.*

We let $C(X)$ denote the convex MP-span of X. The situation here differs with that of convex max-min combinations since for $0 < a, b < 1$, $a \neq a \cdot b \neq b$.

Proposition 3.1.5 *(1) $X \subseteq C(X)$;*
(2) If $X_1 \subseteq X_2$, then $C(X_1) \subseteq C(X_2)$;
(3) $C(C(X)) = C(X)$. ∎

We can think of C as a function of $\mathcal{P}(S)$ into $\mathcal{P}(S)$. When doing so, recall that an important property to consider is the Exchange Property: if $x \in C(X \cup \{y\})$ and $x \notin C(X)$, then $y \in C(X \cup \{x\})$. In the next example, we show that the Exchange Property does not hold for C.

Example 3.1.6 *Let $X = \{(1, \frac{1}{8})\}$, $x = (\frac{1}{8}, 1)$, and $y = (\frac{1}{2}, \frac{1}{4})$. Then $y \notin C(X)$. Since $y = \frac{1}{2} \cdot (1, \frac{1}{8}) \vee \frac{1}{4} \cdot (\frac{1}{8}, 1)$, $y \in C(X \cup \{x\})$. However $x \notin C(X \cup \{y\})$. Thus the Exchange Property does not hold.*

Definition 3.1.7 Y is called a convex **MP-set** if for all $y_1, y_2 \in Y$ all convex MP-combinations of $\{y_1, y_2\}$ are also in Y.

Proposition 3.1.8 Y is a convex MP-set if and only if $Y = C(Y)$.

Proof. Suppose that Y is a convex MP-set. Then $Y = C(X)$ for some subset X of Y. Thus $C(Y) = C(C(X)) = C(X) = Y$. The converse follows by the definitions. ■

Proposition 3.1.9 For every X, $C(X)$ is a convex MP-set. ■

Proof. The proof follows from Propositions 3.1.8 and 3.1.5(2). ■

Definition 3.1.10 Let Y denote a convex MP-set and let $X \subseteq Y$. Then X is called a **set of generators** of Y if $Y = C(X)$. If X does not contain any proper subset which is itself a set of generators of Y, then X is called a **set of vertices** of Y.

Proposition 3.1.11 Let Y denote a convex MP-set and let $X \subseteq Y$. Then X is a set of vertices of Y if and only if
(1) $Y = C(X)$ and
(2) if $x \in X$, then $x \notin C(X \backslash \{x\})$. ■

Proposition 3.1.12 Let Y denote a convex MP-set and let X be a set of generators of Y such that X is finite. Then there exists $X' \subseteq X$ such that X' is a set of vertices of Y.

Proof. The proof follows from Propositions 3.1.11 and the fact that X is finite. ■

We write $x_1 \geq x_2$ if $x_1^k \geq x_2^k$ for all k.

Proposition 3.1.13 Let Y be a convex MP-set. Let X be a set of vertices of Y and X' be a set of generators of Y. Then $X \subseteq X'$.

Proof. Let $x \in X$. By Proposition 3.1.11(1),

$$x = \vee_{i=1}^n a_i x_i', \tag{3.2}$$

where for $i = 1, 2, \ldots, n, x_i' \in X'$ and $0 \leq a_i \leq 1$. Moreover

$$x_i' = \vee_{j=0}^m b_{ij} x_j, \tag{3.3}$$

where for $j = 0, 1, 2, \ldots, m, x_j \in X$ and $0 \leq b_{ij} \leq 1$. Thus

$$x = \vee_{j=0}^{m} c_j x_j,$$

where

$$c_j = \vee_{i=1}^{n} a_i b_{ij}. \tag{3.4}$$

By Proposition 3.1.11(2), x must be one of the x_j, say $x = x_0$. Since for all $c, b \in [0, 1]$, $cb < b$ if $c < 1$ it follows that $c_0 = 1$ else $x = \vee_{i=1}^{n} c_j x_j \in C(X\backslash\{x\})$ which contradicts the assumption that X is a set of vertices of Y. Therefore, by (3.4), there exists k such that $a_k = 1$ and $b_{k0} = 1$. It follows from (3.2) and (3.3) that

$$x \geq x_k' \qquad \text{and} \qquad x_k' \geq x_0 = x.$$

Thus $x = x_k' \in X'$. Consequently, $X \subseteq X'$. ∎

In the proof of Proposition 3.1.13, the conclusion that $c_0 = 1$ in the last paragraph followed partially from the fact that $\forall c, b \in [0, 1]$, $cb < b$ if $c < 1$. This property for product does not hold for minimum, i.e., $c \wedge b < b$ cannot necessarily be concluded if $c < 1$.

Theorem 3.1.14 *Let Y be a convex MP-set. Then the set of vertices of Y is unique if it exists.*

Proof. The proof follows from Proposition 3.1.13. ∎

Proposition 3.1.15 *Let Y be a convex MP-set. Let X be the set of vertices of Y. If X_1 and X_2 are sets of generators of Y such that $|X_1| > |X_2|$, then there exists $x \in X_1$ such that $x \in C(X_1\backslash\{x\})$.*

Proof. From Proposition 3.1.13, $X \subseteq X_1$ and $X \subseteq X_2$. Therefore, $|X| \leq |X_2| < |X_1|$. Thus X is a proper subset of X_1. The desired result follows from Proposition 3.1.11. ∎

Definition 3.1.16 *Let X_1, $X_2 \subseteq Y$. X_1 is called a **basis** of X_2 if every $x \in X_2$ can be expressed uniquely as a convex MP-combination of a finite subset of X_1.*

Proposition 3.1.17 *Let Y be a convex MP-set. Let X be a basis of Y. Then X is a set of vertices of Y.*

Proof. Clearly, $Y = C(X)$, i.e., (1) of Proposition 3.1.11 holds. Suppose there exists $x \in X$ such that $x \in C(X\backslash\{x\})$. Then there exists x_1, $x_2, \ldots, x_n \in X\backslash\{x\}$ such that x is a convex MP-combination of $x_1, x_2, \ldots, x_n$. Now $x = 1 \cdot x$ is another combination of elements of X. However, this contradicts the fact that x has a unique representation as a convex MP-combination of elements of X. Thus (2) of Proposition 3.1.11 holds. ∎

Example 3.1.18 *Let* $X = \{(\frac{1}{4}, \frac{1}{4}), (1, 0)\}$ *and* $X' = \{(\frac{1}{2}, \frac{1}{4}), (1, 0)\}$. *Then* X *and* X' *are sets of vertices for* $C(X)$ *with respect to max-min convex combinations. In fact,* X *is fundamental. However, for max-product convex combinations,* X' *does not generate* $C(X)$. *This follows since* $(\frac{1}{4}, \frac{1}{4}) \notin C(X')$ *with respect to max-product convex combinations.* X *is not a basis for* $C(X)$ *here for suppose* $(a, b) = x \cdot (\frac{1}{4}, \frac{1}{4}) \vee y \cdot (1, 0)$. *Then* $a = \frac{1}{4}x \vee y$ *and* $b = \frac{1}{4}x$. *Thus* $x = 4b$. *Hence* $a = b \vee y$. *Thus either* $y = a > b$ *(in which case the representation of* (a, b) *is unique) or* $a = b$ *and* y *can be any element in* $[0, b]$. *Clearly,* X *is a set of vertices of* $C(X)$.

In this section, we presented a max-product algebra. We note in the Exercises that this algebra can be extended by replacing product with a t-norm, which we now define.

Definition 3.1.19 *Let* T *be a binary operation on* $[0, 1]$. *Then* T *is called a* t-**norm** *if the following conditions hold* $\forall a, b, c \in [0, 1]$:
  (1) $T(a, 1) = a$;
  (2) $b \leq c$ *implies* $T(a, b) \leq T(a, c)$;
  (3) $T(a, b) = T(b, a)$;
  (4) $T(a, T(b, c)) = T(T(a, b), c)$.

3.2   Equivalences

Definition 3.2.1 *A quadruple* (X, Q, Y, μ) *is called a* **max-product sequential-like machine (MPSM)** *if* Q, X, Y *are finite nonempty sets and* μ *is a function from* $Q \times X \times Y \times Q$ *into* $[0, 1]$.

We note that Definition 2.9.1 and Definition 3.2.1 are the same. The use of "max-min" in the first and "max-product" in the second refers to the type of extended transition function μ^*.

The sets X and Y are the input and output alphabets, respectively. Q is the set of internal states and μ is the transition function. We can regard $\mu(q, u, v, q')$ as the grade of membership that the MPSM will enter state q' and produce output v given that the present state is q and the input is u.

The reader should note the differences between the above definition and the one given in [195].

As before we call finite sequences of elements of X **input tapes** or **strings**. Similarly, we call finite sequences of elements of Y **output tapes** or **strings**. The collection of all input (output) tapes will be denoted by $X^*(Y^*)$. Let

$$(X \times Y)^* = \{(x, y) \mid x \in X^*, y \in Y^*, \ |x| = |y|\}.$$

Unless stated otherwise, the symbol M, with or without subscripts, denotes an MPSM. All MPSMs are assumed to have the same input and output sets. Hence for ease of notation, we represent an MPSM by (Q, μ). In addition, we assume that $(X \times Y)^*$ is ordered in such a way that $|x_1| < |x_2|$ implies $(x_1, y_1) < (x_2, y_2)$. The order of $(X \times Y)^*$ will be kept fixed and assumed to be the same for all MPSMs under consideration.

Definition 3.2.2 *Let $M = (Q, \mu)$ be an MPSM. The* **extended transition function** μ^* *of M is a function from $Q \times (X \times Y)^* \times Q$ into $[0,1]$ defined recursively on $|x|$, $x \in X^*$, as follows:*

$$\mu^*(q', \Lambda, \Lambda, q'') = \begin{cases} 1 & \text{if } q' = q'' \\ 0 & \text{if } q' \neq q''; \end{cases}$$

$$\mu^*(q', ux, vy, q'') = \vee_{q \in Q}\{\mu(q', u, v, q) \cdot \mu^*(q, x, y, q'')\},$$

where $u \in X$, $v \in Y$, and $y \in Y^*$, $|x| = |y|$. Moreover, the output function q^M of M is the function from $Q \times (X \times Y)^*$ into $[0,1]$ defined as follows:

$$q^M(q, x, y) = \vee_{q' \in Q}\mu^*(q, x, y, q').$$

If no confusion arises, we sometimes write q for q^M.

Assume that $Q = \{q_1, q_2, \ldots, q_n\}$. Furthermore, let

(1) $P^M(x, y) = [p_{ij}]$ be the matrix, where $p_{ij} = \mu^*(q_i, x, y, q_j)$;

(2) $Q^M(x, y) = [q_i]$ be the column matrix, where the i-th row $q_i = q^M(q_i, x, y)$;

(3) E be the column matrix with all entries 1.

Proposition 3.2.3 *Let $M = (Q, \mu)$ and $(x_1, y_1), (x_2, y_2) \in (X \times Y)^*$. Then*

(1) $P^M(x_1 x_2, y_1 y_2) = P^M(x_1, y_1) \odot P^M(x_2, y_2)$,

(2) $Q^M(x_1 x_2, y_1 y_2) = P^M(x_1, y_1) \odot Q^M(x_2, y_2) = P^M(x_1 x_2, y_1 y_2) \odot E$.

Proof. The proof follows from Proposition 3.1.2. ■

Definition 3.2.4 *Let $M = (Q, \mu)$.*

(1) An **initial distribution** *(**i.d.**) of M is a function ι from Q into $[0,1]$.*

(2) ι is said to be **concentrated** *at $q \in Q$ if $\iota(q) = 1$ and 0 elsewhere.*

We also use the symbol ι to denote the row matrix whose i-th row is $\iota(q_i)$.

Definition 3.2.5 *Let M be an MPSM and ι be an i.d. of M. Then the ordered pair (M, ι) is called an* **initialized max-product sequential-like machine (IMPSM).**

If ι is concentrated at q, we write (M, q) for (M, ι).

Definition 3.2.6 *Let $I = (M, \iota)$ be an IMPSM. The **response function** r^I of I is a function from $(X \times Y)^*$ into $[0, 1]$ defined as follows:*

$$r^I(x, y) = \vee_{q \in Q} \{\iota(q) \cdot q^M(q, x, y)\}.$$

We let A^M denote the semi-infinite matrix whose columns are $Q^M(x, y)$ arranged in the order of $(X \times Y)^*$.

If A is a matrix, we let $\rho(A)$ denote the collection of all rows of A.

If $R^I = \iota \odot A^M$, then R^I is a row matrix whose entries are $r^I(x, y)$. Moreover, R^I is a convex MP-combination of $\rho(A^M)$.

Definition 3.2.7 *Let I_1 and I_2 be IMPSMs. Then I_1 and I_2 are called **equivalent**, denoted by $I_1 \sim I_2$, if $r^{I_1} = r^{I_2}$.*

Definition 3.2.8 *Let ι_1 and ι_2 be i.d.s of an MPSM M. Then ι_1 and ι_2 are called M-**equivalent**, denoted by $\iota_1 \sim \iota_2$, if $(M, \iota_1) \sim (M, \iota_2)$.*

If ι_1 is concentrated at q, then we also write $q \sim \iota_2$ for $\iota_1 \sim \iota_2$.

Proposition 3.2.9 *Let ι_1 and ι_2 be i.d.s of an MPSM M. Then $\iota_1 \sim \iota_2$ if and only if*

$$\iota_1 \odot A^M = \iota_2 \odot A^M. \blacksquare$$

Definition 3.2.10 *Let $M_1 = (Q_1, \mu_1)$ and $M_2 = (Q_2, \mu_2)$:*

*(1) M_1 and M_2 are called **statewise equivalent**, written $M_1 \sim M_2$, if for all $q' \in Q_1$, there exists $q'' \in Q_2$ such that $(M_1, q') \sim (M_2, q'')$ and vice versa.*

*(2) M_1 and M_2 are called **compositewise equivalent**, written $M_1 \simeq M_2$, if for all $q \in Q_1$, there exists an i.d. ι of M_2 such that $(M_1, q) \sim (M_2, \iota)$ and vice versa.*

*(3) M_1 and M_2 are called **distributionwise equivalent**, written $M_1 \approx M_2$, if for every i.d. ι_1 of M_1, there exists i.d. ι_2 of M_2 such that $(M_1, \iota_1) \sim (M_2, \iota_2)$ and vice versa.*

Theorem 3.2.11 *Let M_1 and M_2 be MPSMs. Then the following assertions hold.*

(1) $M_1 \sim M_2$ if and only if $\rho(A^{M_1}) = \rho(A^{M_2})$;

(2) $M_1 \simeq M_2$ if and only if $\rho(A^{M_1}) \subseteq C[\rho(A^{M_2})]$ and $\rho(A^{M_2}) \subseteq C[\rho(A^{M_1})]$;

(3) $M_1 \approx M_2$ if and only if $C[\rho(A^{M_1})] \subseteq C[\rho(A^{M_2})]$. $\blacksquare$

Proposition 3.2.12 *Let M_1 and M_2 be MPSMs. If $M_1 \sim M_2$, then $M_1 \simeq M_2$.*

Proof. The proof follows from Proposition 3.1.5(1) and Theorem 3.2.11.

$\blacksquare$

Proposition 3.2.13 *Let M_1 and M_2 be MPSMs. Then $M_1 \simeq M_2$ if and only if $M_1 \approx M_2$.*

Proof. The proof follows from Proposition 3.1.5 and Theorem 3.2.11.
∎

3.3   Irreducibility and Minimality

In this section, we study the irreducibility and minimality of an MPSM.

Definition 3.3.1 *Let $M = (Q, \mu)$.*

*(1) M is called **statewise irreducible** if for all q', $q'' \in Q$, $q' \sim q''$ implies $q' = q''$.*

*(2) M is called **compositewise irreducible** if for every $q \in Q$ and i.d. ι of M, $q \sim \iota$ implies $\iota(q) > 0$.*

*(3) M is called **distributionwise irreducible** if for every i.d. ι_1 and ι_2 of M, $\iota_1 \sim \iota_2$ implies $\iota_1 = \iota_2$.*

Theorem 3.3.2 *Let M be an MPSM. Then the following assertions hold.*

(1) M is statewise irreducible if and only if no two rows of A^M are identical.

(2) M is compositewise irreducible if and only if $\rho(A^M)$ is a set of vertices of $C[\rho(A^M)]$, i.e., no row of A^M is a convex MP-combination of the other rows of A^M.

(3) M is distributionwise irreducible if and only if $\rho(A^M)$ is a basis of $C[\rho(A^M)]$.

Proof. (1) follows from Proposition 3.2.9, (2) follows from Propositions 3.2.9 and 3.1.11, and (3) follows from Proposition 3.2.9 and the definition of basis. ∎

Example 3.3.3 *Let $M = (Q, X, Y, \rho)$, where $Q, X,$ and Y are defined as in Example 2.10.24 and $\rho = \mu$. Then*

$$q^M(q_1, u, 1) = \tfrac{1}{2},$$
$$q^M(q_1, uu, 10) = \tfrac{1}{2}\tfrac{3}{4},$$

$$\vdots$$

$$q^M(q_1, u^{2n}, (10)^n) = (\tfrac{1}{2})^n(\tfrac{3}{4})^n,$$
$$q^M(q_1, u^{2n+1}, (10)^n 1) = (\tfrac{1}{2})^{n+1}(\tfrac{3}{4})^n,$$

$$q^M(q_2, u, 0) = \tfrac{3}{4},$$
$$q^M(q_2, uu, 01) = \tfrac{3}{4}\tfrac{1}{2},$$

$$\vdots$$

$$q^M(q_2, u^{2n}, (01)^n) = (\tfrac{3}{4})^n(\tfrac{1}{2})^n,$$
$$q^M(q_2, u^{2n+1}, (01)^n 0) = (\tfrac{3}{4})^{n+1}(\tfrac{1}{2})^n.$$

The nonzero columns of A^M are

$$\begin{bmatrix} \frac{1}{2} \\ 0 \end{bmatrix}, \begin{bmatrix} 0 \\ \frac{3}{4} \end{bmatrix}, \begin{bmatrix} \frac{3}{8} \\ 0 \end{bmatrix}, \begin{bmatrix} 0 \\ \frac{3}{8} \end{bmatrix}, \dots, \begin{bmatrix} (\frac{3}{8})^n \\ 0 \end{bmatrix}, \begin{bmatrix} 0 \\ (\frac{3}{8})^n \end{bmatrix},$$

$$\begin{bmatrix} \frac{1}{2}(\frac{3}{8})^n \\ 0 \end{bmatrix}, \begin{bmatrix} 0 \\ \frac{3}{4}(\frac{3}{8})^n \end{bmatrix}, \dots.$$

Clearly, no two rows of A^M are identical. In fact, $\rho(A^M)$ is clearly a set of vertices of $C[\rho(A^M)]$. It also follows that $\rho(A^M)$ is a basis of $C[\rho(A^M)]$.

Proposition 3.3.4 *Let M be an MPSM. If M is distributionwise irreducible, then M is compositewise irreducible.* ■

Proposition 3.3.5 *Let M be an MPSM. If M is compositewise irreducible, then M is statewise irreducible.* ■

Definition 3.3.6 *Let M be an MPSM. Then M is called **statewise (compositewise) minimal** if M is not statewise (compositewise) equivalent to any MPSM with a fewer number of states.*

Let $M = (Q, \mu)$. Then the cardinality of M is defined by $|M| = |Q|$.

Theorem 3.3.7 *Let M be an MPSM. Then M is statewise minimal if and only if M is statewise irreducible.*

Proof. Let $M = (Q, \mu)$. Suppose that M is not statewise minimal. Then there exists an MPSM M' such that $|M'| < |M|$ and $M' \sim M$. By Theorems 3.2.11(1) and 3.3.2(1), M is not statewise irreducible. Conversely, suppose that M is not statewise irreducible. Then there exist $q', q'' \in Q$ such that $q' \sim q''$ and $q' \neq q''$. By renumbering the elements of Q, if necessary, we may assume that $q' = q_{n-1}$ and $q'' = q_n$, where $n = |Q|$. Let $M' = (Q', \mu')$, where $Q' = Q \backslash \{q_n\}$ and for $i = 1, 2, \dots, n-1$,

$$\mu'(q_i, u, v, q_j) = \begin{cases} \mu(q_i, u, v, q_j) & \text{if } j = 1, 2, \dots, n-2 \\ \mu(q_i, u, v, q_{n-1}) \vee \mu(q_i, u, v, q_n) & \text{if } j = n-1. \end{cases}$$

It follows that $\rho(A^M) = \rho(A^{M'})$. Hence by Theorem 3.2.11(1), $M \sim M'$. Thus M is not statewise minimal. ■

Theorem 3.3.8 *Let M be an MPSM. Then M is compositewise minimal if and only if M is compositewise irreducible.*

Proof. Let $M = (Q, \mu)$. Suppose that M is not compositewise minimal. Then there exists an MPSM M' such that $|M'| < |M|$ and $M' \simeq M$. By Theorems 3.2.11 and 3.3.2 and Proposition 3.1.15, M is not compositewise irreducible. Conversely, suppose that M is not compositewise irreducible. Then there exists $q' \in Q$ and an i.d. ι of M such that $q' \sim \iota$ and $\iota(q') = 0$.

By renumbering the elements of Q, if necessary, we may assume that $q' = q_n$, where $n = |Q|$. Since $q_n \sim \iota$ and $\iota(q_n) = 0$,

$$q^M(q_n, x, y) = \vee \{\iota(q_i) \cdot q^M(q_i, x, y) \mid i = 1, 2, \ldots, n\}$$

for all $(x, y) \in (X \times Y)^*$. Let $M' = (Q', \mu')$, where $Q' = Q \backslash \{q_n\}$ and for $i, j = 1, 2, \ldots, n - 1$,

$$\mu'(q_i, u, v, q_j) = \mu(q_i, u, v, q_j) \vee \iota(q_j) \cdot \mu(q_i, u, v, q_n).$$

Since for all $(x, y) \in (X \times Y)^*$ and $i = 1, 2, \ldots, n - 1$, $q^M(q_n, x, y) = q^M(q_i, x, y)$, we have $C[\rho(A^M)] = C[\rho(A^{M'})]$. Hence by Theorem 3.2.11, $M' \simeq M$. Thus M is not compositewise minimal. ∎

Proposition 3.3.9 *Let M_1 and M_2 be MPSMs. Suppose that M_1 is statewise minimal and $M_1 \sim M_2$. Then M_2 is statewise minimal if and only if $|M_1| \leq |M_2|$.* ∎

Proposition 3.3.10 *Let M_1 and M_2 be MPSMs. Suppose that M_1 is statewise minimal and $M_1 \simeq M_2$. Then M_2 is compositewise minimal if and only if $|M_1| \leq |M_2|$.* ∎

Proposition 3.3.11 *Let M_1 and M_2 be MPSMs that are compositewise minimal. Then $M_1 \sim M_2$ if and only if $M_1 \simeq M_2$.*

Proof. The proof follows from Proposition 3.2.12 and Theorems 3.1.14, 3.3.2, and 3.3.8 ∎

Proposition 3.3.12 *Let M_1 and M_2 be MPSMs. Suppose that M_1 and M_2 are statewise minimal and $M_1 \sim M_2$. Then for all $(x, y) \in (X, Y)^*$,*

$$P^{M_1}(x, y) \odot A^{M_1} = P^{M_2}(x, y) \odot A^{M_1}$$

after an appropriate rearrangement of states.

Proof. From Theorems 3.2.11(1), 3.3.2(1), and 3.3.7, $A^{M_1} = A^{M_2}$ after an appropriate rearrangement of states. ∎

Definition 3.3.13 *Let M_1 and M_2 be MPSMs. Then M_1 and M_2 are called **isomorphic** ($\equiv$) if they are equal up to a permutation of states.*

Definition 3.3.14 *Let M be an MPSM. Then M is called **simple** if there exists no MPSM M' such that $|M'| \leq |M|$, $M' \sim M$, but M' and M are not isomorphic.*

Let M be an MPSM. Then we let

$$\rho(M) = \cup_{u \in X} \cup_{v \in Y} \rho[P^M(u,v)]$$

and

$$\bar{\rho}(M) = \{\iota \odot A^M \mid \iota \in \rho(M)\}.$$

Theorem 3.3.15 *Let M be an MPSM such that M is statewise minimal. Then M is simple if and only if $\rho(A^M)$ is a basis of $\bar{\rho}(M)$.*

Proof. Suppose that $\rho(A^M)$ is a basis of $\bar{\rho}(M)$. Let M' be an MPSM such that $|M'| \leq |M|$ and $M' \sim M$. Then by Proposition 3.3.9, M' is statewise minimal. It follows from Proposition 3.3.12 that $P^M(x,y) = P^{M'}(x,y)$ for all $(x,y) \in (X \times Y)^*$ after an appropriate rearrangement of states. Thus $M \equiv M'$. Hence M is simple. Conversely, suppose that $\rho(A^M)$ is not a basis of $\bar{\rho}(M)$. Then there exists $j \in \rho(M)$ such that

$$j \odot A^M = j' \odot A^M \tag{3.5}$$

for some $j' \neq j$. Let j be the i-th row of the matrix $P^M(u,v)$. Construct M' from M by replacing the i-th row of the matrix $P^M(u,v)$ by j' and leaving the rest unaltered. By (3.5), $A^M = A^{M'}$. Thus $|M'| = |M|$ and $M' \sim M$. However, $M' \not\equiv M$. Hence M is not simple. ∎

Proposition 3.3.16 *Let M be an MPSM such that M be statewise (compositewise) minimal. Then there exists no statewise (compositewise) minimal MPSM that is statewise (compositewise) equivalent, but not isomorphic to M if and only if $\rho(A^M)$ is a basis of $\bar{\rho}(M)$.*

Proof. If M is statewise minimal, the desired result follows immediately from Theorem 3.3.15. The case when M is compositewise minimal follows from Theorem 3.3.15 and Propositions 3.3.10 and 3.3.11. ∎

Proposition 3.3.17 *Let M be an MPSM such that M is distributionwise irreducible. Then there exists no statewise (compositewise) minimal MPSM that is statewise (compositewise) equivalent, but not isomorphic to M.*

Proof. The proof follows from Theorem 3.3.2(3) and Proposition 3.3.16. ∎

3.4   Max-Product Grammars and Languages

The remainder of the chapter is based on [208].

In [204], Santos introduced a mathematical formulation of probabilistic grammars and random languages generated by probabilistic grammars. (See Chapter 5 also.) In that formulation, the probability associated with each "derivation" of a probabilistic grammar is taken to be the sum of the probabilities of all distinct "paths" of the "derivation." This leads to the sum interpretation which is customary in the theory of probabilistic automata.

The sum interpretation is one of the two interpretations considered in [188]. The other interpretation is the maximal interpretation, which leads naturally to the max-product operation introduced in [193, 202]. (See Chapter 2 and Sections 3.1 and 3.2.) A new type of grammar arises naturally from the application of the maximal interpretation to the probabilistic grammars defined in [204]. This grammar, which is readily obtained from a probabilistic grammar by replacing the "sum" operation by the "max-min" or "least upper bound" operations, is called a **normalized max-product grammar** here.

Normalized max-product grammars are a special case of max-product grammars. In max-product grammars, the functions involved are allowed to assume any nonnegative value. In view of [81, 255], max-product grammars may be considered as a model of fuzzy grammars. Another formulation of max-product grammars is given from this point of view. It is shown that the two formulations are equivalent as far as the family of fuzzy languages generated are concerned.

By imposing restrictions on the fuzzy productions, certain more restricted types of max-product grammars can be obtained. Two of these are regular and weak regular max-product grammars. It turns out that regular and weak regular max-product grammars are closely related to max-product automata [193, 207] and asynchronous max-product automata. This allows for the establishment of many interesting properties of languages generated by regular and weak regular max-product grammars with cut point.

An interesting family of languages that arises from the consideration of regular and weak regular max-product grammars is the family of languages $\mathcal{L}$ generated by finitary weak regular max-product grammars with cut point. The family $\mathcal{L}$ may also be characterized by regular max-product grammars, asynchronous max-product automata, max-product automata, and type 3 weighted grammars under maximal interpretation [188].

The family $\mathcal{L}$ contains the family of regular languages as a proper subfamily. Moreover, there are context-free and stochastic languages [235] that are not in $\mathcal{L}$.

A formulation of max-product grammars is given in this section. An alternative formulation of max-product grammars is also given. This formulation is motivated by the fact that max-product grammars may be viewed as a model of fuzzy grammars. The two formulations are shown to be equivalent as far as the family of fuzzy languages generated is concerned.

Let S denote a nonempty set. Let $\mathbb{R}^{\geq 0}$ denote the collection of all nonnegative real numbers and $\mathbb{R}^{\geq 0}_{\infty} = \mathbb{R}^{\geq 0} \cup \{\infty\}$.

Definition 3.4.1 *A* **fuzzy production** ρ *over* S *is a function from* $S^* \times S^*$ *into* $\mathbb{R}^{\geq 0}$ *such that the set* $\overline{\rho} = \{s \in S^* \mid \rho(s,t) > 0 \text{ for some } t \in S^*\}$ *is finite and* $\Lambda \notin \overline{\rho}$. *If, in addition, for all* $s \in S^*$, *the set* $\{t \in S^* \mid \rho(s,t) > 0\}$ *is finite, then* ρ *is called* **finitary**.

Definition 3.4.2 *A fuzzy production* ρ *over* S *is* **strict** *if the range of* ρ *is a subset of* $[0,1]$.

The concept of a fuzzy subset requires that the grades of membership function assume values in $[0,1]$. This concept is extended in [81] to allow the grades of membership function to assume values in any partially ordered set. Since $\mathbb{R}^{\geq 0}$, as well as $\mathbb{R}^{\geq 0}_{\infty}$, is a partially ordered set, we may interpret $\rho(s,t)$ as the grade of membership that s will be replaced by t. If ρ is normalized, i.e., $\sum_{t \in S^*} \rho(s,t) \leqslant 1$ for all $s \in S^*$, then $\rho(s,t)$ may also be interpreted as the probability that s will be replaced by t.

Definition 3.4.3 *A* **(finitary) max-product grammar** *is a quadruple* $G = (T, N, P, h)$, *where (1)* T *and* N *are disjoint finite nonempty sets; (2)* P *is a finite collection of (finitary) fuzzy productions over* $T \cup N$ *such that for all* $\rho \in P$, $\rho(s,t) > 0$ *implies* $s \in (T \cup N)^* N (T \cup N)^*$; *and (3)* h *is a function from* N *into* $\mathbb{R}^{\geq 0}$. *If in addition, every* $\rho \in P$ *is strict and* h *is a function from* N *into* $[0,1]$, *then* G *is said to be* **strict**. *Moreover, if every* $\rho \in P$ *is normalized and* $\sum_{A \in N} h(A) \leqslant 1$, *then* G *is said to be* **normalized**.

In the above definition, T and N are, respectively, the terminals and the nonterminals. $h(A)$ is the grade of membership that A is the start symbol of G.

It is important to note that in Definition 3.4.3(2), that s must contain a nonterminal.

The following concepts are needed to define the fuzzy languages generated by max-product grammars.

Recall that $\mathbb{N}$ is the collection of all positive integers.

Definition 3.4.4 *Let* $t, s \in S^*$ *and* $k \in \mathbb{N}$. $m(t,s) = k$ *if there exist* $u_i, v_i \in S^*$, $i = 1, 2, \ldots, k$, *where* $v_i \neq v_j$ *for* $i \neq j$, *such that (1)* $t = u_i s v_i$ *for* $i = 1, 2, \ldots, k$, *and (2)* $t = usv$ *implies* $u = u_i$ *and* $v = v_i$ *for some* i *such that* $1 \leqslant i \leqslant k$. *Define* $m(t,s) = 0$ *if and only if* $t \neq usv$ *for all* $u, v \in S^*$.

Note that $m(t,s) = k$ if and only if t can be expressed in the form $t = usv$ in exactly k distinct ways.

Definition 3.4.5 *Let $G = (T, N, P, h)$ be a max-product grammar. A re-placement function of G is a function δ from $(T \cup N)^*$ into the set $\{s \in (T \cup N)^* \mid \rho(s, t) > 0 \text{ for some } \rho \in P \text{ and } t \in (T \cup N)^*\}$ such that $\delta(r) = s$ implies $m(r, s) > 0$.*

It follows that $\delta(r) = s$ if and only if some occurrence of s in r will be replaced.

Let $R(G)$ denote the collection of all replacement functions of G.

Example 3.4.6 *Let $N = \{s_0, t\}$, $T = \{a, b\}$, and $P = \{s_0 \to bs_0, s_0 \to at, t \to bt, t \to b\}$. Then $m(bt, t) = 1$ since $bt = utv$ implies $u = b$ and $v = \Lambda$. Also, $\delta(bt) = t$ since $\rho(t, bt) > 0$.*

Example 3.4.7 *Let $N = \{s_0, A, B\}$, $T = \{a, b\}$, and $P = \{s_0 \to B, B \to BAB, B \to b, A \to a\}$. Then $m(BABABAB, BAB) = 3$ since $BABABAB = BABA(BAB)\Lambda = BA(BAB)AB = \Lambda(BAB)ABAB$. How-ever, there does not exist $t \in (T \cup N)^*$ and $\rho \in P$ such that $\rho(BAB, t) > 0$.*

Definition 3.4.8 *Let $r, w, s, t \in S^*$ and $k \in \mathbb{N}$. $r \overset{k}{\sim} w \bmod(s, t)$ if there exist $u, v \in S^*$ such that $r = usv$, $w = utv$, and $m(us, s) = k$.*

We have that $r \overset{k}{\sim} w \bmod(s, t)$ if and only if w can be obtained from r by replacing the kth occurrence of s in r by t.

Definition 3.4.9 *Let $G = (T, N, P, h)$ be a max-product grammar and $S = T \cup N$.*

(1) With all $\rho \in P$, $\delta \in R(G)$, and $k \in \mathbb{N}$, associate the function $\lambda_{(\rho, \delta, k)}$ from $S^* \times S^*$ into $\mathbb{R}^{\geq 0}$, where

$$\lambda_{(\rho, \delta, k)}(r, w) = \begin{cases} \rho(\delta(r), t) & \text{if } r \overset{k}{\sim} w \bmod(\delta(r), t) \\ 0 & \text{otherwise.} \end{cases}$$

(2) With all $z \in (P \times R(G) \times \mathbb{N})^*$, associate the function λ_z from $S^* \times S^*$ into $\mathbb{R}^{\geq 0}_\infty$ defined recursively as follows:

$$\lambda_\Lambda(r, w) = \begin{cases} 1 & \text{if } r = w \\ 0 & \text{if } r \neq w \end{cases}$$

and

$$\lambda_{z(\rho, \delta, k)}(r, w) = \vee_{v \in S^*}[\lambda_{z(\rho, \delta, k)}(v, w)\lambda_z(r, v)].$$

(3) Define the function λ_G from T^* into $[0, 1]$ by $\forall r \in T^*$

$$\lambda_G(r) = \vee_{z \in Z^*} \vee_{A \in N} h(A)\lambda_z(A, r),$$

where $Z = P \times R(G) \times \mathbb{N}$.

Example 3.4.10 *Let* $N = \{s_0, A, B, C, D\}$, $T = \{a, b\}$, *and* $P = \{s_0 \to aAC$, $C \to AD$, $A \to B$, $A \to a$, $B \to b$, $D \to b\}$. *Then* $m(aAA, A) = 2$ *since* $aAA = aA(A)\Lambda = a(A)A$. *Now* $\rho(A, B) > 0$ *and so* $\delta(B) = A$. *Also,* $aAAb \overset{2}{\sim} aABb \bmod(A, B)$ *since* $m(uA, A) = 2$, *where* $u = aA$. *(Note* $aAA = aA(A)\Lambda = a(A)A$.*) We also have that* $\delta(aAAb) = A$ *since* $\rho(A, B) > 0$. *Hence*

$$\lambda_{(\rho, \delta, 2)}(aAAb, aABb) = \rho(A, B).$$

Now $m(aAAD, D) = 1$ *and* $\widehat{\rho}(D, b) > 0$. *Hence* $\widehat{\delta}(aAAD) = D$. *We have that*

$$
\begin{aligned}
\lambda_{(\widehat{\rho}, \widehat{\delta}, 1)(\rho, \delta, 2)}(aAAD, aABb) &= \vee\{\lambda_{(\rho, \delta, 2)}(z, aABb)\lambda_{(\widehat{\rho}, \widehat{\delta}, 1)}(aAAD, z) \\
&\quad \mid z \in (T \cup N)^*\} \\
&= \lambda_{(\rho, \delta, 2)}(aAAb, aABb) \\
&\quad \lambda_{(\widehat{\rho}, \widehat{\delta}, 1)}(aAAD, aAAb) \\
&= \rho(A, B)\widehat{\rho}(D, b).
\end{aligned}
$$

Now

$$
\begin{aligned}
\lambda_G(aABb) &= \vee\{h(s_0)\lambda_z(s_0, aABb) \vee h(A)\lambda_z(A, aABb) \vee \\
&\quad h(B)\lambda_z(B, aABb) \vee h(C)\lambda_z(C, aABb) \vee \\
&\quad h(D)\lambda_z(D, aABb) \mid z \in Z^*\} \\
&= h(s_0) \vee_z \lambda_z(s_0, aABb) \\
&= h(s_0)\rho(s_0, aAC)\rho(aAC, aAAD)\rho(aAAD, aAAb) \\
&\quad \rho(aAAb, aABb).
\end{aligned}
$$

Also

$$\lambda_G(b) = h(A)\rho(A, B)\rho(B, b) \vee h(B)\rho(B, b) \vee h(D)\rho(D, b).$$

Note that $m(A, A) = m(B, B) = m(D, D) = 1$, $\delta(A) = A$, $\delta(B) = B$, *and* $\delta(D) = D$ *here.*

In Definition 3.4.9, as well as in the rest of the chapter, we assume the usual arithmetic of the extended real number system, which includes the assumption that $0 \cdot \infty = 0$. Moreover, the concept of supremum is extended in a natural manner to include ∞.

Definition 3.4.11 *A fuzzy language* λ *over* S *is a function from* S^* *into* $\mathbb{R}_\infty^{\geq 0}$.

$\lambda(s)$ is the grade of membership that s is a member of the language, where $s \in S^*$.

Clearly, if G is a max-product grammar with terminal set T, then λ_G is a fuzzy language over T. Thus we refer to λ_G as the fuzzy language generated by G. Moreover, if G is strict, then λ_G is a function from T^* into $[0, 1]$.

In Section 5.12, the concept of random control sets for probabilistic grammars will be defined. A similar concept can be readily introduced for max-product grammars. The max-product grammars defined above are referred to as **type 0 max-product grammars**. Other more restrictive types of max-product grammars can be obtained by imposing certain restrictions on the fuzzy productions. Four such grammars are defined below. They correspond to context-sensitive, context-free, weak regular, and regular probabilistic grammars, Section 5.12.

Definition 3.4.12 *Let* $G = (T, N, P, h)$ *be a max-product grammar.*

(1) G *is* **context-sensitive** *if for every* $\rho \in P$, $s, t \in (T \cup N)^*$, $\rho(s, t) > 0$ *implies* $|s| \leqslant |t|$.

(2) G *is* **context-free** *if for every* $\rho \in P$, $s, t \in (T \cup N)^*$, $\rho(s, t) > 0$ *implies* $s \in N$.

(3) G *is* **weak regular** *if for every* $\rho \in P$ *and* $s, t \in (T \cup N)^*$, $\rho(s, t) > 0$ *implies* $s \in N$, $t = uA$, *where* $u \in T^*$ *and* $A \in N \cup \{\Lambda\}$.

(4) G *is* **regular** *if for every* $\rho \in P$ *and* $s, t \in (T \cup N)^*$, $\rho(s, t) > 0$ *implies* $s \in N$, $t = aA$, *where* $a \in T$ *and* $A \in N \cup \{\Lambda\}$.

It is easy to see that these definitions contain the crisp case.

We will see that most of the relations established in Section 5.12 between the various types of probabilistic grammars and probabilistic automata may be carried over to the max-product case. In fact, much stronger results are valid for the latter case, as we shall see in subsequent sections.

If $\lambda = \lambda_G$ for some max-product grammar G of a particular kind, then we say that λ is a fuzzy language of that kind, and vice versa.

Theorem 3.4.13 *If* λ *is a (finitary) (type 0, context-sensitive, context-free, weak regular, regular)(strict) fuzzy language over* T, *then* $\lambda = \lambda_G$ *for some (finitary)(type 0, context-sensitive, context-free, weak regular, regular)(strict) max-product grammar* $G = (T, N, \{\rho_0\}, h)$, *where there exists* $A_0 \in N$ *such that the following conditions hold:*

(1) $h(A_0) = 1$ *and* $h(A) = 0$ *for* $A \neq A_0$;

(2) $\rho_0(s, t) = 0$ *for all* $s \in (Y \cup N)^*$ *and* $t \in (T \cup N)^*\{A_0\}(T \cup N)^*$;

(3) $\rho_0(s, t) > 0$, *where* $s \in (T \cup N)^*\{A_0\}(T \cup N)^*$ *and* $t \in (T \cup N)^*$ *implies* $s = A_0$.

Proof. Let $\lambda = \lambda_{G'}$, where $G' = (T, N', P, h')$ is a (finitary) (type 0, context-sensitive, context-free, weak regular, regular) (strict) max-product grammar. Let $A_0 \notin T \cup N'$ and $N = N' \cup \{A_0\}$. Define $G = (T, N, \{\rho_0\}, h)$, where

$$\rho_0(s, t) = \begin{cases} \vee_{\rho \in P} \rho(s, t) & \text{if } s, t \in (T \cup N')^* \\ \vee_{\rho \in P} \vee_{A \in N} h(A)\rho(A, t) & \text{if } s = A_0, t \in (T \cup N')^* \\ 0 & \text{otherwise} \end{cases}$$

and

$$h(A) = \begin{cases} 1 & \text{if } A = A_0 \\ 0 & \text{if } A \neq A_0. \end{cases}$$

Clearly, G has the desired properties. Moreover, it can be verified that $\lambda = \lambda_G$. ∎

If G satisfies conditions (1) to (3) of Theorem 3.4.13, then we also write $G = (T, N, \rho_0, A_0)$.

From a result established in [205], it follows that the above theorem does not hold for probabilistic grammars.

It is clear that every fuzzy production ρ over S is completely characterized by the collection $D(\rho)$ of all triples (s, t, p) where $p = \rho(s, t) > 0$. We also write the triple (s, t, p) as $s \xrightarrow{p} t$ and refer to it as a **discrete fuzzy production** over S. Let $D(\rho)$ denote the set of all discrete fuzzy productions over S. We associate with $D(\rho)$ the function ρ_D from $S^* \times S^*$ into $\mathbb{R}^{\geq 0}$ such that $\forall (s, t) \in S^* \times S^*$,

$$\rho_D(s, t) = \begin{cases} \vee \Gamma(s, t) & \text{if } \Gamma(s, t) \neq \varnothing \\ 0 & \text{if } \Gamma(s, t) = \varnothing, \end{cases}$$

where $\Gamma(s, t) = \{p \mid (s \xrightarrow{p} t) \in D\}$. Clearly, if the set $\{s \mid (s \xrightarrow{p} t) \in D\}$ is finite and for every $s, t \in S^*$, the set $\Gamma(s, t)$ is empty or bounded, then ρ_D is a fuzzy production over S. A set D of discrete fuzzy productions over S satisfying these two conditions is said to be admissible.

In terms of discrete fuzzy productions, Theorem 3.4.13 states that there exists a distinguished symbol A_0 such that (1) A_0 is the start symbol of G; (2) A_0 does not occur on the right hand side of any discrete fuzzy productions of G; and (3) no discrete fuzzy productions of G has A_0 as a proper subword on the left hand side.

Notation 3.4.14 *Let D be a collection of discrete fuzzy productions over S. Then $r \xRightarrow{p} w \bmod[(d_1, k_1)(d_2, k_2) \ldots (d_n, k_n)]$, where $r, w \in S^*$, $d_i = (s_i \xrightarrow{p_i} t_i) \in D$, $k_i \in \mathbb{N}$, $i = 1, 2, \ldots, n$, and $p = p_1 p_2 \ldots p_n$ if and only if there exist $v_i \in S^*$, $i = 0, 1, 2, \ldots, n$, such that $v_0 = r, v_n = w$, and $v_{i-1} \overset{k_i}{\sim} v_i \bmod(s_i, t_i)$ for every i such that $1 \leq i \leq n$. Otherwise, $r \xRightarrow{0} w \bmod[(d_1, k_1)(d_2, k_2) \ldots (d_n, k_n)]$.*

Example 3.4.15 *Let N, T, and P be as defined in Example 3.4.10. Let $r = aAAD$ and $w = aABb$. Let $v_0 = aAAD$, $v_1 = aAAb$, and $v_2 = aABb$. Let $d_1 = (s_1 \xrightarrow{p_1} t_1)$ and $d_2 = (s_2 \xrightarrow{p_2} t_2)$, where $s_1 = D$, $t_1 = b$, $s_2 = A$, and $t_2 = B$. Then from Example 3.4.10, we have that $k_1 = 1$ and $k_2 = 2$. Thus $v_0 \overset{k_1}{\sim} v_1 \bmod(s_1, t_1)$ and $v_1 \overset{k_2}{\sim} v_2 \bmod(s_2, t_2)$. Hence $aAAD \xRightarrow{p} ABb \bmod[(D \xrightarrow{p_1} b), (A \xrightarrow{p_2} B)]$, where $p = p_1 p_2$.*

It is clear that, in the above, p is uniquely determined by r, $w \in S^*$ and $\omega \in (D \times \mathbb{N})^*$. Thus we may define a set of functions λ_{zp} from $S^* \times S^*$ into $\mathbb{R}^{\geq 0}$, one for each $z \in (D \times \mathbb{N})^*$ such that $\lambda_z(r, w) = p$ if and only if $r \Longrightarrow w \bmod(z)$.

Theorem 3.4.16 λ *is a type 0 fuzzy language over* T *if and only if there exists a quadruple* (T, N, D, h) *such that the following properties hold:*

(1) T *and* N *are disjoint finite nonempty sets;*

(2) D *is an admissible set of discrete fuzzy productions over* $T \cup N$ *such that* $(s \xrightarrow{p} t) \in D$ *implies* $s \in (T \cup N)^* N (T \cup N)^*$;

(3) h *is a function from* N *into* $\mathbb{R}^{\geq 0}$; *and such that for every* $r \in T^*$,

$$\lambda(r) = \vee \vee_{z \in W^*} \vee_{A \in N} h(A) \lambda_z(r, A), \tag{3.6}$$

where $W = D \times \mathbb{N}$. *In addition,*

(a) λ *is finitary if and only if* D *is finite;*

(b) λ *is strict if and only if every* $d = (s \xrightarrow{p} t) \in D$ *is strict, i.e.,* $p \in [0, 1]$;

(c) λ *is context-sensitive if and only if every* $d = (s \xrightarrow{p} t) \in D$ *is context sensitive, i.e.,* $|s| \leqslant |t|$;

(d) λ *is context-free if and only if every* $d = (s \xrightarrow{p} t) \in D$ *is context free, i.e.,* $s \in N$;

(e) λ *is weak regular if and only if every* $d = (s \xrightarrow{p} t) \in D$ *is weak regular, i.e.,* $s \in N$ *and* $t = uA$, *where* $u \in T^*$ *and* $A \in N \cup \{\Lambda\}$;

(f) λ *is regular if and only if every* $d = (s \xrightarrow{p} t) \in D$ *is regular, i.e.,* $s \in N$ *and* $t = aA$, *where* $a \in T$ *and* $A \in N \cup \{\Lambda\}$.

Proof. Suppose λ is a type 0 fuzzy language over T. By Theorem 3.4.13, $\lambda = \lambda_G$ for some max-product grammar $G = (T, N, \rho, A_0)$. It follows that the quadruple $(T, N, D(\rho), h)$, where $h(A_0) = 1$ and $h(A) = 0$ for $A = A_0$, has the desired properties. Conversely, let (T, N, D, h) be a quadruple satisfying conditions (1) to (2), and λ a fuzzy language over T satisfying (3.6). It follows that $\lambda = \lambda_G$, where $G = (T, N, \{\rho_D\}, h)$ is a max-product grammar. The remainder of the proof follows easily. ∎

Theorem 3.4.16 yields another formulation of max-product grammars and fuzzy languages generated by max-product grammars. This formulation illustrates the fact that max-product grammars may be viewed as a model of fuzzy grammars.

3.5   Weak Regular Max-Product Grammars

In this section, we study weak regular max-product grammars and their relation to asynchronous max-product automata.

Theorem 3.5.1 *If λ is a strict weak regular fuzzy language over T, then $\lambda = \lambda_G$ for some strict weak regular max-product grammar $G = (T, N, \rho, A_0)$, where $\rho(A, B) = 0$ for all $A, B \in N$, and $\rho(\Lambda, A) = 0$ for all $A \in N \backslash \{A_0\}$. If λ is finitary, then for every $A \in N$, $\rho(A, t) > 0$ implies $t = aB$, where $a \in T$ and $B \in N \cup \{\Lambda\}$, or $A = A_0$ and $t = \Lambda$.*

Proof. By Theorem 3.4.13, $\lambda = \lambda_{G_0}$, where $G_0 = (T, N, \rho_0, A_0)$ is a strict weak regular max-product grammar. Let ρ'_n, $n = 0, 1, 2, \ldots$, be functions from $(T \cup N)^* \times (T \cup N)^*$ into $[0, 1]$ defined recursively as follows:

$$\rho'_0(s, t) = \begin{cases} 1 & \text{if } s = t \in N \\ 0 & \text{otherwise} \end{cases}$$

and

$$\rho'_{n+1}(s, t) = \begin{cases} \vee_{A \in N} [\rho'_n(A, t) \rho_0(s, A)] & \text{if } s, t \in N \\ 0 & \text{otherwise.} \end{cases}$$

Let ρ_1 be a function from $(T \cup N)^* \times (T \cup N)^*$ into $[0, 1]$ such that for all $s, t \in (T \cup N)^*$,

$$\rho_1(s, t) = \begin{cases} \vee \{\rho_0(s, t), \ \vee_{n=0}^{\infty} \vee_{A \in N} [\rho_0(A, t) \rho'_n(s, A)]\} & \text{if } s \in N, t \notin N \\ 0 & \text{otherwise.} \end{cases}$$

Define $G = (T, N, \rho, A_0)$ such that for all $s, t \in (T \cup N)^*$,

$$\rho(s, t) = \begin{cases} \rho_1(s, t) & \begin{array}{l} \text{if } s \in N, t \in T^*N \\ \text{or } s = A_0, t = \Lambda \end{array} \\ \vee \{\rho_1(s, t), \ \vee_{A \in N} [\rho_1(A, \Lambda) \rho_1(s, tA)]\}, & \text{if } s \in N, t \in T^+ \\ 0 & \text{otherwise.} \end{cases}$$

It follows that $\lambda = \lambda_G$ and G has the desired properties. The second part of the theorem follows from the fact that every weak regular discrete fuzzy production $A \xrightarrow{p} uaA'$, where $u \in T$, $a \in T$, and $A' \in N \cup \{\Lambda\}$, may be replaced by $A \xrightarrow{p} uA''$ and $A'' \xrightarrow{1} aA'$, where A'' is a new nonterminal symbol. ∎

In terms of discrete fuzzy productions, the above theorem states that every strict weak regular fuzzy language is generated by a strict weak regular max-product grammar containing no discrete fuzzy productions of the form $A \xrightarrow{p} B$, where $A, B \in N$, or $A \xrightarrow{p} \Lambda$, where $A \in N \backslash \{A_0\}$. In addition, every finitary strict weak fuzzy language is generated by a weak regular max-product grammar containing only regular discrete fuzzy productions and $A_0 \xrightarrow{p} \Lambda$.

Theorem 3.5.2 *λ is a finitary strict weak regular fuzzy language if and only if there exists a strict regular fuzzy language λ' over T such that $\lambda(r) = \lambda'(r)$ for all $r \neq \Lambda$.*

Proof. Suppose $\lambda = \lambda_G$, where $G = (T, N, \rho, A_0)$ is a finitary strict weak regular max-product grammar. By Theorem 3.5.1, we may assume that $D(\rho)$ contains only regular discrete fuzzy productions and $A_0 \xrightarrow{p} \Lambda$. Let D be the collection of all regular fuzzy productions in $D(\rho)$, and $G' = (T, N, \rho_D, A_0)$. Clearly, G' is a strict regular max-product grammar, and $\lambda(r) = \lambda_{G'}(r)$ for all $r \neq \Lambda$. Conversely, suppose $\lambda(r) = \lambda_G(r)$ for all $r \neq \Lambda$, where $G = (T, N, \rho, A_0)$ is a strict regular max-product grammar. Let D be the collection of all discrete fuzzy productions in $D(\rho)$, and $A_0^p \longrightarrow \Lambda^p$, where $p = \lambda(r)$. Then $G' = (T, N, \rho_D, A_0)$ is a finitary strict weak regular max-product grammar such that $\lambda = \lambda_{G'}$. ∎

Definition 3.5.3 *An **asynchronous max-product automaton** (**AMA**) is a sextuple $M = (Q, X, Y, p, h, g)$, where Q, X, and Y are finite nonempty sets, p is a function from $Q \times X \times Y^* \times Q$ into $\mathbb{R}^{\geq 0}$, and h and g are functions from Q into $\mathbb{R}^{\geq 0}$. If, in addition, the image of the functions p, h, and g are subsets of $[0, 1]$, then M is called **strict**.*

Definition 3.5.4 *Let $M = (Q, X, Y, p, h, g)$ be an AMA.*

*(1) M is called **finitary** if for all $q, q' \in Q$ and $u \in X$, $p(q, u, y, q') > 0$ except for finitely many $y \in Y^*$.*

*(2) M is called **synchronous** if for all $q, q' \in Q$ and $u \in X$, $p(q, u, y, q') > 0$ implies $y \in Y$.*

*(3) M is called **normalized** if for all $q \in Q$ and $u \in X$,*

$$\sum_{y \in Y^*} \sum_{q' \in Q} p(q, u, y, q') \leqslant 1$$

and $\sum_{q \in Q} h(q) \leqslant 1$.

In the above definitions, Q, X, and Y are, respectively, the state, input, and output sets. $p(q', y, u, q)$ is the grade of membership, or conditional probability in the case of normalized AMA, that the next state of M is q' and output string y is produced, given that the present state of M is q and input symbol u is applied. $h(q)$ and $g(q)$ are, respectively, the grade of membership, or probability in the case of normalized AMA, that q is the initial state of M and that q is a final state of M.

Definition 3.5.5 *Let $M = (Q, X, Y, p, h, g)$ be an AMA.*

(1) p^M is the function from $Q \times X^* \times Y^* \times Q$ into $\mathbb{R}^{\geq 0}_\infty$ such that for all $q', q'' \in Q$, $u \in X$, $x \in X^*$, and $y \in Y^*$,

$$p^M(q', \Lambda, \Lambda, q'') = \begin{cases} 1 & \text{if } q' = q'' \\ 0 & \text{if } q' \neq q'' \end{cases}$$

and

$$p^M(q', ux, y, q'') = \vee\{p(q', u, y_1, q)p^M(q, x, y_2, q'') \mid q \in Q, \\ y = y_1 y_2, \ y_1, y_2 \in Y^*\}.$$

(2) F^M is the function from $X \times Y$ into $\mathbb{R}_\infty^{\geq 0}$ such that for all $x \in X^*$ and $y \in Y^*$,

$$F^M(x,y) = \vee\{h(q)p^M(q,x,y,q')g(q') \mid q,q' \in Q\}.$$

(3) D^M is the function from X^* into $\mathbb{R}_\infty^{\geq 0}$ such that for all $x \in X^*$,

$$D^M(x) = \vee\{F^M(x,y) \mid y \in Y^*\}.$$

(4) R^M is the function from Y^* into $\mathbb{R}_\infty^{\geq 0}$ such that for all $y \in Y^*$,

$$R^M(y) = \vee\{F^M(x,y) \mid x \in X^*\}.$$

Clearly, D^M and R^M are fuzzy languages over X and Y, respectively. Using the terminology given in [216], the class of all F^M is the class of fuzzy functions computable by AMA, the class of all D^M is the class of fuzzy languages acceptable by AMA, and the class of all R^M is the class of fuzzy languages that can be generated by AMA.

It follows from the above definitions that a synchronous strict AMA reduces to a max-product machine if $g(q) = 1$ for all $q \in Q$.

Example 3.5.6 Let $M = (Q, X, Y, p, h, g)$, where $Q = \{q_1, q_2\}$, $X = \{u\}$, $Y = \{0, 1\}$, $h : Q \to [0, 1]$, $g : Q \to [0, 1]$, and $p : Q \times X \times Y^* \times Q \to [0, 1]$ is defined as follows:

$$p(q_1, u, 11, q_2) = \tfrac{1}{2},$$
$$p(q_2, u, 0, q_1) = \tfrac{3}{4},$$

and $p(q, u, y, q') = 0$ for any other (q, u, y, q'). Then

$$p^M(q_1, uu, 110, q_1) = \tfrac{1}{2} \cdot \tfrac{3}{4}$$
$$= \tfrac{3}{8},$$
$$p^M(q_2, uu, 011, q_2) = \tfrac{3}{4} \cdot \tfrac{1}{2}$$
$$= \tfrac{3}{8}.$$

Let y_{nn}, $y_{n,n-1}$, z_{nn}, and $z_{n,n-1}$ denote, respectively, the following alternating sequences of 11's and 0's

$110110\ldots110$,	*where 11 and 0 appear n times,*
$110110\ldots11$,	*where 11 appear n times and*
	0 appear n − 1 times,
$011011\ldots011$,	*where 0 and 11 appear n times,*
$011011\ldots0110$,	*where 0 appears n times and*
	11 appears n − 1 times,

$n \in \mathbb{N}$, $n \geq 2$. Then

$$p^M(q_1, u^{2n}, y_{nn}, q_1) = (\tfrac{1}{2})^n (\tfrac{3}{4})^n,$$
$$p^M(q_1, u^{2n-1}, y_{n,n-1}, q_2) = (\tfrac{1}{2})^n (\tfrac{3}{4})^{n-1},$$
$$p^M(q_2, u^{2n}, z_{nn}, q_2) = (\tfrac{3}{4})^n (\tfrac{1}{2})^n,$$
$$p^M(q_2, u^{2n-1}, z_{n,n-1}, q_1) = (\tfrac{3}{4})^n (\tfrac{1}{2})^{n-1},$$

$$F^M(u^{2n}, y_{nn}) = h(q_1)(\tfrac{1}{2})^n (\tfrac{3}{4})^n g(q_1),$$
$$F^M(u^{2n-1}, y_{n,n-1}) = h(q_1)(\tfrac{1}{2})^n (\tfrac{3}{4})^{n-1} g(q_2),$$
$$F^M(u^{2n}, z_{nn}) = h(q_2)(\tfrac{3}{4})^n (\tfrac{1}{2})^n g(q_2),$$
$$F^M(u^{2n-1}, z_{n,n-1}) = h(q_2)(\tfrac{3}{4})^n (\tfrac{1}{2})^{n-1} g(q_1),$$

$$
\begin{aligned}
D^M(u^{2n}) &= \vee\{F^M(u^{2n}, y) \mid y \in Y^*\} \\
&= (h(q_1)(\tfrac{1}{2})^n (\tfrac{3}{4})^n g(q_1)) \vee (h(q_2)(\tfrac{3}{4})^n (\tfrac{1}{2})^n g(q_2)) \\
D^M(u^{2n-1}) &= \vee\{F^M(u^{2n-1}, y) \mid y \in Y^*\} \\
&= (h(q_1)(\tfrac{1}{2})^n (\tfrac{3}{4})^{n-1} g(q_2)) \vee (h(q_2)(\tfrac{3}{4})^n (\tfrac{1}{2})^{n-1} g(q_1)),
\end{aligned}
$$

$$
\begin{aligned}
R^M(y_{nn}) &= D^M(u^{2n}), \\
R^M(y_{n,n-1}) &= D^M(u^{2n-1}), \\
R^M(z_{nn}) &= D^M(u^{2n}), \\
R^M(z_{n,n-1}) &= D^M(u^{2n-1}).
\end{aligned}
$$

Theorem 3.5.7 λ *is a (finitary)(strict, normalized) weak regular fuzzy language if and only if* $\lambda = R^M$ *for some (finitary)(strict, normalized) AMA M.*

Proof. Suppose $\lambda = \lambda_G$, where $G = (T, N, P, h)$ is a (finitary)(strict, normalized) weak regular max-product grammar. Let $A_0 \notin T \cup N$ and $N_0 = N \cup \{A_0\}$. Define $M = (N_0, P, T, p, h', g)$ so that for all $\rho \in P$, $r \in T^*$ and $A, A' \in N_0$,

$$
p(A, \rho, r, A') = \begin{cases} \rho(A, rA') & \text{if } A, A' \in N \\ \rho(A, r) & \text{if } A \in N,\ A' = A_0 \\ 0 & \text{otherwise} \end{cases}
$$

$$
g(A) = \begin{cases} 1 & \text{if } A = A_0 \\ 0 & \text{if } A \neq A_0 \end{cases}
$$

and

$$
h'(A) = \begin{cases} h(A) & \text{if } A \in N \\ 0 & \text{if } A = A_0. \end{cases}
$$

It follows that $\lambda = R^M$ and M has the desired properties. Conversely, suppose $\lambda = R^M$, where $M = (Q, X, Y, p, h, g)$ is a (finitary) (strict, normalized) AMA M. Without loss of generality, we may assume that $Q \cap Y = \emptyset$. Let $q_0 \notin Q \cup Y$ and $Q' = Q \cup \{q_0\}$. For all $u \in X$, we associate the fuzzy production ρ_u over $Y \cup Q'$, where $\overline{\rho}_u = Q$ and for all $q \in Q$, $y \in Y^*$, and $t \in (Y \cup Q)^*$,

$$\rho_u(q, t) = \begin{cases} p(q, u, y, q') & \text{if } t = yq' \\ 0 & \text{otherwise.} \end{cases}$$

Let ρ_0 be the fuzzy production over $Y \cup Q'$ such that $\overline{\rho}_0 = \{q_0\}$ and for every $t \in (Y \cup Q')^*$,

$$\rho_0(q_0, t) = \begin{cases} h(q) & \text{if } t = q \in Q \\ 0 & \text{otherwise.} \end{cases}$$

Define $G = (Y, Q', P, h')$ such that $P = \{\rho_0\} \cup \{\rho_u \mid u \in X\}$ and

$$h(q) = \begin{cases} 1 & \text{if } q = q_0 \\ 0 & \text{if } q \neq q_0. \end{cases}$$

It follows that $\lambda = \lambda_G$ and G has the desired properties. $\blacksquare$

Theorem 3.5.8 *If λ is a finitary strict weak regular fuzzy language, then $\lambda = R^M$ for some synchronous strict AMA M with a single input.*

Proof. Let $\lambda = \lambda_G$, where $G = (T, N, \rho, A_0)$ is a weak regular max-product grammar satisfying the conditions given in Theorem 3.5.1. Let A_1, $A_2 \notin T \cup N$ and $N' = N \cup \{A_1, A_2\}$. Define $M = (N', \{\rho\}, T, p, h, g)$ so that for every $r \in T$ and $A, A' \in N$,

$$p(A, \rho, r, A') = \begin{cases} \rho(A, rA') & \text{if } A, A' \in N \\ \rho(A, r) & \text{if } A \in N, \ A = A_1, \text{ and } r \neq \Lambda \\ 0 & \text{otherwise} \end{cases}$$

$$g(A) = \begin{cases} 1 & \text{if } A = A_1 \text{ or } A = A_2 \\ 0 & \text{otherwise} \end{cases}$$

and

$$h(A) = \begin{cases} \rho(A_0, \Lambda) & \text{if } A = A_2 \\ 1 & \text{if } A = A_0 \\ 0 & \text{otherwise.} \end{cases}$$

It follows that $\lambda = R^M$ and M has the desired properties. $\blacksquare$

Theorem 3.5.9 $\lambda = R^M$ *for some synchronous (strict) AMA $M = (Q, X, Y, p, h, g)$ if and only if $\lambda = D^{M'}$ for some synchronous (strict) AMA $M' = (Q, Y, X, p', h, g)$.*

Proof. Suppose $\lambda = R^M$, where $M = (Q, X, Y, p, h, g)$ is a synchronous (strict) AMA. Define $M' = (Q, Y, X, p', h, g)$, where for every $q, q' \in Q$, $v \in Y$, and $x \in X^*$,

$$p'(q, v, x, q') = \begin{cases} p(q, x, v, q') & \text{if } x \in X \\ 0 & \text{otherwise.} \end{cases}$$

It follows that $\lambda = D^{M'}$. The converse follows in a similar manner. ∎

Definition 3.5.10 *A **max-product automaton(MA)** is a quintuple $M = (Q, X, p, h, g)$, where Q and X are finite nonempty sets, p is a function from $Q \times X \times Q$ into $\mathbb{R}^{\geq 0}$, and h and g are functions from Q into $\mathbb{R}^{\geq 0}$. If, in addition, the images of p, h, and g are subsets of $[0, 1]$, then M is called* **strict.**

In the above definition, Q and X are, respectively, the state and input sets. $p(q, u, q')$ is the grade of membership that the next state of M is q', given that the present state of M is q and the input symbol u is applied. $h(q)$ and $g(q)$ are the grade of membership that q is the initial state of M and q is a final state of M, respectively.

Definition 3.5.11 *Let $M = (Q, X, p, h, g)$ be an MA.*
(1) Define the function p^M from $Q \times X \times Q$ into $\mathbb{R}^{\geq 0}$ by for every $q', q'' \in Q$, $u \in X$, and $x \in X^*$,

$$p^M(q', \Lambda, q'') = \begin{cases} 1 & \text{if } q' = q'' \\ 0 & \text{if } q' \neq q'' \end{cases}$$

and

$$p^M(q', ux, q'') = \vee\{p(q, u, q'')p^M(q', x, q) \mid q \in Q\}.$$

(2) Define the function λ^M from X^* into $\mathbb{R}^{\geq 0}$ by for every $x \in X^*$,

$$\lambda^M(x) = \vee\{h(q)p^M(q, x, q')g(q') \mid q, q' \in Q\}.$$

λ^M is the fuzzy language accepted by M.

Theorem 3.5.12 *The following statements are equivalent:*
(1) $\lambda = \lambda^M$ for some (strict) MA M;
(2) $\lambda = D^M$ for some synchronous (strict) AMA M with a single output;
(3) $\lambda = D^M$ for some (strict) AMA $M = (Q, X, Y, p, h, g)$, where there exists $m \in \mathbb{R}^{\geq 0}$ such that $p(q, u, y, q') \leqslant m$ for all $q, q' \in Q$, $u \in X$, and $y \in Y^*$.

Proof. Let $\lambda = \lambda^M$, where $M = (Q, X, p, h, g)$ is an MA. Define $M' = (Q, X, Y, p', h, g)$, where $Y = \{v_0\}$ and for all $q, q' \in Q$, $u \in X$, $y \in Y^*$,

$$p'(q, u, y, q') = \begin{cases} p(q, u, q') & \text{if } y = v_0 \\ 0 & \text{otherwise.} \end{cases}$$

Clearly, M' is a synchronous AMA with a single input, and $\lambda = D^{M'}$. Thus (1) implies (2). That (2) implies (3) is immediate. Now, suppose $\lambda = D^M$, where $M = (Q, X, Y, p, h, g)$ is an AMA. Define $M' = (Q, X, p', h, g)$, where for every $q, q' \in Q$ and $u \in X$, $p'(q, u, q') = \vee_{y \in Y^*} p(q, u, y, q')$. It follows that $\lambda = \lambda^{M'}$. Hence (3) implies (1). ∎

Theorem 3.5.13 *The following statements are equivalent:*

(1) λ *is a finitary strict weak regular fuzzy language;*

(2) $\lambda = \lambda_G$ *for some strict max-product grammar* $G = (T, N, \rho, A_0)$, *where* $D(\rho)$ *contains only regular discrete fuzzy productions and* $A_0 \xrightarrow{p} \Lambda$;

(3) there exists a strict regular fuzzy language λ' *such that* $\lambda(r) = \lambda'(r)$ *for all* $r \neq \Lambda$;

(4) $\lambda = R^M$ *for some finitary strict AMA* M;

(5) $\lambda = R^M$ *for some synchronous strict AMA* M *with a single input;*

(6) $\lambda = D^M$ *for some strict AMA* M;

(7) $\lambda = D^M$ *for some synchronous strict AMA* M *with a single output;* and

(8) $\lambda = \lambda^M$ *for some strict MA* M.

Proof. The proof follows from Theorems 3.5.1 and 3.5.12. ∎

Theorem 3.5.14 *The following statements are equivalent:*

(1) $\lambda = \lambda_G$ *for some max-product grammar* $G = (T, N, \rho, A_0)$, *where* $D(\rho)$ *contains only regular discrete fuzzy productions and* $A_0 \xrightarrow{p} \Lambda$;

(2) there exists a regular fuzzy language λ' *such that* $\lambda(r) = \lambda'(r)$ *for all* $r \neq \Lambda$;

(3) $\lambda = R^M$ *for some synchronous AMA* M;

(4) $\lambda = D^M$ *for some synchronous AMA* M *with a single input;*

(5) $\lambda = D^M$ *for some AMA* M *where the range of the transition function* p *is a subset of* $[0, m]$ *for some* $m \in \mathbb{R}^{\geq 0}$;

(6) $\lambda = D^M$ *for some synchronous AMA* M *with a single output; and*

(7) $\lambda = \lambda^M$ *for some MA* M.

By Theorem 3.5.13, many interesting properties of finitary strict weak regular fuzzy languages can be obtained from the results of Section 5.16. Similar results can be established for the family of fuzzy languages characterized by Theorem 3.5.14.

3.6   Weak Regular Max-Product Languages

In this section, we study the family of languages generated by weak regular max-product grammars with cut point.

Definition 3.6.1 *Let λ be a fuzzy language over T and $c \in \mathbb{R}^{\geq 0}_\infty$. Define*

$$L(\lambda, c, >) = \{r \in T^* \mid \lambda(r) > c\}.$$

$L(\lambda, c, \geqslant)$, $L(\lambda, c, =)$, etc. are defined in a similar manner.

Note that if $G = (T, N, \rho, A_0)$ is a max-product grammar, then $r \in L(\lambda_G, c, >)$ if and only if there exists $w \in (D(\rho) \times \mathbb{N})^*$ such that $A_0 \overset{p}{\Longrightarrow} r$ mod(w) and $p > c$.

Theorem 3.6.2 *For every $L \subseteq T^*$, there exists a strict weak regular fuzzy language λ over T such that $L = L(\lambda, 1, =) = L(\lambda, c, >)$ for all c such that $0 \leqslant c < 1$.*

Proof. Let $G = (T, N, \rho, A_0)$ be the max-product grammar such that $N = \{A_0\}$ and $D(\rho)$ contains all discrete fuzzy productions of the form $A_0 \overset{p}{\longrightarrow} r$, where $r \in L$. Then λ_G has the desired properties. ∎

Definition 3.6.3 *A **phrase-structure grammar** is a quadruple $G = (T, N, P, A_0)$, where T is the terminal set, N is the nonterminal set, P is a finite collection of productions, and $A_0 \in N$ is the start symbol. Derivations according to G, the language $L(G)$ generated by G, and the type i $(i = 0, 1, 2, 3,)$ grammars in the Chomsky hierarchy are defined in the usual manner. (See [77] or Chapter 1.)*

Theorem 3.6.4 *L is regular if and only if $L = L(\lambda, 0, >)$ for some finitary weak regular fuzzy language λ.*

Proof. Suppose $\lambda = \lambda_G$, where $G = (T, N, \rho, A_0)$ is a finitary weak regular max-product grammar. Define a phrase structure grammar $G' = (T, N, P, A_0)$, where P contains all productions of the form $s \longrightarrow t$ where $\rho(s, t) > 0$. Clearly, the language $L(G')$ is regular and $L(\lambda, 0, >) = L(G')$. The converse follows easily. ∎

Theorem 3.6.5 *Let $0 \leqslant c < 1$. $L = L(\lambda, c, >)$ for some finitary strict weak regular fuzzy language λ if and only if $L \backslash \{\Lambda\} = L(\lambda', c, >)$ for some strict regular fuzzy language λ'.*

Proof. The proof follows from Theorem 3.5.13 ∎

Theorem 3.6.6 *L is regular if and only if $L = L(\lambda, c, >)$ for some finitary strict weak regular fuzzy language λ and $0 \leqslant c < 1$.*

Proof. Suppose $L = L(\lambda, c, >)$ for some finitary strict weak regular fuzzy language λ and $0 \leqslant c < 1$. By Theorem 3.6.5 $L \backslash \{\Lambda\} = L(\lambda_G, c, >)$, where $G = (T, N, \rho, A_0)$ is a strict regular max-product grammar. Let D' be the collection of all discrete fuzzy productions of $D(\rho)$ of the form $s \xrightarrow{1} t$, $t \notin T^*$. Let D'' be the collection of all discrete fuzzy productions of $D(\rho)$ of the form $s \xrightarrow{p} t$, $t \notin T^*$ and $p < 1$. Clearly,

$$r \in L(\lambda_G, c, >)$$

if and only if there exists $A_{j_i} \in N$, $r_{j_{i-1}j_i} \in T^*$, $i = 1, 2, \ldots, n$, and $u \in T$ such that $r = r_{j_0 j_1} r_{j_1 j_2} \cdots r_{j_{n-1} j_n} u$, $j_0 = 0$, $(A_{j_n} \xrightarrow{p_n} u) \in D(\rho)$, and for every $i = 1, 2, \ldots, n$, $A_{j_{i-1}} \overset{p_i}{\Longrightarrow} r_{j_{i-1} j_i} A_{j_i} \mathrm{mod}(w_i)$ for some $w_i \in (D' \times \mathbb{N})^*$ or $w_i \in (D'' \times \mathbb{N})^*$, $\Pi_{j=1}^n p_j > c$. Let $N = \{A_0, A_1, \ldots, A_m\}$. This shows that $L(\lambda_G, c, >)$ is a finite union of regular languages of the form $r_1 L_2 r_3 L_4 \ldots r_{n-1} L_n r_{n+1}$, where the r's are in T^* and the L's are languages generated by type 3 phrase structure grammars of the form $G_{ij} = (T, N, P_j, A_i)$, $i, j = 0, 1, 2, \ldots, m$, where P_j contains all productions of the form $A_j \longrightarrow \Lambda$ and $s \longrightarrow t$ such that $(s \xrightarrow{1} t) \in D'$. Thus $L(\lambda_G, c, >)$, as well as L, is regular. The converse follows easily. ∎

Theorem 3.6.7 *Let* $L = L(\lambda, c, >)$*, where* $0 \leqslant c < 1$*.* L *is regular if the following conditions hold:*
  (1) $\lambda = R^M$ *for some finitary strict AMA M;*
  (2) $\lambda = D^M$ *for some strict AMA M;*
  (3) $\lambda = \lambda^M$ *for some strict MA M; or*
  (4) $\lambda = \lambda_G$ *for some strict weak regular max-product grammar* $G = (T, N, \rho, A_0)$*, where for each* $A \in N$*,* $\sum_{t \in (T \cup N)^*} \rho(A, t)$ *is finite.*

Proof. By Theorems 3.5.13 and 3.6.6, it suffices to prove (4). Suppose $\lambda = \lambda_G$, where $G = (T, N, \rho, A_0)$ has the desired properties. Let $G' = (T, N, \rho', A_0)$ be such that for every $s, t \in (T \cup N)^*$,

$$\rho'(s, t) = \begin{cases} \rho(s, t) & \text{if } \rho(s, t) > c \\ 0 & \text{otherwise.} \end{cases}$$

Clearly, G' is a finitary strict weak regular max-product grammar. In addition, $L = L(\lambda_{G'}, c, >)$. Thus by Theorem 3.6.6, L is regular. ∎

Theorem 3.6.8 *Let* $G = (T, N, \rho, A_0)$ *be a weak regular max-product grammar where the range of* ρ *is a subset of* $[0, m]$ *with* $m < 1$*. Then* $L(\lambda_G, c, >)$ *is finite for all* $0 \leqslant c < 1$*.*

Proof. It is clear that $r \in L(\lambda_G, c, >)$ implies $|r| \leqslant n$, where n is the largest integer such that $m^n > c$. ∎

Note that Theorems 3.6.5 to 3.6.8 hold if $>$ is replaced by $\geqslant$ or $=$.

Theorem 3.6.9 $L(\lambda, \infty, =)$ *is regular for all finitary weak regular fuzzy languages* λ.

Proof. Let $L = L(\lambda_G, \infty, =)$, where $G = (T, N, \rho, A_0)$ is a finitary weak regular max-product grammar. Without loss of generality, we may assume that $D(\rho)$ contains only discrete fuzzy productions of the forms $A \xrightarrow{p} aB$, $A \xrightarrow{p} B$, and $A \xrightarrow{p} \Lambda$, where $A, B \in N$ and $a \in T$. Let N_0 be the subset of N consisting of all $A \in N$ such that $A \overset{p}{\Longrightarrow} A \bmod(w)$ for some $w \in (D(\rho) \times \mathbb{N})^*$ and $p > 1$. It is clear that $r \in L\backslash\{\Lambda\} \Leftrightarrow$ there exist $A \in N_0$ and $w, z \in T^*$ such that $r = wz$, $A_0 \overset{p'}{\Longrightarrow} wA \bmod(w')$, and $A \overset{p''}{\Longrightarrow} z$ $\bmod(w'')$ for some $w', w'' \in (D(\rho) \times \mathbb{N})^*$ and $p' > 0$, $p'' > 0$. Thus $L\backslash\{\Lambda\}$ is the union of all languages of the form $L(G'_A)L(G''_A)$, where $A \in N_0$. Therefore, for every $A \in N$, $G'_A = (T, N, P', A_0)$ and $G''_A = (T, N, P'', A)$ are phrase structure grammars such that P' contains all productions of the form $s \longrightarrow t$, where $(s \xrightarrow{p} t) \in D(\rho)$, $t \notin T^*$ and $A \longrightarrow \Lambda$; and P'' contains all productions of the form $s \longrightarrow t$, where $(s \xrightarrow{p} t) \in D(\rho)$. Clearly, $L(G'_A)$ and $L(G''_A)$ are regular for all $A \in N$. Thus $L\backslash\{\Lambda\}$, as well as L, is regular. ∎

Corollary 3.6.10 $L(\lambda, \infty, <)$ *is regular for all finitary weak regular fuzzy languages* λ. ∎

Theorem 3.6.11 *If* $L = L(\lambda, c >)$ *for some finitary weak regular fuzzy language* λ *and* $c \geqslant 0$, *then* $L = L(\lambda_G, c, >)$ *for some finitary max-product grammar* $G = (T, N, \rho, A_0)$, *where* $D(\rho)$ *contains only regular discrete fuzzy productions and* $A_0 \xrightarrow{p} \Lambda$.

Proof. Let $L = L(\lambda_{G_1}, c, >)$, where $G_1 = (T, N_1, \rho_1, A_1)$ is a finitary weak regular max-product grammar and $c \geqslant 0$. By Theorem 3.6.5 we may assume that $0 \leqslant c < 1$. It follows from the proof of Theorem 3.5.1 that we may assume that ρ_1 satisfies the conditions given in Theorem 3.5.1 as long as we allow ρ_1 to assume the value ∞. By Theorem 3.6.9, $L(\lambda_{G_1}, \infty, =) = L(G_2)$, where $G_2 = (T, N_2, P, A_2)$ is a phrase structure grammar such that P contains only productions of the form $A_2 \longrightarrow \Lambda$, $A \longrightarrow a$, and $A \longrightarrow aB$, where $A, B \in N_2$ and $a \in T$. Without loss of generality, we assume that $N_1 \cap N_2 = \emptyset$. Let $A_0 \notin T \cup N_1 \cup N_2$ and $N = N_1 \cup N_2 \cup \{A_0\}$. Let D be the collection of discrete fuzzy productions over $T \cup N$ consisting of all discrete fuzzy productions of the following forms:

    (i) $s \xrightarrow{p} t$, where $(s \xrightarrow{p} t) \in D(\rho)$ and $p < \infty$;

    (ii) $A_0 \xrightarrow{p} t$, where $(A_1 \xrightarrow{p} t) \in D(\rho)$ and $p < \infty$;

    (iii) $s \xrightarrow{1} t$, where $(s \longrightarrow t) \in P$;

    (iv) $A_0 \xrightarrow{1} t$, where $(A_2 \longrightarrow t) \in P$.

Clearly, D is finite and contains only regular discrete fuzzy productions, and $A_0 \xrightarrow{p} \Lambda$. Moreover, it can be shown that $L = L(\lambda_G, c, >)$, where

$G = (T, N, \rho_D, A_0)$ is a finitary max-product grammar with the desired properties. ∎

Let $c \geqslant 0$. Let $\pounds_c$ denote the family of all languages of the form $L(\lambda, c, >)$, where λ is a finitary weak regular fuzzy language. Let $\pounds = \cup_{c \in \mathbb{R}^{\geq 0}} \pounds_c$.

Theorem 3.6.12 *The following statements are equivalent:*

   (1) $L \in \pounds_c$;

   (2) $L \backslash \{\Lambda\} = L(\lambda, c, >)$ *for some regular fuzzy language* λ;

   (3) $L = L(R^M, c, >)$ *for some synchronous AMA* M;

   (4) $L = L(R^M, c, >)$ *for some synchronous AMA* M *with a single input;*

   (5) $L = L(D^M, c, >)$ *for some AMA* M *such that the range of the transition function* p *is a subset of* $[0, m]$ *for some* $m \in \mathbb{R}^{\geq 0}$;

   (6) $L = L(D^M, c, >)$ *for some synchronous AMA* M *with a single output;*

   and

   (7) $L = L(\lambda^M, c, >)$ *for some MA* M.

Proof. The proof follows from Theorems 3.5.14 and 3.6.11. ∎

Another characterization of $\pounds_c$ will be given below via languages generated by type 3 weighted grammars under maximal interpretation [188]. (Also, see Chapter 5.)

Definition 3.6.13 *A **weighted grammar of type** i $(i = 0, 1, 2, 3)$ is a triple* $G_w = (G, \delta, \phi)$, *where* $G = (T, N, P, A_0)$ *is a type i phrase structure grammar,* δ *is a function from* P *into* $\mathbb{R}^{\geq 0}$, *and* ϕ *is a function from* $P \times P$ *into* $\mathbb{R}^{\geq 0}$. *The language* $L_m(G_w, c)$ *generated by* G_w *with cut point* $c \in \mathbb{R}^{\geq 0}$ *is said to be under maximal interpretation if there is at least one derivation of* P *with weight greater than* c *(see [188] and Section 5.8).*

Theorem 3.6.14 *Let* $c \geqslant 0$. $L \in \pounds_c$ *if and only if* $L = L_m(G_w, c)$ *for some weighted grammar* G_w *of type 3.*

Proof. Suppose $L = L(\lambda_G, c, >)$ for some finitary weak regular max-product grammar $G = (T, N, \rho, A_0)$. Let $G' = (T, N, P, A_0)$ be a phrase structure grammar such that P contains all productions of the form $s \longrightarrow t$, where $\rho(s, t) > 0$. Clearly, G' is of type 3. Let $G_w = (G', \delta, \phi)$ be the type 3 weighted grammar such that for every $q = (s \longrightarrow t)$, $q' = (s' \longrightarrow t') \in P$, we have $\delta(q) = \rho(s, t)$ and $\phi(q, q') = \rho(s', t')$. It can be shown that $L = L_m(G_w, c)$. Conversely, suppose $L = L_m(G_w, c)$ for some type 3 weighted grammar $G_w = (G, \delta, \phi)$, where $G = (T, N, P, A_0)$. With each $n \in N$ and $q \in P \cup \{\Lambda\}$, we associate the abstract symbol A_{rq}. Let $N' = \{A_{rq} \mid n \in N \cup \{\Lambda\}\}$ and $q \in P \cup \{\Lambda\}$, and D be the collection of all discrete fuzzy productions over $T \cup N'$ of the forms

   (1) $A_{n\Lambda} \xrightarrow{\delta(t)} A_{q\hat{t}}$, where $n \in N$, $q \in N \cup \{\Lambda\}$, $r \in T^*$, and $\hat{t} = (n \longrightarrow rq) \in P$;

(2) $A_{nq} \xrightarrow{\phi(q,\hat{t})} rA_{q\hat{t}}$, where $q \in P$, $n \in N \cup \{\Lambda\}$, $r \in T^*$, and $\hat{t} = (n \longrightarrow rq) \in P$; and

(3) $A_{\Lambda q} \xrightarrow{1} \Lambda$, where $q \in P$.

Let $G' = (T, N', \rho_D, h)$, where $h(A_{nq}) = 1$ if $q = \Lambda$ and $n \neq \Lambda$, and $h(A_{nq}) = 0$ otherwise. Clearly, G' is a finitary weak regular max-product grammar. Furthermore, it follows that $L = L(\lambda_{G'}, c, >)$. Thus $L \in \pounds_c$. ∎

Combining Theorem 3.6.14 and Theorems 10 and 12 of [188] yields:

Theorem 3.6.15 *The family of regular languages is a proper subfamily of $\pounds$ and there exist context-free and stochastic languages* [235] *which do not belong to $\pounds$. Moreover,* $\{a^n b^m \mid n > m \geqslant 1\} \in \pounds$, *while* $\{a^n b^n \mid n \geqslant 1\} \notin \pounds$.

3.7   Properties of $\pounds$

In this section, we examine the properties of $\pounds_c$ and $\pounds$.

Let $\mathcal{R}$ denote the family of all regular languages.

Theorem 3.7.1 $\mathcal{R} \subseteq \pounds_c$ *for all* $c \geqslant 0$.

Proof. The proof follows from Theorem 3.6.6. ∎

Theorem 3.7.2 $\pounds_0 = \mathcal{R}$.

Proof. The proof follows from Theorem 3.6.4. ∎

Theorem 3.7.3 $\pounds = \pounds_c$ *for all* $c > 0$.

Proof. Suppose $c > 0$ and $L \in \pounds$. Then $L = L(\lambda_G, c, >)$ for some finitary max-product grammar $G_1 = (T, N, P, h_1)$ and $c_1 \geqslant 0$. By Theorems 3.7.1 and 3.7.2, it suffices to consider the case where $c_1 > 0$. Let $G = (T, N, P, h)$, where $h(A) = (c/c_1)h_1(A)$ for all $A \in N$. Clearly $L = L(\lambda_G, c, >)$. Therefore, $L \in \pounds_c$. Hence $\pounds \subseteq \pounds_c$. Since $\pounds_c \subseteq \pounds$, it follows that $\pounds = \pounds_c$. ∎

Theorem 3.7.4 *$\pounds$ is closed under union, i.e., $L_1, L_2 \in \pounds$ implies $L_1 \cup L_2 \in \pounds$.*

Proof. Let $L_i = L(\lambda_{G_i}, 1, >)$, where $G_i = (T_i, N_i, \rho_i, A_i)$ is a finitary weak regular max-product grammar for $i = 1, 2$. Without loss of generality, we may assume that $T_1 \cap N_2 = T_2 \cap N_1 = N_1 \cap N_2 = \emptyset$. Let $T = T_1 \cup T_2$, $N_0 = N_1 \cup N_2$, $A_0 \notin T \cup N_0$ and $N = N_0 \cup \{A_0\}$. Let $D = D(\rho_1) \cup D(\rho_2) \cup \{A_0 \xrightarrow{1} A_1, A_0 \xrightarrow{1} A_2\}$. Clearly, $G = (T, N, \rho_D, A_0)$ is a finitary weak regular max-product grammar. Furthermore, it can be shown that $L_1 \cup L_2 = L(\lambda_G, 1, >)$. Thus $L_1 \cup L_2 \in \pounds$. ∎

Theorem 3.7.5 *£ is closed under intersection with regular languages, i.e.,*
$L_1 \in £, L_2 \in \mathcal{R}$ *implies* $L_1 \cap L_2 \in £.$

Proof. By Theorem 3.6.12, $L_i = L(\lambda^{M_i}, \frac{1}{2}, >)$, where $M_i = (X_i, Q_i,$
$p_i, h_i, g_i)$ are max-product automata, $i = 1, 2$. Since $L_2 \in \mathcal{R}$, we may assume that the images of p_2, h_2, and g_2 are subsets of $\{0, 1\}$. Let $M =$
(Q, X, p, h, g), where $Q = Q_1 \times Q_2$, $X = X_1 \cap X_2$, and for every $q_i, q_i' \in Q_i$
with $i = 1, 2$ and $u \in X$, we have $p((q_1', q_2'), u, (q_1, q_2)) = p_1(q_1', u, q_1) p_2(q_2', u,$
$q_2)$, $h(q_1, q_2) = h_1(q_1) h_2(q_2)$, and $g(q_1, q_2) = g_1(q_1) g_2(q_2)$. Clearly, for every $x \in X^*$,

$$\lambda^M(x) = \begin{cases} \lambda^{M_1}(x) & \text{if } x \in L \\ 0 & \text{if } x \notin L. \end{cases}$$

Thus $L_1 \cap L_2 = L(\lambda^M, \frac{1}{2}, >)$. Hence $L_1 \cap L_2 \in £.$ ■

Example 3.7.6 *£ is not closed under intersection, i.e., there exist $L_1, L_2 \in$*
£ such that $L_1 \cap L_2 \notin £$: Let $G_i = (\{a, b\}, \{A_0, A_1, A_2\}, \rho_i, A_0)$ for $i = 1, 2$,
where
(1) $D(\rho_1)$ contains the discrete fuzzy productions $A_0 \xrightarrow{1} A_1, A_1 \xrightarrow{2}$
aA_1, $A_1 \xrightarrow{1} A_2$, $A_2 \xrightarrow{0.5} bA_2$, and $A_2 \xrightarrow{1} \Lambda$;
(2) $D(\rho_2)$ contains the discrete fuzzy productions $A_0 \xrightarrow{1} A_1, A_1 \xrightarrow{0.5}$
aA_1, $A_1 \xrightarrow{1} A_2$, $A_2 \xrightarrow{2} bA_2$, and $A_2 \xrightarrow{1} \Lambda$.
Clearly, G_1 and G_2 are finitary weak regular max-product grammars.
Furthermore,

$$L_1 = L(\lambda_{G_1}, \frac{1}{2}, >) = \{a^n b^m \mid n \geqslant m \geqslant 1\},$$

$$L_2 = L(\lambda_{G_2}, \frac{1}{2}, >) = \{a^n b^m \mid m \geqslant n \geqslant 1\},$$

and

$$L_1 \cap L_2 = \{a^n b^m \mid n \geqslant 1\}.$$

By Theorem 3.6.15, $L_1 \cap L_2 \notin £.$

Theorem 3.7.7 $L_1 \in £$, $L_2 \in \mathcal{R}$ *implies* $L_1 \backslash L_2 \in £.$

Proof. $L_1 \backslash L_2 = L_1 \cap L_2'$, where L_2' is the complement of L_2, which is
also a regular language. Thus by Theorem 3.7.5, $L_1 \backslash L_2 \in £.$ ■

Theorem 3.7.8 *£ is closed under concatenation with regular languages,*
i.e., $L_1 \in £, L_2 \in \mathcal{R}$ implies $L_1 L_2 \in £$ and $L_2 L_1 \in £.$

Proof. Let $L_i = L(\lambda_{G_i}, \frac{1}{2}, >)$, where $G_i = (T_i, N_i, \rho_i, A_i)$ are finitary weak regular max-product grammars, $i = 1, 2$. Since $L_2 \in \mathcal{R}$, we may assume that for all $s, t \in (T_2 \cup N_2)^*$, $\rho(s, t) = 0$ or 1. Moreover, without loss of generality, we may assume that $T_1 \cap N_2 = T_2 \cap N_1 = N_1 \cap N_2 = \emptyset$. Let $D = D(\rho_2) \cup D'$, where D' is the collection of discrete fuzzy productions of the forms

(1) $s \xrightarrow{\rho} t$, where $(s \xrightarrow{\rho} t) \in D(\rho_1)$ and $t \notin T_1^*$, and

(2) $s \xrightarrow{\rho} tA$, where $(s \xrightarrow{\rho} t) \in D(\rho_1)$ and $t \in T_1^*$. Clearly, $L_1 L_2 = L(\lambda_G, \frac{1}{2}, >)$, where $G = (T, N, \rho_D, A)$, $T = T_1 \cup T_2$, and $N = N_1 \cup N_2$. Since G is a finitary weak regular max-product grammar, it follows that $L_1 L_2 \in \mathcal{L}$. In a similar manner, it can be shown that $L_2 L_1 \in \mathcal{L}$. ∎

Definition 3.7.9 *For each $a \in T$, let T_a^* be a finite nonempty set and $\psi(a) \subseteq T_a^*$. Let $\psi(\Lambda) = \{\Lambda\}$ and $\psi(ar) = \psi(a)\psi(r)$ for every $a \in T$ and $r \in T^*$. Then ψ is called a **substitution**. If $L \subseteq T^*$, then $\psi(L) = \cup_{r \in L} \psi(r)$. If $\psi(a)$ consists of a single word in T_a^* for each $a \in T$, then ψ is regarded as a function from T^* into $(\cup_{a \in T} T_a)^*$ and is called a **homomorphism**.*

Theorem 3.7.10 *$\mathcal{L}$ is closed under regular substitution, i.e., if $L \subseteq T^*$ is in $\mathcal{L}$ and ψ is a substitution such that $\psi(a) \in \mathcal{R}$ for all $a \in T$, then $\psi(L) \in \mathcal{L}$.*

Proof. If $\Lambda \in L$, then $\psi(L) = \psi(L \backslash \{\Lambda\}) \cup \{\Lambda\}$. Thus by Theorem 3.7.4, it suffices to consider the case where $\Lambda \notin L$. By Theorem 3.6.12, $L = L(\lambda_G, \frac{1}{2}, >)$ for some regular max-product grammar $G = (T, N, \rho, r_0)$. For every $a \in T$, $\psi(a) = L(G_a)$ for some phrase structure grammar $G_a = (T_a, N_a, P_a, a)$, where P_a contains only productions of the forms $a \longrightarrow \Lambda$, $m \longrightarrow u$, and $m \longrightarrow um'$ with $m, m' \in N_a$ and $u \in T_a$. Without loss of generality, we may assume that $T \cap N_a = \{a\}$ and $T_a \cap N = N_a \cap N_b = \emptyset$ for all $a, b \in T$ and $a \neq b$. Let $\overline{T} = \cup_{a \in T} T_a$, $\overline{N} = \cup_{a \in T} N_a$, and $\overline{P} = \cup_{a \in T} P_a$. For each $m \in \overline{N} \cup \{\Lambda\}$ and $n \in N \cup \{\Lambda\}$, we associate the abstract symbol A_{mn}. Let D be the collection of all discrete fuzzy productions of the forms

(1) $A_{mn} \xrightarrow{1} A_{m'n}$, where $m \in \overline{N}$, $m' \in \overline{N} \cup \{\Lambda\}$, $u \in \overline{T} \cup \{\Lambda\}$, $n \in N \cup \{\Lambda\}$, and $(m \longrightarrow um') \in \overline{P}$;

(2) $A_{\Lambda n} \xrightarrow{p} A_{mn'}$, where $n \in N$, $n' \in N \cup \{\Lambda\}$, $m \in T$, and $(n \xrightarrow{p} mn') \in D(\rho)$;

(3) $A_{\Lambda\Lambda} \xrightarrow{1} \Lambda$.

It follows that $\psi(L) = L(\lambda_{G'}, \frac{1}{2}, >)$, where $G' = (\overline{T}, N', \rho_D, A_{\Lambda r_0})$ and $N' = \{A_{mn} \mid m \in \overline{N} \cup \{\Lambda\}, n \in N \cup \{\Lambda\}\}$. Since G' is a finitary weak regular max-product grammar, it follows that $\psi(L) \in \mathcal{L}$. ∎

Corollary 3.7.11 *$\mathcal{L}$ is closed under homomorphism.* ∎

Definition 3.7.12 *A **sequential transducer** is a sextuple (Q, X, Y, H, q_0, F), where Q, X, and Y are finite nonempty sets of states, input and output symbols, respectively, $q_0 \in Q$ is the start state, $F \subseteq Q$ is the set of accepting states, and H is a finite subset of $Q \times X^* \times Y^* \times Q$.*

Definition 3.7.13 *Let $M = (Q, X, Y, H, q_0, F)$ be a sequential transducer. For each $x \in X^*$, define $M(x)$ to be the set of all $y \in Y^*$ with the property that there exist $x_1, x_2, \ldots, x_k \in X^*$, $y_1, y_2, \ldots, y_k \in Y^*$, and $q_1, q_2, \ldots, q_k \in Q$ such that $x = x_1 x_2 \ldots x_k$, $y = y_1 y_2 \ldots y_k, q_k \in F$, and $(q_{i-1}, x_i, y_i, q_i) \in H$ for each i such that $1 \leqslant i \leqslant k$. Furthermore, for each $L \subseteq X^*$, let $M(L) = \cup_{x \in L} M(x)$, and for each $L \subseteq Y^*$, let*

$$M^{-1}(L) = \{x \in X^* \mid M(x) \cap L \neq \emptyset\}.$$

It follows from the proof of Theorem 3.3.1 in [77, p.92] that any family of languages that contains all regular sets and is closed under union, regular substitution, and intersection with regular sets, is closed under mappings induced by sequential transducers of the form (Q, X, Y, H, q_0, Q). By a slight modification of the proof, it can be shown that the same family is closed under mappings induced by arbitrary sequential transducers. Thus we have the following result.

Theorem 3.7.14 *£ is closed under mappings induced by sequential transducers, i.e., if $L \in £$ and M is a sequential transducer, then $M(L) \in £$.* ∎

Theorem 3.7.15 *£ is closed under pseudo inverse mappings induced by sequential transducers, i.e., if $L \in £$ and M is a sequential transducer, then $M^{-1}(L) \in £$.*

Proof. Let $L \in £$ and $M = (Q, X, Y, H, q_0, F)$ be a sequential transducer. Let $M' = (Q, Y, X, H', q_0, F)$ be the sequential transducer such that $H' = \{(q, y, x, q') \mid (q, x, y, q') \in H\}$. It is clear that $M^{-1}(L) = M'(L)$. Thus by Theorem 3.7.14, $M^{-1}(L) \in £$. ∎

Many of the well-known devices are special cases of sequential transducers, e.g., nondeterministic generalized sequential machines, sequential machines, etc. Theorems 3.7.14 and 3.7.15 show that £ is closed under mappings and pseudo inverse mappings induced by these devices.

Corollary 3.7.16 *£ is closed under inverse homomorphism, i.e., if ψ is a homomorphism from X^* into Y^* and $L \subseteq Y^*$ is in £, then $\psi^{-1}(L) = \{x \in X^* \mid \psi(x) \in L\} \in £$.* ∎

Let $L \subseteq T^*$. Let $L^T = \{r^T \mid r \in L\}$, where $\Lambda^T = \Lambda^T$ and $(rw)^T = w^T r^T$ for all $r, w \in T^*$.

Theorem 3.7.17 *£ is closed under transposition, i.e., $L \in £$ implies $L^T \in £$.*

Proof. Let $L \in £$. By Theorem 3.6.12, $L = L(\lambda^M, 1, >)$ for some MA $M = (Q, X, p, h, g)$. Let $M' = (Q, X, p', g, h)$, where $p'(q', u, q) = p(q, u, q')$

for all $q, q' \in Q$ and $u \in X$.   Clearly, M' is a MA and $L^T = L(\lambda^{M'}, 1, >)$.
Thus $L^T \in \mathcal{L}$.   ∎

The following theorem has been shown [79] to be true for any family
of languages closed under union, homomorphism, inverse homomorphism,
and intersection with regular sets.

Theorem 3.7.18 *Let $L \in \mathcal{L}$ and $R \in \mathcal{R}$.   Then*
(1) $L/R = \{x \mid xy \in L \text{ for some } y \in R\}$,
(2) $R \backslash L = \{x \mid yx \in L \text{ for some } y \in R\}$,
(3) $\text{Init}(L) = \{x \mid xy \in L \text{ for some } y\}$,
(4) $\text{Fin}(L) = \{x \mid yx \in L \text{ for some } y\}$, and
(5) $\text{Sub}(L) = \{x \mid yxz \in L \text{ for some } y \text{ and } z\}$
are all in $\mathcal{L}$.   ∎

Let $L \subseteq T^*$ and $x \in T^*$.   Let $\eth_x(L) = \{y \mid xy \in L\}$ and $\overline{\eth}_x(L) = \{y \mid yx \in L\}$.

Theorem 3.7.19 *If $L \subseteq T^*$ is in $\mathcal{L}$ and $x \in T^*$, then $\eth_x(L) \in \mathcal{L}$ and $\overline{\eth}_x(L) \in \mathcal{L}$.*

Proof. The proof follows from Theorem 3.7.18 and the fact that $\eth_x(L) = \{x\}/L$ and $\overline{\eth}_x(L) = L/\{x\}$.   ∎

Theorem 3.7.20 *Let $L \subseteq T^*$.   Then the following statements are equivalent.*
(1) $L \in \mathcal{L}$.
(2) There exists $k \in \mathbb{N}$ such that $\eth_x(L) \in \mathcal{L}$ for all $x \in T^*$ such that $|x| = k$.
(3) There exists $k \in \mathbb{N}$ such that $\overline{\eth}_x(L) \in \mathcal{L}$ for all $x \in T^*$ such that $|x| = k$.

Proof. By Theorems 3.7.17 and 3.7.19, it suffices to show that (2)
implies (1).   Suppose (2) holds.   It is clear that L is the union of all
languages of the form $x\eth_x(L)$, where $x \in T^*$ and $|x| = k$.   Thus by
Theorems 3.7.4 and 3.7.8, $L \in \mathcal{L}$.   ∎

Example 3.7.21 *$\mathcal{L}$ is not closed under concatenation, i.e., there exists
$L \in \mathcal{L}$ such that $L^* \notin \mathcal{L}$: Let $L = \{a^n b^m \mid m \geqslant n \geqslant 1\} \in \mathcal{L}$. Suppose $L^* \in \mathcal{L}$.   Then, by Theorem 3.6.12, $L = L(\lambda_G, 1, c)$, where $G = (\{a, b\}, N, \rho, A_0)$
is a regular max-product grammar.   Let k_0 be the number of elements in
N and $k > k_0$.   Clearly, $x = (ab)^{2^k} (a^2 b^2)^{2^k} (a^3 b^3)^{2^k} \ldots (a^k b^k)^{2^k} \in L^*$.
Thus there exists $A \in N$ such that $x = x_1 ba^1 b^1 a x_2 ba^2 b^2 a x_3 ba^3 b^3 a x_4$, $0 <
l_1 \leqslant l_2 < l_3$, and $A_0 \overset{p_1}{\Longrightarrow} x_1 ba^{l_1} A \bmod(w_1)$, $A \overset{p_2}{\Longrightarrow} b^{l_2} a x_2 ba^{l_2} A \bmod(w_2)$,
$A \overset{p_3}{\Longrightarrow} b^{l_2} a x_3 ba^{l_2} A \bmod(w_3)$, $A \overset{p_4}{\Longrightarrow} b^{l_2} a x_4 \bmod(w_4)$, $p_1 p_2 p_3 p_4 > 1$, for
some $w_1, w_2, w_3, w_4 \in (D(\rho) \times 1)^*$.   Hence*

$$x_1 ba^{l_1} b^{l_2} a x_3 ba^{l_3} b^{l_1} a x_2 ba^{l_2} b^{l_3} a x_4 \in L,$$

a contradiction.

3.8 Exercises

1. Let T be a norm on $[0,1]$. Prove that $\forall a \in A$, $aTa \leq a$. Prove also that $aTa = a$ $\forall a \in [0,1]$ if and only if $T = \wedge$.

2. Let $V = \{(a_1, \ldots, a_n) \mid a_i \in [0,1], i = 1, \ldots, n\}$, where $n \in \mathbb{N}$. Let T be a t-norm on $[0,1]$. Assume that $\forall a, b \in [0,1]$, $a < 1$ implies $aTb < b$. For all $a \in [0,1]$ and $\forall x, y \in V$, let aTx denote $(aTa_1, \ldots, aTa_n)$, where $x = (a_1, \ldots, a_n)$ and let $x \vee y$ denote $(a_1 \vee b_1, \ldots, a_n \vee b_n)$, where $y = (b_1, \ldots, b_n)$. Define the function $C : \mathcal{P}(V) \to \mathcal{P}(V)$ by $\forall X \in \mathcal{P}(V)$,

   $$C(X) = \{\vee_{i=1}^m a_i T x_i \mid a_i \in [0,1],\ x_i \in X,\ i = 1, \ldots, m;\ m \in \mathbb{N}\}.$$

   Prove that the following assertions hold:

   (a) $\forall X \in \mathcal{P}(V)$, $X \subseteq C(X)$;

   (b) $\forall X, Y \in \mathcal{P}(V)$, $X \subseteq Y$ implies $C(X) \subseteq C(Y)$;

   (c) $\forall X \in \mathcal{P}(V)$, $C(C(X)) = C(X)$.

3. Let $X, Y \in \mathcal{P}(V)$ be such that $X \subseteq Y$. X is said to **generate** Y if $Y = C(X)$, where C is defined in Exercise 2. If X generates Y and there does not exists $X' \subset X$ which generates Y, then X is called a **set of vertices** of Y. Prove that X is a set of vertices of Y if and only if $Y = C(X)$ and $\forall x \in X$, $x \notin C(X \setminus \{x\})$.

4. Let $X, X', Y \in \mathcal{P}(V)$. If X is a set of vertices of Y and X' generates Y, prove that $X \subseteq X'$. Conclude that a set of vertices is unique.

5. Let $X, Y \in \mathcal{P}(V)$ be such that $X \subseteq Y$. If every element of Y can be expressed uniquely in the form $\vee_{i=1}^m a_i T x_i$, $a_i \in [0,1]$, $x_i \in X$, prove that X is a set of vertices of Y.

6. Complete the proof of Theorem 3.4.16.

7. Show that Theorems 3.5.1 and 3.5.2 do not hold in general for weak regular fuzzy languages.

8. Prove that Theorems 3.6.5 to 3.6.8 hold if $>$ is replaced by $\geq$ or $=$.

Chapter 4

Fuzzy Languages and Grammars

4.1 Fuzzy Languages

We introduced in the previous chapter the notion of max-product grammars and languages. Before returning to these, we introduce some basic ideas of fuzzy languages such as certain normal forms. We base our work on [122].

Formal languages are quite precise while natural languages are quite imprecise. To reduce the gap between them, it is natural to introduce randomness into the structure of formal languages. This leads to the concept of stochastic languages [64, 66, 67, 111, 232]. Another possibility lies in the introduction of fuzziness.

It appears that much of the existing theory of formal languages can be extended quite readily to fuzzy languages.

As usual, we let T denote a set of terminals and N a set of non-terminals such that $T \cap N = \emptyset$. As before, a fuzzy language is a fuzzy subset of T^*.

Let λ_1 and λ_2 be two fuzzy languages over T. The **union** of λ_1 and λ_2 is the fuzzy language denoted by $\lambda_1 \cup \lambda_2$ and defined by

$$(\lambda_1 \cup \lambda_2)(x) = \lambda_1(x) \vee \lambda_2(x) \qquad \forall x \in T^*. \tag{4.1}$$

The **intersection** of λ_1 and λ_2 is the fuzzy language denoted by $\lambda_1 \cap \lambda_2$ and defined by

$$(\lambda_1 \cap \lambda_2)(x) = \lambda_1(x) \wedge \lambda_2(x) \qquad \forall x \in T^*. \tag{4.2}$$

The **concatenation** of λ_1 and λ_2 is the fuzzy language denoted by $\lambda_1 \lambda_2$ and is defined by $\forall x \in T^*$,

$$(\lambda_1 \lambda_2)(x) = \vee\{\lambda_1(u) \wedge \lambda_2(v) \mid x = uv, \ u, v \in T^*\}. \tag{4.3}$$

By the distributivity of $\vee$ and $\wedge$, the operation concatenation is associative.

Let λ be a fuzzy language in T. Then the fuzzy subset λ^{∞} of T^* defined by

$$\lambda^{\infty}(x) = \vee\{\lambda^n(x) \mid n = 0, 1, \ldots\}$$

$\forall x \in T^*$ is called the **Kleene closure** of λ.

Informally, a fuzzy grammar may be viewed as a set of rules for generating the elements of a fuzzy subset. Recall that a fuzzy grammar, or simply a grammar, is a quadruple $G = (N, T, P, S)$ in which T is a set of terminals, N is a set of nonterminals $(T \cap N = \emptyset)$, P is a set of fuzzy productions, and $S \in N$. Essentially, the elements of N are labels for certain fuzzy subsets of T^* called **fuzzy syntactic categories**, with S being the label for the syntactic category "sentence." The elements of P define conditioned fuzzy subsets in $(T \cup N)^*$.

More specifically, the elements of P are expressions of the form

$$\mu(r \longrightarrow w) = c, \qquad c > 0, \tag{4.4}$$

where r and w are strings in $(T \cup N)^*$ and c is the grade of membership of w given r. At times, we abbreviate $\mu(r \longrightarrow w) = c$ to $r \xrightarrow{c} w$ or, more simply, $r \longrightarrow w$.

As in the case of nonfuzzy grammars, the expression $r \longrightarrow w$ represents a rewriting rule. Thus if $r \xrightarrow{c} w$ and s and t are arbitrary strings in $(T \cup N)^*$, then we have

$$srt \xrightarrow{c} swt \tag{4.5}$$

and swt is said to be **directly derivable** from srt.

If $r_1, \ldots, r_m$ are strings in $(T \cup N)^*$ and

$$r_1 \xrightarrow{c_2} r_2, \ldots, r_{m-1} \xrightarrow{c_m} r_m, c_2, \ldots, c_m > 0,$$

then r_1 is said to **derive** r_m in grammar G, or, equivalently, r_m is **derivable** from r_1 in grammar G. This is expressed by $r_1 \underset{G}{\Longrightarrow} r_m$ or simply $r_1 \Longrightarrow r_m$. The expression

$$r_1 \xrightarrow{c_2} r_2 \rightarrow \ldots \rightarrow r_{m-1} \xrightarrow{c_m} r_m \tag{4.6}$$

is referred to as a **derivation chain** from r_1 to r_m.

A fuzzy grammar G generates a fuzzy language $L(G)$ in the following manner. A string of terminals x is said to be in $L(G)$ if and only if x is derivable from S. The grade of membership of x in $L(G)$ is given by

$$\mu_G(x) = \vee(\mu(S, r_1) \wedge \mu(r_1, r_2) \wedge \ldots \wedge \mu(r_m, x)), \tag{4.7}$$

where the supremum is taken over all derivation chains from S to x. Thus (4.7) defines $L(G)$ as a fuzzy subset of $(T \cup N)^*$. If $L(G_1) = L(G_2)$ in the sense of equality of fuzzy subsets, then the grammars G_1 and G_2 are said to be **equivalent**.

Equation (4.7) may be expressed as follows:

$\mu_G(x)$ is the grade of membership of x in the language generated by grammar G;

$\mu_G(x)$ is the strength of the strongest derivation chain from S to x.

Let $\mu(S, r_1) = c_1$, $\mu(r_1, r_2) = c_2, \ldots, \mu(r_m, x) = c_{m+1}$. Then, on writing (4.7) in the form

$$\mu_G(x) = \vee(c_1 \wedge c_2 \ldots \wedge c_{m+1}), \tag{4.8}$$

it follows at once from the associativity of $\wedge$ that (4.7) is equivalent to a more general expression in which the successive r's are derivable from their immediate predecessors rather than directly derivable from them, as in (4.7).

Example 4.1.1 *Let* $T = \{0, 1\}$, $N = \{A, B, S\}$, *and let* P *be given by*

$$\begin{array}{llll}
\mu(S, AB) & = & 0.5 & \quad \mu(A, 0) & = & 0.5 \\
\mu(S, A) & = & 0.8 & \quad \mu(A, 1) & = & 0.6 \\
\mu(S, B) & = & 0.8 & \quad \mu(B, A) & = & 0.4 \\
\mu(AB, BA) & = & 0.4 & \quad \mu(B, 0) & = & 0.2
\end{array}$$

Consider the terminal string $x = 0$. *The possible derivation chains for this string are* $S \xrightarrow{0.8} A \xrightarrow{0.5} 0$, $S \xrightarrow{0.8} B \xrightarrow{0.2} 0$, $S \xrightarrow{0.8} B \xrightarrow{0.4} A \xrightarrow{0.5} 0$. *Thus*

$$\mu_G(0) = (0.8 \wedge 0.5) \vee (0.8 \wedge 0.2) \vee (0.8 \wedge 0.4 \wedge 0.5) = 0.5.$$

Similarly, the possible derivation chains for the terminal string $x = 01$ *are* $S \xrightarrow{0.5} AB \xrightarrow{0.5} 0B \xrightarrow{0.4} 0A \xrightarrow{0.6} 01$, $S \xrightarrow{0.5} AB \xrightarrow{0.4} AA \xrightarrow{0.5} 0A \xrightarrow{0.6} 01$, $S \xrightarrow{0.5} AB \xrightarrow{0.4} BA \xrightarrow{0.2} 0A \xrightarrow{0.6} 01$, *and* $S \xrightarrow{0.5} AB \xrightarrow{0.4} BA \xrightarrow{0.4} AA \xrightarrow{0.5} 0A \xrightarrow{0.6} 01$. *Hence*

$$\mu_G(01) = 0.4 \vee 0.4 \vee 0.2 \vee 0.4 = 0.4.$$

Let G be a fuzzy grammar. An important question that arises in connection with the definition of μ_G is whether or not there exists an algorithm for computing $\mu_G(x)$ by the use of the defining equation (4.7). G is said to be **recursive** if such an algorithm exists.

4.2   Types of Grammars

Similar to the usual definitions of non-fuzzy grammars, we define four principal types of fuzzy grammars as follows:

Type 0 grammar

The allowable productions are of the general form $r \xrightarrow{c} w, c > 0$, where r and w are strings in $(T \cup N)^*$.

Type 1 grammar (context-sensitive)

The productions are of the form $r_1 A r_2 \xrightarrow{c} r_1 w r_2, c > 0$, with r_1, r_2, and w in $(T \cup N)^*$, A in N, and $w \neq \Lambda$. The production $S \longrightarrow \Lambda$ is also allowed.

Type 2 grammar (context-free)

Productions here are of the form $A \xrightarrow{c} w, c > 0, A \in N, w \in (T \cup N)^*$, $w \neq \Lambda$, and $S \longrightarrow \Lambda$.

Type 3 grammar (regular)

In this case, productions are of the form $A \xrightarrow{c} aB$ or $A \xrightarrow{c} a, c > 0$, where $a \in T, A, B \in N$. In addition, $S \longrightarrow \Lambda$ is allowed.

We ask the reader to compare the definition of a type 2 grammar here (with $c = 1$) and that of Definition 1.8.7.

In the following, we concentrate on context-free grammars. However, there is a basic property of context-sensitive grammars that needs to be stated. It is easily shown that context-sensitive and hence also context-free and regular grammars are recursive. This can be stated as an extension of [96, Theorem 2.2, p. 26].

Theorem 4.2.1 *If $G = (N, T, P, S)$ is a fuzzy context-sensitive grammar, then G is recursive.*

Proof. First we show that for any type of grammar, the supremum in (4.7) may be taken over a subset of the set of all derivation chains from S to x, namely, the subset of all loop-free derivation chains, i.e., chains in which no $r_i, i = 1, \ldots, m$, occurs more than once.

Suppose that in a derivation chain C,

$$C = S \xrightarrow{c_1} r_1 \xrightarrow{c_2} r_2 \ldots \xrightarrow{c_m} r_m \xrightarrow{c_{m+1}} x,$$

r_i, say, is the same as $r_j, j > i$. Let C' be the chain resulting from replacing the subchain

$$r_i \xrightarrow{c_{i+1}} \ldots \xrightarrow{c_j} r_j \xrightarrow{c_{j+1}} r_{j+1}$$

in C by $r_i \xrightarrow{c_{j+1}} r_{j+1}$. Clearly, if C is a derivation chain from S to x, then C' is also. However,

$$\wedge \{c_1, \ldots, c_i, c_{i+1}, \ldots, c_{j+1}, \ldots, c_{m+1}\} \leq \wedge \{c_1, \ldots, c_i, c_{j+1}, \ldots, c_{m+1}\}$$

and thus C may be deleted without affecting the supremum in (4.7). Hence we can replace the definition (4.7) for $\mu_G(x)$ by

$$\mu_G(x) = \vee \wedge \{\mu(S, r_1), \mu(r_1, r_2), \ldots, \mu(r_m, x)\}, \tag{4.9}$$

where the supremum is taken over all loop-free derivation chains from S to x.

We now show that for context-sensitive grammars, the set over which the supremum is taken in (4.9) can be further restricted to derivation chains of bounded length l_0, where l_0 depends on $|x|$ and the number of symbols in $T \cup N$.

If G is context-sensitive, then due to the noncontracting character of the productions in P, it follows that

$$|r_j| \geq |r_i| \qquad \text{if } j > i. \tag{4.10}$$

Let $|T \cup N| = k$. Since there are at most k' distinct strings in $(T \cup N)^*$ of length l and since the derivation chain is loop-free, it follows from (4.10) that the total length of the chain is bounded by

$$l_0 = 1 + k + \ldots + k^{|x|}.$$

We next provide a method for generating all finite derivation chains from S to x of length $\leq l_0$. We start with S and using P generate the set Q_1 of all strings in $(T \cup N)^*$ of length $\leq |x|$ that are derivable from S in one step. We then construct Q_2, the set of all strings in $(T \cup N)^*$ of length $\leq |x|$, that are derivable from S in two steps. This is accomplished by noting that Q_2 is identical with the set of all strings in $(T \cup N)^*$ of length $\leq |x|$ that are directly derivable from strings in Q_1. Continuing this process, we construct consecutively $Q_3, Q_4, \ldots, Q_k$ until $k = l_0$ or $Q_k = \emptyset$, whichever happens first. Since the Q_i, $i = 1, \ldots, k$, are finite sets, we are able to find in a finite number of steps all loop-free derivation chains from S to x of length $\leq l_0$ and thus to compute $\mu_G(x)$ by the use of (4.9). This, then, constitutes an algorithm, though not necessarily an efficient one, for the computation of $\mu_G(x)$. Hence G is recursive. ∎

4.3 Fuzzy Context-Free Grammars

Many of the basic results in the theory of formal languages can easily be extended to fuzzy languages. As an illustration, we now sketch, without giving proofs, such extensions in the case of the Chomsky and Greibach normal forms for context-free languages.

We first consider the Chomsky normal form for fuzzy context-free languages.

We recall the following result from the theory of crisp grammars. Any context-free language without Λ is generated by a grammar in which all productions are of the form $A \to BC$ or $A \to a$, where A, B, C are nonterminals and a is a terminal, [96, Theorem 4.5, p. 92]. This form is known as Chomsky normal form.

Let G be a fuzzy context-free grammar. Then, any such grammar is equivalent to a grammar G' in which all productions are of the form $A \xrightarrow{c} BC$ or $A \xrightarrow{c} a$, where $c > 0$ and A, B, C are nonterminals and a is a terminal.

We construct G' in three stages.

First, we construct a grammar G_1 equivalent to G in which there are no productions of the form $A \longrightarrow B, A, B \in N$.

Suppose that in G we have productions of the form $A \longrightarrow B$, which lead to derivation chains of the form

$$A \xrightarrow{c_1} B_1 \xrightarrow{c_2} B_2 \xrightarrow{c_3} \ldots \xrightarrow{c_m} B_m \xrightarrow{c_{m+1}} B \xrightarrow{c_{m+2}} r,$$

where $r \notin N$. Then we replace all such productions of the form $A \xrightarrow{c_1} B_1, B_1 \xrightarrow{c_2} B_2, \ldots, B_m \xrightarrow{c_{m+1}} B$ in G by single productions of the form $A \xrightarrow{c} r$, in which

$$c = \mu(A \Longrightarrow B) \wedge \mu(B, r), \tag{4.11}$$

where

$$\mu(A \Longrightarrow B) = \vee\{\mu(A_1, B_1) \wedge \ldots \wedge \mu(B_m, B)\} \tag{4.12}$$

with the supremum taken over all loop-free derivation chains from A to B. It follows that the resultant grammar, G_1, is equivalent to G.

Second, we construct a grammar G_2 equivalent to G_1 in which there are no productions of the form $A \xrightarrow{c} B_1 B_2 \ldots B_m, c > 0, m > 2$, in which one or more of the B's are terminals. Thus suppose that B_i, say, is a terminal a. Then B_i in $B_1 B_2 \ldots B_m$ is replaced by a new nonterminal C_i which does not appear on the right-hand side of any other production. We then set

$$\mu(A, B_1 B_2 \ldots B_i \ldots B_m) = \mu(A, B_1 B_2 \ldots C_i \ldots B_m). \tag{4.13}$$

We add to the productions of G the production $C_i \xrightarrow{1} a$. We do this for all terminals in $B_1 \ldots B_m$ in all productions of the form $A \to B_1 \ldots B_m$. We thus arrive at a grammar G_2 in which all productions are of the form $A \longrightarrow a$ or $A \longrightarrow B_1 \ldots B_m, m \geq 2$, where all B's are nonterminals. Clearly, G_2 is equivalent to G_1.

Third, we construct a grammar G_3 equivalent to G_2 in which all productions are of the form $A \longrightarrow a$ or $A \longrightarrow BC, A, B, C \in N, a \in T$. Consider a typical production in G_2 of the form $A \xrightarrow{c} B_1 \ldots B_m, c > 0, m > 2$. We replace this production by the productions

$$A \xrightarrow{c} B_1 D_1$$
$$D_1 \xrightarrow{1} B_2 D_2$$
$$\vdots$$
$$D_{m-2} \xrightarrow{1} B_{m-1} B_m,$$

(4.14)

where the D's are new nonterminals that do not appear on the right-hand side of any production in G_2. Once such replacements for all productions in G_2 of the form $A \xrightarrow{c} B_1 \ldots B_m$ have been made, a grammar G_3 that is equivalent to G_2 is obtained. Hence G_3, that is in Chomsky normal form, is equivalent to G.

Example 4.3.1 *Consider the following fuzzy grammar, where $T = \{a, b\}$ and $N = \{A, B, S\}$ and where the productions are as follows:*

$$S \xrightarrow{0.8} bA \qquad B \xrightarrow{0.4} b$$
$$S \xrightarrow{0.6} aB \qquad A \xrightarrow{0.3} bSA$$
$$A \xrightarrow{0.2} a \qquad B \xrightarrow{0.5} aSB$$

To find the equivalent grammar in Chomsky normal form, we proceed as follows.

First we replace $S \xrightarrow{0.8} bA$ by $S \xrightarrow{0.8} C_1 A$, $C_1 \xrightarrow{1} b$. Similarly, $S \xrightarrow{0.6} aB$ is replaced by $S \xrightarrow{0.6} C_2 B$ and $C_2 \xrightarrow{1} a$. We replace $A \xrightarrow{0.3} bSA$ by $A \xrightarrow{0.3} C_3 SA$, $C_3 \xrightarrow{1} b$; and $B \xrightarrow{0.5} aSB$ is replaced by $B \xrightarrow{0.5} C_4 SB$ and $C_4 \xrightarrow{1} a$.

Second, the production $A \xrightarrow{0.3} C_3 SA$ is replaced by $A \xrightarrow{0.3} C_3 D_1$, $D_1 \xrightarrow{1} SA$; and the production $B \xrightarrow{0.5} C_4 SB$ is replaced by $B \xrightarrow{0.5} C_4 D_2$, $D_2 \xrightarrow{1} SB$. Thus the productions in the equivalent Chomsky normal form are as follows:

$$S \xrightarrow{0.8} C_1 A \qquad A \xrightarrow{0.3} C_3 D_1$$
$$C_1 \xrightarrow{1} b \qquad D_1 \xrightarrow{1} SA$$
$$S \xrightarrow{0.6} C_2 B \qquad C_3 \xrightarrow{1} b$$
$$C_2 \xrightarrow{1} a \qquad B \xrightarrow{0.5} C_4 D_2$$
$$A \xrightarrow{0.2} a \qquad D_2 \xrightarrow{1} SB$$
$$B \xrightarrow{0.4} b \qquad C_4 \xrightarrow{1} a$$

We now consider Greibach normal form.

Let G be any fuzzy context-free grammar. Then G is equivalent to a fuzzy grammar G_G in which all productions are of the form $A \longrightarrow ar$, where A is a nonterminal, a is a terminal, and r is a string in N^*. The fuzzy grammar G_G is in *Greibach normal form*. See [96, Theorem 4.6, p.65].

Paralleling the approach used in [96], we state two lemmas that are of use in constructing G_G.

Lemma 4.3.2 *Let G be a fuzzy context-free grammar. Let $A \longrightarrow r_1 B r_2$ be a production in P, where $A, B \in N$, and $r_1, r_2 \in (T \cup N)^*$. Let $B \longrightarrow w_1, \ldots, B \longrightarrow w_k$ be the set of all B-productions (that is, all productions with B on the left-hand side). Let G_1 be the grammar resulting from the replacement of each of the productions of the form $A \longrightarrow r_1 B r_2$ with the productions $A \longrightarrow r_1 w_1 r_2, \ldots, A \longrightarrow r_1 w_r r_2$, in which*

$$\mu(A, r_1 w_i r_2) = \mu(A, r_1 w r_2) \wedge \mu(B, w_i), \tag{4.15}$$

$i = 1, \ldots, k$. Then G_1 is equivalent to G. ∎

Lemma 4.3.3 *Let G be a fuzzy context-free grammar. Let $A \longrightarrow A r_i, i = 1, \ldots, k$, be the A-productions for which A is also the left most symbol on the right-hand side. Let $A \longrightarrow w_j, j = 1, \ldots, m$ be the remaining A-productions, with $r_i, w_j \in (T \cup N)^*, i = 1, \ldots, k$. Let G_2 be the grammar resulting from the replacement of the $A \longrightarrow A r_i$ in G with the productions:*

$$A \longrightarrow w_j Z, \qquad j = 1, \ldots, m \tag{4.16}$$

$$Z \longrightarrow r_i, \quad Z \longrightarrow r_i Z, \qquad i = 1, \ldots, k, \tag{4.17}$$

where

$$\begin{aligned}
\mu(Z, w_j Z) &= \mu(A, w_j), \\
\mu(Z, r_i) &= \mu(A, A r_i), \\
\mu(Z, r_i Z) &= \mu(A, A r_i), \qquad i = 1, \ldots, k.
\end{aligned} \tag{4.18}$$

Then G_2 is equivalent to G. ∎

With the use of these lemmas, we now derive the Greibach normal form for G.

We first put G into the Chomsky normal form. Let the nonterminals in this form be denoted by $A_1, \ldots, A_m$.

We then modify the productions of the form $A_i \longrightarrow A_j s, s \in (T \cup N)^*$, in such a way that for all such productions, $j \geqslant i$. This is done in the following stages. Suppose that it has been done for $i \leqslant k$, that is, if

$$A_i \longrightarrow A_j s \tag{4.19}$$

is a production with $i \leqslant k$, then $j > i$. To extend this to A_{k+1}-productions, suppose that $A_{k+1} \longrightarrow A_j s$ is any production with $j < k+1$. Using Lemma 4.3.2 and substituting for A_j the right-hand side of each A_j-production, we obtain by repeated substitution productions of the form

$$A_{k+1} \longrightarrow A_l s, \qquad l \geqslant k + 1. \tag{4.20}$$

In (4.20), those productions in which l is equal to $k+1$ are replaced by the use of Lemma 4.3.3. This results in a new nonterminal Z_{k+1}. Then by repeating this process, all productions are put into the form

$$A_k \longrightarrow A_l s, \qquad l > k, \ s \in (N \cup \{Z_1, \dots, Z_n\})^* \qquad (4.21)$$

$$A_k \longrightarrow as, \qquad a \in T \qquad (4.22)$$

$$Z_k \longrightarrow s \qquad (4.23)$$

with membership grades given by Lemmas 4.3.2 and 4.3.3.

By (4.21) and (4.22), the leftmost symbol on the right-hand side of any production for A_m must be a terminal. Similarly, for A_{m-1}, the leftmost symbol on the right-hand side must be either A_m or a terminal. Substituting for A_m using Lemma 4.3.2, we obtain productions whose right-hand sides start with terminals. Repeating this process for $A_{m-2}, \dots, A_1$, all productions for the $A_i, i = 1, \dots, m$, are put into a form where their right-hand sides start with terminals.

At this stage, only the productions in (4.23) may not be in the desired form. It follows that the leftmost symbol in s in (4.23) may be either a terminal or one of the $A_i, i = 1, \dots, m$. If the latter case holds, application of Lemma 4.3.2 to each Z_i production yields productions of the desired form. This completes the construction.

Example 4.3.4 *We convert into the Greibach normal form the following fuzzy grammar G. Let $T = \{a, b\}, N = \{A_1, A_2, A_3\}$, and let the productions be in the Chomsky normal form:*

$$A_1 \xrightarrow{0.8} A_2 A_3 \qquad A_3 \xrightarrow{0.2} A_1 A_2$$
$$A_2 \xrightarrow{0.7} A_3 A_1 \qquad A_3 \xrightarrow{0.5} a$$
$$A_2 \xrightarrow{0.6} b.$$

Step 1. The right-hand sides of the productions for A_1 and A_2 start with terminals or higher-numbered variables. We thus begin with the production $A_3 \longrightarrow A_1 A_2$ and substitute $A_2 A_3$ for A_1. Note that $A_1 \longrightarrow A_2 A_3$ is the only production with A_1 on the left.

The resulting productions are as follows:

$$A_1 \xrightarrow{0.8} A_2 A_3 \qquad A_2 \xrightarrow{0.7} A_3 A_1$$
$$A_2 \xrightarrow{0.6} b \qquad A_3 \xrightarrow{0.2} A_2 A_3 A_2$$
$$A_3 \xrightarrow{0.6} b$$

Note that in $A_3 \xrightarrow{0.2} A_2 A_3 A_2, 0.2 = 0.8 \wedge 0.2$.

The right-hand side of the production $A_3 \longrightarrow A_2 A_3 A_2$ begins with a lower-numbered variable. We thus substitute for the first occurrence of A_2 either $A_3 A_1$ or b.

The new productions are given below.

$$A_1 \xrightarrow{0.8} A_2 A_3 \qquad A_2 \xrightarrow{0.7} A_3 A_1$$
$$A_2 \xrightarrow{0.6} b \qquad A_3 \xrightarrow{0.2} A_3 A_1 A_3 A_2$$
$$A_3 \xrightarrow{0.2} b A_3 A_2 \qquad A_3 \xrightarrow{0.5} a.$$

We now apply Lemma 4.3.3 to the productions $A_1 \longrightarrow A_3 A_1 A_3 A_2$, $A_3 \longrightarrow bA_3 A_2$, and $A_3 \longrightarrow a$. We introduce Z_3 and replace the production $A_3 \longrightarrow A_3 A_1 A_3 A_2$ by $A_3 \longrightarrow b A_3 A_2 Z_3$, $A_3 \longrightarrow a Z_3$, $Z_3 \longrightarrow A_1 A_3 A_2$, and $Z_3 \longrightarrow A_1 A_3 A_2 Z_3$.

The resulting productions are as follows:

$$A_1 \xrightarrow{0.8} A_2 A_3 \qquad A_2 \xrightarrow{0.7} A_3 A_1 \qquad A_3 \xrightarrow{0.2} b A_3 A_2 Z_3$$
$$A_2 \xrightarrow{0.6} b \qquad A_3 \xrightarrow{0.2} b A_3 A_2$$
$$A_3 \xrightarrow{0.5} a \qquad A_3 \xrightarrow{0.5} a Z_3$$
$$Z_3 \xrightarrow{0.2} A_1 A_3 A_2 Z_3 \qquad Z_3 \xrightarrow{0.2} A_1 A_3 A_2$$

Step 2. Now all productions with A_3 on the left have right-hand sides that start with terminals. These are used to replace A_3 in the production $A_2 \longrightarrow A_3 A_1$ and then the productions with A_2 on the left are used to replace A_2 in the production $A_1 \longrightarrow A_2 A_3$. The resulting productions are as follows:

$$A_3 \xrightarrow{0.2} b A_3 A_2 \qquad\qquad A_3 \xrightarrow{0.2} b A_3 A_2 Z_3$$
$$A_3 \xrightarrow{0.5} a \qquad\qquad A_3 \xrightarrow{0.5} a Z_3$$
$$A_2 \xrightarrow{0.2} b A_3 A_2 A_1 \qquad\qquad A_2 \xrightarrow{0.2} b A_3 A_2 Z_3 A_1$$
$$A_2 \xrightarrow{0.5} a A_1 \qquad\qquad A_2 \xrightarrow{0.5} a Z_3 A_1$$
$$A_2 \xrightarrow{0.6} b \qquad\qquad A_1 \xrightarrow{0.2} b A_3 A_2 A_1 A_3$$
$$A_1 \xrightarrow{0.2} b A_3 A_2 Z_3 A_1 A_3 \qquad\qquad A_1 \xrightarrow{0.5} a A_1 A_3$$
$$A_1 \xrightarrow{0.6} b A_3 \qquad\qquad Z_3 \xrightarrow{0.2} A_1 A_3 A_2 Z$$
$$Z_3 \xrightarrow{0.2} A_1 A_3 A_2$$

Note that in $A_2 \xrightarrow{0.2} b A_3 A_2 A_1$, $0.2 = 0.7 \wedge 0.2$. Similarly, in $A_1 \xrightarrow{0.5} a A_1 A_3$, $0.5 = 0.5 \wedge 0.8$. The membership grades of other productions are determined similarly.

Step 3. The two Z_3 productions, $Z_3 \xrightarrow{0.2} A_1 A_3 A_2$ and $Z_3 \xrightarrow{0.2} A_1 A_3 A_2 Z_3$, are converted to desired form by substituting the right-hand side of each of the five productions with A_1 on the left for the first occurrence of A_1. Hence

$Z_3 \xrightarrow{0.2} A_1 A_3 A_2$ *is replaced by*

$$Z_3 \xrightarrow{0.2} bA_3 A_3 A_2 \qquad\qquad Z_3 \xrightarrow{0.2} bA_3 A_2 A_1 A_3 A_3 A_2$$
$$Z_3 \xrightarrow{0.2} aA_1 A_3 A_3 A_2 \qquad\qquad Z_3 \xrightarrow{0.2} bA_3 A_2 Z_3 A_1 A_3 A_3 A_2$$
$$Z_3 \xrightarrow{0.2} aZ_3 A_1 A_3 A_3 A_2$$

The other production for Z_3 is converted in a similar manner. The final set of productions is given below.

$$A_3 \xrightarrow{0.2} bA_3 A_2 \qquad\qquad A_3 \xrightarrow{0.2} bA_3 A_2 Z_3$$
$$A_3 \xrightarrow{0.5} a \qquad\qquad A_3 \xrightarrow{0.5} aZ_3$$
$$A_2 \xrightarrow{0.2} bA_3 A_2 A_1 \qquad\qquad A_2 \xrightarrow{0.2} bA_3 A_2 Z_3 A_1$$
$$A_2 \xrightarrow{0.5} aA_1 \qquad\qquad A_2 \xrightarrow{0.5} aZ_3 A_1$$
$$A_2 \xrightarrow{0.6} a \qquad\qquad A_1 \xrightarrow{0.2} bA_3 A_2 A_1 A_3$$
$$A_1 \xrightarrow{0.2} bA_3 A_2 Z_3 A_1 A_3 \qquad\qquad A_1 \xrightarrow{0.5} aA_1 A_3$$
$$A_1 \xrightarrow{0.5} aZ_3 A_1 A_3 \qquad\qquad A_1 \xrightarrow{0.6} bA_3$$
$$Z_3 \xrightarrow{0.2} bA_3 A_3 A_2 \qquad\qquad Z_3 \xrightarrow{0.2} bA_3 A_3 A_2 Z_3$$
$$Z_3 \xrightarrow{0.2} bA_3 A_2 A_1 A_3 A_3 A_2 \qquad\qquad Z_3 \xrightarrow{0.2} bA_3 A_2 A_1 A_3 A_3 A_2 Z_3$$
$$Z_3 \xrightarrow{0.2} aA_1 A_3 A_3 A_2 \qquad\qquad Z_3 \xrightarrow{0.2} aA_1 A_3 A_3 A_2 Z_3$$
$$Z_3 \xrightarrow{0.2} bA_3 A_2 Z_3 A_1 A_3 A_3 A_2 \qquad\qquad Z_3 \xrightarrow{0.2} bA_3 A_2 Z_3 A_1 A_3 A_3 A_2 Z_3$$
$$Z_3 \xrightarrow{0.2} aZ_3 A_1 A_3 A_3 A_2 \qquad\qquad Z_3 \xrightarrow{0.2} aZ_3 A_1 A_3 A_3 A_2 Z_3$$

The theory of fuzzy languages may prove to be of use in the construction of better models for natural languages. It may also be of use in a better understanding of the role of fuzzy algorithms and fuzzy automata in decision making, pattern recognition, and other processes involving the manipulation of fuzzy data.

4.4 Context-Free Max-Product Grammars

The results of this and the next section are from [206].

In Chapter 3, two different formulations of context-free max-product grammars were introduced. The first formulation corresponds to the maximal interpretation [188] of probabilistic grammars [204] and the second is an attempt to provide a grammar to generate fuzzy languages [207]. (See Chapter 5 also.) The second formulation is simpler. Consequently, it is adopted here and the next section. Further details can be found in Chapter 3.

Definition 4.4.1 *A **context-free max-product grammar** (**CMG**) is a quadruple $G = (T, N, P, \eta)$ such that the following conditions hold:*
(1) T and N are disjoint finite nonempty sets;

(2) P is a finite collection of fuzzy productions each of which is of the form $A \xrightarrow{p} x$, where $A \in N$, $x \in (T \cup N)^*$, and $p \in \mathbb{R}^{\geq 0}$;

(3) η is a function from N into $\mathbb{R}^{\geq 0}$.

If, in addition, for every $(A \xrightarrow{p} x) \in P$, $p \in [0,1]$, and η is a function from N into $[0,1]$, then G is called a **strict CMG**.

(It is to be understood in Definition 4.4.1 that the strength of a derivation chain is determined by the product operation.)

In Definition 4.4.1, the elements of T are called terminals, the elements of N are called nonterminals, $\eta(A)$ is the grade of membership that A is the start symbol of G, and $A \xrightarrow{p} x$ means the grade of membership is p that A will be replaced by x.

Let $G = (T, N, P, \eta)$ be a CMG. Then we write

$$x \Rrightarrow^p y(\mathrm{mod}\ w),$$

where $w = (r_1, k_1)(r_2, k_2) \cdots (r_n, k_n)$, $x, y \in (T \cup N)^*$, $r_i = (A_i \xrightarrow{p_i} x_i) \in P$, $k_i \in \mathbb{N}$, $i = 1, 2, \ldots, n$, and $p = p_1 p_2 \cdots p_n$ if and only if there exist $z_i \in (T \cup N)^*$, $i = 0, 1, 2, \ldots, n$, such that $z_0 = x$, $z_n = y$, and for each $i = 1, 2, \ldots, n$, z_i is obtained from z_{i-1} by replacing the k_ith occurrence of A_i in z_{i-1} by x_i. Otherwise, we write

$$x \Rrightarrow^0 y(\mathrm{mod}\ w).$$

Also, if for every $i = 1, 2, \ldots, n$, $k_i = 1$ and $z_{i-1} = u A_i v$, where $u \in T^*$, then we write

$$x \Rrightarrow_L^p y(\mathrm{mod}\ w).$$

Otherwise, we write

$$x \Rrightarrow_L^0 y(\mathrm{mod}\ w).$$

Clearly, in the above notation, p is uniquely determined by $x, y \in (T \cup N)^*$ and $w \in (P \times \mathbb{N})^*$. This allows us to define, for each $w \in (P \times \mathbb{N})^*$, functions λ_w and λ_w^L, both from $(T \cup N)^* \times (T \cup N)^*$ into $\mathbb{R}^{\geq 0}$ such that $\lambda_w(x, y) = p$ if and only if $x \Rrightarrow^p y\ (\mathrm{mod}\ w)$ and $\lambda_w^L(x, y) = p$ if and only if $x \Rrightarrow_L^p y\ (\mathrm{mod}\ w)$.

Let $G = (T, N, P, \eta)$ be a CMG. Let $W = P \times \mathbb{N}$. Define $\lambda_G : T^* \to \mathbb{R}^{\geq 0}$ and $\lambda_G^L : T^* \to \mathbb{R}^{\geq 0}$ by $\forall x \in T^*$,

$$\lambda_G(x) = \wedge_{w \in W^*} \vee_{A \in N} \eta(A) \lambda_w(A, x)$$

and

$$\lambda_G^L(x) = \quad \wedge_{w \in W^*} \vee_{A \in N} \eta(A) \lambda_w^L(A, x).$$

In the above definition, as well as in the rest of this section and the next, we assume the usual arithmetic of the extended real number system including the fact that $0 \cdot \infty = 0$. Furthermore, the concept of least upper bound is extended in a natural manner to include ∞.

Example 4.4.2 *Let $G = (N, T, P, \eta)$ be the grammar defined in Example 3.4.10. Let $x = aAAD$ and $y = aABb$. Then $r_3 = (A \xrightarrow{.7} B)$, $k_1 = 2$, $r_6 = (D \xrightarrow{.4} b)$, $k_2 = 1$, and $z_1 = aABB$, and $w = (r_3, 2)(r_6, 1)$. Thus $x \Rightarrow^p y (\mathrm{mod}\, w)$, where $p = (.7)(.4) = .28$. Also, $\lambda_w(x, y) = .28$.*

Definition 4.4.3 *A **fuzzy language** λ over S is a function from S^* into $\mathbb{R}_\infty^{\geq 0}$. A **finitary fuzzy language** over S is a function from S^* into $\mathbb{R}^{\geq 0}$.*

Let G be a CMG. Then λ_G is the fuzzy language generated by G and λ_G^L is the fuzzy language generated by G using left most derivations only.

Theorem 4.4.4 *Let G be a CMG. Then $\lambda_G = \lambda_G^L$.*

Proof. Let $G = (T, N, P, \eta)$. It suffices to show that for all $A \in N$, $x \in T^*$, and $p \in \mathbb{R}^{\geq 0}$, if $A \Rightarrow^p x \ (\mathrm{mod}\ w)$ for some $w \in (P \times N)^*$, then

$$A \Rightarrow_L^p x (\mathrm{mod}\ w')$$

for some $w' \in (P \times \{1\})^*$. This can be accomplished by induction on $|x|$, [96]. ∎

Definition 4.4.5 *Let λ be a (finitary) fuzzy language over T. λ is called a **(finitary) context-free fuzzy language (CFFL)** if $\lambda = \lambda_G$ for some CMG G.*

Proposition 4.4.6 *Let λ be a CFFL over T. Then $\lambda = \lambda_G$ for some CMG $G = (T, N, P, \eta)$ such that the following properties hold:*

(1) There exists $A_0 \in N$ such that $\eta(A_0) = 1$, and $(A \xrightarrow{} x) \in P$ implies A_0 does not occur in x.

(2) For all $A \in N$ and $x \in (T \cup N)^*$, $(A \xrightarrow{p_1} x) \in P$ and $(A \xrightarrow{p_2} x) \in P$ implies $p_1 = p_2 > 0$.

(3) For all $A \in N$, there exist $u, v \in T^*$ and $w \in (P \times N)^*$ such that $\lambda_w(A_0, uAv) > 0$.

(4) For all $A \in N$, there exists $x \in T^*$ and $w \in (P \times N)^*$ such that $\lambda_w(A, x) > 0$. ∎

A CMG satisfying conditions (1) to (4) of Proposition 4.4.6 is said to be **reduced**. In this case, we also write (T, N, P, A_0) for (T, N, P, η).

Example 4.4.7 *Let* $G = (N, T, P, \eta)$ *be the grammar defined in Example* 3.4.10. *Let* $r_1 = (s_0 \xrightarrow{.9} aAC)$, $r_2 = (C \xrightarrow{.8} AD)$, $r_3 = (A \xrightarrow{.7} B)$, $r_4 = (A \xrightarrow{.6} a)$, $r_5 = (B \xrightarrow{.5} b)$, *and* $r_6 = (D \xrightarrow{.4} b)$. *Then* $s_0 = A_0$. *In* (3) *of Proposition* 4.4.6, $\lambda_w(s_0, aAC) = .9$, $u = a$, $v = C$, *where* $w = (r_1, 1)$; $\lambda_w(s_0, aAC) = .9$, $u = aA$, $v = \Lambda$, *where* $w = (r, 1)$; $\lambda_w(s_0, aAAD) = (.9)(.8)$, $u = aAA$, $v = \Lambda$, *where* $w = (r_1, 1)(r_2, 1)$; $\lambda_w(s_0, aBC) = (.9)(.7)$, $u = a$, $v = C$, *where* $w = (r_1, 1)(r_3, 1)$.

Proposition 4.4.8 *Let* $G = (T, N, P, A_0)$ *be a reduced CMG. Then* λ_G *is finitary if and only if for every* $A \in N$ *and* $w \in (P \times \mathbb{N})^*$, $\lambda_w(A, A) \leq 1$. ∎

Lemma 4.4.9 *Let* λ *be a finitary CFFL. Then* $\lambda = \lambda_G$ *for some reduced CMG* $G = (T, N, P, A)$, *where* P *does not contain any fuzzy productions of the form* $A \xrightarrow{p} \Lambda$ *and where* $A \in N \backslash \{A_0\}$.

Proof. Let $\lambda = \lambda_{G_0}$, where $G_0 = (T, N, P_0, A_0)$ is a CMG. For every fuzzy production $A \xrightarrow{p} x_0 A_1 x_1 A_2 x_2 \cdots A_n x_n$ in P_0, where $n \geq 0$, every $A_i \in N$ and $x = x_0 x_1 \cdots x_n \in (T \cup N)^+$, let $A \xrightarrow{p'} x$ be a fuzzy production such that $p' = p \prod_{i=1}^{n} q(A_i) > 0$. Then for each $B \in N$, $q(B) = \wedge_{w \in W^*} \lambda_w(B, \Lambda)$, where $W = P_0 \times \mathbb{N}$. Let $G = (T, N, P, A_0)$ be the CMG, where P is the set of all fuzzy productions obtained in the manner described above plus the fuzzy production $A_0 \xrightarrow{p} \Lambda$, where $p = \lambda_G(\Lambda)$. Clearly, $(A \xrightarrow{p} \Lambda) \in P$ implies $A = A_0$. Furthermore, it can be shown that $\lambda = \lambda_G$. If G is not a reduced CMG, it can be easily placed in reduced form and still keep the desired property. ∎

Theorem 4.4.10 λ *is a finitary CFFL if and only if* $\lambda = \lambda_G$ *for some reduced CMG* $G = (T, N, P, A_0)$, *where* P *does not contain any fuzzy productions of the forms* $A \xrightarrow{p} B$, $A, B \in N$, *and* $A \xrightarrow{p} \Lambda$, $A \in N \backslash \{A_0\}$.

Proof. Suppose that λ is a finitary CFFL. By Lemma 4.4.9, $\lambda = \lambda_G$ for some reduced CMG $G_0 = (T, N, P_0, A_0)$, where $(A \xrightarrow{p} \Lambda) \in P_0$ implies $A = A_0$. Let ξ be the function from $(T \cup N)^* \times (T \cup N)^*$ into $\mathbb{R}^{\geq 0}$ such that for all $s, t \in (T \cup N)^*$,

$$\xi(s, t) = \begin{cases} p & \text{if } (s \xrightarrow{p} t) \in P \\ 0 & \text{otherwise.} \end{cases}$$

Let ξ_n, $n = 0, 1, 2, \ldots$, be functions from $(T \cup N)^* \times (T \cup N)^*$ into $\mathbb{R}^{\geq 0}$ defined recursively as follows

$$\xi_0(s, t) = \begin{cases} 1 & \text{if } s = t \in N, \\ 0 & \text{otherwise,} \end{cases}$$

and

$$\xi_{n+1}(s,t) = \begin{cases} \vee_{A \in N}[\xi_n(A,t)\xi(s,A)] & \text{if } s,t \in N, \\ 0 & \text{otherwise.} \end{cases}$$

Let ξ' be the function from $(T \cup N)^* \times (T \cup N)^*$ into $\mathbb{R}^{\geq 0}$ such that for all $s, t \in (T \cup N)^*$,

$$\xi'(s,t) = \begin{cases} \vee\{\xi(s,t), \ \wedge \vee[\xi(A,t)\xi_n(s,A)]\} & \text{if } s \in N, t \in N, \\ 0 & \text{otherwise.} \end{cases}$$

Let $G = (T, N, P, A_0)$ be the CMG, where P is the set of all fuzzy productions $s \xrightarrow{p} t$ in which $\xi'(s,t) = p > 0$. It follows that G has the desired properties. Let $x \in T^*$. For all $w \in (P \times \mathbb{N})^+$ and $\varepsilon > 0$, there exists $w_0 \in (P_0 \times \mathbb{N})^+$ such that $\lambda_w(A,x) > \lambda_{w_0}(A,x) - \varepsilon$. Moreover, for every $w_0 \in (P_0 \times \mathbb{N})^+$, there exists $w \in (P \times \mathbb{N})^+$ such that $\lambda_{w_0}(A,x) \leq \lambda_w(A,x)$. Thus $\lambda = \lambda_G$. The converse follows from the fact that

$$\lambda_G(x) = \wedge\{\vee\{\eta(A)\lambda_w(A,x) \mid A \in N\} \mid w \in (P \times \mathbb{N})^*, \\ |w| \leq |x|\}. \ \blacksquare$$

By Theorem 4.4.10, it can be seen that the usual procedures, [96], can be modified to obtain the following normal form theorems for context-free max-product grammars.

Theorem 4.4.11 *(Chomsky normal form) λ is a finitary CFFL if and only if $\lambda = \lambda_G$ for some reduced CMG $G = (T, N, P, A_0)$, where P contains only fuzzy productions of the forms $A_0 \xrightarrow{p} \Lambda$, $A \xrightarrow{p} a$, and $A \xrightarrow{p} BC$, where $A, B, C \in N$, and $a \in T$.* $\blacksquare$

Theorem 4.4.12 *(Greibach normal form) λ is finitary CFFL if and only if $\lambda = \lambda_G$ for some reduced CMG $G = (T, N, P, A_0)$, where P contains only fuzzy productions of the forms $A_0 \xrightarrow{p} \Lambda$ and $A \xrightarrow{p} az$, where $A \in N$, $a \in T$, $z \in N^*$, $|z| \leq 2$.* $\blacksquare$

Theorems 4.4.11 and 4.4.12 are, in general, not valid if λ is not finitary, unless we extend the definition of fuzzy productions to include $A \xrightarrow{p} x$, where $p = \infty$.

4.5 Context-Free Fuzzy Languages

In this section, we study the set of languages generated by context-free max-product grammars with cutpoints.

Let λ be a fuzzy language over T and $r \in \mathbb{R}^{\geq 0}_\infty$. Let $L(\lambda, r, >)$ denote the set $\{x \in T^* \mid \lambda(x) > r\}$.

Define $L(\lambda, r, \geq)$ and $L(\lambda, r, =)$ similarly.

Let $G = (T, N, P, A_0)$ be a CMG. Then $x \in L(\lambda_G, r, >)$ if and only if there exists $w \in (P \times \mathbb{N})^*$ such that $A_0 \Rightarrow^p x \pmod{w}$ and $p > r$.

Recall that a phrase structure grammar is a quadruple $G = (T, N, P, A_0)$ where T and N are the terminals and the nonterminals, respectively, such that $T \cap N = \emptyset$, P is a finite collection of productions, and $A_0 \in N$ is the start symbol. Derivations according to G, the language $L(G)$ generated by G, and the various types of grammars in the Chomsky hierarchy are defined in the usual manner.

Theorem 4.5.1 *L is a context-free language (CFL) if and only if $L = L(\lambda, 0, >)$ for some CFFL λ.*

Proof. Suppose $\lambda = \lambda_G$, where $G = (T, N, P, A_0)$ is a CMG. Let $G' = (T, N, P', A_0)$ be a context-free grammar, where $P' = \{s \rightarrow t \mid (s \xrightarrow{p} t) \in P$ and $p > 0\}$. Then it follows that $L = L(G')$. The converse follows easily. ∎

Definition 4.5.2 *For all $a \in T$, let T_a be a finite nonempty set and $\psi : T_a^* \rightarrow \mathcal{P}(T_a^*)$. Suppose that $\psi(\Lambda) = \{\Lambda\}$ and $\psi(ax) = \psi(a)\psi(x)$ for every $a \in T$ and $x \in T^*$. Then ψ is called a **substitution**. If $L \subseteq T^*$, let $\psi(L) = \cup_{x \in L} \psi(x)$.*

Definition 4.5.3 *An a-**transducer** is a 6-tuple $M = (X, Q, Y, H, q_0, F)$ where X, Q, and Y are finite nonempty sets (of input, state, and output symbols, respectively), H is a finite subset $Q \times X^* \times Y^* \times Q$, $q_0 \in Q$ is the initial state, and $F \subseteq Q$ is the set of accepting states.*

Let $M = (X, Q, Y, H, q_0, F)$ be an a-transducer. For each $x \in X^*$, let $M(x) = \{y \in Y^* \mid \exists x_1, x_2, \ldots, x_k \in X^*, y_1, y_2, \ldots, y_k \in Y^*,$ and $q_1, q_2, \ldots, q_k \in Q$ such that $x = x_1 x_2 \cdots x_k$, $y = y_1 y_2 \cdots y_k, q_k \in F$, and $(q_{i-1}, x_i, y_i, q_i) \in H$ for all $i, 1 \leq i \leq k\}$. For each $L \subseteq X^*$, let

$$M(L) = \cup_{x \in L} M(x).$$

It is well known that if L is a CFL and M is an a-transducer, then $M(L)$ is a CFL. It is also well known that if L is a CFL, and ψ is a substitution such that $\psi(a)$ is a CFL for all a, then $\psi(L)$ is also a CFL.

Theorem 4.5.4 *L is a CFL if and only if $L = L(\lambda_G, r, >)$ for some strict CMG G and $r \in \mathbb{R}^{\geq 0}$.*

Proof. Let $L = L(\lambda_G, r, >)$, where $G = (T, N, P, A_0)$ is a strict CMG and $r \in \mathbb{R}^{\geq 0}$. Without loss of generality, we may assume that $N = N_1 \cup N_2$, where

(1) $N_1 \cap N_2 = \emptyset$,

(2) $A_0 \in N_1$,

(3) if $(A \xrightarrow{1} x) \in P$, then $A \in N_1$,

(4) if $(A \xrightarrow{p} x) \in P$ and $p < 1$, then $A \in N_2$.

Let P_1 be the set of all fuzzy productions in P of the form $A \xrightarrow{1} x$, and let $P_2 = P \backslash P_1$. For all $r = (A \xrightarrow{p} s) \in P_2$, let $M_r = (X, Q, Y, H, q_0, \{q_1\})$ be an a-transducer, where $X = T \cup N_2$, $Q = \{q_0, q_1\}$, $Y = T \cup N$, and $H = \{(q, u, u, q) \mid q \in Q, u \in T \cup N_2\} \cup \{(q_0, A, x, q_1)\}$. Note that if $L_0 \subseteq (T \cup N_2)^*$, then $M_r(L_0)$ is obtained from L by first omitting all words in L_0 not containing any occurrence of A, and then replacing exactly one occurrence of A by x in the remaining words of L_0. For all $A \in N_1$, let $L(A) = (\lambda_{G_A}, 0, >)$, where

$$G_A = (T \cup N_2, \ N_1, \ P_1, \ A).$$

Clearly, $L(A)$ is a CFL for all $A \in N_1$. Let ψ be a substitution such that $\psi(A) = L(A)$ if $A \in N_1$, and $\psi(a) = \{a\}$ if $a \in T \cup N_2$. For all $L_1, L_2 \subseteq (T \cup N_2)^*$ and $k \in \mathbb{N}$, define $L_1 \to L_2$ if and only if there exists $r \in P_2$ such that $L_2 = (M_r(L_1))$; and $L_1 \xrightarrow{k} L_2$ if and only if there exists

$$L_0', \ L_1', \ ..., \ L_k' \subseteq (T \cup N_2)^*$$

such that $L_1 = L_0'$, $L_2 = L_k'$, and $L_{i-1} \to L_i$ for $i = 1, 2, \ldots, k$. Define $L_1 \Rightarrow^k L_2$ if and only if there exists $L_3 \subseteq (T \cup N_2)^*$ such that $L_1 \Rightarrow^k L_3$ and $L_2 = L_3 \cap T^*$. Clearly, if L_1 is a CFL and $L_1 \Rightarrow^k L_2$ for some k, then L_2 is also a CFL. Since $r > 0$, there exists $n \in \mathbb{N}$ such that $x \in L$ implies $x \in L_2$ for some $L_2 \subseteq T^*$, where $\psi(A_0) \Rightarrow^k L_2$ and $k \le n$. Let $\Gamma = \{L_2 \subseteq T^* \mid \psi(A_0) \Rightarrow^k L_2 \text{ for some } k \le n \text{ and } L_2 \cap L \ne \emptyset\}$. It follows easily that $L_2 \in \Gamma$ implies $L_2 \subseteq L$. Thus $L = \cup_{L_2 \in \Gamma} L_2$. Therefore, L is a CFL. The converse is straightforward. ∎

Theorem 4.5.5 *Let* $G = (T, N, P, A_0)$ *be a* CMG, *where there exists* $c < 1$ *such that* $(A \xrightarrow{p} x) \in P$ *implies* $p \le c$. *Then* $L(\lambda_G, r, >)$ *is finite for all* $r \in \mathbb{R}^{\ge 0}$.

Proof. Clearly, λ_G is a finitary $CFFL$. Thus by Theorem 4.4.12, $\lambda_G = \lambda_{G_1}$ for some CMG G_1, which is in Greibach normal form. Let n be the largest integer such that $c^n > r$. Then $x \in L(\lambda_{G_1}, r, >)$ implies $|x| \le n$. Thus $L(\lambda_G, r, >) = L(\lambda_{G_1}, r, >)$ is finite. ∎

Note that Theorems 4.5.4 and 4.5.5 remain valid if $>$ is replaced by $\ge$.

Theorem 4.5.6 *Let* λ *be a* $CFFL$. *Then* $L(\lambda, \infty, =)$ *is a* CFL.

Proof. Let $\lambda = \lambda_G$, where $G = (T, N, P, A_0)$ is a CMG. By the remarks following Theorems 4.4.11 and 4.4.12, we may assume that G is in Greibach normal form as long as we allow fuzzy productions of the form $s \xrightarrow{p} t$, where $p = \infty$. For all $a \in T$, let $\bar{a}$ be a new symbol and let $\bar{T} = \{\bar{a} \mid a \in T\}$. For all $x \in (T \cup N)^*$, let $\bar{x} \in (\bar{T} \cup N)^*$, where $\bar{x}$ is obtained from x by replacing every $a \in T$ in x by $\bar{a}$. Let $G = (T \cup \bar{T}, N, P_0, A_0)$ be the context-free grammar, where $P_0 = \{A \rightarrow x \mid (A \xrightarrow{p} x) \in P$ and $p < \infty\} \cup \{A \rightarrow \bar{x} \mid (A \xrightarrow{p} x) \in P$ and $p = \infty\}$. Let $M = (X, Q, Y, H, q_0, \{q_1\})$ be an a-transducer where $X = T \cup \bar{T}$, $Q = \{q_0, q_1\}$, $Y = T$, and

$$H = \{(q, u, u, q) \mid u \in T, \ q \in Q\} \cup \{(q, \bar{a}, a, q_1) \mid a \in T, \ q \in Q\}.$$

For all $L_0 \subseteq (T \cup \bar{T})^*$, $M(L_0)$ is obtained from L_0 by first omitting all those words in L_0 that do not contain any $\bar{a} \in T$, and then replacing all $\bar{a} \in \bar{T}$ by the corresponding $a \in T$ in the remaining words. Since G is in Greibach normal form, $x \in L(\lambda, \infty, =)$ if and only if x can be derived from A by using at least a fuzzy production of the form $(A \xrightarrow{p} x) \in P$ with $p = \infty$. Thus $L = M(L(G_0))$. Hence L is a CFL. ∎

Let $r \in \mathbb{R}^{\geq 0}$. Let

$$\mathcal{L}_r = \{L(\lambda, r, >) \mid \lambda \text{ is a CFFL}\}$$

and

$$\mathcal{L} = \cup_{r \in \mathbb{R}^{\geq 0}} \mathcal{L}_r.$$

Let $\mathcal{C}$ denote the family of all CFL and $\mathcal{R}$ is the family of all regular languages.

Theorem 4.5.7 $\mathcal{L}_0 = \mathcal{C}$.

Proof. The proof follows from Theorem 4.5.1. ∎

Theorem 4.5.8 *Let $r \in \mathbb{R}^{\geq 0}$. Then $\mathcal{C} \subseteq \mathcal{L}_r$.*

Proof. Let $L \in \mathcal{C}$. Then $L = L(G)$ for some context-free grammar $G = (T, N, P, A_0)$. Let $p = 1 + r$ and $G' = (T, N, P', A_0)$ be a CMG, where $P' = \{A \rightarrow x \mid (A \rightarrow x) \in P\}$. It follows that $L = L(\lambda_{G'}, r, >)$. Thus $L \in \mathcal{L}_r$. Hence $\mathcal{C} \subseteq \mathcal{L}_r$. ∎

Theorem 4.5.9 *Let $r \in \mathbb{R}$ and $r > 0$. Then $\mathcal{L} = \mathcal{L}_r$.*

Proof. Let $L \in \mathcal{L}$. Then $L = L(\lambda_{G_1}, r, >)$ for some CMG $G_1 = (T, N, P, \eta_1)$ and $r_1 \geq 0$. By Theorems 4.5.7 and 4.5.8, it suffices to consider the case where $r_1 > 0$. Let

$$G = (T, N, P, \eta),$$

where $\eta(A) = (r/r_1) \eta_1(A)$ for all $A \in N$. Clearly, $L = L(\lambda_G, r, >)$. Thus $L \in \mathcal{L}_r$. Hence $\mathcal{L} \subseteq \mathcal{L}_r$. Since $\mathcal{L}_r \subseteq \mathcal{L}$, $\mathcal{L} = \mathcal{L}_r$. ∎

Theorem 4.5.10 $L \in \mathcal{L}$ *if and only if* $L = L(\lambda, r, >)$ *for some finitary* $CFFL$ λ *and* $r \in \mathbb{R}^{\geq 0}$.

Proof. Let $L \in \mathcal{L}$. By Theorem 4.5.9, $L = L(\lambda_G, 1/2, >)$ for some CMG $G = (T, N_1, P_1, A_1)$. By remarks following Theorems 4.4.11 and 4.4.12, we may assume that G is in Greibach normal form provided we allow fuzzy productions of the form $A \xrightarrow{p} x$, where $p = \infty$. By Theorem 4.5.6,

$$L(\lambda_{G_1}, \infty, =) = L(G_2)$$

for some context-free grammar $G_2 = (T, N_2, P_2, A_2)$. Assume that G_2 is also in Greibach normal form and that $N_1 \cap N_2 = \emptyset$. Let $A_0 \notin T \cup N_1 \cup N_2$. Let $G = (T, N, P, A_0)$ be a CMG, where $N = N_1 \cup N_2 \cup \{A_0\}$, and P consists of all fuzzy productions of the following forms:

(1) $A \xrightarrow{p} x$, where $(A \xrightarrow{p} x) \in P_1$ and $p < \infty$,

(2) $A_0 \xrightarrow{p} x$, where $(A \xrightarrow{p} x) \in P_1$ and $p < \infty$,

(3) $A \xrightarrow{1} x$, where $(A \to x) \in P_2$,

(4) $A_0 \xrightarrow{1} x$, where $(A \to x) \in P_2$.

Clearly, G is in Greibach normal form. Hence by Theorems 4.4.12, λ_G is finitary. Now it follows that $L = L(\lambda_G, 1/2, >)$. The converse is straightforward. ∎

Theorem 4.5.11 $\mathcal{C}$ *is a proper subfamily of* $\mathcal{L}$.

Proof. Let $L = \{a^n b^n c^m \mid n \geq m \geq 1\}$. It is well known that $L \notin \mathcal{C}$. Now let $G = (\{a, b, c\}, \{A_0, A_1, A_2\}, P, A_0)$ be a CMG where P consists of the fuzzy productions $A_0 \xrightarrow{1} A_1 A_2$, $A_1 \xrightarrow{2} a A_1 b$, $A_1 \xrightarrow{1} ab$, $A_2 \xrightarrow{0.5} A_2$, and $A_2 \xrightarrow{1} c$. It follows that $L = L(\lambda_G, b, >)$. Thus $L \in \mathcal{L}$. ∎

We write

$$s \Rightarrow^p t \text{ if } s \Rightarrow^p t \pmod{w}$$

for some w.

Lemma 4.5.12 *For all* L *where* $L = L(\lambda_G, r, >)$ *for some* CMG G *and* $r \in \mathbb{R}^{\geq 0}$, *there exists* $n \in \mathbb{N}$ *such that every word* $x \in L$ *with* $|x| > n$ *is of the form* $s_1 s_2 s_3 s_4 s_5$, *where* $s_2 \neq \Lambda$, *and either* $s_1 s_2^k s_3 s_4^k s_5 \in L$ *for all* $k \in \mathbb{N}$ *or* $s_1 s_3 s_5 \in L$.

Proof. By Theorems 4.4.12, 4.5.9, and 4.5.10, $L = L(\lambda_G, 1, >)$, where $G = (T, N, P, A_0)$ is a CMG in Greibach normal form. Let $|N| = m$. If $x = s_1 s_2 s_3 s_4 s_5 \in L$ and $|x| > m$, then it follows that there exists an $A \in N$ such that $A_0 \Rightarrow^{p_1} s_1 A t_1 \Rightarrow^{p_2} s_1 s_2 A t_2 t_1 \Rightarrow^{p_3} s_1 s_2 s_3 s_4 s_5 = x$, where $s_i \in T^*$, $t_i \in N^*$, $s_2 \neq \Lambda$, $A \Rightarrow^{p_4} s_3$, $t_2 \Rightarrow^{p_5} s_4$, $t_1 \Rightarrow^{p_6} s_5$, $p_4 p_5 p_6 = p_3$, and $p_1 p_2 p_3 > 1$. It is easily verified that (1) if $p_2 p_5 \geq 1$, then $s_1 s_2^k s_3 s_4^k s_5 \in L$ $\forall k \in \mathbb{N}$ and (2) if $p_2 p_5 < 1$, then $s_1 s_3 s_5 \in L$. ∎

Theorem 4.5.13 $\{a^n b^n c^n \mid n \geq 1\} \notin \mathcal{L}$.

Proof. Suppose that $L = \{a^n b^n c^n \mid n \geq 1\} \in \mathcal{L}$. Then by Theorems 4.4.12, 4.5.9, and 4.5.10, $L = L(\lambda_G, 1, >)$, where $G = (T, N, P, A)$ is a CMG in Greibach normal form. Let $|N| = m$ and let $x = s_1 s_2 s_3 s_4 s_5 s_6 s_7 \in L$, where $|x| > m^2$. Then there exists $A \in N$ such that

$$A_0 \xrightarrow{p_1} s_1 A t_1 \xrightarrow{p_2} s_1 s_2 A t_2 t_1 \xrightarrow{p_3} s_1 s_2 s_3 A t_3 t_2 t_1 \xrightarrow{p_4} s_1 s_2 s_3 s_4 s_5 s_6 s_7,$$

where $s_i \in T^*$, $t_i \in N^*$, $s_2, s_3 \neq e$, $A \Rrightarrow^{p_5} s_3$, $t_3 \Rrightarrow^{p_6} s_5$, $t_1 \Rrightarrow^{p_7} s_7$, $p_5 p_6 p_7 p_8 = p_4$, and $p_1 p_2 p_3 p_4 > 1$. It follows from the proof of Lemma 4.5.12 that neither $p_2 p_7 \geq 1$ nor $p_3 p_6 \geq 1$. However, if $p_2 p_7 < 1$ and $p_3 p_6 < 1$, then by Lemma 4.5.12, $s_1 s_2 s_4 s_6 s_7$ and $s_1 s_3 s_4 s_5 s_7$ are in L. This is impossible. Hence $L \notin \mathcal{L}$. ∎

It is well known that $\{a^n b^n c^n \mid n \geq 1\}$ is a context-sensitive language. Hence the following result holds.

Theorem 4.5.14 *There exists a context-sensitive language that is not in* $\mathcal{L}$. ∎

It can be shown by Theorem 4.5.13 and the proof of Theorem 4.5.11 that $\mathcal{L}$ is not closed under intersection. However, it can be shown that $\mathcal{L}$ is closed under union, intersection with regular sets, concatenation with CFL, substitution by CFL, homomorphism, inverse homomorphism, reversal, a-transducer mapping and so on. This can be accomplished by modifying existing proofs of the closure properties of CFLs.

4.6   On the Description of the Fuzzy Meaning of Context-Free Languages

The material in the remainder of the chapter is essentially from [99].

It is stated in [99] that if there is a connection between context-free grammars and grammars of natural languages, it is undoubtedly, as Chomsky proposes, through some stronger concept like that of transformational grammar. In this framework, it is not the context-free language itself that is of interest, but, rather, the set of derivation trees, i.e., the structural descriptions or markers. From the viewpoint of the syntax directed description of fuzzy meanings, sets of trees rather than the sets of strings are of prime importance.

Thus we are motivated to study systems to manipulate fuzzy sets of trees. The purpose of the remainder of the chapter is to propose three systems to manipulate fuzzy sets of trees, namely, generators, acceptors, and transducers.

We show that the set of derivation trees of any fuzzy context-free grammar is a fuzzy set of trees generated by a fuzzy context-free dendrolanguage
generating system and that it is recognizable by a fuzzy tree automaton.

We also show that the fuzzy tree transducer is able to describe the fuzzy
meanings of fuzzy context-free languages at the level of syntax structure
in the sense that it can fuzzily associate each fuzzy derivation of the fuzzy
language with a tree representation of the computation process of its fuzzy
meaning.

4.7 Trees and Pseudoterms[1]

As before, let $\mathbb{N}$ be the set of natural numbers and $\mathbb{N}^*$ be the set of all
strings on $\mathbb{N}$ including the null string Λ. A finite closed subset U of $\mathbb{N}^*$ is
called a **finite tree domain** if the following conditions hold:

(1) $w \in U$ and $w = uv$ implies $u \in U$, where $u, v, w \in \mathbb{N}^*$;

(2) $wn \in U$ and $m \leq n$ implies $wm \in U$, where $w \in \mathbb{N}^*$, $m, n \in \mathbb{N}$.

Let U be a finite tree domain. Then the subset $\overline{U} = \{w \in U \mid w \cdot 1 \notin U\}$
is called the **leaf node set**. A pair $(N; T)$ of finite alphabets N and T,
where $N \cap T = \emptyset$, is called a **partially ranked alphabet**. A **tree** t on a
partially ranked alphabet $(N; T)$ is a function from a finite tree domain U
into $N \cup T$, written $t : U \longrightarrow (N; T)$, such that

$t(w) \in N$ for $w \in U \backslash \overline{U}$,

$t(w) \in T$ for $w \in \overline{U}$.

Of course, a finite tree $t : U \longrightarrow (N; T)$ can be represented by a finite
set of pairs $(w, t(w))$, i.e., $\{(w, t(w)) \mid w \in U\}$.

Trees on $(N; T)$ can be represented graphically by constructing a rooted
tree (where the successors of each node are ordered), representing the domain of the mapping, and labeling the nodes with elements of $N \cup T$,
representing the values of the function. Thus in the following figures
there are two examples; as a mapping, the left-hand tree has the domain
$U = \{\Lambda, 1, 2, 11, 12\}$ and the value at 11 is a. Note also that $\overline{U} = \{2, 11,$
$12\}$.

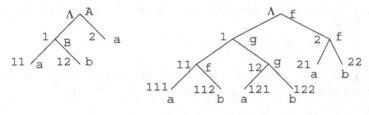

Figure 4.1

The definition of a tree and the corresponding pictorial representation
provide a good basis for intuition for considering tree manipulating systems.

---

[1] Figures 4.1-4.26 are from [99], reprinted with permission of Academic Press.

However, the development of the theory is simpler if the familiar linear representation of such trees is considered. Thus we define the set $D^p_{(N;T)}$ of **pseudoterms** on $N \cup T$ as the smallest subset of $[N \cup T \cup \{(,)\}]^*$ satisfying the following conditions:

(1) $T \subset D^p_{(N;T)}$.

(2) If $n > 0$ and $A \in N$ and $t_1, t_2, \ldots, t_n \in D^p_{(N;T)}$, then $A(t_1 \, t_2 \ldots t_n) \in D^p_{(N;T)}$.

(We note that parentheses are not symbols of $N \cup T$.)

We consider trees and pseudoterms to be equivalent formalizations. The translation between the two is the usual one. By way of example, the trees of the above figure correspond to the following pseudoterms:

$A(B(a\ b)a),\ f(g(f(ab)\ g(ab))\ f(ab))$.

This correspondence can be made precise in the following manner:

(1) If a pseudoterm $t^p \in D^p_{(N;T)}$ is atomic, i.e., $t^p = a \in T$, then the corresponding tree t has domain $\{\Lambda\}$ and $t\{\Lambda\} = a$.

(2) If $t^p = A(t^p_1 \ldots t^p_m)$, then t has domain $\cup_{i \le m}\{iw \mid w \in \text{domain } (t_i)\} \cup \{\Lambda\}$, $t(\Lambda) = A$, and for $w = iw'$ in the domain of t, $t(w) = t_i(w')$.

We denote the set of trees on $(N; T)$ by $D_{(N;T)}$, its element by t, and the pseudoterm corresponding to a tree t by $p(t)$ or t^p.

A **fuzzy set T of trees** is defined by a membership function $\mu_T :$ $D_{(N;T)} \to [0, 1]$. The set of all fuzzy set of trees is denoted by $F(D_{(N;T)})$.

4.8 Fuzzy Dendrolanguage Generating Systems

We present a fuzzy system that generates fuzzy sets of trees, as an extension of the dendrolanguage generating system of [101].

Definition 4.8.1 *A **fuzzy context-free dendrolanguage generating system (F-CFDS)** is 5-tuple, $S = (N_0, N, T, P, \lambda_0)$, such that the following conditions hold:*

*(1) N_0 is a finite set of symbols whose elements are called **nonterminal node symbols**,*

*(2) N is a finite set of symbols whose elements are called **node symbols**,*

*(3) T is a finite set of symbols whose elements are called **leaf symbols**,*

*(4) P is a finite set of **fuzzy rewriting rules** of the form*

$$\mu(\lambda, t) = c$$

which are also represented by

$$\lambda \xrightarrow{c} t;$$

$t \in D_{(N;N_0 \cup T)}$; $c \in [0,1]$ *or equivalently by*

$$\lambda \xrightarrow{c} p(t);$$

$p(t)$ is a pseudoterm corresponding to a tree t.

(5) $\lambda_0 \in N_0$ *is an **initial nonterminal node symbol**.*

Define the fuzzy relation $\xRightarrow{c}$ on the set $D_{(N;N_0 \cup T)}$ of trees as follows: For all $\alpha, \beta \in D_{(N;N_0 \cup T)}$ $\alpha \xRightarrow{c} \beta$ if and only if (i) $p(\alpha) = x\lambda y$, (ii) $p(\beta) = xp(t)y$, and (iii) $\lambda \xrightarrow{c} t$ is in P, where $x, y \in [N \cup N_0 \cup T \cup \{(,)\}]^*$, $\lambda \in N_0$ and $t \in D_{(N;N_0 \cup T)}$. Moreover, define the transitive closure $\xRightarrow{*c}$ of fuzzy relation $\xRightarrow{c}$ as follows:

(1) $\alpha \xRightarrow{*1} \alpha$ for all $\alpha \in D_{(N;N_0 \cup T)}$,

(2) $\alpha \xRightarrow{*c} \beta$ if and only if $c = \vee_{c \in D_{(N;N_0 \cup T)}} \{c' \wedge c'' \mid \alpha \xRightarrow{*c'} \gamma, \gamma \xRightarrow{c''} \beta\}$.

Definition 4.8.2 *The fuzzy set $D(S) = \{(t;c) \mid \lambda_0 \xRightarrow{*c} t \in D_{(N;T)}\}$ is called a **fuzzy context-free dendrolanguage (F-CFDL)** generated by the F-CFDS, S.*

Example 4.8.3 *Suppose that $N_0 = \{\lambda, \xi, \eta\}$, $N = \{A\}$, $T = \{a\}$, and P is given as follows:*

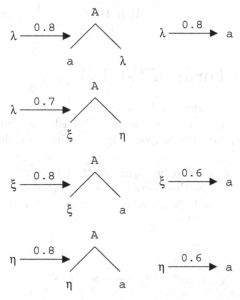

Figure 4.2

Then the F-CFDS, $S = (N_0, N, T, P, \lambda)$ generates F-CFDS, $D(S) = \{(t;$

$0.7)\mid p(t) = A \overbrace{(a \dots A(a}^{n} A(aa) \overbrace{) \dots)}^{n}, n \geq 0\} \cup \{(t; 0.6)\mid p(t) = \overbrace{A(\dots A}^{n}$

$$(A(aa)\overbrace{a)\ldots}^{n}), \quad n \geq 1\}.$$

For example, as a derivation, we have

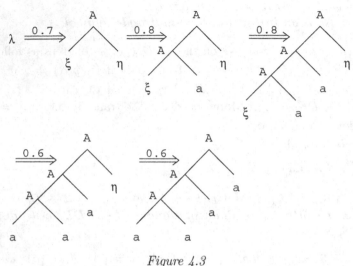

Figure 4.3

or equivalently,

$$\lambda \overset{0.7}{\Longrightarrow} A(\xi\eta) \overset{0.8}{\Longrightarrow} A(A(\xi a)\eta) \overset{0.8}{\Longrightarrow} A(A(A(\xi a)a)\eta) \overset{0.6}{\Longrightarrow} A(A(A(A(aa)a)\eta) \overset{0.6}{\Longrightarrow}$$
$$A(A(A(A(aa)a)a).$$

4.9   Normal Form of F-CFDS

The **depth** of tree t with domain U_t is defined by $d(t) = \vee\{|w| \mid w \in U_t\}$, where $|w|$ is the length of w. The order of F-CFDS is defined to be the maximum value of the depths of trees appearing in the right-hand side of the rules.

Two F-CFDS's are said to be **equivalent** if they generate the same fuzzy dendrolanguage. In this section, we show that for any F-CFDS's there is an equivalent F-CFDS of order 1, i.e., one whose rules are of the form

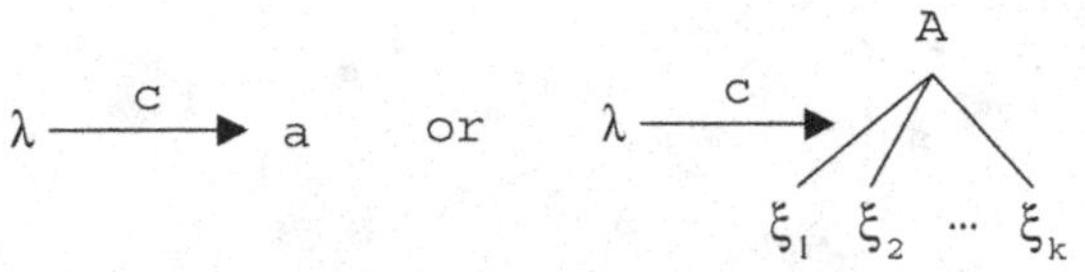

$$\lambda \overset{c}{\longrightarrow} a \quad or \quad \lambda \overset{c}{\longrightarrow} \overset{A}{\underset{\xi_1 \ \xi_2 \ \cdots \ \xi_k}{\bigwedge}} \quad k \geq 1,$$

Figure 4.4

where $\lambda, \xi_i \in N_0 \ (i = 1, \ldots, k)$, $A \in N$, and $a \in T$.

Lemma 4.9.1 *Let $S = (N_0, N, T, P, \lambda_0)$ be a F-CFDS of order n $(n \geq 2)$. Then there exists an equivalent F-CFDS of order $n - 1$.*

Proof. We determine a new F-CFDS, $S' = (N_0', N, T, P', \lambda_0)$ from a given F-CFDS, S. P' is defined as follows: For each rule

$$\lambda \xrightarrow{c} t$$

in P, (1) if $d(t) < n$, then $\lambda \xrightarrow{c} t$ is in P'; (2) if $d(t) = n$ and $p(t) = X(p(t_1) \ldots p(t_k))$, then

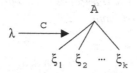

Figure 4.5

and $\xi_i \xrightarrow{1} t_i$ for all i such that $p(t_i) \notin N_0$ is in P', where ξ_i is a new distinct nonterminal node symbol if $p(t_i) \notin N_0$ and $\xi_i = p(t_i)$ if $p(t_i) \in N_0$.

Clearly, N_0' is the union of N_0 and the set of all new nonterminal node symbols introduced by applying the above rule (2).

Suppose $\alpha \overset{c}{\Longrightarrow} \beta$ under S. Then $p(\alpha) = x\lambda y$, $p(\beta) = xp(t)y$, and $\lambda \xrightarrow{c} t$ is in P. If $d(t) < n$, the above construction shows that $\alpha \overset{c}{\Longrightarrow} \beta$ under S' since $\lambda \xrightarrow{c} t$ is also contained in P'. If $d(t) = n$, we have by the above construction that

$$p(\alpha) = x\lambda y \overset{c}{\Longrightarrow} xX(\xi_1 \ldots \xi_k)y \overset{*1}{\Longrightarrow} xX(p(t_1) \ldots p(t_k))y = p(\beta).$$

Conversely, if $x\lambda y \overset{c}{\Longrightarrow} xX(\xi_1 \ldots \xi_k)y$ under S', then $\xi_i \xrightarrow{1} t_i$; $i = 1, \ldots, k$ should be applied since nonterminal symbols ξ_i's can be rewritten only by them. Hence $x\lambda y \overset{c}{\Longrightarrow} xX(\xi_1 \ldots \xi_k)y \overset{*1}{\Longrightarrow} xX(p(t_1) \ldots p(t_k))y$. For this derivation, we have

$$x\lambda y \overset{c}{\Longrightarrow} xp(t)y$$

under S.

Thus $D(S) = D(S')$. From the construction procedure of S', it is clear that S' is of order $n - 1$. ∎

By repeated application of Lemma 4.9.1, we obtain the following lemma.

Lemma 4.9.2 *For any F-CFDS, S, there exists an equivalent F-CFDS, S' of order 1.*

Theorem 4.9.3 *For any F-CFDS there is an equivalent F-CFDS whose rules are in the form of*

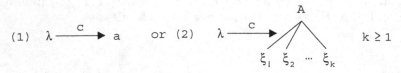

Figure 4.6

where λ, ξ_i, $i = 1, 2, \ldots, k$, *are nonterminal node symbols, a is a leaf symbol, and A is a terminal node symbol.*

Proof. Let S be an F-CFDS. By Lemma 4.9.2, there is an equivalent F-CFDS whose rules are in the following form:

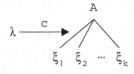

Figure 4.7

Here, if we replace the rule of type (2) by a rule

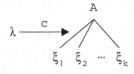

Figure 4.8

where $\xi_i = X_i$ if $X_i \in N_0$ and ξ_i is a new symbol if $X_i \in T$, and rules

$$\xi_i \xrightarrow{1} X_i$$

for all $X_i \in T$, then we obtain the desired F-CFDS. ∎

An F-CFDS whose rules are in the form of (1) or (2) of Theorem 4.9.3 is said to be **normal**.

Example 4.9.4 *Consider the following set of rules:*

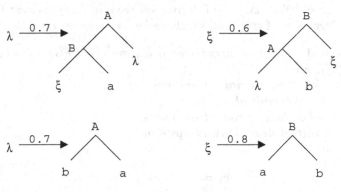

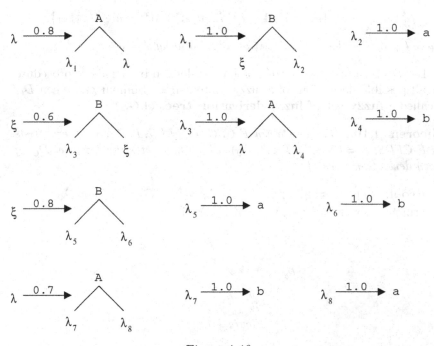

Figure 4.9

This gives an F-CFDS of order 2. The normal form for this F-CFDS is given by the following rules:

Figure 4.10

4.10   Sets of Derivation Trees of Fuzzy Context-Free Grammars

In this section, we define the sets of derivation trees of fuzzy context-free grammars as fuzzy sets of trees and we characterize them by F-CFDS's.

Definition 4.10.1 *A* **fuzzy context-free grammar** *(F-CFG) is a 4-tuple* $G = (N, T, P, S)$*, where*
   (1) N *is a set of* **nonterminal symbols,**
   (2) T *is a set of* **terminal symbols,**
   (3) P *is a set of* **fuzzy production rules,**
   (4) S *is an* **initial nonterminal symbol.**
For a derivation

$$w_0(= S) \xRightarrow{c_1} w_1 \xRightarrow{c_2} \ldots \xRightarrow{c_m} w_m(= w)$$

under a fuzzy context-free grammar G*, we define a derivation tree with a degree of membership as follows:*
   (1) For $w_0(= S)$*,* $(\alpha^{w_0}; 1) = (\{(\Lambda, S)\}; 1)$*.*
   (2) Suppose that $(\alpha^{w_{i-1}}; c)$ *is given for some* i *and that* $w_{i-1} \xRightarrow{c_i} w_i$ *is realized by* $A \xrightarrow{c_i} Y_1 Y_2 \ldots Y_k (Y_i \in N \cup T)$ *with* $w_{i-1} = xAy$ *and* $w_i = xY_1 Y_2 \ldots Y_k y$*. (It is assumed that the symbol* A *replaced by* $Y_1 Y_2 \ldots Y_k$ *corresponds to a leaf node* u *in* $\overline{U}_{\alpha^{w_{i-1}}}$*.) Then* $(\alpha^{w_i}; c')$ *is given by*

$$\alpha^{w_i} = \alpha^{w_{i-1}} \cup \{(u \cdot i, Y_i) | 1 \le i \le k, (u, A) \in \alpha^{xAy}, u \in \overline{U}_{\alpha^{w_{i-1}}}\},$$

where $\overline{U}_{\alpha^{w_{i-1}}}$ *is the leaf node set of* $\alpha^{w_{i-1}}$ *and by* $c' = c \wedge c_i$*.*

Let D_G be a fuzzy set of trees on $(N; T)$ defined by the above procedure for all possible derivations of a fuzzy context-free grammar G. Then D_G is called a **fuzzy set of fuzzy derivation trees** of G.

Theorem 4.10.2 *For any given F-CFG,* $G = (N, T, P_G, S)$*, there exists an F-CFDS,* $S = (N_0, N_S, T_S, P_S, \lambda_0)$*, which generates the fuzzy set* D_G *of fuzzy derivation trees of* G*.*

    Proof. Set $N_0 = \{\lambda_X | X \in N\}$, $N_S = N$, $T_S = T$, and $\lambda_0 = \lambda_S$. Determine P_S as follows: If $X \xrightarrow{c} Y_1 Y_2 ... Y_k$ is in P, then

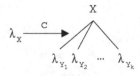

Figure 4.11

is contained in P_S if $Y_i = a \in T$, then $\lambda_{Y_i} = a$.

Since the process of obtaining (α^{w_i}, c') from $(\alpha^{w_{i-1}}, c)$ in the definition of D_G corresponds to the application of the rule

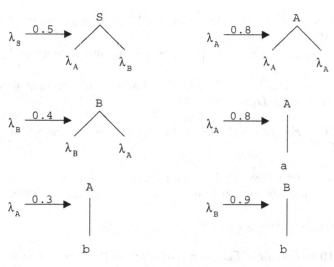

Figure 4.12

in F-CFDS, S, it follows that $D_G = D(S)$. ∎

Example 4.10.3 *Consider the F-CFG given by the following rules:*

$$S \xrightarrow{0.5} AB, A \xrightarrow{0.8} AB, B \xrightarrow{0.4} BA, A \xrightarrow{0.8} a, A \xrightarrow{0.3} b, B \xrightarrow{0.9} b.$$

For this F-CFG, construct an F-CFDS determined by the following rules:

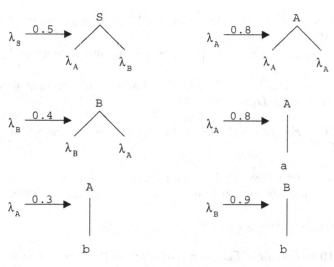

Figure 4.13

For the derivation of F-CFG,

$$S \xRightarrow{0.5} AB \xRightarrow{0.4} ABA \xRightarrow{0.8} aBA \xRightarrow{0.9} abA \xRightarrow{0.3} abb$$

its derivation tree is generated by the F-CFDS as follows:

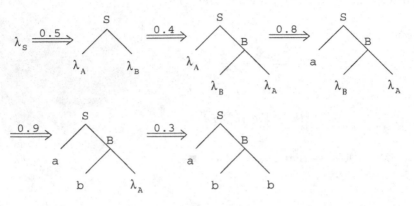

Figure 4.14

We now consider the converse of Theorem 4.10.2. We prove that for any F-CFDS, S, there exists a F-CFG G corresponding to S in the sense of Theorem 4.10.5 below. Let $h : [N \cup T \cup \{(,)\}]^* \longrightarrow T$ be a homomorphism defined by $h(a) = a$ for a in T and $h(X) = \Lambda$ for $X \notin T$.

Lemma 4.10.4 *Let S be an F-CFDS. Then the fuzzy set $p(D(S))$ of pseudoterms of $D(S)$ is a fuzzy context-free language.*

Proof. Let $S = (N_0, N, T, P, \lambda_0)$ be an F-CFDS. Construct an F-CFG, $G = (N_G, T_G, P_G, S_G)$ as follows:

Set $N_G = N_0$, $T_G = N \cup T \cup \{(,)\}$, $S_G = \lambda_0$ and determine P_G as follows:

If $\lambda \xrightarrow{c} t$ is in P, then $\lambda \xrightarrow{c} p(t)$ is in P_G. From the above construction, it follows that $L(G) = p(D(S))$. ∎

Theorem 4.10.5 *For any F-CFDS, S, $h(p(D(S)))$ is a fuzzy context-free language on T.*

Proof. The proof follows by Lemma 4.10.4 and the fact that a homomorphic image of a fuzzy context-free language is also a fuzzy context-free language. ∎

The next result is stronger.

Theorem 4.10.6 *Every F-CFDL is a projection of the fuzzy set of derivation trees of an F-CFG.*

Proof. Let $S = (N_0, N, T, P, \lambda_0)$ be a normal F-CFDS. Consider the F-CFG, $G = (N_G, T_G, P_G, S_G)$, where $N_G = N_0 \times (N \cup T)$, $T_G = \{\langle f, a \rangle \mid a \in T\}$ and where f is a new symbol not in N_0, $S_G = \{\langle \lambda_0, X \rangle \mid X \in N \cup T\}$ and P_G is defined as follows:

(1) If

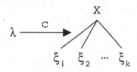

Figure 4.15

is in P, then

$$\langle \lambda, X \rangle \xrightarrow{c} \langle \xi_1, X_1 \rangle \langle \xi_2, X_2 \rangle \ldots \langle \xi_k, X_k \rangle$$

is in P_G, where $X_i \in N \cup T$.

(2) If $\lambda \xrightarrow{c} a$ is in P, then the rule

$$\langle \lambda, a \rangle \xrightarrow{c} \langle f, a \rangle$$

is in P_G.

It follows that if $(t; c)$ is a fuzzy derivation tree of this grammar, $(\pi(t); c)$, a projection of $(t; c)$, is a fuzzy tree generated by the F-CFDS, S, where $\pi(t)$ is defined, in terms of pseudoterms, as follows:

(1) $\pi[\langle \lambda, a \rangle (\langle f, a \rangle)] = a$

(2) $\pi[\langle \lambda, X \rangle (p(t_1) \ldots p(t_k))] = X (\pi[p(t_1)] \ldots \pi[p(t_k)])$. ∎

Example 4.10.7 *Consider the F-CFDS given by the following rules:*

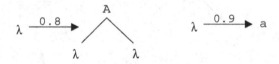

Figure 4.16

For this F-CFDS, define a F-CFG by the following rules:

$$\langle \lambda, A \rangle \xrightarrow{0.8} \langle \lambda, A \rangle \langle \lambda, A \rangle$$

$$\langle \lambda, A \rangle \xrightarrow{0.8} \langle \lambda, A \rangle \langle \lambda, a \rangle$$

$$\langle \lambda, A \rangle \xrightarrow{0.8} \langle \lambda, a \rangle \langle \lambda, A \rangle$$

$$\langle \lambda, A \rangle \xrightarrow{0.8} \langle \lambda, a \rangle \langle \lambda, a \rangle$$

$$\langle \lambda, a \rangle \xrightarrow{0.9} \langle f, a \rangle$$

For example, the fuzzy derivation

$$\langle \lambda, a \rangle \stackrel{0.8}{\Longrightarrow} \langle \lambda, A \rangle \langle \lambda, a \rangle \stackrel{0.8}{\Longrightarrow} \langle \lambda, a \rangle \langle \lambda, a \rangle \langle \lambda, a \rangle \stackrel{*0.9}{\Longrightarrow} \langle f, a \rangle \langle f, a \rangle \langle f, a \rangle$$

has a derivation tree

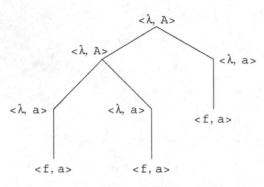

Figure 4.17

and its degree of membership is 0.8.
The projection of this tree defined in the proof of Theorem 4.10.6 is

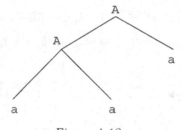

Figure 4.18

which is contained in the F-CFDL generated by the given F-CFDS.

4.11   Fuzzy Tree Automaton

In Section 4.9, we presented a fuzzy dendrolanguage generating system that was used in Section 4.10 to characterize the fuzzy set of derivation trees of a fuzzy context-free grammar. In this section, we define a fuzzy tree automaton as an acceptor of a fuzzy dendrolanguage.

Definition 4.11.1 *A **fuzzy tree automaton** (F-TA) is a 5-tuple $A = (S, N, T, \alpha, F)$, where*
   *(1) S is a finite set of **state symbols**,*
   *(2) N is a finite set of **terminal node symbols**,*
   *(3) T is a finite set of **leaf symbols**, where $N \cap T = \emptyset$,*

(4) $\alpha : (N \cup T) \longrightarrow \{f \mid f : S \times S \to [0,1]\}$, *where S is a finite subset of S^* containing the null string Λ. For $X \in N$, $\alpha(X) = a_X$ is a function from $(S \backslash \{\Lambda\}) \times S$ into $[0,1]$. It is called a **fuzzy direct transition function**. For $a \in T$, $\alpha(a) = \alpha_a$ is a function from $\{\Lambda\} \times S$ into $[0,1]$. α_a defines a fuzzy subset on S that is assigned to the node of a. $\alpha_X(s_1 s_2 \ldots s_k, s) = c$ means that when a node of X has k sons with states $s_1, s_2, \ldots, s_k$, the state s is assigned to the node with degree c. This may be represented graphically as follows:*

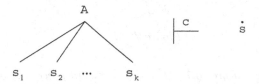

Figure 4.19

*(5) F is a distinct subset of S, called a set of **final states**.*

For a tree $t \in D_{(N;T)}$, define a fuzzy transition function

$$\alpha_t : \{f \mid f : S^* \times S \longrightarrow [0,1]\} \longrightarrow [0,1]$$

as follows: Let t be $X(t_1 t_2 \ldots t_k)$ in terms of pseudoterm. Then for $f = (s_1 s_2 \ldots s_n, s)$,

$$
\begin{aligned}
\alpha_t(s_1 s_2 \ldots s_n, s) &= \alpha_{X(t_1 \ldots t_k)}(s_1 s_2 \ldots s_n, s) \\
&= \vee_{\overline{s}_i \in S} \wedge \{\alpha_x(\overline{s}_1 \overline{s}_2 \ldots \overline{s}_k, s), i = 1, \ldots, k \\
&\quad \alpha_{t_1}(s_1 s_2 \ldots s_n, s_1), \ldots, \alpha_{t_k}(s_1 s_2 \ldots s_n, s_k)\}.
\end{aligned}
$$

If $t = a \in T$, then $\alpha_t = \alpha_a$.

Using α_t, we can assign a fuzzy subset of S to the root node of the tree t. The mapping α_t can also define a fuzzy set of trees, i.e., a fuzzy set on $D_{(N;T)}$, by

$$D(A) = \{(t;c) \mid c = \vee_{s \in F} \{\alpha_t(\Lambda, s)\}\},$$

which is called a **fuzzy set of trees recognized by a fuzzy tree automaton A.**

Example 4.11.2 Let $S = \{s_\lambda, s_\xi, s_\eta\}$, $N = \{A\}$, $T = \{a\}$, and $F = \{s_\eta\}$. Define α as follows:

$\alpha(a)$	s_λ	s_ξ	s_η
Λ	0.8	0.7	1.0

$\alpha(A)$	s_λ	s_ξ	s_η
s_λ	1.0	0	0
s_ξ	0	1.0	0
s_η	0	0	1.0
$s_\lambda s_\lambda$	0.2	0.8	0
$s_\lambda s_\xi s_\eta$	0	0	1.0

For the F-TA, $A = (S, N, T, \alpha, F)$, $D(A)$ contains the following t:

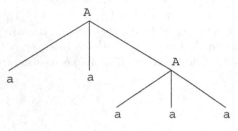

Figure 4.20

with the membership degree 0.7. The computation process of $\alpha_t(\Lambda, s_\eta)$ can be represented as follows:

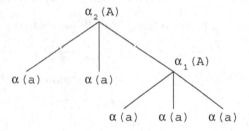

Figure 4.21

where

$$\alpha(a) = [0.8s_\lambda, 0.7s_\xi, 1.0s_\eta]$$

(i.e., $\alpha_a(s_\lambda) = 0.8, \alpha_a(s_\xi) = 0.7, \alpha_a(s_\eta) = 1.0$),

$$\alpha_1(A) = [0s_\lambda, 0s_\xi, 0.7s_\eta]$$

and

$$\alpha_2(A) = [0s_\lambda, 0s_\xi, 0.7s_\eta].$$

Hence we have that $\alpha_t(\Lambda, s_\eta) = 0.7$. Alternatively, $\alpha_t(\Lambda, s_\eta)$ can also be determined by enumerating all the fuzzy reductions from t to s_η such as the *following one:*

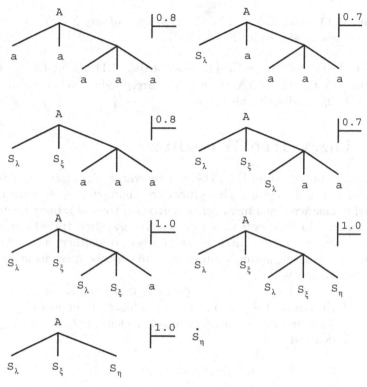

Figure 4.22

Theorem 4.11.3 *Any F-CFDL is recognizable by an F-TA.*

Proof. By Theorem 4.9.3, we can assume that for any given F-CFDL, D, there exists a normal F-CFDS, $S = (N_0, N, T, P, \lambda_0)$ with $D(S) = D$. From the F-CFDS, S, we construct an F-TA, $A = (S, N, T, \alpha, F)$ as follows:

Set $S = \{s_\lambda \mid \lambda \in N_0\}$, $F = \{s_{\lambda_0}\}$.

Then α is determined as follows:

(1) If $\lambda \xrightarrow{c} X(\lambda_1 \lambda_2 \ldots \lambda_k)$ is in P, then
$$\alpha(X)(s_{\lambda_1} s_{\lambda_2} \ldots s_{\lambda_k}, s_\lambda) = c$$

(2) If $\lambda \xrightarrow{c} a$ is in P, then
$$\alpha(a)(\Lambda, s_\lambda) = c.$$

Clearly, the set S in the Definition 4.11.1 is
$$\{s_{\lambda_1} s_{\lambda_2} \ldots s_{\lambda_k} \mid \lambda \xrightarrow{c} X(\lambda_1 \ldots \lambda_k) \text{ is in } P\} \cup \{\Lambda\}.$$

The above construction of A yields $D(A) = D(S)$. ∎

Theorem 4.11.4 *Any fuzzy set of trees recognizable by an F-TA is an F-CFDL.*

Proof. The converse of the construction in the proof of Theorem 4.11.3 proves the theorem. ∎

Corollary 4.11.5 *A fuzzy set of derivation trees of any F-CFG is a fuzzy set of trees recognizable by an F-TA.*

Proof. The proof follows by Theorems 4.10.2, 4.11.3, and 4.11.4. ∎

It follows that, for an F-TA, the results corresponding to Lemma 4.10.4, Theorems 4.10.5, and 4.10.6 also hold.

4.12   Fuzzy Tree Transducer

Previously, we presented an F-CFDS as a generator of a fuzzy set of trees and an F-TA as an acceptor. These fuzzy tree manipulating systems have been used to characterize a fuzzy set of derivation trees of a fuzzy context-free grammar. In this section, we present a fuzzy tree transducer that can define a fuzzy function from a set of trees to another one. These three tree manipulating systems can be used to describe fuzzy meanings of context-free languages.

Let $(N_1; T_1)$ and $(N_2; T_2)$ be finite partially ranked alphabets. A **fuzzy tree translation** from $D_{(N_1;T_1)}$ to $D_{(N_2;T_2)}$ is a fuzzy subset Φ of $D_{(N_1;T_1)} \times D_{(N_2;T_2)}$ for which the grade of membership of an element (t_1, t_2) of $D_{(N_1;T_1)} \times D_{(N_2;T_2)}$ is denoted by

$$\mu_\Phi(t_1, t_2) = c \in [0, 1].$$

We also denote it by a triple (t_1, t_2, c) and then the fuzzy subset Φ can be considered to be a set of such triples. The **domain of a fuzzy tree translation** Φ is $\{t_1|$ for some t_2 and some $c > 0$, (t_1, t_2, c) is in $\Phi\} = \{t_1|$ for some $t_2, \mu_\Phi(t_1, t_2) > 0\}$, which is denoted by dom Φ. The **range of a fuzzy tree translation** Φ is $\{t_2|$ for some t_1 and some $c > 0$, (t_1, t_2, c) is in $\Phi\} = \{t_2|$ for some $t_1, \mu_\Phi(t_1, t_2) > 0\}$, which is denoted by range Φ. We also define two underlying fuzzy dendrolanguages; the one is that of the domain that is defined to be a fuzzy subset $ufd\Phi$ of $D_{(N_1;T_1)}$ for which the grade of membership of an element t_1 of $D_{(N_1;T_1)}$ is given by

$$\mu_{ufd\Phi}(t_1) = \vee\{c \mid \mu_\Phi(t_1, t_2) = c, t_2 \in D_{(N_2;T_2)}\}.$$

The other is that of the range that is a fuzzy subset $ufr\Phi$ of $D_{(N_2;T_2)}$. The membership function is given by

$$\mu_{ufr\Phi}(t_2) = \vee\{c \mid \mu(t_1, t_2) = c, t_1 \in D_{(N_1;T_1)}\}.$$

We next introduce a relatively simple system in order to define a fuzzy tree translation.

Definition 4.12.1 *A **fuzzy simple tree transducer**, (F-STT), is a 7-tuple*

$$M = (N_0, N_1, T_1, N_2, T_2, R, \lambda_0)$$

such that the following conditions hold:

*(1) N_0 is a finite set of symbols whose elements are called **nonterminal node symbols**,*

*(2) N_1, N_2 are finite sets of symbols, called **node symbols**,*

*(3) T_1, T_2 are finite sets of symbols, called **leaf symbols**,*

*(4) $R: N_0 \times D_{(N_1;N_0 \cup T_1)} \times D_{(N_2;N_0 \cup T_2)} \longrightarrow [0,1]$; if $\mu(\lambda, t_1, t_2) = c$, then we write $\lambda \xrightarrow{c} (t_1, t_2)$, where t_1, t_2 contain the same nonterminal symbols in the same order. We call R a **fuzzy translation rule**.*

(5) λ_0 is the initial nonterminal node symbol.

A form of M is a pair (t_1, t_2), where t_1 is in $D_{(N_1;N_0 \cup T_1)}$ and t_2 is in $D_{(N_2;N_0 \cup T_2)}$. If (1) $\lambda \xrightarrow{c} (t_1, t_2)$ is a fuzzy translation rule, (2) (α_1, α_2) and (β_1, β_2) are forms such that $p(\alpha_1) = x_1 \lambda y_1$, $p(\alpha_2) = x_2 \lambda y_2$, $p(\beta_1) = x_1 p(t_1) y_1$, $p(\beta_2) = x_2 p(t_2) y_2$, and (3) if λ is the k-th nonterminal node symbol in $p(\alpha_1)$, then λ of $p(\alpha_2) = x_2 \lambda y_2$ is also the k-th one in it and we write

$$(\alpha_1, \alpha_2) \xRightarrow{c} (\beta_1, \beta_2).$$

We also define the relation $\xRightarrow{*c}$ as follows:

$$(\alpha_1, \alpha_2) \xRightarrow{*1} (\alpha_1, \alpha_2)$$

and if $(\alpha_1, \alpha_2) \xRightarrow{*c} (\beta_1, \beta_2)$ and $(\beta_1, \beta_2) \xRightarrow{c_2} (\gamma_1, \gamma_2)$, then $(\alpha_1, \alpha_2) \xRightarrow{*c} (\gamma_1, \gamma_2)$, where

$$c = \vee_{(\beta_1, \beta_2)} \{c_1 \wedge c_2\}.$$

The fuzzy tree translation defined by M, written $\Phi(M)$, is

$$\{(t_1, t_2, c) \,|\, (\lambda, \lambda) \xRightarrow{*c} (t_1, t_2) \in D_{(N_1;T_1)} \times D_{(N_2;T_2)}\},$$

i.e., $\Phi(M)$ is a fuzzy subset of $D_{(N_1;T_1)} \times D_{(N_2;T_2)}$ for which the grade of membership of an element (t_1, t_2) of $D_{(N_1;T_1)} \times D_{(N_2;T_2)}$ is given by

$$\mu_{\Phi(M)}(t_1, t_2) = c$$

or $(\lambda, \lambda) \xRightarrow{*c} (t_1, t_2)$.

Example 4.12.2 *Consider the following set of rules:*

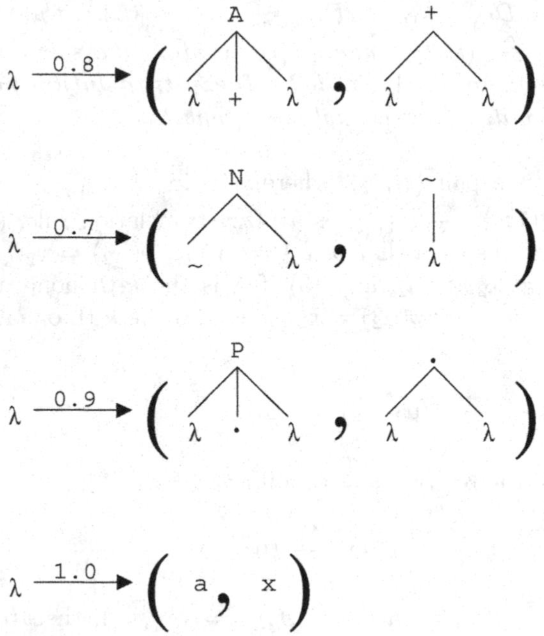

Figure 4.23

From these rules, we have a fuzzy tree translation whose elements can be determined as follows:

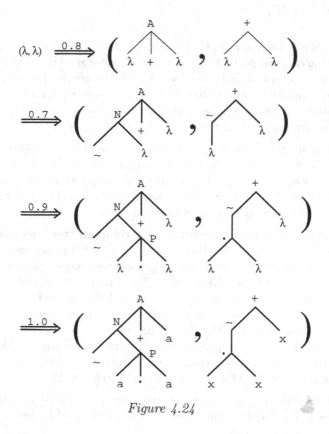

Figure 4.24

Theorem 4.12.3 *For any F-STT, M, both $ufd\Phi(M)$ and $ufr\Phi(M)$ are F-CFDL's.*

Proof. From a given F-STT, $M = (N_0, N_1, T_1, N_2, T_2, R, \lambda_0)$, we construct an F-CFDS, $S = (N_0, N_1, T_1, P, \lambda_0)$, by defining P as follows: If $\lambda \xrightarrow{c} (t_1, t_2)$ is in R, then P contains $\lambda \xrightarrow{c} t_1$.

By noting that $(\lambda_0, \lambda_0) \overset{*c}{\Longrightarrow} (t_1, t_2)$ if and only if $\lambda_0 \overset{*c}{\Longrightarrow} t_1$, it follows that $ufd\Phi(M) = D(S)$.

The remaining part of the theorem can be proved in a similar manner.

∎

Similar arguments as used in the proof of Theorem 4.12.3 can be used in the proof of the following theorem.

Theorem 4.12.4 *For any F-STT, M, both $\mathrm{dom}\,\Phi(M)$ and range $\Phi(M)$ are context-free dendrolanguages.* ∎

4.13 Fuzzy Meaning of Context-Free Languages

Specifying a set of semantic rules that can serve as an algorithm for computing the meaning of a composite term from the knowledge of the meanings of
its components is a central problem of semantics. However, the complexity
of natural languages is so great that it is not even clear what the form of
the rules should be. Hence it is natural to start with a few relatively simple
cases involving fragments of natural or artificial languages.

It has been suggested by Zadeh that a possible start to approach the
problem concerning semantics is by proposing a quantitative theory of semantics: The meaning of a term is defined to be a fuzzy subset of a universe
of discourse and an approach similar to that described in [116] can be used
to compute the meaning of a composite term.

This method of assigning the meanings to a composite term is essentially
considered to be a syntax-directed one. A fuzzy subset that is assigned to a
composite term is considered to be an image of some composite function of
fuzzy subsets on the universe of discourse. Then, it should be fairly natural
to define the semantic domain as the universe of discourse and (composite)
functions. A syntax structure in an expression representing a function
can also be recognized here. That is, we consider as the semantic domain
only the set of all functions representable by using some syntax rules and
we consider that assigning the meaning to a composite term as assigning a
syntax tree representation of a corresponding function to it.

We are thus led to apply our fuzzy tree transducer to describing the
fuzzy meaning of context-free languages. This is illustrated in the following
examples.

Example 4.13.1 *We construct a fuzzy tree transducer for a slightly modified example of the one described in [261, 260] as follows:*

$$(1) \quad \lambda_S \xrightarrow{\ 1.0\ } \left(\begin{array}{c} S \\ | \\ \lambda_A \end{array} \ , \ \lambda_A \right)$$

$$(2) \quad \lambda_A \xrightarrow{\ 1.0\ } \left(\begin{array}{c} A \\ | \\ \lambda_B \end{array} \ , \ \lambda_B \right)$$

$$(3) \quad \lambda_B \xrightarrow{\ 1.0\ } \left(\ \begin{array}{c} B \\ | \\ \lambda_C \end{array} \ , \ \lambda_C \ \right)$$

$$(4) \quad \lambda_S \xrightarrow{\ 0.9\ } \left(\ \overset{S}{\underset{\lambda_S \ \text{or} \ \lambda_A}{\bigwedge}} \ , \ \overset{f_v}{\underset{\lambda_S \quad \lambda_A}{\bigwedge}} \ \right)$$

$$(5) \quad \lambda_A \xrightarrow{\ 0.9\ } \left(\ \overset{A}{\underset{\lambda_A \ \text{and} \ \lambda_B}{\bigwedge}} \ , \ \overset{f_\Lambda}{\underset{\lambda_S \quad \lambda_A}{\bigwedge}} \ \right)$$

$$(6) \quad \lambda_B \xrightarrow{\ 0.8\ } \left(\ \overset{B}{\underset{\text{not} \quad \lambda_C}{\bigwedge}} \ , \ \begin{array}{c} f_\sim \\ | \\ \lambda_C \end{array} \ \right)$$

$$(7) \quad \lambda_O \xrightarrow{\ 0.9\ } \left(\ \overset{O}{\underset{\text{very} \quad \lambda_C}{\bigwedge}} \ , \ \begin{array}{c} f_v \\ | \\ \lambda_O \end{array} \ \right)$$

$$(8) \quad \lambda_Y \xrightarrow{\ 0.9\ } \left(\ \overset{Y}{\underset{\text{very} \quad \lambda_Y}{\bigwedge}} \ , \ \begin{array}{c} f_v \\ | \\ \lambda_Y \end{array} \ \right)$$

$$(9) \quad \lambda_C \xrightarrow{\ 1.0\ } \left(\ \begin{array}{c} C \\ | \\ \lambda_O \end{array} \ , \ \lambda_O \ \right)$$

$$(10) \quad \lambda_C \xrightarrow{1.0} \left(\begin{array}{c} C \\ | \\ \lambda_Y \end{array} , \ \lambda_Y \right)$$

$$(11) \quad \lambda_C \xrightarrow{0.6} \left(\begin{array}{c} C \\ | \\ \lambda_S \end{array} , \ \lambda_S \right)$$

$$(12) \quad \lambda_O \xrightarrow{1.0} \left(\begin{array}{c} O \\ | \\ old \end{array} , \ f_{old} \right)$$

$$(13) \quad \lambda_Y \xrightarrow{1.0} \left(\begin{array}{c} Y \\ | \\ young \end{array} , \ f_{young} \right)$$

Let $f_\vee$, $f_\wedge$, and $f_\sim$ be the fuzzy set operations union, intersection, and complement and let f_v be the concentrating function defined as follows: if $f_v(A) = B$, then the membership function $\mu_B(x)$ of B is given by $\mu_B(x) = \mu_A^2(x)$. Moreover, assume that f_{old} and f_{young} are constant functions whose values are, for example, fuzzy subsets of the set of the first one-hundred positive integers $K = \{k \mid k = 1, 2, \ldots, 100\}$ and that are characterized by the following membership functions:

$$\mu_{N_0}(old, k) = \begin{cases} 0 & \text{for } k < 50 \\ [1 + (\frac{k-50}{5})^{-2}]^{-1} & \text{for } k \geq 50 \end{cases}$$

and

$$\mu_{N_0}(young, k) = \begin{cases} 1 & \text{for } k < 25 \\ [1 + (\frac{k-25}{5})^2]^{-1} & \text{for } k \geq 25, \end{cases}$$

respectively, where $k \in K$.

We now consider a composite term $x = $ old or young and not very old. For the term x, the translation is given by (A) and (B):

(A)

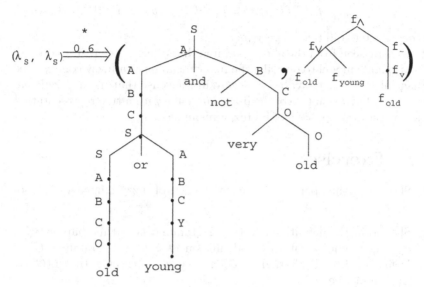

Figure 4.25

which is realized by applications of rules (1), (5), (2), (3), (11), (4), (1), (2), (3), (9), (12), (2), (3), (10), (13), (6), (9), (7), and (12) in this order, and

(B)

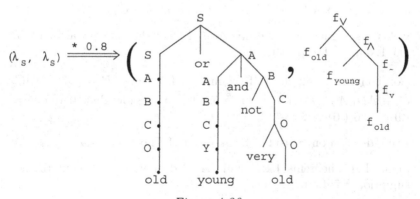

Figure 4.26

which is realized by applications of rules (4), (1), (2), (3), (9), (12), (5), (2), (3), (10), (13), (6), (9), (7), and (12).

Hence we have that in the system under consideration, the composite term x = old or young and not very old has two meanings

$$f_\wedge(f_\vee(f_{old}, f_{young}), f_\sim(f_v(f(_{old})))) = [\mu(old) \vee \mu(young)] \wedge [1 - \mu^2(old)]$$

and

$$f_\vee(f_{old}, f_\wedge(f_{young}, f_\sim(f_v(f_{old})))) = \mu(old) \vee [\mu(young) \wedge [1 - \mu^2(old)]]$$

with degrees 0.6 and 0.8, respectively.

It is easily seen from the above example that a fuzzy tree transducer can be a reasonable model to describe the fuzzy meaning of a fuzzy context-free language at the level of syntax structure in the sense that it can fuzzily associate each derivation tree of the fuzzy language with a tree representation of the computation process of its fuzzy meaning.

4.14 Exercises

1. Show that the operation of concatenation of fuzzy languages is associative.

2. Show that the definition of a type 2 grammar G in this chapter (with $c = 1$) is equivalent to the definition of a type 2 grammar G' of Definition 1.8.7 in that, given G, there exists a G' such that $L(G') = L(G)$ and vice versa.

3. Consider the grammar $G = (N, T, P, S)$, where $N = (S, A, B)$, $T = \{a, b\}$, and $P = \{S \rightarrow ab, S \rightarrow aB, A \rightarrow bAA, A \rightarrow aS, A \rightarrow a, B \rightarrow aBB, B \rightarrow bS, B \rightarrow b\}$. Show that

$$
\begin{array}{llll}
S \rightarrow C_b A, & S \rightarrow C_a B, & A \rightarrow C_a S, & A \rightarrow C_b D_1, \\
A \rightarrow a, & B \rightarrow C_b S, & B \rightarrow C_a D_2, & B \rightarrow b, \\
D_1 \rightarrow AA, & D_2 \rightarrow BB, & C_a \rightarrow a, & C_a \rightarrow b,
\end{array}
$$

yields a grammar in Chomsky normal form that is equivalent to G, [96, Example 4.9, p. 93].

4. Consider the grammar G of Example 4.4.2, where $P = \{s_0 \xrightarrow{.9} aAC, C \xrightarrow{.8} AD, A \xrightarrow{.7} B, A \xrightarrow{.6} a, B \xrightarrow{.5} b, D \xrightarrow{.4} b\}$. Show that $\lambda_G(aaab) = (.9)(.8)(.6)(.6)(.4)$.

5. Use induction on $|x|$ to finish the proof of Theorem 4.4.4.

6. Show that Theorems 4.4.11 and 4.4.12 do not hold without the assumption λ is finitary.

7. Prove that Theorems 4.5.4 and 4.5.5 hold if $>$ is replaced by $\geq$.

8. Use Theorem 4.5.13 and the proof of Theorem 4.5.11 to show that $\mathcal{L}$ is not closed under intersection.

9. Show that $\mathcal{L}$ is closed under union, intersection with regular sets, concatenation with CFL, substitution by CFL, homomorphism, inverse homomorphism, reversal, and a-transducer.

10. Prove for an F-TA that results corresponding to Lemma 4.10.4 and Theorems 4.10.5 and 4.10.6 hold.

Chapter 5

Probabilistic Automata and Grammars

5.1 Probabilistic Automata and Their Approximation

The presentation in Sections 5.1–5.7 is essentially from [166]. We consider probabilistic automata and their approximation by nonprobabilistic automata.

Let X be a finite alphabet and let X^* be the set of all words over X. Suppose that a preassigned function $f : X^* \to [0,1]$ is given. We consider the construction of a physical device (black box) that is capable of reading words fed into it and such that after a word x is read, an output $q(x)$ is produced such that the following properties hold $\forall x \in X^*$:

(1) $q(x)$ provides full information as to the exact value $f(x)$;

(2) $q(x)$ provides full information as to whether $f(x) > c$ or $f(x) \leq c$ for a given real number c, $0 \leq c < 1$;

(3) $q(x)$ provides enough information to approximate $f(x)$ within any preassigned ϵ;

(4) $q(x)$ provides enough information to approximate within any preassigned ϵ whether $f(x) > c$ or $f(x) \leq c$ for a given c, $0 \leq c < 1$.

The solution to the above problems has potential applications in areas such as pattern recognition, communication of information, and analysis of noise processes in sequential networks. The study of functions f of the above form, but over general spaces (not necessarily X^*), has been introduced in [255] with a similar purpose. Functions over X^*, which may result from noise processes, have been studied in [21] while the study of special kinds of functions of the above type, namely, functions induced by probabilistic automata, have been studied in [165, 180].

The opinion is expressed in [166] that the solution to problems (1) and

(2) above involves requirements that are too stringent. The following reasons are given.

In connection with the second problem, it was proved in [180, Theorem 2] that even the set of functions f over X^* induced by probabilistic automata is an uncountable set. However, the set of Turing machines (which is the most powerful set of sequential devices having a finite number of internal states) is a countable set. It was also shown in [180] that if a function induced by a probabilistic machine is such that there is a number $d > 0$ with $|f(x) - c| > d$ for all $x \in X^*$, then f can be realized by a finite state automaton, but the above condition (that $|f(x) - c| > d$) seems to hold for only degenerate cases. In addition to the above-mentioned theorem in [180], there are other theorems on probabilistic automata that seem to imply that the nature of the cutpoint c and its relation to the range of the function $f(x)$ is related to the equivalence or nonequivalence of probabilistic automata to nonprobabilistic automata. The following are such theorems.

Theorem 5.1.1 (165) *There exists a probabilistic automata A such that the set of words $T(A, c)$ (e.g., the set of all words x such that $f(x) > c$, where $f(x)$ is the function induced by A) is regular if and only if c is a rational number.* ∎

Theorem 5.1.2 (165) *There exists a probabilistic automata A such that the set of words $T(A, c)$, with a given rational number c, is not a regular set.* ∎

Theorem 5.1.3 (186) *For any probabilistic automata A with a single input letter, there are only finitely numbers c such that $T(A, c)$ is not regular.* ∎

In view of the above discussion, we conclude that a new and weaker criterion for comparison between probabilistic automata and deterministic automata is needed in order to overcome the cardinality gap between the two kind of automata and that will loosen the stringent quality of the cutpoint c.

Consequently, we consider only problems (3) and (4). We introduce the concept of ϵ-approximation of functions from X^* into the interval $[0, 1]$ by sequential machines. Problems (1) and (2) have been considered in [180], [165], [164].

We give a characterization of functions for which Problems (3) and (4) are solvable by using finite automata as the approximating device. We also show that Problems (3) and (4) are not equivalent.

We show that a noncountable class of functions can be approximated by finite automata whose realization is given explicitly.

Functions induced by probabilistic automata are shown to have a property that is necessary, but not sufficient for the existence of a solution to Problems (3) and (4).

We show that there are functions induced by probabilistic automata that cannot be approximated by finite automata in the sense of Problems (3) and (4). Hence the class or probabilistic automata is stronger than the class of nonprobabilistic automata because of the intrinsic nature of the probabilistic automata and not because of the actual properties of the cutpoint c or of the relation of c to the range of the function $f(x)$.

We begin with some definitions. Let $x, y \in X^*$. Then y is a k-**suffix** (**prefix**) of x if $x = uy$ ($x = yu$) for some (possible empty) word u and $|y| = k$.

Definition 5.1.4 *A function from X^* into $[0,1]$ is called a **fuzzy star function (fsf)**.*

Definition 5.1.5 *A **fuzzy star acceptor (fsac)** is a pair (ψ, c), where ψ is a fsf and $0 \leq c < 1$.*

Definition 5.1.6 *If (ψ, c) is an fsac, then the set of tapes accepted or defined by (ψ, c), written $T((\psi, c))$, is defined as follows:*

$$T((\psi, c)) = \{x \mid x \in X^*, \ \psi(x) > c\}.$$

We now define and make distinctions between probabilistic, deterministic, automata acceptors, and machines.

Definition 5.1.7 *A **finite automaton (fa)** over the alphabet X is a triple $A = (Q, \mu, q_0)$, where Q is a finite nonempty set (the internal states of A), q_0 is an element of Q (the initial state), and μ is a function from $Q \times X$ into Q (the transition function).*

Define the extension $\mu^* : Q \times X^* \to Q$ of μ as follows: $\forall q \in Q$, $\mu^*(q, \Lambda) = q$ and $\forall a \in X$, $x \in X^*$, $\mu^*(q, ax) = \mu^*(\mu(q, a), x)$. For the purpose of distinguishing an fa $A = (Q, \mu, q_0)$, we use the notation $A(q_j, x) = \mu(q_j, x)$.

Definition 5.1.8 *A **finite acceptor (fac)** is an fa together with a subset F of Q (the set of final states). The set of words **defined** or **accepted** by an fac A is defined to be the set $T(A) = \{x \mid A(q_0, x) \in F\}$.*

Definition 5.1.9 *If a set of words U is equal to $T(A)$ for some fac A, then U is called a **regular set**.*

Other nonprobabilistic devices of a more complex structure such as push down automata and linear bounded automata, [77], can be defined in the same form. When referring to machines rather than automata, we refer to devices that, at each instant of time, receive an input and yield an output.

Recall that an $n \times n$ matrix whose entries are nonnegative real numbers is referred to as a **stochastic matrix** if the sum of the elements in each of its rows is equal to one.

We let P_n denote the set of all n-dimensional probabilistic vectors, i.e., vectors whose entries are nonnegative real numbers that add 1.

Definition 5.1.10 *A **probabilistic automaton** (**pa**) is a 4-tuple $A = (Q, \pi, \{A(u)\}, F)$, where Q is a finite set (the set of states of A), $\pi \in P_n$, where $|Q| = n$, $\{A(u)\}$ is a set of $|X|$ stochastic matrices for each $u \in X$ such that $A(u) = [a_{ij}(u)]$, where $a_{ij} : X \rightarrow [0,1]$, and F is a subset of Q (the set of final states).*

In Definition 5.1.10, π represents the "initial distribution" of A, and $a_{ij}(u)$ is the probability that the automaton will enter the state q_j starting from state q_i after scanning the symbol u of X.

$A(x) = [a_{ij}(x)]$ denotes the matrix $A(x_1) \ldots A(x_k)$, where $x = x_1 \ldots x_k$, $x_i \in X$, $i = 1, 2, \ldots, n$, and $\pi(x)$ denotes the vector $\pi A(x)$. It follows that $\pi(xy) = \pi(x)A(y)$ for $x, y \in X^*$.

It follows that $a_{ij}(x)$ is the probability that the automaton will enter the state q_j starting from state q_i after scanning the word x, and $\pi(x)$ is the final distribution over the states, starting with initial distribution π and after scanning the word x.

It is assumed throughout that the values $a_{ij}(u)$ are computable.

Example 5.1.11 *Let $X = \{a, b\}$, $Q = \{q_1, q_2\}$, $\pi = \left[\begin{array}{cc} \frac{1}{4} & \frac{3}{4} \end{array}\right]$, and*

$$A(a) = \begin{bmatrix} \frac{1}{2} & \frac{1}{2} \\ \frac{1}{3} & \frac{2}{3} \end{bmatrix}, \quad A(b) = \begin{bmatrix} \frac{2}{3} & \frac{1}{3} \\ \frac{1}{2} & \frac{1}{2} \end{bmatrix}.$$

Then $\pi(a) = \pi A(a) = \left[\begin{array}{cc} \frac{3}{8} & \frac{5}{8} \end{array}\right]$ and $\pi(b) = \pi A(b) = \left[\begin{array}{cc} \frac{13}{24} & \frac{11}{24} \end{array}\right]$. Also, $\pi(ab) = \pi(a)A(b) = \pi A(a)A(b) = \left[\begin{array}{cc} \frac{9}{16} & \frac{7}{16} \end{array}\right].$

Let η^F be an n-dimensional column vector $\eta^F = (\eta_i^F)$ defined as follows:

$$\eta_i^F = \begin{cases} 1 & \text{if } q_i \in F \\ 0 & \text{otherwise.} \end{cases}$$

Then $\eta^F(x) = A(x)\eta^F$ denotes a column vector whose ith entry is the probability of entering a state in F when beginning in state q_i and after scanning the word x.

Thus $p(x)$, the probability of entering a state in F when beginning with initial distribution π and after scanning the word x, is given by

$$p(x) = \pi A(x)\eta^F = \pi(x)\eta^F = \pi\eta^F(x).$$

It follows that

$$p(xy) = \pi(x)\eta^F(y),$$

where $x, y \in X^*$.

Let $p(A, x)$ denote the value $p(x)$ above, for a word x as related to a given pa A.

Example 5.1.12 *Let X, Q, and π be defined as in Example 5.1.11. Let $F = \{q_2\}$. Then $\eta^F = \begin{bmatrix} 0 \\ 1 \end{bmatrix}$. Also*

$$
\begin{aligned}
\eta^F(ab) &= A(ab)\eta^F \\
&= A(a)A(b)\eta^F \\
&= \begin{bmatrix} \frac{7}{12} & \frac{5}{12} \\ \frac{5}{9} & \frac{4}{9} \end{bmatrix} \begin{bmatrix} 0 \\ 1 \end{bmatrix} \\
&= \begin{bmatrix} \frac{5}{12} \\ \frac{4}{9} \end{bmatrix}.
\end{aligned}
$$

Furthermore,

$$
p(ab) = \begin{bmatrix} \frac{1}{4} & \frac{3}{4} \end{bmatrix} \begin{bmatrix} \frac{5}{12} \\ \frac{4}{9} \end{bmatrix} = \frac{7}{16}.
$$

Definition 5.1.13 *A **probabilistic acceptor** (pac) is a pair (A, c), where A is a pa and $0 \le c < 1$. The set of words **defined** or **accepted** by a pac (A, c) is the set*

$$
T(A, c) = \{x \mid p(A, x) > c, \ x \in X^*\}.
$$

Clearly, a pac gives rise to an fsac.

Definition 5.1.14 *Two acceptors are said to be **equivalent** if they accept the same set of words.*

Definition 5.1.15 *An fa B is said to be **equivalent** to a pa A (to an fsf ψ) if there is a function ϕ from the set of states Q of B into the interval $[0, 1]$ such that for all $x \in X^*$, $\phi(B(q_0, x)) = p(A, x)$, $(\phi(B(q_0, x)) = \psi(x))$.*

If B is a fa that is equivalent to a pa A (fsf ψ), then for any cutpoint c, it is possible to transform B into an fac that is equivalent to the pac (A, c) (fsac (ψ, c)).

5.2   ε-Approximating by Nonprobability Devices

The importance of Turing machines follows from the widely held belief that any function that can be computed by a digital computer can be computed by a Turing machine. This assertion is known as Turing's hypothesis or Church's thesis. Church's thesis implies that a Turing machine is the correct abstract model of a digital computer. The formal definition of an algorithm follows from these ideas, namely, an algorithm is a Turing machine that eventually stops after given an input string.

Definition 5.2.1 *A Turing automaton over the alphabet X is a triple $A = (Q, \mu, q_0)$, where Q is a finite nonempty set (the internal states of A), $q_0 \in Q$ (the initial state), and μ is a function from $Q \times X$ into $Q \times X \times \{-1, 0, 1\}$ (the transition function).*

The operation of a Turing machine is described as follows: An infinite tape is fed into the machine. The machine begins scanning the input word on the tape, starting at the left-most input symbol in its initial state and reading a symbol at a time. If the scanned symbol is x_i and the machine is in state q_j, then it replaces x_i by another symbol x_k, changes its state into a new state q_l, and moves the tape either left or right or keeps it stationary depending on the value m in $\mu(q_j, x_i) = (q_l, x_k, m)$, where $m = 1$ means left move, $m = -1$ means right move, and $m = 0$ means no move. In particular, the input string can be changed. A Turing machine A accepts a string x if A halts in an accepting state, where x is input. It can be shown that a language L is generated by a phase-structure grammar if and only if there is a Turing machine that accepts L.

Example 5.2.2 *Consider the Turing automaton $A = (Q, \mu, q_0)$ over $X = \{x_1, x_2, x_3, x_4, \Box\}$, where $Q = \{q_1, q_2, q_3, q_4, q_5, q_6, q_7\}$ and where μ is defined as follows:*

$$\mu(q_0, \Box) = (\Box, q_4, 1) \qquad \mu(q_4, \Box) = (\Box, q_7, -1)$$
$$\mu(q_0, x_1) = (x_1, q_1, -1) \qquad \mu(q_4, x_1) = (x_1, q_4, 1)$$
$$\mu(q_0, x_2) = (x_2, q_0, -1) \qquad \mu(q_4, x_2) = (x_2, q_4, 1)$$
$$\mu(q_0, x_3) = (x_3, q_0, -1) \qquad \mu(q_4, x_3) = (x_3, q_4, 1)$$
$$\mu(q_0, x_4) = (x_4, q_0, -1) \qquad \mu(q_4, x_4) = (x_4, q_4, 1)$$
$$\mu(q_1, \Box) = (\Box, q_3, 1) \qquad \mu(q_5, \Box) = (x_2, q_6, -1)$$
$$\mu(q_1, x_1) = (x_1, q_1, -1) \qquad \mu(q_5, x_1) = (x_1, q_4, 1)$$
$$\mu(q_1, x_2) = (x_2, q_0, -1) \qquad \mu(q_5, x_2) = (x_2, q_4, 1)$$
$$\mu(q_1, x_3) = (x_3, q_0, -1) \qquad \mu(q_5, x_3) = (x_3, q_4, 1)$$
$$\mu(q_1, x_4) = (x_3, q_2, 1) \qquad \mu(q_5, x_4) = (x_4, q_4, 1)$$
$$\mu(q_2, x_1) = (x_4, q_7, 1) \qquad \mu(q_6, x_1) = (x_3, q_6, 1)$$
$$\mu(q_3, x_1) = (x_1, q_5, 1) \qquad \mu(q_6, x_2) = (x_2, q_7, 1)$$

*and for any other $(q, x) \in Q \times X$, $\mu(q, x) = (q, x, 0)$. The symbol $\Box$ is a special symbol called a **blank**.*

Let p^* and q^* be arbitrary expressions. Suppose $\mu(q_i, x_j) = (x_k, q_l, -1)$. Then expressions of the form $p^* q_i x_j s q^*$ are transformed into expressions of the form $p^* x_k q_l s q^*$. Expressions of the form $p^* q_i x_j$ are transformed into expressions of the form $p^* x_k q_l \Box$. Other expressions are not transformed.

Suppose $\mu(q_i, x_j) = (x_k, q_l, 1)$. *Then expressions of the form* $p^*x'q_ix_jq^*$ *are transformed into expressions of the form* $p^*q_lx'x_kq^*$. *Expressions of the form* $q_ix_jq^*$ *are transformed into expressions of the form* $q_l\square x_kq^*$. *Other expressions are not transformed.*

Note that in the definition of μ, *when* $\mu(q, x) \neq (q, x, 0)$, q *is never* q_7. *Hence* q_7 *is, in effect, a "stop" state. Consider the expression* $q_0x_2x_3x_1x_4x_3$. *The machine will transform this successively into*

$$x_2q_0x_3x_1x_4x_3, \quad x_2x_3q_0x_1x_4x_3, \quad x_2x_3x_1q_1x_4x_3,$$

$$x_2x_3q_2x_1x_3x_3, \quad \text{and} \quad x_2q_7x_3x_4x_3x_3,$$

at which point it will stop.

Definition 5.2.3 *A **linear bounded automaton** is the same as a Turing automaton, but its input tape is finite. The length of the input tape of a linear bounded automaton is a linear function of the length of the input word printed on the tape.*

It is known that a language is context sensitive just if it is recognized by some nondeterministic linear-bounded acceptor. However, it is not known whether each such language can be recognized by a deterministic linear bounded acceptor. It is worth noting that the theory of context-sensitive languages has had little impact on programming languages.

Any tape of nonprobabilistic automata having finitely many states and capable of reading tapes is called a **nonprobabilistic (np)** device.

Definition 5.2.4 *An np device B is said to ϵ-**approximate** an fsf ψ if there is a function ϕ from the states of B into the interval $[0, 1]$ such that for all $x \in X^*$,*

$$|\psi(x) - \phi(B(q_0, x))| \leq \epsilon.$$

If for given ψ and ϵ there is a B satisfying the above properties, then ψ can be ϵ-approximated.

The motivation of this criterion can be seen by the proposition and corollary below.

Definition 5.2.5 *A Turing automaton (or a linear bounded automaton) B is said to ϵ-**approximate** the fsf ψ if B when fed with a tape x performs a finite number of operations, at the end of which a number d is printed on the tape such that $|d - \psi(x)| \leq \epsilon$.*

We assume here that the symbols X are numerals so that a word $x = x_1, \ldots, x_k$ represents the number $x_1 \ldots x_k$.

Proposition 5.2.6 *Let an fsf ψ and ϵ be given. Then ψ can be ϵ-approximated by a Turing automaton (or linear bounded automaton) B if and only if ψ is $\frac{1}{2}\epsilon$-computable by B.* ∎

The proof of this proposition follows easily from the definitions.

Corollary 5.2.7 *The fsf induced by pa's can be ϵ-approximated by Turing automata.* ∎

Clearly, Turing automata, having infinite tape, can compute the fsf's induced by pa's for any tape x within any arbitrary given ϵ. However, this may not be true for linear bounded automata. The set of fsf's induced by pa's is not a countable set, but the set of Turing automata is countable.

5.3 ϵ-Approximating by Finite Automata

In this section, we characterize fsf's that can be ϵ-approximated by finite automata.

Definition 5.3.1 *Let ψ be an fsf and $\epsilon > 0$. An ϵ-**cover** induced by ψ is a finite set $\{C_i\}_{i=0}^k$, where the C_i are subsets of $[0, 1]$ satisfying the following conditions:*

(1) $\cup_{i=0}^k C_i = \{d \mid \psi(x) = d, \ x \in X^*\}$.
(2) $d_1, d_2 \in C_i \Rightarrow |d_1 - d_2| \leq \epsilon, \ i = 0, 1, 2, \ldots, k$.
(3) Let $C_i z$ for $z \in X^*$ be defined as

$$C_i z = \{d \mid \psi(xz) = d, \ \psi(x) \in C_i\}.$$

It follows that for any i and any z there exists a j such that $C_i z \subseteq C_j$.

Theorem 5.3.2 *Let ψ be an fsf and $\epsilon > 0$. Then ψ can be ϵ-approximated by a fa B if and only if there is an 2ϵ-cover induced by ψ.*

Proof. Suppose there is an ϵ-cover induced by ψ, where the fa B is as follows: The states of B are $C_0, \ldots, C_k$ (the elements of the ϵ-cover). Let C_0 be the first set such that $\psi(\Lambda) \in C_0$. Then the initial state of B is C_0. Define the transition function of B by the relation

$$B(C_i, u) = C_j \text{ if } C_i u \subseteq C_j,$$

where j is the smallest index satisfying this relation.
    Set

$$\phi(C_i) = \frac{1}{2}[\vee_{d \in C_i} d + \wedge_{d \in C_i} d].$$

We prove by induction that for any $x \in X^*$, $\psi(x) \in B(q_0, x)$. For $x = \Lambda$, the statement follows from the definition of B. Assume that $\psi(x) \in B(q_0, x) = C_i$. Then $\psi(xu) \in C_i u = B(q_0, xu)$ by the definitions of $C_i u$ and B. Thus the statement holds.

Hence by Definition 5.3.1(2), we have for any $x \in X^*$ that

$$|\psi(x) - \phi(B(q_0, x))| = |\psi(x) - \frac{1}{2}[\vee_{d \in C_i} d + \wedge_{d \in C_i} d]| \leq \epsilon$$

since $\psi(x) \in C_i$. Suppose that ψ can be ϵ-approximated by an npa B. Let the states of B be $q_0, q_1, \ldots, q_k$ and define the sets

$$C_i = \{d \mid \psi(x) = d, \ B(q_0, x) = q_i\},$$

$i = 0, 1, \ldots, k$. It follows that the set $\{C_i \mid i = 0, 1, \ldots, k\}$ is a 2ϵ-cover. ∎

5.4 Applications

Definition 5.4.1 *An fsf ψ is said to be **quasi-definite** if for all $\epsilon > 0$ there is $k(\epsilon)$ such that for any x with $|x| \geq k(\epsilon)$ the inequality $|\psi(x) - \psi(y)| \leq \epsilon$ holds, where y is the $k(\epsilon)$-suffix of x.*

Quasi-definite pa's have been introduced in [165] by a similar definition. In that paper, a decision procedure was given for determining whether a given pa is quasi-definite. A theorem similar to the next theorem has been proved in that paper for pac's.

Theorem 5.4.2 *Let $\epsilon > 0$. Any quasi-definite fsf ψ can be ϵ-approximated by an fa.*

Proof. Given ψ and ϵ, we define the following ϵ-cover induced by ψ.
Let $y_1, y_2, \ldots, y_n$ be all the words such that $|y_i| < k(\frac{1}{2}\epsilon)$ where $k(\epsilon)$ is as in Definition 5.4.1, $i = 1, 2, \ldots, n$. Let $z_1, z_2, \ldots, z_m$ be all the words such that $|z_i| = k(\frac{1}{2}\epsilon)$, $i = 1, 2, \ldots, m$. Define the sets C_i as follows:

$$\begin{array}{llll} C_i & = & \{d \mid \psi(y_i) = d\} & i = 1, 2, \ldots, n, \\ C_{t+i} & = & \{d \mid \psi(xz_i) = d, \ x \in X^*\} & i = 1, 2, \ldots, m. \end{array}$$

Clearly,

$$\cup_{i=1}^{n+m} C_i = \{d \mid \psi(x) = d, \ x \in X^*\}.$$

If $\psi(u)$ and $\psi(v)$ are in the same set C_{n+i} (the sets C_i for $i \leq n$ are one-sets and hence are out of consideration), then $u = u_1 z_i$ and $v = v_1 z_i$. Thus

$$|\psi(u_1 z_i) - \psi(z_i)| \leq \frac{1}{2}\epsilon \quad \text{and} \quad |\psi(v_1 z_i) - \psi(z_i)| \leq \frac{1}{2}\epsilon$$

since ψ is quasi-definite. Hence

$$|\psi(u) - \psi(v)| = |\psi(u_1 z_i) - \psi(v_1 z_i)| \leq \epsilon.$$

Finally, it follows from the definitions that

$$
\begin{aligned}
C_{n+i}w &= \{d \mid \psi(uw) = d,\ \psi(u) \in C_{n+i}\} \\
&= \{d \mid \psi(xz_iw) = d,\ x \in X^*\} \\
&= \{d \mid \psi(xw_1z_j) = d,\ x \in X^*\} \\
&\subseteq C_{n+j},
\end{aligned}
$$

where z_j is the $k(\frac{1}{2}\epsilon)$ suffix of z_iw.

The set $\{C_{n+i}\}_{i=1}^{n+m}$ is thus an ϵ-cover. Hence by Theorem 5.3.2, the desired result follows. ∎

A definite fa can be defined as an automaton B such that there is an integer k with the property that $B(q_0, xy) = B(q_0, y)$ for all y with $|y| \geq k$ and for all x. Then from the proof of Theorem 5.4.2, it follows that if A is a quasi-definite automaton, then A can be ϵ-approximated by a definite automaton B. The converse is also true. The following example shows that there are fsf's that can be ϵ-approximated by npa's, but are not quasi-definite.

Example 5.4.3 *Consider the fsf ψ induced by the pa A defined as follows:*

$$
X = \{a, b\}, \quad Q = \{q_0, q_1\},
$$

the initial vector is $\pi = \begin{bmatrix} 0 & 1 \end{bmatrix}$, the final vector is $\eta = \begin{bmatrix} 1 \\ 0 \end{bmatrix}$, and the transition matrices are

$$
P(a) = \begin{bmatrix} 1 & 0 \\ 0 & 1 \end{bmatrix} \quad and \quad P(b) = \begin{bmatrix} \frac{1}{2} & \frac{1}{2} \\ 1 & 0 \end{bmatrix}.
$$

Then $\psi(x)$ is the $(1,1)$ entry in the matrix $P(x)$. Clearly, ψ is not quasi-definite, for $\psi(x) = \psi(b^k)$, where k is the number of b's in the word x (e.g., $\psi(ba^n) = \psi(b)$ for any n). Therefore, for any n, $|\psi(ba^n) - \psi(a^n)| = \frac{1}{2}$ contradicting Definition 5.4.1.

To show that ψ can be ϵ-approximated for any ϵ, consider the following cover induced by ψ.

Divide the interval $[0,1]$ by $k-1$ points $d_1, d_2, \ldots, d_{k-1}$ ($d_0 = 0$, $d_k = 1$) so that $d_i - d_{i-1} = \frac{1}{2}\epsilon$, $i = 1, 2, \ldots, k-1$ and $1 - d_{k-1} \leq \frac{1}{2}\epsilon$.

Define the cover $\{C_i\}_{i=0}^{k-2}$ as follows:

$$
C_i = \{d \mid d_i \leq d \leq d_{i+2}\} \cap \{d \mid \psi(x) = d,\ x \in X^*\},
$$

$i = 1, 2, \ldots, k-2$. Then (1) and (2) of Definition 5.3.1 clearly hold. We now show that (3) of Definition 5.3.1 holds. Let $\psi(x), \psi(y) \in C_i$. Then $\psi(xa), \psi(ya) \in C_i$. If

$$
\pi(x) = (u, 1-u) \quad and \quad \pi(y) = (v, 1-v),
$$

then

$$u - v \leq \epsilon,$$

for $u = \psi(x)$ and $v = \psi(y)$ with $\psi(x), \psi(y) \in C_i$, and

$$\begin{array}{rcl}
\pi(xb) & = & (\frac{1}{2}u + 1 - u, \frac{1}{2}u) = (1 - \frac{1}{2}u, \frac{1}{2}u), \\
\pi(yb) & = & (\frac{1}{2}v + 1 - v, \frac{1}{2}v) = (1 - \frac{1}{2}v, \frac{1}{2}v).
\end{array}$$

Hence

$$\begin{array}{rcl}
|\psi(xb) - \psi(yb)| & = & \left|(1 - \frac{1}{2}u) - (1 - \frac{1}{2}v)\right| \\
& = & \left|\frac{1}{2}(v - u)\right| \\
& \leq & \frac{1}{2}\epsilon.
\end{array}$$

By induction, it follows that for all z, either

$$\psi(xz) = \psi(x) \text{ and } \psi(yz) = \psi(y)$$

or

$$|\psi(xz) - \psi(yz)| \leq \frac{1}{2}\epsilon$$

provided that

$$|\psi(x) - \psi(y)| \leq \epsilon.$$

Thus for any z and i, either $C_i z = C_i$ or the maximal distance between any two points in $C_i z$ is not greater than $\frac{1}{2}\epsilon$. Now let c and d be two points in C_i such that $c < d$. Suppose that $d_j \leq c \leq d_{j+1}$. Then since $|c - d| \leq \frac{1}{2}\epsilon$, we have that $d_j \leq d \leq d_{j+2}$ and, therefore, c and d are both in C_j. Hence $C_i z \subseteq C_j$ for some j. Thus the function ψ can be ϵ-approximated.

5.5   The $\overline{P}_\epsilon$ Relation

Definition 5.5.1 *Let ψ be an fsf and $\epsilon > 0$. A P_ϵ-relation induced by ψ is a relation over X^* satisfying the following properties:*

(1) P_ϵ is symmetric and reflexive;

(2) P_ϵ is right invariant, i.e., $xP_\epsilon y \Rightarrow xzP_\epsilon yz$ for $x, y, z \in X^*$;

(3) $xP_\epsilon y \Rightarrow |\psi(x) - \psi(y)| \leq \epsilon$.

Definition 5.5.2 *A relation over a set X is said to be of **finite index** if there is an integer k such that in any subset of $k + 1$ elements of X there are at least two elements that are related.*

The following lemma is easily proved.

Lemma 5.5.3 *Given* $\psi : X^* \to [0,1]$ *and* $\epsilon > 0$. *Define the relation* $\overline{P}_\epsilon$ *on* X^* *by* $\forall x, y \in X^*$,

$$x\overline{P}_\epsilon y \Leftrightarrow \forall z \ |\psi(xz) - \psi(yz)| \leq \epsilon.$$

Then any P_ϵ-*relation induced by* ψ *is a refinement of* $\overline{P}_\epsilon$. ∎

Lemma 5.5.4 *For given fsf* ψ *and* ϵ, *if* ψ *can be* ϵ-*approximated by an fa, then there is a* $P_{2\epsilon}$ *relation of finite index induced by* ψ.

   Proof. Consider the machine B that ϵ-approximates P. As in the proof of Theorem 5.3.2, let $\{C_i\}_{i=0}^k$ be the ϵ-cover defined by B. Then $\psi(x)$ and $\psi(y)$ are in the same set C_i if $B(q_0, x) = B(q_0, y) = q_i$.
   Define the relation P_ϵ over X^* by

$$xP_\epsilon y \Leftrightarrow B(q_0, x) = B(q_0, y).$$

Then P_ϵ is clearly symmetric, reflexive, right-invariant, and of finite index, by definition. Condition (3) in Definition 5.5.1 is also satisfied since $xP_\epsilon y$ implies that $\psi(x)$ and $\psi(y)$ are in the same set C_i, and as in the proof of Theorem 5.3.2, this implies that

$$|\psi(x) - \psi(y)| \leq 2\epsilon.$$

∎

   Combining Lemma 5.5.3 and Lemma 5.5.4, we have the following result.

Proposition 5.5.5 *Let* ψ *be an fsf and* $\epsilon > 0$. *If* ψ *can be* ϵ-*approximated by an fa, then the explicit relation* $\overline{P}_{2\epsilon}$ *defined in Lemma 5.5.3 is of finite index.* ∎

   Consider the set

$$P_n = \{\overline{c} = (c_1, \dots, c_n), \ c_i \geq 0, \ \sum_{i=1}^{n} c_i = 1\}.$$

Let $\epsilon > 0$. Let U_ϵ be any set such that $U_\epsilon \subset P_n$ for any pair of vectors $\overline{c}$ and $\overline{d} = (d_1, \dots, d_n)$ in U_ϵ,

$$\sum_{i=1}^{n} |c_i - d_i| \geq \epsilon.$$

Lemma 5.5.6 *If* P_n *and* U_ϵ *are as above, then* U_ϵ *is a finite set containing at most* $k(\epsilon)$ *elements, where*

$$k(\epsilon) = (1 + \frac{2}{\epsilon})^{n-1} \ for \ n > 1.$$

Proof. The proof of this lemma is implicit in the proof of Theorem 3 in [180]. ■

The bound in Lemma 5.5.6 is not sharp and from a practical point of view, it would be desirable to have a sharp bound.

We can prove the following result.

Theorem 5.5.7 *Let A be a pa and $\epsilon > 0$. If $\overline{P}_\epsilon$ is induced by A, then P_ϵ is of finite index k, where*

$$k \le (1 + \frac{1}{\epsilon})^{n-1}$$

and n here is the number of states of A.

Proof. Let $x_1, \ldots, x_k$ be tapes that are pairwise nonrelatives by $\overline{P}_\epsilon$. Then for all $1 \le i < j \le k$, there is a tape y such that

$$|p(x_i y) - p(x_j y)| \ge \epsilon \tag{5.1}$$

or

$$|\pi(x_i)\eta(y) - \pi(x_j)\eta(y)| = |(\pi(x_i) - \pi(x_j))\eta(y)| \ge \epsilon.$$

Now

$$\sum_{t=1}^{n}(\pi_t(x_i) - \pi_t(x_j)) = \sum_{t=1}^{n} \pi_t(x_i) - \sum_{t=1}^{n} \pi_t(x_j) = 1 - 1 = 0.$$

Therefore,

$$X_t^+(\pi_t(x_i) - \pi_t(x_j)) = -X_t^-(\pi_t(x_i) - \pi_t(x_j)), \tag{5.2}$$

where X_t^+ and X_t^- are summations over indices t for which $\pi_t(x_i) - \pi_t(x_j)$ is nonnegative or negative, respectively.

Combining (5.1) and (5.2), we have that

$$
\begin{aligned}
\epsilon &\le \left|\sum_t(\pi_t(x_i) - \pi_t(x_j))\eta_t(y)\right| \\
&\le \left|\sum\nolimits^+(\pi_t(x_i) - \pi_t(x_j)) \vee_t \eta_t(y) + \right. \\
&\qquad \left. \sum\nolimits^-(\pi_t(x_i) - \pi_t(x_j)) \wedge_t \eta_t(y)\right| \\
&= \sum\nolimits^+(\pi_t(x_i) - \pi_t(x_j))(\vee_t\eta_t(y) - \wedge_t\eta_t(y)) \\
&\le \sum\nolimits^+ \pi_t(x_i) - \pi_t(x_j).
\end{aligned}
$$

Applying (5.2), we have

$$2\epsilon \le \sum_{t=1}^{n}|\pi_t(x_i) - \pi_t(x_j)|$$

and this inequality implies by Lemma 5.5.6 that the set of vectors

$$\pi(x_1), \pi(x_2), \ldots, \pi(x_k)$$

is finite with

$$k \le (1 + \frac{1}{\epsilon})^{n-1}.$$

■

5.6 Fuzzy Stars Acceptors and Probabilistic Acceptors

Definition 5.6.1 *Let A be an fsac, B an fac, and $T(A)$ and $T(B)$ the languages defined by A and B, respectively. Let $\epsilon > 0$. Then $T(B)$ ϵ-approxi-mates $T(A)$ if*

$$(T(B)\backslash T(A)) \cup (\overline{T(B)}\backslash\overline{T(A)}) \subseteq \{x \mid x \in X^*,\ |\psi(x) - c| \leq \epsilon\},$$

where $\overline{T(A)}$ and $\overline{T(B)}$ are the complements of $T(A)$ and $T(B)$, respectively, in X^*.

Proposition 5.6.2 *Let ψ be an fsf and B an fa. If B ϵ-approximates ψ, then for any c, B can be transformed into an fac that ϵ-approximates the fsac (ψ, c).*

Proof. Let the final states of B be the states such that $\phi(q_i) > c$. If $x \in T(B)$, then $B(q_0, x) = q_i$ with $\phi(q_i) > c$ and since

$$|\psi(x) - \phi(q_i)| \leq \epsilon,$$

it follows that

$$\psi(x) > c - \epsilon.$$

If $x \notin T(B)$, then $B(q_0, x) = q_j$ with $\phi(q_j) \leq c$ and thus $\psi(x) \leq c + \epsilon$. Now $x \in T(A)$ implies that $\psi(x) > c$ and $x \in T(B)$ implies that $\psi(x) \leq c$. Hence the desired result follows. ∎

The following proposition is a converse of Proposition 5.6.2 and is related to Theorem 5 in [180].

Proposition 5.6.3 *Let ψ be an fsf such that for any c, (ψ, c) is an fsac that can be ϵ-approximated by some fac B. Then there is an fa that 2ϵ-approximates ψ.*

Proof. Divide the interval $[0, 1]$ into k parts $c_1, c_2, \ldots, c_{k-1}$ ($c_0 = 0$, $c_k = 1$) so that $c_i - c_{i-1} \leq \epsilon$, $i = 1, 2, \ldots, k$ and let B_{c_i}, $i = 1, 2, \ldots, k-1$ be the corresponding ϵ-approximating acceptors for (ψ, c_i). Define the machine B as follows: $B = (Q, q_0, \mu)$, where

$$Q = \{(q_{i_1}(c_0),\ q_{i_2}(c_1), \ldots, q_{i_k}(c_{k-1})) \mid q_{i_k}(c_j) \in Q_{c_j}\},$$

$$q_0 = (q_0(c_0),\ q_0(c_1), \ldots, q_0(c_{k-1})),$$

$$\mu((q_{i_1}(c_0),\ q_{i_2}(c_1), \ldots, q_{i_k}(c_{k-1})), a) = (\mu_{c_0}(q_{i_1}, a),\ \mu_{c_1}(q_{i_2}, a), \ldots, \mu_{c_{k-1}}(q_{i_k}, a)),$$

with

$$B_{c_i} = (Q_{c_i}, q_0(c_i), \mu_{c_i}, F_{c_i}).$$

Set

$$\phi(q) = \phi(q_{i_1}(c_0),\ q_{i_2}(c_1), \dots, q_{i_k}(c_{k-1})) = \vee_j\{c_j \mid q_{i_{j-1}}(c_j) \in F_{c_j}\}$$

so that

$$\phi(B(q_0, x)) = c_j \Rightarrow x \in T(B_{c_j}) \text{ and } x \notin T(B_{c_{j+1}}).$$

Then

$$\phi(B(q_0, x)) = c_j$$

implies that

$$\psi(x) > c_j - \epsilon$$

and

$$\psi(x) \le c_{j+1} + \epsilon \le c_j + 3\epsilon.$$

Thus

$$|\phi(B(q_0, x)) - \psi(x)| \le 2\epsilon.$$

∎

It follows from Propositions 5.6.2 and 5.6.3 that ϵ-approximation of an fsf by an fa is possible if and only if ϵ-approximation of the fsac derived from that function, with any given c, is possible.

We shall show, however, by an example that there are fsf's that cannot be ϵ-approximated by fa's, but the derived fsac with some c can be ϵ-approximated by an fac.

Corollary 5.6.4 *The pac's are ϵ-approximable by Turing acceptors.*

Proof. The proof follows from Corollary 5.2.7 and Proposition 5.6.2.

∎

It was shown in [180, Theorem 2] and [165, Theorem 12] that the set of pac's is not a countable set even if the transition matrices have only rational entries. However, the set of Turing machines is countable.

5.7 Characterizations and the $\overline{R}_\epsilon$-Relation

Definition 5.7.1 *Let (ψ, c) be an fsac and $\epsilon > 0$. An ϵ-**cover** induced by (ψ, c) is a finite set $\{C_i \mid i = 0, \dots, k\}$, where the C_i are subsets of $[0, 1]$ satisfying the following conditions:*

(1) $\cup_{i=0}^{k} C_i = \{d \mid \psi(x) = d, \ x \in X^*\}$;
(2) either

$$C_i \subseteq \{d \mid d \geq c - \epsilon\}$$

or

$$C_i \subseteq \{d \mid d \leq c + \epsilon\},$$

$i = 0, 1, \ldots, k$;
(3) for any i and z there exists a j such that $C_i z \subseteq C_j$, where $C_i z$ is defined as in Definition 5.3.1.

Theorem 5.7.2 *Let (ψ, c) be an fsac and $\epsilon > 0$. Then (ψ, c) can be ϵ-approximated by an fac if and only if there is an ϵ-cover induced by the fsac (ψ, c). If there exists an ϵ-cover induced by ψ, then there exists an ϵ-cover induced by (ψ, c).* ■

The proof of the first assertion of Theorem 5.7.2 is similar to the proof of Theorem 5.3.2 and is omitted. The machine B here would be defined as in Theorem 5.3.2 and the final states of B would be those C_i that satisfy the relation

$$C_i \subseteq \{d \mid d \geq c - \epsilon\}.$$

Clearly, any ϵ-cover satisfying the conditions of Definition 5.3.1 also satisfies the conditions of Definition 5.7.1.

Definition 5.7.3 *The relation $\overline{R}_\epsilon$ on X^* induced by an fsac (ψ, c) is defined as follows:* $\forall x, y \in X^*$,

$$x\overline{R}_\epsilon y \ \Leftrightarrow \forall z([|\psi(xz) - c| > \epsilon \text{ and } |\psi(yz) - c| > \epsilon]$$
$$\Rightarrow [\psi(xz) > c \Leftrightarrow \psi(yz) > c]).$$

It follows that the relation $\overline{P}_\epsilon$ defined in Lemma 5.5.3 is a refinement of the relation $\overline{R}_\epsilon$ here. Thus $\overline{P}_\epsilon$ of finite index implies that $\overline{R}_\epsilon$ is of finite index. The following result thus holds.

Corollary 5.7.4 *The relation $\overline{R}_\epsilon$ induced by a pac is of finite index.*

Proof. The proof follows from Theorem 5.5.7 and the above remark.
■

Corollary 5.7.5 *If the relation $\overline{R}_\epsilon$ is induced by an fsac such that the corresponding fsf can be ϵ-approximated, then $\overline{R}_{2\epsilon}$ is of finite index.*

Proof. The proof follows from Proposition 5.5.5 and the above remark.
■

The next result follows by using the same kind of reasoning as that used in the proof of Proposition 5.5.5.

Proposition 5.7.6 *Let (ψ, c) be an fsac and $\epsilon > 0$. If (ψ, c) can be ϵ-approximated by an fac, then the induced relation $\overline{R}_\epsilon$ is of finite index.* ∎

Definition 5.7.7 *An fsac (ψ, c) is called **quasi-definite** if for any $\epsilon > 0$ there is $k(\epsilon)$ such that for all x with $|x| \geq k(\epsilon)$, the following holds for all $y \in X^*$:*

$$\psi(x) > c \; (\psi(x) \in T(\psi, c)) \; \Rightarrow \psi(yx) > c - \epsilon;$$

$$\psi(x) \leq c \; (\psi(x) \notin T(\psi, c)) \; \Rightarrow \psi(yx) \leq c + \epsilon.$$

Clearly, if ψ is a quasi-definite fsf, then (ψ, c) is a quasi-definite fsac. The converse is, however, not true. For example, suppose that the matrices of a pa are

$$A(a) = \begin{bmatrix} 1 & 0 \\ 0 & 1 \end{bmatrix} \quad \text{and} \quad A(b) = \begin{bmatrix} \frac{3}{4} & \frac{1}{4} \\ \frac{1}{4} & \frac{3}{4} \end{bmatrix}$$

and $\psi(x)$ is the $(1,1)$ entry in $A(x)$. Then with $c = \frac{1}{2}$, $T((\psi, c)) = X^*$. Therefore, (ψ, c) is a quasi-definite acceptor. However, for any n,

$$|\psi(ba^n) - \psi(a^n)| = \frac{1}{4}.$$

Hence ψ is not quasi-definite.

We thus have the following corollary.

Corollary 5.7.8 *If ψ is a quasi-definite fsf, then for any c and ϵ, then the fsac (ψ, c) can be ϵ-approximated and the approximating acceptor may be chosen to be a definite acceptor.*

Proof. The proof follows by the remarks following Theorem 5.4.2 and the above remark. ∎

Corollary 5.7.8 follows, also, from Theorem 5.7.9 below, which can be proved in the same manner as Theorem 5.4.2.

Theorem 5.7.9 *Let (ψ, c) be a quasi-definite acceptor and $\epsilon > 0$. Then (ψ, c) can be ϵ-approximated by a definite finite acceptor.* ∎

In [180, Theorem 2] and [165, Theorem 12] the fsac's are quasi-definite so that the set of quasi-definite acceptors is not a countable set.

We now present a counterexample based on an example of H. Kesten and on an idea of R.E. Stearns (private communication to A. Paz).

Consider the following pa $A = (Q, \pi, \{A(\sigma)\}, F)$ over $X = \{0, 1\}$ with $Q = \{q_0, q_1, q_2, q_3\}$, $\pi = \begin{bmatrix} 1 & 0 & 0 & 0 \end{bmatrix}$,

$$\eta^F = \begin{bmatrix} 1 \\ 0 \\ 0 \\ 0 \end{bmatrix}, \; A(0) = \begin{bmatrix} \frac{1}{2} & 0 & \frac{1}{2} & 0 \\ 0 & \frac{1}{2} & 0 & \frac{1}{2} \\ 0 & 0 & 1 & 0 \\ 0 & 0 & 0 & 1 \end{bmatrix}, \; \text{and } A(1) = \begin{bmatrix} \frac{1}{2} & \frac{1}{2} & 0 & 0 \\ \frac{1}{2} & \frac{1}{2} & 0 & 0 \\ 1 & 0 & 0 & 0 \\ 0 & 1 & 0 & 0 \end{bmatrix}.$$

By straightforward computations, it can be shown that the following relations hold:

$$p(x) = \begin{cases} (\frac{1}{2})^n & \text{if } x = 0^n & n = 0, 1, \ldots, 0^0 = \Lambda \\ = \frac{1}{2} & \text{if } x = 0^{n_1} 10^{n_2} 1 \ldots 0^{n_k} 1 & n_j \geq 0, \; j = 1, 2, \ldots, k \\ & \text{and there is } i \text{ with } n_i = 0 \\ > \frac{1}{2} & \text{if } x = 0^{n_1} 10^{n_2} 1 \ldots 0^{n_k} 1 & n_j > 0, \; j = 1, 2, \ldots, k \\ < \frac{1}{2} & \text{if } x = 0^{n_1} 10^{n_2} 1 \ldots 0^{n_k} 10^{n_{k+1}} & n_j \geq 0, \; j = 1, 2, \ldots, k \\ & & n_{k+1} > 0, \end{cases}$$

where $p(x)$ is the $(1, 1)$ entry in $A(x)$.

Consider now the fsf defined by A, P_A, and let (P_A, c) be the pac with $c = \frac{1}{2}$. Then

$$T((P_A, c)) = \{x \mid P_A(x) > \frac{1}{2}\}.$$

It follows from the above inequalities that $T((P_A, c))$ for $c = \frac{1}{2}$ is the set of tapes x such that $x = \Lambda$ or x begins with a zero, ends with a one, and contains no subtapes of two or more consecutive ones. It follows that this set of tapes is a regular set since there exists an fac accepting it. Therefore, it can be ϵ-approximated (even for $\epsilon = 0$) by an fac.

We now show that by the use of Proposition 5.6.2 there exists c such that (P_A, c) cannot be ϵ-approximated by an fac. Hence it is for that c with the result that P_A is an fsf that cannot be approximated by an npa.

Let x_n^m be the word $x_n^m = (0^n 1)^m$. It follows easily that

$$p(x_n^m) = \frac{1 + [1 - (\frac{1}{2})^n]^m}{2}.$$

Thus $\lim_{n \to \infty} p(x_n^m) = 1$ for fixed $m > 0$, while $\lim_{m \to \infty} p(x_n^m) = \frac{1}{2}$ for fixed $n > 0$. Now let c be a real number such that $\frac{1}{2} < c < 1$, say $c = \frac{3}{4}$, and let $\epsilon > 0$ be a real number such that $\epsilon < \frac{1}{4}$. Suppose that (P_A, c) can be ϵ-approximated for the given c and ϵ. Let the approximating machine have k states. Choose n_0 sufficiently large so that

$$p(x_{n_0}^m) > c + \epsilon, \; m = 1, 2, \ldots, k + 1.$$

The first $k + 1$ applications of the input sequence x must send the approximating machine B through a sequence of states $q_0, q_1, \ldots, q_{k+1}$, which are all final states of B. However, B has only k states. Hence $q_{k+1} = q_i$ for $i < k + 1$. Consequently, all the tapes of the form $x_{n_0}^m$, $m = 1, 2, \ldots$, are in $T(B)$. Thus B cannot ϵ-approximate P_A since there is an m_0 with $p(x_{n_0}^{m_0}) < c - \epsilon$, i.e.,

$$\left| p(q_{n_0}^{m_0}) - c \right| > \epsilon,$$

while $x_{n_0}^{m_0} \in T(B)$ and $x_{n_0}^{m_0} \notin T(A)$. The following are direct consequences of the above example.

(1) There is a pac that cannot be approximated by an fac.

(2) There is an fsf that cannot be approximated by an fa. This follows from the example and Proposition 5.6.2.

(3) There is a pa that cannot be approximated by an fa, but the acceptor defined by the pa with some c ($c = \frac{1}{2}$ in the example) can be approximated by a fa. Therefore, the two concepts of approximation are not equivalent.

(4) The class of pa's is stronger than the class of fa's. This is a consequence of the intrinsic nature of the probabilistic automata and not of the actual properties of the cut point.

(5) There is an fsf and $\epsilon > 0$ such that there is no ϵ-cover induced by this fsf. (See Definition 5.3.1.)

(6) There is a fsac and $\epsilon > 0$ such that there is no ϵ-cover induced by this fsac. (See Definition 5.7.1.)

We have shown that fsf's that are computable (e.g., pa's) can be approximated by Turing automata. However, as shown by the previous example, there are fsf's defined by pa's that cannot be approximated by fa's.

We pose in the Exercises some rather difficult problems concerning the ability of certain types of automata to approximate fsf's and pc's.

We have shown that Turing automata can approximate pa's, but fa's cannot. We ask in the Exercises if push-down automata can approximate pa's.

We have shown in Section 5.5 that the condition that the relation $\overline{P}_\epsilon$ defined in Lemma 5.5.3 induced by a given fsf ψ is of finite index is a necessary condition for the function ψ to be approximated by a fa. This condition is however not sufficient since, by Theorem 5.5.7, that relation is of finite index for any pa, but the previous example is a pa that cannot be approximated by a fa.

The previous example also shows that there is a pa which cannot be approximated by fa's, but the acceptor defined by that pa with some c can be approximated by a fa. The two concepts are therefore not identical and the above remarks and the problems posed in the Exercises may be stated and posed also for the case of approximating fuzzy star acceptors separately.

5.8 Probabilistic and Weighted Grammars

In the next several sections, we consider the work in [188].

The customary Chomsky hierarchy of formal languages is obtained by imposing restrictions on the form of the rewriting rules, i.e., productions. In certain grammars, an application of some production determines which productions are applicable on the next step. These are called programmed grammars. In ordered grammars, some productions can never be applied if some others are applicable. In matrix grammars, only certain previously specified strings of productions can be applied or, more generally, in a grammar with a control set, the string of productions corresponding to a derivation must belong to a set of strings previously specified. See, for example, [185, 61, 1, 80]. We consider time variant and probabilistic grammars.

In a *probabilistic grammar*, there is given together with each production a stochastic vector whose i-th component indicates the probability that the i-th production is applied after the production. Furthermore, there is given an initial probability distribution over the set of productions. Thus each derivation is assigned a probability. The language generated by a probabilistic grammar consists of all words P generated by the grammar such that the probability assigned to the derivations of P is greater than some previously chosen cutpoint. We consider the following two interpretations. In the *maximal interpretation*, it is required that there is at least one derivation of P with probability greater than the cutpoint. In the *sum interpretation*, it is required that the sum of the probabilities assigned to the distinct derivations of P is greater than the cutpoint. The latter interpretation is customary in connection with probabilistic automata.

A probabilistic grammar is a special case of a weighted grammar. In a weighted grammar, vectors with arbitrary nonnegative components are considered instead of stochastic vectors. This situation can be viewed as having a reward or punishment associated with the application of each production. Some restrictions on probabilistic as well as on weighted grammars may be imposed in order to guarantee the effectiveness of the procedures. For example, one may assume that the probabilities and the weights are rational.

Let $G = (N, T, X_o, F)$ be a phrase structure grammar, where N is the set of nonterminals, T is the set of terminals, X_0 is the initial symbol, and F is the set of productions. Derivations according to G, the language $L(G)$ generated by G, as well as type $i(i = 0, 1, 2, 3)$ grammars in the Chomsky hierarchy obtained by imposing restrictions on F, are defined in the usual fashion. Let $\{f_1, \ldots, f_k\}$ be a set of distinct labels for the productions in F. Let

$$D : X_0 = P_0 \Longrightarrow_{f_{j(1)}} P_1 \Longrightarrow_{f_{j(2)}} P_2 \Longrightarrow_{f_{j(3)}} \cdots \Longrightarrow_{f_{j(r)}} P_T \qquad (5.3)$$

be a derivation according to G, where in the transition from P_i to $P_{i+1}(0 \leq i < r)$ the production labeled by $f_{j(i+1)}$ with $1 \leq j(i+1) \leq k$ is applied. Then the word

$$f_{j(1)} f_{j(2)} \cdots f_{j(r)}$$

over the alphabet $\{f_1, \ldots, f_k\}$ is termed a **control word** of the derivation (5.3). If $r = 0$, then the control word is defined to be the empty word Λ. A derivation from X_0 determines a unique control word, provided the productions in F are distinct. However, the existence of two identical productions in F is not excluded in the following discussion.

By a language, we mean any set of words. Let C be a language over the alphabet $\{f_1, \ldots, f_k\}$. Then the language $L_C(G)$ is defined to be the subset of $L(G)$ consisting of words that possess at least one derivation whose control word is in C. $L_C(G)$ is called the **language generated by** G **with control language** C.

Example 5.8.1 *Consider the grammar*

$$G = (\{X, Y, Z\}, \{x, y, z\}, X, F),$$

where the productions of F are as follows:

$$
\begin{aligned}
f_1 &: X \longrightarrow XYZ \\
f_2 &: X \longrightarrow xX \\
f_3 &: Y \longrightarrow yY \\
f_4 &: Z \longrightarrow zZ \\
f_5 &: X \longrightarrow x \\
f_6 &: Y \longrightarrow y \\
f_7 &: Z \longrightarrow z
\end{aligned}
$$

Assume that C consists of all words of the form

$$f_1(f_2 f_3 f_3)^i f_5 f_6 f_7, \qquad i = 0, 1, 2, \ldots .$$

Then

$$L_C(G) = \{x^i y^i z^i \mid i \geq 1\}. \tag{5.4}$$

The language (5.4) is also generated by the grammar

$$G = (\{X, Z\}, \{x, y, z\}, X, F),$$

where the productions of F are as follows:

$$
\begin{aligned}
f_1 &: X \longrightarrow XZ \\
f_2 &: X \longrightarrow xXy \\
f_3 &: Z \longrightarrow zZ \\
f_4 &: X \longrightarrow xy \\
f_5 &: Z \longrightarrow z,
\end{aligned}
$$

with the control language consisting of words of the form

$$f_1(f_2 f_3)^i f_4 f_5, \qquad i = 0, 1, 2, \ldots .$$

Control languages provide a uniform way of describing grammars with restrictions on the use of productions. The notion of a control language used here differs from the notion of a control set used in [80], where only leftmost derivations were considered.

Let G be a type i grammar $(i = 0, 1, 2, 3)$ whose productions are labeled by the labels in the set $\{f_1, \ldots, f_k\}$. (See Definitions 1.8.3 and 1.8.7.) Let φ be a function of the set $\{f_1, \ldots, f_k\}$ into the set of k-dimensional row vectors with nonnegative components, and let δ be a k-dimensional row vector with nonnegative components. Then the triple (G, δ, φ) is called a **weighted grammar of type** i. If, in addition, δ as well as the values of φ are stochastic vectors, then (G, δ, φ) is a probabilistic grammar of type i. In a probabilistic grammar, δ is referred to as the initial distribution of the productions. The u-th component of the vector $\varphi(f_v)$, where $1 \leq u, v \leq k$, is referred to as the probability of applying the production labeled by f_u after applying the production labeled by f_v.

Consider a weighted grammar $G_u = (G, \delta, \varphi)$. We assign a numerical value $\psi(D)$ to each derivation (5.3), where $r > 0$. If $r = 1$ then $\psi(D)$ is defined to be the $j(1)$-th component of δ. Suppose that $\psi(D)$ has been defined for the derivation (5.3), where $r \geq 1$. Then, for the derivation

$$D_1 : X_0 = P_0 \Longrightarrow_{j(1)} P_1 \Longrightarrow_{f_{j(2)}} \cdots \Longrightarrow_{f_j(r)} P_r \Longrightarrow_{f_{j(r+1)}} P_{r+1},$$

$\psi(D_1)$ is defined to be $\varphi(D)[\varphi(f_{j(r)})]_{j(r+1)}$, where the second factor is the $j(r + 1)$-th component of the vector $\varphi(f_{j(r)})$.

Hence for a probabilistic grammar $G_p = (G, \delta, \varphi)$, the number $\psi(D)$ can be interpreted as the probability of the derivation D.

Let c be a nonnegative number. We now define two languages $L_m(G_w, c)$ and $L_s(G_w, c)$ generated by the weighted grammar $G_w = (G, \delta, \varphi)$ with cutpoint c. The former is defined to be the subset of $L(G)$ consisting of all words that possess at least one derivation D such that

$$\psi(D) > c. \tag{5.5}$$

The latter is defined to be the subset of $L(G)$ consisting of all words P such that

$$\sum_D \psi(D) > c, \tag{5.6}$$

where D ranges over all distinct derivations of P. Thus two derivations are distinct if they possess distinct control words.

Suppose that $L = L_m(G_w, c)$, for some weighted grammar $G_w = (G, \delta, \varphi)$, where G is a type i $(i = 0, 1, 2, 3)$ grammar and c is some cutpoint. Then L is said to be generated by a weighted grammar of type i under *maximal interpretation* or, merely, L is *w.i.m.* Similarly, if $L = L_s(G_w, c)$, L is said to be generated by a weighted grammar of type i under *sum interpretation* or, merely, L is *w.i.s.* Moreover, if G_w is a probabilistic grammar we say that L is *p.i.m.* or *p.i.s.*, respectively. Clearly, for a probabilistic grammar G_p, both $L_m(G_p, c)$ and $L_s(G_p, c)$ are empty whenever $c \geq 1$.

By a **rational weighted (probabilistic) grammar**, we mean a weighted (probabilistic) grammar, where the components of δ as well as the components of each value of φ are rational. If a language L is *w.i.m.* and also the corresponding weighted grammar and cutpoint are rational, then L is said to be $r.w.i.m.$ The abbreviations $r.w.i.s.$, $r.p.i.m.$ and $r.p.i.s.$ are defined similarly.

It will be seen that the sum interpretation corresponds to the interpretation customary in connection with probabilistic automata. Following the customary definition in automata theory, we have assumed a strict inequality in (5.5) and (5.6).   A more difficult problem arises in characterizing the language families if in (5.5) and (5.6) the symbol $>$ is replaced by the symbol $\geq$.

Example 5.8.2 *Consider the type 2 probabilistic grammar G_p with nonterminals X and Z, terminals x, y, and z, initial symbol X, and the following labeled productions:*

$$
\begin{aligned}
f_1 &: X \longrightarrow XZ \quad & [\; 0 \;\; \tfrac{1}{2} \;\; 0 \;\; \tfrac{1}{2} \;\; 0 \;] \\
f_2 &: X \longrightarrow xXy \quad & [\; 0 \;\; 0 \;\; 1 \;\; 0 \;\; 0 \;] \\
f_3 &: Z \longrightarrow zZ \quad & [\; 0 \;\; \tfrac{1}{2} \;\; 0 \;\; \tfrac{1}{2} \;\; 0 \;] \\
f_4 &: X \longrightarrow xy \quad & [\; 0 \;\; 0 \;\; 0 \;\; 0 \;\; 1 \;] \\
f_5 &: Z \longrightarrow z \quad & [\; 0 \;\; 0 \;\; 0 \;\; 0 \;\; 1 \;]
\end{aligned}
$$

The values of φ are given together with the productions. Let the initial distribution be $[\; 1 \;\; 0 \;\; 0 \;\; 0 \;\; 0 \;]$. It is easy to see that (5.4) is the language generated by G_p with cutpoint 0 under both maximal and sum interpretation. Thus (5.4) is p.2.m. and p.2.s. and, hence, also w.2.m. and w.2.s. Since all components involved are rational, the language (5.4) is also r.p.2.m, r.p.2.s., r.w.2.m., and r.w.2.s.

In a probabilistic grammar, the probabilities tend to 0 with the length of derivations.   Consequently, $c = 0$ is the only interesting cutpoint for probabilistic grammars, whereas the structure of weighted grammars is much richer.

We first prove that the family of type i languages in the Chomsky hierarchy is contained in each of the families involving i introduced above, where $i = 0, 1, 2, 3$.

Theorem 5.8.3 *For $i = 0, 1, 2, 3$, any language L of type i is r.p.i.m, and r.p.i.s. Furthermore, L is r.w.i.m., r.w.i.s., p.i.m., p.i.s., w.i.m., and w.i.s.*

Proof. Assume that $L = L(G)$, where G is a type i grammar whose productions are labeled by the elements of the set $\{f_1, \ldots, f_k\}$. Define

$$\delta = \varphi(f_j) = [\; \tfrac{1}{k} \;\; \cdots \;\; \tfrac{1}{k} \;], \qquad j = 1, \ldots, k.$$

Consider the probabilistic grammar $G_p = (G, \delta, \varphi)$. Clearly,

$$L = L_s(G_p, 0) = L_m(G_p, 0).$$

Thus L is r.p.i.m. and r.p.i.s. That L satisfies the remaining properties follows easily from definitions. ■

In the statement of the next theorem, PQ^* denotes the language consisting of all words $PQ^i, i = 0, 1, 2, \ldots$. Thereby, P and Q are words.

Theorem 5.8.4 *Let G be a grammar. If the control language C is a finite sum of languages of the form PQ^*, then $L_C(G)$ is finite.*

Proof. Since clearly

$$L_{C+D}(G) = L_C(G) + L_D(G),$$

it suffices to consider the case where the control language C is of the form PQ^*. If Q is the empty word, then the desired result holds. Hence assume that Q is not empty.

Clearly, only a finite number of derivations from the initial symbol possess the control word P. Thus there is only a finite number of last words in these derivations. It suffices to consider one such last word R and show that, starting from R, control words in Q^* do not lead to an infinite number of terminal words.

Consider the productions labeled by the letters of Q. Let $u(v)$ be the total number of nonterminals appearing on the left (right) sides of these productions, where each nonterminal is counted as many times as it occurs. If $u \leq v$, then the control words $Q^j, j \geq 1$, do not lead to any terminal words. If $u > v$ and t is the number of nonterminals in R, then the control words Q^j, where $j > t$, are not applicable. Hence the desired result holds. ■

As was seen in the examples above, Theorem 5.8.4 does not remain valid for control languages of the form $P_1 Q^* P_2$. It is clearly not valid even for control languages of the form $Q^* P$.

Theorem 5.8.5 *Let G_p be a probabilistic grammar of type i, $i = 0, 1, 2, 3$, and $c > 0$. Then the language $L_m(G_p, c)$ is finite.*

Proof. Let $G_p = (G, \delta, \varphi)$. Then we show that

$$L_m(G_p, c) = L_C(G), \tag{5.7}$$

where the control language C is a finite sum of languages of the form PQ^*. Since $c > 0$, the derivation of any word belonging to the left side of (5.7) contains at most u transitions with probability < 1, where the bound u depends on c and on the greatest probability < 1 occurring in G_p. On the other hand, if a sequence of transitions with probability 1 does not constitute a loop, then this sequence cannot contain more productions than the total number of productions. It is clear that if a loop (with probability 1) is entered, it is impossible in the derivation to leave this loop. Thus if there are k distinctly labeled productions in G, then the control words of the derivations of the words in the language on the left side of (5.7) are of the form PQ^j, $j = 0, 1, 2, \ldots$, where the lengths of P and Q possess a finite upper bound. In fact, the length of P does not exceed $ku + k + u$, and the length of Q does not exceed k. Hence (5.7) holds with C of the form mentioned and the desired conclusion follows by Theorem 5.8.4. ∎

Theorem 5.8.6 *For* $i = 0, 1, 2, 3$, *the family of p.i.m. languages is a subset of the family of p.i.s. languages. Similarly, the family of r.p.i.m. languages is a subset of the family of r.p.i.s. languages.*

Proof. For any probabilistic grammar G_p,

$$L_m(G_p, 0) = L_s(G_p, 0).$$

By Theorem 5.8.3, all finite languages are r.p.3.s. The desired result now follows by Theorem 5.8.5. ∎

We ask in the Exercises whether or not the family of w.i.m. languages is a subset of the family of w.i.s. languages. The same problem can be stated also for the corresponding rational families.

We show next how one of the decidability results concerning ordinary grammars can be extended to weighted grammars. For simplicity, we restrict ourselves to the rational case. By length-increasing productions, we mean productions of the form $P \longrightarrow Q$, where the length of P is less than or equal to the length of Q.

Theorem 5.8.7 *Let* G_w *be a rational weighted grammar whose productions are length-increasing and c a rational number. Then there is an algorithm for deciding whether or not a word P belongs to the language* $L_m(G_w, c)$ *and an algorithm for deciding whether or not a word* $P \in L_s(G_u, n)$.

Proof. An algorithm can be obtained by modifying the well-known algorithm for length-increasing grammars, [187, pp. 171-172]. In fact, the difference is that the occurrence of loops

$$X_0 \Longrightarrow P_1 \Longrightarrow \ldots \Longrightarrow P_u \Longrightarrow \ldots \Longrightarrow P_u \Longrightarrow \ldots \Longrightarrow P$$

cannot be ignored. The existence of such a loop with weight > 1 guarantees that P belongs to both languages under consideration. Loops with weight ≤ 1 can be ignored in the case of maximal interpretation. Under sum interpretation, the existence of a loop with weight 1 guarantees that P belongs to the language, and the effect of a loop with weight < 1 can be determined through summation. ∎

5.9 Probabilistic and Weighted Grammars of Type 3

A **stochastic language** is a language that is acceptable by a finite probabilistic automaton with some cutpoint. See [189, pp.73-77]. The automaton considered may possess an initial state or an initial distribution of states. This does not affect the family of stochastic languages. A language is called **rational stochastic** if it is acceptable by a finite probabilistic automaton, where all of the probabilities involved are rational, with some rational cutpoint. In this section, interrelations between stochastic languages and languages generated by weighted and probabilistic grammars of type 3 are studied.

Theorem 5.9.1 *Let L be a stochastic language. Then L is w.3.s.*

Proof. Let L be accepted with cutpoint c by the finite probabilistic automaton $A = (Q, X, q_0, Q_1, M)$, where Q is the set of states, X the set of input symbols, q_0 the initial state, Q_1 the final state set, and M the set of transition matrices. Consider the type 3 grammar $G = (Q, X, q_0, F)$, where F consists of the following productions:

$$q_u \longrightarrow x q_v, \qquad q_u \in Q, \qquad q_v \in Q, \qquad x \in X; \qquad (5.8)$$

$$q_u \longrightarrow \Lambda, \qquad q_u \in Q_1. \qquad (5.9)$$

Let k denote the number of these productions. The productions are labeled by the elements of the set $\{f_1, \ldots, f_k\}$. The grammar G is extended to a weighted grammar $G_w = (G, \delta, \varphi)$ in the following manner. The vector δ consists of 0's and 1's with 1's in the positions indicating the productions with q_0 on the left side. The vector associated by φ to productions of the form (5.9) consists of 0's only. The vector $\overline{v}$ associated to a production of the form (5.8) is defined in the following manner. A component of $\overline{v}$ corresponding to a production with q_v on the left side equals the probability of A entering q_v, after being in q_u and receiving the input x. The other components of $\overline{v}$ equal 0. Then the transition probabilities of A are preserved in the grammar G_w and

$$L = L_s(G_w, c).$$

■

We call a language L **stochastic under maximal interpretation** if there is a finite probabilistic automaton A and a cutpoint c such that L consists of words that move A from the initial state to a final state through at least one path whose probability is greater than c. Then every language stochastic under maximal interpretation is $w.3.m.$ This is established exactly as in Theorem 5.9.1.

Theorem 5.9.2 *Let G_w be a type 3 weighted grammar. If G_w does not contain productions of the form $X \longrightarrow Y$, where X and Y are nonterminals, then for any c, the language $L_s(G_w, c)$ is stochastic.*

Proof. We may assume, without loss of generality, that the productions of G_w are of the form

$$X \longrightarrow xY \tag{5.10}$$

and

$$X \longrightarrow \Lambda, \tag{5.11}$$

where X and Y are nonterminals and x is a terminal. This can be seen as follows. A production of the form

$$X \longrightarrow x_1 x_2 \ldots x_r, \qquad r \geq 1, \tag{5.12}$$

where the x_i's are terminals, is replaced by the sequence of productions

$$X \longrightarrow x_1 X, \ X \longrightarrow x_2 X, \ldots, \ X \longrightarrow x_r X, \ X \longrightarrow \Lambda. \tag{5.13}$$

The weights are adjusted so that in the sequence (5.13), the transition to the next production is given weight 1 and transitions elsewhere are given weight 0. Furthermore, the transition to the first production of (5.13) is given the weight originally associated with the transition to the production (5.12). A production of the form

$$X \longrightarrow x_1 x_2 \ldots x_r Y, \qquad r \geq 2, \tag{5.14}$$

is replaced by the sequence of productions

$$X \longrightarrow x_1 X, \ X \longrightarrow x_2 X, \ldots, \ X \longrightarrow x_{r-1} X, \ X \longrightarrow x_r Y, \tag{5.15}$$

where the weights are adjusted as above, the vector associated with the last production of (5.15) corresponding to the vector originally associated with (5.14). (Note that in a weighted grammar of type 3, the vector associated with a production $X \longrightarrow P$, where P is a word over the terminal alphabet, does not influence the language generated.) Clearly, these changes do not affect the language $L_s(G_w, c)$.

Suppose that the productions of $G_w = (G, \delta, \varphi)$ are of the forms (5.10) and (5.11). We construct a "generalized probabilistic automaton" A, in the following manner. The states of A are the labels of the productions of G. The initial distribution is δ. The final state set consists of the labels of the productions (5.11). Consider a production of the form (5.10) whose label is f_u. Then the "probability" of A entering the state f_v, after being in f_u and receiving the input x, equals the v-th component of the vector associated with f_u or 0, depending on whether Y or some other nonterminal appears on the left side of the production labeled by f_v. The probability of all other transitions equals 0. Hence

$$L_s(G_w, c) = L(A, c),$$

where the right side denotes the language accepted by A with cutpoint c. Note that this language is defined for A by exactly the same matrix product for ordinary probabilistic automata. Although A is not a probabilistic automaton, it follows by a result in [235] that $L(A, c)$ is stochastic. Thus the desired result holds. ∎

It follows from Theorems 5.9.1 and 5.9.2 that the family of stochastic languages equals the family of languages generated under sum interpretation by type 3 weighted grammars that do not contain productions of the form $X \longrightarrow Y$. It may be possible to remove this restriction on the form of the productions. To accomplish this, it would suffice to prove that a language obtained from a stochastic language by deleting all occurrences of one letter is stochastic. This again is a special case of the conjecture that the family of stochastic languages is closed under homomorphism.

It follows from Theorem 5.9.1 that the family of $w.i.s.$ languages is nondenumerable and therefore contains languages that are not of type 0. This reflects the fact that no computability assumptions are made in the definitions about the real numbers involved. The following theorem is established exactly as are Theorems 5.9.1 and 5.9.2.

Theorem 5.9.3 *Every rational stochastic language is r.w.3.s. Let G_w be a type 3 rational weighted grammar that does not contain productions of the form $X \longrightarrow Y$, where X and Y are nonterminals. Then, for any rational number c, the language $L_s(G_w, c)$ is rational stochastic.* ∎

Theorem 5.9.4 *For a type 3 probabilistic grammar G_p and $c > 0$, the language $L_s(G_p, c)$ is finite. The family of p.3.s. languages, as well as the family of p.3.m. languages, equals the family of type 3 languages.*

Proof. First replace the grammar G_p by a grammar whose rules are of the forms (5.10), (5.11), and $X \longrightarrow Y$, where X and Y are nonterminals. The new grammar is then rewritten as an automaton, exactly as was done in the proof of Theorem 5.9.2. Productions of the form $X \longrightarrow Y$ correspond to transitions caused by the empty word. The first assertion of the theorem now follows since every loop with an exit to a final state possesses a probability less than 1. The second assertion follows from Theorems 5.8.3 and 5.8.5 and the fact that the languages $L_s(G_p, 0)$ and $L_m(G_p, 0)$ are of type 3. This holds since these languages are accepted by finite nondeterministic automata. ∎

Clearly, the families of r.p.3.s. and r.p.3.m. languages also equal the family of type 3 languages. However, as was seen in Section 5.8, the family of r.p.2.s. languages, as well as the family of r.p.2.m. languages, properly includes the family of type 2 languages.

The families of p.3.m., p.3.s., and w.3.s. languages have been considered. We now consider the remaining family of w.3.m. languages.

Theorem 5.9.5 *The family of r.w.3.m. languages properly includes the family of type 3 languages.*

Proof. The inclusion follows, by Theorem 5.8.3. Consider the rational weighted grammar G_w of type 3 with the productions

$$
\begin{array}{ll}
f_1 : X \longrightarrow xX & \left[\begin{array}{ccc} 2 & 1 & 0 \end{array}\right] \\
f_2 : X \longrightarrow yX & \left[\begin{array}{ccc} 0 & \frac{1}{2} & 1 \end{array}\right] \\
f_3 : X \longrightarrow \Lambda & \left[\begin{array}{ccc} 0 & 0 & 0 \end{array}\right]
\end{array}
$$

and with $\delta = \left[\begin{array}{ccc} 1 & 0 & 0 \end{array}\right]$. Clearly,

$$
L_m(G_w, 1) = \{x^i y^j \mid i > j \geq 1\}.
$$

Consequently, the inclusion is proper. ∎

Theorem 5.9.6 *There is an algorithm for deciding whether or not the language $L_m(G_w, c)$ is empty, where G_w is a rational weighted grammar of type 3 and c is a rational number.*

Proof. We first determine all of the finitely many parts of the derivations according to G_w whose control word begins and ends with the same letter, but does not have proper subwords with this property. Suppose there exists such a loop with weight > 1 that is also a part of a terminating derivation with weight > 0. Then the language under consideration is not empty. Suppose no such loop exists with the described properties. Then the emptiness can be decided by checking all of the (finitely many) derivations without loops since, in this case, loops do not increase the total weight of a derivation. ∎

It can be shown that there is also an algorithm for deciding whether or not the language $L_m(G_w, c)$ is infinite. The same problems are undecidable for the languages $L_s(G_w, c)$, with G_w and c as above since the existence of a decision method would imply the decidability of the emptiness and infinity problems for rational stochastic languages. Consequently, not every r.w.3.s. language is r.w.3.m. An example is given in our next theorem.

Theorem 5.9.7 *The language*

$$
\{a^i b^i \mid i \geq 1\} \tag{5.16}
$$

is r.w.3.s., but not r.w.3.m.

Proof. The first assertion follows, by Theorem 5.9.3, and results in [234]. To prove the second assertion, we assume on the contrary that $\{a^i b^i \mid i \geq 1\}$ equals $L_m(G_w, c)$, for some rational weighted grammar G_w of type 3 and rational cutpoint c. Without loss of generality, we may assume that the productions of G_w are of the forms (5.10), (5.11), and $X \longrightarrow Y$. Consider a word $a^j b^j$, where j exceeds the number of productions of G_w. There is a derivation of this word with weight greater than c. Furthermore, in this derivation there is a loop that begins and ends with the same production and possesses $r \geq 1$ occurrences of productions of the form $X \longrightarrow aY$. If the weight associated with this loop is > 1, then the word $a^{j+r}b^j$ belongs to the language $L_m(G_\omega, c)$. If the weight is ≤ 1, then the word $a^{j-r}b^j$ belongs to the language. Thus we have a contradiction in both cases. Hence the desired result holds. ∎

5.10 Interrelations with Programmed and Time-Variant Grammars

The family of *w.i.s.* languages is a subset of the family of *w.j.s.* languages, for $i > j$, by definition. By Theorem 5.9.1, the family of *w.3.s.* languages contains all stochastic languages. It is very difficult to give examples of languages that are not stochastic and so it is also difficult to solve the problem of whether the family of *w.i.s. languages, $i = 0, 1, 2, 3$,* constitute a proper hierarchy.

A detailed definition of a *programmed grammar* can be found in [185]. In a programmed grammar, the productions are labeled and together with each production there are given two sets of labels, namely, the *success field* and the *failure field*. After the application of some production f, only productions with labels in the success field of f are applicable on the next step of the derivation. If f is not applicable, the next production applied must have its label in the failure field of f. It is shown in [185] that all recursively enumerable (i.e., type 0) languages are generated by programmed grammars with context-free (i.e., type 2) core productions.

From the point of view of weighted (and time-variant) grammars, programmed grammars with context-free core productions and with *empty failure fields* are of special interest.

Theorem 5.10.1 *Every language generated by a programmed grammar G with context-free core productions and with empty failure fields is r.p.2.m. and r.p.2.s.*

Proof. Let G be a programmed grammar. Then G is transformed into a rational probabilistic grammar G_p of type 2 in the following manner. If the success field of a production f contains $r \geq 1$ labels, then the transition from f to each of these labels is given probability $1/r$, and the transition from f to all other labels is given probability 0. If the success field of f is empty then a "sink" production $X \longrightarrow X$ is added to the grammar and the transition from f to this sink production is given probability 1. (A common sink production may be used for all productions of G with empty success fields.) The initial distribution δ is defined similarly. Then the languages $L_s(G_p, 0)$ and $L_m(G_p, 0)$ equal the language generated by G and hence the desired result holds. ∎

Consider a type i $(i = 0, 1, 2, 3)$ grammar

$$G = (N, T, X_o, F).$$

The grammar G together with an infinite sequence

$$F_1, F_2, F_3, \ldots \tag{5.17}$$

of subsets of F forms a *time-variant grammar* G_{iv} of type i. (See [187] for a corresponding recognition device of type 3.) The language $L(G_{iv})$ generated by G_{iv} is defined to be $L_C(G)$, where the control language C is the union of all languages of the form $F_1 F_2 \ldots F_u, u \geq 1$, where juxtaposition denotes catenation. Such type 2 time-variant grammars, where the sequence (5.17) is periodic, are of special interest. Languages generated by these grammars are called **periodically time-variant context-free languages (ptvcf)**. Actually, it can be shown that the same languages are obtained by assuming that the sequence (5.17) is almost periodic. For an example, consider the type 2 grammar with the initial symbol X_0 and the productions

$$f_1 : X_0 \longrightarrow xX_1yX_2$$
$$f_2 : X_1 \longrightarrow xX_1y$$
$$f_3 : X_1 \longrightarrow \Lambda$$
$$f_4 : Y_2 \longrightarrow \Lambda$$
$$f_5 : X_2 \longrightarrow zX_2$$
$$f_6 : X_2 \longrightarrow Y_2.$$

Time-variance is specified in such a way that, for $u = 1, 2, \ldots$, the productions f_1, f_2, f_3, f_4 belong to the set F_{2u-1} and f_5, f_6 belong to the set F_{2u}. Then the language generated is that of (5.4) and thus is ptvcf. (Note that Y_2 is introduced to prevent the derivation of words $x^{i+1}y^{i+1}z^i$.) Derivations according to ptvcf grammars are easy to describe and perform. Nevertheless the generative power of these grammars is remarkable. It follows that the family of ptvcf languages includes all languages generated by context-free matrix grammars. Thus the family of ptvcf languages contains all context-sensitive languages, provided the result in [1] is correct. On the other hand, ptvcf languages form a subset of the family of languages generated by programmed grammars with context-free core productions and empty failure fields. This subset is obtained by imposing on programmed grammars the additional restriction that whenever two labels f_1 and f_2 are in the success field of a production, then the productions labeled by f_1 and f_2 possess identical success fields.

Throughout, we have assumed in considering a step

$$P_1 \Longrightarrow_f P_2 \tag{5.18}$$

of a derivation that the production f is actually applied, i.e., $P_1 = Q_1 Q Q_2$, $P_2 = Q_1 R Q_2$, and f is the label of the production $Q \longrightarrow R$, for some Q_1

and Q_2. Another possibility is to specify a subset F_1 of productions such that the notation (5.18) may be used also in case $f \in F_1$ is not applicable, i.e., P_1 does not contain an occurrence of Q and $P_1 = P_2$. We assume that this possibility is included in (5.3) when control words are defined. Everything concerning weighted and probabilistic grammars is defined now as it was defined before using this new interpretation, the so-called *checking interpretation*, of control words. Then Theorem 5.10.1 can be strengthened as shown in the next result.

Theorem 5.10.2 *Every recursively enumerable language is r.p.2.m. and r.p.2.s., under checking interpretation.*

Proof. By the result in [185], any recursively enumerable language L is generated by a programmed grammar G with context-free core productions. G is first written so that for every production f, either the success field of f or the failure field of f is empty. This is accomplished by making two copies of f, one with the success field and the other with the failure field of the original f, and putting the labels of both copies into all fields that contain the label of the original production f. All productions $X \longrightarrow P$ are then replaced with an empty success field by productions $X \longrightarrow Y$, where Y is a new nonterminal (which, thus, does not appear on the left side of any production). The remainder of the proof proceeds like the proof of Theorem 5.10.1. ∎

It is always possible in Theorem 5.10.2 to choose the required rational probabilistic type 2 grammar $G_p = (G, \delta, \varphi)$ in such a way that the language $L(G)$ generated by the basic grammar G is of type 3.

5.11 Probabilistic Grammars and Automata

The material in the remainder of the chapter is from [204] and [207].

In Sections 5.8–5.10, the concept of a probabilistic grammar is formulated and random languages generated by probabilistic grammars are defined. See also [188] and [57]. According to Santos, none of these formulations were broad enough to encompass the conventional deterministic grammars and still preserve its probabilistic character.

In Sections 5.11–5.15, another formulation of probabilistic grammars is presented. This formulation reduces to the conventional deterministic grammars if all functions involved are deterministic.

Various types of probabilistic grammars are considered, namely, type-0, context-sensitive, context-free, weakly regular, and regular probabilistic grammars. They correspond to the various types of grammars in the conventional theory.

Most of the study here concerns the relations between the various types of probabilistic grammars and the corresponding types of probabilistic automata.

In Section 5.13, the concept of asynchronous probabilistic automata (APA) is defined. It is shown that every weakly regular random language can be generated by an APA, and vice versa. It is also shown that every regular random language can be generated by a synchronous APA, and vice versa.

In Sections 5.14 and 5.15, we define the concepts of probabilistic Turing machines (PTM) and probabilistic pushdown automata (PPA). It is shown that every bounded type-0 (leftmost-bounded context-free) random language can be generated by a bounded PTM (PPA), and vice versa.

It is shown in [136] that every language generated by any grammar using only leftmost derivations is context free. A similar result for random languages is given in Section 5.15.

5.12 Probabilistic Grammars

In this section, we present a mathematical formulation of probabilistic grammar and we define the class of random languages generated by pr-obabilistic grammars.

Definition 5.12.1 *Let X and Y be nonempty sets. A **random function** from X into Y is a function F from $X \times Y$ into $[0,1]$ such that $\sum_{y \in Y} F(x,y) \leq 1$ for all $x \in X$. If $\sum_{y \in Y} F(x,y) = 1$ for all $x \in X$, then F is called a **total random function**.*

$F(x,y)$ is the probability that the value of the function at x is y.
Let S be a nonempty set.

Definition 5.12.2 *A (total) **probabilistic production** ρ over S is a (total) random function from S^* into S^* such that $\rho(\Lambda, \Lambda) = 1$ and the set*

$$\overline{\rho} = \{s \in S^* \mid \rho(s,t) > 0 \text{ for some } t \in S^*, \text{ where } t \neq s\}$$

*is finite. If, in addition, for all $s \in S^*$, the set $\{t \in S^* \mid \rho(s,t) > 0\}$ is finite, then ρ is called **bounded**. An element of $\overline{\rho}$ is called a **genetrix** of ρ.*

$\rho(s,t)$ is the probability that s will be replaced by t.

Definition 5.12.3 *A (bounded) **probabilistic grammar** is a quadruple $G = (T, N, P, h)$ such that the following conditions hold:*
(1) T and N are disjoint finite nonempty sets;
(2) P is a finite collection of (bounded) probabilistic productions over $T \cup N$ such that $s \in \overline{P} = U_{\rho \in P} \overline{\rho}$ implies $s \in (T \cup N)^* N (T \cup N)^*$;
(3) h is a function from N into $[0,1]$ such that $\sum_{A \in N} h(A) \leqslant 1$.

In the above definition, T and N are, respectively, the sets of terminal and nonterminal symbols. Furthermore, $h(A)$ is the probability that A is the start symbol of G.

Recall that $\mathbb{N}$ is the collection of all positive integers.

We now restate some definitions and observations quite similar to those in Section 3.4 for max-product grammars and languages.

Definition 5.12.4 *Let* r, $s \in S^*$ *and* $k \in \mathbb{N}$. *Then* $m(r, s) = k$ *if there exist* $u_i, v_i \in S^*$, $i = 1, 2, \dots, k$, *where* $v_i \neq v_j$ *for* $i \neq j$, *such that* (1) $r = u_i s v_i$ *for all* $i = 1, 2, \dots, k$, *and* (2) $r = usv$ *implies* $u = u_i$ *and* $v = v_i$ *for some* i, *where* $i = 1, 2, \dots, k$. *Define* $m(r, s) = 0$ *if* $r \neq usv$ *for all* $u, v \in S^*$.

Note that $m(r, s) = k$ if and only if r can be expressed in the form usv in exactly k distinct ways.

Definition 5.12.5 *Let* $G = (T, N, P, h)$ *be a probabilistic grammar. A* **replacement function** *of* G *is a function* δ *from* $(T \cup N)^*$ *into* $\overline{P}$ *such that* $\delta(r) = s$ *implies* $m(r, s) > 0$. *If, in addition,* $\delta(r) = s$ *implies* $r = usv$, *where* $u \in T^*$ *and* $v \in (T \cup N)^*$, *then* δ *is called* **leftmost**.

Let $D(G)$ denote the collection of all replacement functions of G, and let $D_L(G)$ denote the collection of all leftmost replacement functions of G.

Note that $\delta(r) = s$ if and only if some occurrence of s in r will be replaced.

Definition 5.12.6 *Let* $r, w, s, t, \in S^*$ *and* $k \in \mathbb{N}$. *Then define* $r \sim^k w$ $mod(s, t)$ *if there exist* $u, v \in S^*$ *such that* $r = usv$, $w = utv$, *and* $m(us, s) = k$.

Note that $r \sim^k w \ mod(s, t)$ if and only if w can be obtained from r by replacing the k-th occurrence of s in r by t.

Definition 5.12.7 *Let* $G = (T, N, P, h)$ *be a probabilistic grammar and* $S = T \cup N$.

(1) *For all* $\rho \in P$, $\delta \in D(G)$, *and* $k \in \mathbb{N}$, *associate the function* $f_{(\rho, \delta, k)}$ *from* $S^* \times S^*$ *into* $[0, 1]$, *where*

$$f_{(\rho, \delta, k)}(r, w) = \begin{cases} \rho(\delta(r), t) & \text{if } r \sim^k w \ mod(\delta(r), t) \\ 0 & \text{otherwise.} \end{cases}$$

(2) *For all* $z \in (P \times D(G) \times \mathbb{N})^*$, *associate the function* f_z *from* $S^* \times S^*$ *into* $[0, 1]$ *defined recursively as follows:*

$$f_\Lambda(r, w) = \begin{cases} 1 & \text{if } r = w \\ 0 & \text{if } r \neq w \end{cases}$$

and

$$f_{z(\rho, \delta, k)}(r, w) = \sum_{x \in S^*} f_z(r, x) f_{(\rho, \delta, k)}(x, w).$$

It is clear that f_z is a random function from $(T \cup N)^*$ into $(T \cup N)^*$. Furthermore, $f_z(r, w)$ is the probability that w can be obtained from r by the "derivation" z.

Definition 5.12.8 *A **random language** f over S is a function from S^* into $[0, 1]$.*

$f(r)$ is the probability that r is a member of the language.

Definition 5.12.9 *A **random control set** π of a probabilistic grammar $G = (T, N, P, h)$ is a random language over $P \times D(G) \times \mathbb{N}$.*

$\pi(z)$ is the probability that the derivation z will be applied. This concept of a control set is similar to, but distinct from that introduced in [80].

Definition 5.12.10 *Let $G = (T, N, P, h)$ be a probabilistic grammar and π a random control set of G. The **random language** $f_{G,\pi}$ generated by G under π is the random language over T defined by $\forall r \in T^*$,*

$$f_{G,\pi}(r) = \vee_{z \in Z^*} \pi(z) \{ \sum_{A \in N} h(A) f_z(A, r) \},$$

where $Z = P \times D(G) \times \mathbb{N}$. If $\pi(z) = 1$ for all $z \in Z^*$, then we write f_G for $f_{G,\pi}$. In this case, f_G is said to be the **random language generated** by G. Furthermore, if

$$\pi(z) = \begin{cases} 1 & \text{if } z \in (P \times D_L(G) \times \{1\})^* \\ 0 & \text{otherwise,} \end{cases}$$

then we write $f_{G,L}$ for $f_{G,\pi}$. In this case, $f_{G,L}$ is said to be the **random language generated** by G with leftmost derivations only.

$f_{G,\pi}(r)$ is the probability that r will be generated by G under π.

Example 5.12.11 *Let $G = (N, T, \rho, \eta)$ be the grammar defined in Example 3.4.10. Let $r_1 = (s_0 \xrightarrow{.9} aAC)$, $r_2 = (C \xrightarrow{.8} AD)$, $r_3 = (A \xrightarrow{.7} B)$, $r_4 = (A \xrightarrow{.6} a)$, $r_5 = (B \xrightarrow{.5} b)$, and $r_6 = (D \xrightarrow{.4} b)$. Then*

$$\begin{aligned} f_{G,\pi}(aABb) &= \vee_{z \in Z} \pi(z) \cdot h(A) f_z(aAAb, aABb) \\ &= \pi(\rho, \delta, n) h(A) f_{(\rho,\delta,n)}(aAAb, aABb) \\ &= \pi(\rho, \delta, n) h(A) \rho(A, B) \\ &= \pi(\rho, \delta, n) h(A)(.7). \end{aligned}$$

The probabilistic grammars defined above are referred to as type-0 probabilistic grammars. Other more restrictive types of probabilistic grammars can be obtained by imposing certain restrictions on the probabilistic productions. We define four such probabilistic grammars below. Three of them correspond to the conventional context-sensitive, context-free, and regular grammars.

Definition 5.12.12 *Let $G = (T, N, P, h)$ be a probabilistic grammar.*

*(1) G is called **context-sensitive** if for all $\rho \in P$, $\rho(s, t) > 0$ implies $|s| \leqslant |t|$.*

*(2) G is called **context-free** if $\overline{P} \subseteq N$.*

*(3) G is called **weakly regular** if for all $\rho \in P$, $\overline{\rho} \subseteq N$ and for all $s \in \overline{\rho}$, $\rho(s, t) > 0$ implies $t = uA$, where $u \in T^*$ and $A \in N \cup \{\Lambda\}$.*

*(4) G is called **regular** if for all $\rho \in P$, $\overline{\rho} = N$ and for all $s \in \overline{\rho}$, $\rho(s, t) > 0$ implies $t = aA$, where $a \in T$ and $A \in N \cup \{\Lambda\}$.*

The grammar in Example 5.12.11 is context-free.

Clearly, every language generated by a conventional type-0, context-sensitive, context-free, or regular grammar in the conventional manner may be associated with f_G for some type-0, context-sensitive, context-free, or regular probabilistic grammar $G = (T, N, P, h)$, where h and every $\rho \in P$, are deterministic, and vice versa.

From the above definition, it follows that every regular probabilistic grammar is weakly regular and every weakly regular probabilistic grammar is context-free. Furthermore, if G is a context-free probabilistic grammar, then $D_L(G)$ contains exactly one replacement function of G.

If $f = f_G$ for some specified (type-0), context-sensitive, context-free, weak-regular, and regular probabilistic grammar G, then we say that f is that specific random language, and vice versa.

Theorem 5.12.13 *Let f be a (bounded) (type-0, context-sensitive, context-free, weakly regular, regular) random language over T. Then $f = f_G$ for some (bounded) (type-0, context-sensitive, context-free, weakly regular, regular) probabilistic grammar $G = (T, N, P, h)$ such that (1) every $\rho \in P$ is a total probabilistic production, (2) $h(A_0) = 1$ for some $A_0 \in N$, and (3) $N \subseteq \overline{P}$.*

Proof. Let $f = f_G$, where $G_0 = (T, N_0, P_0, h_0)$ is a (bounded) (type-0, context-sensitive, context-free, weakly regular, regular) probabilistic grammar. Let A_0, $A_1 \notin T \cup N_0$ and a_0 an arbitrarily fixed element of T. Let $N = N_0 \cup \{A_0, A_1\}$ and $N_1 = N \backslash \overline{P}$. For every $\rho \in P_0$, we associate the probabilistic productions ρ' and ρ'' over $T \cup N$ such that for all $s, t \in (T \cup N)^*$,

$$
\rho'(s, t) = \begin{cases}
\rho(s, t) & \text{if } s \in \overline{P},\ t \in (T \cup N_0)^* \\
1 - \sum_{w \in (T \cup N_0)^*} \rho(s, w) & \text{if } s \in \overline{P},\ t = a_0 A_1 \\
1 & \text{if } s = t \notin \overline{P} \cup N \\
  & \quad \text{or } s \in N_1,\ t = a_0 A_1 \\
0 & \text{otherwise}
\end{cases}
$$

and

$$
\rho''(s,t) = \begin{cases} \sum_{A \in N_0} h(A)\rho(A,t) & \text{if } s = A_0,\ t \in (T \cup N_0)^* \\ 1 - \sum_{A \in N_0} h(A)\rho(A,t) & \text{if } s = A_0,\ t = a_0 A_1 \\ 1 & \text{if } s = t \neq A_0 \\ 0 & \text{otherwise.} \end{cases}
$$

Let $G = (T, N, P, h)$, where P is the collection of all ρ' and ρ'' defined above and

$$
h(A) = \begin{cases} 1 & \text{if } A = A_0 \\ 0 & \text{if } A \neq A_0. \end{cases}
$$

It follows that G has the desired properties. ∎

We also write $G = (T, N, P, A_0)$ if it satisfies condition (2) above. Moreover, we say that G is **total** if it satisfies conditions (1)-(3) above.

We conclude this section by introducing the concepts of random domains and ranges of random functions, which are needed later.

Definition 5.12.14 *Let F be a random function from X into Y.*

(1) *The **random domain** $D(F)$ of F is the function from X into $[0,1]$ such that $D(F)(x) = \sum_{y \in Y} F(x,y)$ for all $x \in X$.*

(2) *The **random range** $R(F)$ of F is the function from Y into $[0,1]$ such that $R(F)(y) = \bigvee_{x \in X} F(x,y)$ for all $y \in Y$.*

5.13 Weakly Regular Grammars and Asynchronous Automata

In this section, we study weakly regular probabilistic grammars and their relationship with asynchronous probabilistic automata.

Proposition 5.13.1 *If G is a weakly regular probabilistic grammar, then $f_G = f_{G,L}$.* ∎

Proposition 5.13.2 *If f is a bounded weakly regular random language, then $f = f_G$ for some total weakly regular probabilistic grammar $G = (T, N, P, h)$ such that for all $\rho \in P$ and $A \in N$, $\rho(A,t) > 0$ implies $t = aB$, where $a \in T \cup \{\Lambda\}$ and $B \in N \cup \{\Lambda\}$.* ∎

Definition 5.13.3 *An **asynchronous probabilistic automaton (APA)** is a sextuple $M = (Q, X, Y, p, h, g)$, where $Q, X,$ and Y are finite nonempty sets, p is a random function from $Q \times X$ into $Y^* \times Q$, and h and g are functions from Q into $[0,1]$ such that $\sum_{q \in Q} h(q) \leqslant 1$.*

In Definition 5.13.3, $Q, X,$ and Y are, respectively, the state, input, and outputs sets. $p(q, u, y, q')$ is the conditional probability that the next state of M is q' and output string y is produced given that the present state of M is q and input symbol u is applied. $h(q)$ and $g(q)$ are, respectively, the probabilities that q is the initial state of M and q is a final state of M.

Definition 5.13.4 *Let $M = (Q, X, Y, p, h, g)$ be an APA.*

(1) Define the function p^M from $Q \times X^* \times Y^* \times Q$ into $[0, 1]$ by for all q', $q'' \in Q$, $u \in X$, $x \in X^*$, and $y \in Y^*$,

$$p^M(q', \Lambda, \Lambda, q'') = \begin{cases} 1 & \text{if } q' = q'' \\ 0 & \text{if } q' \neq q'' \end{cases}$$

and

$$p^M(q', ux, y, q'') = \sum p(q', u, y_1, q) p^M(q, x, y_2, q''),$$

where the summation ranges over all $q \in Q$ and $y_1, y_2 \in Y^*$ such that $y = y_1 y_2$.

(2) Define the function F^M from $X^* \times Y^*$ into $[0, 1]$ by for all $x \in X^*$ and $y \in Y^*$,

$$F^M(x, y) = \sum_{q, q' \in Q} h(q) p^M(q, x, y, q') g(q').$$

The function p^M has the following interpretation. $p^M(q', x, y, q'')$ is the conditional probability that M will be in state q'' and produce output string y given that the present state of M is q' and input string x is applied. $F^M(x, y)$ is the probability that M will produce y when x is applied.

The above model of an APA is a generalization of the model introduced in [221].

Clearly, F^M is a random function from X^* into Y^*, and $D(F^M)$ and $R(F^M)$ are random languages over X and Y, respectively. Using the terminology introduced in [216], the class of all F^M is the class of all random functions computable by an APA, the class of all $D(F^M)$ is the class of all random languages acceptable by an APA, and the class of all $R(F^M)$ is the class of all random languages that can be generated by an APA.

Definition 5.13.5 *Let $M = (Q, X, Y, p, h, g)$ be an APA.*

*(1) M is called **bounded** if for all $q, q' \in Q$ and $u \in X$, $p(q, u, y, q') = 0$ except for finitely many $y \in Y^*$.*

*(2) M is called **synchronous** if for all $q, q' \in Q$ and $u \in X$,*

$$p(q, u, y, q') > 0$$

implies $y \in Y$.

A synchronous total APA reduces to a stochastic sequential machine if $g(q) = 1$ for all $q \in Q$. It follows that a random language is acceptable by a synchronous APA if and only if it is realizable by a conventional probabilistic automaton, Section 5.16 or [207].

Theorem 5.13.6 *The following statements are equivalent.*

(1) f is a weakly regular random language;

(2) $f = R\left(F^M\right)$ for some APA $M = (Q, X, Y, p, h, g)$, where

    (i) p is a total random function from $Q \times X$ into $Y^* \times Q$,

    (ii) $h(q_0) = 1$ for some $q_0 \in Q$, and

    (iii) there exists $q_1 \in Q$ such that $g(q_1) = 1$ and $g(q) = 0$ for all $q \neq q_1$;

(3) $f = R(F^M)$ for some APA M.

Proof. Suppose f is a weakly regular random language. Then, by Theorem 5.12.13, $f = f_G$ for some total weakly regular probabilistic grammar $G = (T, N, P, A_0)$.

Let $A_1, A_2 \notin T \cup N$ and $N_0 = N \cup \{A_1, A_2\}$. Define $M = (P, N_0, T, p, h, g)$ such that for all $\rho \in P$, $r \in T^*$, and $A, A' \in N_0$,

$$p(A, \rho, r, A') = \begin{cases} \rho(A, aA') & \text{if } A, A' \in N \\ \rho(A, a) & \text{if } A \in N, A' = A_1 \\ 1 & \text{if } A' = A_2 \text{ and } A \in \{A_1, A_2\} \\ 0 & \text{otherwise} \end{cases}$$

$$g(A) = \begin{cases} 1 & \text{if } A = A_1 \\ 0 & \text{if } A \neq A_1 \end{cases}$$

and

$$h(A) = \begin{cases} 1 & \text{if } A = A_0 \\ 0 & \text{if } A \neq A_0. \end{cases}$$

It follows that M has the desired properties. Thus (1) implies (2).

That (2) implies (3) is immediate.

Now suppose that $f = R(F^M)$ for some APA $M = (Q, X, Y, p, h, g)$. We may assume without loss of generality that $Q \cap Y = \emptyset$. For all $u \in X$, we associate the probabilistic production ρ_u over $Y \cup Q$, where $\overline{\rho}_u = Q$ and

$$\rho_u(q, t) = \begin{cases} p(q, u, y, q') & \text{if } t = yq \\ 0 & \text{otherwise.} \end{cases}$$

Moreover, let ρ_0 be the probabilistic production over $Y \cup Q$, where $\overline{\rho}_0 = Q$ and

$$\rho_0(q, t) = \begin{cases} g(q) & \text{if } t = \Lambda \\ 0 & \text{otherwise.} \end{cases}$$

Define $G = (Q, Y, P, h)$, where $P = \{\rho_0\} \cup \{\rho_u \mid u \in X\}$. It follows that $f = f_G$. Thus (3) implies (1). ∎

Theorem 5.13.7 *The following statements are equivalent:*
(1) f is a bounded weakly regular random language;
(2) $f = R(F^M)$ for some APA $M = (Q, X, Y, p, h, g)$ satisfying conditions (i)-(iii) of Theorem 5.13.6(2) and (iv) for all $q, q' \in Q$ and $u \in X, p(q, u, y, q') > 0$ implies $y \in Y \cup \{\Lambda\}$;
(3) $f = R(F^M)$ for some bounded APA M.

Proof. The proof follows from Proposition 5.13.2 and the proof of Theorem 5.13.6. ■

Theorem 5.13.8 *(1) If f is a regular random language, then $f = R(F^M)$ for some synchronous APA M satisfying conditions (i)-(iii) of Theorem 5.13.6.*
(2) If $f = R(F^M)$ for some synchronous APA M, then there exists a regular probabilistic grammar G such that $f(r) = f_G(r)$ for all $r \neq \Lambda$.

Proof. Assertion (1) follows from the proof of Theorem 5.13.6. Now, suppose $f = R(F^M)$, where $M = (Q, X, Y, p, h, g)$ is a synchronous APA. We assume without loss of generality that $Q \cap Y = \emptyset$. For all $u \in X$, we associate the probabilistic productions ρ_u^1 and ρ_u^2, where $\overline{\rho}_u^1 = \overline{\rho}_u^2 = Q$ and

$$\rho_u^1(q, t) = \begin{cases} p(q, u, v, q') & \text{if } t = vq' \\ 0 & \text{otherwise} \end{cases}$$

$$\rho_u^2(q, t) = \begin{cases} \sum_{q' \in Q} p(q, u, v, q') g(q') & \text{if } t = v \\ 0 & \text{otherwise} \end{cases}$$

for all $q \in Q$, $v \in Y$, and $t \in (Y \cup Q)^*$. Let $G = (Q, Y, P, h)$, where $P = \{\rho_u^1 \mid u \in X\} \cup \{\rho_u^2 \mid u \in X\}$. It follows that $f(r) = f_g(r)$ for all $r \in Y^+$. Hence assertion (2) holds. ■

The above result states that every regular random language can be generated by a synchronous APA, and vice versa. For deterministic languages, it is known that every regular language is also acceptable by a finite automaton, and vice versa. This does not hold in general for random languages. However, a sufficient condition is given below.

Theorem 5.13.9 *If $f = f_G$ for some regular probabilistic grammar $G = (T, N, P, h)$, where P contains exactly one element, then f is acceptable by some synchronous APA.*

Proof. It follows from the proof of Theorem 5.13.6 that $f = R(F^M)$ for some synchronous APA $M = (Q, X, T, p, h, g)$, where $X = \{u_0\}$. Let $M' = (Q, T, \{u_0\}, p', h, g)$, where for all $q, q' \in Q$ and $v \in T$,

$$p'(q, v, u_0, q') = p(q, v_0, v, q').$$

Clearly, M' is a synchronous APA. Moreover, it follows that $f = D(F^{M'})$. ■

5.14 Type-0 Probabilistic Grammars and Probabilistic Turing Machines

In this section, we study type-0 probabilistic grammars and their relationship to probabilistic Turing machines.

Definition 5.14.1 *A **probabilistic Turing machine (PTM)** is a sextuple $M = (Q, X, Y, W, p, h)$, where $Q, X, Y,$ and W are finite nonempty sets, $X \cup Y \subseteq W, Q \cap W = \emptyset, p$ is a random function from $Q \times W$ into $(W^* \cup \{+, -\}) \times Q$, where $+, - \notin W$, and h is a function from Q into $[0, 1]$ such that $\sum_{q \in Q} h(q) \leqslant 1$.*

In Definition 5.14.1, $X, Y,$ and W are, respectively, the set of input, output, and tape symbols. Q denotes the set of states. $h(q)$ is the probability that $q \in Q$ is the initial state. $p(q, w, z, q')$ is the conditional probability of the "next act" of the PTM given that its present state is q and the tape symbol w is scanned. Like the conventional Turing machines, the "next act" of the PTM may be one of the following:

(1) $z \in W^*$: replace w by z and go to state q'.

(2) $z = +$: move one square to the right and go to state q'.

(3) $z = -$: move one square to the left and go to state q'.

In the following, if $M = (Q, X, Y, W, p, h)$ is a PTM, then we assume that $b \in W$ and $b \notin X \cup Y$. The symbol b stands for blank.

Definition 5.14.2 *Let $M = (Q, X, Y, W, p, h)$ be a PTM and $r \in (W \cup Q)^*$. r is called an **instantaneous description** of M if (1) r contains exactly one $q \in Q$ and q is not the rightmost symbol of r, (2) the leftmost symbol of r is not b, and (3) the rightmost symbol of r is not b unless it is the symbol immediately to the right of q.*

Let $I(M)$ denote the set of all instantaneous descriptions of M.

Definition 5.14.3 *Let $M = (Q, X, Y, W, p, h)$ be a PTM.*

(1) Define the function p^M from $I(M) \times I(M)$ into $[0, 1]$ by for all $r, s \in I(M)$,

$$
p^M(r, s) = \begin{cases}
p(q, w, z, q') & \text{if } r = uqwv,\ s = uq'zv,\ zv \neq \Lambda \\
& \text{or } r = uqw,\ s = uq'b,\ z = \Lambda \\
p(q, w, +, q') & \text{if } r = uqww'v,\ s = uwq'w'v,\ uw \neq b \\
& \text{or } r = qww'v,\ s = q'w'v,\ w = b \\
& \text{or } r = uqw,\ s = uwq'b,\ uw \neq b \\
& \text{or } r = qw,\ s = q'b,\ w = b \\
p(q, w, -, q') & \text{if } r = uw'qwv,\ s = uq'w'wv,\ wv \neq b \\
& \text{or } r = uw'qw,\ s = u'q'w',\ w = b \\
& \text{or } r = qwv,\ s = q'bwv,\ wv \neq b \\
& \text{or } r = qw,\ s = q'b,\ w = b \\
0 & \text{otherwise,}
\end{cases}
$$

where $u, v \in W^*, q, q' \in Q, w, w' \in W$, and $z \in W^*$.

(2) For all $k \in \mathbb{N} \cup \{0\}$, define the function p_k^M from $I(M) \times I(M)$ into $[0, 1]$ by for all $r, s \in I(M)$,

$$p_0^M(r, s) = \begin{cases} 1 & if \quad r = s \\ 0 & if \quad r \neq s \end{cases}$$

$$p_{k+1}^M(r, s) = \sum_{u \in I(M)} p_k^M(u, s) p^M(r, u).$$

(3) For all $k \in \mathbb{N}$, define the function q_k^M from $I(M) \times I(M)$ into $[0, 1]$ by for all $r, s \in I(M)$,

$$q_k^M(r, s) = p_{k-1}^M(r, s)[1 - \sum_{q' \in Q} \sum_{z \in Z} p(q, w, z, q')],$$

where $Z = W^* \cup \{+, -\}, q$ is the state symbol contained in s, and w is the symbol contained in s, which is immediately to the right of q.

(4) Define the function F^M from $X^* \times Y^*$ into $[0, 1]$ by for all $x \in X^*$ and $y \in Y^*$,

$$F^M(x, y) = \sum_{<r>=y} \sum_{q \in Q} h(q) \left[\sum_{k=1}^{\infty} q_k^M(qx, r) \right],$$

where $< r >$ is the output string obtained from r by deleting all symbols in r not belonging to Y.

The functions in Definition 5.14.3 have the following interpretations.

(1) $p^M(r, s)$ is the conditional probability that the "next" instantaneous description is s given that M "starts" with instantaneous description r.

(2) $q_k^M(r, s)$ is the conditional probability that the instantaneous description of M is s "after k steps" given that M "starts" with r.

(3) $q_k^M(r, s)$ is the conditional probability that M will "terminate" with s "after k steps" given that M "starts" with r.

(4) $F^M(x, y)$ is the conditional probability that the output of M is y given that input x is applied.

It follows that F^M is a random function from X^* into Y^*.

Definition 5.14.4 *Let $M = (Q, X, Y, W, p, h)$ be a PTM.*

*(1) M is called **bounded** if for all $q, q' \in Q$ and $w \in W$, $p(q, w, z, q') = 0$ except for finitely many $z \in W^* \cup \{+, -\}$.*

*(2) M is called **synchronous** if for all $q, q' \in Q$ and $w \in W$, $p(q, w, z, q') > 0$ implies $z \in W \cup \{+, -\}$.*

The above model of a PTM differs slightly from those given in [198] and [201].

Proposition 5.14.5 *Every random function that is computable by a bounded PTM is computable by a synchronous PTM.* ∎

Proposition 5.14.6 *If F is a random function computable by a PTM, then $F = F^M$ for some PTM $M = (Q, X, Y, W, p, h)$ satisfying the following conditions:*

(1) There exists $q_1 \in Q$ such that $h(q_1) = 1$.

(2) For all $q \in Q$ and $w \in W$, $p(q, w, z, q') > 0$ implies $z \in W^*$, $p(q, w, z, q') > 0$ implies $z = +$, or $p(q, w, z, q') > 0$ implies $z = -$.

(3) There exists $q_2 \in Q$ such that

(i) $p(q_2, w, z, q) = 0$ for all $q \in Q$, $w \in W$, and $z \in W^* \cup \{+, -\}$,

(ii) for all $w \in W$ and $q \in Q$, where $q \neq q_2$,

$$\sum_{q' \in Q} \sum_{z \in Z} p(q, w, z, q') = 1,$$

where $Z = W^* \cup \{+, -\}$, and

(iii) for all $x \in X^*$, $\sum_{k=1}^{\infty} q_k^M(q, x, r) > 0$ implies $r = q_2 y$, where $y \in Y^*$. ∎

Theorem 5.14.7 *If f can be generated by a (bounded) PTM, then f is a (bounded) type-0 random language.*

Proof. Let $f = R(F^M)$, where $M = (Q, X, Y, W, p, h)$ is a (bounded) PTM satisfying conditions $(1)-(3)$ of Proposition 5.14.6. Let A_0, A_1, $\phi, \$ \notin W$ and $W_0 = W \cup \{A_0, A_1, \phi, \$\}$. Let $G = (Y, N, P, h')$, where $N = W_0 \backslash Y$, $h'(A_0) = 1$, $P = \{\rho_0, \rho_1, \rho_2, \rho_3, \rho_4\} \cup \{\rho_u \mid u \in X\}, \overline{\rho}_0 = \{A_0\}, \overline{\rho}_u = \{A_1\}$ for all $u \in X$, $\overline{\rho}_1 = \{A_1\}, \overline{\rho}_2 = \{q_2\}, \overline{\rho}_3 = \{\phi, \$\}, \overline{\rho}_4 \subseteq \{qw \mid q \in Q, w \in W_0\} \cup \{w'qw \mid q \in Q$ and $w, w' \in W_0\}, \overline{\rho}_0(A_0, \phi A_1 \$) = 1$, $\rho_u(A_1, A_1 u) = 1$ for all $u \in X$, $\rho_1(A_1, q_1) = 1, \rho_2(q_2, \Lambda) = 1, \rho_3(\phi, \Lambda) = \rho_3(\$, \Lambda) = 1$, and

$$\rho_4(s, t) = \begin{cases} p(q, w, z, q') & \text{if } s = qw, \ t = q'z, \ w \neq \$ \\ p(q, w, +, q') & \text{if } s = qw, t = wq, \ w \neq \$ \\ p(q, w, -, q') & \text{if } s = w'qw, \ t = qw'w, \ w' \neq \phi \\ 1 & \text{if } s = q\$, \ t = qb\$ \\ & \text{or } s = \phi qw, \ t = \phi bqw \\ 0 & \text{otherwise.} \end{cases}$$

Clearly, G is a (bounded) type-0 probabilistic grammar. It also follows that $f = f_G$. ∎

Theorem 5.14.8 *If f is a bounded type-0 random language, then f can be generated by a synchronous PTM.*

Proof. Let $f = f_G$, where $G = (T, N, P, h)$ is a bounded type-0 probabilistic grammar. We may assume without loss of generality that G is total and $h(A_0) = 1$ for some $A_0 \in N$. For all $s \in (T \cup N)^*$, we associate the abstract symbol $\bar{s}$. Let $Q_1 = \{\bar{s} \mid s \in \overline{P}\}$ and $Q_2 = \{\bar{t} \mid \rho(s, t) > 0$ for some $\rho \in P$ and $s \in \overline{\rho}\}$. Let $M = (Q, X, Y, W, p, h')$ be a synchronous PTM, where

$$X = P \cup Q_1 \cup \{1\}, P \cup Q_1 \cup Q_2 \cup \{1, b\} \subseteq W$$

and $h(q_0) = 1$ for some $q_0 \in Q$. We now describe informally the behavior of M.

(a) Suppose M has the instantaneous description $q_0 x \in Q(P \cup Q_1 \cup \{1\})^*$. Then M will go to state q_1 if $x \notin (PQ_1^*\{1\}^*)^+$, where $p(q_1, w, w, q_1) = 1$ for all $w \in W$, i.e., M loops. Otherwise, M will have the instantaneous description $q_2 x b b A_0$, $q_2 \in Q$.

(b) Suppose M has the instantaneous description $q_2 z$, where

$$z = \rho \bar{s}_1 \bar{s}_2 \ldots \bar{s}_m k z_0 b z_1 b r_1 z_2 b r_2 \ldots z_n b r_n,$$

$n \geqslant 0$, $\rho \in P$, $\bar{s}_i \in Q_1$, $i = 1, 2, \ldots, m$, $z_0 \in (PQ_1^*\{1\}^*)^*$, $z_i \in (PQ_1 Q_2)^*$, $r_i \in (T \cup N)^*$, $i = 1, 2, \ldots, n$, and k stands for $11 \ldots 1$ (k times). Then M will go to state q_1 if (1) $n = 0$, (2) $m < n$, or (3) for some i and j, where $i, j = 1, 2, \ldots, n$, $r_i = r_j$, but $s_i \neq s_j$. Otherwise, M will have the instantaneous description $q_3 z$, $q_3 \in Q$.

(c) Suppose M has the instantaneous description $q_3 z$, where z is the same as in (b) satisfying none of the conditions (1)–(3). Then M will have instantaneous description $q_2 z'$, where z' is obtained from z by (1) erasing $\rho \bar{s}_1 \bar{s}_2 \ldots \overline{\rho}_m k$, (2) erasing $z_i b r_i$ if $m(r_i, s_i) < k$, and (3) replacing $z_i b r_i$ by

$$z_i \rho \bar{s}_i \bar{t}_{i1} b w_{i1} z_i \rho \bar{s}_i \bar{t}_{i2} b w_{i2} \ldots z_i \rho \bar{s}_i \bar{t}_{il} b w_{il} \qquad \text{if } m(r_i, s_i) \geqslant k.$$

Here, $w_{ir} \sim^k r_i \bmod(s_i, t_{ij})$, $j = 1, 2, \ldots, l$, and $\{t_{i1}, t_{i2}, \ldots, t_{in}\} = \{t \mid \rho(s, t_i) > 0\}$. Thereby, we assume that for all $\rho \in P$ and $s \in \overline{\rho}$, the set $\{t \mid \rho(s, t) > 0\}$ is well ordered.

(d) Suppose M has the instantaneous description $q_3 b z$, where $z \in W^*$. Then M will go to state q_1 if $z = \Lambda$. Otherwise, M will have the instantaneous description $q_4 z$, where $q_4 \in Q$.

(e) Suppose M has the instantaneous description $q_4 z$, where

$$z = z_1 b r_1 z_2 b r_2 \ldots z_n b r_n,$$

$n \geqslant 1$, $z_i \in (PQ_1 Q_2)^*$, $r_i \in (T \cup N)^*$, $i = 1, 2, \ldots, n$. If $z_1 \neq \Lambda$ and for all i, where $i = 1, 2, \ldots, n$, z_i starts with $\rho \bar{s}$ where $\rho \in P$ and $\bar{s} \in Q_1$, then M will have the instantaneous description $q_{\rho,s} z$, where $q_{\rho,s} \in Q$. If $z_1 = \Lambda$, then M will have the instantaneous description $q_5 z'$, where z' is obtained from z by erasing b, and $q_5 \in Q$. If $z = \Lambda$, then M will go to state q_1.

(f) Suppose M has the instantaneous description $q_{\rho,s} z$, where z is defined as in (e) and for all i, $i = 1, 2, \ldots, n$, z_i starts with $\rho \bar{s}$. Then, with probability $\rho(s,t)$, M will have the instantaneous description $q_4 z'$, where z' is obtained from z by erasing all $z_i br_i$ where z_i does not start with $\rho \overline{st}$, and erasing $\rho \overline{st}$ from all z_i that starts with $\rho \overline{st}$.

(g) Suppose M has the instantaneous description $q_5 r$, where $r \in W^*$. Then M will go to state q_1 if $z \notin T^*$. Otherwise, M will have the instantaneous description $q_6 r$, where for all $q \in Q$, $w \in W$, and $z \in W^* \cup \{+,-\}$, $p(q_6, w, z, q) = 0$.

Note that M acts probabilistically only when case (f) occurs. Otherwise, it acts deterministically. It follows that $f = R(F^M)$. ∎

Combining Theorem 5.14.7 and 5.14.8 yields the following result.

Theorem 5.14.9 *Every bounded type-0 random language can be generated by a synchronous PTM, and vice versa.* ∎

In view of Theorem 5.14.9, many interesting properties of bounded type-0 random languages can be obtained from the results given in [201].

5.15 Context-Free Probabilistic Grammars and Pushdown Automata

In this section, we study context-free probabilistic grammars and their relationship to probabilistic pushdown automata. We also examine the relationship between leftmost random languages and leftmost context-free random languages.

Proposition 5.15.1 *If f is a context-free random language, then $f = f_G$ for some total context-free probabilistic grammar $G = (T, N, P, h)$ such that $\overline{P} = N, h(A_0) = 1$ for some $A_0 \in N$, and for all $\rho \in P$ and $s \in \overline{\rho}, \rho(s,t) > 0$ implies $t = av$, where $a \in T \cup \{\Lambda\}$ and $v \in N^*$. If, in addition, f is bounded, then for all $\rho \in P$ and $s \in \overline{\rho}, \rho(s,t) > 0$ implies $t = av$, where $a \in T \cup \{\Lambda\}$ and $v \in NN$.*

In the remainder of the section, if $f = f_{G,L}$ for some specified (type-0, context-sensitive, context-free, weakly regular, and regular) probabilistic grammar G, then f is said to be a leftmost (type-0, context-sensitive, context-free, weakly regular, and regular) random language.

Theorem 5.15.2 *If $G = (T, N, P, h)$ is a context-free probabilistic grammar such that for all $r \in T^*$,*

$$f_G(r) = \bigvee_{z \in Z_1^*} \sum_{A \in N} h(A) f_z(A, r),$$

where $Z_1 = P \times D_L(G) \times \mathbb{N}$, then f_G is leftmost context free. In fact, $f_G = f_{G,L}$.

Proof. By a previous remark, $D_L(G)$ contains exactly one replacement function, say δ_0. For every

$$z = z_1(\rho, \delta_0, k)z_2 \in (P \times \{\delta_0\} \times \mathbb{N})^*,$$

where $z_2 \in (P \times \{\delta_0\} \times \{1\})^*$ and $k > 1$, let $z' = z_1(\rho, \delta_0, 1)z_2$. It can be shown that $f_z(A, r) \leqslant f_{z'}(A, r)$ for all $A \in N$ and $r \in T^*$. Thus for all $z \in (P \times \{\delta_0\} \times \mathbb{N})^*$, there exists $z_0 \in (P \times \{\delta_0\} \times \{1\})^*$ such that $f_z(A, r) \leqslant f_{z_0}(A, r)$ for all $A \in N$ and $r \in T^*$. Hence $f_G = f_{G,L}$. $\blacksquare$

Theorem 5.15.3 *If $G = (T, N, P, h)$ is a context-free probabilistic grammar such that P contains exactly one element, then f_G is leftmost context free. In fact, $f_G = f_{G,L}$.*

Proof. Let $P = \{\rho\}$ and $D_L(G) = \{\delta_0\}$. It follows by induction on $|z|$, $z \in (P \times D(G) \times \mathbb{N})^*$, that for all $z \in (\{\rho\} \times D(G) \times \mathbb{N})^*$, there exists $z' \in (\{\rho\} \times \{\delta_0\} \times \{1\})^*$ such that $f_z(A, r) \leqslant f_{z'}(A, r)$ for all $r \in T^*$ and $A \in N$. Thus $f_G = f_{G,L}$. $\blacksquare$

Definition 5.15.4 *A (total) probabilistic pushdown automaton (PPA) is a septuple $M = (Q, X, Y, W, p, h, g)$, where Q, X, Y, and W are finite nonempty sets, p is a (total) random function from $Q \times W \times X$ into $Y^* \times W^* \times Q$, h is a function from $Q \times W$ into $[0, 1]$ such that*

$$\sum_{q \in Q} \sum_{w \in W} h(q, w) = 1,$$

*and g is a function from Q into $[0, 1]$. If, in addition, for all $q, q' \in Q, u \in X$, and $w \in W$, $p(q, w, u, y, z, q') = 0$ except for finitely many $z \in W^*$ and $y \in Y^*$, then M is said to be **bounded**.*

In the Definition 5.15.4, Q, X, Y, and W are, respectively, the state, input, output, and pushdown alphabets. $p(q, w, u, y, z, q')$ is the conditional probability that the next state of M is q', the leftmost symbol w in the pushdown list is replaced by z and output string y is produced, given that the present state of M is q, the leftmost symbol in the pushdown list is w and input u is applied. $h(q, w)$ is the probability that q is the initial state of M and w is the initial symbol in the pushdown list. $g(q)$ is the probability that q is a final state of M.

Definition 5.15.5 *Let $M = (Q, X, Y, W, p, h, g)$ be a PPA.*
(1) Define the function p^M from $(Q \times W^* \times X^* \times Y^*) \times (Q \times W^* \times X^* \times Y^*)$ into $[0, 1]$ by for all $r, s \in Q \times W^* \times X^* \times Y^*$,

$$p^M(r, s) = \begin{cases} p(q, w, u, y_0, z_0, q') & \text{if } r = (q, wz, ux, y), \ s = (q', z_0z, x, yy_0) \\ 1 & \text{if } r = (q, \Lambda, ux, y), \ s = (q, \Lambda, x, y) \\ & \text{or } r = s = (q, z, \Lambda, y) \\ 0 & \text{otherwise,} \end{cases}$$

where $q, q' \in Q, u \in X, x \in X^*, y_0, y \in Y^*, w \in W,$ and $z_0, z \in W^*.$

(2) For all $k \in \mathbb{N} \cup \{0\}$, define the function p_k^M from $(Q \times W^* \times X^* \times Y^*) \times (Q \times W^* \times X^* \times Y^*)$ into $[0,1]$ by for all $r, s \in Q \times W^* \times X^* \times Y^*$,

$$p_0^M(r,s) = \begin{cases} 1 & if \ r = s \\ 0 & if \ r \neq s \end{cases}$$

$$p_{k+1}^M(r,s) = \sum_{v \in \Gamma} p_k^M(v,s) p^M(r,v),$$

where $\Gamma = Q \times W^* \times X^* \times Y^*.$

(3) Define the function F^M from $X^* \times Y^*$ into $[0,1]$ by for all $x \in X^*$ and $y \in Y^*,$

$$F^M(x,y) = \sum_{q,q' \in Q} \sum_{w \in W} h(q,w) \left[\sum_{k=0}^{\infty} p_k^M(q,w,x,\Lambda,q',\Lambda,\Lambda,y) \right].$$

(4) Define the function G^M from $X^* \times Y^*$ into $[0,1]$ by for all $x \in X^*$ and $y \in Y^*,$

$$G^M(x,y) = \sum_{q,q' \in Q} \sum_{w \in W} \sum_{z \in W^*} h(q,w)g(q') \left[\sum_{k=0}^{\infty} p_k^M(q,w,x,\Lambda,q',z,\Lambda,y) \right].$$

It follows that F^M and G^M are both random functions from X into Y^*. Moreover, for all $x \in X^*$ and $y \in Y^*$, where $|x| = k$,

$$F^M(x,y) = \sum_{q,q' \in Q} \sum_{w \in W^*} h(q,w) p_k^M(q,w,x,\Lambda,q',\Lambda,\Lambda,y)$$

and

$$G^M(x,y) = \sum_{q,q' \in Q} \sum_{w \in W} \sum_{z \in W} h(q,w)g(q') p_k^M(q,w,x,\Lambda,q',z,\Lambda,y).$$

It follows from the above definition that PPAs are stochastic generalizations of conventional pushdown automata. F^M is the random function computed by M with empty store, while G^M is the random function computed by M with final states.

Theorem 5.15.6 *If f is a leftmost(bounded) context-free random language, then f can be generated with empty store by some total (bounded) PPA having a single state and such that $p(q,w,u,y,z,q') > 0$ implies $y \in T \cup \{\Lambda\}.$*

Proof. Let $f = f_{G,L}$, where $G = (T, N, P, h)$ is a (bounded) context-free probabilistic grammar. By Proposition 5.15.1, we may assume that G is total and for all $\rho \in P$ and $s \in \overline{\rho}$, $\rho(s, t) > 0$ implies $t = av$, where $a \in T \cup \{\Lambda\}$ and $v \in N^*$. Let $M = (P, \{q_0\}, T, N, p, h', g)$, where $h'(q_0, A) = h(A)$ and $p(q_0, w, y, \rho, z, q_0) = \rho(w, yz)$ for all $z \in N^*$ and $y \in T^*$, $\rho \in P$, and $w \in N$. Clearly M is a PPA with the desired properties. Moreover, it can be shown that $f = R(F^M)$. ■

Theorem 5.15.7 *If f is a leftmost (bounded) context-free random language, then f can be generated with final states by some total (bounded) PPA having two states and such that $p(q, w, u, y, z, q') > 0$ implies $y \in T \cup \{\Lambda\}$.*

Proof. Let $f = f_{G,L}$, where $G = (T, N, P, h)$ is a (bounded) context-free probabilistic grammar. By Proposition 5.15.1, we may assume that G is total and for all $\rho \in P$ and $s \in \overline{\rho}$, $\rho(s, t) > 0$ implies $t = av$, where $a \in T \cup \{\Lambda\}$ and $v \in N^*$. Moreover, we may assume, without loss of generality, that there exist uniquely $A_0, A_1, A_2 \in N$ and $\rho_0, \rho_1 \in P$ such that (1) $h(A_0) = 1$, (2) $\overline{\rho}_0 = \{A_0\}$ and $\rho_0(A_0, A_2 A_1) = 1$, (3) $\overline{\rho}_1 = \{A_1\}$ and $\rho_1(A_1, \Lambda) = 1$, (4) for $i = 0, 1$, $A_i \in \overline{\rho}$ implies $\rho = \rho_i$, (5) for all $\rho \in P$ and $s \in \overline{\rho}$, $\rho(s, t) > 0$ implies $t \notin (T \cup N)^*\{A_0\}(T \cup N)^*$, and (6) for all $\rho \in P$, $s \in \overline{\rho}$, and $\rho \neq \rho_0$, $\rho(s, t) > 0$ implies $t \notin (T \cup N)^*\{A_1\}(T \cup N)^*$. That is, we modify G in such a way that A_1 serves as an endmarker. Let $M = (Q, P, T, N, p, h', g)$ where $Q = \{q_1, q_2\}$, $h'(q_1, A_0) = 1$, $g(q_2) = 1$ and for all $q, q' \in Q$, $\rho \in P$, $w \in N$, $z \in N^*$, and $y \in T^*$,

$$
p(q, w, \rho, y, z, q') = \begin{cases} \rho(w, zy) & \text{if } q = q' = q_1, \ w \neq A_1 \text{ or } \rho \neq \rho_1 \\ 1 & \text{if } q = q_1, \ q' = q_2, \ w = z = A_1, \\ & \phantom{\text{if }} \rho = \rho_1, \ y = \Lambda \\ & \phantom{\text{if }} \text{or } q = q' = q_2, \ w = z, \ y = \Lambda \\ 0 & \text{otherwise.} \end{cases}
$$

Clearly, M is a PPA with the desired properties. Moreover, it can be shown that $f = R(G^M)$. ■

Let S be a nonempty set and $k \in \mathbb{N}$. Let $S^k = \{r \in S^* \mid |r| \leqslant k\}$.

Theorem 5.15.8 *If $f = R(F^M)$ for some bounded PPA M, then f is a leftmost bounded context-free random language.*

Proof. Let $M = (Q, X, Y, W, p, h, g)$. Since M is bounded, there exists $k \in \mathbb{N}$ such that for all $q, q' \in Q$, $u \in X$, $w \in W$, $y \in Y^*$, and $z \in W^*$, $p(q, w, u, y, z, q') > 0$ implies $|z| \leqslant k$. Let $N_0 = \{(q, w, q') \mid q, q' \in Q$ and $w \in W\}$ and $N = N_0 \cup \{A_0\}$, where $A_0 \notin N_0$. For ease of notation, we write $(q_1 q_2 \ldots q_n, w_1 w_2 \ldots w_n, q'_1 q'_2 \ldots q'_n)$ for $(q_1, w_1, q'_1)(q_2, w_2, q'_2) \ldots (q_n, w_n, q'_n)$, where $(q_i, w_i, q'_i) \in N_0, i = 1, 2, \ldots, n$. Let $\mathcal{F}$ be the set of all functions f from W^k into Q^k such that $f(z) = r$ implies $|r| = |z| - 1$. For all $u \in X$ and

$f \in \mathcal{F}$, let $\rho_{u,f}$ be the probabilistic production over $Y \cup N$, where $\overline{\rho}_{u,f} = N_0$ and for all $A \in N_0$, $t \in (Y \cup N)^*$,

$$\rho_{u,f}(A,t) = \begin{cases} p(q,w,u,y,z,q') & \text{if } A = (q,w,q') \text{ and} \\ & t = y(qf(z),z,f(z)q') \\ 0 & \text{otherwise.} \end{cases}$$

Moreover, for all $q \in Q$, let ρ_q be the probabilistic production over $Y \cup N$ where $\overline{\rho}_q = \{A_0\}$ and

$$\rho_q(A,t_0) = \begin{cases} h(q,w) & \text{if } t = (q,w,q') \\ 0 & \text{otherwise.} \end{cases}$$

Let $G = (Y, N, P, h')$, where $h'(A_0) = 1$ and $P = \{\rho_{u,f} \mid u \in X, f \in \mathcal{F}\} \cup \{\rho_q \mid q \in Q\}$. Clearly, G is a bounded context-free probabilistic grammar. Furthermore, it can be shown that $f = f_{G,L}$. ∎

Theorem 5.15.9 *If* $f = R(G^M)$ *for some bounded PPA* M, *then* f *is a leftmost-bounded context-free random language.*

Proof. By Theorem 5.15.6, it suffices to show that $f = R(F^{M_0})$ for some bounded PPA M_0. Let $M = (Q, X, Y, W, p, h, g)$ and $M_0 = (Q_0, X_0, Y, W_0, p_0, h_0, g_0)$, where $Q_0 = Q \cup \{q_0, q_1\}$, $q_0, q_1 \notin Q$, $X_0 = X \cup \{u_0, u_1\}$, $u_0, u_1 \notin X$, $W_0 = W \cup \{w_0, \$\}$, w_0, $\$ \notin W$, $h_0(q_0, w_0) = 1$, and for all $q, q' \in Q_0$, $u \in X_0$, $w \in W_0$, $y \in Y_0^*$, $z \in W_0^*$,

$$p_0(q,w,u,y,z,q') = \begin{cases} h(q,w') & \text{if } q = q_0, \ w = w_0, \ u = u_0, \\ & q' \in Q, \ z = w'\$, \ y = \Lambda \\ p(q,w,u,y,z,q') & \text{if } q, \ q' \in Q, \ w \in W, \ u \in X, \\ & y \in Y^*, \ z \in W^* \\ g(q) & \text{if } q \in Q, \ q' = q_1, \ w \in W_0, \\ & u = u_1, \ z = y = \Lambda \\ 1 & \text{if } q = q' = q_1, \ w \in W_0, \\ & u = u_1, \ z = y = \Lambda \\ 0 & \text{otherwise.} \end{cases}$$

Clearly, M_0 is a bounded PPA. Furthermore, it can be shown that $f = F(F^{M_0})$. ∎

Combining Theorems 5.15.6, 5.15.7, 5.15.8, and 5.15.9 yields the following result.

Theorem 5.15.10 *The following statements are equivalent:*

(1) f is a leftmost-bounded context-free random language.

(2) $f = R(F^M)$ for some bounded PPA M.

(3) $f = R(G^M)$ for some bounded PPA M.

(4) $f = R(F^M)$ for some total bounded PPA M, where $p(q,w,u,y,z,q') > 0$ implies $y \in Y \cup \{\Lambda\}$.

(5) $f = R(G^M)$ for some total bounded PPA M, where $p(q,w,u,y,z,q') > 0$ implies $y \in Y \cup \{\Lambda\}$. ∎

It is known, [136], that languages generated by any grammar using only leftmost derivations are context free. We show next that a similar result holds for random languages.

Theorem 5.15.11 *If $f = f_{G,L}$ for some bounded probabilistic grammar $G = (T, N, P, h)$ such that $P = \{\rho\}$, then f is leftmost-bounded context-free.*

Proof. For all $s \in \overline{\rho}$, let A_i^s denote the i-th symbol of s. Moreover, for all $s \in \overline{\rho}$, we associate the symbols u_i^s, q_i^s, $i = 1, 2, \ldots, |s| - 1$. Let $M = (Q, X, T, W, p, h', g)$, where $X = \{\rho\} \cup \{u_i^s \mid s \in \overline{\rho} \text{ and } i = 1, 2, \ldots, |s| - 1\}$, $Q = \{q_0\} \cup \{q_i^s \mid s \in \overline{\rho} \text{ and } i = 1, 2, \ldots, |s| - 1\}$, $W = T \cup N$, and for all $q, q' \in Q, u \in X, w \in W, y \in T^*$, and $z \in W^*$,

$$p(q, w, u, y, z, q') = \begin{cases} \rho(s, t) & \text{if } q = q_{k-1}^s, \ w = A_k^s, \ u = \rho, \\ & q' = q_0, \ z = t, \ y = \Lambda, \ k = |s| \\ 1 & \text{if } (u, w, q) = (u_k^s, A_k^s, q_{k-1}^s), \ q' = q_k^s, \\ & z = y = \Lambda, \ k < |s| \\ & \text{or } q = q' = q_0, \ w \in T \text{ and } w \neq A_1^s \\ & \text{for any } s \in \overline{P}, \ z = \Lambda, \ y = w \\ & \text{or } (u, w, q) \text{ is not equal to any of those} \\ & \text{given above, } q' = q, \ z = w, \ y = \Lambda \\ 0 & \text{otherwise,} \end{cases}$$

$$h'(q, w) = \begin{cases} h(w) & \text{if } q = q_0, \ w \in N \\ 0 & \text{otherwise.} \end{cases}$$

Clearly, M is a bounded PPA. Moreover, it can be shown that $f = R(F^M)$. Hence by Theorem 5.15.10, f is leftmost-bounded context-free. ∎

5.16   Realization of Fuzzy Languages by Various Automata

The remainder of the chapter concerns the results in [207].

Consider a probabilistic automaton A over an input set X. For all $x \in X^*$, we associate a number $r^A(x)$ representing the probability that x is accepted by A. Then r^A may be viewed as a fuzzy language realized by A. In a similar manner, one may consider fuzzy languages realized by max-product and max-min automata.

In the remainder of the chapter, we examine the class of fuzzy languages realized by probabilistic, max-product, and max-min automata, denoted by $L_1, L_2,$ and L_3, respectively. Related results for L_1 and L_3 can be found in [158] and [196].

We give characterizations of L_k, $k = 1,2,3$. They reduce to Nerode's Theorem [181] in the case of deterministic automata.

We show that neither L_1 is a subclass of L_2, nor L_2 is a subclass of L_1. We also show that L_3 is generated by the class of all probabilistic, max-product, or max-min automata with deterministic transition function. This leads to the interesting result that $L_3 = L_1 \cap FL = L_2 \cap FL$, where FL is the class of all fuzzy languages with a finite image.

The results obtained in the next two sections can be applied to the study of input-output relations [30] or sequential functions [196] realized by stochastic, max-product, and max-min machines and to the study of languages accepted by probabilistic, max-product, and max-min automata.

Definition 5.16.1 *Let X be a finite nonempty set. A **pseudoautomaton** over X is a quadruple $A = (Q,p,h,g)$, where Q is a finite nonempty set, p is a function from $Q \times X \times Q$ into $[0,1]$, and h and g are functions from Q into $[0,1]$. If, in addition, $\sum_{q \in Q} h(q) = 1$ and $\sum_{q \in Q} p(q,u,q') = 1$ for all $q \in Q$ and $u \in X$, then A is called a **probabilistic pseudoautomaton**.*

In Definition 5.16.1, Q is the set of states, X is the set of input symbols, and $p(q,u,q')$ is the grade of membership that the next state of A is q' given that the present state of A is q and input u is applied. $h(q)$ is the grade of membership that q is the initial state of A and $g(q)$ is the grade of membership that q belongs to the final states of A. For probabilistic pseudoautomaton, grade of membership is equivalent to probability.

For ease of notation, the input set X is assumed fixed throughout the remainder of the chapter. Hence any pseudoautomaton is understood to be a pseudoautomaton over X.

As usual X^* is the free monoid with identity Λ generated by X.

Given a pseudoautomaton $A = (Q,p,h,g)$, various response functions r^A from X^* into $[0,1]$ may be associated with A. For $x \in X^*$, $r^A(x)$ is the grade of membership that x is accepted by A. The manner in which r^A is related to p, h, and g determines the type of automaton generated by A.

Definition 5.16.2 *Let $A = (Q,p,h,g)$ be a probabilistic pseudoautomaton.*

(1) Define the function p_1^A from $Q \times X^* \times Q$ into $[0,1]$ by for all $q',q'' \in Q$, $u \in X$, $x \in X^*$,

$$p_1^A(q',\Lambda,q'') = \begin{cases} 1 & \text{if } q' = q'' \\ 0 & \text{if } q' \neq q'' \end{cases}$$

$$p_1^A(q',xu,q'') = \sum_{q \in Q} p(q,u,q'')p_1^A(q',x,q).$$

(2) Define the function q_1^A from $Q \times X^*$ into $[0,1]$ by for all $q \in Q$ and $x \in X^*$,

$$q_1^A(q,x) = \sum_{q' \in Q} p_1^A(q,x,q')g(q').$$

(3) Define the function r_1^A from X^* into $[0,1]$ by for all $x \in X^*$,

$$r_1^A(x) = \sum_{q \in Q} h(q)q_1^A(q,x).$$

A probabilistic pseudoautomaton A together with the response function r_1^A defined above constitute a probabilistic automaton. This model of probabilistic automata reduces to the conventional model [189] if $\text{Im}(g) \subseteq \{0,1\}$.

Definition 5.16.3 *Let $A = (Q, p, h, g)$ be a pseudoautomaton.*

(1) The functions p_2^A, q_2^A, and r_2^A are defined in exactly the same manner as p_1^A, q_1^A, and r_1^A, respectively, with summation replaced by maximum. (For example,

$$p_2^A(q', xu, q'') = \vee\{p(q, u, q'')p_2^A(q', x, q) \mid q \in Q\}.)$$

(2) The functions p_3^A, q_3^A, and r_3^A are defined in exactly the same manner as p_2^A, q_2^A, and r_2^A, respectively, with product replaced by minimum. (For example,

$$p_3^A(q', xu, q'') = \vee\{p(q, u, q'') \wedge p_3^A(q', x, q) \mid q \in Q\}.)$$

A pseudoautomaton A together with the response function r_2^A defined above constitute a max-product automaton. Similarly, a pseudoautomaton A together with the response function r_3^A defined above constitute a max-min automaton.

As usual, a fuzzy language (over X) is a function from X^* into $[0,1]$.

Let L denote the set of all fuzzy languages over X.

Let L_1 denote the set of all response functions r_1^A, where A is a probabilistic pseudoautomaton.

Let L_k denote the set of all response functions r_k^A, where A is a pseudoautomaton, $k = 2, 3$.

The following operations of fuzzy languages are considered in the remainder of the chapter.

If $\lambda_1, \lambda_2 \in L$, then $\lambda_1\lambda_2, \lambda_1 \cup \lambda_2$, and $\lambda_1 \cap \lambda_2$ are fuzzy languages such that for all $x \in X^*$, $(\lambda_1\lambda_2)(x) = \lambda_1(x) \cdot \lambda_2(x)$, $\lambda_1 \cup \lambda_2(x) = \lambda_1(x) \vee \lambda_2(x)$, and $\lambda_1 \cap \lambda_2(x) = \lambda_1(x) \wedge \lambda_2(x)$. In general, if $\lambda_i \in L$, for $i = 1, 2, \ldots, n$, then $\Pi_{i=1}^n \lambda_i = (\lambda_1 \ldots \lambda_{n-1})\lambda_n$, $\cup_{i=1}^n \lambda_i = (\lambda_1 \cup \ldots \cup \lambda_{n-1}) \cup \lambda_n$, and $\cap_{i=1}^n \lambda_i = (\lambda_1 \cap \ldots \cap \lambda_{n-1}) \cap \lambda_n$. Moreover, $\lambda_1 + \lambda_2$ and $\sum_{i=1}^n \lambda_i$ are defined similarly provided $\lambda_1 + \lambda_2$ and $\sum_{i=1}^n \lambda_i$ are fuzzy languages.

If $\lambda_0 \in L$ is constant, i.e., there exists $0 \leq c \leq 1$ such that $\lambda_0(x) = c$ for all $x \in X^*$, then we write $c\lambda$, $c \cup \lambda$, and $c \cap \lambda$ for $\lambda_0\lambda$, $\lambda_0 \cup \lambda$, and $\lambda_0 \cap \lambda$, respectively, where $\lambda \in L$.

Definition 5.16.4 *Let $\lambda_i \in L$, $i = 0, 1, 2, \ldots, n$.*

(1) λ_0 *is a convex combination or* 1-*combination of* $\{\lambda_1, \lambda_2, \ldots, \lambda_n\}$ *if there exists* $0 \leq c_i \leq 1$, $i = 1, 2, \ldots, n$ *such that* $\sum_{i=1}^{n} c_i = 1$ *and* $\lambda_0 = \sum_{i=1}^{n} c_i \lambda_i$.

(2) λ_0 *is a max-product combination or* 2-*combination of*

$$\{\lambda_1, \lambda_2, \ldots, \lambda_n\}$$

if there exists $0 \leq c_i \leq 1$, $i = 1, 2, \ldots, n$ *such that* $\lambda_0 = \cup_{i=1}^{n} c_i \lambda_i$.

(3) λ_0 *is a max-min combination or* 3-*combination of* $\{\lambda_1, \lambda_2, \ldots, \lambda_n\}$ *if there exists* $0 \leq c_i \leq 1$, $i = 1, 2, \ldots, n$ *such that* $\lambda_0 = \cup_{i=1}^{n}(c_i \cap \lambda_i)$.

Let $\lambda \in L$ and $x \in X^*$. Let λ^x denote the fuzzy language over X such that $\lambda^x(y) = \lambda(xy)$ for all $y \in X^*$. Moreover, let $T(\lambda) = \{\lambda^x \mid x \in X^*\}$.

Definition 5.16.5 *Let* $k = 1, 2, 3$, *and* $\lambda_i \in L$, $i = 1, 2, \ldots, n$. *Then* $\{\lambda_1, \lambda_2, \ldots, \lambda_n\}$ *is called a* k-***admissible set*** *if for all* $i = 1, 2, \ldots, n$ *and* $u \in X$, λ_i^u *is a* k-*combination of* $\{\lambda_1, \lambda_2, \ldots, \lambda_n\}$. *Moreover, let* $\lambda \in L$. *Then* λ *is said to have a* k-*admissible set if there exists a* k-*admissible set* $\{\lambda_1, \lambda_2, \ldots, \lambda_n\}$ *such that* λ *is a* k-*combination of* $\{\lambda_1, \lambda_2, \ldots, \lambda_n\}$. *In this case,* $\{\lambda_1, \lambda_2, \ldots, \lambda_n\}$ *is called a* k-***admissible*** *set of* λ.

Proposition 5.16.6 *Let* $k = 1, 2, 3$. *If* $\{\lambda_1, \lambda_2, \ldots, \lambda_n\}$ *is a* k-*admissible set of* $\lambda \in L$, *then for all* $x \in X^*$, $\{\lambda_1, \lambda_2, \ldots, \lambda_n\}$ *is a* k-*admissible set of* λ^x. ∎

In the remainder of the chapter, the symbol A denotes a probabilistic pseudoautomaton if the case $k = 1$ is under consideration and a pseudoautomaton if the case $k = 2$ or 3 is under consideration.

Theorem 5.16.7 *Let* $\lambda \in L$. *Then for* $k = 1, 2, 3$, $\lambda \in L_k$ *if and only if* λ *has a* k-*admissible set.*

Proof. We prove the result for the case $k = 1$ only. The other cases can be proved in a similar manner. Suppose $\lambda \in L_1$. Then $\lambda = r_1^A$ for some $A = (Q, p, h, g)$, where $Q = \{q_1, q_2, \ldots, q_n\}$. For all $i = 1, 2, \ldots, n$ and $x \in X^*$, let $\lambda_i(x) = q_1^A(q_i, x)$. Then it follows that $\{\lambda_1, \lambda_2, \ldots, \lambda_n\}$ is a 1-admissible set of λ. Conversely, suppose $\{\lambda_1, \lambda_2, \ldots, \lambda_n\}$ is a 1-admissible set of λ. Then for all $i = 1, 2, \ldots, n$ and $u \in X$, $\lambda_i^u = \sum_{j=1}^{n} \alpha_{ij}(u)\lambda_j$ for some $\alpha_{ij}(u) \geq 0$ and $\sum_{j=1}^{n} \alpha_{ij}(u) = 1$. Moreover $\lambda = \sum_{i=1}^{n} c_i \lambda_i$ for some $c_i \geq 0$ and $\sum_{i=1}^{n} c_i = 1$. Let $A = (Q, p, h, g)$, where $Q = \{q_1, q_2, \ldots, q_n\}$ and for all $1 \leq i, j \leq n$ and $u \in X$, $p(q_j, u, q_i) = \alpha_{ij}(u)$, $h(q_i) = c_i$, and $g(q_i) = \lambda_i(\Lambda)$. It follows that $\lambda = r_1^A$. ∎

We show in Section 5.17 that the above characterization for L_k, $k = 1, 2, 3$, reduces to Nerode's Theorem [181] in the case of deterministic automata.

Corollary 5.16.8 *If* $\lambda \in L_1$, *then the vector space* $V(\lambda)$ *spanned by* $T(\lambda)$ *is finite-dimensional.* ∎

The converse of Corollary 5.16.8 does not hold in general. A counter-example may be constructed by adopting a method described in [45]. Nevertheless, it follows from the proof of Theorem 5.16.7 that $V(\lambda)$ is finite dimensional if and only if λ is realized by a generalized automaton [196]. Moreover, the dimension of $V(\lambda)$ is a lower bound for the number of states of any probabilistic automaton realizing λ.

Let $\lambda \in L$ and $x \in X^*$. Let $\lambda^{(x)}$ denote the fuzzy language over X such that $\lambda^{(x)}(y) = \lambda(yx)$ for all $y \in X^*$. Moreover, let $T_0(\lambda) = \{\lambda^{(x)} \mid x \in X^*\}$.

It is clear that, for $k = 2$ or 3, a result similar to Theorem 5.16.7 can be established using $\lambda^{(x)}$ instead of λ^x. However, for $k = 1$, some modifications are needed.

Theorem 5.16.9 *Let $\lambda \in L$. Then $\lambda \in L_1$ if and only if there exist $\lambda_i \in L$, $i = 1, 2, \ldots, n$, such that the following conditions hold:*

(1) for all $x \in X^*$, $\sum_{i=1}^{n} \lambda_i(x) = 1$;

(2) for all $1 \leq i \leq n$, and $u \in X$, $\lambda_i^{(u)} = \sum_{j=1}^{n} \alpha_{ij}(u)\lambda_j$, where $\alpha_{ij}(u) \geq 0$ and $\sum_{j=1}^{n} \alpha_{ij}(u) = 1$;

(3) $\lambda = \sum_{i=1}^{n} c_i \lambda_i$, where for all $0 \leq c_i \leq 1$, $i = 1, 2, \ldots, n$.

Proof. Suppose $\lambda = r_1^A$, where $A = (Q, p, h, g)$ and $Q = \{q_1, q_2, \ldots, q_n\}$. For all $i = 1, 2, \ldots, n$ and $x \in X^*$, let $\lambda_i(x) = \sum_{j=1}^{n} h(q_j)p^A(q_j, x, q_i)$. It follows that $\{\lambda_1, \lambda_2, \ldots, \lambda_n\}$ has the desired properties. Conversely, suppose there exists $\lambda_i \in L$, $i = 1, 2, \ldots, n$, such that conditions (1), (2), and (3) hold. Let $A = (Q, p, h, g)$, where $Q = \{q_1, q_2, \ldots, q_n\}$ and for all $i = 1, 2, \ldots, n$ and $u \in X$, $p(q_i, u, q_j) = \alpha_{ji}(u)$, $h(q_i) = \lambda_i(\Lambda)$, and $g(q_i) = c_i$. It follows that $\lambda = r_1^A$. ∎

Corollary 5.16.10 *If $\lambda \in L_1$, then the vector space $V_0(\lambda)$ spanned by $T_0(\lambda)$ is finite dimensional.* ∎

Suppose that X^* is well ordered. Consider the infinite matrix $M(\lambda)$ whose (i, j)-th entry is $\lambda(x_i x_j)$ where x_i is the i-th element of X^*. Then the rows and columns of $M(\lambda)$ are precisely $T(\lambda)$ and $T_0(\lambda)$, respectively. Thus, the dimension of $V(\lambda)$, the dimension of $V_0(\lambda)$, and the rank of $M(\lambda)$ are equal. Suppose the dimension of $V(\lambda)$ is n. Then there exist $x_i', x_j'' \in X^*$, $1 \leq i, j \leq n$, such that the matrix M_0, whose (i, j)-th entry is $\lambda(x_i' x_j'')$, is nonsingular and has rank n. It follows from the proof of Theorem 5.16.7 that a generalized automaton that realizes λ can be constructed from the knowledge of M and $\lambda(x_i' u x_j'')$, $1 \leq i, j \leq n$, and $u \in X$. Since x_i', x_i'' can always be chosen [30] such that $|x_i'| < n$ and $|x_i''| < n$ for all $i = 1, 2, \ldots, n$, we have proved the following theorem.

Theorem 5.16.11 *If the dimension of $V(\lambda)$ is $\leq n$, then λ is uniquely determined by $\lambda(x)$, where $|x| < 2n - 1$.* ∎

An elaborate procedure, which is essentially the same as that given above, can be found in [30] for stochastic sequential machines. A similar procedure was given in [196] for max-min sequential machines.

We conclude this section by showing that for L_k, $k = 1, 2, 3$, there is no loss of generality by considering only certain more restrictive classes of pseudo automata, e.g., the conventional probabilistic, max-product, and max-min automata, respectively. For $k = 3$, additional results along this direction can be found in [196].

Theorem 5.16.12 *Let $k = 1, 2, 3$. If $\lambda \in L_k$, then $\lambda = r_k^A$ for some $A = (Q, p, h, g)$, where $\text{Im}(g) \subseteq \{0, 1\}$.*

Proof. By Theorem 5.16.7, λ has a k-admissible set $\{\lambda_1, \lambda_2, \ldots, \lambda_n\}$. For all $i = 1, 2, \ldots, n$ and $j = 1, 2$, define

$$\lambda_{ij}(x) = \begin{cases} \lambda_i(x) & \text{if } x \neq \Lambda \\ 1 & \text{if } x = \Lambda \text{ and } j = 1 \\ 0 & \text{if } x = \Lambda \text{ and } j = 2, \end{cases}$$

$\forall x \in X^*$. It is clear that

(a) if $\lambda \in L$ and $\lambda = \sum_{i=1}^n c_i \lambda_i$, then

$$\lambda = \sum_{i=1}^n c_i \lambda_i(\Lambda) \lambda_{i1} + \sum_{i=1}^n c_i [1 - \lambda_i(\Lambda)] \lambda_{i2},$$

(b) if $\lambda \in L$ and $\lambda = \cup_{i=1}^n c_i \lambda_i$, then

$$\lambda = (\cup_{i=1}^n c_i \lambda_i(\Lambda) \lambda_{i1}) \cup (\cup_{i=1}^n c_i \lambda_{i2}),$$

and

(c) if $\lambda \in L$ and $\lambda = \cup_{i=1}^n (c_i \cap \lambda_i)$, then

$$\lambda = (\cup_{i=1}^n (c_i \cap \lambda_i(\Lambda) \cap \lambda_{i1})) \cup (\cup_{i=1}^n (c_i \cap \lambda_{i2})).$$

Thus $\{\lambda_{ij} \mid i = 1, 2, \ldots, n \text{ and } j = 1, 2\}$ is a k-admissible set of λ. The desired result follows from Theorem 5.16.7. ∎

Theorem 5.16.13 *For $k = 1, 2, 3$, if $\lambda \in L_k$, then $\lambda = r_k^A$ for some $A = (Q, p, h, g)$ such that $\text{Im}(h) \subseteq \{0, 1\}$.*

Proof. The proof follows from Theorem 5.16.7 and the fact that if $\{\lambda_1, \lambda_2, \ldots, \lambda_n\}$ is a k-admissible set of λ, then so is $\{\lambda, \lambda_1, \lambda_2, \ldots, \lambda_n\}$. ∎

Theorem 5.16.14 *For $k = 2, 3$, if $\lambda \in L_k$, then $\lambda = r_k^A$ for some $A = (Q, p, h, g)$, where $\vee\{h(q) \mid q \in Q\} = 1$, $\vee\{g(q) \mid q \in Q\} = 1$, and $\vee\{p(q, u, q') \mid q' \in Q\} = 1$ for all $q \in Q$ and $u \in X$.*

Proof. By Theorem 5.16.7, λ has a k-admissible set $\{\lambda_1, \lambda_2, \ldots, \lambda_n\}$. For all $x \in X^*$, let $\lambda_0(x) = 0$ and $\lambda_{n+1}(x) = 1$. Since $\{\lambda_0, \lambda_1, \lambda_2, \ldots, \lambda_n, \lambda_{n+1}\}$ is a k-admissible set of λ, the conclusion follows from the proof of Theorem 5.16.7. ∎

5.17 Properties of L_k, $k = 1, 2, 3$

In this section, properties of L_k, $k = 1, 2, 3$ are derived using the characterizations established in the previous section.

Theorem 5.17.1 *If $\lambda \in L$ is constant, then $\lambda \in L_k$ for $k = 1, 2, 3$.*

Proof. The proof follows from Theorem 5.16.7 and the fact that if λ is constant, then $\{\lambda\}$ is a k-admissible set of λ. ∎

Theorem 5.17.2 *For $k = 1, 2, 3$, if $\lambda \in L_k$, then for all $x \in X^*$, $\lambda^x \in L_k$.*

Proof. The proof follows from Proposition 5.16.6 and Theorem 5.16.7. ∎

The next theorem gives a converse of Theorem 5.17.2.

Theorem 5.17.3 *Let $\lambda \in L$. For $k = 1, 2, 3$, if there exists a positive integer m such that $\lambda^x \in L_k$ for all $x \in X^*$ such that $|x| = m$, then $\lambda \in L_k$.*

Proof. For all $x \in X^*$ such that $|x| = m$, let $\Gamma(x)$ be a k-admissible set of λ^x. Let $\Gamma_0 = \cup\{\Gamma(x) \mid x \in X^*$ and $|x| = m\}$ and $\Gamma = \Gamma_0 \cup \{\lambda^x \mid x \in X^*$ and $|x| < m\}$. It follows that Γ is a k-admissible set of λ. Thus $\lambda \in L_k$. ∎

Theorem 5.17.4 *Let $\lambda \in L$. For $k = 1, 2, 3$, if $\lambda_0 \in L_k$ and $\{x \in X^* \mid \lambda(x) \neq \lambda_0(x)\}$ is a finite set, then $\lambda \in L_k$.*

Proof. There exists a positive integer m such that $\lambda(x) = \lambda_0(x)$ for all $x \in X^*$ and $|x| \geq m$. Thus by Theorem 5.17.2, for all $x \in X^*$ where $|x| = m$, $\lambda_0^x \in L_k$. By Theorem 5.17.3, $\lambda \in L_k$. ∎

Definition 5.17.5 *A **Rabin (probabilistic) pseudoautomaton** is a (probabilistic) pseudoautomaton $A = (Q, \rho, h, g)$ such that $\mathrm{Im}(h) \cup \mathrm{Im}(g) \subseteq \{0, 1\}$.*

Theorem 5.17.6 *Let $\lambda \in L$. For $k = 1, 2, 3$, $\lambda = r_k^A$ for some Rabin pseudoautomaton A if and only if (1) $\lambda(\Lambda) \in \{0, 1\}$ and (2) there exists $\lambda_0 \in L_k$ such that $\lambda(x) = \lambda_0(x)$ for all $x \in X^* \backslash \{\Lambda\}$.*

Proof. The proof follows from Theorems 5.16.12, 5.16.13, and 5.17.4. ∎

Theorem 5.17.7 *Let $\lambda \in L$. For $k = 1, 2, 3$, if λ is a k-combination of $\{\lambda_1, \lambda_2, \ldots, \lambda_n\}$, where $\lambda_i \in L_k$, $i = 1, 2, \ldots, n$, then $\lambda \in L_k$.*

Proof. For all $i = 1, 2, \ldots, n$, let Γ_i be a k-admissible set of λ_i. Let $\Gamma = \cup_{i=1}^n \Gamma_i$. Then it follows that Γ is a k-admissible set of λ. Thus $\lambda \in L_k$. ∎

Corollary 5.17.8 Let $k = 1, 2$. If $\lambda \in L_k$ and $0 \leq c \leq 1$, then $c\lambda \in L_k$. ∎

Corollary 5.17.9 Let $k = 2, 3$. If $\lambda_1, \lambda_2 \in L_k$, then $\lambda_1 \cup \lambda_2 \in L_k$. ∎

Theorem 5.17.10 Let $k = 1, 2$. If $\lambda_1, \lambda_2 \in L_k$, then $\lambda_1 \lambda_2 \in L_k$.

Proof. Let Γ_i be a k-admissible set of λ_i, $i = 1, 2$ and $\Gamma = \{\lambda_1' \lambda_2' \mid \lambda_i' \in \Gamma_i, i = 1, 2\}$. It follows that Γ is a k-admissible set of $\lambda_1 \lambda_2$. Thus $\lambda_1 \lambda_2 \in L_k$. ∎

Theorem 5.17.11 $\lambda \in L_1$ if and only if $\overline{\lambda} \in L_1$, where $\overline{\lambda} = 1 - \lambda$.

Proof. The proof follows from Theorem 5.16.7 and the fact that if $\{\lambda_1, \lambda_2, \ldots, \lambda_n\}$ is a 1-admissible set of λ, then $\{\overline{\lambda_1}, \overline{\lambda_2}, \ldots, \overline{\lambda_n}\}$ is a 1-admissi-ble set of $\overline{\lambda}$. ∎

Theorem 5.17.12 If $\lambda_1, \lambda_2, \lambda_3 \in L_1$, then $\lambda_1 \lambda_2 + \overline{\lambda_1} \lambda_3 \in L_1$.

Proof. Let Γ_i be a 1-admissible set of λ_i, $i = 1, 2, 3$. It can be shown that $\{\lambda\lambda' + \overline{\lambda}\lambda'' \mid \lambda \in \Gamma_1, \lambda' \in \Gamma_2, \lambda'' \in \Gamma_3\}$ is a 1-admissible set of $\lambda_1 \lambda_2 + \overline{\lambda_1} \lambda_3$. Thus $\lambda_1 \lambda_2 + \overline{\lambda_1} \lambda_3 \in L_1$. ∎

Let $B \subseteq X^*$. Recall that χ_B is the characteristic function of B.

Proposition 5.17.13 If $\lambda_1, \lambda_2 \in L$ and

$$\{x \in X^* \mid \lambda_1(x) > \lambda_2(x)\} \subseteq B \subseteq \{x \in X^* \mid \lambda_1(x) \geq \lambda_2(x)\},$$

then $\lambda_1 \cup \lambda_2 = \chi_B \lambda_1 + \overline{\chi_B} \lambda_2$ and $\lambda_1 \cap \lambda_2 = \overline{\chi_B} \lambda_1 + \chi_B \lambda_2$. ∎

Theorem 5.17.14 Let $\lambda_1, \lambda_2 \in L_1$. If $\chi_B \in L_1$ for some B such that

$$\{x \in X^* \mid \lambda_1(x) > \lambda_2(x)\} \subseteq B \subseteq \{x \in X^* \mid \lambda_1(x) \geq \lambda_2(x)\},$$

then $\lambda_1 \cup \lambda_2, \lambda_1 \cap \lambda_2 \in L_1$.

Proof. The proof follows from Proposition 5.17.13 and Theorems 5.17.11 and 5.17.12. ∎

Theorem 5.17.15 If $\lambda_1, \lambda_2 \in L_1$ and $\mathrm{Im}(\lambda_2) \subseteq \{0, 1\}$, then $\lambda_1 \cup \lambda_2$ and $\lambda_1 \cap \lambda_2 \in L_1$.

Proof. Let $B = \{x \in X^* \mid \lambda_2(x) = 0\}$. Clearly,

$$\{x \in X^* \mid \lambda_1(x) > \lambda_2(x)\} \subseteq B \subseteq \{x \in X^* \mid \lambda_1(x) \geq \lambda_2(x)\}.$$

Thus by Theorem 5.17.14, $\lambda_1 \cup \lambda_2$ and $\lambda_1 \cap \lambda_2 \in L_1$. ∎

Let $\lambda \in L$. Let λ^T denote the fuzzy language over X such that for all $x \in X^*$, $\lambda^T(x) = \lambda(x^T)$, where $\Lambda^T = \Lambda$ and $(xu)^T = ux^T$ for all $u \in X$ and $x \in X^*$.

Theorem 5.17.16 *Let $k = 2, 3$. Then $\lambda \in L_k$ if and only if $\lambda^T \in L_k$.*

Proof. The proof follows from Theorem 5.16.7 and a remark made in the previous section. ■

The following operations of functions are needed in the proof of the next theorem. The reader is referred to [158] for a detailed treatment.

Definition 5.17.17 *Let g be a real-valued function with domain Q, a finite nonempty set.*

*(1) A real-valued function g^c with domain $\mathcal{P}(Q)$ is called a **column recomposition** of g if for all $Q' \in \mathcal{P}(Q)$, $g^c(Q') = \sum_{q \in Q'} g(q)$.*

*(2) A real-valued function g^r with domain $\mathcal{P}(Q)$ is called a **row recomposition** of g if for all $q \in Q$, $g(q) = \sum_{Q' \in \Omega(q)} g^r(Q')$, where $\Omega(q) = \{Q' \in \mathcal{P}(Q) \mid q \in Q'\}$.*

The following lemma was proved in [158].

Lemma 5.17.18 *(1) If g_1 and g_2 are real-valued functions with domain Q, then*

$$\sum_{q \in Q} g_1(q) \cdot g_2(q) = \sum_{Q' \in \mathcal{P}(Q)} g_1^r(Q') \cdot g_2^c(Q').$$

(2) If g is a function from Q into $[0, 1]$, then a row recomposing g^r of g may be constructed from g such that $\sum_{Q' \in \mathcal{P}(Q)} g^r(Q') = 1$ and $g^r(Q') \geq 0$ for all $Q' \subseteq Q$. ■

Theorem 5.17.19 *$\lambda \in L_1$ if and only if $\lambda^T \in L_1$.*

Proof. There exists $\lambda_i \in L$, $i = 1, 2, \ldots, n$, satisfying conditions (1), (2), and (3) of Theorem 5.16.9. For all $x \in X^*$, define $g_x(i) = \lambda_i(x)$ for all $i = 1, 2, \ldots, n$. Then g_x is a function from $Q_0 = \{1, 2, \ldots, n\}$ into $[0, 1]$. For all $Q \in \mathcal{P}(Q_0)$ and $x \in X^*$, let $\lambda_Q(x) = g_x^c(Q)$. It follows from Lemma 5.17.18 and conditions (1), (2), and (3) of Theorem 5.16.9 that $\{\lambda_Q^T \mid Q \in \mathcal{P}(Q_0)\}$ is a 1-admissible set of λ^T. Thus $\lambda^T \in L$. ■

We conclude this section by constructing two examples showing that neither L_1 is a subclass of L_2 nor L_2 is a subclass of L_1.

Example 5.17.20 *Let $X = \{u_0, u_1\}$ and for all $x \in X^*$,*

$$\lambda(x) = .b_n b_{n-1} \ldots b_2 b_1$$

written in binary expansion, where $x = u_{b_1} u_{b_2} \ldots u_{b_n}$. It is well known [180] that $\lambda \in L_1$. Suppose $\lambda \in L_2$. By Theorem 5.16.7, λ has a 2-admissible set $\{\lambda_1, \lambda_2, \ldots, \lambda_n\}$. Hence for all $x \in X^*$, there exists $0 \leq \beta_i(x) \leq 1$, $i = 1, 2, \ldots, n$, such that for all $x' \in X^*$, $\lambda(xx') = \vee_{i=1}^n \beta_i(x)\lambda_i(x')$. For all $x \in X^*$, let $I_x = \{i \mid \beta_i(x) = 1\}$ and $\beta(x) = \vee\{\beta_i(x) \mid i \notin I_x\} < 1$.

Clearly, there exists an $I \subseteq \{1, 2, \ldots, n\}$ such that $X^*(I) = \{x \in X^* \mid I_x = I\}$ is infinite. Let $X_0^*(I)$ be a subset of $X^*(I)$ containing exactly $n + 1$ elements. By the definition of λ, there exists $y_0 \in X^*$ such that $\lambda(y_0) > \vee\{\beta(x) \mid x \in X^*(I)\}$. Thus for all $x \in X_0^*(I)$, $\lambda(xy_0) = \vee_{i \in I} \lambda(y_0)$. This is impossible since $X_0^*(I)$ contains more than n elements and λ is one-one. Hence $\lambda \notin L$. Thus $L_1 \nsubseteq L_2$.

Example 5.17.21 *Let $X = \{u_1, u_2\}$ and for all $x \in X^*$, let $\lambda(x) = 2^{-m(x)}$, where $m(x) = m_1(x) \wedge m_2(x)$ and $m_1(x)$ and $m_2(x)$ are, respectively, the number of occurrences of u_1 and u_2 in x. First we show that $\lambda \in L_2$. Consider $A = (Q, p, h, g)$, where $Q = \{q_1, q_2\}$ and*

$$p(q_i, u_k, q_j) = \begin{cases} 1 & \text{if } i = j = k \\ \frac{1}{2} & \text{if } i = j \neq k \\ 0 & \text{otherwise,} \end{cases}$$

$$h(q_i) = \begin{cases} 1 & \text{if } i = 1 \\ 0 & \text{if } i = 2, \end{cases}$$

$$g(q_i) = 1 \text{ for } i = 1, 2.$$

It follows that $\lambda = r_2^A$. Suppose $\lambda \in L_1$. By Corollary 5.16.8, $V(\lambda)$ is finite-dimensional. For all $x \in X^*$, let $m_0(x) = m_1(x) - m_2(x)$. Clearly, $m_0(x') = m_0(x'')$ implies $2^{m_2(x')}\lambda^{x'} = 2^{m_2(x'')}\lambda^{x''}$. Thus there exist positive integers k_i, $i = 1, 2, \ldots, m$, and l_i, $i = 1, 2, \ldots, n$, such that $\lambda^{x_1}, \lambda^{x_2}, \ldots, \lambda^{x_m}, \lambda^{x'_1}, \lambda^{x'_2}, \ldots, \lambda^{x'_n}$, is a basis of $V(\lambda)$, where $x_i = u_1^{k_i}$, $i = 1, 2, \ldots, m$, $x'_i = u_2^{l_i}$, $i = 1, 2, \ldots, n$. Consequently, for all $x \in X^*$, $x^0 = \Lambda$ and $x^{i+1} = xx^i$. We assume that $k_{i'}$ as well as $l_{i'}$ are arranged in ascending order. Hence there exists constants a_i, $i = 1, 2, \ldots, r$, b_i, $i = 1, 2, \ldots, t$, where $r \leq m$, $t \leq n$, and $a_r \neq 0$, $b_t \neq 0$ such that for all $x \in X^*$, $\lambda(x) = \sum_{i=1}^{r} a_i \lambda(x_i x) + \sum_{i=1}^{t} b_i \lambda(x'_i x)$. Clearly, not both r and t are 0. Suppose $r > 0$. Let $x'_0 = u^{k_r}$ and $x''_0 = u^{k_{r-1}}$. Then $\lambda(x_i x'_0) = \lambda(x_i x''_0)$ for all $i = 1, 2, \ldots, r$ and $\lambda(x'_i x'_0) = \lambda(x'_i x''_0) = 1$ for all $i = 1, 2, \ldots, t$. Moreover, $\lambda(x'_0) = \lambda(x''_0) = 1$ and $\lambda(x_r x'_0) = \frac{1}{2}\lambda(x_r x''_0)$. Thus $a_r = 0$, a contradiction. Suppose $r = 0$. Then $t > 0$. By a similar argument, it can be shown that $b_t = 0$, a contradcition. Hence $\lambda \notin L_2$. Consequently, $L_2 \nsubseteq L_1$.

5.18 Further Properties of L_3

In this section, we show that L_3 is generated by the classes of all probabilistic, max-product, or max-min automata with deterministic or nondeterministic transition functions.

For $k = 1, 2, 3$, let $NL_k = \{\lambda \in L_k \mid \lambda = r_k^A$ for some $A = (Q, p, h, g)$, where $\text{Im}(p) \subseteq \{0, 1\}\}$.

Theorem 5.18.1 *Let $\lambda \in L$. For $k = 1, 2, 3$, $\lambda \in NL_k$ if and only if $T(\lambda)$ is finite.*

Proof. We only prove the case for $k = 1$. The other two cases can be proved similarly. Suppose $\lambda \in NL_1$. It follows from the proof of Theorem 5.16.7 that there exists a 1-admissible set $\Gamma = \{\lambda_1, \lambda_2, \ldots, \lambda_n\}$ of λ such that for all $i = 1, 2, \ldots, n$ and $u \in X$, $\lambda_i^u \in \Gamma$. Since $\lambda = \sum_{i=1}^{n} c_i \lambda_i'$ for some $c_i \geq 0$ and $\sum_{i=1}^{n} c_i = 1$, we have for all $x \in X^*$ that $\lambda^x = \sum_{i=1}^{n} c_i \lambda_i'$, where $\lambda_i' \in \Gamma$ for all $i = 1, 2, \ldots, n$. Thus $T(\lambda)$ is finite. Conversely, suppose that $T(\lambda)$ is finite. Clearly, $T(\lambda)$ is a k-admissible set of λ. The desired follows immediately from the proof of Theorem 5.16.7. ∎

For $k = 3$, Theorem 5.18.1 can be strengthened as follows:

Theorem 5.18.2 *Let $\lambda \in L$. Then $\lambda \in L_3$ if and only if $T(\lambda)$ is finite.*

Proof. Suppose $\lambda = r_3^A$, where $A = (Q, p, h, g)$ and $Q = \{q_1, q_2, \ldots, q_n\}$. For all $i = 1, 2, \ldots, n$ and $x \in X^*$, define $\lambda_i(x) = q_3^A(x, q_i)$ and $\beta_i(x) = \vee_{j=1}^{n}(h(q_j) \wedge p_3^A(q_j, x, q_i))$. Then for all $x \in X^*$, $\lambda^x = \cup_{i=1}^{n}(\beta_i(x) \cap \lambda_i)$. Let $R = \mathrm{Im}(p) \cup \mathrm{Im}(h)$. Clearly, R is finite and for all $i = 1, 2, \ldots, n$ and $x \in X^*$, $\beta_i(x) \in R$. Thus $T(\lambda)$ is finite. The converse follows from Theorem 5.16.7. ∎

Theorem 5.18.3 *Let $k = 1, 2, 3$. Then $\lambda \in NL_k$ if and only if $\lambda = r_k^A$ for some $A = (Q, p, h, g)$, where (1) $\mathrm{Im}(p) \subseteq \{0, 1\}$, and for all $q \in Q$ and $u \in X$, there exists a unique $q' \in Q$ such that $p(q, u, q') = 1$, and (2) $\mathrm{Im}(g) \subseteq \{0, 1\}$ or $\mathrm{Im}(h) \subseteq \{0, 1\}$ and there exists a unique $q \in Q$ such that $h(q) = 1$.*

Proof. The proof follows from Theorems 5.16.12, 5.16.13, and 5.18.1. ∎

Theorem 5.18.4 *Let $k = 1, 2$. If $\lambda \in NL_k$, then $\mathrm{Im}(\lambda)$ is finite.*

Proof. We prove only the case $k = 1$. The other case can be proved in a similar manner. Let $\lambda = r_1^A$, where $A = (Q, p, h, g)$ and $\mathrm{Im}(p) \subseteq \{0, 1\}$. Then for all $q, q' \in Q$ and $x \in X^*$, $p_1^A(q, x, q') \in \{0, 1\}$. Since $r_1^A(x) = \sum_{q,q' \in Q} h(q) p_1^A(q, x, q') g(q')$ for all $x \in X^*$, we have that $\mathrm{Im}(\lambda) = \mathrm{Im}(r_1^A)$ is finite. ∎

Let $FL = \{\lambda \in L \mid \mathrm{Im}(\lambda) \text{ is finite}\}$.

Theorem 5.18.5 $NL_1 = L_1 \cap FL$.

Proof. Let $\lambda \in L_1 \cap FL$. By Corollary 5.16.8, $V(\lambda)$ is finite-dimensional. Let $x_i \in X^*$, $i = 1, 2, \ldots, n$, be such that $\{\lambda^{x_1}, \lambda^{x_2}, \ldots, \lambda^{x_n}\}$ is a basis of $V(\lambda)$. Then for all $x \in X^*$, there exist unique $\beta_i(x)$, $i = 1, 2, \ldots, n$, such that $\lambda^x = \sum_{i=1}^{n} \beta_i(x) \lambda^{x_i}$. Clearly there exists a positive integer m such that for all $x \in X^*$, $\beta_i(x)$, $i = 1, 2, \ldots, n$, is completely determined by

$\lambda^x(y)$, where $y \in X^*$ and $|y| \le m$. Since $\text{Im}(\lambda)$ is finite, we have that $T(\lambda)$ is finite. Thus by Theorem 5.18.1, $\lambda \in NL_1$. Therefore, $L_1 \cap FL \subseteq NL_1$. By Theorem 5.18.4, $NL_1 \subseteq L_1 \cap FL$. Hence $NL_1 = L_1 \cap FL$. ∎

Theorem 5.18.6 *Let $\lambda \in L_2$. If there exists $c > 0$ such that for all $x \in X^*$, $\lambda(x) = 0$ or $\lambda(x) > c$, then $T(\lambda)$ is finite.*

Proof. Let $\lambda = r_2^A$, where $A = (Q, p, h, g)$ and $Q = \{q_1, q_2, \ldots, q_n\}$. For all $i = 1, 2, \ldots, n$ and $x \in X^*$, define $\lambda_i(x) = q_2^A(q_i, x)$ and $\beta_i(x) = \vee_{j=1}^{n}(h(q_j)p_2^A(q_j, x, q_i))$. Then for all $x \in X^*$, $\lambda^x = \cup_{i=1}^{n}(\beta_i'(x)\lambda_i)$, where

$$
\beta_i'(x) = \begin{cases} \beta_i(x) & \text{if there exists } y \in X^* \text{ such that} \\ & 0 \ne \lambda^x(y) = \beta_i(x)\lambda_i(y) \\ 0 & \text{otherwise.} \end{cases}
$$

Thus for all $i = 1, 2, \ldots, n$ and $x \in X^*$, $\beta_i'(x) = 0$ or $\beta_i'(x) > c$. Let $R = \text{Im}(p) \cup \text{Im}(h) \cup \text{Im}(g)$, $F_n = \{\prod_{i=0}^{n} c_i \in R, \ i = 0, 1, 2, \ldots, n, \text{ and } \prod_{i=0}^{n} c_i > c\}$ and $F = (\cup_{n=0}^{\infty} F_n) \cup \{0\}$. Clearly, F is finite and for all $i = 1, 2, \ldots, n$ and $x \in X^*$, $\beta_i'(x) \in F$. Thus $T(\lambda)$ is finite. ∎

Theorem 5.18.7 $NL_2 = L_2 \cap FL$.

Proof. The proof follows from Theorems 5.18.1, 5.18.4, and 5.18.6. ∎
The next result follows from Theorems 5.18.1, 5.18.2, 5.18.5, and 5.18.7.

Theorem 5.18.8 $L_3 = L_1 \cap FL = L_2 \cap FL$. ∎

The following closure property of L_3 follows from the results given in the previous section and Theorem 5.18.8.

Theorem 5.18.9 *(1) If $\lambda \in L_3$, then $\overline{\lambda}$ and $\lambda^T \in L_3$.*
(2) If $\lambda_1, \lambda_2 \in L_3$, then $\lambda_1\lambda_2, \lambda_1 \cup \lambda_2$ and $\lambda_1 \cap \lambda_2 \in L_3$.
(3) If $\lambda_1, \lambda_2, \lambda_3 \in L_3$, then $\lambda_1\lambda_2 + \overline{\lambda_1}\lambda_3 \in L_3$.
(4) Let $k = 1, 2, 3$. If $\lambda_i \in L_3$, $i = 1, 2, \ldots, n$, and λ is a k-combination of $\{\lambda_1, \lambda_2, \ldots, \lambda_n\}$, then $\lambda \in L_3$. ∎

Closure properties of L_3 with respect to other operations can be found in [196].

Definition 5.18.10 *A deterministic pseudoautomaton is a probabilistic automaton (Q, p, h, g), where $\text{Im}(p) \cup \text{Im}(h) \cup \text{Im}(g) \subseteq \{0, 1\}$.*

Note that if A is a deterministic pseudoautomaton, then $r_1^A = r_2^A = r_3^A$. For simplicity, we write $r^A = r_k^A$ for $k = 1, 2, 3$.
Let $DL = \{r^A \mid A \text{ is a deterministic pseudoautomaton}\}$.
Clearly, $\lambda \in DL$ if and only if $\text{Im}(\lambda) \subseteq \{0, 1\}$ and $\{x \in X^* \mid \lambda(x) = 1\}$ is a regular language [189].

Theorem 5.18.11 *Let $\lambda \in L$. Then $\lambda \in DL$ if and only if $\mathrm{Im}(\lambda) \subseteq \{0,1\}$ and $T(\lambda)$ is finite.*

Proof. The proof follows from Theorem 5.18.2. ∎

The above theorem is Nerode's Theorem [181].

Theorem 5.18.12 *Let $k = 1, 2, 3$. Then $\lambda \in DL$ if and only if $\lambda \in L_k$ and $\mathrm{Im}(\lambda) \subseteq \{0,1\}$.*

Proof. The proof follows from Theorem 5.18.8. ∎

Theorem 5.18.13 *For $k = 1, 2, 3$, L_k is a proper subclass of L.*

Proof. Let $\lambda \in L$ be such that $\mathrm{Im}(\lambda) \subseteq \{0,1\}$ and $\{x \in X^* \mid \lambda(x) = 1\}$ is a nonregular language. By Theorem 5.18.12, $\lambda \notin L_k$, $k = 1, 2, 3$. Thus L_k, $k = 1, 2, 3$, is a proper subclass of L. ∎

Theorem 5.18.14 *Let $\lambda \in FL$. Then $\lambda \in L_3$ if and only if all $0 \le c < 1$, $\{x \in X^* \mid \lambda(x) > c\}$ is a regular language.*

Proof. Suppose $\lambda \in L_3$. It follows from Theorems 5.18.2 and 5.18.11 that for all $0 \le c < 1$, $\{x \in X^* \mid \lambda(x) > c\}$ is a regular language. Conversely, suppose for all $0 \le c < 1$, $\{x \in X^* \mid \lambda(x) > c\}$ is a regular language. Let $\mathrm{Im}(\lambda) = \{c_1, c_2, \ldots, c_n\}$. Then $\lambda = \cup_{i=1}^{n} c_i \lambda_i$, where for all $i = 1, 2, \ldots, n$,

$$\lambda_i(x) = \left\{ \begin{array}{ll} 1 & \text{if } \lambda(x) > c_i \\ 0 & \text{otherwise.} \end{array} \right.$$

By hypothesis, $\lambda_i \in L_3$ for every i, $i = 1, 2, \ldots, n$. Hence by Theorem 5.18.9(4), $\lambda \in L_3$. ∎

Theorem 5.18.14 was first proved in [195]. Clearly, $\{x \in X^* \mid \lambda(x) > c\}$ may be replaced by $\{x \in X^* \mid \lambda(x) = c\}$, $\{x \in X^* \mid \lambda(x) \ge c\}$, $\{x \in X^* \mid \lambda(x) < c\}$, or $\{x \in X^* \mid \lambda(x) \le c\}$ in Theorem 5.18.14 .

5.19   Exercises

1. Let $A = (Q, \pi, \{A(u)\}, F)$ be a probabilistic automaton. Prove that $\pi(xy) = \pi(x)A(y)$, where $x, y \in X^*$.

2. Prove that $\rho(xy) = \pi(x)\eta^F(y)$, where $x, y \in X^*$.

3. Prove Proposition 5.2.6.

4. Let A be a probabilistic automaton. If A can be ϵ-approximated by a definite automaton, prove that A is quasi-definite.

5. Characterize the fsf's (the pa's) that can be approximated by

   (a) linear bounded automata,

   (b) pushdown automata,

   (c) sequential automata.

6. Determine the most powerful class of np devices that suffices for approximating the pa's.

7. Determine whether pushdown automata can approximate pa's.

8. Determine whether or not the family of w.i.m. languages is a subset of the family of w.i.s. languages.

9. Prove that every stochastic language under maximal interpretation is $w.3.m.$

10. Prove Theorem 5.16.7 for the cases $k = 2, 3$.

11. For $k = 2, 3$, state and prove a result similar to Theorem 5.16.7 replacing λ^x with $\lambda^{(x)}$.

12. Prove Theorem 5.18.1 for the cases $k = 2, 3$.

13. Prove Theorem 5.18.4 for the case $k = 2$.

14. Prove Theorem 5.18.14 with $>$ in $\{x \in X^* \mid \lambda(x) > c\}$ replaced by $=$, $\geq$, and $<$.

Chapter 6

Algebraic Fuzzy Automata Theory

6.1 Fuzzy Finite State Machines

A sequential machine consists of two main structures, the transition structure and the output structure. The transition structure is an internal part of the machine while the output structure is the external part. Consequently, the output structure is of more interest for practical applications than the input structure. The output structure is dependent on the transition structure while the transition structure is independent of the output structure. Hence the input structure can be studied separately. In this chapter, we study the transition structure of fuzzy machines.

A **fuzzy finite state machine** (**ffsm**) is a triple $M = (Q, X, \mu)$, where Q and X are finite nonempty sets and μ is a fuzzy subset of $Q \times X \times Q$, i.e., $\mu : Q \times X \times Q \to [0, 1]$. As usual X^* denotes the set of all words of elements of X of finite length.

Q is called the set of states and X is called the set of input symbols. Let Λ denote the empty word in X^* and $|x|$ denote the length of x $\forall x \in X^*$. X^* is a free semigroup with identity Λ with respect to the binary operation concatenation of two words.

6.2 Semigroups of Fuzzy Finite State Machines

Definition 6.2.1 *Let $M = (Q, X, \mu)$ be a ffsm. Define $\mu^* : Q \times X^* \times Q \to [0, 1]$ by*

$$\mu^*(q, \Lambda, p) = \begin{cases} 1 & \text{if } q = p \\ 0 & \text{if } q \neq p \end{cases}$$

237

and

$$\mu^*(q, xa, p) = \vee\{\mu^*(q, x, r) \wedge \mu(r, a, p) | r \in Q\}$$

$\forall\ x \in X^*, a \in X.$

Let $X^+ = X^* \backslash \{\Lambda\}$. Then X^+ is a semigroup. For μ^* given in Definition 6.2.1, we let $\mu^+ = \mu^*$ restricted to $Q \times X^+ \times Q$.

Lemma 6.2.2 *Let* $M = (Q, X, \mu)$ *be a ffsm. Then*

$$\mu^*(q, xy, p) = \vee\{\mu^*(q, x, r) \wedge \mu^*(r, y, p) | r \in Q\}$$

$\forall q, p \in Q$ *and* $\forall x, y \in X^*.$

Proof. Let $q, p \in Q$ and $x, y \in X^*$. We prove the result by induction on $|y| = n$. If $n = 0$, then $y = \Lambda$ and hence $xy = x\Lambda = x$. Thus $\vee\{\mu^*(q, x, r) \wedge \mu^*(r, y, p) | r \in Q\} = \vee\{\mu^*(q, x, r) \wedge \mu^*(r, \Lambda, p) | r \in Q\} = \mu^*(q, x, p) = \mu^*(q, xy, p)$ by the definition of μ^*. Thus the result is true for $n = 0$. Suppose the result is true for all $u \in X^*$ such that $|u| = n - 1, n > 0$. Let $y = ua$ where $u \in X^*$, $a \in X$, and $|u| = n - 1, n > 0$. Now $\mu^*(q, xy, p) = \mu^*(q, xua, p) = \vee\{\mu^*(q, xu, r) \wedge \mu(r, a, p) | r \in Q\} = \vee\{(\vee\{\mu^*(q, x, s) \wedge \mu^*(s, u, r) | s \in Q\}) \wedge \mu(r, a, p) | r \in Q\} = \vee\{\vee\{\mu^*(q, x, s) \wedge \mu^*(s, u, r) \wedge \mu(r, a, p)\} | r, s \in Q\} = \vee\{\mu^*(q, x, s) \wedge (\vee\{\mu^*(s, u, r) \wedge \mu(r, a, p) | r \in Q\}) | s \in Q\} = \vee\{\mu^*(q, x, s) \wedge \mu^*(s, ua, p) | s \in Q\} = \vee\{\mu^*(q, x, s) \wedge \mu^*(s, y, r) | s \in Q\}$. Thus the result is true for $|y| = n$. ∎

Define a relation $\equiv$ on X^* by $\forall x, y \in X^*$, $x \equiv y$ if and only if $\mu^*(q, x, p) = \mu^*(q, y, p)\ \forall\ q, p \in Q$.

Clearly $\equiv$ is an equivalence relation on X^*. Let $z \in X^*$ and let $x \equiv y$. Then $\forall p, q \in Q$, $\mu^*(q, xz, p) = \vee\{\mu^*(q, x, r) \wedge \mu^*(r, z, p) | r \in Q\} = \vee\{\mu^*(q, y, r) \wedge \mu^*(r, z, p) | r \in Q\} = \mu^*(q, yz, p)$. Thus $xz \equiv yz$. Similarly $zx \equiv zy$. Thus $\equiv$ is a congruence relation on the semigroup X^*. We have thus proved the following result.

Theorem 6.2.3 *Let* $M = (Q, X, \mu)$ *be a ffsm. Define a relation* $\equiv$ *on* X^* *by* $\forall x, y \in X^*$, $x \equiv y$ *if and only if* $\mu^*(q, x, p) = \mu^*(q, y, p)\ \forall\ q, p \in Q$. *Then* $\equiv$ *is a congruence relation on* X^*. ∎

Let $x \in X^*$, $[x] = \{y \in X^* | x \equiv y\}$, and $E(M) = \{[x] | x \in X^*\}$.

Theorem 6.2.4 *Let* $M = (Q, X, \mu)$ *be a ffsm. Define a binary operation* $*$ *on* $E(M)$ *by* $\forall\ [x], [y] \in E(M), [x] * [y] = [xy]$. *Then* $(E(M), *)$ *is a finite semigroup with identity.*

Proof. Clearly $*$ is well defined and associative. Now $[x] * [\Lambda] = [x\Lambda] = [x] = [\Lambda x] = [\Lambda] * [x]\ \forall\ [x] \in E(M)$. Thus $[\Lambda]$ is the identity of $(E(M), *)$.

Hence $(E(M), *)$ is a semigroup with identity. Let $x \in X^*$ and let $x = x_1 x_2 \ldots x_n$, where $x_1, x_2, \ldots, x_n \in X$. Then $\forall q, p \in Q$,

$$\mu^*(q, x, p) = \vee \{\mu(q, x_1, q_1) \wedge \mu(q_1, x_2, q_2) \wedge \ldots \wedge \mu(q_{n-1}, x_n, p) \mid q_1, q_2, \ldots, q_{n-1} \in Q\}.$$

Hence since $\mathrm{Im}(\mu)$ is finite, $\mathrm{Im}(\mu^*)$ is finite. Thus $(E(M), *)$ is a finite semigroup with identity. ∎

Example 6.2.5 Let $M = (Q, X, \mu)$ be a ffsm, where $Q = \{q\}$ and $X = \{a, b\}$. Define $\mu : Q \times X \times Q \to [0, 1]$ by $\mu(q, a, q) = \frac{1}{2} = \mu(q, b, q)$. Then $\forall x, y \in X^+$, $x \equiv y$ since $\mu^*(q, x, q) = \frac{1}{2} = \mu^*(q, y, q)$. Hence $E(M) = \{[\Lambda], [x]\}$, where $x \in X^+$. Clearly, $[\Lambda]$ is the identity of $E(M)$ and $[x]^2 = [x]$.

Example 6.2.6 Let $M = (Q, X, \mu)$ be a ffsm, where $Q = \{q\}$ and $X = \{a, b\}$. Define $\mu : Q \times X \times Q \to [0, 1]$ by $\mu(q, a, q) = \frac{2}{3}$ and $\mu(q, b, q) = \frac{1}{3}$. Then $\forall x \in X^*$, $\mu^*(q, x, q) = \frac{1}{3}$ if and only if x contains a b. Hence $\forall x, y \in X^+$, $x \equiv y$ if and only if x and y both contain a b. Hence $E(M) = \{[\Lambda], [x], [y]\}$, where x contains a b and y does not contain a b, $x, y \in X^+$. Here $[\Lambda]$ is the identity of $E(M)$, $[x]^2 = [x]$, $[y]^2 = [y]$, and $[x] * [y] = [x] = [y] * [x]$.

We now define another type of congruence relation on X^*. Let $x, y \in X^*$. Define $x \approx y$ if and only if $(\forall s, t \in Q, \mu^*(s, x, t) > 0 \Leftrightarrow \mu^*(s, y, t) > 0)$. Clearly $\approx$ is an equivalence relation on X^*. Let $z \in X^*$ and $x \approx y$. Then $\forall s, t \in Q$, $\mu^*(s, zx, t) = \vee \{\mu^*(s, z, r) \wedge \mu^*(r, x, t) \mid r \in Q\} > 0$ if and only if $\exists r \in Q$ such that $\mu^*(s, z, r) \wedge \mu^*(r, x, t) > 0$ if and only if $\exists r \in Q$ such that $\mu^*(s, z, r) \wedge \mu^*(r, y, t) > 0$ if and only if $\mu^*(s, zy, t) = \vee \{\mu^*(s, z, r) \wedge \mu^*(r, y, t) \mid r \in Q\} > 0$. Hence $zx \approx zy$. Similarly $xz \approx yz$. Thus $\approx$ is a congruence relation on X^*. Hence, we have the following theorem.

Theorem 6.2.7 Let $M = (Q, X, \mu)$ be a ffsm. Let $x, y \in X^*$. Define a relation $\approx$ on X^* by $x \approx y$ if and only if $\forall s, t \in Q, \mu^*(s, x, t) > 0 \Leftrightarrow \mu^*(s, y, t) > 0$. Then $\approx$ is a congruence relation on X^*. ∎

Let $x \in X^*$ and let $[\![x]\!] = \{y \in X^* \mid x \approx y\}$. Let $\widetilde{E(M)} = \{[\![x]\!] \mid x \in X^*\}$.

Theorem 6.2.8 Let $M = (Q, X, \mu)$ be a ffsm. Define a binary operation $\widetilde{*}$ on $\widetilde{E(M)}$ by $\forall [\![x]\!], [\![y]\!] \in \widetilde{E(M)}$, $[\![x]\!] \widetilde{*} [\![y]\!] = [\![xy]\!]$. Then $(\widetilde{E(M)}, \widetilde{*})$ is a finite semigroup with identity and $[x] \to [\![x]\!]$ is a homomorphism of $E(M)$ onto $\widetilde{E(M)}$.

Proof. Clearly $(\widetilde{E(M)}, \widetilde{*})$ is a semigroup with identity. Define $f : E(M) \to \widetilde{E(M)}$ by $f([x]) = [\![x]\!] \forall [x] \in E(M)$. Let $x, y \in X^*$ and $[x] = [y]$. Then $\forall s, t \in Q$, $\mu^*(s, x, t) = \mu^*(s, y, t)$. Thus $\forall s, t \in Q, \mu^*(s, x, t) > 0 \Leftrightarrow \mu^*(s, y, t) > 0$. Hence $x \approx y$ or $[\![x]\!] = [\![y]\!]$. Thus f is well defined. Clearly f is an onto homomorphism. Now since $E(M)$ is finite $\widetilde{E(M)}$ is finite. ∎

Example 6.2.9 *Let $M = (Q, X, \mu)$ be the ffsm as defined in Example 6.2.5. Then $\forall x, y \in X^*$, $x \approx y$. Thus $\widetilde{E(M)} = \{[[x]]\}$, where $x \in X^+$. Now $[[x]]^2 = [[x]]$.*

Definition 6.2.10 *Let $M = (Q, X, \mu)$ be a ffsm. For all $x \in X^*$ define the fuzzy subset x^M of $Q \times Q$ by $x^M(s, t) = \mu^*(s, x, t)$ $\forall s, t \in Q$.*

Theorem 6.2.11 *Let $M = (Q, X, \mu)$ be a ffsm. Let $S_M = \{x^M | x \in X^*\}$. Then*

(1) $x^M \circ y^M = (xy)^M$ $\forall x, y \in X^*$,

(2) $(S_M, \circ)$ is a finite semigroup with identity, where $\circ$ is defined as in Definition 1.4.4.

Proof. (1) Let $s, t \in Q$. Then $(xy)^M(s, t) = \mu^*(s, xy, t) = \vee\{\mu^*(s, x, q) \wedge \mu^*(q, y, t) | q \in Q\} = \vee\{x^M(s, q) \wedge y^M(q, t) | q \in Q\} = (x^M \circ y^M)(s, t)$. Thus $(xy)^M = x^M \circ y^M$.

(2) Clearly $(S_M, \circ)$ is a finite semigroup with identity, where Λ^M is the identity element. S_M is finite since Q and $\text{Im}(\mu)$ are finite. ∎

Theorem 6.2.12 *Let $M = (Q, X, \mu)$ be a ffsm. Then $S_M \simeq E(M)$, i.e., S_M and $E(M)$ are isomorphic as semigroups.*

Proof. Define $f : S_M \to E(M)$ by $f(x^M) = [x]$ $\forall x^M \in S_M$. Let $x^M, y^M \in S_M$. Then $x^M = y^M$ if and only if $x^M(s, t) = y^M(s, t)$ $\forall s, t \in Q$ if and only if $\mu^*(s, x, t) = \mu^*(s, y, t)$ $\forall s, t \in Q$ if and only if $[x] = [y]$. Thus f is single valued and one-one. Now

$$f(x^M \circ y^M) = f((xy)^M) = [xy] = [x] * [y] = f(x^M) * f(y^M).$$

Thus f is a homomorphism. Clearly f is onto. Hence $S_M \simeq E(M)$. ∎

Let $M = (Q, X, \mu)$ be a ffsm. The index of an equivalence relation is the number of distinct equivalence classes. Let $\sim$ be a congruence relation of finite index on X^*. Let $x \in X^*$ and $\prec x \succ = \{y \in X^* | x \sim y\}$. Let $\widetilde{Q} = \{\prec x \succ \, | \, x \in X^*\}$. Define $\sigma : \widetilde{Q} \times X \times \widetilde{Q} \to [0, 1]$ by $\forall \prec x \succ \in \widetilde{Q}$ and $\forall a \in X$, $\sigma(\prec x \succ, a, \prec xa \succ)$ an arbitrary fixed element in $(0, 1]$ and $\forall \prec x \succ, \prec w \succ \in \widetilde{Q}$,

$$\sigma(\prec x \succ, a, \prec w \succ) = \begin{cases} \sigma(\prec x \succ, a, \prec xa \succ) & \text{if } w \sim xa \\ 0 & \text{otherwise.} \end{cases}$$

Let $\prec x \succ, \prec y \succ, \prec u \succ, \prec v \succ \in \widetilde{Q}$ and $a, b \in X$. Suppose that

$$(\prec x \succ, a, \prec u \succ) = (\prec y \succ, b, \prec v \succ).$$

Then $\prec x \succ = \prec y \succ$, $a = b$, $\prec u \succ = \prec v \succ$. Now $u \sim xa$ if and only if $v \sim ya$. Thus

$$\sigma(\prec x \succ, a, \prec u \succ) = \sigma(\prec y \succ, b, \prec v \succ).$$

Hence σ is single valued. Thus $\widetilde{M} = (\widetilde{Q}, X, \sigma)$ is ffsm. Now extend σ to σ^* as μ was extended to μ^* in Definition 6.2.1.

Lemma 6.2.13 *Let $\widetilde{M}$ be as above and let $\prec z \succ, \prec w \succ \in \widetilde{Q}$. Then the following assertions hold.*

(1) $\forall x \in X^*$, *if* $\sigma^*(\prec z \succ, x, \prec w \succ) > 0$, *then* $\prec zx \succ = \prec w \succ$.

(2) $\sigma^*(\prec z \succ, x, \prec zx \succ) > 0 \ \forall \ z, x \in X^*$.

Proof. (1) Let $x \in X^*$ and $|x| = n$. If $n = 0$, then $x = \Lambda$. Hence if $\sigma^*(\prec z \succ, x, \prec w \succ) > 0$, then $\prec z \succ = \prec w \succ$ or $\prec zx \succ = \prec w \succ$. Suppose the result is true $\forall \ y \in X^*$ such that $|y| = n - 1, n > 0$. Let $x = ya$ where $y \in X^*$, $a \in X$, and $|y| = n - 1$. Let

$$\sigma^*(\prec z \succ, ya, \prec w \succ) = \sigma^*(\prec z \succ, x, \prec w \succ) > 0.$$

Now

$$\begin{aligned}
\sigma^*(\prec z \succ, ya, \prec w \succ) &= \vee\{\sigma^*(\prec z \succ, y, \prec q \succ) \wedge \sigma(\prec q \succ, a, \prec w \succ) \\
&\quad | \prec q \succ \in \widetilde{Q}\} \\
&> 0.
\end{aligned}$$

Hence $\sigma^*(\prec z \succ, y, \prec q \succ) > 0$ and $\sigma^*(\prec q \succ, a, \prec w \succ) > 0$ for some $\prec q \succ \in \widetilde{Q}$. Thus by the induction hypothesis, $\prec zy \succ = \prec q \succ$ and $\prec qa \succ = \prec w \succ$. Hence $\prec zx \succ = \prec zya \succ = \prec qa \succ = \prec w \succ$. The result now follows by induction.

(2) Let $z, x \in X^*$ and $|x| = n$. If $n = 0$, then $x = \Lambda$. Hence

$$\sigma^*(\prec z \succ, x, \prec zx \succ) > 0.$$

Suppose the result is true $\forall \ y \in X^*$ such that $|y| = n - 1, n > 0$. Let $x = ya$ where $y \in X^*$, $a \in X$, and $|y| = n - 1$. Now

$$\begin{aligned}
\sigma^*(\prec z \succ, x, \prec zx \succ) &= \sigma^*(\prec z \succ, ya, \prec zya \succ) \\
&= \vee\{\sigma^*(\prec z \succ, y, \prec q \succ) \wedge \sigma(\prec q \succ, a, \prec zya \succ) \\
&\quad | \prec q \succ \in \widetilde{Q}\} \\
&\geq \sigma^*(\prec z \succ, y, \prec zy \succ) \wedge \sigma(\prec zy \succ, a, \prec zya \succ) \\
&> 0.
\end{aligned}$$

The result now follows by induction. ∎

Let $x, y \in X^*$. Suppose $x \sim y$. Let $\prec z \succ, \prec w \succ \in \widetilde{Q}$. Suppose

$$\sigma^*(\prec z \succ, x, \prec w \succ) > 0.$$

Then $\prec zx \succ = \prec w \succ$. Since $x \sim y$ and $\sim$ is a congruence relation, $\prec zx \succ = \prec zy \succ$. Thus $\prec zy \succ = \prec w \succ$. Hence $\sigma^*(\prec z \succ, y, \prec w \succ) > 0$. Similarly if $\sigma^*(\prec z \succ, y, \prec w \succ) > 0$, then

$$\sigma^*(\prec z \succ, x, \prec w \succ) > 0.$$

Hence $x \approx y$. Conversely, suppose $x \approx y$. Let $\prec z \succ \in \widetilde{Q}$. Now $\sigma^*(\prec z \succ, x, \prec zx \succ) > 0$. Hence $\sigma^*(\prec z \succ, y, \prec zx \succ) > 0$. Thus $\prec zy \succ = \prec zx \succ$. Choose $z = \Lambda$. Then $\prec x \succ = \prec y \succ$ or $x \sim y$. Thus $x \sim y$ if and only if $x \approx y$. We summarize the above discussion in the following theorem.

Theorem 6.2.14 *A fuzzy finite state machine M can be constructed from a given congruence relation $\sim$ on X^* of finite index in such a way that $\sim$ is the same congruence relation $\approx$ as on M.* ∎

In [113], a triple (Q, X, τ) is called a **generalized state machine** if Q, X are finite sets and $\tau : Q \times X \times Q \to [0, 1]$ is such that $\sum_{q \in Q} \tau(p, a, q) \le 1$ for all $p \in Q$ and $a \in X$. Let M be a generalized machine. Then the condition $\sum_{q \in Q} \tau(p, a, q) \le 1$ contains the crisp case in the sense that τ can be considered a partial function, i.e., if $\mathrm{Im}(\tau) \subseteq \{0, 1\}$, then τ is a partial function if there exists $p \in Q$ such that $\tau(p, a, q) = 0$ for all $q \in Q$ and $a \in X$. M is called **complete** if $\sum_{q \in Q} \tau(p, a, q) = 1$ for all $p \in Q$ and $a \in X$. If M is not complete, it can be completed in the following manner: Let $Q' = Q \cup \{z\}$, where $z \notin Q$. Let $M^c = (Q', X, \tau')$, where

$$\tau'(p', a, q') = \begin{cases} \tau(p', a, q') & \text{if } p', q' \in Q, \\ 1 - \sum_{q \in Q} \tau(p', a, q) & \text{if } p' \in Q \text{ and } q' = z, \\ 0 & \text{if } p' = z \text{ and } q' \in Q, \\ 1 & \text{if } p' = z \text{ and } q' = z, \end{cases}$$

for all $a \in X$. Let $X^+ = X^* \backslash \{\Lambda\}$. Let T be a t-norm on $[0, 1]$. Define $\tau^+ : Q \times X^+ \times Q \to [0, 1]$ by

$$\tau^+(p, a_1 \ldots a_n, q) = \vee \{\tau(p, a_1, r_1) T \tau(r_1, a_2, r_2) T \ldots T \tau(r_{n-1}, a_n, q) \mid r_i \in Q, \; i = 1, 2, \ldots, n-1\},$$

where $p, q \in Q$ and $a_1, \ldots, a_n \in X$. M is called a T-**generalized state machine**, when τ^+ is defined in terms of T.

A theory for T-generalized state machines is developed in [113]. We present the main results of [113] in the Exercises.

6.3   Homomorphisms

Definition 6.3.1 *Let $M_1 = (Q_1, X_1, \mu_1)$ and $M_2 = (Q_2, X_2, \mu_2)$ be ff-sms. A pair (α, β) of mappings, $\alpha : Q_1 \to Q_2$ and $\beta : X_1 \to X_2$, is called a **homomorphism**, written $(\alpha, \beta) : M_1 \to M_2$, if $\mu_1(q, x, p) \le \mu_2(\alpha(q), \beta(x), \alpha(p)) \; \forall \; q, p \in Q_1$ and $\forall \; x \in X_1$.*

*The pair (α, β) is called a **strong homomorphism** if*

$$\mu_2(\alpha(q), \beta(x), \alpha(p)) = \vee \{\mu_1(q, x, t) \mid t \in Q_1, \; \alpha(t) = \alpha(p)\}$$

$\forall \; q, p \in Q_1$ *and* $\forall \; x \in X_1$.

A homomorphism (strong homomorphism) $(\alpha, \beta) : M_1 \to M_2$ is called an **isomorphism** (**strong isomorphism**) if α and β are both one-one and onto.

Example 6.3.2 *Let $M_1 = (Q_1, X_1, \mu_1)$ and $M_2 = (Q_2, X_2, \mu_2)$ be ffsms, where $Q_1 = \{q_1, q_2, q_3\}$, $X_1 = \{a, b\}$, $Q_2 = \{q_1', q_2', q_3'\}$, $X_2 = \{a, b\}$, and μ_1 and μ_2 are defined as follows:*

$$
\begin{array}{llll}
\mu_1(q_1, a, q_1) & = & \tfrac{1}{3} & \quad \mu_2(q_1', a, q_1') & = & \tfrac{1}{3} \\
\mu_1(q_1, b, q_2) & = & \tfrac{2}{3} & \quad \mu_2(q_1', b, q_2') & = & \tfrac{2}{3} \\
\mu_1(q_2, a, q_1) & = & \tfrac{1}{3} & \quad \mu_2(q_2', a, q_1') & = & \tfrac{1}{3} \\
\mu_1(q_2, b, q_3) & = & \tfrac{2}{3} & \quad \mu_2(q_2', b, q_1') & = & \tfrac{2}{3} \\
\mu_1(q_3, a, q_3) & = & \tfrac{1}{3} & \quad \mu_2(q_3', a, q_1') & = & \tfrac{1}{2} \\
\mu_1(q_3, b, q_2) & = & \tfrac{2}{3} & \quad \mu_2(q_3', b, q_2') & = & \tfrac{1}{2}.
\end{array}
$$

For all other ordered triples, (q, x, s) and (q', x', s') define $\mu_1(q, x, s) = 0$ and $\mu_2(q', x', s') = 0$. Define $\alpha : Q_1 \to Q_2$ and $\beta : X_1 \to X_2$ as follows: $\alpha(q_1) = \alpha(q_3) = q_1'$, $\alpha(q_2) = q_2'$, $\beta(a) = a$, and $\beta(b) = b$. We note that (α, β) is a strong homomorphism of Q_1 into Q_2. This follows from the following equalities.

$$
\begin{array}{lllllll}
\mu_2(\alpha(q_1), \beta(a), \alpha(q_1)) & = & \mu_2(q_1', a, q_1') & = & \tfrac{1}{3} & = & \mu_1(q_1, a, q_1) \\
\mu_2(\alpha(q_1), \beta(b), \alpha(q_2)) & = & \mu_2(q_1', b, q_2') & = & \tfrac{2}{3} & = & \mu_1(q_1, b, q_2) \\
\mu_2(\alpha(q_2), \beta(a), \alpha(q_1)) & = & \mu_2(q_2', a, q_1') & = & \tfrac{1}{3} & = & \mu_1(q_2, a, q_1) \\
\mu_2(\alpha(q_2), \beta(b), \alpha(q_3)) & = & \mu_2(q_2', b, q_1') & = & \tfrac{2}{3} & = & \mu_1(q_2, b, q_3) \\
\mu_2(\alpha(q_3), \beta(a), \alpha(q_3)) & = & \mu_2(q_1', a, q_1') & = & \tfrac{1}{3} & = & \mu_1(q_3, a, q_3) \\
\mu_2(\alpha(q_3), \beta(b), \alpha(q_2)) & = & \mu_2(q_1', b, q_2') & = & \tfrac{2}{3} & = & \mu_1(q_3, b, q_2).
\end{array}
$$

(1) In Definition 6.3.1, if $X_1 = X_2$ and β is the identity map, then we simply write $\alpha : M_1 \to M_2$ and say that α is a **homomorphism** or **strong homomorphism** accordingly.

(2) If (α, β) is a strong homomorphism with α one-one, then

$$\mu_2(\alpha(q), \beta(x), \alpha(p)) = \mu_1(q, x, p)$$

$\forall\, q, p \in Q_1$ and $\forall\, x \in X_1$.

Lemma 6.3.3 *Let $M_1 = (Q_1, X_1, \mu_1)$ and $M_2 = (Q_2, X_2, \mu_2)$ be two ffsms. Let $(\alpha, \beta) : M_1 \to M_2$ be a strong homomorphism. Then $\forall q, r \in Q_1$, $\forall x \in X_1$, if $\mu_2(\alpha(q), \beta(x), \alpha(r)) > 0$, then $\exists\, t \in Q_1$ such that $\mu_1(q, x, t) > 0$ and $\alpha(t) = \alpha(r)$. Furthermore, $\forall p \in Q$ if $\alpha(p) = \alpha(q)$, then $\mu_1(q, x, t) \geq \mu_1(p, x, r)$.*

Proof. Let $p, q, r \in Q_1$, $x \in X_1$, and $\mu_2(\alpha(q), \beta(x), \alpha(r)) > 0$. Then

$$\vee\{\mu_1(q, x, s) | s \in Q_1, \ \alpha(s) = \alpha(r)\} > 0.$$

Since Q_1 is finite, $\exists \ t \in Q_1$ such that $\alpha(t) = \alpha(r)$ and $\mu_1(q, x, t) = \vee\{\mu_1(q, x, s) | s \in Q_1, \alpha(s) = \alpha(r)\} > 0$. Suppose $\alpha(p) = \alpha(q)$. Then

$$\mu_1(q, x, t) = \mu_2(\alpha(q), \beta(x), \alpha(r)) = \mu_2(\alpha(p), \beta(x), \alpha(r)) \geq \mu_1(p, x, r). \blacksquare$$

Definition 6.3.4 Let $M_1 = (Q_1, X_1, \mu_1)$ and $M_2 = (Q_2, X_2, \mu_2)$ be two ffsms. Let $(\alpha, \beta) : M_1 \to M_2$ be a homomorphism. Define $\beta^* : X_1^* \to X_2^*$ by $\beta^*(\Lambda) = \Lambda$ and $\beta^*(ua) = \beta^*(u)\beta(a) \ \forall \ u \in X_1^*$, $a \in X_1$.

Lemma 6.3.5 Let M_1, $M_2, (\alpha, \beta)$, and β^* be as above. Then $\beta^*(uv) = \beta^*(u)\beta^*(v) \ \forall \ u, v \in X_1^*$.

Proof. Let $u, v \in X_1^*$ and $|v| = n$. If $n = 0$, then $v = \Lambda$ and hence $\beta^*(uv) = \beta^*(u) = \beta^*(u)\beta^*(v)$. Suppose now the result is true $\forall \ y \in X_1^*$ such that $|y| = n - 1, n > 0$. Let $v = ya$ where $y \in X_1^*, a \in X_1$, and $|y| = n - 1$. Then $\beta^*(uv) = \beta^*(uya) = \beta^*(uy)\beta(a) = \beta^*(u)\beta^*(y)\beta(a) = \beta^*(u)\beta^*(ya) = \beta^*(u)\beta^*(v)$. The result now follows by induction. $\blacksquare$

Theorem 6.3.6 Let M_1, M_2 be as above. Let $(\alpha, \beta) : M_1 \to M_2$ be a homomorphism. Then $\mu_1^*(q, x, p) \leq \mu_2^*(\alpha(q), \beta^*(x), \alpha(p)) \ \forall \ q, p \in Q_1$ and $x \in X_1^*$.

Proof. Let $q, p \in Q_1$ and $x \in X_1^*$. We prove the result by induction on $|x| = n$. If $n = 0$, then $x = \Lambda$ and $\beta^*(x) = \beta^*(\Lambda) = \Lambda$. Now if $q = p$, then $\mu_1^*(q, \Lambda, p) = 1 = \mu_2^*(\alpha(q), \Lambda, \alpha(p))$. If $q \neq p$, then $\mu_1^*(q, \Lambda, p) = 0 \leq \mu_2^*(\alpha(q), \Lambda, \alpha(p))$. Suppose now the result is true $\forall \ y \in X^*$ such that $|y| = n - 1, n > 0$. Let $x = ya$ where $y \in X_1^*$, $a \in X_1$, and $|y| = n - 1$. Now

$$
\begin{aligned}
\mu_1^*(q, x, p) &= \mu_1^*(q, ya, p) \\
&= \vee\{\mu_1^*(q, y, r) \wedge \mu_1^*(r, a, p) | r \in Q_1\} \\
&\leq \vee\{\mu_2^*(\alpha(q), \beta^*(y), \alpha(r)) \wedge \mu_2^*(\alpha(r), \beta(a), \alpha(p)) | r \in Q_1\} \\
&\leq \vee\{\mu_2^*(\alpha(q), \ \beta^*(y), \ r') \wedge \mu_2^*(r', \ \beta^*(a), \ \alpha(p) \ | \ r' \in Q_2\} \\
&= \mu_2^*(\alpha(q), \beta^*(y)\beta(a), \alpha(p)) \\
&= \mu_2^*(\alpha(q), \beta^*(ya), \alpha(p)) \\
&= \mu_2^*(\alpha(q), \beta^*(x), \ \alpha(p)). \blacksquare
\end{aligned}
$$

Theorem 6.3.7 Let M_1, M_2 be as above. Let $(\alpha, \beta) : M_1 \to M_2$ be a strong homomorphism. Then α is one-one if and only if $\mu_1^*(q, x, p) = \mu_2^*(\alpha(q), \beta^*(x), \alpha(p)) \ \forall \ q, p \in Q_1$ and $x \in X_1^*$.

Proof. Suppose α is one-one. Let $p, q \in Q_1$ and $x \in X_1^*$. Let $|x| = n$. We prove the result by induction on n. Let $n = 0$. Then $x = \Lambda$ and $\beta^*(\Lambda) =$

Λ. Now $\alpha(q) = \alpha(p)$ if and only if $q = p$. Hence $\mu_1^*(q, \Lambda, p) = 1$ if and only if $\mu_2^*(\alpha(q), \beta^*(\Lambda), \alpha(p)) = 1$. Suppose the result is true $\forall\, y \in X_1^*$, $|y| = n - 1$, $n > 0$. Let $x = ya$, $|y| = n - 1$, $y \in X_1^*$, $a \in X_1$. Then

$$
\begin{aligned}
\mu_2^*(\alpha(q), \beta^*(x), \alpha(p)) &= \mu_2^*(\alpha(q), \beta^*(ya), \alpha(p)) \\
&= \mu_2^*(\alpha(q), \beta^*(y)\beta(a), \alpha(p)) \\
&= \vee\{\mu_2^*(\alpha(q), \beta^*(y), \alpha(r)) \wedge \mu_2(\alpha(r), \beta(a), \alpha(p)) \\
&\qquad \mid r \in Q_1\} \mid r \in Q_1\} \\
&= \vee\{\mu_1^*(q, y, r) \wedge \mu_1(r, a, p) \mid r \in Q_1\} \\
&= \mu_1^*(q, ya, p) \\
&= \mu_1^*(q, x, p).
\end{aligned}
$$

Conversely, let $q, p \in Q_1$ and let $\alpha(q) = \alpha(p)$. Then $1 = \mu_2^*(\alpha(q), \Lambda, \alpha(p))$ $= \mu_1^*(q, \Lambda, p)$. Hence $q = p$, i.e., α is one-one. ∎

6.4 Admissible Relations

Definition 6.4.1 *Let $M = (Q, X, \mu)$ be a ffsm and let $\sim$ be an equivalence relation on Q. Then $\sim$ is called an **admissible relation** if and only if $\forall$ $p, q, r \in Q, \forall a \in X$, if $p \sim q$ and $\mu(p, a, r) > 0$, then $\exists\, t \in Q$ such that $\mu(q, a, t) \geq \mu(p, a, r)$ and $t \sim r$.*

Theorem 6.4.2 *Let $M = (Q, X, \mu)$ be a ffsm and let $\sim$ be an equivalence relation on Q. Then $\sim$ is an admissible relation if and only if $\forall$ $p, q, r \in Q, \forall x \in X^*$, if $p \sim q$ and $\mu^*(p, x, r) > 0$, then $\exists\, t \in Q$ such that $\mu^*(q, x, t) \geq \mu^*(p, x, r)$ and $t \sim r$.*

Proof. Suppose $\sim$ is admissible. Let $p, q \in Q$ be such that $p \sim q$. Let $x \in X^*$, $r \in Q$ be such that $\mu^*(p, x, r) > 0$. Suppose $|x| = n$. If $n = 0$, then $x = \Lambda$. Thus $\mu^*(p, x, r) > 0 \implies p = r$ and $\mu^*(p, x, p) = 1$. Now $\mu^*(q, x, q) = 1 = \mu^*(p, x, p)$ and $q \sim p$. Thus the result is true for $n = 0$. Suppose now the result is true $\forall\, y \in X^*$ such that $|y| = n - 1$, $n > 0$. Let $x = ya$ where $y \in X_1^*$, $a \in X_1$, and $|y| = n - 1$. Now $\mu^*(p, x, r) = \mu^*(p, ya, r) = \vee\{\mu^*(p, y, q_1) \wedge \mu^*(q_1, a, r) | q_1 \in Q\} > 0$. Let $s \in Q$ be such that $\mu^*(p, y, s) \wedge \mu^*(s, a, r) = \vee\{\mu^*(p, y, q_1) \wedge \mu^*(q_1, a, r) \mid q_1 \in Q\}$. Then $\mu^*(p, y, s) > 0$ and $\mu^*(s, a, r) > 0$. By the induction hypothesis, $\exists\, t_s \in Q$ such that $\mu^*(q, y, t_s) \geq \mu^*(p, y, s)$ and $t_s \sim s$. Now $\mu(s, a, r) > 0$ and $t_s \sim s$. Since $\sim$ is admissible, $\exists\, t \in Q$ such that $\mu(t_s, a, t) \geq \mu(s, a, r)$ and $t \sim r$. Thus $\exists\, t \in Q$ such that $t \sim r$ and $\mu^*(q, y, t_s) \wedge \mu(t_s, a, t) \geq \mu^*(p, y, s) \wedge \mu^*(s, a, r)$. Thus

$$
\begin{aligned}
\mu^*(p, x, r) &= \mu^*(p, y, s) \wedge \mu^*(s, a, r) \\
&\leq \mu^*(q, y, t_s) \wedge \mu(t_s, a, t) \\
&\leq \vee\{\mu^*(q, y, r_1) \wedge \mu(r_1, a, t) | r_1 \in Q\} \\
&= \mu^*(q, ya, t) \\
&= \mu^*(q, x, t)
\end{aligned}
$$

and $t \sim r$. The result now follows by induction. The converse is trivial. ∎

Let $M = (Q, X, \mu)$ be a ffsm and let $\sim$ be an admissible relation on Q. For $q \in Q$, let $[q]$ denote the equivalence class of q. Let $\widetilde{Q} = Q/\sim$ $= \{[q]|q \in Q\}$. Define the fuzzy subset $\widetilde{\mu}$ of $\widetilde{Q} \times X \times \widetilde{Q}$ by

$$\widetilde{\mu}([q], x, [p]) = \vee\{\mu(q, x, t)|t \in [p]\}$$

$\forall q, p \in Q, x \in X$. Suppose that $[q] = [q']$, $x = y$, and $[p] = [p']$, $q, q', p, p' \in Q$ and $x, y \in X$. Then $q \sim q'$. Now

$$\widetilde{\mu}([q], x, [p]) = \vee\{\mu(q, x, r)|r \in [p]\}$$

and

$$\widetilde{\mu}([q'], y, [p']) = \widetilde{\mu}([q'], x, [p']) = \vee\{\mu(q', x, t)|t \in [p']\}.$$

Let $r \in [p]$ be such that $\mu(q, x, r) > 0$. Then since $\sim$ is admissible, $\exists$ $t \in Q$ such that $\mu(q', x, t) \geq \mu(q, x, r) > 0$ and $t \sim r$. Now since $t \sim r$, $t \in [p] = [p']$. Thus $\exists t \in [p']$ such that $\mu(q', x, t) \geq \mu(q, x, r) > 0$. Similarly if $\mu(q', x, t) > 0$ for some $t \in [p']$, then $\exists r \in [p]$ such that $\mu(q, x, r) \geq \mu(q', x, t) > 0$. Hence

$$\widetilde{\mu}([q], x, [p]) = \widetilde{\mu}([q'], x, [p']).$$

Thus $\widetilde{\mu}$ is single-valued. Hence $(\widetilde{Q}, X, \widetilde{\mu})$ is a ffsm. Define $\underline{\alpha} : Q \to \widetilde{Q}$ by $\underline{\alpha}(q) = [q] \; \forall \; q \in Q$. Clearly, $\underline{\alpha}$ maps Q onto $\widetilde{Q}$. Let $\beta : X \to X$ be the identity map. Let $q, t \in Q$ and $x \in X$. Then $\widetilde{\mu}(\underline{\alpha}(q), x, \underline{\alpha}(t)) = \widetilde{\mu}([q], x, [t])$ $= \vee\{\mu(q, x, r)|r \in [t]\} \geq \mu(q, x, t)$. Hence $(\underline{\alpha}, \beta)$ is a homomorphism.

Definition 6.4.3 Let $M_1 = (Q_1, X, \mu_1)$ and $M_2 = (Q_2, X, \mu_2)$ be two ff-sms. Let $\alpha : (Q_1, X, \mu_1) \to (Q_2, X, \mu_2)$ be a strong homomorphism. The kernel of α, denoted Ker α, is defined to be the set

$$Ker \; \alpha = \{(p, q)|\alpha(p) = \alpha(q)\}.$$

Lemma 6.4.4 Let α be as defined in Definition 6.4.3. Then Ker α is an admissible relation.

Proof. Now clearly Ker α is an equivalence relation. Let $p, q \in Q_1$ and $(p, q) \in$ Ker α. Then $\alpha(p) = \alpha(q)$. Let $a \in X$, $r \in Q_1$, and $\mu_1(p, a, r) > 0$. Then $\mu_2(\alpha(q), a, \alpha(r)) = \mu_2(\alpha(p), a, \alpha(r)) \geq \mu_1(p, a, r) > 0$. By Lemma 6.3.3, $\exists t \in Q_1$ such that $\mu_1(q, a, t) \geq \mu_1(p, a, r) > 0$ and $\alpha(t) = \alpha(r)$. Since $\alpha(t) = \alpha(r)$, $(t, r) \in$ Ker α. Thus Ker α is admissible. ∎

Theorem 6.4.5 Let $M_1 = (Q_1, X, \mu_1)$ and $M_2 = (Q_2, X, \mu_2)$ be two ffsms and let $\alpha : (Q_1, X, \mu_1) \to (Q_2, X, \mu_2)$ be an onto strong homomorphism. Then $\exists$ an isomorphism

$$\gamma : (Q_1/(Ker \; \alpha), \; X, \; \widetilde{\mu_1}) \to (Q_2, X, \mu_2)$$

such that $\alpha = \gamma \circ \underline{\alpha}$.

Proof. Define $\gamma : Q_1/(Ker\ \alpha) \to Q_2$ by $\gamma([q]) = \alpha(q)$. Let $p, q \in Q_1$ be such that $[p] = [q]$. Then $(p, q) \in Ker\ \alpha$ and hence $\alpha(p) = \alpha(q)$ or $\gamma([q]) = \gamma([p])$. Now, let $q, p \in Q_1$ and $x \in X$. Then

$$
\begin{aligned}
\widetilde{\mu_1}([q], x, [p]) &= \vee\{\mu_1(q, x, r) | r \in [p]\} \\
&= \vee\{\mu_1(q, x, r) | \alpha(r) = \alpha(p),\ r \in Q_1\} \\
&= \mu_2(\alpha(q), x, \alpha(p)) \\
&= \mu_2(\gamma([q]), x, \gamma([p])).
\end{aligned}
$$

Thus γ is a homomorphism. Clearly γ maps $(Q_1/(Ker\ \alpha), X, \widetilde{\mu_1})$ one-to-one onto (Q_2, X, μ_2). ∎

Example 6.4.6 Let $M_i = (Q_i, X_i, \mu_i)$ be the ffsm of Example 6.3.2, $i = 1, 2, 3$. Let $M'_2 = (Q'_2, X'_2, \mu'_2)$ be the ffsm, where $Q'_2 = \{q'_1, q'_2\}$, $X'_2 = \{a, b\}$, and $\mu'_2 = \mu_2|_{Q'_2 \times X_2 \times Q'_2}$. Let α and β be defined as in Example 6.3.2. Then (α, β) is a strong homomorphism of M_1 onto M'_2. $Ker\ \alpha = \{(q_1, q_1), (q_2, q_2), (q_3, q_3), (q_1, q_3), (q_3, q_1)\}$. Then $[q_1] = \{q_1, q_3\} = [q_3]$ and $[q_2] = \{q_2\}$. Also $\bar{Q}_1 = Q_1/Ker\ \alpha = \{[q_1], [q_2]\}$ and

$$
\begin{aligned}
\widetilde{\mu_1}([q_1], a, [q_1]) &= \tfrac{1}{3} \\
\widetilde{\mu_1}([q_1], b, [q_2]) &= \tfrac{2}{3} \\
\widetilde{\mu_1}([q_2], a, [q_1]) &= \tfrac{1}{3} \\
\widetilde{\mu_1}([q_2], b, [q_1]) &= \tfrac{2}{3}.
\end{aligned}
$$

It is easily seen that there exists an isomorphism $\gamma : (Q_1/Ker\ \alpha, X, \widetilde{\mu_1}) \to (Q'_2, X'_2, \mu'_2)$ such that $\alpha = \gamma \circ \underline{\alpha}$, where $X = X_2$.

6.5 Fuzzy Transformation Semigroups

A **transformation semigroup** is a pair (Q, S), where Q is a finite nonempty set and S is a finite semigroup with an **action** δ of S on Q, i.e., a partial function δ of $Q \times S$ into Q such that
   (1) $\delta(\delta(q, s), s') = \delta(q, ss')\ \forall q \in Q$, $s, s' \in S$, and
   (2) $\delta(q, s) = \delta(q, s')\ \forall q \in Q$ implies $s = s'$, where $s, s' \in S$, [92, p. 33].

Definition 6.5.1 A *fuzzy transformation semigroup (fts)* is a triple (Q, S, ρ), where Q is a finite nonempty set, S is a finite semigroup, and ρ is a fuzzy subset of $Q \times S \times Q$ such that
   (1) $\rho(q, uv, p) = \vee\{\rho(q, u, r) \wedge \rho(r, v, p) | r \in Q\}\ \forall u, v \in S$ and $\forall q, p \in Q$;
   (2) If S contains the identity e, then $\rho(q, e, p) = 1$ if $q = p$ and $\rho(q, e, p) = 0$ if $q \neq p$, $\forall\ q, p \in Q$.
   If, in addition, the following property holds, then (Q, S, ρ) is called **faithful**.
   (3) Let $u, v \in S$. If $\rho(q, u, p) = \rho(q, v, p)\ \forall q, p \in Q$, then $u = v$.

Let $M = (Q, S, \rho)$ be a fts. This fts may not be faithful. Define a relation R on S by $\forall\ u, v \in S$, uRv if and only if $\forall\ q, p \in Q$, $\rho(q, u, p) = \rho(q, v, p)$. Clearly R is an equivalence relation on S. Suppose that $u, v, x \in S$ and uRv. Then

$$
\begin{aligned}
\rho(q, ux, p) &= \vee\{\rho(q, u, r) \wedge \rho(r, x, p) \mid r \in Q\} \\
&= \vee\{\rho(q, v, r) \wedge \rho(r, x, p) \mid r \in Q\} \\
&= \rho(q, vx, p)
\end{aligned}
$$

$\forall\ q, p \in Q$. Similarly, $\rho(q, xu, p) = \rho(q, xv, p)\ \forall\ q, p \in Q$. Hence R is a congruence relation on S. Let $[u]$ denote the equivalence class of R induced by u. Let $S/R = \{[u] \mid u \in S\}$. Define

$$
\overline{\rho} : Q \times S/R \times Q \to [0, 1]
$$

by

$$
\overline{\rho}(q, [x], p) = \rho(q, x, p)
$$

$\forall\ q, p \in Q$ and $\forall\ [x] \in S/R$. Clearly, $\overline{\rho}$ is single-valued. Now

$$
\begin{aligned}
\overline{\rho}(q, [x][y], p) &= \overline{\rho}(q, [xy], p) \\
&= \rho(q, xy, p) \\
&= \vee\{\rho(q, x, r) \wedge \rho(r, y, p) \mid r \in Q\} \\
&= \vee\{\overline{\rho}(q, [x], r) \wedge \overline{\rho}(r, [y], p) \mid r \in Q\}\ .
\end{aligned}
$$

Also

$$
\overline{\rho}(q, [e], p) = \begin{cases} 1 \text{ if } p = q \\ 0 \text{ otherwise.} \end{cases}
$$

Suppose that $\overline{\rho}(q, [x], p) = \overline{\rho}(q, [y], p)\ \forall q, p \in Q$. Then $\rho(q, x, p) = \rho(q, y, p)\ \forall q, p \in Q$. Hence xRy and so $[x] = [y]$. Thus $(Q, S/R, \overline{\rho})$ is a faithful fts. We call $(Q, S/R, \overline{\rho})$ the **faithful fuzzy transformation semigroup represented** by the triple (Q, S, ρ).

Theorem 6.5.2 *Let $M = (Q, X, \mu)$ be a ffsm. Let $E(M)$ be defined as before. Then $(Q, E(M), \rho)$ is a faithful fts where $\rho(q, [x], p) = \mu^*(q, x, p)\ \forall q, p \in Q$, $x \in X^*$.*

Proof. By Theorem 6.2.4, $E(M)$ is a finite semigroup with identity $[\Lambda]$. Clearly ρ is single-valued. Let $q, p \in Q$ and $[x], [y] \in E(M)$. Then

$$
\begin{aligned}
\rho(q, [x] * [y], p) &= \rho(q, [xy], p) \\
&= \mu^*(q, xy, p) \\
&= \vee\{\mu^*(q, x, r) \wedge \mu^*(r, y, p) \mid r \in Q\} \\
&= \vee\{\rho(q, [x], r) \wedge \rho(r, [y], p) \mid r \in Q\}.
\end{aligned}
$$

Now

$$\rho(q,[\Lambda],p) = \left\{ \begin{array}{l} \mu^*(q,\Lambda,p) = 1 \ \text{ if } p = q \\ \mu^*(q,\Lambda,p) = 0 \ \text{ if } q \neq p, \end{array} \right.$$

by the definition of μ^*. Suppose $\rho(q,[x],p) = \rho(q,[y],p) \ \forall \ q,p \in Q$. Then $\mu^*(q,x,p) = \mu^*(q,y,p) \ \forall q,p \in Q$. Thus $x \equiv y$ or $[x] = [y]$. Hence $(Q,E(M),\rho)$ is a faithful fuzzy transformation semigroup. ∎

Let $M = (Q,X,\mu)$ be a ffsm. Then by Theorem 6.5.2 $(Q,E(M),\rho)$ is a fuzzy transformation semigroup that we denote by $FTS(M)$. We call $FTS(M)$ the **fuzzy transformation semigroup associated** with M.

Let $M = (Q,X,\mu)$ be a ffsm. Define the relation $\equiv^+$ on X^+ by $\forall x,y \in X^+$, $x \equiv^+ y$ if and only if $\mu^+(q,x,p) = \mu^+(q,y,p) \ \forall q,p \in Q$. Then $\equiv^+$ is the restriction of $\equiv$ to $X^+ \times X^+$. Let $S(M)$ denote the set of all equivalence classes induced by $\equiv^+$. Then $S(M) = E(M) \backslash \{\Lambda\}$ and $S(M)$ is a subsemigroup of $E(M)$. In view of the crisp case, it would have been reasonable to define $(Q,S(M),\rho)$ as the fuzzy transformation semigroup associated with M.

Example 6.5.3 *Let $M = (Q,X,\mu)$ be the ffsm such that $Q = \{q_1,q_2,q_3\}$, $X = \{a,b\}$, and $\mu : Q \times X \times Q \to [0,1]$ is defined as follows:*

$$\begin{array}{rcl} \mu(q_1,a,q_1) & = & \frac{1}{3} \\ \mu(q_1,b,q_2) & = & \frac{2}{3} \\ \mu(q_2,a,q_1) & = & \frac{1}{3} \\ \mu(q_2,b,q_j) & = & 0 \\ \mu(q_3,a,q_3) & = & \frac{1}{3} \\ \mu(q_3,b,q_2) & = & \frac{2}{3} \end{array}$$

for $j = 1,2,3$ and $\mu(q,x,q') = 0$ for any other triple $(q,x,q') \in Q \times X \times Q$. Then

$$\widetilde{E(M)} = \{[[\Lambda]], \ [[a]], \ [[b]], \ [[ab]], \ [[ba]], \ [[aba]]\} \cup \{[[b^2]]\}$$

and

$$E(M) = \{[\Lambda], \ [a], \ [b], \ [ab], \ [ba], \ [aba], \ [bab]\} \cup \{[[b^2]]\}.$$

In the crisp case, μ would be considered a "partial" function since $\mu(q_2,b,q_j) = 0$, for $j = 1,2,3$ and the equivalence classes $[[b^2]]$ and $[b^2]$ would be designated as an empty relation θ. Hence if we set $\theta = [[b^2]]$ and then $\theta = [b^2]$, we have that the following tables give the semigroup operations for $\widetilde{E(M)} \backslash \{\theta\}$ and $E(M) \backslash \{\theta\}$, respectively. Note that $\theta \tilde{*} x = x \tilde{*} \theta = \theta$ for all

$x \in \widetilde{E(M)}$ *and* $\theta * x = x * \theta = \theta$ *for all* $x \in E(M)$.

$\widetilde{*}$	$[[\Lambda]]$	$[[a]]$	$[[b]]$	$[[ab]]$	$[[ba]]$	$[[aba]]$
$[[\Lambda]]$	$[[\Lambda]]$	$[[a]]$	$[[b]]$	$[[ab]]$	$[[ba]]$	$[[aba]]$
$[[a]]$	$[[a]]$	$[[a]]$	$[[ab]]$	$[[ab]]$	$[[aba]]$	$[[aba]]$
$[[b]]$	$[[b]]$	$[[ba]]$	θ	$[[b]]$	θ	$[[ba]]$
$[[ab]]$	$[[ab]]$	$[[aba]]$	θ	$[[ab]]$	θ	$[[aba]]$
$[[ba]]$	$[[ba]]$	$[[ba]]$	$[[b]]$	$[[b]]$	$[[ba]]$	$[[ba]]$
$[[aba]]$	$[[aba]]$	$[[aba]]$	$[[ab]]$	$[[ab]]$	$[[aba]]$	$[[aba]]$

$\widetilde{*}$	$[\Lambda]$	$[a]$	$[b]$	$[ab]$	$[ba]$	$[aba]$	$[bab]$
$[\Lambda]$	$[\Lambda]$	$[a]$	$[b]$	$[ab]$	$[ba]$	$[aba]$	$[bab]$
$[a]$	$[a]$	$[a]$	$[ab]$	$[ab]$	$[aba]$	$[aba]$	$[ba]$
$[b]$	$[b]$	$[ba]$	θ	$[b]$	θ	$[ba]$	θ
$[ab]$	$[ab]$	$[aba]$	θ	$[ab]$	θ	$[aba]$	$[bab]$
$[ba]$	$[ba]$	$[ba]$	$[b]$	$[b]$	$[ba]$	$[ba]$	θ
$[aba]$	$[aba]$	$[aba]$	$[ab]$	$[ab]$	$[aba]$	$[aba]$	$[ba]$
$[bab]$	$[bab]$	$[ba]$	θ	$[bab]$	θ	$[ba]$	θ

The fts $(Q, E(M), \rho)$ *is now easily determined.*

Definition 6.5.4 *Let* (Q, S, ρ) *be a fts. Let* $\sim$ *be an equivalence relation on* Q*. Then* $\sim$ *is called an admissible relation if and only if* $\forall p, q, r \in Q, \forall u \in S$*, if* $p \sim q$ *and* $\rho(p, u, r) > 0$*, then* $\exists t \in Q$ *such that* $\rho(q, u, t) \geq \rho(p, u, r)$ *and* $t \sim r$*.*

Theorem 6.5.5 *Let* $M = (Q, X, \mu)$ *be a ffsm and let* $\sim$ *be an equivalence relation on* Q*. Then* $\sim$ *is an admissible relation for* M *if and only if* $\sim$ *is an admissible relation for the fuzzy transformation semigroup,* $FTS(M) = (Q, E(M), \rho)$*.*

Proof. Suppose $\sim$ is admissible for M. Let $p, q \in Q$ be such that $p \sim q$ and $[u] \in E(M)$. Let $\rho(p, [u], r) > 0$ for some $r \in Q$. Then $\mu^*(p, u, r) > 0$. Hence by Theorem 6.4.2, $\exists t \in Q$ such that $\mu^*(q, u, t) \geq \mu^*(p, u, r)$ and $t \sim r$. Thus $\rho(q, [u], t) = \mu^*(q, u, t) \geq \mu^*(p, u, r) = \rho(p, [u], r)$. Hence $\sim$ is admissible for $FTS(M)$. Conversely, suppose that $\sim$ is admissible for $FTS(M)$. Let $p, q \in Q$ be such that $p \sim q$ and $u \in X$. Let $\mu^*(p, u, r) > 0$ for some $r \in Q$. Then $\rho(p, [u], r) > 0$. Then $\exists t \in Q$ such that $\rho(q, [u], t) \geq \rho(p, [u], r)$ and $t \sim r$. Now $\mu^*(q, u, t) = \rho(q, [u], t) \geq \rho(p, [u], r) = \mu^*(p, u, r)$ and $t \sim r$. Hence $\sim$ is admissible for M. ∎

Definition 6.5.6 *Let* (Q_1, S_1, ρ_1) *and* (Q_2, S_2, ρ_2) *be two fts's. A pair* (f, g) *of mappings, where* $f : Q_1 \rightarrow Q_2$ *and* $g : S_1 \rightarrow S_2$*, is said to be a* **homomorphism** *from* (Q_1, S_1, ρ_1) *to* (Q_2, S_2, ρ_2) *if*
    (1) $g(xy) = g(x)g(y) \; \forall \, x, y \in S_1$,

(2) If e_1 is the identity of S_1 and e_2 is the identity of S_2, then $g(e_1) = e_2$,
(3) $\rho_1(q, x, p) \leq \rho_2(f(q), g(x), f(p))$ $\forall q, p \in Q_1, x \in S_1$.
*(f, g) is called a **strong homomorphism** if it satisfies (1), (2), and*

$$\rho_2(f(q), g(x), f(p)) = \vee\{\rho_1(q, x, t) | t \in Q_1, \ f(t) = f(p)\}$$

$\forall q, p \in Q_1, x \in S_1$.
A *homomorphism (strong homomorphism) $(f, g) : (Q_1, S_1, \rho_1) \to (Q_2, S_2, \rho_2)$ is called an **isomorphism (strong isomorphism)** if f and g are both one-one and onto.*

Let S be a semigroup with identity. Let $\mathcal{A} = (Q, S, \delta)$ be a faithful fts. Define the ffsm $M = (Q, S, \mu)$ by taking $\mu = \delta$. Consider FTS$(M) = (Q, E(M), \rho)$, where $E(M) = S^*/\sim$ and $\rho(q, [u], p) = \mu^*(q, u, p)$. Let e be the identity element of S and Λ the empty word in S^*. Now $\rho(q, [e], p) = \mu^*(q, e, p) = \delta(q, e, p) = 1$ if $p = q$ and 0 if $p \neq q$. Hence $\rho(q, [e], p) = \rho(q, [\Lambda], p)$ $\forall p, q \in Q$. Thus $[e] = [\Lambda]$.

Theorem 6.5.7 *Let S be a semigroup with identity. Then FTS(M) is isomorphic to $\mathcal{A} = (Q, S, \delta)$.*

Proof. Define $f : Q \to Q$ by $f(q) = q$ $\forall q \in Q$ and $g : S \to E(M)$ by $g(x) = [x]$ $\forall x \in S$. Let $x, y \in S$ be such that $g(x) = g(y)$. Then $[x] = [y]$. Thus $\mu^*(q, x, p) = \mu^*(q, y, p)$ $\forall q, p \in Q$. Hence $\mu(q, x, p) = \mu(q, y, p)$ $\forall q, p \in Q$. This implies that $\delta(q, x, p) = \delta(q, y, p)$ $\forall q, p \in Q$. Since $\mathcal{A}$ is faithful, we find that $x = y$. Hence g is injective. Let $\cdot$ denote the binary operation of the semigroup S. Let $a, b \in S$. Then $a \cdot b \in S$ and $ab \in S^*$. Let $q, p \in Q$. Now

$$
\begin{aligned}
\mu^*(q, a \cdot b, p) &= \mu(q, a \cdot b, p) \\
&= \delta(q, a \cdot b, p) \\
&= \vee\{\delta(q, a, r) \wedge \delta(r, b, p) | r \in Q\} \\
&= \vee\{\mu(q, a, r) \wedge \mu(r, b, p) | r \in Q\} \\
&= \mu^*(q, ab, p).
\end{aligned}
$$

Hence $[a \cdot b] = [ab]$. Thus $g(ab) = [a \cdot b] = [ab] = [a][b] = g(a)g(b)$. By induction it can be shown that if $c_i \in S$, $1 \leq i \leq n$, then $[c_1 \cdot c_2 \cdot \ldots \cdot c_n] = [c_1 c_2 \ldots c_n]$. Let $[u] \in E(M)$. If $u = \Lambda$, then $[\Lambda] = [e]$ and $g(e) = [\Lambda]$. Suppose $u = a_1 a_2 \ldots a_n$, $a_i \in S$, $1 \leq i \leq n$. Then $g(a_1 \cdot a_2 \cdot \ldots \cdot a_n) = [a_1 \cdot a_2 \cdot \ldots \cdot a_n] = [a_1 a_2 \ldots a_n] = [u]$. Thus g is surjective. Finally, $\rho(f(q), g(x), f(p)) = \rho(q, [x], p) = \mu^*(q, x, p) = \mu(q, x, p) = \delta(q, x, p)$. ∎

Theorem 6.5.8 *Let $M_1 = (Q_1, X_1, \mu_1)$ and $M_2 = (Q_2, X_2, \mu_2)$ be two ffsms and let $(\alpha, \beta) : M_1 \to M_2$ be a strong homomorphism with α one-one and onto. Then $\exists$ a strong homomorphism (f_α, g_β) from FTS(M_1) to FTS(M_2).*

Proof. Define $f_\alpha : Q_1 \to Q_2$ by $f_\alpha(q) = \alpha(q) \; \forall \; q \in Q_1$ and $g_\beta : E(M_1) \to E(M_2)$ by $g_\beta([x]) = [\beta^*(x)] \; \forall \; [x] \in E(M_1)$. Let $[x], [y] \in E(M_1)$ and $[x] = [y]$. Then $\mu^*(q, x, p) = \mu^*(q, y, p) \; \forall \; q, p \in Q$. Now

$$
\begin{aligned}
\mu_2^*(\alpha(q), \beta^*(x), \alpha(p)) &= \mu_1^*(q, x, p) \\
&= \mu_1^*(q, y, p) \\
&= \mu_2^*(\alpha(q), \beta^*(y), \alpha(p))
\end{aligned}
$$

$\forall \; q, p \in Q_1$. Thus since α is onto, $[\beta^*(x)] = [\beta^*(y)]$. Hence g_β is well defined. Now $g_\beta([x] * [y]) = g_\beta([xy]) = [\beta^*(xy)] = [\beta^*(x)\beta^*(y)] = [\beta^*(x)] * [\beta^*(y)] = g_\beta([x]) * g_\beta([y])$ and $g_\beta([\Lambda]) = [\beta^*(\Lambda)] = [\Lambda]$. Also

$$
\begin{aligned}
\rho_1(q, [x], p) &= \mu_1^*(q, x, p) \\
&= \mu_2^*(\alpha(q), \beta^*(x), \alpha(p)) \\
&= \rho_2(f_\alpha(q), \; g_\beta([x]), f_\alpha(p)).
\end{aligned}
$$

Hence by definition (f_α, g_β) is a strong homomorphism. ∎

Definition 6.5.9 *A **polytransformation semigroup** is a triple (Q, S, ν), where Q is a finite nonempty set, S is a finite semigroup, and $\nu : Q \times S \to \mathcal{P}(Q) \setminus \{\emptyset\}$ such that*

(1) $\nu(\nu(q, u), v) = \nu(q, uv) \; \forall \; q \in Q, u, v \in S$. It is being understood for any subset P of Q, $\nu(P, u) = \cup_{u \in P} \nu(p, u)$.

(2) If S contains identity e, then $\nu(q, e) = \{q\} \; \forall \; q \in Q$.

*If, in addition, the following holds, then (Q, S, ν) is called **faithful**.*

(3) Let $u, v \in S$. If $\nu(q, u) = \nu(q, v) \; \forall \; q \in Q$, then $u = v$.

Let $M = (Q, X, \mu)$ be a ffsm. For $q \in Q$ and $x \in X^*$, we define

$$
S_x(q) = \{p \in Q | \mu^*(q, x, p) > 0\}.
$$

Theorem 6.5.10 *Let $M = (Q, X, \mu)$ be a ffsm. Let $\widetilde{E(M)}$ be defined as before. If $\forall x \in X^*$ and $\forall q \in Q$, $S_x(q) \neq \emptyset$, then $(Q, \widetilde{E(M)}, \nu)$ is a faithful polytransformation semigroup with identity.*

Proof. Define $\nu : Q \times \widetilde{E(M)} \to \mathcal{P}(Q) \setminus \{\emptyset\}$ by $\nu(q, \prec x \succ) = \{p \in Q \mid \mu^*(q, x, p) > 0\}$. Suppose that $\prec x \succ = \prec y \succ$. Then $x \approx y$. Thus

$$
\begin{aligned}
\nu(q, \prec x \succ) &= \{p \in Q | \mu^*(q, x, p) > 0\} \\
&= \{p \in Q | \mu^*(q, y, p) > 0\} \\
&= \nu(q, \prec y \succ).
\end{aligned}
$$

Thus ν is single-valued. We write $q \prec x \succ$ for $\nu(q, \prec x \succ)$. Then, by definition, $(q \prec x \succ) \prec y \succ = \cup_{p \in q \prec x \succ} p \prec y \succ$. Let $p \in q \prec xy \succ = \{p \in Q | \mu^*(q, xy, p) > 0\}$. Now $\mu^*(q, xy, p) > 0 \Rightarrow \mu^*(q, x, r) \wedge \mu^*(r, y, p) > 0$ for some $r \in Q$. Thus $\mu^*(q, x, r) > 0$ and $\mu^*(r, y, p) > 0$. Thus

$r \in q \prec x \succ$ and $p \in r \prec y \succ$. Hence $p \in \cup_{t \in q \prec x \succ} t \prec y \succ$. Thus $q \prec xy \succ \subseteq (q \prec x \succ)(\prec y \succ)$. Let

$$p \in (q \prec x \succ)(\prec y \succ) = \cup_{t \in q \prec x \succ} t \prec y \succ .$$

Then $p \in t \prec y \succ$ for some $t \in q \prec x \succ$. Thus $\mu^*(t, y, p) > 0$ and $\mu^*(q, x, t) > 0$. Hence $\mu^*(q, xy, p) > 0$. Thus $p \in q \prec xy \succ$. Hence $(q \prec x \succ)(\prec y \succ) \subseteq q \prec xy \succ$. Therefore, $q \prec xy \succ = (q \prec x \succ)(\prec y \succ)$. Hence (1) of Definition 6.5.9 holds. It follows easily that $q \prec \Lambda \succ = \{q\}$ and so (2) of Definition 6.5.9 holds. Suppose that $q \prec x \succ = q \prec y \succ \forall q \in Q$. Then $\mu^*(q, x, p) > 0$ iff $\mu^*(q, y, p) > 0 \forall p \in Q$. Thus $\prec x \succ = \prec y \succ$. Hence (3) of Definition 6.5.9 holds. ∎

6.6 Products of Fuzzy Finite State Machines

The concepts of transformation semigroup, covering, cascade product, and wreath product play a prominent role in the study of automata [92]. In this chapter, we examine these ideas for fuzzy finite state machines. Some severe and interesting complications arise when introducing these ideas to the fuzzy setting. One of the concepts we introduce to overcome some of the problems that arise is that of a polysemigroup. Let P be a nonempty set and $\mathcal{P}(P)$ be the power set of P. Let $*$ be a function of $P \times P$ into $\mathcal{P}(P)\backslash\{\emptyset\}$. Then $(P, *)$ is called a **polysemigroup**, [39, 102], if and only if $x * (y * z) = (x * y) * z \; \forall x, y, z \in P$. The obvious abuse of notation is explained as follows: If $x \in P$ and $A, B \subseteq P$, then $x * A$ denotes $\{x\} * A$, $A * x$ denotes $A * \{x\}$, and $A * B = \cup_{a \in A, b \in B} a * b$.

Definition 6.6.1 *Let* $M_i = (Q_i, X_i, \mu_i)$ *be a ffsm,* $i = 1, 2$. *Let* η *be a function of* Q_2 *onto* Q_1 *and let* ξ *be a function of* X_1 *into* X_2. *Extend* ξ *to a function* ξ^* *of* X_1^* *into* X_2^* *by* $\xi^*(\Lambda) = \Lambda$ *and* $\forall \; x \in X_1^*$, $\xi^*(x) = \xi(x_1)\xi(x_2)\ldots\xi(x_n)$ *where* $x = x_1 x_2 \ldots x_n$ *and* $x_i \in X_1$, $i = 1, 2, \ldots, n$. *Then* (η, ξ) *is called a **covering** of* M_1 *by* M_2, *written* $M_1 \leq M_2$, *if and only if* $\forall \; q_2 \in Q_2$, $q_1 \in Q_1$, *and* $x \in X_1^*$,

$$\mu_1^*(\eta(q_2), x, q_1) = \vee\{\mu_2^*(q_2, \xi^*(x), r_2) | \eta(r_2) = q_1, r_2 \in Q_2\}.$$

Clearly (η, ξ) is a covering of M_1 by M_2 if and only if $\forall \; q_2 \in Q_2$, $q_1 \in Q_1$, and $x \in X_1^*$, $\mu_1^*(\eta(q_2), x, q_1) \geq \mu_2^*(q_2, \xi^*(x), r_2) \; \forall \; r_2 \in Q_2$ such that $\eta(r_2) = q_1$ and $\exists \; r_2 \in Q_2$ such that $\eta(r_2) = q_1$ and $\mu_1^*(\eta(q_2), x, q_1) = \mu_2^*(q_2, \xi^*(x), r_2)$.

Example 6.6.2 *Let* $M_1 = (Q_1, X_1, \mu_1)$ *and* $M_2 = (Q_2, X_2, \mu_2)$ *be ffsms such that* $Q_1 = \{q_1, q_2\}$, $X_1 = \{a, b\}$, $Q_2 = \{q_1', q_2', q_3'\}$, $X_2 = \{a, b\}$, *and*

μ_1 and μ_2 are defined as follows:

$$\mu_1(q_1, a, q_1) = \tfrac{1}{3} \qquad \mu_2(q_1', b, q_2') = \tfrac{2}{3}$$

$$\mu_1(q_1, b, q_2) = \tfrac{2}{3} \qquad \mu_2(q_2', a, q_2') = \tfrac{1}{3}$$

$$\mu_1(q_2, a, q_2) = \tfrac{1}{3} \qquad \mu_2(q_2', b, q_3') = \tfrac{2}{3}$$

$$\mu_1(q_2, b, q_1) = \tfrac{2}{3} \qquad \mu_2(q_3', a, q_1') = \tfrac{1}{3}$$

$$\mu_2(q_1', a, q_3') = \tfrac{1}{3} \qquad \mu_2(q_3', b, q_2') = \tfrac{1}{4}$$

and $\mu_1(q, x, p) = 0$ for all other $(q, x, p) \in Q_1 \times X_1 \times Q_1$ and $\mu_2(q', x, p') = 0$ for all other $(q', x, p') \in Q_2 \times X_2 \times Q_2$. Define $\eta : Q_2 \to Q_1$ by $\eta(q_1') = \eta(q_3') = q_1$ and $\eta(q_2') = q_2$. Let ξ be the identity map on $X_1 = X_2$. Now for all $x \in X_1^* = X_2^*$

$$\mu_1^*(q_1, x, q_1) = \mu_2^*(q_1', x, q_1') \vee \mu_2^*(q_1', x, q_3')$$
$$\mu_1^*(q_1, x, q_2) = \mu_2^*(q_1', x, q_2')$$
$$\mu_1^*(q_2, x, q_1) = \mu_2^*(q_2', x, q_1') \vee \mu_2^*(q_2', x, q_3')$$
$$\mu_1^*(q_2, x, q_2) = \mu_2^*(q_2', x, q_2').$$

Thus (η, ξ) is a covering of M_1 by M_2.

Let $M_i = (Q_i, X_i, \mu_i)$ be a ffsm, $i = 1, 2$. Let $\overline{X}$ be a finite set and f a function from $\overline{X}$ into $X_1 \times X_2$. Let π_i be the projection map of $X_1 \times X_2$ onto X_i, $i = 1, 2$.

Definition 6.6.3 Define $\mu_f : (Q_1 \times Q_2) \times \overline{X} \times (Q_1 \times Q_2) \to [0, 1]$ as follows: $\forall (q_1, q_2), (p_1, p_2) \in Q_1 \times Q_2$ and $\forall a \in \overline{X}$,

$$\mu_f((q_1, q_2), a, (p_1, p_2)) = \mu_1 \times \mu_2((q_1, q_2), (\pi_1(f(a)), \pi_2(f(a))), (p_1, p_2)).$$

Then $(Q_1 \times Q_2, \overline{X}, \mu_f)$ is called the **general direct product** of M_1 and M_2 and we write $M_1 * M_2$ for $(Q_1 \times Q_2, \overline{X}, \mu_f)$.

Recall

$$\mu_1 \times \mu_2((q_1, q_2), (a_1, a_2), (p_1, p_2)) = \mu_1(q_1, a_1, p_1) \wedge \mu_2(q_2, a_2, p_2)$$

$\forall (q_1, q_2), (p_1, p_2) \in Q_1 \times Q_2 \ \forall (a_1, a_2) \in X_1 \times X_2$.

If $\overline{X} = X_1 \times X_2$ and f is the identity map, then $M_1 * M_2$ is called the **full direct product** of M_1 and M_2 and we write $M_1 \times M_2$ for $M_1 * M_2$. If $X_1 = X_2, \overline{X} = \{(a_1, a_2) | a_i \in X_i, i = 1, 2, \ a_1 = a_2\}$, and f is the identity map, then $M_1 * M_2$ is called the **restricted direct product** of M_1 and M_2 and we write $M_1 \wedge M_2$ for $M_1 * M_2$. (We could also let $\overline{X} = X_1 = X_2$ and $f : \overline{X} \to \{(a_1, a_2) \mid a_i \in X_i, i = 1, 2, a_1 = a_2\}$ where $f(a) = (a, a)$ to obtain the restricted direct product.)

For every result concerning $M_1 * M_2$, there is a corresponding result for $M_1 \wedge M_2$. We see this by making the identifications $(a, a) \to a \ \forall a \in X_1 = X_2$ and $(x_1, x_1) \ldots (x_n, x_n) \to x_1 \ldots x_n$ for $x_i \in X_1, i = 1, \ldots, n$.

Example 6.6.4 *Let* $M_1 = (Q_1, X_1, \mu_1)$ *and* $M_2 = (Q_2, X_2, \mu_2)$ *be ffsms, where* $Q_1 = \{q_1, q_2\}$, $X_1 = \{a\}$, $Q_2 = \{q'_1, q'_2\}$, $X_2 = \{a\}$, *and* μ_1 *and* μ_2 *are defined as follows:*

$$
\begin{aligned}
\mu_1(q_1, a, q_1) &= 0 & \mu_2(q'_1, a, q'_1) &= 0 \\
\mu_1(q_1, a, q_2) &= c_1 > 0 & \mu_2(q'_1, a, q'_2) &= d_1 > 0 \\
\mu_1(q_2, a, q_1) &= c_2 > 0 & \mu_2(q'_2, a, q'_1) &= 0 \\
\mu_1(q_2, a, q_2) &= 0 & \mu_2(q'_2, a, q'_2) &= d_2 > 0.
\end{aligned}
$$

Then

$$
\begin{aligned}
\mu_1 \times \mu_2((q_1, q'_1), (a, a), (q_2, q'_2)) &= \mu_1(q_1, a, q_2) \wedge \mu_2(q'_1, a, q'_2) \\
&= c_1 \wedge d_1 \\
\mu_1 \times \mu_2((q_2, q'_1), (a, a), (q_1, q'_2)) &= \mu_1(q_2, a, q_1) \wedge \mu_2(q'_1, a, q'_2) \\
&= c_2 \wedge d_1 \\
\mu_1 \times \mu_2((q_1, q'_2), (a, a), (q_2, q'_2)) &= \mu_1(q_1, a, q_2) \wedge \mu_2(q'_2, a, q'_2) \\
&= c_1 \wedge d_2 \\
\mu_1 \times \mu_2((q_2, q'_2), (a, a), (q_1, q'_2)) &= \mu_1(q_2, a, q_1) \wedge \mu_2(q'_2, a, q'_2) \\
&= c_2 \wedge d_2.
\end{aligned}
$$

Suppose that $c_1 = c_2$ *and* $d_1 = d_2$. *Then* $a^n \equiv_1 a^m$ *if and only if* n *and* m *are both even or both odd. Hence* $E(M_1) = \{[\Lambda], [a], [a^2]\}$. *For* M_2, $a^n \equiv_2 a^m$ $\forall n, m \in \mathbb{N}$. *Thus* $E(M_2) = \{[\Lambda], [a]\}$. *(The reader is asked to determine* $E(M_1)$ *and* $E(M_2)$ *in the Exercises when* $c_1 \neq c_2$ *and* $d_1 \neq d_2$.*)*

Since $c_1 = c_2$ *and* $d_1 = d_2$, $c_1 \wedge d_1 = c_2 \wedge d_1 = c_1 \wedge d_2 = c_2 \wedge d_2$. *Hence* $a^n \equiv_{12} a^m$ *if and only if* n *and* m *are both even or both odd. Thus* $E(M_1 \times M_2) = \{[\Lambda], [a], [a^2]\}$, *where* $[a^3] = [a]$. *In this example,* $E(M_1 \wedge M_2) = E(M_1 \times M_2)$.

Example 6.6.5 *Let* $M_1 = (Q_1, X_1, \mu_1)$ *and* $M_2 = (Q_2, X_2, \mu_2)$ *be ffsms, where* $Q_1 = \{q_1, q_2\}$, $X_1 = \{a\}$, $Q_2 = \{q'_1, q'_2\}$, $X_2 = \{a, b\}$, *and* μ_1 *and* μ_2 *are defined as follows:*

$$
\begin{aligned}
\mu_1(q_1, a, q_1) &= 0 & \mu_2(q'_1, b, q'_1) &= d_2 > 0 \\
\mu_1(q_1, a, q_2) &= c_1 > 0 & \mu_2(q'_1, b, q'_2) &= 0 \\
\mu_1(q_2, a, q_1) &= c_2 > 0 & \mu_2(q'_2, a, q'_1) &= 0 \\
\mu_1(q_2, a, q_2) &= 0 & \mu_2(q'_2, a, q'_2) &= d_3 > 0 \\
\mu_2(q'_1, a, q'_1) &= 0 & \mu_2(q'_2, b, q'_1) &= d_4 > 0 \\
\mu_2(q'_1, a, q'_2) &= d_1 > 0 & \mu_2(q'_2, b, q'_2) &= 0.
\end{aligned}
$$

Then

$$
\begin{aligned}
\mu_1 \times \mu_2((q_1, q'_1), (a, b), (q_2, q'_2)) &= c_1 \wedge d_2 \\
\mu_1 \times \mu_2((q_2, q'_1), (a, b), (q_1, q'_1)) &= c_2 \wedge d_2 \\
\mu_1 \times \mu_2((q_1, q'_1), (a, a), (q_2, q'_2)) &= c_1 \wedge d_1 \\
\mu_1 \times \mu_2((q_2, q'_2), (a, b), (q_1, q'_1)) &= c_2 \wedge d_4 \\
\mu_1 \times \mu_2((q_2, q'_1), (a, a), (q_1, q'_2)) &= c_2 \wedge d_1 \\
\mu_1 \times \mu_2((q_1, q'_2), (a, b), (q_2, q'_1)) &= c_1 \wedge d_4 \\
\mu_1 \times \mu_2((q_1, q'_2), (a, a), (q_2, q'_2)) &= c_1 \wedge d_3 \\
\mu_1 \times \mu_2((q_2, q'_2), (a, a), (q_1, q'_2)) &= c_2 \wedge d_3.
\end{aligned}
$$

*Suppose that $c_1 = c_2$ and $d_1 = d_2 = d_3 = d_4$. Then $E(M_1) = \{[\Lambda], [a],$
$[a^2]\}$ and $(\{[a], [a^2]\}, *)$ is a group with identity $[a^2]$. $E(M_2) = \{[\Lambda], [a],$
$[b]\}$, where $[a] = [a^2]$, $[b] = [b^2]$, $[ab] = [b]$, and $[ba] = [a]$. Thus $S(M_1) =$
$\{[a], [a^2]\}$ and $S(M_2) = \{[a], [b]\}$. $S(M_1) \times S(M_2) = \{([a], [a]), ([a], [b]),$
$([a^2], [a]), ([a^2], [b])\}$. It follows that $S(M_1 \times M_2) = \{[(a, a)], [(a, a)^2], [(a, b)],$
$[(a, b)^2]\}$. The operation tables of $S(M_1) \times S(M_2)$ and $S(M_1 \times M_2)$ are given
below.*

$$S(M_1) \times S(M_2)$$

	$([a], [a])$	$([a^2], [a])$	$([a], [b])$	$([a^2], [b])$
$([a], [a])$	$([a^2], [a])$	$([a], [a])$	$([a^2], [b])$	$([a], [b])$
$([a^2], [a])$	$([a], [a])$	$([a^2], [a])$	$([a], [b])$	$([a^2], [b])$
$([a], [b])$	$([a^2], [a])$	$([a], [a])$	$([a^2], [b])$	$([a], [b])$
$([a^2], [b])$	$([a], [a])$	$([a^2], [a])$	$([a], [b])$	$([a^2], [b])$

$$S(M_1 \times M_2)$$

	$[(a, a)]$	$[(a, a)^2]$	$[(a, b)]$	$[(a, b)^2]$
$[(a, a)]$	$[(a, a)^2]$	$[(a, a)]$	$[(a, b)^2]$	$[(a, b)]$
$[(a, a)^2]$	$[(a, a)]$	$[(a, a)^2]$	$[(a, b)]$	$[(a, b)^2]$
$[(a, b)]$	$[(a, a)^2]$	$[(a, a)]$	$[(a, b)^2]$	$[(a, b)]$
$[(a, b)^2]$	$[(a, a)]$	$[(a, a)^2]$	$[(a, b)]$	$[(a, b)^2]$

*We see that $S(M_1) \times S(M_2) \cong S(M_1 \times M_2)$ under $([a], [a]) \rightarrow [(a, a)]$,
$([a^2], [a]) \rightarrow [(a, a)^2]$, $([a], [b]) \rightarrow [(a, b)]$, $([a^2], [b]) \rightarrow [(a, b)^2]$.*

Lemma 6.6.6 *Let $M_i = (Q_i, X_i, \mu_i)$ be a ffsm, $i = 1, 2$. Consider the
general direct product $M_1 * M_2$. Then*

(1)

$$\mu_f^*((q_1, q_2), \Lambda, (p_1, p_2)) = \mu_1^*(q_1, \Lambda, p_1) \wedge \mu_2^*(q_2, \Lambda, p_2)$$

$\forall q_1, p_1 \in Q_1$, $\forall q_2, p_2 \in Q_2$.

(2)

$$\mu_f^*((q_1, q_2), \overline{a_1}...\overline{a_n}, (p_1, p_2)) = \mu_1^*(q_1, \pi_1(f(\overline{a_1}))...\pi_1(f(\overline{a_n})), p_1) \wedge$$
$$\mu_2^*(q_2, \pi_2(f(\overline{a_1}))...\pi_2(f(\overline{a_n})), p_2))$$

where $\overline{a_i} \in \overline{X}$, $i = 1, ..., n$, $\forall q_1, p_1 \in Q_1$, $\forall q_2, p_2 \in Q_2$.

Proof. (1) The proof is straightforward.

(2) Then $\mu_f^*((q_1, q_2), \overline{a_1}...\overline{a_n}, (p_1, p_2)) = \vee\{\mu_f((q_1, q_2), \overline{a_1}, (r_1^{(1)}, r_2^{(1)})) \wedge$
$\mu_f((r_1^{(1)}, r_2^{(1)}), a_2, (r_1^{(2)}, r_2^{(2)})) \wedge ... \wedge \mu_f((r_1^{(n-1)}, r_2^{(n-1)}), \overline{a_n}, (p_1, p_2)) \mid (r_1^{(1)},$
$r_2^{(1)}) \in Q_1 \times Q_2, i = 1, ..., n-1\} = \vee\{\mu_1 \times \mu_2((q_1, q_2), (\pi_1(f(\overline{a_1})), \pi_2(f(\overline{a_1})),$
$(r_1^{(1)}, r_2^{(1)})) \wedge \mu_1 \times \mu_2((r_1^{(1)}, r_2^{(1)}), (\pi_1(f(\overline{a_2})), \pi_2(f(\overline{a_2})), (r_1^{(2)}, r_2^{(2)})) \wedge ... \wedge$
$\mu_1 \times \mu_2((r_1^{(n-1)}, r_2^{(n-1)}), (\pi_1(f(\overline{a_n})), \pi_2(f(\overline{a_n}), (p_1, p_2)) \mid (r_1^{(1)}, r_2^{(1)}) \in Q_1 \times$

$Q_2, i = 1, \ldots, n-1\} = \vee\{\mu_1(q_1, \pi_1(f(\overline{a_1})), r_1^{(1)}) \wedge \mu_2(q_2, \pi_2(f(\overline{a_1}), r_2^{(1)}) \wedge$
$\mu_1(r_1^{(1)}, \pi_1(f(\overline{a_2})), r_1^{(2)}) \wedge \mu_2(r_2^{(1)}, \pi_2(f(\overline{a_2}), r_2^{(2)}) \wedge \ldots \wedge \mu_1(r_1^{(n-1)}, \pi_1(f(\overline{a_n})),$
$p_1) \wedge \mu_2(r_2^{(n-1)}, \pi_2(f(\overline{a_n})), p_2) \mid r_1^{(1)} \in Q_1, r_2^{(1)} \in Q_2, i = 1, \ldots, n-1\}$
$= \vee\{\mu_1(q_1, \pi_1(f(\overline{a_1})), r_1^{(1)}) \wedge \mu_1(r_1^{(1)}, \pi_1(f(\overline{a_2})), r_1^{(2)}) \wedge \ldots \wedge \mu_1(r_1^{(n-1)},$
$\pi_1(f(\overline{a_n})), p_1) \mid r_1^{(1)} \in Q_1, i = 1, \ldots, n-1\} \wedge \vee\{\mu_2(q_2, \pi_2(f(\overline{a_1})), r_2^{(1)}) \wedge$
$\mu_2(r_2^{(1)}, \pi_2(f(\overline{a_2})), r_2^{(2)}) \wedge \ldots \wedge \mu_2(r_2^{(n-1)}, \pi_2(f(\overline{a_n})), p_2) \mid r_2^{(1)} \in Q_2, i =$
$1, \ldots, n-1\} = \mu_1^*(q_1, \pi_1(f(\overline{a_1})) \ldots \pi_1(f(\overline{a_n})), p_1) \wedge \mu_2^*(q_2, \pi_2(f(\overline{a_1})) \ldots$
$\pi_2(f(\overline{a_n})), p_2).$ ∎

Corollary 6.6.7 *Let $M_i = (Q_i, X_i, \mu_i)$ be a ffsm, $i = 1, 2$. Then*
   (1) $(\mu_1 \times \mu_2)^* = \mu_1^* \times \mu_2^*$,
   (2) $(\mu_1 \wedge \mu_2)^* = \mu_1^* \wedge \mu_2^*$. ∎

Proposition 6.6.8 *Let $M_i = (Q_i, X, \mu_i)$ be a ffsm, $i = 1, 2$. Define the relation $\equiv_{12}$ on X^* by $x \equiv_{12} y$ if and only if $x \equiv_1 y$ and $x \equiv_2 y$ where $\equiv_i$ is the congruence relation on M_i defined in Theorem 6.2.3, $i = 1, 2$. Then $\equiv_{12}$ is a congruence relation on X^*.* ∎

Definition 6.6.9 *Let $M_i = (Q_i, X, \mu_i)$ be a ffsm, $i = 1, 2$. $\forall x \in X^*$, let $< x > = \{y \in X^* \mid y \equiv_{12} x\}$ and let $T = \{< x > \mid x \in X^*\}$. Let*

$$FTS(M_1) \wedge FTS(M_2) = (Q_1 \times Q_2, T, \rho_1 \wedge \rho_2)$$

where

$$\rho_1 \wedge \rho_2((q_1, q_2), < x >, (p_1, p_2)) = \rho(q_1, [x]_1, p_1) \wedge \rho(q_2, [x]_2, p_2).$$

We wish to point out that the coverings of the fuzzy transformation semigroups that appear in Theorems 6.6.10 and 6.6.11 are coverings as fuzzy finite state machines as stated in Definition 6.6.1.

Theorem 6.6.10 *Let $M_i = (Q_i, X, \mu_i)$ be a ffsm, $i = 1, 2$. Then the following assertions hold.*

$$FTS(M_1 \wedge M_2) \geq FTS(M_1) \wedge FTS(M_2).$$

Proof. Let $x \in X^*$ and $(q_1, q_2), (p_1, p_2) \in Q_1 \times Q_2$. Then

$$\begin{aligned}
\rho((q_1, q_2), [x], (p_1, p_2)) &= (\mu_1 \wedge \mu_2)^*((q_1, q_2), x, (p_1, p_2)) \\
&= \mu_1^*(q_1, x, p_1) \wedge \mu_2^*(q_2, x, p_2) \\
&= \rho_1(q_1, [x]_1, p_1) \wedge \rho_2(q_2, [x]_2, p_2) \\
&= (\rho_1 \wedge \rho_2)((q_1, q_2), < x >, (p_1, p_2)).
\end{aligned}$$

Let T be as in Definition 6.6.9. Define $\zeta : T \to E(M_1 \wedge M_2)$ by

$$\zeta(< x >) = [x] \forall x \in X^*.$$

Then ζ is single-valued since $< x > = < y > \Leftrightarrow x \equiv_{12} y \Leftrightarrow x \equiv_1 y$ and $x \equiv_2 y \Rightarrow x \equiv y$ (w. r. t. $M_1 \wedge M_2$) $\Leftrightarrow [x] = [y] \Leftrightarrow \zeta(< x >) = \zeta(< y >)$ where the implication holds since $x \equiv y \Leftrightarrow \forall (q_1, q_2), (p_1, p_2) \in Q_1 \times Q_2, (\mu_1 \wedge \mu_2)^*((q_1, q_2), x, (p_1, p_2)) = (\mu_1 \wedge \mu_2)^*((q_1, q_2), y, (p_1, p_2)) \Leftrightarrow \forall (q_1, q_2), (p_1, p_2) \in Q_1 \times Q_2, \mu_1^*(q_1, x, p_1) \wedge \mu_2^*(q_2, x, p_2) = \mu_1^*(q_1, y, p_1) \wedge \mu_2^*(q_2, y, p_2)$. (The latter does not imply $x \equiv_1 y$ and $x \equiv_2 y$.)

If we take η to be the identity map of $Q_1 \times Q_2$, then we have (η, ζ) is a covering of $\mathrm{FTS}(M_1) \wedge \mathrm{FTS}(M_2)$ by $\mathrm{FTS}(M_1 \wedge M_2)$. ∎

Theorem 6.6.11 *Let $M_i = (Q_i, X, \mu_i)$ be a ffsm, $i = 1, 2$. Then the following assertions hold:*

$$FTS(M_1 \times M_2) \geq FTS(M_1) \times FTS(M_2) \geq FTS(M_1) \wedge FTS(M_2).$$

Proof. Let $a_i \in X_1$ and $b_i \in X_2$, $i = 1, \dots, n$. Then $(a_1, b_1)\dots(a_n, b_n) = (a_1 \dots a_n, b_1 \dots b_n) = (x_1, x_2)$ where $x_1 = a_1 \dots a_n$ and $x_2 = b_1 \dots b_n$. Now $M_1 \times M_2 = (Q_1 \times Q_2, X_1 \times X_2, \mu_1 \wedge \mu_2)$ and so

$$\mathrm{FTS}(M_1 \times M_2) = (Q_1 \times Q_2, E(M_1 \times M_2), \rho)$$

and

$$\mathrm{FTS}(M_1) \times \mathrm{FTS}(M_2) = (Q_1 \times Q_2, E(M_1) \times E(M_2), \rho_1 \wedge \rho_2).$$

Let $x_1 \in X_1^*$ and $x_2 \in X_2^*$. Then

$$
\begin{aligned}
\rho((q_1, q_2), [(x_1, x_2)], (p_1, p_2)) &= (\mu_1 \times \mu_2)^*((q_1, q_2), (x_1, x_2), (p_1, p_2)) \\
&= \mu_1^*(q_1, x_1, p_1) \wedge \mu_2^*(q_2, x_2, p_2) \\
&= \rho_1(q_1, [x_1], p_1) \wedge \rho_2(q_2, [x_2], p_2) \\
&= (\rho_1 \wedge \rho_2)((q_1, q_2), ([x_1], [x_2]), (p_1, p_2)).
\end{aligned}
$$

Now define

$$\zeta : E(M_1) \times E(M_2) \to E(M_1 \times M_2)$$

by

$$\zeta(([x_1], [x_2])) = [(x_1, x_2)]$$

$\forall x_1, x_2 \in X^*$. Then $([x_1], [x_2]) = ([y_1], [y_2]) \Leftrightarrow [x_1] = [y_1]$ and $[x_2] = [y_2] \Leftrightarrow \mu_1^*(q_1, x_1, p_1) = \mu_1^*(q_1, y_1, p_1) \ \forall q_1, p_1 \in Q_1$ and $\mu_2^*(q_2, x_2, p_2) = \mu_2^*(q_2, y_2, p_2) \ \forall q_2, p_2 \in Q_2 \Rightarrow \mu_1^*(q_1, x_1, p_1) \wedge \mu_2^*(q_2, x_2, p_2) = \mu_1^*(q_1, y_1, p_1) \wedge \mu_2^*(q_2, y_2, p_2) \ \forall q_1, p_1 \in Q_1, \forall q_2, p_2 \in Q_2 \Leftrightarrow (\mu_1 \times \mu_2)^*((q_1, q_2), (x_1, x_2), (p_1, p_2)) = (\mu_1 \times \mu_2)^*((q_1, q_2), (y_1, y_2), (p_1, p_2)) \ \forall (q_1, q_2), (p_1, p_2) \in Q_1 \times Q_2 \Leftrightarrow [(x_1, x_2)] = [(y_1, y_2)] \Leftrightarrow \zeta(([x_1], [x_2])) = \zeta(([y_1], [y_2]))$. Thus ζ is single-valued.

Let η be the identity map of $Q_1 \times Q_2$. Then (η, ζ) is a covering of $\text{FTS}(M_1) \times \text{FTS}(M_2)$ by $\text{FTS}(M_1 \times M_2)$. Let T be the set of all equivalence classes of $E(M_1 \times M_2)$ given by the relation $\equiv_{12}$ on $(X_1 \times X_2)^*$ where $(x_1, x_2) \equiv_{12} (y_1, y_2) \Leftrightarrow x_1 \equiv_1 y_1$ and $x_2 \equiv_2 y_2$. Define

$$\tau : T \to E(M_1) \times E(M_2)$$

by

$$\tau(< (x_1, x_2) >) = ([x_1], [x_2])$$

$\forall < (x_1, x_2) > \in T$. Then $< (x_1, x_2) > = < (y_1, y_2) > \Leftrightarrow (x_1, x_2) \equiv_{12} (y_1, y_2) \Leftrightarrow x_1 \equiv_1 y_1$ and $x_2 \equiv_2 y_2 \Leftrightarrow [x_1] = [y_1]$ and $[x_2] = [y_2] \Leftrightarrow ([x_1], [x_2]) = ([y_1], [y_2])$. Thus τ is single-valued and one-to-one. It follows that (η, τ) is a covering of $\text{FTS}(M_1) \wedge \text{FTS}(M_2)$ by $\text{FTS}(M_1) \times \text{FTS}(M_2)$.
∎

Let $M_1 = (Q_1, X_1, \mu_1)$ and $M_2 = (Q_2, X_2, \mu_2)$ be fuzzy finite state machines. Let ω be a function of $Q_2 \times X_2$ into X_1. Let $Q = Q_1 \times Q_2$. Define

$$\mu^\omega : Q \times X_2 \times Q \to [0, 1]$$

as follows: $\forall((q_1, q_2), b, (p_1, p_2)) \in Q \times X_2 \times Q,$

$$\mu^\omega((q_1, q_2), b, (p_1, p_2)) = \mu_1(q_1, \omega(q_2, b), p_1) \wedge \mu_2(q_2, b, p_2).$$

Then $M = (Q, X_2, \mu^\omega)$ is a ffsm. M is called the **cascade product** of M_1 and M_2 and we write $M = M_1 \omega M_2$. μ^ω is called **separable** if $\forall$ $(q_1, q_2), (p_1, p_2) \in Q$ and $\forall y = y_1 \ldots y_n \in X_2^*$, $\mu^{\omega*}((q_1, q_2), y, (p_1, p_2)) = \mu_1^*(q_1, \omega(q_2, y_1)\omega(q_2^{(1)}, y_2) \ldots \omega(q_2^{(n-1)}, y_n), p_1) \wedge \mu_2^*(q_2, y, p_2)$ for some $q_2^{(i)} \in Q_2$, $i = 1, 2, \ldots, n-1$.

Example 6.6.12 *Let $M_1 = (Q_1, X_1, \mu_1)$ and $M_2 = (Q_2, X_2, \mu_2)$ be ffsms, where $Q_1 = \{q_1, q_2\}$, $X_1 = \{a, b\}$, $Q_2 = \{q_1', q_2'\}$, $X_2 = \{a\}$, and μ_1 and μ_2 are defined as follows:*

$$
\begin{array}{llll}
\mu_1(q_1, a, q_1) & = & 0 & \quad \mu_1(q_2, b, q_1) = c_4 > 0 \\
\mu_1(q_1, a, q_2) & = & c_1 > 0 & \quad \mu_1(q_2, b, q_2) = 0 \\
\mu_1(q_1, b, q_1) & = & c_2 > 0 & \quad \mu_2(q_1', a, q_1') = 0 \\
\mu_1(q_1, b, q_2) & = & 0 & \quad \mu_2(q_1', a, q_2') = d_1 > 0 \\
\mu_1(q_2, a, q_1) & = & 0 & \quad \mu_2(q_2', a, q_1') = d_2 > 0 \\
\mu_1(q_2, a, q_2) & = & c_3 > 0 & \quad \mu_2(q_2', a, q_2') = 0.
\end{array}
$$

Let $\omega : Q_2 \times X_2 \to X_1$ be defined as follows:

$$\omega(q_1', a) = a, \qquad \omega(q_2', a) = b.$$

Let $Q = Q_1 \times Q_2$. Then $\mu^\omega : Q \times X_2 \times Q \to [0,1]$ is such that

$$
\begin{aligned}
\mu^\omega((q_1, q_1'), a, (q_2, q_2')) &= \mu_1(q_1, \omega(q_1', a), q_2) \wedge \mu_2(q_1', a, q_2') &= c_1 \wedge d_1, \\
\mu^\omega((q_2, q_2'), a, (q_1, q_1')) &= \mu_1(q_2, \omega(q_2', a), q_1) \wedge \mu_2(q_2', a, q_1') &= c_4 \wedge d_2, \\
\mu^\omega((q_1, q_2'), a, (q_1, q_1')) &= \mu_1(q_1, \omega(q_2', a), q_1) \wedge \mu_2(q_2', a, q_1') &= c_2 \wedge d_2, \\
\mu^\omega((q_2, q_1'), a, (q_2, q_2')) &= \mu_1(q_2, \omega(q_1', a), q_2) \wedge \mu_2(q_1', a, q_2') &= c_3 \wedge d_1,
\end{aligned}
$$

and μ^ω is 0 elsewhere.
Note that

$$
\begin{aligned}
(\mu^\omega)^*((q_1, q_2'), aa, (q_2, q_2')) &= \mu_1^*(q_1, \omega(q_2', a)\omega(q_1', a), q_2) \wedge \mu_2^*(q_2', aa, q_2') \\
&= \mu_1^*(q_1, ba, q_2) \wedge \mu_2^*(q_2', aa, q_2') \\
&= \mu_1(q_1, b, q_1) \wedge \mu_1(q_1, a, q_2) \wedge \mu_2^*(q_2', aa, q_2') \\
&= c_2 \wedge c_1 \wedge d_2 \wedge d_1.
\end{aligned}
$$

It follows that μ^ω is separable.

Proposition 6.6.13 Let $M_i = (Q_i, X_i, \mu_i)$ be a ffsm, $i = 1, 2$. Let $M = M_1 \omega M_2$ for some ω. Then $\forall\ (q_1, q_2), (p_1, p_2) \in Q$ and $\forall\ y = y_1 \ldots y_n \in X_2^*$,
$\mu^{\omega*}((q_1, q_2), y, (p_1, p_2)) = \vee\{\mu_1^*(q_1, \omega(q_2, y_1)\omega(q_2^{(1)}, y_2) \ldots \omega(q_2^{(n-1)}, y_n), p_1)$
$\wedge \mu_2(q_2,\ y_1, q_2^{(1)}) \wedge \mu_2(q_2^{(1)},\ y_2, q_2^{(2)}) \wedge \ldots \wedge \mu_2(q_2^{(n-1)},\ y_n, p_2) \mid q_2^{(i)} \in Q_2,$
$i = 1, 2, \ldots, n - 1\}$.

Proof. $\mu^{\omega*}((q_1,\ q_2),\ y,\ (p_1, p_2)) = \vee\{\mu^\omega((q_1,\ q_2),\ y_1,\ (q_1^{(1)}, q_2^{(1)})) \wedge$
$\mu^\omega((q_1^{(1)},\ q_2^{(1)}),\ y_2,\ (q_1^{(2)},\ q_2^{(2)})) \wedge \ldots \wedge \mu^\omega((q_1^{(n-1)},\ q_2^{(n-1)}),\ y_n, (p_1, p_2)) \mid$
$(q_1^{(1)}, q_2^{(1)}), \ldots, (q_1^{(n-1)}, q_2^{(n-1)}) \in Q\} = \vee\{\mu_1(q_1, \omega(q_2, y_1), q_1^{(1)}) \wedge \mu_2(q_2, y_1,$
$q_2^{(1)}) \wedge \mu_1(q_1^{(1)}, \omega(q_2^{(1)}, y_2), q_1^{(2)}) \wedge \mu_2(q_2^{(1)}, y_2, q_2^{(2)}) \ldots \wedge \mu_1(q_1^{(n-1)},\ \omega(q_2^{(n-1)},$
$y_n),\ p_1) \wedge \mu_2(q_2^{(n-1)},\ y_n,\ p_2) \mid (q_1^{(1)},\ q_2^{(1)}), \ldots, (q_1^{(n-1)},\ q_2^{(n-1)}) \in Q\}$
$= \vee\{\mu_1(q_1, \omega(q_2, y_1), q_1^{(1)}) \wedge \mu_1(q_1^{(1)}, \omega(q_2^{(1)}, y_2), q_1^{(2)}) \wedge \ldots \wedge \mu_1(q_1^{(n-1)},$
$\omega(q_2^{(n-1)}, y_n), p_1) \wedge \mu_2(q_2, y_1, q_2^{(1)}) \wedge \mu_2(q_2^{(1)}, y_2, q_2^{(2)}) \wedge \ldots \wedge \mu_2(q_2^{(n-1)},$
$y_n, p_2) \mid (q_1^{(1)}, q_2^{(1)}), \ldots, (q_1^{(n-1)}, q_2^{(n-1)}) \in Q\} = \vee\{\mu_1^*(q_1, \omega(q_2, y_1)\ \omega(q_2^{(1)},$
$y_2) \ldots \omega(q_2^{(n-1)}, y_n), p_1) \wedge \mu_2(q_2, y_1, q_2^{(1)}) \wedge \mu_2(q_2^{(1)}, y_2, q_2^{(2)}) \wedge \ldots \wedge \mu_2(q_2^{(n-1)},$
$y_1,\ p_2) \mid q_2^{(i)} \in Q_2, i = 1, 2, \ldots, n - 1\}$. ∎

Proposition 6.6.14 Let $M_i = (Q_i, X_i, \mu_i)$ be a ffsm, $i = 1, 2$. If $Im(\mu_2) = \{0, 1\}$, then μ^ω is separable.

Proof. Let $(q_1, q_2), (p_1, p_2) \in Q$ and $y = y_1 \ldots y_n \in X_2^*$. Now $\mu^{\omega*}((q_1,$
$q_2), y, (p_1, p_2)) = \vee\{\mu_1^*(q_1, \omega(q_2, y_1)\omega(q_2^{(1)}, y_2) \ldots \omega(q_2^{(n-1)}, y_n), p_1) \wedge \mu_2(q_2,$
$y_1, q_2^{(1)}) \wedge \mu_2(q_2^{(1)}, y_2, q_2^{(2)}) \wedge \ldots \wedge \mu_2(q_2^{(n-1)}, y_1, p_2) \mid q_2^{(i)} \in Q_2, i = 1, 2, \ldots, n -$
$1\}$ by Proposition 6.6.13.
    Case 1: $\mu_2^*(q_2, y, p_2) = 1$.
    Now $\vee\{\mu_2(q_2, y_1, q_2^{(1)}) \wedge \mu_2(q_2^{(1)}, y_2, q_2^{(2)}) \wedge \ldots \wedge \mu_2(q_2^{(n-1)}, y_1, p_2) \mid q_2^{(i)} \in$
$Q_2, i = 1, 2, \ldots, n - 1\} = \mu_2^*(q_2, y, p_2) = 1$. Hence $\exists\ q_2^{(1)}, \ldots, q_2^{(n-1)} \in$

Q_2 such that $\mu_2(q_2, y_1, q_2^{(1)}) \wedge \mu_2(q_2^{(1)}, y_2, q_2^{(2)}) \wedge \ldots \wedge \mu_2(q_2^{(n-1)}, y_1, p_2) = 1$. Thus $\mu^{\omega*}((q_1 q_2), y, (p_1, p_2)) = \vee\{\mu_1^*(q_1, \omega(q_2, y_1)\omega(q_2^{(1)}, y_2) \ldots \omega(q_2^{(n-1)}, y_n), p_1) \mid \mu_2(q_2, y_1, q_2^{(1)}) \wedge \mu_2(q_2^{(1)}, y_2, q_2^{(2)}) \wedge \ldots \wedge \mu_2(q_2^{(n-1)}, y_1, p_2) = 1, q_2^{(1)}, \ldots, q_2^{(n-1)} \in Q_2\}$. Since μ_1^* is finite valued, the supremum is attained at some $q_2^{(1)}, \ldots, q_2^{(n-1)} \in Q_2$. Hence $\exists\, q_2^{(1)}, \ldots, q_2^{(n-1)} \in Q_2$ such that $\mu^{\omega*}((q_1 q_2), y, (p_1, p_2)) = \mu_1^*(q_1, \omega(q_2, y_1)\ \omega(q_2^{(1)}, y_2)\ \ldots\ \omega(q_2^{(n-1)}, y_n), p_1) = \mu_1^*(q_1, \omega(q_2, y_1)\ \omega(q_2^{(1)}, y_2) \ldots \omega(q_2^{(n-1)}, y_n), p_1) \wedge 1 = \mu_1^*(q_1, \omega(q_2, y_1)\omega(q_2^{(1)}, y_2) \ldots \omega(q_2^{(n-1)}, y_n), p_1) \wedge \mu_2^*(q_2, y, p_2)$.

Case 2: $\mu_2^*(q_2, y, p_2) = 0$. Then $\forall\, q_2^{(1)}, \ldots, q_2^{(n-1)} \in Q_2$, $\mu_2(q_2, y_1, q_2^{(1)}) \wedge \mu_2(q_2^{(1)}, y_2, q_2^{(2)}) \wedge \ldots \wedge \mu_2(q_2^{(n-1)}, y_1, p_2) = 0$. Thus $\mu^{\omega*}((q_1 q_2), y, (p_1, p_2)) = 0 = \mu_1^*(q_1, \omega(q_2, y_1)\omega(q_2^{(1)}, y_2) \ldots \omega(q_2^{(n-1)}, y_n), p_1) \wedge 0 = \mu_1^*(q_1, \omega(q_2, y_1) \omega(q_2^{(1)}, y_2) \ldots \omega(q_2^{(n-1)}, y_n), p_1) \wedge \mu_2^*(q_2, y, p_2)\ \forall\, q_2^{(1)}, \ldots, q_2^{(n-1)} \in Q_2$. ∎

Let $M_1 = (Q_1, X_1, \mu_1)$ and $M_2 = (Q_2, X_2, \mu_2)$ be fuzzy finite state machines. Let f be a function from Q_2 into X_1. Define $\mu^0 : Q \times (X_1^{Q_2} \times X_2) \times Q \to [0, 1]$ as follows: $\forall\, ((q_1, q_2), (f, b), (p_1, p_2)) \in Q \times (X_1^{Q_2} \times X_2) \times Q$,

$$\mu^0((q_1, q_2), (f, b), (p_1, p_2)) = \mu_1(q_1, f(q_2), p_1) \wedge \mu_2(q_2, b, p_2).$$

Then $M = (Q, X_1^{Q_2} \times X_2, \mu^0)$ is a ffsm. $M = M_1 \circ M_2$ is called the **wreath product** of M_1 and M_2. μ^0 is called **separable** if $\forall\, (q_1, q_2), (p_1, p_2) \in Q$ and $\forall\, (f_1, b_1) \ldots (f_n, b_n) \in (X_1^{Q_2} \times X_2)^*$, $\mu^{0*}((q_1, q_2), (f_1, b_1) \ldots (f_n, b_n), (p_1, p_2)) = \mu_1^*(q_1, f_1(q_2)f_2(q_2^{(1)}) \ldots f_n(q_2^{(n-1)}), p_1) \wedge \mu_2^*(q_2, b_1 \ldots b_n, p_2)$ for some $q_2^{(1)} \in Q_2$, $i = 1, 2, \ldots, n-1$.

Example 6.6.15 *Let M_1 and M_2 be the ffsms defined in Example 6.6.12. Let $X_1^{Q_2} = \{f_1, f_2, f_3, f_4\}$, where $f_1(q_1') = f_1(q_2') = a$, $f_2(q_1') = a$, $f_2(q_2') = b$, $f_3(q_1') = b$, $f_3(q_2') = a$, and $f_4(q_1') = f_4(q_2') = b$. Then*

$$
\begin{aligned}
\mu^0((q_1, q_1'), (f_1, a), (q_2, q_2')) &= \mu_1(q_1, f_1(q_1'), q_2) \wedge \mu_2(q_1', a, q_2') \\
&= c_1 \wedge d_1 \\
\mu^0((q_2, q_1'), (f_1, a), (q_2, q_2')) &= \mu_1(q_2, f_1(q_1'), q_2) \wedge \mu_2(q_1', a, q_2') \\
&= c_3 \wedge d_1 \\
\mu^0((q_1, q_2'), (f_1, a), (q_2, q_1')) &= \mu_1(q_1, f_1(q_2'), q_2) \wedge \mu_2(q_2', a, q_1') \\
&= c_1 \wedge d_2 \\
\mu^0((q_2, q_2'), (f_1, a), (q_2, q_1')) &= \mu_1(q_2, f_1(q_2'), q_2) \wedge \mu_2(q_2', a, q_1') \\
&= c_3 \wedge d_2 \\
\mu^0((q_1, q_1'), (f_2, a), (q_2, q_2')) &= \mu_1(q_1, f_2(q_1'), q_2) \wedge \mu_2(q_1', a, q_2') \\
&= c_1 \wedge d_1 \\
\mu^0((q_2, q_1'), (f_2, a), (q_2, q_2')) &= \mu_1(q_2, f_2(q_1'), q_2) \wedge \mu_2(q_1', a, q_2') \\
&= c_3 \wedge d_1 \\
\mu^0((q_1, q_2'), (f_2, a), (q_1, q_1')) &= \mu_1(q_1, f_2(q_2'), q_1) \wedge \mu_2(q_2', a, q_1') \\
&= c_2 \wedge d_2 \\
\mu^0((q_2, q_2'), (f_2, a), (q_1, q_1')) &= \mu_1(q_2, f_2(q_2'), q_1) \wedge \mu_2(q_2', a, q_1') \\
&= c_4 \wedge d_2
\end{aligned}
$$

$$\mu^0((q_1,q_1'),(f_3,a),(q_1,q_2')) = \mu_1(q_1,f_3(q_1'),q_1) \wedge \mu_2(q_1',a,q_2')$$
$$= c_2 \wedge d_1$$
$$\mu^0((q_2,q_1'),(f_3,a),(q_1,q_2')) = \mu_1(q_2,f_3(q_1'),q_1) \wedge \mu_2(q_1',a,q_2')$$
$$= c_4 \wedge d_1$$
$$\mu^0((q_1,q_2'),(f_3,a),(q_2,q_1')) = \mu_1(q_1,f_3(q_2'),q_2) \wedge \mu_2(q_2',a,q_1')$$
$$= c_1 \wedge d_2$$
$$\mu^0((q_2,q_2'),(f_3,a),(q_2,q_1')) = \mu_1(q_2,f_3(q_2'),q_2) \wedge \mu_2(q_2',a,q_1')$$
$$= c_3 \wedge d_2$$
$$\mu^0((q_1,q_1'),(f_4,a),(q_1,q_2')) = \mu_1(q_1,f_4(q_1'),q_1) \wedge \mu_2(q_1',a,q_2')$$
$$= c_2 \wedge d_1$$
$$\mu^0((q_2,q_1'),(f_4,a),(q_1,q_2')) = \mu_1(q_2,f_4(q_1'),q_1) \wedge \mu_2(q_1',a,q_2')$$
$$= c_4 \wedge d_1$$
$$\mu^0((q_1,q_2'),(f_4,a),(q_1,q_1')) = \mu_1(q_1,f_4(q_2'),q_1) \wedge \mu_2(q_2',a,q_1')$$
$$= c_2 \wedge d_2$$
$$\mu^0((q_2,q_2'),(f_4,a),(q_1,q_1')) = \mu_1(q_2,f_4(q_2'),q_1) \wedge \mu_2(q_2',a,q_1')$$
$$= c_4 \wedge d_2.$$

In the case that $c_1 = c_2 = c_3 = c_4$ and $d_1 = d_2$, it follows that the semigroup of this ffsm has eight elements and the transformation semigroup is isomorphic to the wreath product $\overline{2} \circ \mathbb{Z}_2$, [92, p. 59].

Example 6.6.16 Let M_1 and M_2 be the ffsms defined in Example 6.6.15. Let $X_2^{Q_1} = \{f\}$, where $f(q_1) = f(q_2) = a$. Then

$$\mu^0((q_1',q_1),(f,a),(q_2',q_2)) = \mu_2(q_1',a,q_2') \wedge \mu_1(q_1,a,q_2) = d_1 \wedge c_1$$
$$\mu^0((q_2',q_1),(f,a),(q_1',q_2)) = \mu_2(q_2',a,q_1') \wedge \mu_1(q_1,a,q_2) = d_2 \wedge c_1$$
$$\mu^0((q_1',q_2),(f,a),(q_2',q_2)) = \mu_2(q_1',a,q_2') \wedge \mu_1(q_2,a,q_2) = d_1 \wedge c_3$$
$$\mu^0((q_2',q_2),(f,a),(q_1',q_2)) = \mu_2(q_2',a,q_1') \wedge \mu_1(q_2,a,q_2) = d_2 \wedge c_3$$
$$\mu^0((q_1',q_1),(f,b),(q_2',q_1)) = \mu_2(q_1',a,q_2') \wedge \mu_1(q_1,b,q_1) = d_1 \wedge c_2$$
$$\mu^0((q_2',q_1),(f,b),(q_1',q_1)) = \mu_2(q_2',a,q_1') \wedge \mu_1(q_1,b,q_1) = d_2 \wedge c_2$$
$$\mu^0((q_1',q_2),(f,b),(q_2',q_1)) = \mu_2(q_1',a,q_2') \wedge \mu_1(q_2,b,q_1) = d_1 \wedge c_4$$
$$\mu^0((q_2',q_2),(f,b),(q_1',q_1)) = \mu_2(q_2',a,q_1') \wedge \mu_1(q_2,b,q_1) = d_2 \wedge c_4.$$

Suppose that $d_1 = d_2$ and $c_1 = c_2 = c_3 = c_4$. Then it follows that $S(M_1 \circ M_2)$ has four elements while $S(M_1) \circ S(M_2)$ has eight elements, [92, p.59].

Proposition 6.6.17 $\forall (q_1,q_2),\ (p_1,p_2) \in Q$ and $\forall\ (f_1,b_1)\ldots(f_n,\ b_n) \in (X_1^{Q_2} \times X_2)^*$, $\mu^{0*}((q_1,q_2),(f_1,b_1)\ldots(f_n,b_n),(p_1,p_2)) = \vee\{\mu_1^*(q_1,f_1(q_2),\ f_2(q_2^{(1)})\ldots f_n\ (q_2^{(n-1)})\ ,p_1) \wedge\ \mu_2(q_2,\ b_1,\ q_2^{(1)}) \wedge \ldots \wedge \mu_2(q_2^{(n-1)},\ b_n,\ p_2)\ |\ q_2^{(1)},\ldots,q_2^{(n-1)} \in Q_2\}$.

Proof. $\mu^{0*}((q_1\ q_2),(f_1\ b_1)\ldots(f_n,b_n),(p_1,p_2)) = \vee\{\mu_1(q_1,\ f_1(q_2),q_1^{(1)}) \wedge \mu_2(q_2,\ b_1,\ q_2^{(1)}) \wedge \mu_1(q_1^{(1)},\ f_2(q_2^{(1)}),\ q_1^{(2)}) \wedge \mu_2(q_2^{(1)},\ b_2,\ q_2^{(2)}) \wedge \ldots \wedge \mu_1(q_1^{(n-1)},\ f_n(q_2^{(n-1)}),\ p_1) \wedge \mu_2(q_2^{(n-1)},\ b_n,\ p_2)\ |\ q_1^{(1)},\ldots,\ q_1^{(n-1)} \in Q_1;\ q_2^{(1)},\ldots,\ q_2^{(n-1)} \in Q_2\} = \vee\{\mu_1(q_1,\ f_1(q_2),\ q_1^{(1)}) \wedge \mu_1(q_1^{(1)},\ f_2(q_2^{(1)}),\ q_1^{(2)}) \wedge \ldots \wedge \mu_1(q_1^{(n-1)},$

$f_n(q_2^{(n-1)}), p_1) \wedge \mu_2(q_2, b_1, q_2^{(1)}) \wedge \mu_2(q_2^{(1)}, b_2, q_2^{(2)}) \wedge \ldots \wedge \mu_2(q_2^{(n-1)}, b_n, p_2) \mid q_1^{(1)}, \ldots, q_1^{(n-1)} \in Q_1; q_2^{(1)}, \ldots, q_2^{(n-1)} \in Q_2\} = \vee\{\mu_1{}^*(q_1, f_1(q_2)f_2(q_2^{(1)})) \ldots f_n(q_2^{(n-1)}), p_1) \wedge \mu_2(q_2, b_1, q_2^{(1)}) \wedge \mu_2(q_2^{(1)}, b_2, q_2^{(2)}) \wedge \ldots \wedge \mu_2(q_2^{(n-1)}, b_n, p_2) \mid q_2^{(1)}, \ldots, q_2^{(n-1)} \in Q_2\}.$ ∎

Proposition 6.6.18 *Let $M_i = (Q_i, X_i, \mu_i)$ be a ffsm, $i = 1, 2$. If $Im(\mu_2) = \{0, 1\}$, then μ^0 is separable.*

Proof. Let $(q_1, q_2), (p_1, p_2) \in Q$ and $(f_1, b_1) \ldots (f_n, b_n) \in (X_1^{Q_2} \times X_2)^*$. Now $\mu^{0*}((q_1 q_2), (f_1, b_1) \ldots (f_n, b_n), (p_1, p_2)) = \vee\{\mu_1^*(q_1, f_1(q_2), f_2(q_2^{(1)})) \ldots f_n(q_2^{(n-1)}), p_1) \wedge \mu_2(q_2, b_1, q_2^{(1)}) \wedge \ldots \wedge \mu_2(q_2^{(n-1)}, b_n, p_2) \mid q_2^{(1)}, \ldots, q_2^{(n-1)} \in Q_2\}$ by Proposition 6.6.17.

Case 1: $\mu_2^*(q_2, b_1 \ldots b_n, p_2) = 1$.

Now $\vee\{\mu_2(q_2, b_1, q_2^{(1)}) \wedge \ldots \wedge \mu_2(q_2^{(n-1)}, b_n, p_2) \mid q_2^{(1)}, \ldots, q_2^{(n-1)} \in Q_2\} = \mu_2^*(q_2, b_1 \ldots b_n, p_2) = 1$. Hence $\exists \, q_2^{(1)}, \ldots, q_2^{(n-1)} \in Q_2$ such that $\mu_2(q_2, b_1, q_2^{(1)}) \wedge \ldots \wedge \mu_2(q_2^{(n-1)}, b_n, p_2) = 1$. Thus $\mu^{0*}((q_1 q_2), (f_1, b_1) \ldots (f_n, b_n), (p_1, p_2)) = \vee\{\mu_1^*(q_1, f_1(q_2) f_2(q_2^{(1)})) \ldots f_n(q_2^{(n-1)}), p_1) \mid \mu_2(q_2, b_1, q_2^{(1)}) \wedge \ldots \wedge \mu_2(q_2^{(n-1)}, b_n, p_2) = 1, q_2^{(1)}, \ldots, q_2^{(n-1)} \in Q_2\}$. Since μ_1^* is finite valued, the supremum is attained at some $q_2^{(1)}, \ldots, q_2^{(n-1)} \in Q_2$. Hence $\exists \, q_2^{(1)}, \ldots, q_2^{(n-1)} \in Q_2$ such that $\mu^{0*}((q_1, q_2), (f_1, b_1) \ldots (f_n, b_n), (p_1, p_2)) = \mu_1^*(q_1, f_1(q_2), f_2(q_2^{(1)})) \ldots f_n(q_2^{(n-1)}), p_1) = \mu_1^*(q_1, f_1(q_2) f_2(q_2^{(1)})) \ldots f_n(q_2^{(n-1)}), p_1) \wedge 1 = \mu_1^*(q_1, f_1(q_2) f_2(q_2^{(1)})) \ldots f_n(q_2^{(n-1)}), p_1) \wedge \mu_2^*(q_2, b_1 \ldots b_n, p_2)$.

Case 2: $\mu_2^*(q_2, b_1 \ldots b_n, p_2) = 0$. Then $\forall \, q_2^{(1)}, \ldots, q_2^{(n-1)} \in Q_2, \mu_2(q_2, b_1, q_2^{(1)}) \wedge \ldots \wedge \mu_2(q_2^{(n-1)}, b_n, p_2) = 0$. Thus $\mu^{0*}((q_1, q_2), (f_1, b_1) \ldots (f_n, b_n), (p_1, p_2)) = 0 = \mu_1^*(q_1, f_1(q_2) f_2(q_2^{(1)})) \ldots f_n(q_2^{(n-1)}), p_1) \wedge 0 = \mu_1^*(q_1, f_1(q_2) f_2(q_2^{(1)})) \ldots f_n(q_2^{(n-1)}), p_1) \wedge \mu_2^*(q_2, b_1 \ldots b_n, p_2)$. ∎

We are now interested in obtaining covering results involving the cascade and wreath product of fuzzy finite state machines. The method that is immediately suggested is to suitably relate elements of $E(M)$ with those of $E(M_1)^{Q_2} \times E(M_2)$. However it turns out that these relations are not necessarily single-valued. Hence we replace $E(M)$ with X^* in the following definition.

Definition 6.6.19 *Let $M = (Q, X, \mu)$ and $M_i = (Q_i, X_i, \mu_i)$, $i = 1, 2$, be ffsms. Let η be a function of $Q_1 \times Q_2$ onto Q and ζ a function of X^* into $E(M_1)^{Q_2} \times E(M_2)$. Then (η, ζ) is said to be a **weak covering** of $FTS(M)$ by $FTS(M_1) \circ FTS(M_2)$ if*

$$\rho(\eta(q_1, q_2), [x], \eta(p_1, p_2)) = \vee\{\widehat{\mu}^0((q_1, q_2), \zeta(x), (r_1, r_2)) \mid \eta(r_1, r_2) (p_1, p_2), (r_1, r_2) \in Q_1 \times Q_2\}$$

$\forall \, x \in X^*$ and $(q_1, q_2), (p_1, p_2) \in Q_1 \times Q_2$.

Theorem 6.6.20 *Let $M_i = (Q_i, X_i, \mu_i)$ be a ffsm, $i = 1, 2$. If μ^0 is separable, then*

$$FTS(M_1 \circ M_2) \leq (weakly) \ FTS(M_1) \circ FTS(M_2).$$

Proof. Now $M_1 \circ M_2 = (Q_1 \times Q_2, X_1^{Q_2} \times X_2, \mu^0)$, where

$$\mu^0((q_1, q_2), (g, b), (p_1, p_2)) = \mu_1(q_1, g(q_2), p_1) \wedge \mu_2(q_2, b, p_2)$$

and

$$FTS(M_1 \circ M_2) = (Q_1 \times Q_2, E(M_1 \circ M_2), \rho^0),$$

where $\rho^0((q_1, q_2), [(g_1, b_1)...(g_k, b_k)], (p_1, p_2)) = \mu^{0*}((q_1, q_2), (g_1, b_1) \dots (g_k, b_k), (p_1, p_2))$. Also,

$$FTS(M_1) \circ FTS(M_2) = (Q_1 \times Q_2, E(M_1)^{Q_2} \times E(M_2), \widehat{\mu}^0),$$

where $\widehat{\mu}^0((q_1, q_2), (f, [b_1...b_k]), (p_1, p_2)) = \rho_1(q_1, f(q_2), p_1) \wedge \rho_2(q_2, [b_1...b_k], p_2) = \mu_1^*(q_1, x_1, p_1) \wedge \mu_2^*(q_2, b_1 \dots b_k, p_2)$ where $f(q_2) = [x_1]$ and $x_1 \in X_1^*$ is selected below.

Let η be the identity map of $Q_1 \times Q_2$. Define

$$\zeta : (X_1^{Q_2} \times X_2)^* \to E(M_1)^{Q_2} \times E(M_2)$$

as follows:

$$\zeta((g_1, b_1) \dots (g_k, b_k)) = (f, [b_1 \dots b_k]).$$

Now

$$(g_1, b_1) \dots (g_k, b_k) = (h_1, a_1) \dots (h_j, a_j)$$

if and only if $k = j$ and $(g_i, b_i) = (h_i, a_i)$ $i = 1, \dots, k$ if and only if $k = j$ and $g_i = h_i, b_i = a_i, i = 1, \dots, k$. Thus ζ is single-valued. $\rho^0((q_1, q_2), [(g_1, b_1) \dots (g_k, b_k)], (p_1, p_2)) = \mu^{0*}((q_1, q_2), (g_1, b_1) \dots (g_k, b_k), (p_1, p_2)) = \vee\{\mu_1^*(q_1, g_1(q_2)g_2(q_2^{(1)}) \dots g_k(q_2^{(k-1)}), p_1) \wedge \mu_2(q_2, b_1, q_2^{(1)}) \wedge \dots \wedge \mu_2(q_2^{(k-1)}, b_k, p_2) \mid q_2^{(1)}, \dots, q_2^{(k-1)} \in Q_2\}$ " $=$ " $\mu_1^*(q_1, x_1, p_1) \wedge \mu_2^*(q_2, x_2, p_2) = \rho_1(q_1, f(q_2), p_1) \wedge \rho_2(q_2, [x_2], p_2) = \widehat{\mu}^0((q_1, q_2), (f, [x_2]), (p_1, p_2)) = \vee\{\widehat{\mu}^0((q_1, q_2), \zeta((g_1, b_1) \dots (g_k, b_k)), (r_1, r_2)) \mid \eta(r_1, r_2) = (p_1, p_2), (r_1, r_2) \in Q_1 \times Q_2\}$ since η is the identity map. Select $x_1 = g_1(q_2)g_2(q_2^{(1)}) \dots g_k(q_2^{(k-1)})$ so that $q_2^{(1)}, \dots, q_2^{(k-1)}$ gives " $=$ ". Hence (η, ζ) is a weak covering of $FTS(M_1 \circ M_2)$ by $FTS(M_1) \circ FTS(M_2)$. ∎

Theorem 6.6.21 *Let $M_i = (Q_i, X_i, \mu_i)$ be a ffsm, $i = 1, 2$. If μ^ω is separable, then*

$$FTS(M_1 \omega M_2) \leq (weakly) \ FTS(M_1) \circ FTS(M_2).$$

Proof. Now $M_1 \omega M_2 = (Q_1 \times Q_2, X_2, \mu^\omega)$, where $\omega : Q_2 \times X_2 \to X_1$ and

$$\mu^\omega((q_1, q_2), b, (p_1, p_2)) = \mu_1(q_1, \omega(q_2, b), p_1) \wedge \mu_2(q_2, b, p_2)$$

$\forall (q_1, q_2), (p_1, p_2) \in Q_1 \times Q_2, b \in X_2^*$ and

$$\mathrm{FTS}(M_1 \omega M_2) = (Q_1 \times Q_2, E(M_1 \omega M_2), \rho),$$

where

$$\rho((q_1, q_2), [x_2], (p_1, p_2)) = \mu^{\omega *}((q_1, q_2), x_2, (p_1, p_2)),$$

$\forall (q_1, q_2), (p_1, p_2) \in Q_1 \times Q_2, \forall x_2 \in X_2^*$. Also

$$\mathrm{FTS}(M_1) \circ \mathrm{FTS}(M_2) = (Q_1 \times Q_2, E(M_1)^{Q_2} \times E(M_2), \widehat{\mu}^0),$$

where

$$\widehat{\mu}^0((q_1, q_2), (f, [x_2]), (p_1, p_2)) = \rho_1(q_1, f(q_2), p_1) \wedge \rho_2(q_2, [x_2], p_2)$$

$\forall (q_1, q_2), (p_1, p_2) \in Q_1 \times Q_2, \forall x_2 \in X_2^*$. Let η be the identity map of $Q_1 \times Q_2$. Since μ^ω is separable we can define ζ in the following manner. Define

$$\zeta : X_2^* \to E(M_1)^{Q_2} \times E(M_2)$$

by

$$\zeta(x_2) = (f_{x_2}, [x_2]) \; \forall x_2 \in X_2^*,$$

where $f_{x_2}(q_2) = [x_1]$ for $x_2 = b_1 ... b_n$ and $x_1 = a_1 ... a_n$ and $\omega(q_1, b_1) = a_1$ and $\omega(q_1^{(i-1)}, b_i) = a_i$, $i = 2, \ldots, n$ for those $q_1^{(i-1)}$ so that :

$\vee \{\mu_1^*(q_1, \omega(q_2, b_1) \omega(q_2^{(1)}, b_2) \ldots \omega(q_2^{(n-1)}, b_n), p_1) \wedge \mu_2(q_2, b_1, q_2^{(1)}) \wedge \mu_2(q_2^{(1)}, b_2, q_2^{(2)}) \wedge \ldots \wedge \mu_2(q_2^{(n-1)}, b_n, p_2) \mid q_2^{(1)} \in Q_2, i = 1, 2, \ldots, n-1\}$
$= \mu_1^*(q_1, a_1 \ldots a_n, p_1) \wedge \mu_2^*(q_2, x_2, p_2)$.

Now ζ is single-valued since for $y_2 = d_1 \ldots d_j$, $x_2 = y_2$ if and only if $n = j$ and $d_i = b_i$ for $i = 1, \ldots, n$.

Then $\rho((q_1, q_2), [x_2], (p_1, p_2)) = \mu^{\omega *}((q_1, q_2), x_2, (p_1, p_2)) = \vee \{\mu_1^*(q_1, \omega(q_2, b_1) \omega(q_2^{(1)}, b_2) \ldots \omega(q_2^{(n-1)}, b_n), p_1) \wedge \mu_2(q_2, b_1, q_2^{(1)}) \wedge \mu_2(q_2^{(1)}, b_2, q_2^{(2)}) \wedge \ldots \wedge \mu_2(q_2^{(n-1)}, b_n, p_2) \mid q_2^{(1)} \in Q_2, i = 1, 2, \ldots, n-1\} = \mu_1^*(q_1, a_1 \ldots a_n, p_1) \wedge \mu_2^*(q_2, x_2, p_2) = \rho_1(q_1, f(q_2), p_1) \wedge \rho_2(q_2, [x_2], p_2) = \widehat{\mu}^0((q_1, q_2), (f, [x_2]), (p_1, p_2)) = \widehat{\mu}^0((q_1, q_2), \zeta(x_2), (p_1, p_2))$. ∎

If $\mathcal{A} = (Q, S)$ and $\mathcal{A}' = (Q', S')$ are transformation semigroups, Section 6.5, then in the wreath product $\mathcal{A} \circ \mathcal{A}' = (Q \times Q', S^{Q'} \times S')$ one has that $S^{Q'} \times S'$ is a semigroup where we recall that S and S' are finite semigroups.

We now consider the fuzzy case. Here the difficulty that arises is that the natural way to define a binary operation on $S^{Q'} \times S'$ fails since $\vee\{\mu'(q', s', r') \mid r' \in Q'\}$ may not attain its maximum uniquely.

Let (Q, S, μ) and (Q', S', μ') be fuzzy transformation semigroups. $\forall$ $s' \in S'$, $\forall q' \in Q'$, let $\mathcal{M}(q', s') = \{m' \in Q' \mid \mu'(q', s', m') = \vee\{\mu'(q', s', r') \mid r' \in Q'\}$. Let $Q' = \{q'_1, \ldots, q'_k\}$. Define $\cdot : (S^{Q'} \times S') \times (S^{Q'} \times S') \to \mathcal{P}(S^{Q'} \times S')\backslash\{\emptyset\}$ as follows: $\forall (f, s), (g, s') \in S^{Q'} \times S'$, $(f, s) \cdot (g, s') = \{(h_{s,m}, ss') \mid h_{s,m}(q'_i) = f(q'_i)g(m'_i), i = 1, \ldots, k, m = (m'_1, \ldots, m'_k) \in \mathcal{M}(q'_1, s) \times \cdots \times \mathcal{M}(q'_k, s)\}$.

Theorem 6.6.22 *Suppose that* $\forall q' \in Q'$, $\forall s, s' \in S'$,

$$\mathcal{M}(q', ss') = \cup_{m' \in \mathcal{M}(q', s)} \mathcal{M}(m', s').$$

Then $S^{Q'} \times S'$ is polysemigroup.

Proof. Let $(f, s), (g, s'), (k, s'') \in S^{Q'} \times S'$. Then

$$
\begin{aligned}
(f, s) \cdot (g, s') &= \{(h_{s,m}, ss') \mid h_{s,m}(q'_i) = f(q'_i)g(m'_i), i = 1, \ldots, k, \\
&\quad m = (m'_1, \ldots, m'_k) \in \mathcal{M}(q'_1, s) \times \cdots \times \mathcal{M}(q'_k, s)\}.
\end{aligned}
$$

Thus

$$
\begin{aligned}
((f, s) \cdot (g, s')) \cdot (k, s'') &= \cup_{m \in \mathcal{M}(q'_1, s) \times \ldots \times \mathcal{M}(q'_k, s)} (h_{s,m}, ss') \cdot (k, s'') \\
&= \cup_{m \in \mathcal{M}(q'_1, s) \times \ldots \times \mathcal{M}(q'_k, s)} \{(_{s,m}l_{ss',r}, (ss')s'') \mid \\
&\quad {}_{s,m}l_{ss',r}(q'_i) = h_{s,m}(q'_i)k(r'_i), \\
&\quad r = (r'_1, \ldots, r'_k) \in \\
&\quad \mathcal{M}(q'_1, ss') \times \ldots \times \mathcal{M}(q'_k, ss')\}
\end{aligned}
$$

$$\tag{6.1}$$

and so

$${}_{s,m}l_{ss',r}(q'_i) = h_{s,m}(q'_i)k(r'_i) = f(q'_i)g(m'_i)k(r'_i), i = 1, \ldots, k.$$

Now

$$
\begin{aligned}
(g, s') \cdot (k, s'') &= \{(j_{s',u}, s's'') \mid j_{s',u}(q'_i) = g(q'_i)k(u'_i), i = 1, \ldots, k, \\
&\quad u = (u'_1, \ldots, u'_k) \in \mathcal{M}(q'_1, s') \times \ldots \times \mathcal{M}(q'_k, s')\}.
\end{aligned}
$$

Thus

$$
\begin{aligned}
(f, s) \cdot ((g, s') \cdot (k, s'')) &= \cup_{u \in \mathcal{M}(q'_1, s') \times \ldots \times \mathcal{M}(q'_k, s')} (f, s)(j_{s',u}, s's'') \\
&= \cup_{u \in \mathcal{M}(q'_1, s') \times \ldots \times \mathcal{M}(q'_k, s')} \{(_{s',u}w_{s,v}, s(s's'')) \mid \\
&\quad {}_{s',u}w_{s,v}(q'_i) = f(q'_i)j_{s',u}(v'_i), \\
&\quad v = (v'_1, \ldots, v'_k) \in \\
&\quad \mathcal{M}(q'_1, s) \times \ldots \times \mathcal{M}(q'_k, s)\}
\end{aligned}
$$

$$\tag{6.2}$$

and so

$$_{s',u}w_{s,v}(q'_i) = f(q'_i)j_{s',u}(v'_i) = f(q'_i)g(v'_i)k(u(v'_i)),$$

where $u(v'_i) \in \mathcal{M}(v'_i, s'), i = 1, \ldots, k$. Now $m'_i, v'_i \in \mathcal{M}(q'_i, s), r'_i \in \mathcal{M}(q'_i, ss')$, $u(v'_i) \in \mathcal{M}(v'_i, s')$ for $i = 1, \ldots, k$. Since

$$\mathcal{M}(q', ss') = \cup_{m' \in \mathcal{M}(q',s)} \mathcal{M}(m', s'), i = 1, \ldots, k,$$

by hypothesis, the sets in (6.1) and (6.2) are equal. Thus $\cdot$ is associative and so $S^{Q'} \times S'$ is polysemigroup. ∎

Definition 6.6.23 *Let (Q, X, μ) be a ffsm. Define the fuzzy subset $\widehat{\mu}$ of $Q \times (\mathcal{P}(X) \backslash \{\emptyset\}) \times Q$ as follows: $\forall q, p \in Q$ and $\forall Y \in \mathcal{P}(X) \backslash \{\emptyset\}$,*

$$\widehat{\mu}(q, Y, p) = \vee \{\mu(q, y, p) | y \in Y\}.$$

Definition 6.6.24 *Let (Q, X, μ) be a ffsm. By $\mu^0((q, q'), \{(h_{s,m}, ss') \mid h_{s,m}(q'_i) = f(q'_i)h(m'_i), i = 1, \ldots, k, m = (m'_1, \ldots, m'_k) \in \mathcal{M}(q'_1, s) \times \ldots \times \mathcal{M}(q'_k, s)\}, (p, p'))$, we mean*

$$\mu(q, \{h_{s,m}(q'_i) \mid m \in \mathcal{M}(q'_1, s) \times \ldots \times \mathcal{M}(q'_k, s)\}, p) \wedge \mu'(q', ss', p').$$

Let (Q, S, μ) and (Q', S', μ') be fuzzy transformation semigroups. Assume $(S^{Q'} \times S', \cdot)$ is a polysemigroup. Recall that we have the ffsm $(Q \times Q', S^{Q'} \times S', \rho)$ where $\forall (q, q'), (p, p') \in Q \times Q'$ and $\forall (f, s) \in S^{Q'} \times S'$,

$$\rho((q, q'), (f, s), (p, p')) = \mu(q, f(q'), p) \wedge \mu'(q', s, p').$$

The following result examines condition (1) of Definition 6.5.1.

Theorem 6.6.25 $\forall (q, q'), (p, p') \in Q \times Q'$ and $\forall (f, s), (g, s') \in S^{Q'} \times S'$, $\vee \{\mu^0((q, q'), (f, s), (r, r')) \wedge \rho((r, r'), (g, s'), (p, p')) \mid (r, r') \in Q \times Q'\} \geq \mu^0((q, q'), (f, s) \cdot (g, s'), (p, p'))$.

Proof. $\mu^0((q, q'), (f, s) \cdot (g, s'), (p, p')) = \mu^0((q, q'), \{(h_{s,m}, ss') \mid h_{s,m}(q'_i) = f(q'_i)h(m'_i), i = 1, \ldots, k, m = (m'_1, \ldots, m'_k) \in \mathcal{M}(q'_1, s) \times \ldots \times \mathcal{M}(q'_k, s)\}, (p, p')) = \mu(q, \{h_{s,m}(q'_i) \mid m \in \mathcal{M}(q'_1, s) \times \ldots \times \mathcal{M}(q'_k, s)\}, p) \wedge \mu'(q', ss', p')$ (by Definition 6.6.24) $= \mu(q, \{f(q')g(m') , p) \mid m' \in \mathcal{M}(q', s')\} \wedge \mu'(q', ss', p') \leq \vee \{\mu(q, f(q')g(r'), p) \mid r' \in Q'\} \wedge \mu'(q', ss', p')$ (by Definition 6.6.23) $= \vee \{\mu(q, f(q'), r) \wedge \mu(r, g(r'), p) \mid r \in Q, r' \in Q'\} \wedge \vee \{\mu'(q', s, r') \wedge \mu'(r', s', p') \mid r' \in Q'\} = \vee \{\mu(q, f(q'), r) \wedge \mu'(q', s, r') \wedge \mu(r, g(r'), p) \wedge \mu'(r', s', p') \mid r \in Q, r' \in Q'\} = \vee \{\mu^0((q, q'), (f, s), (r, r')) \wedge \mu^0((r, r'), (g, s'), (p, p')) \mid r \in Q, r' \in Q'\}$. ∎

6.7 Submachines of a Fuzzy Finite State Machine

In this section, we continue our study of a fuzzy finite state machine utilizing algebraic techniques.

Definition 6.7.1 *Let $M = (Q, X, \mu)$ be a ffsm. Let $p, q \in Q$. p is called an* **immediate successor** *of q if $\exists a \in X$ such that $\mu(q, a, p) > 0$. p is called a* **successor** *of q if $\exists x \in X^*$ such that $\mu^*(q, x, p) > 0$.*

Proposition 6.7.2 *Let $M = (Q, X, \mu)$ be a ffsm. Let $q, p, r \in Q$. Then the following assertions hold.*

(1) q is a successor of q.

(2) if p is a successor of q and r is a successor of p, then r is a successor of q.

Proof. (1) Since $\mu^*(q, \Lambda, q) = 1 > 0$, q is a successor of q.

(2) Now $\exists x, y \in X$ such that $\mu^*(q, x, p) > 0$ and $\mu^*(p, y, r) > 0$. Thus

$$\mu^*(q, xy, r) \geq \mu^*(q, x, p) \wedge \mu^*(p, y, r) > 0,$$

by Lemma 6.2.2. Hence r is a successor of q. ∎

Definition 6.7.3 *Let $M = (Q, X, \mu)$ be a ffsm and let $q \in Q$. We denote by $S(q)$ the set of all successors of q.*

Definition 6.7.4 *Let $M = (Q, X, \mu)$ be a ffsm and let $T \subseteq Q$. The set of all successors of T, denoted by $S_Q(T)$ in Q, is defined to be the set*

$$S_Q(T) = \cup\{S(q) | q \in T\}.$$

If no confusion arises, then we write $S(T)$ for $S_Q(T)$.

Theorem 6.7.5 *Let $M = (Q, X, \mu)$ be a ffsm. Let $A, B \subseteq Q$. Then the following assertions hold.*

(1) If $A \subseteq B$, then $S(A) \subseteq S(B)$.

(2) $A \subseteq S(A)$.

(3) $S(S(A)) = S(A)$.

(4) $S(A \cup B) = S(A) \cup S(B)$.

(5) $S(A \cap B) \subseteq S(A) \cap S(B)$.

Proof. The proofs of (1), (2), (4), and (5) are straightforward.

(3) Clearly $S(A) \subseteq S(S(A))$. Let $q \in S(S(A))$. Then $q \in S(p)$ for some $p \in S(A)$. Thus $p \in S(r)$ for some $r \in A$. Now q is a successor of p and p is a successor of r. Hence by Proposition 6.7.2 , q is a successor of r. Thus $q \in S(r) \subseteq S(A)$. Hence $S(S(A)) = S(A)$. ∎

Exchange Property: Let $M = (Q, X, \mu)$ be a ffsm. Let $q, p \in Q$ and let $T \subseteq Q$. Suppose that if $p \in S(T \cup \{q\})$, $p \notin S(T)$, then $q \in S(T \cup \{p\})$. Then we say that M satisfies the Exchange Property.

Proposition 6.7.6 *Let $M = (Q, X, \mu)$ be a ffsm. Then the following assertions are equivalent.*

(1) M satisfies the Exchange Property.

(2) $\forall p, q \in Q$, $q \in S(p)$ if and only if $p \in S(q)$.

Proof. (1)$\Longrightarrow$(2): Let $p, q \in Q$ and $p \in S(q)$. Now $p \notin S(\phi)$. Hence $q \in S(p)$. Similarly if $q \in S(p)$ then $p \in S(q)$.

(2)$\Longrightarrow$(1): Let $T \subseteq Q$, $p, q \in Q$. Suppose $p \in S(T \cup \{q\})$, $p \notin S(T)$. Then $p \in S(q)$. Hence $q \in S(p) \subseteq S(T \cup \{p\})$. ∎

Definition 6.7.7 *Let $M = (Q, X, \mu)$ be a ffsm. Let $T \subseteq Q$. Let ν be a fuzzy subset of $T \times X \times T$ and let $N = (T, X, \nu)$. The fuzzy finite state machine N is called a **submachine** of M if*

(1) $\mu|_{T \times X \times T} = \nu$ and

(2) $S_Q(T) \subseteq T$.

We assume that $\phi = (\phi, X, \nu)$ is a submachine of M. Clearly, if K is a submachine of N and N is a submachine of M, then K is a submachine of M.

Theorem 6.7.8 *Let $M = (Q, X, \mu)$ be a ffsm. Let $M_i = (Q_i, X, \mu_i)$, $i \in I$, be a family of submachines of M, where $Q_i \subseteq Q$. Then the following assertions hold.*

(1) $\cap_{i \in I} M_i = (\cap_{i \in I} Q_i, X, \cap_{i \in I} \mu_i)$ is a submachine of M.

(2) $\cup_{i \in I} M_i = (\cup_{i \in I} Q_i, X, \nu)$ is a submachine of M, where

$$\nu = \mu|_{\cup_{i \in I} Q_i \times X \times \cup_{i \in I} Q_i}.$$

Proof. (1) Let $(q, x, p) \in \cap_{i \in I} Q_i \times X \times \cap_{i \in I} Q_i$. Then

$$(\cap_{i \in I} \mu_i)(q, x, p) = \wedge_{i \in I} \mu_i(q, x, p) = \wedge_{i \in I} \mu(q, x, p) = \mu(q, x, p).$$

Also

$$S(\cap_{i \in I} Q_i) \subseteq \cap_{i \in I} S(Q_i) \subseteq \cap_{i \in I} Q_i.$$

Thus $\cap_{i \in I} M_i$ is a submachine of M.

(2) Since

$$S(\cup_{i \in I} Q_i) = \cup_{i \in I} S(Q_i) = \cup_{i \in I} Q_i,$$

$\cup_{i \in I} M_i$ is a submachine of M. ∎

Definition 6.7.9 *Let $M = (Q, X, \mu)$ be a ffsm. Then M is called **strongly connected** if $\forall p, q \in Q$, $p \in S(q)$.*

Definition 6.7.10 *Let $M = (Q, X, \mu)$ be a ffsm and let $N = (T, X, \nu)$ be a submachine of M. N is called* **proper** *if $T \neq Q$ and $T \neq \emptyset$.*

Theorem 6.7.11 *Let $M = (Q, X, \mu)$ be a ffsm. Then M is strongly connected if and only if M has no proper submachines.*

Proof. Suppose M is strongly connected. Let $N = (T, X, \nu)$ be a submachine of such that $T \neq \emptyset$. Then $\exists q \in T$. Let $p \in Q$. Since M is strongly connected $p \in S(q)$. Hence $p \in S(q) \subseteq S(T) \subseteq T$. Thus $T = Q$ and so $M = N$. Conversely, suppose M has no proper submachine. Let $p, q \in Q$ and let $N = (S(q), X, \nu)$ where $\nu = \mu|_{S(q) \times X \times S(q)}$. Then N is a submachine of M and $S(q) \neq \emptyset$. Hence $S(q) = Q$. Thus $p \in S(q)$. Hence M is strongly connected. ∎

Proposition 6.7.12 *Let $M = (Q, X, \mu)$ be a ffsm. Let $R \subseteq Q$. Then $N = (S(R), X, \mu_R)$ is a submachine of M where $\mu_R = \mu|_{S(R) \times X \times S(R)}$.* ∎

Definition 6.7.13 *Let $M = (Q, X, \mu)$ be a ffsm. Let $R \subseteq Q$ and $\{N_i \mid i \in \Lambda\}$ be the collection of all submachines of M whose state set contains R. Define $< R > = \cap_{i \in \Lambda}\{N_i \mid i \in \Lambda\}$. Then $< R >$ is called the* **submachine generated** *by R.*

In Definition 6.7.13, it is clear that $< R >$ is the smallest submachine of M whose state set contains R.

Proposition 6.7.14 *Let $M = (Q, X, \mu)$ be a ffsm. Let $R \subseteq Q$. Then*

$$< R > \ = (S(R), X, \mu_R).$$

Proof. Now $< R > = (\cap_{i \in \Lambda}Q_i, X, \cap_{i \in \Lambda}\mu_i)$, where $\{N_i \mid i \in \Lambda\}$ is the collection of all submachines of M whose state set contains R and $N_i = (Q_i, X, \mu_i)$, $i \in \Lambda$. It suffices to show that $S(R) = \cap_{i \in \Lambda}Q_i$. Since $(S(R), X, \mu_R)$ is a submachine of M such that $R \subseteq S(R)$, we have that $S(R) \supseteq \cap_{i \in \Lambda}Q_i$. Let $p \in S(R)$. Then $\exists\, r \in R$ and $x \in X^*$ such that $\mu^*(r, x, p) > 0$. Now $r \in \cap_{i \in \Lambda}Q_i$, and since $< R >$ is a submachine of M, $p \in \cap_{i \in \Lambda}Q_i$. Thus $S(R) \subseteq \cap_{i \in \Lambda}Q_i$. Hence $S(R) = \cap_{i \in \Lambda}Q_i$. ∎

Definition 6.7.15 *Let $M = (Q, X, \mu)$ be a ffsm. M is called* **singly generated** *if $\exists q \in Q$ such that $M =< \{q\} >$. In this case q is called a* **generator** *of M and we say that M is* **generated** *by q.*

Theorem 6.7.16 *Let $M = (Q, X, \mu)$ be a ffsm. Let $R, T \subseteq Q$. Then the following assertions hold.*
(1) $< R \cup T > \ = \ < R > \cup < T >$.
(2) $< R \cap T > \ \subseteq \ < R > \cap < T >$.

Proof. (1) By Theorem 6.7.5, $S(R \cup T) = S(R) \cup S(T)$. Now

$$
\begin{aligned}
\mu_{S(R) \cup S(T)} &= \mu|_{(S(R) \cup S(T)) \times X \times (S(R) \cup S(T))} \\
&= \mu|_{S(R \cup T) \times X \times S(R \cup T)} \\
&= \mu_{S(R \cup T)}.
\end{aligned}
$$

Hence $< R \cup T > = < R > \cup < T >$.

(2) By Theorem 6.7.5, $S(R \cap T) \subseteq S(R) \cap S(T)$. Now

$$
\mu_{S(R \cap T)} = \mu|_{S(R \cap T) \times X \times S(R \cap T)}
$$

and

$$
\mu_{S(R) \cap S(T)} = \mu|_{(S(R) \cap S(T)) \times X \times (S(R) \cap S(T))}.
$$

Hence

$$
\mu_{S(R \cap T)} = \mu_{S(R) \cap S(T)}|_{S(R \cap T) \times X \times S(R \cap T)}.
$$

Hence $< R \cap T > \subseteq < R > \cap < T >$. ∎

Definition 6.7.17 Let $M = (Q, X, \mu)$ be a ffsm. Let $M_i = (Q_i, X, \mu_i)$, $i = 1, 2$, be submachines of M. If $M = < Q_1 \cup Q_2 >$, then we say that M is the **union** of M_1 and M_2 and we write $M = M_1 \cup M_2$. If $M = M_1 \cup M_2$ and $Q_1 \cap Q_2 = \emptyset$, then we say that M is the **(internal) direct union** of M_1 and M_2 and we write $M = M_1 \dot{\cup} M_2$.

Suppose $M = M_1 \cup M_2$. Then $S(Q_i) = Q_i$ in M, $i = 1, 2$, since M_i is a submachine of M, $i = 1, 2$. Now $S(Q_1 \cup Q_2) = S(Q_1) \cup S(Q_2) = Q_1 \cup Q_2$.

Definition 6.7.18 Let $M = (Q, X, \mu)$ be a ffsm. Let $T \subseteq Q$. T is called **free** if $\forall t \in T$, $t \notin S(T \backslash \{t\})$.

Definition 6.7.19 Let $M = (Q, X, \mu)$ be a ffsm. Let $T \subseteq Q$. If T is free and $M = < T >$, then T is called a **basis** of M.

Theorem 6.7.20 Let $M = (Q, X, \mu)$ be a ffsm. Let $T \subseteq Q$. Then the following assertions are equivalent.

(1) T is a minimal system of generators of M.
(2) T is a maximally free subset of Q.
(2) T is a basis of M. ∎

Theorem 6.7.21 Let $M = (Q, X, \mu)$ be a ffsm. Suppose that M satisfies the exchange property. Then M has a basis and the cardinality of a basis is unique. ∎

Theorem 6.7.22 Let $M = (Q, X, \mu)$ be a ffsm. Suppose that M satisfies the exchange property. Let $\{q_1, q_2, \ldots, q_n\}$ be a basis of M. Then $M = < q_1 > \dot{\cup} < q_2 > \dot{\cup} \cdots \dot{\cup} < q_n >$.

Proof. We have $< q_i > = (S(q_i), X, \mu_i)$, where $\mu_i = \mu|_{S(q_i) \times X \times S(q_i)}$. Now if $i \neq j$, then $S(q_i) \cap S(q_j) = \emptyset$ since the exchange property is equivalent to the statement that $\forall p, q \in Q$, $p \in S(q)$ if and only if $q \in S(p)$. Since $M = < q_1, q_2, \ldots, q_n >$, it follows that $M = < q_1 > \dot{\cup} < q_2 > \dot{\cup} \ldots \dot{\cup} < q_n >$. ∎

6.8   Retrievability, Separability, and Connectivity

Definition 6.8.1 *Let $M = (Q, X, \mu)$ be a ffsm. M is said to be **retrievable** if $\forall q \in Q$, $\forall y \in X^*$ if $\exists t \in Q$ such that $\mu^*(q, y, t) > 0$, then $\exists x \in X^*$ such that $\mu^*(t, x, q) > 0$.*

Definition 6.8.2 *Let $M = (Q, X, \mu)$ be a ffsm. M is said to be **quasi-retrievable** if $\forall q \in Q$, $\forall y \in X^*$ if $\exists t \in Q$ such that $\mu^*(q, y, t) > 0$, then $\exists x \in X^*$ such that $\mu^*(q, yx, q) > 0$.*

Definition 6.8.3 *Let $M = (Q, X, \mu)$ be a ffsm. Let $q, r, s \in Q$. Then r and s are said to be q-**related** if $\exists y \in X^*$ such that $\mu^*(q, y, r) > 0$ and $\mu^*(q, y, s) > 0$. If r and s are q-related, then r and s are said to be q-**twins** if $S(s) = S(r)$.*

Remark 6.8.4 *Let $M = (Q, X, \mu)$ be a ffsm. The following assertions are equivalent.*

(1) $\forall q, r, p \in Q$, $\forall x, y \in X^*$ if $\mu^*(q, y, r) > 0$ and $\mu^*(q, yx, p) > 0$, then $p \in S(r)$.

(2) $\forall q, r, s \in Q$, if r and s are q-related, then r and s are q-twins.

Proof. (1)$\Rightarrow$(2): Let $q, r, s \in Q$ be such that r and s are q-related. Then $\exists y \in X^*$ such that $\mu^*(q, y, r) > 0$ and $\mu^*(q, y, s) > 0$. Let $p \in S(s)$. Then $\exists x \in X^*$ such that $\mu^*(s, y, p) > 0$. Then $\mu^*(q, yx, p) > 0$. Thus by the hypothesis $p \in S(r)$. Similarly if $p \in S(r)$ then $p \in S(s)$.

(2)$\Rightarrow$(1): Let $q, r, p \in Q$, $x, y \in X^*$ be such that $\mu^*(q, y, r) > 0$ and $\mu^*(q, yx, p) > 0$. Now $\mu^*(q, yx, p) = \vee\{\mu^*(q, y, s) \wedge \mu^*(s, x, p) | s \in Q\} > 0$. Hence $\exists s \in Q$ such that $\mu^*(q, y, s) > 0$ and $\mu^*(s, x, p) > 0$. Then r and s are q-twins and $p \in S(s)$. Thus by the hypothesis $p \in S(r)$. ∎

Proposition 6.8.5 *Let $M = (Q, X, \mu)$ be a ffsm. Then the following assertions are equivalent:*

(1) M is retrievable.

(2) M is quasi-retrievable and $\forall\ q, r, s \in Q$, if r and s are q-related, then r and s are q-twins.

Proof. (1)$\Rightarrow$(2): It is immediate that retrievability implies quasi-retrieva-bility. Let $q, r, p \in Q$ and $x, y \in X^*$. Suppose that $\mu^*(q, y, r) > 0$ and $\mu^*(q, yx, p) > 0$. Since M is retrievable, $\exists z \in X^*$ such that $\mu^*(p, z, q) > 0$. Thus $p \in S(r)$. Hence (2) holds by Remark 6.8.4.

(2)$\Rightarrow$(1): Let $q \in Q$ and $y \in X^*$. Suppose $\exists t \in Q$ such that $\mu^*(q, y, t) > 0$. Then $\exists x \in X^*$ such that $\mu^*(q, yx, q) > 0$ since M is quasi-retrievable. By Remark 6.8.4, $q \in S(t)$. ∎

Theorem 6.8.6 *Let $M = (Q, X, \mu)$ be a ffsm. The following assertions are equivalent.*

(1) M satisfies the Exchange Property,

(2) M is the union of strongly connected submachines,

(3) M is retrievable.

Proof. $(1) \Rightarrow (2)$: By (1), $M = \cup_{i=1}^{n} < q_i >$, where $\{q_1, q_2, \ldots, q_n\}$ is a basis of M. Also $S(q_i) \cap S(q_j) = \emptyset$ if $i \neq j$. Let $p, q \in S(q_i)$. Then $q_i \in S(p)$ and so $q \in S(p)$. Thus $< q_i >$ is strongly connected.

$(2) \Rightarrow (1)$: Now $M = \cup_{i=1}^{n} M_i$, where each $M_i = (Q_i, X, \mu_i)$ is strongly connected. Let $p, q \in Q$. Suppose $p \in S(q)$. Now $\exists i$ such that $q \in Q_i$. Then $p \in S(q) \subseteq S(Q_i) = Q_i$. Thus $p, q \in Q_i$. Since M_i is strongly connected, $q \in S(p)$. Hence M satisfies the Exchange Property by Proposition 6.7.6.

$(2) \Rightarrow (3)$: Now $M = \cup_{i=1}^{n} M_i$, where each $M_i = (Q_i, X, \mu_i)$ is strongly connected. Let $q \in Q$, $y \in X^*$ be such that $\mu^*(q, y, t) > 0$ for some $t \in Q$. Now $q \in Q_i$ for some i. Thus $t \in S(q) \subseteq S(Q_i)$. Since M_i is strongly connected, $q \in S(t)$. Hence $\exists x \in X^*$ such that $\mu^*(t, x, q) > 0$. Thus M is retrievable.

$(3) \Rightarrow (2)$: Let $q \in Q$ and let $r, t \in S(q)$. Then $\exists y, z \in X^*$ such that $\mu^*(q, y, r) > 0$ and $\mu^*(q, z, t) > 0$. Since M is retrievable $\exists x \in X^*$ such that $\mu^*(r, x, q) > 0$. Hence $q \in S(r)$. Thus $t \in S(r)$. Hence $< q >$ is strongly connected. Now $M = \cup_{q \in Q} < q >$. ∎

Definition 6.8.7 *Let $M = (Q, X, \mu)$ be a ffsm. Let $N = (T, X, \nu) \neq \emptyset$ be a submachine of M. Then N is said to be **separated** if $S(Q \backslash T) \cap T = \emptyset$.*

Theorem 6.8.8 *Let $M = (Q, X, \mu)$ be a ffsm. Let $N = (T, X, \nu) \neq \emptyset$ be a submachine of M. Then N is separated if and only if $S(Q \backslash T) = Q \backslash T$.*

Proof. Suppose N is separated. Let $q \in S(Q \backslash T)$. Now $S(Q \backslash T) \cap T = \emptyset$. Hence $q \notin T$. Thus $q \in Q \backslash T$. Hence $S(Q \backslash T) \subseteq Q \backslash T$. Thus $S(Q \backslash T) = Q \backslash T$. Conversely, suppose that $S(Q \backslash T) = Q \backslash T$. Clearly then $S(Q \backslash T) \cap T = \emptyset$. Thus N is separated. ∎

Theorem 6.8.9 *Let $M = (Q, X, \mu)$ be a ffsm. Let $N = (T, X, \nu) \neq \emptyset$ be a submachine of M. If N is separated, then so is $C = (Q \backslash T, X, \sigma)$ where $\sigma = \mu|_{(Q \backslash T) \times X \times (Q \backslash T)}$.*

Proof. Now $\emptyset \neq T \neq Q$ and so $Q \backslash T \neq \emptyset$. By Theorem 6.8.8, $S(Q \backslash T) = Q \backslash T$. Hence C is a submachine of M. Now $S(Q \backslash (Q \backslash T)) = S(T) = T$. Thus $S(Q \backslash (Q \backslash T)) \cap (Q \backslash T) = T \cap (Q \backslash T) = \emptyset$. Hence C is separated. ∎

Definition 6.8.10 *Let $M = (Q, X, \mu)$ be a ffsm. Then M is said to be **connected** if M has no separated proper submachine.*

Theorem 6.8.11 *Let $M = (Q, X, \mu)$ be a ffsm. Then the following asser-tions are equivalent.*

(1) *M is strongly connected.*

(2) *M is connected and retrievable.*

(3) *Every submachine of M is strongly connected.*

Proof. (1)$\Rightarrow$(2): By Theorem 6.7.11, M does not have any proper submachines and so M has no proper separated submachines. Thus M is connected. We now show that M is retrievable. Let $q, t \in Q$ and $y \in X^*$ be such that $\mu^*(q, y, t) > 0$. Since M is strongly connected, $q \in S(t)$. Then $\exists x \in X^*$ such that $\mu^*(t, x, q) > 0$. Hence M is retrievable.

(2)$\Rightarrow$(3): Let $N = (T, X, \nu)$ be a submachine of M. Suppose $p, q \in T$ are such that $p \notin S(q)$. Then $S(q) \neq Q$ and so $K = (S(q), X, \mu|_{S(q) \times X \times S(q)})$ is a proper submachine of M. Since M is connected, $S(Q \backslash S(q)) \cap S(q) \neq \emptyset$. Let $r \in S(Q \backslash S(q)) \cap S(q)$. Then $r \in S(t)$ for some $t \in Q \backslash S(q)$ and $r \in S(q)$. Now $\exists y \in X^*$ such that $\mu^*(t, y, r) > 0$. Since M is retrievable, $\exists z \in X^*$ such that $\mu^*(r, z, t) > 0$. Thus $t \in S(r)$. Hence $t \in S(r) \subseteq S(q)$, a contradiction. Thus $p \in S(q)$ $\forall p, q \in T$. Hence N is strongly connected.

(3)$\Rightarrow$(1): Obvious. ∎

6.9   Decomposition of Fuzzy Finite State Machines

Definition 6.9.1 *Let $M = (Q, X, \mu)$ be a ffsm. Let P be a submachine of M. Then P is called a **primary submachine** of M if*

(1) $\exists q \in Q$ *such that* $P = <q>$;

(2) $\forall s \in Q$ *if* $P \subseteq <s>$, *then* $P = <s>$.

Theorem 6.9.2 (Decomposition Theorem) *Let $M = (Q, X, \mu)$ be a ffsm. Let $\wp = \{P_1, P_2, \ldots, P_n\}$ be the set of all distinct primary subma-chines of M. Then*

(1) $M = \cup_{i=1}^n P_i$;

(2) $M \neq \cup_{i=1, i \neq j}^n P_i$ *for any* $j \in \{1, 2, \ldots, n\}$.

Proof. (1) Let $q_0 \in Q$. Now $\forall q_i \in Q$, either (a) $<q_i> \in \wp$ or (b) $\exists q_{i+1} \in Q \backslash S(q_i)$ such that $<q_i> \subset <q_{i+1}>$. Since Q is finite, either $<q_0> \in \wp$ or there exists a positive integer k such that $<q_0> \subset <q_k> \in \wp$. Thus $Q = \cup_{i=1}^n S(p_i)$ where $P_i = <p_i>$, $i = 1, 2, \ldots, n$. Hence $M = \cup_{i=1}^n P_i$.

(2) Let $N = \cup_{i=1, i \neq j}^n P_i$ and let $P_j = <p_j>$. If $p_j \in \cup_{i=1, i \neq j}^n S(p_i)$, then $p_j \in S(p_i)$ for some $i \neq j$. Hence $P_j = <p_j> \subset P_i$. However this contradicts the maximality of P_j since $P_j \neq P_i$. Thus $p_j \notin \cup_{i=1, i \neq j}^n S(p_i)$. Hence $M \neq N$. ∎

Corollary 6.9.3 *Let $M = (Q, X, \mu)$ be a ffsm. Then every singly generated submachine of $M \neq \emptyset$ is a submachine of a primary submachine of M.* ∎

Corollary 6.9.4 *Let* $M = (Q, X, \mu)$ *be a ffsm. Then* $P_1, P_2, \ldots, P_n$ *in Theorem 6.9.2 are unique.* ■

Definition 6.9.5 *Let* $M = (Q, X, \mu)$ *be a ffsm. Then* **rank** *of* M, *rank*(M), *is the number of distinct primary submachines of* M.

Theorem 6.9.6 *Let* $M = (Q, X, \mu)$ *be a ffsm. The following assertions are equivalent.*
  (1) M *is retrievable.*
  (2) Every primary submachine of M *is strongly connected.*

  Proof. (1)$\Longrightarrow$(2): Let P be a primary submachine of M. Then $P = <p>$ for some $p \in Q$. Then as in the proof of (3)$\Longrightarrow$(2) of Theorem 6.8.6, $<p>$ is strongly connected.
  (2)$\Longrightarrow$(1): Now $M = \cup_{i=1}^{n} P_i$ where P_i are primary submachines of M. Then the P_i are strongly connected. Thus M is the union of strongly connected submachines. By Theorem 6.8.6, (1) holds. ■

Lemma 6.9.7 *Let* $M = (Q, X, \mu)$ *be a ffsm. Then* M *has a strongly connected submachine.*

  Proof. We prove the result by induction on $|Q| = n$. If $n = 1$, then the result is obvious. Suppose the result is true for all ffsms $N = (T, X, \nu)$ such that $|T| < n$, $n > 1$. Let $q \in Q$. Then $M' = (S(q), X, \mu|_{S(q) \times X \times S(q)})$ is a submachine of M. If M' is strongly connected, then the result follows. Suppose that M' is not strongly connected. Then $\exists p \in S(q)$ such that $q \notin S(p)$ and hence $S(p) \subset S(q)$. Now $|S(p)| < n$. Hence by the induction hypothesis the ffsm $M'' = (S(p), X, \mu|_{S(p) \times X \times S(p)})$ has a strongly connected submachine. Since M'' is a submachine of M, M has a strongly connected submachine. ■

Theorem 6.9.8 *Let* $M = (Q, X, \mu)$ *be a ffsm. The following assertions are equivalent.*
  (1) M *is retrievable.*
  (2) Every singly generated submachine of M *is primary.*
  (3) Every nonempty connected submachine of M *is primary.*

  Proof. (1)$\Longrightarrow$(2): Now $M = \cup_{i=1}^{n} P_i$ where the P_i are primary submachines of M. By Theorem 6.7.11, the P_i are strongly connected. Let $N = <q>$ be a singly generated submachine of M. Then $<q> \subseteq P_i$ for some i. Hence $<q> = P_i$ by Theorem 6.7.11. Thus N is primary.
  (2)$\Longrightarrow$(1): Since every singly generated submachine of M is primary, every singly generated submachine of M is strongly connected. Thus every primary submachine of M is strongly connected. By Theorem 6.9.6, (1) holds.
  (2)$\Longrightarrow$(3): Let $N = (T, X, \nu)$ be a nonempty connected submachine of M. Let $q \in T$. Suppose $S(q) \neq T$. Since N is connected, $S(T \backslash S(q)) \cap S(q) \neq$

$\emptyset$. Let $r \in S(T\backslash S(q)) \cap S(q)$. Then $r \in S(t)$ for some $t \in T\backslash S(q)$ and $r \in S(q)$. Now $< r > \subseteq < t >$ and $< r > \subseteq < q >$. Since $< r >$ is primary, $< t > = < r > = < q >$. Hence $t \in S(q)$, which is a contradiction. Hence $N = < q >$ and so N is primary.

(3)$\Longrightarrow$(2): Let $N = < s >$ be a singly generated submachine. By Lemma 6.9.7, N has a strongly connected submachine $B = < r >$, say. Then B is connected and hence primary. Thus $< r > = < s > = N$. Hence N is primary. ∎

Lemma 6.9.9 *Let $M = (Q, X, \mu)$ be a ffsm and let $N = (T, X, \nu)$ be a separated submachine of M. Then every primary submachine of N is also a primary submachine of M.*

Proof. Let $< q >$ be a primary submachine of N. Suppose $< q >$ is not a primary submachine of M. Then $\exists p \in Q\backslash S(q)$ such that $< q > \subset < p >$. Clearly $p \notin T$. Thus $p \in Q\backslash T$. Since $q \in S(p)$, $q \in S(Q\backslash T)$. Thus $q \in S(Q\backslash T) \cap T$, which is a contradiction since N is separated. Hence $< q >$ is a primary submachine of M. ∎

Theorem 6.9.10 *Let $M = (Q, X, \mu)$ be a ffsm and let $N_i = (T_i, X, \nu_i), i = 1, 2, \dots, n$, be the primary submachines of M. Then a proper submachine $N = (T, X, \nu)$ of M is separated if and only if for some $J \subseteq \{1, 2, \dots, n\}$, $J \neq \emptyset$, $Q\backslash T = \cup_{i \in J} T_i$.*

Proof. Suppose $N = (T, X, \nu)$ be a proper separated submachine of M. Then $S(Q\backslash T) = Q\backslash T$. Since N is proper, the submachine $< Q\backslash T >$ is nonempty. Thus $< Q\backslash T >$ is the union of all its primary submachines. Since $< Q\backslash T >$ is separated every primary submachine of $< Q\backslash T >$ is a primary submachine of M. Thus $S(Q\backslash T) = \cup_{i \in J} T_i$ for some $J \subseteq \{1, 2, \dots, n\}$, $J \neq \emptyset$. Since $Q\backslash T = S(Q\backslash T)$, $Q\backslash T = \cup_{i \in J} T_i$ for some $J \subseteq \{1, 2, \dots, n\}$, $J \neq \emptyset$. Conversely, let $N = (T, X, \nu)$ be a proper submachine of M such that $Q\backslash T = \cup_{i \in J} T_i$ for some $J \subseteq \{1, 2, \dots, n\}$, $J \neq \emptyset$. Then $S(Q\backslash T) = S(\cup_{i \in J} T_i) = \cup_{i \in J} S(T_i) = \cup_{i \in J} T_i = Q\backslash T$. Hence N is separated. ∎

Corollary 6.9.11 *Let $M = (Q, X, \mu)$ be a ffsm. Then M is connected if and only if M has no proper submachine $N = (T, X, \nu)$ such that $Q\backslash T$ is the union of the sets of states of all primary submachines of M.* ∎

Definition 6.9.12 *Let $M = (Q, X, \mu)$ be a ffsm and let $N = (T, X, \nu)$ be a submachine of M. A subset $R \subseteq Q$ is called a **generating set** of N, and is said to **generate** N if $N = < R >$.*

Lemma 6.9.13 *Let $M = (Q, X, \mu)$ be a ffsm and let $N_i = (T_i, X, \nu_i), i = 1, 2, \dots, n$ be the primary submachines of M. Let $R \subseteq Q$. Then R generates M if and only if $\forall i, 1 \leq i \leq n, \exists r_i \in R$ such that $N_i = < r_i >$.*

Proof. Suppose that R generates M. Then $M = <R> = \cup_{r \in R} <r>$. Let $q_i \in T_i$ be such that $T_i = <q_i>$, $i = 1, 2, \ldots, n$. Then $q_i \in \cup_{r \in R} <r>$ and so $q_i \in <r>$ for some $r \in R$. Thus $<q_i> \subseteq <r>$. Since $<q_i>$ is primary, $<q_i> = <r>$. The converse is immediate. ∎

Definition 6.9.14 *Let $M = (Q, X, \mu)$ be a ffsm. Let $R \subseteq Q$ be a generating set of M. Then R is said to be a **minimal generating set** of M if*

(1) $M = <R>$, and
(2) $\forall r \in R, <R\backslash\{r\}> \neq M$.

Theorem 6.9.15 *Let $M = (Q, X, \mu)$ be a ffsm. Let $R \subseteq Q$ be a generating set of M. Then R is a minimal generating set of M if and only if $|R| = rank(M)$.*

Proof. Let $n = \text{rank}(M)$ and let $N_i = (T_i, X, \nu_i)$ be a primary submachine of M, $i = 1, 2, \ldots, n$. By Lemma 6.9.13, since R is a generating set of M, $\exists r_i \in R$ such that $<r_i> = T_i$, $i = 1, 2, \ldots, n$. Since the T_i are distinct, the r_i are distinct. Thus $|R| \geq \text{rank}(M)$. Now assume that R is a minimal generating set. Suppose $|R| > \text{rank}(M)$. Then $\exists r \in R$ such that $r \notin \{r_1, r_2, \ldots, r_n\}$. Thus $<R\backslash\{r\}> = Q$. Hence R is not minimal, a contradiction. Thus $|R| = \text{rank}(M)$. Conversely suppose that $|R| = \text{rank}(M)$. Then $R = \{r_1, r_2, \ldots, r_n\}$. Hence $<R\backslash\{r_i\}> = \cup_{j \neq i} N_j \neq M$. Thus R is minimal. ∎

Theorem 6.9.16 *Let $M = (Q, X, \mu)$ be a ffsm. Then M is not connected if and only if $\exists$ a generating set R of M with nonempty subsets R_1 and R_2 such that $<R_1> \cap <R_2> = \emptyset$ and $M = <R_1> \cup <R_2>$.*

Proof. Suppose that M is not connected. Then $\exists$ a proper submachine $N = (T, X, \nu)$ of M that is separated., i.e., $S(Q\backslash T) \cap T = \emptyset$. Let R_1 be a generating set of $<S(Q\backslash T)>$ and let R_2 be a generating set of N. Then $<R_1> \cap <R_2> = \emptyset$ and $M = <R_1> \cup <R_2>$. Conversely, suppose that R_1 and R_2 exists. Let $N = <R_2>$. Then $<S(Q\backslash T)> = <R_1>$. Hence $N = (T, X, \mu|_{T \times X \times T})$ is a proper submachine of M that is separated. ∎

6.10 Subsystems of Fuzzy Finite State Machines

In this and the next section, we introduce the notion of subsystems and strong subsystems of a ffsm in order to consider state membership as fuzzy.

Definition 6.10.1 *Let $M = (Q, X, \mu)$ be a ffsm. Let δ be a fuzzy subset of Q. Then (Q, δ, X, μ) is called a **subsystem** of M if $\forall p, q \in Q$, $\forall a \in X$,*

$$\delta(q) \geq \delta(p) \wedge \mu(p, a, q).$$

If (Q, δ, X, μ) is a subsystem of M, then we simply write δ for (Q, δ, X, μ).

Theorem 6.10.2 *Let $M = (Q, X, \mu)$ be a ffsm and let δ be a fuzzy subset of Q. Then δ is a subsystem of M if and only if $\forall p, q \in Q$, $\forall x \in X^*$,*

$$\delta(q) \geq \delta(p) \wedge \mu^*(p, x, q).$$

Proof. Suppose δ is a subsystem. Let $q, p \in Q$, and $x \in X^*$. We prove the result by induction on $|x| = n$. If $n = 0$, then $x = \Lambda$. Now if $p = q$, then $\delta(q) \wedge \mu^*(q, \Lambda, q) = \delta(q)$. If $q \neq p$, then $\delta(p) \wedge \mu^*(p, \Lambda, q) = 0 \leq \delta(q)$. Thus the result is true if $n = 0$. Suppose the result is true $\forall y \in X^*$ such that $|y| = n - 1$, $n > 0$. Let $x = ya$, $|y| = n - 1$, $y \in X^*$, $a \in X$. Then

$$\begin{aligned}
\delta(p) \wedge \mu^*(p, x, q) &= \delta(p) \wedge \mu^*(p, ya, q) \\
&= \delta(p) \wedge (\vee\{\mu^*(p, y, r) \wedge \mu(r, a, q) | r \in Q\}) \\
&= \vee\{\delta(p) \wedge \mu^*(p, y, r) \wedge \mu(r, a, q) | r \in Q\} \\
&\leq \vee\{\delta(r) \wedge \mu(r, a, q) | r \in Q\} \\
&\leq \delta(q).
\end{aligned}$$

Hence $\delta(q) \geq \delta(p) \wedge \mu^*(p, x, q)$. The converse is trivial. ∎

Theorem 6.10.3 *Let $M = (Q, X, \mu)$ be a ffsm. Let δ, δ_1, and δ_2 be subsystems of M. Then the following assertions hold.*
(1) $\delta_1 \cap \delta_2$ is a subsystem of M.
(2) $\delta_1 \cup \delta_2$ is a subsystem of M.
(3) $N = (Supp(\delta), X, \nu)$ is a submachine of M, where

$$\nu = \mu|_{Supp(\delta) \times X \times Supp(\delta)}.$$

(4) Let $N_t = (\delta_t, X, \nu_{(t)})$ where $\nu_{(t)} = \mu|_{\delta_t \times X \times \delta_t}$, $t \in [0, 1]$. If $\forall t \in [0, 1]$, N_t is a submachine of M, then δ is a subsystem of M.

Proof. The proofs of (1) and (2) are straightforward.
(3) Let $p \in S(Supp(\delta))$. Then $p \in S(q)$ for some $q \in Supp(\delta)$. Then $\delta(q) > 0$. Since $p \in S(q)$, $\exists x \in X^*$ such that $\mu^*(q, x, p) > 0$. Hence since δ is a subsystem, $\delta(p) \geq \delta(q) \wedge \mu^*(q, x, p) > 0$. Thus $p \in Supp(\delta)$. Hence $S(Supp(\delta)) \subseteq Supp(\delta)$. Thus N is a submachine of M.
(4) Let $q, p \in Q$, $x \in X^*$. If $\delta(p) = 0$ or $\mu^*(p, x, q) = 0$, then $\delta(q) \geq 0 = \delta(p) \wedge \mu^*(p, x, q)$. Suppose $\delta(p) > 0$ and $\mu^*(p, x, q) > 0$. Let $\delta(p) \wedge \mu^*(p, x, q) = t$. Then $p \in \delta_t$. Since N_t is a submachine of M, $S(\delta_t) = \delta_t$. Hence $q \in S(p) \subseteq S(\delta_t) = \delta_t$. Hence $\delta(q) \geq t = \delta(p) \wedge \mu^*(p, x, q)$. Thus δ is a subsystem. ∎

Example 6.10.4 *Let $Q = \{p, q\}$, $X = \{a\}$, $\mu(r, a, t) = \frac{1}{2}$ $\forall r, t \in Q$. Let $\delta(q) = \frac{3}{4}$ and $\delta(p) = \frac{1}{2}$. Then*

$$\delta(q) \wedge \mu(q, a, p) = \frac{1}{2} = \delta(p)$$

and

$$\delta(p) \wedge \mu(p, a, q) = \frac{1}{2} < \frac{3}{4} = \delta(q).$$

Thus δ is a subsystem. Let $\frac{1}{2} < t \leq \frac{3}{4}$. Let $N_t = (\delta_t, X, \nu_{(t)})$ where $\nu_{(t)} = \mu|_{\delta_t \times X \times \delta_t}$. Now $\delta(q) \geq t$. Thus $q \in \delta_t$. Also $\mu(q, a, p) = \frac{1}{2} > 0$. Hence $p \in S(q)$. Thus $p \in S(\delta_t)$. But $\delta(p) = \frac{1}{2} < t$. Hence $p \notin \delta_t$. Thus N_t is not a submachine of M.

Definition 6.10.5 *Let $M = (Q, X, \mu)$ be a ffsm and let δ be a fuzzy subset of Q. For all $x \in X^*$, define the fuzzy subset δx of Q by*

$$(\delta x)(q) = \vee\{\delta(p) \wedge \mu^*(p, x, q)|p \in Q\}$$

$\forall q \in Q$.

Proposition 6.10.6 *Let $M = (Q, X, \mu)$ be a ffsm. Then $\forall$ fuzzy subsets δ of Q and $\forall x, y \in X^*$,*

$$(\delta x)y = \delta(xy).$$

Proof. Let δ be a fuzzy subset of Q and let $x, y \in X^*$. We prove the result by induction on $|y| = n$. If $n = 0$, then $y = \Lambda$. Let $q \in Q$. Now

$$\begin{aligned}((\delta x)\Lambda)(q) &= \vee\{(\delta x)(p) \wedge \mu^*(p, \Lambda, q) \mid p \in Q\} \\ &= (\delta x)(q).\end{aligned}$$

Hence $(\delta x)\Lambda = \delta x = \delta(x\Lambda)$. Suppose now the result is true for all $u \in X^*$ such that $|u| = n - 1$, $n > 0$ and for all δ. Let $y = ua$, where $a \in X$, $u \in X^*$, and $|u| = n - 1$. Let $q \in Q$. Then

$$\begin{aligned}(\delta(xy))(q) &= (\delta(xua))(q) \\ &= (\delta((xu)a))(q) \\ &= \vee\{(\delta(xu))(r) \wedge \mu^*(r, a, q)|r \in Q\} \\ &= \vee\{(\vee\{(\delta x)(p) \wedge \mu^*(p, u, r)|p \in Q\}) \wedge \mu^*(r, a, q)|r \in Q\} \\ &= \vee\{(\delta x)(p) \wedge (\vee\{\mu^*(p, u, r) \wedge \mu^*(r, a, q)|r \in Q\})|p \in Q\} \\ &= \vee\{(\delta x)(p) \wedge \mu^*(p, ua, q)|p \in Q\} \\ &= ((\delta x)y)(q).\end{aligned}$$

Hence $\delta(xy) = (\delta x)y$. The result now follows by induction. ∎

Theorem 6.10.7 *Let $M = (Q, X, \mu)$ be a ffsm and let δ be a fuzzy subset of Q. Then δ is a subsystem of M if and only if $\delta x \subseteq \delta \ \forall x \in X^*$.*

Proof. Let δ be a subsystem of M. Let $x \in X^*$ and $q \in Q$. Then $(\delta x)(q) = \vee\{\delta(p) \wedge \mu^*(p, x, q)|p \in Q\} \leq \delta(q)$. Hence $\delta x \subseteq \delta$. Conversely, suppose $\delta x \subseteq \delta \ \forall x \in X^*$. Let $q \in Q$ and $x \in X^*$. Now

$$\delta(q) \geq (\delta x)(q) = \vee\{\delta(p) \wedge \mu^*(p, x, q) \mid p \in Q\} \geq \delta(p) \wedge \mu^*(p, x, q)$$

$\forall p \in Q$. Hence δ is a subsystem of M. ∎

Definition 6.10.8 Let $M = (Q, X, \mu)$ be a ffsm. Let $t \in (0, 1]$ and $q \in Q$. Define the fuzzy subset $q_t X$ of Q by

$$(q_t X)(p) = \vee \{t \wedge \mu(q, a, p) | a \in X\}$$

$\forall p \in Q$.

Definition 6.10.9 Let $M = (Q, X, \mu)$ be a ffsm. Let $t \in (0, 1]$ and $q \in Q$. Define the fuzzy subset $q_t X^*$ of Q by

$$(q_t X^*)(p) = \vee \{t \wedge \mu^*(q, y, p) | y \in X^*\}$$

$\forall p \in Q$.

Theorem 6.10.10 Let $M = (Q, X, \mu)$ be a ffsm. Let $t \in (0, 1], q \in Q$. Then the following assertions hold.
  (1) $q_t X^*$ is a subsystem of M.
  (2) $Supp(q_t X^*) = S(q)$.

Proof. (1) Let $r, s \in Q$ and $x \in X^*$. Now

$$
\begin{aligned}
((q_t X^*) x)(r) &= \vee \{(q_t X^*)(p) \wedge \mu^*(p, x, r) | p \in Q\} \\
&= \vee \{(\vee \{\mu^*(q, y, p) \wedge t | y \in X^*\}) \wedge \mu^*(p, x, r) | p \in Q\} \\
&= \vee \{\mu^*(q, y, p) \wedge \mu^*(p, x, r) \wedge t | y \in X^*, p \in Q\} \\
&= \vee \{\mu^*(q, yx, r) \wedge t | y \in X^*\} \\
&\leq \{\mu^*(q, u, r) \wedge t | u \in X^*\} \\
&= (q_t X^*)(r).
\end{aligned}
$$

Hence $(q_t X^*) x \subseteq q_t X^*$. Thus $q_t X^*$ is a subsystem by Theorem 6.10.7.
  (2) $p \in S(q) \iff \exists x \in X^*$ such that $\mu^*(q, x, p) > 0 \iff \vee \{t \wedge \mu^*(q, x, p) | x \in X^*\} > 0 \iff (q_t X^*)(p) > 0 \iff p \in Supp(q_t X^*)$. ∎

Theorem 6.10.11 Let $M = (Q, X, \mu)$ be a ffsm and let δ be a fuzzy subset of Q. The following assertions are equivalent.
  (1) δ is a subsystem of M.
  (2) $q_t X^* \subseteq \delta, \forall q_t \subseteq \delta, q \in Q, t \in (0, 1]$.
  (3) $q_t X \subseteq \delta, \forall q_t \subseteq \delta, q \in Q, t \in (0, 1]$.

Proof. (1)$\Longrightarrow$(2): Let $q_t \subseteq \delta, q \in Q, t \in (0, 1]$. Let $p \in Q$ and $y \in X^*$. Then $\mu^*(q, y, p) \wedge t = \mu^*(q, y, p) \wedge q_t(q) \leq \mu^*(q, y, p) \wedge \delta(q) \leq \delta(p)$ since δ is a subsystem. Hence $q_t X^* \subseteq \delta$.
  (2)$\Longrightarrow$(3): Obvious.
  (3)$\Longrightarrow$(1): Let $p, q \in Q$ and $a \in X$. If $\delta(q) = 0$ or $\mu(q, a, p) = 0$ then $\delta(p) \geq 0 = \delta(q) \wedge \mu(q, a, p)$. Suppose $\delta(q) \neq 0$ and $\mu(q, a, p) \neq 0$. Let $\delta(q) = t$. Then $q_t \subseteq \delta$. Thus by the hypothesis, $q_t X \subseteq \delta$. Thus $\delta(p) \geq (q_t X)(p) = \vee \{t \wedge \mu(q, y, p) | y \in X\} \geq t \wedge \mu(q, a, p) = \delta(q) \wedge \mu(q, a, p)$. Hence δ is a subsystem of M. ∎

Definition 6.10.12 *Let $M_1 = (Q_1, X_1, \mu_1)$ and $M_2 = (Q_2, X_2, \mu_2)$ be two ffsms. Let $(f, g) : M_1 \to M_2$ be a homomorphism. Let δ be a fuzzy subset of Q_1. Define the fuzzy subset $f(\delta)$ of Q_2 by*

$$f(\delta)(q') = \begin{cases} \vee\{\delta(q) \mid q \in Q_1, \ f(q) = q'\} & \text{if } f^{-1}(q') \neq \emptyset \\ 0 & \text{if } f^{-1}(q') = \emptyset \end{cases}$$

$\forall q' \in Q_2$.

Theorem 6.10.13 *Let $M_1 = (Q_1, X, \mu_1)$ and $M_2 = (Q_2, X, \mu_2)$ be two ffsms. Let $f : M_1 \longrightarrow M_2$ be an onto strong homomorphism. If δ is a subsystem of Q_1, then $f(\delta)$ is a subsystem of Q_2.*

Proof. Let $p', q' \in Q_2$ and $a \in X$. Then

$$\begin{aligned} f(\delta)(p') \wedge \mu_2(p', a, q') &= (\vee\{\delta(p) | p \in Q_1, \ f(p) = p'\}) \wedge \mu_2(p', a, q') \\ &= \vee\{\delta(p) \wedge \mu_2(p', a, q') | p \in Q_1, \ f(p) = p'\}. \end{aligned}$$

Let $p, q \in Q_1$ be such that $f(p) = p'$ and $f(q) = q'$. Then

$$\begin{aligned} \delta(p) \wedge \mu_2(p', a, q') &= \delta(p) \wedge \mu_2(f(p), a, f(q)) \\ &= \delta(p) \wedge (\vee\{\mu_1(p, a, r) \mid r \in Q_1, \ f(r) = f(q) = q'\}) \\ &= \vee\{\delta(p) \wedge \mu_1(p, a, r) \mid r \in Q_1, \ f(r) = f(q) = q'\} \\ &\leq \vee\{\delta(r) \mid r \in Q_1, \ f(r) = q'\} \\ &= f(\delta)(q'). \end{aligned}$$

Hence

$$\begin{aligned} f(\delta)(p') \wedge \mu_2(p', a, q') &\leq \vee\{f(\delta)(q') \mid p \in Q_1, \ f(p) = p'\} \\ &= f(\delta)(q'). \end{aligned}$$

Thus $f(\delta)$ is a subsystem of M_2. ∎

The following example shows that the above result need not be true if f is not onto.

Example 6.10.14 *Let $Q = \{p, q\} = Q_1 = Q_2$, $X = \{a\}$, and $\mu = \mu_1 = \mu_2$, where $\mu(r, a, s) = 1 \ \forall r, s \in Q$. Then $M = (Q, X, \mu)$ is a fuzzy finite state machine. Let $f : Q \longrightarrow Q$ be a mapping such that $f(p) = f(q) = p$. Then f is not onto. Clearly, f is a strong homomorphism. Let δ be a fuzzy subset of Q such that $\delta(p) = \delta(q) = \frac{1}{2}$. Then $\delta(r) = \frac{1}{2} = \delta(s) \wedge \mu(s, a, r) \ \forall r, s \in Q$. Thus δ is a subsystem of M. Now $f(\delta)(p) \wedge \mu(p, a, q) = \frac{1}{2} > 0 = f(\delta)(q)$. Thus $f(\delta)$ is not a subsystem of Q.*

Definition 6.10.15 *Let $M = (Q, X, \mu)$ be a ffsm and let δ be a subsystem of M. Then δ is called **cyclic** if $\exists q_t \subseteq \delta, q \in Q, t \in (0, 1]$ such that $\delta = q_t X^*$. In this case we call q_t a **generator** of δ.*

Theorem 6.10.16 *Let $M = (Q, X, \mu)$ be a ffsm and let δ be a subsystem of M. Suppose $\exists q \in Q, t \in (0,1]$ such that $\delta = q_t X^*$. Then*

(1) $\delta(q) = t$,

(2) $\forall\, p \in Q, \; \delta(q) \geq \delta(p)$,

(3) $\forall$ subsystems δ' of M such that $\delta' \subseteq \delta$, if $\delta'(q) \geq \delta'(p) \; \forall p \in Q$, then $\delta' = q_{\delta'(q)} X^*$.

Proof. (1) Now

$$\delta(q) = (q_t X^*)(q) = (\vee\{\mu^*(q, x, q)|x \in X^*\}) \wedge t = 1 \wedge t = t.$$

(2) Let $p \in Q$. Then

$$\delta(p) = (q_{\delta(q)} X^*)(p) = (\vee\{\mu^*(q, x, p)|x \in X^*\}) \wedge \delta(q) \leq \delta(q).$$

(3) Let $p \in Q$. Then

$$
\begin{aligned}
\delta'(p) &= \delta'(p) \wedge \delta(p) \\
&= \vee\{\delta'(p) \wedge \delta(q) \wedge \mu^*(q, x, p)|x \in X^*\} \\
&= \vee\{\delta'(p) \wedge \mu^*(q, x, p)|x \in X^*\} \\
&\leq \vee\{\delta'(q) \wedge \mu^*(q, x, p)|x \in X^*\} \\
&= (q_{\delta'(q)} X^*)(p).
\end{aligned}
$$

Hence $\delta' \subseteq q_{\delta'(q)} X^*$. Thus $\delta' = q_{\delta'(q)} X^*$. ∎

Definition 6.10.17 *Let $M = (Q, X, \mu)$ be a ffsm and let δ be a subsystem of M. Then δ is called **super cyclic** if and only if $\delta = q_{\delta(q)} X^* \; \forall q \in Q$.*

By Theorem 6.10.16, if δ is super cyclic, then δ is constant.

Example 6.10.18 *Let $Q = \{p, q\}$, $X = \{a\}$, $\delta(q) = \delta(p) = 1$ and $\mu(q, a, q) = \mu(p, a, p) = 1$ and $\mu(q, a, p) = \mu(p, a, q) = \frac{1}{2}$. Then δ is a subsystem of $M = (Q, X, \mu)$ and δ is constant. Now*

$$(q_1 X^*)(p) = \vee\{1 \wedge \mu^*(q, x, p)|x \in X^*\} = \frac{1}{2} < 1 = \delta(p).$$

Similarly $(p_1 X^*)(q) < \delta(q)$. Hence δ is not cyclic.

Theorem 6.10.19 *Let $M = (Q, X, \mu)$ be a ffsm and let δ be a subsystem of M. Suppose $\text{Supp}(\delta) = Q$. If δ is super cyclic, then M is strongly connected.*

Proof. Let $p, q \in Q$. Then

$$(q_{\delta(q)} X^*)(p) = \vee\{\delta(q) \wedge \mu^*(q, x, p)|x \in X^*\} > 0$$

since $\delta = q_{\delta(q)} X^*$ and $\text{Supp}(\delta) = Q$. Hence $\mu^*(q, x, p) > 0$ for some $x \in X^*$. Thus $p \in S(q)$. Hence M is strongly connected. ∎

Theorem 6.10.20 *Let $M = (Q, X, \mu)$ be a ffsm and let δ be a subsystem of M. Then δ is super cyclic if and only if $\forall p, q \in Q$, $\exists x \in X^*$ such that $\mu^*(p, x, q) \geq \delta(p)$.*

Proof. Suppose that δ is super cyclic. Then δ is constant by Theorem 6.10.16. Suppose $\exists p, q \in Q$ such that $\forall x \in X^*$, $\mu^*(p, x, q) < t$, where $t = \delta(r)$ $\forall r \in Q$. Then

$$(p_{\delta(p)} X^*)(q) = \vee \{\delta(p) \wedge \mu^*(p, x, q) \mid x \in X^*\} < t = \delta(q).$$

Hence $p_{\delta(p)} X^* \neq \delta$. Thus δ is not super cyclic, a contradiction. Conversely, suppose that $\forall p, q \in Q$, $\exists x \in X^*$ such that $\mu^*(p, x, q) \geq \delta(p)$. Then $\forall p, q \in Q$, $\exists x \in X^*$ such that $\delta(q) \geq \delta(p) \wedge \mu^*(p, x, q) = \delta(p)$. Similarly $\delta(p) \geq \delta(q)$. Hence δ is constant. Now

$$(p_{\delta(p)} X^*)(q) = \vee \{\delta(p) \wedge \mu^*(p, x, q) \mid x \in X^*\} = \delta(p) = \delta(q).$$

Thus $p_{\delta(p)} X^* = \delta$. Hence δ is super cyclic. ∎

6.11 Strong Subsystems

Definition 6.11.1 *Let $M = (Q, X, \mu)$ be a ffsm. Let δ be a fuzzy subset of Q. Then (Q, δ, X, μ) is called a **strong subsystem** of M if and only if $\forall p, q \in Q$, if $\exists a \in X$ such that $\mu(p, a, q) > 0$, then $\delta(q) \geq \delta(p)$.*

If (Q, δ, X, μ) is a strong subsystem of M, then we simply write δ for (Q, δ, X, μ).

Theorem 6.11.2 *Let $M = (Q, X, \mu)$ be a ffsm and let δ be a fuzzy subset of Q. Then δ is a strong subsystem of M if and only if $\forall p, q \in Q$, if $\exists x \in X^*$ such that $\mu^*(p, x, q) > 0$, then $\delta(q) \geq \delta(p)$.*

Proof. Suppose δ is a subsystem. We prove the result by induction on $|x| = n$. If $n = 0$, then $x = \Lambda$. Now if $p = q$, then $\mu^*(q, \Lambda, q) = 1$ and $\delta(q) = \delta(q)$. If $q \neq p$, then $\mu^*(q, \Lambda, p) = 0$. Thus the result is true if $n = 0$. Suppose the result is true $\forall y \in X^*$ such that $|y| = n - 1$, $n > 0$. Let $x = ya$, $|y| = n - 1$, $y \in X^*$, $a \in X$. Suppose that $\mu^*(p, x, q) > 0$. Then

$$\begin{aligned}
\vee \{\mu^*(p, y, r) \wedge \mu(r, a, q) \mid r \in Q\} &= \mu^*(p, ya, q) \\
&= \mu^*(p, x, q) \\
&> 0.
\end{aligned}$$

Thus $\exists r \in Q$ such that $\mu^*(p, y, r) > 0$ and $\mu(r, a, q) > 0$. Hence $\delta(q) \geq \delta(r)$ and $\delta(r) \geq \delta(p)$. Thus $\delta(q) \geq \delta(p)$. The converse is trivial. ∎

Theorem 6.11.3 *Let $M = (Q, X, \mu)$ be a ffsm and let δ be a fuzzy subset of Q. If δ is a strong subsystem of M, then δ is a subsystem of M.* ∎

The following example shows that in general, the converse of the above theorem is not true.

Example 6.11.4 *Let δ, Q, X, μ be defined as in Example 6.10.4. Then $\mu(q, a, p) = \frac{1}{2} > 0$, but $\delta(q) = \frac{3}{4} > \frac{1}{2} = \delta(p)$. Thus δ is subsystem of M, which is not a strong subsystem.*

Theorem 6.11.5 *Let $M = (Q, X, \mu)$ be a ffsm. Let δ_1 and δ_2 be strong subsystems of M. Then the following assertions hold.*
(1) $\delta_1 \cap \delta_2$ is a strong subsystem of M.
(2) $\delta_1 \cup \delta_2$ is a strong subsystem of M.
(3) Let δ be a strong subsystem of M. Then $N = (Supp(\delta), X, \nu)$ is a submachine of M, where $\nu = \mu|_{Supp(\delta) \times X \times Supp(\delta)}$.
(4) Let δ be a strong subsystem of M. Let $N_t = (\delta_t, X, \nu_{(t)})$ where $\nu_{(t)} = \mu|_{\delta_t \times X \times \delta_t}$, $t \in [0, 1]$. Then $\forall t \in [0, 1]$, N_t is a submachine of M.
(5) Let δ be fuzzy subset of Q. Let $N_t = (\delta_t, X, \nu_{(t)})$ where $\nu_{(t)} = \mu|_{\delta_t \times X \times \delta_t}$, $t \in [0, 1]$. If $\forall t \in [0, 1]$, N_t is a submachine of M, then δ is a strong subsystem of M.

Proof. The proofs of (1) and (2) are straightforward.
(3) Let $p \in S(Supp(\delta))$. Then $p \in S(q)$ for some $q \in Supp(\delta)$. Then $\delta(q) > 0$. Since $p \in S(q)$, $\exists x \in X^*$ such that $\mu^*(q, x, p) > 0$. Hence since δ is a strong subsystem, $\delta(p) \geq \delta(q) > 0$. Thus $p \in Supp(\delta)$. Hence $S(Supp(\delta)) \subseteq Supp(\delta)$. Thus N is a submachine of M.
(4) Let $q \in S(\delta_t)$. Then $q \in S(p)$ for some $p \in \delta_t$. Thus $\delta(p) \geq t$. Now $\exists x \in X^*$ such that $\mu^*(p, x, q) > 0$. Then $\delta(q) \geq \delta(p) \geq t$. Thus $q \in \delta_t$. Hence N_t is a submachine of M.
(5) Let $q, p \in Q$, $x \in X^*$ be such that $\mu^*(p, x, q) > 0$. Suppose $\delta(p) > 0$. Let $\delta(p) = t$. Then $p \in \delta_t$. Since N_t is a submachine of M, $S(\delta_t) = \delta_t$. Thus $q \in S(p) \subseteq S(\delta_t) = \delta_t$. Hence $\delta(q) \geq t$. Thus δ is a strong subsystem. ∎

Theorem 6.11.6 *Let $M = (Q, X, \mu)$ be a ffsm. Let $N = (T, X, \nu)$ be a submachine of M. Then χ_T is a strong subsystem of M.*

Proof. Let $p, q \in Q$, $a \in X$, and $\mu(p, a, q) > 0$. Then $q \in S(p)$. If $p \in T$, then $q \in S(p) \subseteq S(T) \subseteq T$. Hence $\chi_T(q) = 1 = \chi_T(p)$. If $p \notin T$, then $\chi_T(p) = 0 \leq \chi_T(q)$. Thus χ_T is a strong subsystem of M. ∎

Theorem 6.11.7 *Let $M = (Q, X, \mu)$ be a ffsm. Then M is strongly connected if and only if every strong subsystem of M is constant.*

Proof. Suppose M is strongly connected. Let δ be a strong subsystem of M. Let $p, q \in Q$. Then $p \in S(q)$ and $q \in S(p)$. Hence $\delta(p) \geq \delta(q)$ and $\delta(q) \geq \delta(p)$. Thus $\delta(p) = \delta(q)$. Hence δ is constant. Conversely, suppose that every strong subsystem is constant. Let $p, q \in Q$. Suppose $q \notin S(p)$. Then $\emptyset \neq S(p) \neq Q$. Let δ be a fuzzy subset of Q such that $\delta(r) = 1$ if $r \in S(p)$ and $\delta(r) = 0$ if $r \notin S(p)$. Let $r, s \in Q$ be such that $\mu^*(r, x, s) > 0$

for some $x \in X^*$. If $\delta(r) = 0$, then $\delta(s) \geq 0 = \delta(r)$. Let $\delta(r) = 1$. Then $r \in S(p)$. Hence $s \in S(r) \subseteq S(p)$. Thus $\delta(s) = 1 = \delta(r)$. Hence δ is a strong subsystem. Now $\delta(p) = 1$ and $\delta(q) = 0$. Thus δ is not constant, which is a contradiction. Hence $q \in S(p)$. Thus M is strongly connected. ∎

Definition 6.11.8 Let $M = (Q, X, \mu)$ be a ffsm and let δ be a strong subsystem of M. Suppose Q has at least two elements. Then δ is called **simple** if

(1) δ is not constant, and

(2) for all strong subsystems σ of M, $\chi_\emptyset \neq \sigma \subseteq \delta \implies Supp(\sigma) = Supp(\delta)$.

Theorem 6.11.9 Let $M = (Q, X, \mu)$ be a ffsm and let δ be a strong subsystem of M. Suppose $|Q| \geq 2$. If δ is simple, then $Im(\delta) = \{0, t\}$, where $0 < t \leq 1$.

Proof. Since δ is not constant, $|Im(\delta)| \geq 2$. Suppose $|Im(\delta)| > 2$. Then $\exists t_1, t_2, t_3 \in [0, 1]$, $0 \leq t_1 < t_2 < t_3 \leq 1$, and $\exists r_1, r_2, r_3 \in Q$ such that $\delta(r_i) = t_i$, $i = 1, 2, 3$. Let $m \in [0, 1]$ be such that $t_2 < m \leq t_3$. Let σ be a fuzzy subset of Q such that $\forall r \in Q$,

$$\sigma(r) = \begin{cases} m \text{ if } \delta(r) \geq m \\ 0 \text{ if } \delta(r) < m. \end{cases}$$

Let $p, q \in Q$ be such that $\mu^*(p, x, q) > 0$ for some $x \in X^*$. Then $\delta(q) \geq \delta(p)$ since δ is a strong subsystem. If $\delta(p) \geq m$, then $\delta(q) \geq m$. Hence $\sigma(q) = m = \sigma(p)$. Suppose $\delta(p) < m$. Then $\sigma(q) \geq 0 = \sigma(p)$. Thus σ is a strong subsystem. Clearly, $\sigma \subseteq \delta$. Now $\delta(r_2) = t_2 \neq 0$ and $\sigma(r_2) = 0$. Thus $\chi_\emptyset \neq \sigma \subseteq \delta$ and $Supp(\sigma) \neq Supp(\delta)$, which is a contradiction since δ is simple. Hence $|Im(\delta)| = 2$. Let $Im(\delta) = \{t_2, t_3\}$, $0 \leq t_2 < t_3 \leq 1$. Suppose $t_2 \neq 0$. Let σ be as defined previously. Then $\chi_\emptyset \neq \sigma \subseteq \delta$ and $Supp(\sigma) \neq Supp(\delta)$, which is a contradiction. Hence $t_2 = 0$. ∎

Theorem 6.11.10 Let $M = (Q, X, \mu)$ be a ffsm and let δ be a fuzzy subset of Q. Suppose $Im(\delta) = \{0, t\}$, $0 < t \leq 1$. Then δ is a simple strong subsystem if and only if $N = (Supp(\delta), X, \nu)$, where $\nu = \mu|_{Supp(\delta) \times X \times Supp(\delta)}$, is a strongly connected submachine of M.

Proof. Suppose N is strongly connected. Let $p, q \in Q$ be such that $\mu^*(p, x, q) > 0$. If $\delta(p) = 0$, then $\delta(q) \geq 0 = \delta(p)$. Suppose $\delta(p) > 0$. Then $p \in Supp(\delta)$. Now $q \in S(p) \subseteq S(Supp(\delta)) = Supp(\delta)$, since N is a submachine. Hence $\delta(q) = t = \delta(p)$. Thus δ is a strong subsystem. Let σ be a strong subsystem of M such that $\chi_\emptyset \neq \sigma \subseteq \delta$. Then by Theorem 6.11.5(3), $K = (Supp(\sigma), X, \mu|_{Supp(\sigma) \times X \times Supp(\sigma)})$ is a submachine. Now $K \subseteq N$. Since N is strongly connected, N has no proper submachine. Hence $K = N$. Thus $Supp(\sigma) = Supp(\delta)$. Hence δ is simple. Conversely, suppose δ is simple. Let $K = (T, X, \eta) \subseteq N$, $\emptyset \neq T \subseteq \mathrm{Supp}(\delta) \subseteq Q$, be a submachine. Let σ be a fuzzy subset of Q such that $\forall r \in Q$,

$$\sigma(r) = \left\{ \begin{array}{l} \delta(r) \text{ if } r \in T \\ 0 \text{ if } r \notin T. \end{array} \right.$$

Then $\sigma \subseteq \delta$. Let $p, q \in Q$ be such that $\mu^*(p, x, q) > 0$. Then $q \in S(p)$. If $\sigma(p) = 0$, then $\sigma(q) \geq 0 = \sigma(p)$. Let $\sigma(p) > 0$. Then $p \in T$. Thus $q \in S(p) \subseteq S(T) = T$. Hence $\sigma(q) = \delta(q) \geq \delta(p) \geq \sigma(p)$. Thus σ is a strong subsystem. By the definition of σ, $\sigma \neq \chi_\emptyset$. Since δ is simple, $\mathrm{Supp}(\sigma) = \mathrm{Supp}(\delta)$. Hence $\mathrm{Supp}(\sigma) \subseteq T \subseteq \mathrm{Supp}(\delta) = \mathrm{Supp}(\sigma)$. Thus $T = \mathrm{Supp}(\delta)$. Hence $K = N$. Thus N has no proper submachine. Hence N is strongly connected. ∎

Theorem 6.11.11 *Let* $M_1 = (Q_1, X, \mu_1)$ *and* $M_2 = (Q_2, X, \mu_2)$ *be two ffsms. Let* $f : M_1 \longrightarrow M_2$ *be an onto strong homomorphism. If* δ *is a strong subsystem of* Q_1, *then* $f(\delta)$ *is a strong subsystem of* Q_2.

Proof. Let $p, q \in Q_1$ and $a \in X$ be such that $\mu_2(f(p), a, f(q)) > 0$. Now

$$f(\delta)(f(q)) = \vee\{\delta(r) \mid r \in Q_1, f(r) = f(q)\}$$

and

$$f(\delta)(f(p)) = \vee\{\delta(s) \mid s \in Q_1, f(s) = f(p)\}.$$

Let $s \in Q_1$ be such that $\delta(s) > 0$ and $f(s) = f(p)$. Now

$$\mu_2(f(s), a, f(q)) = \mu_2(f(p), a, f(q)) > 0.$$

Hence

$$\vee\{\mu_1(s, a, r) | r \in Q_1, f(r) = f(q)\} > 0.$$

Thus $\exists r \in Q_1$ such that $\mu_1(s, a, r) > 0$ and $f(r) = f(q)$. Since δ is a strong subsystem, $\delta(r) \geq \delta(s) > 0$. Hence $f(\delta)(f(q)) \geq f(\delta)(f(p))$. Thus $f(\delta)$ is a strong subsystem. ∎

The following example shows that the above result need not be true if f is not onto.

Example 6.11.12 *Let* Q, X, μ, δ, *and* f *be defined as in Example 6.10.14. Then* δ *is a strong subsystem and* f *is a strong homomorphism such that* f *is not onto. Now* $\mu(p, a, q) = 1 > 0$, *but* $f(\delta)(p) = \frac{1}{2} > 0 = f(\delta)(q)$. *Hence* $f(\delta)$ *is not a strong subsystem.*

6.12   Cartesian Composition of Fuzzy Finite State Machines

Here and the next section, we study a new product of two fuzzy finite state machines M_1 and M_2, written $M_1 \cdot M_2$, and called the Cartesian composition of M_1 and M_2, as in [48]. We show that M_1, M_2, and $M_1 \cdot M_2$ share many similar structural properties, e.g., those of singly generated, retrievability, connectedness, strongly connectedness, commutativity, perfectness, and state independence. This is important since a fuzzy finite state machine, which is a Cartesian composition of submachines can thus be studied in terms of smaller machines.

Definition 6.12.1 *Let $M = (Q, X, \mu)$ be a ffsm. Then M is said to be **connected** if and only if M has no separated proper submachine.*

Theorem 6.12.2 *Let $M = (Q, X, \mu)$ be a ffsm. Then M is connected if and only if $\forall$ proper submachines $N = (T, X, \nu) \; \exists \, s \in Q \backslash T$ and $t \in T$ such that $S(s) \cap S(t) \neq \emptyset$.*

   Proof. Suppose M is connected. Let $N = (T, X, \nu)$ be a proper submachine. Then $S(Q \backslash T) \cap T \neq \emptyset$ since M has no separated proper submachine. Thus $\exists \, r \in S(Q \backslash T) \cap T$. Now $T = S(T)$. Hence $r \in S(s)$ for some $s \in Q \backslash T$ and $r \in S(t)$ for some $t \in T$. Thus $S(s) \cap S(t) \neq \emptyset$. Conversely, let $N = (T, X, \nu)$ be a proper submachine. Then $\exists \, s \in Q \backslash T$ and $t \in T$ such that $S(s) \cap S(t) \neq \emptyset$. Hence $\emptyset \neq S(s) \cap S(t) \subseteq S(Q \backslash T) \cap S(T) = S(Q \backslash T) \cap T$. Thus N is not separated. Hence M has no proper separated submachine. Thus M is connected. ∎

Definition 6.12.3 *Let $M = (Q, X, \mu)$ be a ffsm. Let $p, q \in Q$. Then q and p are called **connected** if either $q = p$ or $\exists \, q_0, q_1, \ldots, q_k \in Q$, $q = q_0$, $p = q_k$ and $\exists \, a_1, a_2, \ldots, a_k \in X$ such that $\forall \, i = 1, 2, \ldots, k$ either $\mu(q_{i-1}, a_i, q_i) > 0$ or $\mu(q_i, a_i, q_{i-1}) > 0$.*

   Clearly, if q and p are connected and p and r are connected, then q and r are connected.

Definition 6.12.4 *Let $M = (Q, X, \mu)$ be a ffsm. For all $q \in Q$, let*

$$C(q) = \{p \in Q \mid p \text{ and } q \text{ are connected}\}$$

$\forall q \in Q$.

   Clearly, $\forall q, p \in Q$ if $p \in C(q)$, then $C(p) = C(q)$.

   Let $M = (Q, X, \mu)$ be a ffsm. For all $T \subseteq Q$, let

$$C(T) = \cup_{p \in T} C(p).$$

Lemma 6.12.5 *Let* $M = (Q, X, \mu)$ *be a ffsm. Let* $U, V \subseteq Q$. *Then the following properties hold.*

(1) *If* $U \subseteq V$, *then* $C(U) \subseteq C(V)$.

(2) $U \subseteq C(U)$.

(3) $C(C(U)) = C(U)$.

(4) $C(U \cup V) = C(U) \cup C(V)$.

(5) $C(U \cap V) \subseteq C(U) \cap C(V)$.

(6) *Let* $q, p \in Q$. *If* $q \in C(U \cup \{p\})$ *and* $q \notin C(U)$, *then* $p \in C(U \cup \{q\})$.

(7) $S(Q) \subseteq C(Q)$.

(8) $S(C(U)) = C(U)$.

Proof. The proofs of (1), (2), (4), (5), and (7) are straightforward. Consider (3). Now $C(U) \subseteq C(C(U))$ by (2). Let $q \in C(C(U))$. Then $\exists\, p \in C(U)$ such that $q \in C(p)$. Hence $q \in C(r)$ for some $r \in U$. Thus $q \in C(r) \subseteq C(U)$. Consider (6). Suppose that $q \in C(U \cup \{p\})$ and $q \notin C(U)$. Since $C(U \cup \{p\}) = C(U) \cup C(p)$, $q \in C(p)$. Thus $p \in C(q) \subseteq C(U \cup \{q\})$. Consider (8). Let $p \in S(C(U))$. Then $p \in S(q)$ for some $q \in C(U)$. Now $q \in C(r)$ for some $r \in U$. Hence $p \in C(r) \subseteq C(U)$. Thus $S(C(U)) \subseteq C(U)$ and so $S(C(U)) = C(U)$. ■

Since Q is a finite set, it is clear that $\forall\, q \in Q$ and $U \subseteq Q$, if $q \in C(U)$, then $q \in C(U')$ for some finite subset U' of U. This fact together with properties (1), (2), (3), and (6) give C the basic spanning properties in [264, p. 50] that are used for various algebraic structures to obtain the existence of bases and the uniqueness of their cardinalities.

Definition 6.12.6 *Let* $M = (Q, X, \mu)$ *be a ffsm and let* $T \subseteq Q$. *Then* T *is called a* **connected component** *if* $\forall\, s, t \in T$, *s and t are connected.* T *is called a* **maximal connected component** *when* $\forall\, p \in Q$, *if p is connected to t for some* $t \in T$, *then* $p \in T$.

Theorem 6.12.7 *Let* $M = (Q, X, \mu)$ *be a ffsm and let* $q \in Q$. *Then* $C(q)$ *is a maximal connected component of* Q.

Proof. The proof is obvious. ■

Theorem 6.12.8 *Let* $M = (Q, X, \mu)$ *be a ffsm and let* $q \in Q$. *Let* $N = (C(q), X, \mu|_{C(q) \times X \times C(q)})$. *Then* N *is a submachine of* M. ■

Theorem 6.12.9 *Let* $M = (Q, X, \mu)$ *be a ffsm. Then* M *is connected if and only if* $\forall q \in Q$, $C(q) = Q$.

Proof. Suppose M is connected and let $q \in Q$. Suppose $\exists p \in Q$ such that $p \notin C(q)$. Then $N = (C(q), X, \mu|_{C(q) \times X \times C(q)})$ is a proper submachine of M. Hence by Theorem 6.12.2, $\exists s \in Q \backslash C(q)$ and $t \in C(q)$ such that $S(s) \cap S(t) \neq \emptyset$. Let $r \in S(s) \cap S(t)$. Then s and r are connected and r and t are connected. Hence s and t are connected. Thus $s \in C(t) = C(q)$, which is a contradiction. Hence $C(q) = Q$. Conversely, suppose that $\forall q \in Q$,

$C(q) = Q$. Let $N = (T, X, \nu)$ be a proper submachine of M. Suppose that N is separated. Then $S(Q \backslash T) \cap T = \emptyset$ and $S(Q \backslash T) = Q \backslash T$. Let $q \in Q \backslash T$ and $t \in T$. Then $C(t) = Q = C(q)$. Hence q and t are connected. Thus $\exists \, q_0, q_1, \ldots, q_k \in Q$, $q = q_0$, $t = q_k$, and $\exists \, a_1, a_2, \ldots, a_k \in X$ such that $\forall \, i = 1, 2, \ldots, k$ either $\mu(q_{i-1}, a_i, q_i) > 0$ or $\mu(q_i, a_i, q_{i-1}) > 0$. Now $\exists \, i$ such that $q_{i-1} \in Q \backslash T$ and $q_i \in T$. Hence either $q_i \in S(q_{i-1}) \subseteq S(Q \backslash T)$ or $q_{i-1} \in S(q_i) \subseteq T$, which is a contradiction. Thus N is not separated. Hence M is connected. ∎

Corollary 6.12.10 *Let* $M = (Q, X, \mu)$ *be a ffsm. Then* M *is connected if and only if* $\forall \, p, q \in Q$, p *and* q *are connected.* ∎

6.13 Cartesian Composition

Definition 6.13.1 *Let* $M_i = (Q_i, X_i, \mu_i)$ *be a ffsm,* $i = 1, 2$ *and let* $X_1 \cap X_2 = \emptyset$. *Let*

$$M_1 \cdot M_2 = (Q_1 \times Q_2, X_1 \cup X_2, \mu_1 \cdot \mu_2),$$

where

$$(\mu_1 \cdot \mu_2)((p_1, p_2), a, (q_1, q_2)) = \begin{cases} \mu_1(p_1, a, q_1) \text{ if } a \in X_1 \text{ and } p_2 = q_2 \\ \mu_2(p_2, a, q_2) \text{ if } a \in X_2 \text{ and } p_1 = q_1 \\ 0 \text{ otherwise,} \end{cases}$$

$\forall (p_1, p_2), (q_1, q_2) \in Q_1 \times Q_2$, $a \in X_1 \cup X_2$. *Then* $M_1 \cdot M_2$ *is a ffsm, called the* **Cartesian composition** *of* M_1 *and* M_2.

Theorem 6.13.2 *Let* $M_i = (Q_i, X_i, \mu_i)$ *be a ffsm,* $i = 1, 2$ *and let* $X_1 \cap X_2 = \emptyset$. *Let* $M_1 \cdot M_2 = (Q_1 \times Q_2, X_1 \cup X_2, \mu_1.\mu_2)$ *be the Cartesian composition of* M_1 *and* M_2. *Then* $\forall x \in X_1^* \cup X_2^*$, $x \neq \Lambda$,

$$(\mu_1 \cdot \mu_2)^*((p_1, p_2), x, (q_1, q_2)) = \begin{cases} \mu_1^*(p_1, x, q_1) \text{ if } x \in X_1^* \text{ and } p_2 = q_2 \\ \mu_2^*(p_2, x, q_2) \text{ if } x \in X_2^* \text{ and } p_1 = q_1 \\ 0 \text{ otherwise,} \end{cases}$$

$\forall (p_1, p_2), (q_1, q_2) \in Q_1 \times Q_2$.

   Proof. Let $x \in X_1^* \cup X_2^*$, $x \neq \Lambda$ and let $|x| = n$. Suppose that $x \in X_1^*$. Clearly the result is true if $n = 1$. Suppose the result is true $\forall y \in X_1^*$,

$|y| = n - 1$, $n > 1$. Let $x = ay$ where $a \in X_1$ and $y \in X_1^*$. Now

$$
\begin{aligned}
(\mu_1 \cdot \mu_2)^*((p_1, p_2), ay, (q_1, q_2)) &= \vee\{(\mu_1 \cdot \mu_2)((p_1, p_2), a, (r_1, r_2)) \\
&\quad \wedge (\mu_1 \cdot \mu_2)^*((r_1, r_2), y, (q_1, q_2))| \\
&\quad (r_1, r_2) \in Q_1 \times Q_2\} \\
&= \vee\{\mu_1(p_1, a, r_1) \wedge (\mu_1 \cdot \mu_2)^*((r_1, p_2), y, \\
&\quad (q_1, q_2))| \; r_1 \in Q_1\} \\
&= \begin{cases} \vee\{\mu_1(p_1, a, r_1) \wedge \mu_1^*(r_1, y, q_1) \mid \\ \quad r_1 \in Q_1\} \text{ if } p_2 = q_2 \\ 0 \text{ otherwise} \end{cases} \\
&= \begin{cases} \mu_1^*(p_1, ay, q_1) \text{ if } p_2 = q_2 \\ 0 \text{ otherwise.} \end{cases}
\end{aligned}
$$

The result now follows by induction. The proof is similar if $x \in X_2^*$. ∎

Theorem 6.13.3 Let $M_i = (Q_i, X_i, \mu_i)$ be a ffsm, $i = 1, 2$ and let $X_1 \cap X_2 = \emptyset$. Then $\forall x \in X_1^*$, $\forall y \in X_2^*$

$$
\begin{aligned}
(\mu_1 \cdot \mu_2)^*((p_1, p_2), xy, (q_1, q_2)) &= \mu_1^*(p_1, x, q_1) \wedge \mu_2^*(p_2, y, q_2) \\
&= (\mu_1 \cdot \mu_2)^*((p_1, p_2), yx, (q_1, q_2))
\end{aligned}
$$

$\forall (p_1, p_2), (q_1, q_2) \in Q_1 \times Q_2$.

Proof. Let $x \in X_1^*$, $y \in X_2^*$, $(p_1, p_2), (q_1, q_2) \in Q_1 \times Q_2$. If $x = \Lambda = y$, then $xy = \Lambda$. Suppose $(p_1, p_2) = (q_1, q_2)$. Then $p_1 = q_1$ and $p_2 = q_2$. Hence $(\mu_1 \cdot \mu_2)^*((p_1, p_2), xy, (q_1, q_2)) = 1 = 1 \wedge 1 = \mu_1^*(p_1, x, q_1) \wedge \mu_2^*(p_2, y, q_2)$. If $(p_1, p_2) \neq (q_1, q_2)$, then either $p_1 \neq q_1$ or $p_2 \neq q_2$. Thus $\mu_1^*(p_1, x, q_1) \wedge \mu_2^*(p_2, y, q_2) = 0$. Hence $(\mu_1 \cdot \mu_2)^*((p_1, p_2), xy, (q_1, q_2)) = 0 = \mu_1^*(p_1, x, q_1) \wedge \mu_2^*(p_2, y, q_2)$. If $x = \Lambda$ and $y \neq \Lambda$ or $x \neq \Lambda$ and $y = \Lambda$, then the result follows by Theorem 6.13.2. Suppose $x \neq \Lambda$ and $y \neq \Lambda$. Now

$$
\begin{aligned}
(\mu_1 \cdot \mu_2)^*((p_1, p_2), xy, (q_1, q_2)) &= \vee\{(\mu_1 \cdot \mu_2)^*((p_1, p_2), x, (r_1, r_2)) \wedge \\
&\quad (\mu_1 \cdot \mu_2)^*((r_1, r_2), y, (q_1, q_2))| \\
&\quad (r_1, r_2) \in Q_1 \times Q_2\} \\
&= \vee\{\vee\{(\mu_1 \cdot \mu_2)^*((p_1, p_2), x, (r_1, r_2)) \wedge \\
&\quad (\mu_1 \cdot \mu_2)^* ((r_1, r_2), y, (q_1, q_2)) \\
&\quad | \; r_2 \in Q_2\} \mid r_1 \in Q_1\} \\
&= \vee\{(\mu_1 \cdot \mu_2)^*((p_1, p_2), x, (r_1, p_2)) \wedge \\
&\quad (\mu_1 \cdot \mu_2)^*((r_1, p_2), y, (q_1, q_2))| \\
&\quad r_1 \in Q_1\} \\
&= \mu_1^*(p_1, x, q_1) \wedge \mu_2^*(p_2, y, q_2).
\end{aligned}
$$

Similarly $(\mu_1 \cdot \mu_2)^*((p_1, p_2), yx, (q_1, q_2)) = \mu_1^*(p_1, x, q_1) \wedge \mu_2^*(p_2, y, q_2)$. ∎

Theorem 6.13.4 Let $M_i = (Q_i, X_i, \mu_i)$ be a ffsm, $i = 1, 2$ and let $X_1 \cap X_2 = \emptyset$. Then $\forall w \in (X_1 \cup X_2)^*$ $\exists u \in X_1^*$, $v \in X_2^*$ such that

$$
(\mu_1 \cdot \mu_2)^*((p_1, p_2), w, (q_1, q_2)) = (\mu_1 \cdot \mu_2)^*((p_1, p_2), uv, (q_1, q_2))
$$

$\forall (p_1, p_2), (q_1, q_2) \in Q_1 \times Q_2$.

Proof. Let $w \in (X_1 \cup X_2)^*$ and $(p_1, p_2), (q_1, q_2) \in Q_1 \times Q_2$. If $w = \Lambda$, then we can choose $u = \Lambda = v$. In this case the result is trivially true. Suppose $w \neq \Lambda$. If $w \in X_1^*$ or $w \in X_2^*$, then again the result is trivially true. Suppose $w \notin X_1^*$ and $w \notin X_2^*$.

Case 1: If $w = xy$, $x \in X_1^+$, $y \in X_2^+$, then the result follows by Theorem 6.13.3.

Case 2: Suppose $w = x_1 y x_2$, $x_1, x_2 \in X_1^*$ and $y \in X_2^*$, x_i and y are nonempty strings, $i = 1, 2$. Let $u = x_1 x_2 \in X_1^*$ and $v = y$. Now by Theorem 6.13.3,

$$(\mu_1 \cdot \mu_2)^*((r_1, r_2), x_2 y, (q_1, q_2)) = (\mu_1 \cdot \mu_2)^*((r_1, r_2), y x_2, (q_1, q_2))$$

$\forall (r_1, r_2), (q_1, q_2) \in Q_1 \times Q_2$. Thus $(\mu_1 \cdot \mu_2)^*((p_1, p_2), x_1 y x_2, (q_1, q_2)) = \vee\{(\mu_1 \cdot \mu_2)^*((p_1, p_2), x_1, (r_1, r_2)) \wedge (\mu_1 \cdot \mu_2)^*((r_1, r_2), y x_2, (q_1, q_2)) \mid (r_1, r_2) \in Q_1 \times Q_2\} = \vee\{(\mu_1 \cdot \mu_2)^*((p_1, p_2), x_1, (r_1, r_2)) \wedge (\mu_1 \cdot \mu_2)^*((r_1, r_2), x_2 y, (q_1, q_2)) \mid (r_1, r_2) \in Q_1 \times Q_2\} = (\mu_1 \cdot \mu_2)^*((p_1, p_2), x_1 x_2 y, (q_1, q_2))$.

Case 3: Suppose $w = y_1 x y_2$, $y_1, y_2 \in X_2^*$ and $x \in X_1^*$, y_i and x are nonempty strings, $i = 1, 2$. Let $v = y_1 y_2 \in X_2^*$ and $u = x$. The proof of this case is similar to Case 2.

Case 4: Suppose $w = x_1 y_1 x_2 y_2$, $x_1, x_2 \in X_1^*$, $y_1, y_2 \in X_2^*$, x_i and y_i are nonempty strings. Let $u = x_1 x_2 \in X_1^*$ and $v = y_1 y_2 \in X_2^*$. Then $(\mu_1 \cdot \mu_2)^*((p_1, p_2), x_1 y_1 x_2 y_2, (q_1, q_2)) = \vee\{(\mu_1 \cdot \mu_2)^*((p_1, p_2), x_1, (r_1, r_2)) \wedge (\mu_1 \cdot \mu_2)^*((r_1, r_2), y_1 x_2 y_2, (q_1, q_2)) \mid (r_1, r_2) \in Q_1 \times Q_2\} = \vee\{(\mu_1 \cdot \mu_2)^*((p_1, p_2), x_1, (r_1, r_2)) \wedge (\mu_1 \cdot \mu_2)^*((r_1, r_2), x_2 y_1 y_2, (q_1, q_2)) \mid (r_1, r_2) \in Q_1 \times Q_2\}$ (by Case 3) $= (\mu_1 \cdot \mu_2)^*((p_1, p_2), x_1 x_2 y_1 y_2, (q_1, q_2)) = (\mu_1 \cdot \mu_2)^*((p_1, p_2), uv, (q_1, q_2))$.

Case 5: Suppose $w = y_1 x_1 y_2 x_2$, $x_1, x_2 \in X_1^*$, $y_1, y_2 \in X_2^*$. Let $u = x_1 x_2 \in X_1^*$ and $v = y_1 y_2 \in X_2^*$. The proof of this case is similar to Case 4.

Case 6: Let $w \in (X_1 \cup X_2)^*$. Then $w = x_1 y_1 x_2 y_2 \ldots x_n y_n$ or $w = y_1 x_1 y_2 x_2 \ldots y_n x_n$, $x_i \in X_1^*$, $y_i \in X_2^*$, x_i and y_i are nonempty strings, $i = 1, 2, \ldots, n - 2$. To be specific, let $w = x_1 y_1 x_2 y_2 \ldots x_n y_n$. The proof of the second case is similar. If $n = 0, 1,$ or 2, then the result is true by the previous cases. Suppose the result for all $z = x_1 y_1 x_2 y_2 \ldots x_{n-1} y_{n-1} \in (X_1 \cup X_2)^*$, $n \geq 2$. Let $u_1 = x_1 x_2 \ldots x_{n-1}$, $v_1 = y_1 y_2 \ldots y_{n-1}$, $u = u_1 x_n$, and $v = v_1 y_n$. Now $(\mu_1 \cdot \mu_2)^*((p_1, p_2), x_1 y_1 x_2 y_2 \ldots x_n y_n, (q_1, q_2)) = \vee\{(\mu_1 \cdot \mu_2)^*((p_1, p_2), x_1 y_1 x_2 y_2 \ldots x_{n-1} y_{n-1}, (r_1, r_2)) \wedge (\mu_1 \cdot \mu_2)^*((r_1, r_2), x_n y_n, (q_1, q_2)) \mid (r_1, r_2) \in Q_1 \times Q_2\} = \vee\{(\mu_1 \cdot \mu_2)^*((p_1, p_2), u_1 v_1, (r_1, r_2)) \wedge (\mu_1 \cdot \mu_2)^*((r_1, r_2), x_n y_n, (q_1, q_2)) \mid (r_1, r_2) \in Q_1 \times Q_2\} = (\mu_1 \cdot \mu_2)^*((p_1, p_2), u_1 v_1 x_n y_n, (q_1, q_2)) = (\mu_1 \cdot \mu_2)^*((p_1, p_2), uv, (q_1, q_2))$. The result now follows by induction. ∎

In Theorem 6.13.4, u consists of all the elements from w that are in X_1 (in the given order) and v consists of all the elements from w that are in X_2 (in the given order). We write $w^* = uv$ and call w^* the **standard form** of w.

Theorem 6.13.5 Let $M_i = (Q_i, X_i, \mu_i)$ be a ffsm, $i = 1, 2$ and let $X_1 \cap X_2 = \emptyset$. Then the Cartesian composition $M_1 \cdot M_2$ is cyclic if and only if M_1 and M_2 are cyclic.

Proof. Suppose M_1 and M_2 are cyclic, say $Q_1 = S(q_0)$ and $Q_2 = S(p_0)$ for some $q_0 \in Q_1$, $p_0 \in Q_2$. Let $(q, p) \in Q_1 \times Q_2$. Then $\exists\ x \in X_1^*$ and $y \in X_2^*$ such that $\mu_1^*(q_0, x, q) > 0$ and $\mu_2^*(p_0, y, p) > 0$. Thus $(\mu_1 \cdot \mu_2)^*((q_0, p_0), xy, (q, p)) = \mu_1^*(q_0, x, q) \wedge \mu_2^*(p_0, y, p) > 0$. Hence $(q, p) \in S((q_0, p_0))$. Thus $Q_1 \times Q_2 = S((q_0, p_0))$. Hence $M_1 \cdot M_2$ is cyclic.

Conversely, suppose $M_1 \cdot M_2$ is cyclic. Let $Q_1 \times Q_2 = S((q_0, p_0))$ for some $(q_0, p_0) \in Q_1 \times Q_2$. Let $q \in Q_1$ and $p \in Q_2$. Then $\exists\ w \in (X_1 \cup X_2)^*$ such that $(\mu_1 \cdot \mu_2)^*((q_0, p_0), w, (q, p)) > 0$. By Theorem 6.13.4, $\exists u \in X_1^*$ and $v \in X_2^*$ such that $\mu_1^*(q_0, u, q) \wedge \mu_2^*(p_0, v, p) = (\mu_1 \cdot \mu_2)^*((q_0, p_0), w, (q, p)) > 0$. Hence $\exists u \in X_1^*$ and $v \in X_2^*$ such that $\mu_1^*(q_0, u, q) > 0$ and $\mu_2^*(p_0, v, p) > 0$. Thus $q \in S(q_0)$ and $p \in S(p_0)$. Hence $Q_1 = S(q_0)$ and $Q_2 = S(p_0)$. Thus M_1 and M_2 are cyclic. ∎

Theorem 6.13.6 Let $M_i = (Q_i, X_i, \mu_i)$ be a ffsm, $i = 1, 2$ and let $X_1 \cap X_2 = \emptyset$. Then the Cartesian composition $M_1 \cdot M_2$ is retrievable if and only if M_1 and M_2 are retrievable.

Proof. Suppose that M_1 and M_2 are retrievable. Let $(q, p), (t, s) \in Q_1 \times Q_2$, and $w \in (X_1 \cup X_2)^*$ be such that $(\mu_1 \cdot \mu_2)^*((q, p), w, (t, s)) > 0$. Let $w^* = uv$ be the standard form of w, $u \in X_1^*$, $v \in X_2^*$. Then $(\mu_1 \cdot \mu_2)^*((q, p), w, (t, s)) = (\mu_1 \cdot \mu_2)^*((q, p), uv, (t, s)) = \mu_1^*(q, u, t) \wedge \mu_2^*(p, v, s)$. Thus $\mu_1^*(q, u, t) > 0$ and $\mu_2^*(p, v, s) > 0$. Since M_1 and M_2 are retrievable, $\exists\ u' \in X_1^*$, $v' \in X_2^*$ such that $\mu_1^*(t, u', q) > 0$ and $\mu_2^*(s, v', p) > 0$. Thus $(\mu_1 \cdot \mu_2)^*((t, s), u'v', (q, p)) > 0$. Hence $M_1 \cdot M_2$ is retrievable.

Conversely, suppose that $M_1 \cdot M_2$ is retrievable. Let $q, t \in Q_1$ and $y \in X_1^*$ be such that $\mu_1^*(q, y, t) > 0$. Then $\forall\ s \in Q_2$, $(\mu_1 \cdot \mu_2)^*((q, s), y, (t, s)) = \mu_1^*(q, y, t) > 0$. Thus $\exists\ w \in (X_1 \cup X_2)^*$ such that $(\mu_1 \cdot \mu_2)^*((t, s), w, (q, s)) > 0$. Let $w^* = uv$ be the standard form of w, $u \in X_1^*$, $v \in X_2^*$. Then $0 < (\mu_1 \cdot \mu_2)^*((t, s), w, (q, s)) = \mu_1^*(t, u, q) \wedge \mu_2^*(s, v, s)$. Thus $\mu_1^*(t, u, q) > 0$. Hence M_1 is retrievable. Similarly M_2 is retrievable. ∎

The following corollaries follow from Theorems 6.8.6 and 6.13.6.

Corollary 6.13.7 Let $M_i = (Q_i, X_i, \mu_i)$ be a ffsm, $i = 1, 2$, and let $X_1 \cap X_2 = \emptyset$. Then the Cartesian composition $M_1 \cdot M_2$ is the union of strongly connected submachines if and only if M_1 and M_2 are the union of strongly connected submachines. ∎

Corollary 6.13.8 Let $M_i = (Q_i, X_i, \mu_i)$ be a ffsm, $i = 1, 2$, and let $X_1 \cap X_2 = \emptyset$. Then the Cartesian composition $M_1 \cdot M_2$ satisfies the Exchange Property if and only if M_1 and M_2 satisfies the Exchange Property. ∎

Theorem 6.13.9 *Let $M_i = (Q_i, X_i, \mu_i)$ be a ffsm, $i = 1, 2$, and let $X_1 \cap X_2 = \emptyset$. Then the Cartesian composition $M_1 \cdot M_2$ is connected if and only if M_1 and M_2 are connected.*

Proof. Suppose that M_1 and M_2 are connected. Let $(q, q'), (p, p') \in Q_1 \times Q_2$. Now $\exists\, q_0, q_1, \ldots, q_n \in Q_1, q = q_0, p = q_n$ and $\exists\, a_1, a_2, \ldots, a_n \in X_1$ such that $\forall\, i = 1, 2, \ldots, n$ either $\mu_1(q_{i-1}, a_i, q_i) > 0$ or $\mu_1(q_i, a_i, q_{i-1}) > 0$ and $\exists\, q_0', q_1', \ldots, q_m' \in Q_2, q' = q_0', p' = q_m'$ and $\exists\, b_1, b_2, \ldots, b_m \in X_2$ such that $\forall\, i = 1, 2, \ldots, m$ either $\mu_2(q_{i-1}', b_i, q_i') > 0$ or $\mu_2(q_i', b_i, q_{i-1}') > 0$. Consider the sequence of states $(q, q') = (q_0, q_0'), (q_1, q_0'), \ldots, (q_n, q_0'), (q_n, q_1'), \ldots, (q_n, q_m') = (p, p') \in Q_1 \times Q_2$ and the sequence $a_1, a_2, \ldots, a_n, b_1, b_2, \ldots, b_m \in X_1 \cup X_2$. Then $\forall i = 1, 2, \ldots, n$ either $(\mu_1 \cdot \mu_2)((q_{i-1}, q_0'), a_i, (q_i, q_0')) > 0$ or $(\mu_1 \cdot \mu_2)((q_i, q_0'), a_i, (q_{i-1}, q_0')) > 0$ and $\forall\, j = 1, 2, \ldots, m$ either $(\mu_1 \cdot \mu_2)((q_n, q_{j-1}'), b_j, (q_n, q_j')) > 0$ or $(\mu_1 \cdot \mu_2)((q_n, q_j'), b_j, (q_n, q_{j-1}')) > 0$. Hence (q, q') and (p, p') are connected.

Conversely, suppose that $M_1 \cdot M_2$ is connected. Let $q, p \in Q_1$ and let $r \in Q_2$. If $p = q$ then p and q are connected. Suppose $p \neq q$. Then $\exists\, (q, r) = (q_0, p_0), (q_1, p_1), \ldots, (q_n, p_n) = (p, r) \in Q_1 \times Q_2$ and $a_1, a_2, \ldots, a_n \in X_1 \cup X_2$ such that $\forall i = 1, 2, \ldots, n$ either $(\mu_1 \cdot \mu_2)((q_{i-1}, p_{i-1}), a_i, (q_i, p_i)) > 0$ or $(\mu_1 \cdot \mu_2)((q_i, p_i), a_i, (q_{i-1}, p_{i-1})) > 0$. Clearly, if $q_{i-1} \neq q_i$, then $p_{i-1} = p_i$ and if $p_{i-1} \neq p_i$, then $q_{i-1} = q_i \,\forall\, i = 1, 2, \ldots, n$. Let $\{q = q_{i_1}, q_{i_2}, \ldots, q_{i_k} = p\}$ be the set of all distinct $q_i \in \{q_0, q_1, \ldots, q_n\}$ and let $a_{i_1}, a_{i_2}, \ldots, a_{i_k}$ be the corresponding a_i's. Then $a_{i_1}, a_{i_2}, \ldots, a_{i_k} \in X_1$ and $\forall\, j = 1, 2, \ldots, k$ either $\mu_1(q_{i_{j-1}}, a_{i_j}, q_{i_j}) > 0$ or $\mu_1(q_{i_j}, a_{i_j}, q_{i_{j-1}}) > 0$. Thus q and p are connected. Hence M_1 is connected. Similarly, M_2 is connected. ∎

The following theorem follows from Theorems 6.8.11, 6.13.6, and 6.13.9.

Theorem 6.13.10 *Let $M_i = (Q_i, X_i, \mu_i)$ be a ffsm, $i = 1, 2$, and let $X_1 \cap X_2 = \emptyset$. Then the Cartesian composition $M_1 \cdot M_2$ is strongly connected if and only if M_1 and M_2 are strongly connected.* ∎

Definition 6.13.11 *Let $M = (Q, X, \mu)$ be a ffsm. Then M is said to be **commutative** if $\forall a, b \in X$ and $\forall q, p \in Q$,*

$$\mu^*(q, ab, p) = \mu^*(q, ba, p).$$

The following result is immediate from Theorem 6.13.4.

Theorem 6.13.12 *Let $M_i = (Q_i, X_i, \mu_i)$ be a ffsm, $i = 1, 2$, and let $X_1 \cap X_2 = \emptyset$. Then the Cartesian composition $M_1 \cdot M_2$ is commutative if and only if M_1 and M_2 are commutative.* ∎

Definition 6.13.13 *Let $M = (Q, X, \mu)$ be a ffsm. If M is commutative and strongly connected, then M is said to be **perfect**.*

Theorem 6.13.14 *Let $M_i = (Q_i, X_i, \mu_i)$ be a ffsm, $i = 1, 2$, and let $X_1 \cap X_2 = \emptyset$. Then the Cartesian composition $M_1 \cdot M_2$ is perfect if and only if M_1 and M_2 are perfect.* ∎

Definition 6.13.15 Let $M = (Q, X, \mu)$ be a ffsm. Then M is said to be **state independent** if $\forall q', p' \in Q$, $\forall x, y \in X^+$, $(\mu^*(q', x, p') > 0$ and $\mu^*(q', y, p') > 0) \implies (\forall q, p \in Q, \mu^*(q, x, p) > 0 \Leftrightarrow \mu^*(q, y, p) > 0)$.

Theorem 6.13.16 Let $M_i = (Q_i, X_i, \mu_i)$ be a ffsm, $i = 1, 2$, and let $X_1 \cap X_2 = \emptyset$. Then the Cartesian composition $M_1 \cdot M_2$ is state independent if and only if M_1 and M_2 are state independent.

Proof. Suppose that M_1 and M_2 are state independent. Suppose that $(\mu_1 \cdot \mu_2)^*((q'_1, q'_2), w_1, (p'_1, p'_2)) > 0$ and $(\mu_1 \cdot \mu_2)^*((q'_1, q'_2), w_2, (p'_1, p'_2)) > 0$, where $(q'_1, q'_2), (p'_1, p'_2) \in Q_1 \times Q_2$, and $w_1, w_2 \in (X_1 \cup X_2)^*$. Now $\exists$ $u_1, u_2 \in X_1^+$ and $v_1, v_2 \in X_2^+$ such that $(\mu_1 \cdot \mu_2)^*((q'_1, q'_2), w_1, (p'_1, p'_2)) = \mu_1^*(q'_1, u_1, p'_1) \wedge \mu_2^*(q'_2, v_1, p'_2)$ and $(\mu_1 \cdot \mu_2)^*((q'_1, q'_2), w_2, (p'_1, p'_2)) = \mu_1^*(q'_1, u_2, p'_1) \wedge \mu_2^*(q'_2, v_2, p'_2)$. Thus $\mu_1^*(q'_1, u_1, p'_1) > 0$, $\mu_2^*(q'_2, v_1, p'_2) > 0$, $\mu_1^*(q'_1, u_2, p'_1) > 0$, and $\mu_2^*(q'_2, v_2, p'_2) > 0$. Hence $\forall q_1, p_1 \in Q_1$, $\mu_1^*(q_1, u_1, p_1) > 0 \iff \mu_1^*(q_1, u_2, p_1) > 0$ and $\forall q_2, p_2 \in Q_2$, $\mu_2^*(q_2, v_1, p_2) > 0 \iff \mu_2^*(q_2, v_2, p_2) > 0$. Hence $\mu_1^*(q_1, u_1, p_1) \wedge \mu_2^*(q_2, v_1, p_2) > 0 \iff \mu_1^*(q_1, u_2, p_1) \wedge \mu_2^*(q_2, v_2, p_2) > 0$, $\forall q_1, p_1 \in Q_1$, $\forall q_2, p_2 \in Q_2$. Thus $(\mu_1 \cdot \mu_2)^*((q_1, q_2), w_1, (p_1, p_2)) > 0 \iff (\mu_1 \cdot \mu_2)^*((q_1, q_2), w_2, (p_1, p_2)) > 0$, $\forall q_1, p_1 \in Q_1$, $\forall q_2, p_2 \in Q_2$. Hence $M_1 \cdot M_2$ is state independent.

Conversely, suppose that $M_1 \cdot M_2$ is state independent. Suppose that $\mu_1^*(q'_1, u_1, p'_1) > 0$ and $\mu_1^*(q'_1, u_2, p'_1) > 0$ where $u_1, u_2 \in X_1^+$ and $q'_1, p'_1 \in Q_1$. Then $\forall s \in Q_2$, $(\mu_1 \cdot \mu_2)^*((q'_1, s), u_1, (p'_1, s)) = \mu_1^*(q'_1, u_1, p'_1) > 0$ and $(\mu_1 \cdot \mu_2)^*((q'_1, s), u_2, (p'_1, s)) = \mu_1^*(q'_1, u_2, p'_1) > 0$. Thus $\forall q, p \in Q_1$, $s \in Q_2$, $(\mu_1 \cdot \mu_2)^*((q, s), u_1, (p, s)) > 0 \iff (\mu_1 \cdot \mu_2)^*((q, s), u_2, (p, s)) > 0$. Hence $\forall q, p \in Q_1$, $\mu_1^*(q, u_1, p) > 0 \iff \mu_1^*(q, u_2, p) > 0$. Thus M_1 is state independent. Similarly M_2 is state independent. ∎

Theorem 6.13.17 Let $M_i = (Q_i, X_i, \mu_i)$ be a ffsm, $i = 1, 2$, and let $X_1 \cap X_2 = \emptyset$. Let $N_i = (T_i, X_i, \nu_i)$ be a submachine of M_i, $i = 1, 2$. Then $N_1 \cdot N_2$ is a submachine of $M_1 \cdot M_2$. Conversely, if $N = (T_1 \times T_2, X_1 \cup X_1, \nu)$ is a submachine of $M_1 \cdot M_2$, then there exist submachines N_1 of M_1 and N_2 of M_2 such that $N = N_1 \cdot N_2$.

Proof. Let $N_i = (T_i, X_i, \nu_i)$ be a submachine of M_i, $i = 1, 2$. Now $N_1 \cdot N_2 = (T_1 \times T_2, X_1 \cup X_2, \nu_1 \cdot \nu_2)$. Let $(r, s) \in S(T_1 \times T_2)$. Then $\exists w \in (X_1 \cup X_2)^*$, $(p, q) \in T_1 \times T_2$ such that $(\mu_1 \cdot \mu_2)^*((p, q), w, (r, s)) > 0$. Let $w^* = uv$ be the standard form of w, $u \in X_1^*$, $v \in X_2^*$. Now $\mu_1^*(p, u, r) \wedge \mu_2^*(q, v, s) = (\mu_1 \cdot \mu_2)^*((p, q), w, (r, s)) > 0$. Thus $\mu_1^*(p, u, r) > 0$ and $\mu_2^*(q, v, s) > 0$. Hence $r \in S(p) \subseteq S(T_1) = T_1$ and $s \in S(q) \subseteq S(T_2) = T_2$. Thus $(r, s) \in T_1 \times T_2$.

Hence $S(T_1 \times T_2) \subseteq T_1 \times T_2$. Let $(p,q),(r,s) \in T_1 \times T_2$, $a \in X_1 \cup X_2$. Now

$$(\nu_1 \cdot \nu_2)((p,q),a,(r,s)) = \begin{cases} \nu_1(p,a,r) \text{ if } a \in X_1, \ q = s \\ \nu_2(q,a,s) \text{ if } a \in X_2, \ p = r \\ 0 \text{ otherwise} \end{cases}$$
$$= \begin{cases} \mu_1(p,a,r) \text{ if } a \in X_1, \ q = s \\ \mu_2(q,a,s) \text{ if } a \in X_2, \ p = r \\ 0 \text{ otherwise} \end{cases}$$
$$= (\mu_1 \cdot \mu_2)((p,q),a,(r,s)).$$

Hence $(\mu_1 \cdot \mu_2)|_{(T_1 \times T_2) \times (X_1 \cup X_2) \times (T_1 \times T_2)} = \nu_1 \cdot \nu_2$. Thus $N_1 \cdot N_2$ is a submachine of $M_1 \cdot M_2$.

Conversely, let $N = (T_1 \times T_2, X_1 \cup X_1, \nu)$ be a submachine of $M_1 \cdot M_2$. Let $\nu_1 = \mu_1|_{T_1 \times X_1 \times T_1}$, $\nu_2 = \mu_2|_{T_2 \times X_2 \times T_2}$, $N_1 = (T_1, X_1, \nu_1)$, and $N_2 = (T_2, X_2, \nu_2)$. Let $p \in T_1$, $x \in X_1^*$, $r \in Q_1$ be such that $\mu_1^*(p,x,r) > 0$. Let $t \in T_2$. Then $(\mu_1 \cdot \mu_2)^*((p,t),x,(r,t)) = \mu_1^*(p,x,r) > 0$. Thus $(r,t) \in S(T_1 \times T_2) = T_1 \times T_2$. Hence $r \in T_1$ and so $S(T_1) \subseteq T_1$. Thus N_1 is a submachine of M_1. Similarly N_2 is a submachine of M_2. Let $(p,q),(r,s) \in T_1 \times T_2$, $a \in X_1 \cup X_2$. Now

$$\nu((p,q),a,(r,s)) = (\mu_1 \cdot \mu_2)((p,q),a,(r,s))$$
$$= \begin{cases} \mu_1(p,a,r) \text{ if } a \in X_1, \ q = s \\ \mu_2(q,a,s) \text{ if } a \in X_2, \ p = r \\ 0 \text{ otherwise} \end{cases}$$
$$= \begin{cases} \nu_1(p,a,r) \text{ if } a \in X_1, \ q = s \\ \nu_2(q,a,s) \text{ if } a \in X_2, \ p = r \\ 0 \text{ otherwise} \end{cases}$$
$$= (\nu_1 \cdot \nu_2)((p,q),a,(r,s)).$$

Hence $N = N_1 \cdot N_2$. ∎

Let $M = (Q,X,\mu)$ be a ffsm and let $\sim$ be an equivalence relation on Q. Recall from Definition 6.4.1 that $\sim$ is called an admissible relation if and only if $\forall\, p,q,r \in Q, \forall a \in X$, if $p \sim q$ and $\mu(p,a,r) > 0$, then $\exists\, t \in Q$ such that $\mu(q,a,t) \geq \mu(p,a,r)$ and $t \sim r$.

Let $M = (Q,X,\mu)$ be a ffsm and let $\sim$ be an equivalence relation on Q. By Theorem 6.4.2, $\sim$ is an admissible relation if and only if $\forall\, p,q,r \in Q, \forall x \in X^*$, if $p \sim q$ and $\mu^*(p,x,r) > 0$, then $\exists\, t \in Q$ such that $\mu^*(q,x,t) \geq \mu^*(p,x,r)$ and $t \sim r$.

Let $M_1 = (Q_1,X_1,\mu_1)$ and $M_2 = (Q_2,X_2,\mu_2)$ be two ffsms and let $X_1 \cap X_2 = \emptyset$. Let ρ_i be an admissible relation on M_i, $i = 1,2$. Define a relation $\rho_1 \cdot \rho_2$ on $M_1 \cdot M_2$ by

$$(p_1,p_2)\rho_1 \cdot \rho_2(q_1,q_2) \text{ if and only if } p_1\rho_1q_1 \text{ and } p_2\rho_2q_2$$

$\forall (p_1,p_2),(q_1,q_2) \in Q_1 \times Q_2$.

Clearly $\rho_1 \cdot \rho_2$ is an equivalence relation on $M_1 \cdot M_2$. Let $(p_1,p_2),(q_1,q_2) \in Q_1 \times Q_2$ be such that $(p_1,p_2)\rho_1 \cdot \rho_2(q_1,q_2)$. Let $a \in X_1 \cup X_2$ and $(\mu_1 \cdot$

$\mu_2)((p_1, p_2), a, (r_1, r_2)) > 0$ for some $(r_1, r_2) \in Q_1 \times Q_2$. Suppose that $a \in X_1$. Then $\mu_1(p_1, a, r_1) = (\mu_1 \cdot \mu_2)((p_1, p_2), a, (r_1, r_2)) > 0$. Thus $p_2 = r_2$. Now $p_1 \rho_1 q_1$ and $\mu_1(p_1, a, r_1) > 0$. Since ρ_1 is admissible, $\exists \, t_1 \in Q_1$ such that $\mu_1(q_1, a, t_1) \geq \mu_1(p_1, a, r_1)$ and $t_1 \rho_1 r_1$. Hence $(\mu_1 \cdot \mu_2)((q_1, q_2), a, (t_1, q_2)) = \mu_1(q_1, a, t_1) \geq \mu_1(p_1, a, r_1) = (\mu_1 \cdot \mu_2)((p_1, p_2), a, (r_1, p_2))$. Also, since $t_1 \rho_1 r_1$ and $q_2 \rho_2 p_2$, $(t_1, q_2) \rho_1 \cdot \rho_2 (r_1, p_2)$. Thus $\rho_1 \cdot \rho_2$ is an admissible relation on $M_1 \cdot M_2$. We have thus proved the following theorem.

Theorem 6.13.18 *Let $M_i = (Q_i, X_i, \mu_i)$ be a ffsm, $i = 1, 2$, and let $X_1 \cap X_2 = \emptyset$. Let ρ_i be an admissible relation on M_i, $i = 1, 2$. Then $\rho_1 \cdot \rho_2$ is an admissible relation on $M_1 \cdot M_2$.* ∎

6.14   Admissible Partitions

In this section, we introduce the concept of a covering of a ffsm by another, admissible partitions and relations of a ffsm, μ-orthogonality of admissible partitions, irreducible ffsm, and the quotient of a ffsm induced by an admissible partition of the state set. We derive results concerning μ-orthogonality and covering, Theorems 6.14.14 and 6.14.15. We show that an admissible partition π of Q is maximal if and only if the quotient M/π is irreducible, Theorem 6.14.20. The paper culminates with a result showing that a ffsm can be covered by irreducible ffsms, Theorem 6.14.21. These results allow us to study fuzzy finite state machines via coverings of products of simpler machines. Irreducible finite state machines seem to arise naturally in some applications, e.g., biology. An example is given in [92] of a finite state machine arising from a metabolic pathway.

Recall from Definition 6.6.1 that (η, ξ) is a covering of M_1 by M_2, written $M_1 \leq M_2$, if and only if $\forall \, q_2 \in Q_2$, $q_1 \in Q_1$, and $x \in X_1^*$, $\mu_1^*(\eta(q_2), x, q_1) \geq \mu_2^*(q_2, \xi^*(x), r_2) \, \forall \, r_2 \in Q_2$ such that $\eta(r_2) = q_1$ and $\exists \, r_2 \in Q_2$ such that $\eta(r_2) = q_1$ and $\mu_1^*(\eta(q_2), x, q_1) = \mu_2^*(q_2, \xi^*(x), r_2)$.

Let $M_1 = (Q_1, X_1, \mu_1)$ and $M_2 = (Q_2, X_2, \mu_2)$ be fuzzy finite state machines. Let ω be a function of $Q_2 \times X_2$ into X_1. Let $Q = Q_1 \times Q_2$. Define

$$\mu^\omega : Q \times X_2 \times Q \to [0, 1]$$

as follows: $\forall ((q_1, q_2), b, (p_1, p_2)) \in Q \times X_2 \times Q$,

$$\mu^\omega((q_1, q_2), b, (p_1, p_2)) = \mu_1(q_1, \omega(q_2, b), p_1) \wedge \mu_2(q_2, b, p_2).$$

Then $M = (Q, X_2, \mu^\omega)$ is a ffsm. M is called the **cascade product** of M_1 and M_2 and we write $M = M_1 \omega M_2$.

Let $M = (Q, X, \mu)$ be a ffsm and let $\sim$ be an equivalence relation on Q. Recall that $\sim$ is called an admissible relation on Q if and only if $\forall \, p, q, r \in Q, \forall a \in X$, if $p \sim q$ and $\mu(p, a, r) > 0$, then $\exists \, t \in Q$ such that $\mu(q, a, t) \geq \mu(p, a, r)$ and $t \sim r$.

Theorem 6.14.1 *[3] Let $M = (Q, X, \mu)$ be a ffsm and let $\sim$ be an equivalence relation on Q. Then $\sim$ is an admissible relation on Q if and only if $\forall\, p, q, r \in Q, \forall x \in X^*$, if $p \sim q$ and $\mu^*(p, x, r) > 0$, then $\exists\, t \in Q$ such that $\mu^*(q, x, t) \geq \mu^*(p, x, r)$ and $t \sim r$.* ∎

Definition 6.14.2 *Let $M = (Q, X, \mu)$ be a ffsm and $\mathcal{P} = \{Q_1, Q_2, \ldots, Q_k\}$ be a partition of Q. Then $\mathcal{P}$ is called an **admissible partition** of Q if the following holds: Let $a \in X$, then $\forall i, \exists j, 1 \leq i, j \leq k$ such that $\forall p_1, p_2 \in Q_i$, if $\mu(p_1, a, r) > 0$ for some $r \in Q$, then $\exists\, t \in Q$ such that $\mu(p_2, a, t) \geq \mu(p_1, a, r)$ and $t, r \in Q_j$.*

Proposition 6.14.3 *Let $M = (Q, X, \mu)$ be a ffsm.*
(1) Let $1_Q = \{\{q\} \mid q \in Q\}$. Then 1_Q is an admissible partition of Q.
(2) $\{Q\}$ is an admissible partition of Q. ∎

Theorem 6.14.4 *Let $M = (Q, X, \mu)$ be a ffsm and $\mathcal{P} = \{Q_1, Q_2, \ldots, Q_k\}$ be a partition of Q. The following are equivalent:*
(1) $\mathcal{P}$ is an admissible partition of Q.
(2) Let $x \in X^*$. Then $\forall i, \exists j, 1 \leq i, j \leq k$ such that $\forall p_1, p_2 \in Q_i$, if $\mu^*(p_1, x, r) > 0$ for some $r \in Q$, then $\exists\, t \in Q$ such that $\mu^*(p_2, x, t) \geq \mu^*(p_1, x, r)$ and $t, r \in Q_j$.

Proof. (1)$\Rightarrow$(2): Let $x \in X^*$ and $|x| = n$. Let p_1, $p_2 \in Q_i$ and $\mu^*(p_1, x, r) > 0$ for some $r \in Q$. If $n = 0$, then $x = \Lambda$ and $\mu^*(p_1, x, r) > 0$ implies that $p_1 = r$. Thus $\mu^*(p_2, x, p_2) = 1 = \mu^*(p_1, x, p_1)$. In this case $i = j$. Hence the result is true for $n = 0$. Suppose the result is true for all $y \in X^*$ such that $|y| = n - 1$, where $n > 0$. Let $x = ya$, where $a \in X$. Now $\mu^*(p_1, x, r) = \mu^*(p_1, ya, r) = \vee\{\mu^*(p_1, y, s) \wedge \mu^*(s, a, r) \mid s \in Q\} > 0$. Since Q is finite, there exists $t \in Q$ such that $\mu^*(p_1, y, t) \wedge \mu^*(t, a, r) = \mu^*(p_1, ya, r)$. Thus $\mu^*(p_1, y, t) > 0$ and $\mu(t, a, r) = \mu^*(t, a, r) > 0$. By the induction hypothesis, there exists j and there exists $s \in Q$ such that $\mu^*(p_2, y, s) \geq \mu^*(p_1, y, t)$ and $s, t \in Q_j$. Now $s, t \in Q_j$ and $\mu(t, a, r) > 0$. Hence by (1), there exists l and there exists $q \in Q$ such that $\mu(s, a, q) \geq \mu(t, a, r)$ and $r, q \in Q_l$. Now $\mu^*(p_2, x, q) = \mu^*(p_2, ya, q) \geq \mu^*(p_2, y, s) \wedge \mu(s, a, q) \geq \mu^*(p_1, y, t) \wedge \mu(t, a, r) = \mu^*(p_1, ya, r) = \mu^*(p_1, x, r)$ and $q, r \in Q_l$. The result now follows by induction. ∎

Corollary 6.14.5 *Let $M = (Q, X, \mu)$ be a ffsm. Then every admissible partition $\mathcal{P}$ of Q induces an admissible relation $\sim$ on Q such that the set of all equivalence classes of $\sim$ is $\mathcal{P}$. Conversely, the set of all equivalence classes of an admissible relation on Q is an admissible partition of Q.* ∎

Lemma 6.14.6 *Let $M = (Q, X, \mu)$ be a ffsm and $\pi = \{H_i \mid i \in I\}$ be an admissible partition of Q. Let $i, j \in I$. Then $\forall q, q' \in H_i, \forall a \in X$,*

$$\vee\{\mu(q, a, r) \mid r \in H_j\} = \vee\{\mu(q', a, r) \mid r \in H_j\}.$$

Proof. Let $q, q' \in H_i$, $a \in X$, $A = \{\mu(q, a, r) \mid r \in H_j\}$, and $B = \{\mu(q', a, r) \mid r \in H_j\}$. Suppose $\mu(q, a, r) > 0$ for some $r \in H_j$. Since π is an admissible partition, there exists $r' \in Q$ such that $\mu(q', a, r') \geq \mu(q, a, r)$. Again by the admissibility of π, $r' \in H_j$. Similarly if $\mu(q', a, p) > 0$ for some $p \in H_j$, then there exists $p' \in H_j$ such that $\mu(q, a, p') \geq \mu(q', a, p)$. Also, by the admissibility of π, it follows that $\mu(q, a, r) = 0 \ \forall r \in H_j$ if and only if $\mu(q', a, r) = 0 \ \forall r \in H_j$. Hence

$$\vee\{\mu(q, a, r) \mid r \in H_j\} = \vee\{\mu(q', a, r) \mid r \in H_j\}. \ \blacksquare$$

Theorem 6.14.7 *Let $M = (Q, X, \mu)$ be a ffsm. Let $\pi = \{H_i \mid i \in I\}$ be an admissible partition of Q. Define*

$$\mu^\pi : \pi \times X \times \pi \to [0, 1]$$

by

$$\mu^\pi(H_i, a, H_j) = \vee\{\mu(q, a, r) \mid r \in H_j\}$$

$\forall H_i, H_j \in \pi$ *and* $a \in X$, *where* $q \in H_i$. *Then $M/\pi = (\pi, X, \mu^\pi)$ is a ffsm, called the **quotient fuzzy finite state machine** with respect to π.*

Proof. By Lemma 6.14.6, μ^π is well defined. $\blacksquare$

Proposition 6.14.8 *Let $M = (Q, X, \mu)$ be a ffsm. Let $\pi = \{H_i \mid i \in I\}$ be an admissible partition of Q. Then for $q \in H_i$*

$$\mu^{\pi*}(H_i, x, H_j) \leq \vee\{\mu^*(q, x, r) \mid r \in H_j\}$$

$\forall H_i, H_j \in \pi$ *and* $x \in X^*$.

Proof. Let $H_i, H_j \in \pi$ and $x \in X^*$. Let $|x| = n$. If $n = 0$, then $x = \Lambda$. If $H_i = H_j$ then $\mu^{\pi*}(H_i, x, H_j) = 1$ and $\vee\{\mu^*(q, x, r) \mid r \in H_j\} = \vee\{\mu^*(q, x, r) \mid r \in H_i\} = \mu^*(q, x, q) = 1$, where $q \in H_i$. If $H_i \neq H_j$, then $\mu^{\pi*}(H_i, x, H_j) = 0$ and $H_i \cap H_j = \emptyset$. Since $H_i \cap H_j = \emptyset$ and $q \in H_i$, $\vee\{\mu^*(q, x, r) \mid r \in H_j\} = 0$. Hence $\mu^{\pi*}(H_i, x, H_j) = \vee\{\mu^*(q, x, r) \mid r \in H_j\}$. Suppose that the result is true $\forall y \in X^*$ such that $|y| = n - 1$, $n > 0$. Let $n > 0$ and $x = ya$, where $y \in X^*$, $a \in X$, and $|y| = n - 1$. Now for $q \in H_i$

and $s \in H_k$,

$$
\begin{aligned}
\mu^{\pi*}(H_i, x, H_j) &= \mu^{\pi*}(H_i, ya, H_j) \\
&= \vee\{\mu^{\pi*}(H_i, y, H_k) \wedge \mu^{\pi*}(H_k, a, H_j) \mid H_k \in \pi\} \\
&= \vee\{(\vee\{\mu^*(q, y, r) \mid r \in H_k\}) \wedge (\vee\{\mu^*(s, a, p) \mid \\
&\quad p \in H_j\}) \mid H_k \in \pi\} \\
&= \vee\{\vee\{\mu^*(q, y, r) \wedge \mu^*(s, a, p) \mid r \in H_k, p \in H_j\} \\
&\quad \mid H_k \in \pi\} \\
&\leq \vee\{\vee\{\mu^*(q, y, r) \wedge \mu^*(r, a, p) \mid r \in H_k, p \in H_j\} \\
&\quad \mid H_k \in \pi\} \\
&= \vee\{\vee\{\mu^*(q, y, r) \wedge \mu^*(r, a, p) \mid r \in H_k, H_k \in \pi\} \\
&\quad \mid p \in H_j\} \\
&= \vee\{\vee\{\mu^*(q, y, r) \wedge \mu^*(r, a, p) \mid r \in Q\} \mid p \in H_j\} \\
&= \vee\{\mu^*(q, ya, p) \mid p \in H_j\} \\
&= \vee\{\mu^*(q, x, p) \mid p \in H_j\}. \quad\blacksquare
\end{aligned}
$$

Proposition 6.14.9 *Let* $M = (Q, X, \mu)$ *be a ffsm. Let* $\pi = \{H_i \mid i \in I\}$ *be an admissible partition of* Q. *Then for all* $x \in X^*$

$$
\mu^*(q, x, p) \leq \mu^{\pi*}(H_i, x, H_j),
$$

where $q \in H_i$, $p \in H_j$.

Proof. Let $q \in H_i$, $p \in H_j$. Let $x \in X^*$ and $|x| = n$. If $n = 0$, then $x = \Lambda$. If $q = p$, then $H_i = H_j$ and $\mu^*(q, x, p) = 1 = \mu^{\pi*}(H_i, x, H_i)$. Suppose $q \neq p$. Then $\mu^*(q, x, p) = 0 \leq \mu^{\pi*}(H_i, x, H_j)$. Suppose $n = 1$. Then $x = a \in X$ and

$$
\begin{aligned}
\mu^*(q, a, p) &= \mu(q, a, p) \\
&\leq \vee\{\mu(q, a, r) \mid r \in H_j\} \\
&= \mu^\pi(H_i, a, H_j) \\
&= \mu^{\pi*}(H_i, a, H_j).
\end{aligned}
$$

Hence the result is true for $n = 0$ and $n = 1$. Suppose that the result is true $\forall y \in X^*$ such that $|y| = n - 1$, $n > 0$. Let $n > 0$ and $x = ya$, where $y \in X^*$, $a \in X$, and $|y| = n - 1$. Then

$$
\begin{aligned}
\mu^*(q, x, p) &= \mu^*(q, ya, p) \\
&= \vee\{\mu^*(q, y, r) \wedge \mu(r, a, p) \mid r \in Q\} \\
&= \vee\{\vee\{\mu^*(q, y, r) \wedge \mu(r, a, p) \mid r \in H_k\} \mid H_k \in \pi\} \\
&\leq \vee\{\vee\{\mu^{\pi*}(H_i, y, H_k) \wedge \mu^\pi(H_k, a, H_j) \mid r \in H_k\} \mid H_k \in \pi\} \\
&\quad \text{(by induction and } n = 1 \text{ case)} \\
&= \vee\{\mu^{\pi*}(H_i, y, H_k) \wedge \mu^\pi(H_k, a, H_j) \mid H_k \in \pi\} \\
&= \mu^{\pi*}(H_i, x, H_j).
\end{aligned}
$$

The result now follows by induction. $\blacksquare$

Corollary 6.14.10 Let $M = (Q, X, \mu)$ be a ffsm. Let $\pi = \{H_i \mid i \in I\}$ be an admissible partition of Q. Then for all $x \in X^*$, for all $H_i, H_j \in \pi$,

$$\vee\{\mu^*(q, x, p) \mid p \in H_j\} \leq \mu^{\pi*}(H_i, x, H_j),$$

where $q \in H_i$. ∎

Theorem 6.14.11 Let $M = (Q, X, \mu)$ be a ffsm. Let $\pi = \{H_i \mid i \in I\}$ be an admissible partition of Q. Then for $q \in H_i$

$$\mu^{\pi*}(H_i, x, H_j) = \vee\{\mu^*(q, x, r) \mid r \in H_j\}$$

$\forall H_i, H_j \in \pi$ and $x \in X^*$. ∎

Definition 6.14.12 Let $M = (Q, X, \mu)$ be a ffsm. Let π and τ be admissible partitions of Q. π and τ are called **μ-orthogonal** if
(1) $\pi \cap \tau = 1_Q$, and
(2) $\forall H_i, H_u \in \pi$, $\forall K_j, K_v \in \tau$, $\forall a \in X$, if $H_i \cap K_j = \{q_0\}$ and $H_u \cap K_v = \{p_0\}$, then

$$\mu(q_0, a, p_0) = \vee\{\mu(q_0, a, p) \wedge \mu(q_0, a, p') \mid p \in H_u, p' \in K_v\}.$$

Theorem 6.14.13 Let $M = (Q, X, \mu)$ be a ffsm. Let π and τ be admissible partitions of Q. Then π and τ are μ-orthogonal if and only if
(1) $\pi \cap \tau = 1_Q$, and
(2) $\forall H_i, H_u \in \pi$, $\forall K_j, K_v \in \tau$, $\forall x \in X^*$, if $H_i \cap K_j = \{q_0\}$ and $H_u \cap K_v = \{p_0\}$, then

$$\mu^*(q_0, x, p_0) = \vee\{\mu^*(q_0, x, p) \wedge \mu^*(q_0, x, p') \mid p \in H_u, p' \in K_v\}.$$

Proof. Suppose π and τ are called μ-orthogonal. Then clearly (1) holds.

(2) Let $H_i, H_u \in \pi$, $K_j, K_v \in \tau$, $x \in X^*$, $H_i \cap K_j = \{q_0\}$, and $H_u \cap K_v = \{p_0\}$. Suppose $|x| = n$. If $n = 0$, then $x = \Lambda$. Now $\mu^*(q_0, x, p_0) = 0$ if $q_0 \neq p_0$ and $\mu^*(q_0, x, p_0) = 1$ if $q_0 = p_0$. Suppose $q_0 \neq p_0$. Then either $H_i \cap H_u = \emptyset$ or $K_j \cap K_v = \emptyset$, say, $H_i \cap H_u = \emptyset$. Thus $q_0 \notin H_u$ and $q_0 \notin K_v$. Hence $\vee\{\mu^*(q_0, x, p) \wedge \mu^*(q_0, x, p') \mid p \in H_u, p' \in K_v\} = \vee\{0 \wedge \mu^*(q_0, x, p') \mid p \in H_u, p' \in K_v\} = 0$. Suppose $q_0 = p_0$. Then $H_i = H_u$ and $K_j = K_v$. Thus $\vee\{\mu^*(q_0, x, p) \wedge \mu^*(q_0, x, p') \mid p \in H_u, p' \in K_v\} = \mu^*(q_0, x, p_0) \wedge \mu^*(q_0, x, p_0) = 1$. Hence if $n = 0$, then $\mu^*(q_0, x, p_0) = \vee\{\mu^*(q_0, x, p) \wedge \mu^*(q_0, x, p') \mid p \in H_u, p' \in K_v\}$. Suppose that the result is true $\forall y \in X^*$ such that $|y| = n - 1$, $n > 0$. Let $n > 0$ and $x = ya$, where $y \in X^*$, $a \in X$,

and $|y| = n - 1$. Then

$$
\begin{aligned}
\mu^*(q_0, x, p_0) &= \mu^*(q_0, ya, p_0) \\
&= \vee\{\mu^*(q_0, y, r) \wedge \mu(r, a, p_0) \mid r \in Q\} \\
&= \vee\{\mu^*(q_0, y, r) \wedge (\vee\{\mu(r, a, p) \wedge \mu(r, a, p') \mid p \in H_u, \\
&\quad\ p' \in K_v\}) \mid r \in Q\} \\
&= \vee\{(\vee(\mu^*(q_0, y, r) \wedge \mu(r, a, p))) \wedge (\vee(\mu^*(q_0, y, r) \wedge \\
&\quad\ \mu(r, a, p'))) \mid p \in H_u, p' \in K_v,\ r \in Q\} \\
&= \vee\{(\vee\{\mu^*(q_0, y, r) \wedge \mu(r, a, p) \mid r \in Q\}) \wedge \\
&\quad\ (\vee\{\mu^*(q_0, y, r) \wedge \mu(r, a, p') \mid r \in Q\}) \mid p \in H_u, p' \in K_v\} \\
&= \vee\{\mu^*(q_0, ya, p) \wedge \mu^*(q_0, ya, p') \mid p \in H_u, p' \in K_v\} \\
&= \vee\{\mu^*(q_0, x, p) \wedge \mu^*(q_0, x, p') \mid p \in H_u, p' \in K_v\}.
\end{aligned}
$$

The result now follows by induction. The converse is trivial. ∎

Let $M = (Q, X, \mu)$ be a ffsm. Let π and τ be admissible partitions of Q. Consider the ffsms $M/\pi = (\pi, X, \mu^\pi)$ and $M/\tau = (\tau, X, \mu^\tau)$. Define $\mu^\wedge : (\pi \times \tau) \times X \times (\pi \times \tau) \to [0, 1]$ by

$$
\mu^\wedge((H_i, K_j), a, (H_u, K_v)) = \mu^\pi(H_i, a, H_u) \wedge \mu^\tau(K_j, a, K_v)
$$

$\forall H_i, H_u \in \pi, K_j, K_v \in \tau$, and $a \in X$. Then $M/\pi \wedge M/\tau = (\pi \times \tau, X, \mu^\wedge)$ is a fuzzy finite state machine. Note that

$$
\mu^{\wedge*}((H_i, K_j), x, (H_u, K_v)) = \mu^{\pi*}(H_i, x, H_u) \wedge \mu^{\tau*}(K_j, x, K_v)
$$

$\forall H_i, H_u \in \pi, K_j, K_v \in \tau$, and $x \in X^*$.

Theorem 6.14.14 *Let $M = (Q, X, \mu)$ be a ffsm. Let π and τ be admissible partitions of Q that are μ-orthogonal. Then $M \leq M/\pi \wedge M/\tau$.*

Proof. Define $\eta : \pi \times \tau \to Q$ by $\eta((H_i, K_j)) = q_0$, where $H_i \cap K_j = \{q_0\}$. Since π and τ are μ-orthogonal, η is one-to-one. Let ζ be the identity map on X. Let $H_i, H_u \in \pi, K_j, K_v \in \tau$, and $x \in X^*$. Suppose $H_i \cap K_j = \{q_0\}$ and $H_u \cap K_v = \{p_0\}$. Then

$$
\mu^*(\eta((H_i, K_j)), x, \eta((H_u, K_v))) = \mu^*(q_0, x, p_0).
$$

Also,

$$
\begin{aligned}
\mu^{\wedge*}((H_i, K_j), x, (H_u, K_v)) &= \mu^{\pi*}(H_i, a, H_u) \wedge \mu^{\tau*}(K_j, x, K_v) \\
&= (\vee\{\mu^*(q_0, x, p) \mid p \in H_u\}) \wedge \\
&\quad\ (\vee\{\mu^*(q_0, x, p') \mid p' \in K_v\}) \\
&= \vee\{\mu^*(q_0, x, p) \wedge \mu^*(q_0, x, p') \mid p \in H_u, \\
&\quad\ p' \in K_v\} \\
&= \mu^*(q_0, x, p_0),
\end{aligned}
$$

where the last inequality holds since π and τ are μ-orthogonal. Thus

$$
\mu^*(\eta((H_i, K_j)), x, \eta((H_u, K_v))) = \mu^{\wedge*}((H_i, K_j), x, (H_u, K_v)).
$$

Now $\mu^{\wedge*}((H_i, K_j), x, (H_u, K_v)) = \vee\{\mu^{\wedge*}((H_i, K_j), x, (H_r, K_s)) \mid \eta((H_r, K_s))$
$= \eta((H_u, K_v)), (H_r, K_s) \in \pi \times \tau\}$ since η is one-to-one. Hence $\mu^*(\eta((H_i, K_j)),$
$x, \eta((H_u, K_v))) = \vee\{\mu^{\wedge*}((H_i, K_j), x, (H_r, K_s)) \mid \eta((H_r, K_s)) = \eta((H_u,$
$K_v)), (H_r, K_s) \in \pi \times \tau\}$. Consequently, $M \leq M/\pi \wedge M/\tau$. ∎

Theorem 6.14.15 *Let* $M = (Q, X, \mu)$ *be a ffsm. Let* π *be an admissible partition of* Q. *If there exists a partition* τ *of* Q *such that* π *and* τ *are* μ-*orthogonal, then there exists a ffsm* N *such that* $M \leq N\omega M/\pi$.

Proof. Let $\pi = \{H_i\}_{i \in I}$ and $\tau = \{K_j\}_{j \in J}$ be μ-orthogonal partitions of Q. Let $N = (\tau, \pi \times X, \mu')$, where

$$\mu'(K_j, (H_i, a), K_v) = \vee\{\mu(q_0, a, p) \mid p \in K_v\},$$

and $\{q_0\} = H_i \cap K_j$. Since τ is admissible, μ' is well defined. Define $\omega : \pi \times X \to \pi \times X$ to be the identity map. Define $\eta : \tau \times \pi \to Q$ by $\eta((K_j, H_i)) = q_0$, where $\{q_0\} = H_i \cap K_j$. Then η is one-to-one and onto. Let ζ be the identity map on X. Then for $\{p_0\} = H_u \cap K_v$,

$$
\begin{aligned}
\mu(\eta((K_j, H_i)), a, \eta((K_v, H_u))) &= \mu(q_0, a, p_0) \\
&= \vee\{\mu(q_0, a, p') \wedge \mu(q_0, a, p) \\
&\quad \mid (p', p) \in K_v \times H_u\} \\
&= (\vee\{\mu(q_0, a, p') \mid p' \in K_v\}) \wedge \\
&\quad (\vee\{\mu(q_0, a, p) \mid p \in H_u\}) \\
&= \mu'(K_j, \omega(H_i, a), K_v) \wedge \mu^\pi(H_i, a, H_u) \\
&= \mu^\omega((K_j, H_i), a, (K_v, H_u)).
\end{aligned}
$$

Thus

$$\mu^*(\eta((K_j, H_i)), x, \eta((K_v, H_u))) = \mu^{\omega*}((K_j, H_i), x, (K_v, H_u)) \qquad (6.3)$$

for $x \in X^*$ such that $|x| = 1$. Suppose that (6.3) is true if $|x| = n-1$, $n > 0$, where $x \in X^*$. Now $\mu^*(\eta((K_j, H_i)), xa, \eta((K_v, H_u))) = \vee\{\mu^*(\eta((K_j, H_i)), x, \eta((K, H))) \wedge \mu(\eta((K, H)), a, \eta((K_v, H_u))) \mid \eta(K, H) \in \eta(\tau \times \pi)\}$ (since η is onto) $= \vee\{\mu^{\omega*}((K_j, H_i), x, (K, H)) \wedge \mu^\omega((K, H), a, (K_v, H_u)) \mid (K, H) \in (\tau \times \pi)\} = \mu^{\omega*}((K_j, H_i), xa, (K_v, H_u))$. Now $\mu^*(\eta((K_j, H_i)), \Lambda, \eta((K_v, H_u))) = 1$ if and only if $\eta((K_j, H_i)) = \eta((K_v, H_u))$ if and only if $(K_j, H_i) = (K_v, H_u)$ (since η is one-to-one) if and only if $\mu^{\omega*}((K_j, H_i), \Lambda, (K_v, H_u)) = 1$. From this, it follows that (6.3) holds for $x = \Lambda$. Hence (6.3) holds $\forall x \in X^*$. Thus by induction, (η, ζ) is a covering of M by $N\omega M/\pi$. ∎

Definition 6.14.16 *Let* Q *be a nonempty set and* π *and* τ *be partitions of* Q. *Then* $\pi \leq \tau$ *if* $\forall A \in \pi$, *there exists* $B \in \tau$ *such that* $A \subseteq B$.

The proof of the following lemma is straightforward.

Lemma 6.14.17 *Let* Q *be a finite nonempty set. Let* $\pi = \{H_i\}_{i=1}^n$ *and* $\tau = \{K_j\}_{j=1}^m$ *be partitions of* Q *such that* $\pi \leq \tau$. *Then* $\forall j$, $1 \leq j \leq m$, $K_j = H_{j_1} \cup H_{j_2} \cup \ldots \cup H_{j_{r_j}}$ *for some* $H_{j_1}, H_{j_2}, \ldots, H_{j_{r_j}} \in \pi$ *and* $m \leq n$. *If* $m = n$, *then* $\pi = \tau$. ∎

Definition 6.14.18 Let $M = (Q, X, \mu)$ be a ffsm. Let π be an admissible partition of Q. Then π is called **maximal** if π is nontrivial and if τ is any admissible partition of Q with $\pi \leq \tau \leq \{Q\}$, then either $\tau = \pi$ or $\tau = \{Q\}$.

Definition 6.14.19 Let $M = (Q, X, \mu)$ be a ffsm. Then M is called **irreducible** if $|Q| > 1$ and 1_Q and $\{Q\}$ are the only admissible partitions of Q.

Theorem 6.14.20 Let $M = (Q, X, \mu)$ be a ffsm. Let $\pi = \{H_i\}_{i=1}^{n}$ be an admissible partition of Q. Then π is maximal if and only if M/π is irreducible.

Proof. Suppose π is maximal. Now $M/\pi = (\pi, X, \mu^{\pi})$ where $\mu^{\pi}(H_i, a, H_j) = \vee\{\mu(q, a, r) \mid r \in H_j\} \forall H_i, H_j \in \pi$ and $a \in X$, where $q \in H_i$. Since π is maximal, $\pi \neq \{Q\}$. Thus $|\pi| > 1$. Let $\bar{\pi}$ be an admissible partition of π. Suppose $\bar{\pi} \neq 1_{\pi}$. Then there exists $\tau \subseteq \pi$ such that $\tau \in \bar{\pi}$ and $|\tau| > 1$. Suppose that $\tau \neq \pi$. Without loss of generality, we may assume that $\tau = \{H_1, \ldots, H_m\}$, where $1 < m < n$. Let $\pi' = \{H_1 \cup \ldots \cup H_m, H_{m+1} \cup \ldots \cup H_n\}$. Then $\pi < \pi'$ and π' is a partition of Q. We now show that π' is admissible. It suffices to consider $q, p \in H_1 \cup \ldots \cup H_m$ with $q \in H_1$ and $p \in H_2$. Suppose $\mu(p, a, r) > 0$ where $r \in H_i$. Then $\mu^{\pi}(H_2, a, H_i) = \vee\{\mu(p, a, r') \mid r' \in H_i\} > 0$. Since $\bar{\pi}$ is admissible, there exists $K \in \bar{\pi}$ such that $\mu^{\pi}(H_1, a, K) \geq \mu^{\pi}(H_2, a, H_i)$ and K and H_i belong to the same element of $\bar{\pi}$. Hence $\vee\{\mu(q, a, t') \mid t' \in K\} \geq \vee\{\mu(p, a, r') \mid r' \in H_i\}$. This implies that there exists $t \in K$ such that $\mu(q, a, t) \geq \mu(p, a, r) > 0$. Now $H_i \in \tau$ if and only if $K \in \tau$ since K and H_i belong to the same element of $\bar{\pi}$. If $K, H_i \in \tau$, then $t, r \in H_1 \cup \ldots \cup H_m$ and if $K, H_i \notin \tau$, then $t, r \in H_{m+1} \cup \ldots \cup H_n$, i.e., t and r belong to the same element of π'. Hence π' is admissible. Since π is maximal, it follows that $\pi' = \{Q\}$ and so $\tau = \{Q\}$. This implies that $\bar{\pi} = \{\pi\}$. Thus M/π is irreducible.

Conversely, suppose that M/π is irreducible. Let τ be an admissible partition of Q such that $\pi \leq \tau \leq \{Q\}$. Suppose that $\pi \neq \tau$. By Lemma 6.14.17, there is no loss of generality in assuming that $\tau = \{H_1 \cup \ldots \cup H_m, H_{m+1}, \ldots, H_n\}$, $1 < m \leq n$. Since $\tau \neq \pi$,

$$\bar{\tau} = \{\{H_1, \ldots, H_m\}, \{H_{m+1}\}, \ldots, \{H_n\}\} \neq 1_{\pi}.$$

We now show that $\bar{\tau}$ is admissible. It suffices to consider H_1, H_2. Suppose $\mu^{\pi}(H_1, a, H_i) > 0$ for some $H_i \in \pi$. Then $\vee\{\mu(q, a, r') \mid r' \in H_i\} > 0$ where $q \in H_1$. Let $r \in H_i$ be such that $\mu(q, a, r) = \vee\{\mu(q, a, r') \mid r' \in H_i\} > 0$. Since τ is admissible, $\forall p \in H_2$, there exists $t_p \in Q$ such that $\mu(p, a, t_p) \geq \mu(q, a, r)$ and r, t_p are in the same element of τ $\forall p \in H_2$. Now if $H_i \notin \{H_1, \ldots, H_m\}$, then $r, t_p \in H_i$ $\forall p \in H_2$ and $\mu^{\pi}(H_2, a, H_i) \geq \mu(p, a, t_p) \geq \mu(q, a, r) = \mu^{\pi}(H_1, a, H_i)$. Suppose $H_i \in \{H_1, \ldots, H_m\}$. Then $r, t_p \in H_1 \cup \ldots \cup H_m$ for all $p \in H_2$. Let $\mu(p, a, t_{p'}) = \vee\{\mu(p, a, t_p) \mid p \in H_2\}$. Since $t_{p'} \in H_1 \cup \ldots \cup H_m$, $t_{p'} \in H_k$ for some k, $1 \leq k \leq m$.

Hence $\mu^\pi(H_2, a, H_k) = \vee\{\mu(q, a, r') \mid r' \in H_k\} \geq \mu(p, a, t_{p'}) \geq \mu(q, a, r) = \mu^\pi(H_1, a, H_i)$ and $H_k, H_i \in \{H_1, \ldots, H_m\}$. Consequently $\bar{\tau}$ is an admissible partition of π. Since M/π is irreducible, $\bar{\tau} = \{\pi\}$ and so $\tau = \{Q\}$. Hence π is maximal. ∎

Theorem 6.14.21 Let $M = (Q, X, \mu)$ be a ffsm and $|Q| = n \geq 2$. Then

$$M \leq N_1\omega_1 N_2\omega_2 \ldots \omega_{m-1}N_m,$$

where $N_1, N_2, \ldots, N_m$ are irreducible ffsms and the state sets Q_i of N_i are such that $|Q_i| < n$.

Proof. Since $|Q| \geq 2$, we can choose a maximal admissible partition π of Q. Clearly $|\pi| < |Q|$. By Theorem 6.14.15, there exists a ffsm N such that $M \leq N\omega M/\pi$ for some suitable ω. Since π is maximal, M/π is irreducible. Also π is the state set of M/π and $|\pi| < |Q|$. If R is the state set of N, then $|R| < |Q|$ by the construction of N. We can now apply Theorem 6.14.15 to N, to obtain $N \leq N'\omega'N/\pi'$ such that N/π' is irreducible and the number of states in N' is less than the number of states in N. Thus, we have

$$M \leq N'\omega'N/\pi'\omega M/\pi.$$

Continue this process and apply Theorem 6.14.15 to N'. Since $|Q|$ is finite, this process must stop after a finite number of steps. Hence $M \leq N_1\omega_1 N_2\omega_2 \ldots \omega_{m-1}N_m$, where $N_1, N_2, \ldots, N_m$ are irreducible ffsms and the state sets Q_i of N_i are such that $|Q_i| < n$. ∎

6.15 Coverings of Products of Fuzzy Finite State Machines

We are now interested in the notion of a covering that is more in line with the crisp notion. The next three sections are based on [119]. We present several ways of constructing products of fuzzy finite state machines and their relationship through isomorphisms and coverings.

We prove that the wreath product of two fuzzy finite state machines covers their cascade product. Similar relationships hold for the direct sum and sum of two fuzzy finite state machines. A distributive property of the cascade product over the sum of fuzzy finite state machines in relation to coverings of fuzzy finite state machines is established.

Let $M_1 = (Q_1, X_1, \mu_1)$ and $M_2 = (Q_2, X_2, \mu_2)$ be fuzzy finite state machines. Let $\alpha : Q_1 \to Q_2$ and $\beta : X_1 \to X_2$ be functions.

The homomorphism $(\alpha, \beta) : M_1 \to M_2$ (Definition 6.3.1) is called a **monomorphism (epimorphism, isomorphism)**, if both the functions α, β are injective (surjective, bijective, respectively). If (α, β) is an isomorphism, we write $M_1 \cong M_2$.

Definition 6.15.1 *Let $M_1 = (Q_1, X_1, \mu_1)$ and $M_2 = (Q_2, X_2, \mu_2)$ be fuzzy finite state machines. Let $\eta : Q_2 \to Q_1$ be a surjective partial function and $\xi : X_1 \to X_2$ be a function. Then the pair (η, ξ) is called a **strong covering** of M_1 by M_2, written $M_1 \preceq M_2$, if*

$$\mu_1^+(\eta(p'), x, \eta(q')) \leq \mu_2^+(p', \xi^+(x), q')$$

for all $x \in X_1^+$ and p', q' belong to the domain of η, where $\xi^+ : X_1^+ \to X_2^+$ is defined by $\xi^+(x_1 x_2 \ldots x_n) = \xi(x_1)\xi(x_2)\ldots\xi(x_n)$ for $x_1, x_2, \ldots, x_n \in X_1$.

Clearly, the strong covering relation is reflexive and transitive, but not symmetric.

Definition 6.15.2 *Let $M = (Q, X, \mu)$ be a fuzzy finite state machine. A fuzzy finite state machine $M^s = (Q^s, X^s, \mu^s)$ is called a **subfuzzy finite state machine** of M, written $M^s \subseteq M$, if*
(1) $Q^s \subseteq Q$, $X^s \subseteq X$, and
(2) $\mu^s = \mu|_{Q^s \times X^s \times \mu^s}$.

We point out that a subfuzzy finite state machine of M and a submachine of M (Definition 6.7.7) differ in that for the latter, the set of input symbols remains X.

Identifying X with the diagonal elements of $X \times X$, the restricted direct product of M and M' may be considered as the subfuzzy finite state machine of the direct product of M and M'.

The restricted direct product of M and M' is a special case of their cascade product, where $X = X'$ and $\omega : Q' \times X \to X$ is a projection mapping.

We now introduce two more ways of connecting fuzzy finite state machines.

Definition 6.15.3 *Let $M = (Q, X, \mu)$ and $M' = (Q', X', \mu')$ be fuzzy finite state machines such that $Q \cap Q' = \emptyset$ and $X \cap X' = \emptyset$. Then the fuzzy finite state machine $M \oplus M' = (Q \cup Q', X \cup X', \mu \oplus \mu')$ is called the **direct sum** of M and M', where $\mu \oplus \mu' : (Q \cup Q') \times (X \cup X') \times (Q \cup Q') \to [0,1]$ is defined as follows:*

$$(\mu \oplus \mu')(p, x, q) = \begin{cases} \mu(p, x, q) & \text{if } p, q \in Q \text{ and } x \in X \\ \mu'(p, x, q) & \text{if } p, q \in Q' \text{ and } x \in X' \\ 1 & \text{if either } (p, x) \in Q \times X \text{ and } q \in Q' \\ & \text{or } (p, x) \in Q' \times X' \text{ and } q \in Q \\ 0 & \text{otherwise.} \end{cases}$$

Definition 6.15.4 *Let $M = (Q, X, \mu)$ and $M' = (Q', X', \mu')$ be fuzzy finite state machines such that $Q \cap Q' = \emptyset$ and $X \cap X' = \emptyset$. Then the fuzzy finite state machine $M + M' = (Q \cup Q', X \cup X', \mu + \mu')$ is called the **sum** of M*

and M', where $\mu + \mu' : (Q \cup Q') \times (X \cup X') \times (Q \cup Q') \to [0,1]$ is defined as follows:

$$(\mu + \mu')(p,x,q) = \begin{cases} \mu(p,x,q) & \text{if } p,q \in Q \text{ and } x \in X \\ \mu'(p,x,q) & \text{if } p,q \in Q' \text{ and } x \in X' \\ 0 & \text{otherwise.} \end{cases}$$

6.16 Associative Properties of Products

The direct and restricted direct products of fuzzy finite state machines are associative. As proved in the following theorem, the wreath product, sum, and cascade products of fuzzy finite state machines are also associative. However, it can easily be shown that the direct sum of fuzzy finite state machines is not associative.

Theorem 6.16.1 *Let M, M', and M'' be fuzzy finite state machines. Then the following properties hold:*
(1) $(M \circ M') \circ M'' \cong M \circ (M' \circ M'')$
(2) $(M + M') + M'' \cong M + (M' + M'')$ and
(3) $(M\omega_1 M')\omega_2 M'' \cong M\omega_3(M'\omega_4 M'')$, where ω_3 and ω_4 are defined naturally in terms of ω_1 and ω_2.

Proof. Let $M = (Q, X, \mu)$, $M' = (Q', X', \mu')$, and $M'' = (Q'', X'', \mu'')$.
(1) Then

$$(M \circ M') \circ M'' = ((Q \times Q') \times Q'', (X^{Q'} \times X')^{Q''} \times X'', (\mu \circ \mu') \circ \mu'')$$

and

$$M \circ (M' \circ M'') = (Q \times (Q' \times Q''), (X^{Q' \times Q''}) \times X'^{Q''} \times X'', \mu \circ (\mu' \circ \mu'')).$$

Let $\alpha : (Q \times Q') \times Q'' \to Q \times (Q' \times Q'')$ be the natural map. Let $p_1 : X^{Q'} \times X' \to X^{Q'}$ and $p_2 : X^{Q'} \times X' \to X'$ be the natural projection mappings.

Given a function $f : Q'' \to X^{Q'} \times X'$, let $f_1 = p_1 \circ f$ and $f_2 = p_2 \circ f$. Define $\overline{f_1} : Q' \times Q'' \to X$ by $\overline{f_1}((q',q'')) = f_1(q'')(q')$. Define $\beta : (X^{Q'} \times X')^{Q''} \times X'' \to X^{Q' \times Q''} \times (X'^{Q''} \times X'')$ by $\beta((f,x'')) = (\overline{f_1}, (f_2, x''))$. Now

$$\beta((f,x'')) = \beta((g,y''))$$
$$\Rightarrow \quad (\overline{f_1}, (f_2, x'')) = (\overline{g_1}, (g_2, y''))$$
$$\Rightarrow \quad \overline{f_1} = \overline{g_1}, f_2 = g_2, \text{ and } x'' = y''.$$

Thus $f_1(q'')(q') = g_1(q'')(q')$, $f_2 = g_2$, and $x'' = y''$. This implies that $f_1 = g_1$, $f_2 = g_2$, and $x'' = y''$, which implies that $p_1 \circ f = p_1 \circ g$, $p_2 \circ f = p_2 \circ g$, and $x'' = y''$. Therefore, $f = g$ and $x'' = y''$ and so $(f,x) = (g,y)$. Thus β is injective.

Let $(g, (h, x'')) \in X^{Q' \times Q''} \times (X'^{Q''} \times X'')$. Define $f : Q'' \to X^{Q'} \times X'$ by $f(q'') = (g_{q''}, h(q''))$, where $g_{q''}(q') = g(q', q'')$. Then $\beta((f, x'')) = (g, (h, x''))$. Thus β is surjective.

It follows that (α, β) is a required isomorphism.

(2) Let α and β be identity mappings on $Q \cup Q' \cup Q''$ and $X \cup X' \cup X''$, respectively.

(3) Consider $M\omega_1 M' = (Q \times Q', X', \mu\omega_1\mu')$, where $\omega_1 : Q' \times X' \to X$ and

$$\mu\omega_1\mu'((p, p'), x', (q, q')) = \mu(p, \omega_1(p', x'), q) \wedge \mu'(p', x', q')$$

and $(M\omega_1 M')\omega_2 M'' = ((Q \times Q') \times Q'', X'', (\mu\omega_1\mu')\omega_2\mu'')$, where $\omega_2 : Q'' \times X'' \to X'$ and

$$
\begin{aligned}
(\mu\omega_1\mu')\omega_2\mu''(((p, p'), p''), x'', ((q, q'), q'')) &= \mu\omega_1\mu'((p, p'), \omega_2(p'', x''), \\
&\quad (q, q')) \wedge \mu''(p'', x'', q'') \\
&= \mu(p, \omega_1(p', \omega_2(p'', x'')), q) \\
&\quad \wedge \mu'(p', \omega_2(p'', x''), q) \\
&\quad \wedge \mu''(p'', x'', q'').
\end{aligned}
$$

Define $\omega_3 : (Q' \times Q'') \times X'' \to X$ by $\omega_3((p', p''), x'') = \omega_1(p', \omega_2(p'', x''))$ and set $\omega_4 = \omega_2$.

It follows immediately that $M'\omega_4 M'' = (Q' \times Q'', X'', \mu'\omega_4\mu'')$ and

$$\mu'\omega_4\mu''((p', p''), x'', (q', q'')) = \mu'(p', \omega_4(p'', x''), q') \wedge \mu''(p'', x'', q'').$$

Moreover, $M\omega_3(M'\omega_4 M'') = (Q \times (Q' \times Q''), X'', \mu\omega_3(\mu'\omega_4\mu''))$ and

$$
\begin{aligned}
\mu\omega_3(\mu'\omega_4\mu'')((p, (p', p'')), x'', (q, (q', q''))) &= \mu(p, \omega_3((p', p''), x''), q) \wedge \\
&\quad (\mu'\omega_4\mu''((p', p''), x'', \\
&\quad (q', q''))) \\
&= \mu(p, \omega_3((p', p''), x''), q) \wedge \\
&\quad (\mu'(p', \omega_4(p'', x''), q') \\
&\quad \wedge \mu''(p'', x'', q'')).
\end{aligned}
$$

Let $\alpha : (Q \times Q') \times Q'' \to Q \times (Q' \times Q'')$ be the natural mapping and β be the identity mapping on X''. Then

$$
\begin{aligned}
(\mu\omega_1\mu')\omega_2\mu'')(((p,p'),p''),x'',((q,q'),q'')) &= \mu(p,\omega_1(p',\omega_2(p'',x''))), \\
&\quad q) \wedge \mu'(p',\omega_2(p'',x''), \\
&\quad q') \wedge \mu''(p'',x'',q'') \\
&= \mu(p,\omega_1(p',\omega_2(p'',x''))), \\
&\quad q) \wedge (\mu'(p',\omega_2(p'',x''), \\
&\quad q') \wedge \mu''(p'',x'',q'')) \\
&= \mu(p,\omega_3((p',p''),x''),q) \\
&\quad \wedge(\mu'(p',\omega_4(p'',x''),q') \\
&\quad \wedge\mu''(p'',x'',q'') \\
&= \mu\omega_3(\mu'\omega_4\mu'')((p,(p', \\
&\quad p'')),x'',(q,(q',q''))) \\
&= \mu\omega_3(\mu'\omega_4\mu'')(\alpha(((p,p'), \\
&\quad p'')),\beta(x''),\alpha(((q,q'), \\
&\quad q''))).
\end{aligned}
$$

Thus $(M\omega_1 M')\omega_2 M'' \cong M\omega_3(M'\omega_4 M'')$. ∎

6.17   Covering Properties of Products

The following theorem is a direct consequence of the definition of the direct sum and sum of two fuzzy finite state machines.

Theorem 6.17.1 *Let M and M' be fuzzy finite state machines. Then the following properties hold:*
(1) $M \preceq M + M'$,
(2) $M' \preceq M + M'$,
(3) $M \preceq M \oplus M'$,
(4) $M' \preceq M \oplus M'$.

Proof. We only prove (1). Let $\eta : Q \cup Q' \to Q$ be a partial surjective function defined by $\eta(q) = q$ for all $q \in Q$ and let $\xi : X \to X \cup X'$ be the inclusion function. Clearly, (η, ξ) is a required strong covering of M by $M + M'$. ∎

Theorem 6.17.2 *Let $M = (Q, X, \mu)$ and $M' = (Q', X', \mu')$ be fuzzy finite state machines. Then the following properties hold:*
(1) $M\omega M' \preceq M \circ M'$,
(2) $M + M' \preceq M \oplus M'$.

Proof. (1) Let $\omega_{x'} : Q' \to X$ be the function defined by $\omega_{x'}(p') = \omega(p',x')$ for all $p' \in Q'$ and $x' \in X'$. Define $\xi : X' \to X^{Q'} \times X'$ by $\xi(x') = (\omega_{x'}, x')$ and let η be the identity map on $Q \times Q'$.

(2) Let η and ξ be the identity mappings on $Q \cup Q'$ and $X \cup X'$, respectively.

If $p, q \in Q$ and $x \in X$, then $(\mu + \mu')(\eta(p), x, \eta(q)) = (\mu \oplus \mu')(p, \xi(x), q) = \mu(p, x, q)$.

If $p, q \in Q'$ and $x \in X'$, then $(\mu + \mu')(\eta(p), x, \eta(q)) = (\mu \oplus \mu')(p, \xi(x), q) = \mu'(p, x, q)$.

If $(p, x) \in Q \times X$ and $q \in Q'$ or $(p, x) \in Q' \times X'$ and $q \in Q$, then $(\mu + \mu')(\eta(p), x, \eta(q)) = 0 < 1 = (\mu \oplus \mu')(p, \xi(x), q)$. In all other cases, the values of $\mu + \mu'$ and $\mu \oplus \mu'$ are 0. ∎

Theorem 6.17.3 *Let* $M = (Q, X, \mu)$, $M' = (Q', X', \mu')$, *and* $M'' = (Q'', X'', \mu'')$ *be fuzzy finite state machines such that* $M \preceq M'$. *Then the following assertions hold:*

(1) $M \times M'' \preceq M' \times M''$;

(2) $M'' \times M \preceq M'' \times M'$;

(3) Given $\omega_1 : Q'' \times X'' \to X$, *there exists* $\omega_2 : Q'' \times X'' \to X'$ *such that* $M\omega_1 M'' \preceq M'\omega_2 M''$;

(4) If (η, ξ) *is a covering of* M *by* M' *and* ξ *is surjective, then for all* $\omega_1 : Q \times X \to X''$, *there exists* $\omega_2 : Q' \times X' \to X''$ *such that* $M''\omega_1 M \preceq M''\omega_2 M'$;

(5) $M \circ M'' \preceq M' \circ M''$;

(6) $M + M'' \preceq M' + M''$;

(7) $M'' + M \preceq M'' + M'$;

(8) $M \oplus M'' \preceq M' \oplus M''$;

(9) $M'' \oplus M \preceq M'' \oplus M'$.

Moreover, if $X = X' = X''$, *then*

(10) $M \wedge M'' \preceq M' \wedge M''$;

(11) $M'' \wedge M \preceq M'' \wedge M'$.

Proof. Since $M \preceq M'$, there exists a partial surjective function $\eta : Q' \to Q$ and a function $\xi : X \to X'$ such that $\mu^+(\eta(p), x, \eta(p')) \leq \mu'^+(p, \xi(x), p')$, where $x \in X^+$.

(1) Define $\eta' : Q' \times Q'' \to Q \times Q''$ by $\eta'(p', p'') = (\eta(p'), p'')$ and $\xi' : X \times X' \to X' \times X''$ by $\xi'(x, x'') = (\xi(x), x'')$. Clearly (η', ξ') is the required covering.

(2) The proof is similar to that of (1).

(3) Given $\omega_1 : Q'' \times X'' \to X$, set $\omega_2 = \xi \circ \omega_1$ and let ξ' be the identity map on X''.

(4) Given $\omega_1 : Q \times X \to X''$, let $\omega_2 : Q' \times X' \to X''$ be such that $\omega_2(p', \xi(x)) = \omega_1(\eta(p'), x)$. Since ξ is surjective and X is finite, such ω_2 exists. Clearly, ω_2 is not unique. Define $\eta' : Q'' \times Q' \to Q'' \times Q$ by $\eta'(p'', p') = (p'', \eta(p'))$ and set $\xi' = \xi$.

(5) Define $\eta' : Q' \times Q'' \to Q \times Q''$ by $\eta'(p', p'') = (\eta(p'), p'')$ and $\xi' : X^{Q''} \times X'' \to (X')^{Q''} \times X''$ by $\xi'(f, x'') = (\xi \circ f, x'')$.

(6) Recall that $M + M'' = (Q \cup Q'', X \cup X'', \mu + \mu'')$, where

$$(\mu + \mu'')(p, x, q) = \begin{cases} \mu(p, x, q) & \text{if } p, q \in Q \text{ and } x \in X \\ \mu''(p, x, q) & \text{if } p, q \in Q'' \text{ and } x \in X'' \\ 0 & \text{otherwise} \end{cases}$$

and $M' + M'' = (Q' \cup Q'', X' \cup X'', \mu' + \mu'')$, where

$$(\mu' + \mu'')(p, x, q) = \begin{cases} \mu'(p, x, q) & \text{if } p, q \in Q' \text{ and } x \in X' \\ \mu''(p, x, q) & \text{if } p, q \in Q'' \text{ and } x \in X'' \\ 0 & \text{otherwise.} \end{cases}$$

Define $\eta' : Q' \cup Q'' \to Q \cup Q''$ by

$$\eta'(p') = \begin{cases} \eta(p') & \text{if } p' \in Q' \\ p' & \text{otherwise} \end{cases}$$

and $\xi' : X \cup X'' \to X' \cup X''$ by

$$\xi'(x) = \begin{cases} \xi(x) & \text{if } x \in X \\ x & \text{otherwise.} \end{cases}$$

Since η is a partial surjective function, so is η'.

We now note that $(\mu + \mu'')^+(\eta'(p'), x, \eta'(q')) \leq (\mu' + \mu'')^+(p', (\xi')^+(x), q')$. If $p', q' \in Q'$ and $x \in X$ or $p', q' \in Q''$ and $x \in X''$, then clearly

$$(\mu + \mu'')^+(\eta'(p'), x, \eta'(q')) \leq (\mu' + \mu'')^+(p', (\zeta')^+(x), q').$$

In all other cases, $(\mu + \mu'')^+(\eta'(p'), x, \eta'(q')) = 0$.

The proofs of (7), (8), and (9) are now obvious.

(10) Define $\eta' : Q' \times Q'' \to Q \times Q''$ by $\eta'(p', p'') = (\eta(p'), p'')$ and set $\xi' = \xi$ the identity map on X. Then (η', ξ') is the required strong covering.

(11) Follows easily. ∎

It follows that $M'' \circ M'$, in general, does not strongly cover $M'' \circ M$, even though $M \preceq M'$. Hence we introduce a weaker notion of covering.

Definition 6.17.4 Let $M = (Q, X, \mu)$ and $M' = (Q', X', \mu')$ be fuzzy finite state machines. Let $\eta : Q \to Q'$ be a partial surjective function and $\xi : X \to X'$ be a partial function. The ordered pair (η, ξ) is called a **weak covering** of M by M', written $M \preceq_w M'$, if

$$\mu(\eta(p'), x, \eta(q')) \leq \mu'(p', \xi(x), q')$$

for all p', q' in the domain of η and x in the domain of ξ.

A weak covering differs from a strong covering only in that ξ in Definition 6.17.4 is a partial function, while ξ in Definition 6.15.1 is a function. Thus every strong covering is a weak covering.

Theorem 6.17.5 *Let M, M', and M'' be fuzzy finite state machines and $M \preceq M'$. Then $M'' \circ M \preceq_w M'' \circ M'$.*

Proof. Since $M \preceq M'$, there exists a partial surjective function $\eta :$ $Q' \to Q$ and a function $\xi : X \to X'$ such that $\mu^+(\eta(p'), x, \eta(q')) \leq$ $\mu'^+(p', \xi^+(x), q')$. We have $M'' \circ M = (Q'' \times Q, (X'')^Q \times X, \mu'' \circ \mu)$ and $M'' \circ M' = (Q'' \times Q', (X'')^{Q'} \times X', \mu'' \circ \mu')$. Define $\eta' : Q'' \times Q' \to$ $Q'' \times Q$ by $\eta'(p'', p') = (p'', \eta(p'))$ and $\xi' : (X'')^Q \times X \to (X'')^{Q'} \times X'$ by $\xi'(f, x) = (f \circ \eta, \xi(x))$. Clearly, η' is a partial surjective function and ξ' is a partial function. Now $(\mu'' \circ \mu)(\eta(p'', p'), (f, x), \eta'(q'', q')) = (\mu'' \circ \mu)((p'', \eta(p')), (f, x), (q'', \eta(q'))) = \mu''(p'', f(\eta(p')), q'') \wedge \mu(\eta(p'), x, \eta(q'))$. However, since $\mu^+(\eta(p'), x, \eta(q')) \leq \mu'^+(p', \xi^+(x), q')$, it follows that

$$
\begin{aligned}
(\mu'' \circ \mu)^+(\eta(p'', p'), (f, x), \eta'(q'', q')) \ &\leq \ (\mu'')^+(p'', f(\eta(p')), q'') \wedge \\
&\quad \mu'^+(p', \xi^+(x), q') \\
&= \ (\mu'' \circ \mu')^+((p'', p'), \\
&\quad (f \circ \eta, \xi^+(x)), (q'', q')) \\
&= \ (\mu'' \circ \mu')^+((p'', p'), \xi'^+(f, x), \\
&\quad (q'', q')).
\end{aligned}
$$

∎

The following theorems are easy consequences of the transitive property of strong coverings of fuzzy finite state machines and Theorem 6.17.3.

Theorem 6.17.6 *Let M, M', and M'' be fuzzy finite state machines and $M \preceq M'$. Then*
  (1) $M \wedge M'' \preceq M' \wedge M''$,
  (2) $M'' \wedge M \preceq M'' \wedge M'$,
  (3) $M \omega M'' \preceq M' \circ M''$,
  (4) $M'' \omega M \preceq_w M'' \circ M'$,
  (5) $M + M'' \preceq M' \oplus M''$,
  (6) $M'' + M \preceq M'' \oplus M'$. ∎

We next prove a distributive property for strong coverings.

Theorem 6.17.7 *Let M, M', and M'' be fuzzy finite state machines and $M \preceq M'$. Then*

$$M \circ (M' + M'') \preceq M \circ M' + M \circ M''.$$

Proof. Let $M = (Q, X, \mu)$, $M' = (Q', X', \mu')$, and $M'' = (Q'', X'', \mu'')$. Define

$$\eta : (Q \times Q') \cup (Q \times Q'') \to Q \times (Q' \cup Q'')$$

by $\eta(p, p') = (p, p')$ and

$$\xi : X^{Q' \cup Q''} \times (X' \cup X'') \to (X^{Q'} \times X') \cup (X^{Q''} \times X'')$$

by

$$\xi(g, x') = \begin{cases} (g|_{Q'}, x') & \text{if } x' \in X' \\ (g|_{Q''}, x') & \text{if } x' \in X''. \end{cases}$$

∎

In the case of finite state machines, different types of products and their coverings play a crucial role in their decomposition. We have established fuzzy analogs of different products and their coverings. Therefore, these results may be useful in the decomposition of fuzzy finite state machines.

6.18   Fuzzy Semiautomaton over a Finite Group

In the remainder of the chapter, we assume that the reader is familiar with the basic results of group theory.

The group semiautomaton has been studied extensively in [90] and [60]. We now consider fuzzy semiautomata over a finite group, [41]. A group semiautomaton is a quadruple $(Q, *, X, \delta)$ such that $(Q, *)$ is a finite group, where Q is called the set of states, X is a finite set, called the set of inputs, and δ is a function from $Q \times X$ into Q. We present the notion of a fuzzy semiautomaton over a finite group. The fuzzy kernel and fuzzy subsemiautomaton of a fuzzy semiautomaton over a finite group are defined using the notions of a fuzzy normal subgroup and a fuzzy subgroup [184] of a group.

We let $(G, *)$ denote a group. We sometimes write G for $(G, *)$ when the operation $*$ is understood.

Definition 6.18.1 *A fuzzy subset λ of G is called a **fuzzy subgroup** of G if the following properties hold:*
(1) $\lambda(x * y) \geq \lambda(x) \wedge \lambda(y)$,
(2) $\lambda(x) = \lambda(x^{-1})$
for all $x, y \in G$.

Definition 6.18.2 *A fuzzy subgroup λ of G is called a **fuzzy normal subgroup** of G if*

$$\lambda(x * y * x^{-1}) \geq \lambda(y)$$

for all $x, y \in G$.

Definition 6.18.3 *Let λ and μ be fuzzy subsets of G. The **product** $\lambda * \mu$ of λ and μ is defined by*

$$(\lambda * \mu)(x) = \vee\{\lambda(y) \wedge \mu(z) \mid y, z \in G \text{ such that } x = y * z\}$$

for all $x \in G$.

Definition 6.18.4 *Let λ and μ be fuzzy subgroups of G such that $\lambda \subseteq \mu$. Then λ is called a **fuzzy normal subgroup** of μ if*

$$\lambda(x * y * x^{-1}) \geq \lambda(y) \wedge \mu(x)$$

for all $x, y \in G$.

Let λ and μ be fuzzy subgroups of G such that λ is a fuzzy normal subgroup of μ. Then $\text{Supp}(\lambda)$ is a normal subgroup of the group $\text{Supp}(\mu)$. Define the fuzzy subset μ/λ of the quotient group $\text{Supp}(\mu)/\text{Supp}(\lambda)$ by $\forall$ cosets $q\text{Supp}(\lambda) \in \text{Supp}(\mu)/\text{Supp}(\lambda)$, $\mu/\lambda(q\,\text{Supp}(\lambda)) = \vee\{\mu(p) \mid p \in q\,\text{Supp}(\lambda)\}$. Then μ/λ is a fuzzy subgroup of $\text{Supp}(\mu)/\text{Supp}(\lambda)$.

Definition 6.18.5 *Let λ and μ be fuzzy subsets of the groups G and H, respectively. Let $f : G \to H$ be a group homomorphism. The fuzzy subsets $f(\lambda)$ of H and $f^{-1}(\mu)$ of G are defined as follows:*

$$f(\lambda)(h) = \begin{cases} \vee\{\lambda(g) \mid g \in G \text{ such that } f(g) = h\} & \text{if } f^{-1}(H) \neq \emptyset \\ 0 & \text{if } f^{-1}(H) = \emptyset, \end{cases}$$

for all $h \in H$, and

$$f^{-1}(\mu)(g) = \mu(f(g))$$

for all $g \in G$.

Definition 6.18.6 *Let λ and μ be fuzzy subgroups of the groups G and H, respectively. A function $f : G \to H$ is called a **weak homomorphism** from λ into μ if the following conditions hold:*
(1) f is an epimorphism,
(2) $f(\lambda) \subseteq \mu$.
*If f is an isomorphism from G onto H, then f is called a **weak isomorphism** from λ into μ.*

Definition 6.18.7 *A **fuzzy semiautomaton (fsa)** over a finite group $(Q, *)$ is a triple (Q, X, μ), where X is a finite set and μ is a fuzzy subset of $Q \times X \times Q$.*

Let μ^* be the extension of μ to $Q \times X^* \times Q$ as defined in Definition 6.2.1.

Definition 6.18.8 *Let $S = (Q, X, \mu)$ and $T = (Q_1, X_1, \mu_1)$ be fsa's over a finite group. A pair of functions (f, g), where $f : Q \to Q_1$, $g : X \to X_1$, is called a **homomorphism** from S into T, written $(f, g) : S \to T$ if the following conditions hold:*
(1) f is a group homomorphism,
(2) $\mu(p, x, q) \leq \mu_1(f(p), g(x), f(q))$ for all $p, q \in Q$, $x \in X$.

The pair (f, g) is called a **strong homomorphism** from S into T if it satisfies (1) of Definition 6.18.8 and the added condition,

$$\mu_1(f(p), g(x), f(q)) = \vee\{\mu(p, x, r) \mid r \in Q, \ f(r) = f(q)\}$$

for all $p, q \in Q$, $x \in X$.

In Definition 6.18.8, if $X = X_1$ and g is the identity map of X, then we write $f : S \to T$ and say that f is a homomorphism or strong homomorphism from S into T.

Let $S = (Q, X, \mu)$ be a fsa over a finite group in the remainder of the chapter.

Definition 6.18.9 *A fuzzy subset λ of Q is called a* **fuzzy kernel** *of S if the following conditions hold:*

(1) λ is a fuzzy normal subgroup of Q,

(2) $\lambda(p * r^{-1}) \geq \mu(q * k, x, p) \wedge \mu(q, x, r) \wedge \lambda(k)$ for all $p, q, k, r \in Q$, $x \in X$.

Definition 6.18.10 *A fuzzy subset λ of Q is called a* **fuzzy subsemiautomaton** *of S if the following conditions hold:*

(1) λ is a fuzzy subgroup of Q,

(2) $\lambda(p) \geq \mu(q, x, p) \wedge \lambda(q)$ for all $p, q \in Q$, $x \in X$.

Theorem 6.18.11 *A fuzzy normal subgroup λ of Q is a fuzzy kernel of S if and only if*

$$\lambda(p * r^{-1}) \geq \mu^*(q * k, x, p) \wedge \mu^*(q, x, r) \wedge \lambda(k)$$

for all $p, q, k, r \in Q$, $x \in X^*$.

Proof. Let λ be a fuzzy kernel of S. We prove the theorem by induction on $|x| = n$. Let $n = 0$. Then $x = \Lambda$. Let $p, q, k, r \in Q$. If $p = q * k$, $r = q$, then $\mu^*(q*k, x, p) \wedge \mu^*(q, x, r) \wedge \lambda(k) = \lambda(k) \leq \lambda(q*k*q^{-1})$ since λ is a fuzzy normal subgroup. If $p \neq q*k$ or $r \neq q$, then $\mu^*(q*k, x, p) \wedge \mu^*(q, x, r) \wedge \lambda(k) = 0 \leq \lambda(p * r^{-1})$. Thus the result holds for $n = 0$. Suppose that the result holds for all $y \in X^*$, where $|y| = n - 1$, $n > 0$. Let $x \in X^*$ be such that

$x = ya, \ y \in X^*, a \in X, \ |y| = n - 1, n > 0.$ Then

$$
\begin{aligned}
\mu^*(q * k, x, p) \wedge \mu^*(q, x, r) \wedge \lambda(k) \ &= \ (\vee\{\mu^*(q * k, y, u) \wedge \mu(u, a, p) \\
&\quad \ | \ u \in Q\}) \wedge (\vee\{\mu^*(q, y, v) \wedge \\
&\quad \ \mu(v, a, r) \ | \ v \in Q\}\}) \wedge \lambda(k) \\
&= \ \vee\{\vee\{\mu^*(q * k, y, u) \wedge \mu(u, a, p) \wedge \\
&\quad \ \mu^*(q, y, v) \wedge \mu(v, a, r) \wedge \lambda(k) \\
&\quad \ | \ u \in Q\} \ | \ v \in Q\} \\
&\leq \ \vee\{\vee\{\lambda(u * v^{-1}) \wedge \mu(u, a, p) \\
&\quad \ \wedge \mu(v, a, r)| \ u \in Q\} \ | \ v \in Q\} \\
&\leq \ \vee\{\vee\{\lambda(v^{-1} * u) \wedge \mu(v * v^{-1} * u, \\
&\quad \ a, p) \wedge \mu(v, a, r)| \ u \in Q\} \ | \ v \in Q\} \\
&\quad \ (\text{since } \lambda \text{ is a normal fuzzy} \\
&\quad \ \text{subgroup}) \\
&\leq \ \lambda(p * r^{-1}).
\end{aligned}
$$

Hence the desired condition holds. The converse follows easily. ∎

Theorem 6.18.12 *A fuzzy subgroup λ of Q is a fuzzy subsemiautomaton of S if and only if*

$$
\lambda(p) \geq \mu^*(q, x, p) \wedge \lambda(q)
$$

for all $p, q \in Q, \ x \in X^*$.

Proof. The proof is similar to that of Theorem 6.18.11. ∎

Theorem 6.18.13 *Let $T = (Q_1, X, \mu_1)$ be an fsa over a finite group and let f be a homomorphism from S into T. If λ is a fuzzy subsemiautomaton (fuzzy kernel) of T, then $f^{-1}(\lambda)$ is a fuzzy subsemiautomaton (fuzzy kernel) of S.*

Proof. The proof follows easily. ∎

Theorem 6.18.14 *Let $T = (Q_1, X, \mu_1)$ be an fsa over a finite group and let f be a strong homomorphism from S onto T. If λ is a fuzzy subsemiautomaton (fuzzy kernel) of S, then $f(\lambda)$ is a fuzzy subsemiautomaton (fuzzy kernel) of T.*

Proof. Let λ be a fuzzy kernel of S. Then λ is a fuzzy normal subgroup of Q. Since f is an epimorphism from Q onto Q_1, $f(\lambda)$ is a fuzzy normal subgroup of Q_1. Let $p_1, q_1, r_1, k_1 \in Q_1, \ x \in X$. Then

$$
\begin{aligned}
\mu_1(q_1 * k_1, x, p_1) \wedge \mu_1(q_1, x, r_1) \wedge f(\lambda)(k_1) \ &= \ \mu_1(q_1 * k_1, x, p_1) \wedge \\
&\quad \ \mu_1(q_1, x, r_1) \wedge (\vee\{\lambda(k) \\
&\quad \ | \ k \in Q, \ f(k) = k_1\}) \\
&= \ \vee\{\mu_1(q_1 * k_1, x, p_1) \wedge \\
&\quad \ \mu_1(q_1, x, r_1) \wedge \lambda(k) \\
&\quad \ | \ k \in Q, \ f(k) = k_1\}.
\end{aligned}
$$

Now let $p, q, r, k \in Q$ be such that $f(p) = p_1$, $f(q) = q_1$, $f(r) = r_1$, and $f(k) = k_1$. Then

$$
\begin{aligned}
\mu_1(q_1 * k_1, x, p_1) \wedge \mu_1(q_1, x, r_1) \wedge \lambda(k) &= \mu_1(f(q * k), x, f(p)) \wedge \\
&\quad \mu_1(f(q), x, f(r)) \wedge \lambda(k) \\
&= (\vee\{\mu(q * k, x, a) \mid a \in Q, f(a) \\
&\quad = f(p)\}) \wedge (\vee\{\mu(q, x, b)| \\
&\quad b \in Q, f(b) = f(r)\}) \wedge \lambda(k) \\
&= \vee\{\vee\{\mu(q * k, x, a) \wedge \mu(q, x, b) \\
&\quad \wedge \lambda(k) \,|a \in Q, \ f(a) = f(p)\} \\
&\quad \mid b \in Q, f(b) = f(r)\} \\
&\leq \vee\{\vee\{\lambda(a * b^{-1}) \mid a \in Q, \ f(a) \\
&\quad = f(p)\} \mid b \in Q, f(b) = f(r)\} \\
&\leq f(\lambda)(p_1 * r_1^{-1}).
\end{aligned}
$$

Hence

$$
\begin{aligned}
f(\lambda)(p_1 * r_1^{-1}) &\geq \vee\{\mu_1(q_1 * k_1, x, p_1) \wedge \mu_1(q_1, x, r_1) \wedge \lambda(k) \\
&\quad \mid k \in Q, \ f(k) = k_1\} \\
&= \mu_1(q_1 * k_1, x, p_1) \wedge \mu_1(q_1, x, r_1) \wedge f(\lambda)(k_1).
\end{aligned}
$$

Thus $f(\lambda)$ is a fuzzy kernel of T. ∎

As in Theorem 6.18.14, it can be shown that $f(\lambda)$ is a fuzzy subsemiautomaton of T if λ is a fuzzy subsemiautomaton of S.

For the remainder of the chapter, unless otherwise stated, $S = (Q, X, \mu)$ denotes an fsa over a finite group Q for which the following condition is satisfied:

$$
\mu(p * q, x, r) \leq \mu(p, x, k) \wedge \mu(q, x, k)
$$

for all $p, q, r, k \in Q$, $x \in X$.

Also, e will denote the identity element of the group $(Q, *)$.

Theorem 6.18.15 *Let λ be a fuzzy kernel of S. Then λ is a fuzzy subsemiautomaton of S if and only if*

$$
\lambda(p) \geq \mu(e, x, p) \wedge \lambda(e)
$$

for all $p \in Q$, $x \in X$.

Proof. Suppose that the given condition is satisfied. Then for all $p, q, r \in Q$, $x \in X$,

$$
\begin{aligned}
\lambda(p) &= \lambda(p * r^{-1} * r) \\
&\geq \lambda(p * r^{-1}) \wedge \lambda(r) \\
&\geq \mu(q, x, p) \wedge \mu(e, x, r) \wedge \lambda(q) \wedge \lambda(r) \\
&\geq \mu(q, x, p) \wedge \mu(e, x, r) \wedge \lambda(e) \wedge \lambda(q) \\
&\quad \text{(by the given condition)} \\
&= \mu(q, x, p) \wedge \lambda(q) \\
&\quad \text{(since } \lambda(e) \geq \lambda(q) \text{ and } \mu(e, x, r) \geq \mu(e * q, x, p)).
\end{aligned}
$$

Thus λ is a fuzzy subsemiautomaton of S. The converse is immediate. ∎

Corollary 6.18.16 *If λ is a fuzzy kernel and ν is a fuzzy subsemiautomaton of S such that $\nu \subseteq \lambda$ and $\lambda(e) = \nu(e)$, then λ is a fuzzy subsemiautomaton of S.* ■

Theorem 6.18.17 *Let λ be a fuzzy kernel and ν be a fuzzy subsemiautomaton of S. Then $\lambda * \nu$ is a fuzzy subsemiautomaton of S.*

Proof. Since λ is a fuzzy normal subgroup and ν is a fuzzy subgroup of Q, it follows that $\lambda * \nu$ is a fuzzy subgroup of Q and $\lambda * \nu = \nu * \lambda$.
Now

$$
\begin{aligned}
(\lambda * \nu)(p) &\geq \lambda(p * r^{-1}) \wedge \nu(r) \\
&\geq (\mu(a * b, x, p) \wedge \mu(a, x, r) \wedge \lambda(b)) \wedge \\
&\quad (\mu(a, x, r) \wedge \nu(a)) \\
&= \mu(a * b, x, p) \wedge \lambda(b) \wedge \nu(a) \\
&\quad (\text{since } \mu(a * b, x, p) \leq \mu(a, x, r))
\end{aligned}
$$

for all $a, b, p \in Q, x \in X$.
Thus for all $p, q \in Q, x \in X$,

$$
\begin{aligned}
(\lambda * \nu)(p) &\geq \vee\{\mu(a * b, x, p) \wedge \lambda(b) \wedge \nu(a) \mid a, b \in Q, \ a * b = q\} \\
&= \mu(q, x, p) \wedge (\vee\{\lambda(b) \wedge \nu(a) \mid a, b \in Q, \ a * b = q\}) \\
&= \mu(q, x, p) \wedge (\nu * \lambda)(q) \\
&= \mu(q, x, p) \wedge (\lambda * \nu)(q).
\end{aligned}
$$

Hence $\lambda * \nu$ is a fuzzy subsemiautomaton of S. ■

Theorem 6.18.18 *If λ and ν are fuzzy kernels of S, then $\lambda * \nu$ is a fuzzy kernel of S.*

Proof. Since λ and ν are fuzzy normal subgroups of Q, it follows that $\lambda * \nu$ is a fuzzy normal subgroup of Q and $\lambda * \nu = \nu * \lambda$. Now

$$
\begin{aligned}
(\lambda * \nu)(p * r^{-1}) &\geq \lambda(p * q^{-1}) \wedge \nu(q * r^{-1}) \\
&\geq (\mu(a * b * c, x, p) \wedge \mu(a * b, x, q) \wedge \lambda(c)) \wedge \\
&\quad (\mu(a * b, x, p) \wedge \mu(a, x, r) \wedge \nu(b)) \\
&= \mu(a * b * c, x, p) \wedge \mu(a, x, r) \wedge \lambda(c) \wedge \nu(b) \\
&\quad (\text{since } \mu(a * b * c, x, p) \leq \mu(a * b, x, p))
\end{aligned}
$$

for all $a, b, c, p, r \in Q, \ x \in X$. Thus for all $p, q, r, k \in Q, \ x \in X$,

$$
\begin{aligned}
(\lambda * \nu)(p * r^{-1}) &\geq \vee\{\mu(q * b * c, x, p) \wedge \mu(q, x, r) \wedge \lambda(c) \wedge \nu(b) \\
&\quad \mid b, c \in Q, \ b * c = k\} \\
&= (\mu(q * k, x, p) \wedge \mu(q, x, r)) \wedge \\
&\quad (\vee\{\lambda(c) \wedge \nu(b) \mid b, c \in Q, \ b * c = k\}) \\
&= \mu(q * k, x, p) \wedge \mu(q, x, r) \wedge (\lambda * \nu)(k).
\end{aligned}
$$

Thus $\lambda * \nu$ is a fuzzy kernel of S. ■
Let λ be a fuzzy kernel of S. Let $C(\lambda) = \{\nu \mid \nu$ is a fuzzy kernel of S such that $\lambda(e) = \nu(e)\}$.

Theorem 6.18.19 $(C(\lambda), \subseteq)$ *is a lattice.*

Proof. Let $\nu, \delta \in C(\lambda)$. Then $\nu \cap \delta \in C(\lambda)$ and so $\nu \cap \delta$ is the glb of ν and δ for the inclusion relation " $\subseteq$." By Theorem 6.18.18, $\nu * \delta \in C(\lambda)$, and so $\nu * \delta$ is the lub of ν and δ for the inclusion relation " $\subseteq$." Thus $(C(\lambda), \subseteq)$ is a lattice. ■

Definition 6.18.20 *Let λ be a fuzzy subsemiautomaton of S. A fuzzy subset ν of Q is called a **fuzzy kernel** of λ if*

(1) $\nu \subseteq \lambda$ and ν is a fuzzy normal subgroup of λ and

(2) $\nu(p * r^{-1}) \geq \mu(q * k, x, p) \wedge \mu(q, x, r) \wedge \nu(k)$ for all $p, r, k \in Q$, $q \in$ $Supp(\lambda)$, $x \in X$.

Theorem 6.18.21 *Let λ be a fuzzy subsemiautomaton of S and let ν be a fuzzy kernel of S such that $\lambda(e) = \nu(e)$. Then $\lambda * \nu$ is a fuzzy subsemiautomaton of S, ν is a fuzzy kernel of $\lambda * \nu$, and $\lambda \cap \nu$ is a fuzzy kernel of λ. Moreover, the weak isomorphism from $\lambda/\lambda \cap \nu$ into $\lambda * \nu/\nu$ gives an into homomorphism from the fsa over a finite group (G, X, μ_1) into the fsa over a finite group (H, X, μ_2), where $G = Supp(\lambda/\lambda \cap \nu)$ and $H = Supp(\lambda * \nu/\nu)$.*

Proof. By Theorem 6.18.17, $\lambda * \nu$ is a fuzzy subsemiautomaton of S. Since $\lambda(e) = \nu(e)$, it follows that $\nu \subseteq \lambda * \nu$ and ν is a fuzzy normal subgroup of $\lambda * \nu$. It also follows that λ is a fuzzy kernel of $\lambda * \nu$ since λ is a fuzzy kernel of S. In addition, $\lambda \cap \nu \subseteq \lambda$ and $\lambda \cap \nu$ is a fuzzy normal subgroup of λ. Now for $p, r, k \in Q$, $q \in Supp(\lambda)$, $x \in X$,

$$
\begin{aligned}
(\lambda \cap \nu)(p * r^{-1}) \geq\ & (\mu(q * k, x, p) \wedge \mu(q, x, r) \wedge \nu(k)) \wedge \\
& (\mu(k, x, p * r^{-1}) \wedge \lambda(k)) \\
=\ & \mu(q * k, x, p) \wedge \mu(q, x, r) \wedge (\lambda \cap \nu)(k) \\
& (\text{since } \mu(q * k, x, p) \leq \mu(k, x, p * r^{-1})).
\end{aligned}
$$

Thus $\lambda \cap \nu$ is a fuzzy kernel of λ.

Now, since λ and ν are, respectively, a fuzzy subgroup and a fuzzy normal subgroup of Q, $Supp(\lambda)$ and $Supp(\nu)$ are, respectively, a subgroup and a normal subgroup of Q. Also $Supp(\lambda \cap \nu) = Supp(\lambda) \cap Supp(\nu)$ and $Supp(\lambda * \nu) = Supp(\lambda) * Supp(\nu)$. Thus by the second isomorphism theorem of groups, there is an isomorphism

$$ f : Supp(\lambda)/Supp(\lambda \cap \nu) \rightarrow Supp(\lambda * \nu)/Supp(\nu) $$

given by

$$ f(qSupp(\lambda \cap \nu)) = qSupp(\nu) $$

for all $qSupp(\lambda \cap \nu) \in Supp(\lambda)/Supp(\lambda \cap \nu)$.
    Define

$$ \lambda/\lambda \cap \nu : Supp(\lambda)/Supp(\lambda \cap \nu) \rightarrow [0, 1] $$

by

$$(\lambda/\lambda \cap \nu)(q\text{Supp}(\lambda \cap \nu)) = \vee\{\lambda(p) \mid p \in q\text{Supp}(\lambda \cap \nu)\}$$

for all $q\text{Supp}(\lambda \cap \nu) \in \text{Supp}(\lambda)/\text{Supp}(\lambda \cap \nu)$ and

$$\lambda * \nu/\nu : \text{Supp}(\lambda * \nu)/\text{Supp}(\nu) \to [0,1]$$

by

$$(\lambda * \nu/\nu)(q\text{Supp}(\nu)) = \vee\{(\lambda * \nu)(p) \mid p \in q\text{Supp}(\nu)\}$$

for all $q\text{Supp}(\nu) \in \text{Supp}(\lambda * \nu)/\text{Supp}(\nu)$. Then $\lambda/\lambda \cap \nu$ and $\lambda * \nu/\nu$ are fuzzy subgroups of $\text{Supp}(\lambda)/\text{Supp}(\lambda \cap \nu)$ and $\text{Supp}(\lambda * \nu)/\text{Supp}(\nu)$, respectively. Also, f is a weak isomorphism from $\lambda/\lambda \cap \nu$ into $\lambda * \nu/\nu$.

Now, let $G = \text{Supp}(\lambda)/\text{Supp}(\lambda \cap \nu)$ and $H = \text{Supp}(\lambda * \nu)/\text{Supp}(\nu)$. Then G and H are subgroups of $\text{Supp}(\lambda)/\text{Supp}(\lambda \cap \nu)$ and $\text{Supp}(\lambda * \nu)/\text{Supp}(\nu)$, respectively.

Define $\mu_1 : G \times X \times G \to [0,1]$ by

$$\mu_1(q\text{Supp}(\lambda \cap \nu), x, p\text{Supp}(\lambda \cap \nu)) = \vee\{\mu(a,x,b) \mid a \in q\text{Supp}(\lambda \cap \nu),$$
$$b \in p\text{Supp}(\lambda \cap \nu)\}$$

for all $q\text{Supp}(\lambda \cap \nu), p\text{Supp}(\lambda \cap \nu) \in G$, and $x \in X$. Clearly, μ_1 is well defined and (G, X, μ_1) is an fsa over a finite group. Define $\mu_2 : H \times X \times H \to [0,1]$ by

$$\mu_2(q\text{Supp}(\nu), x, p\text{Supp}(\nu)) = \vee\{\mu(a,x,b) \mid a \in q\text{Supp}(\nu), b \in p\text{Supp}(\nu)\}$$

for all $q\text{Supp}(\nu), p\text{Supp}(\nu) \in H$, and $x \in X$. Then (H, X, μ_2) is an fsa over a finite group. Since f is a weak isomorphism from $\lambda/\lambda \cap \nu$ into $\lambda * \nu/\nu$, $f(\lambda/\lambda \cap \nu) \subseteq \lambda * \nu/\nu$. Let $q\text{Supp}(\lambda \cap \nu) \in G$. Then

$$\begin{aligned}(\lambda * \nu/\nu)(q\text{Supp}(\nu)) &\geq f(\lambda/\lambda \cap \nu)(q\text{Supp}(\nu)) \\ &= (\lambda/\lambda \cap \nu)(q\text{Supp}(\lambda \cap \nu)) \\ &> 0.\end{aligned}$$

Hence $q\text{Supp}(\nu) = f(q\text{Supp}(\lambda \cap \nu)) \in H$. Thus f induces a monomorphism $g : G \to H$ given by $g(q\text{Supp}(\lambda \cap \nu)) = q\text{Supp}(\nu)$ for all $q\text{Supp}(\lambda \cap \nu) \in G$. It follows that g is a homomorphism from (G, X, μ_1) into (H, X, μ_2). ∎

Theorem 6.18.22 *Let $T = (Q, X, \delta)$ be an fsa over a finite group and let λ be a fuzzy kernel of T with $Supp(\lambda) = Q$. Then there is an onto homomorphism from T into the fsa over a finite group (G, X, σ), where $G = Supp(Q/\lambda)$.*

Proof. Now λ is a fuzzy normal subgroup of Q. Define $\lambda_p : Q \to [0,1]$ by $\lambda_p(q) = \lambda(p * q^{-1})$ for all $q \in Q$. Let $H = \{\lambda_p \mid p \in Q\}$. Then H is a group with respect to the binary operation $\lambda_p * \lambda_q = \lambda_{p*q}$ for all $\lambda_p, \lambda_q \in H$. Define $Q/\lambda : H \to [0,1]$ by

$$(Q/\lambda)(\lambda_p) = \lambda(p)$$

for all $\lambda_p \in H$. Then Q/λ is a fuzzy subgroup of H, called the **quotient subgroup** of Q by λ. Let $G = \mathrm{Supp}(Q/\lambda)$. Define $\sigma : G \times X \times G \to [0,1]$ by

$$\sigma(\lambda_p, x, \lambda_q) = \vee\{\delta(a, x, b) \mid a, b \in Q, \ \lambda_a = \lambda_p, \ \lambda_b = \lambda_q\}$$

for all $\lambda_p, \lambda_q \in G$, $x \in X$. Clearly, σ is well defined and (G, X, σ) is an fsa over a finite group. Define $f : Q \to G$ by $f(q) = \lambda_q$ for all $q \in Q$. Then it follows that f is an onto homomorphism from T into (G, X, σ). ∎

Let $T = (Q, X, \delta)$ be an fsa over a finite group. An element $x_0 \in X$ is called an **e-input** if $\delta(e, x_0, e) > 0$.

Definition 6.18.23 *An fsa $T = (Q, X, \delta)$ over a finite group is called **multiplicative** if there exists an e-input $x_0 \in X$ having the following properties:*
(1) $\delta(q, x, p * r) = \delta(q, x_0, p) \wedge \delta(e, x, r)$ for all $p, q, r \in Q$, $x \in X$.
(2) $\delta(p_1 * p_2^{-1}, x_0, q_1 * q_2^{-1}) = \delta(p_1, x_0, q_1) \wedge \delta(p_2, x_0, q_2)$ for all p_1, p_2, $q_1, q_2 \in Q$.

Theorem 6.18.24 *Let $T = (Q, X, \delta)$ be a multiplicative fsa over a finite group with e-input $x_0 \in X$. Let λ be a fuzzy subsemiautomaton of T and let ν be a fuzzy kernel of λ. If ν is a fuzzy normal subgroup of Q, then ν is a fuzzy kernel of T.*

Proof. Since T is multiplicative, for all $p, q, r, k \in Q$, $x \in X$,

$$
\begin{aligned}
\delta(q * k, x, p) \wedge \delta(q, x, r) &= (\delta(q * k, x_0, p) \wedge \delta(e, x, e)) \wedge \\
&\quad (\delta(q, x_0, r) \wedge \delta(e, x, e)) \\
&= \delta(q, x_0, p) \wedge \delta(k^{-1}, x_0, e) \wedge \\
&\quad \delta(e, x, e) \wedge \delta(q, x_0, r).
\end{aligned}
$$

Now for any $q \in Q$, $b \in \mathrm{Supp}(\lambda)$, $q = b * (b * q'^{-1})^{-1}$ for some $q' \in Q$. Thus we have

$$
\begin{aligned}
\delta(q * k, x, p) \wedge \delta(q, x, r) &= \delta(b, x_0, p) \wedge \delta(b * q'^{-1}, x_0, e) \wedge \\
&\quad \delta(k^{-1}, x_0, e) \wedge \delta(e, x, e) \wedge \delta(b, x_0, r).
\end{aligned}
\tag{6.4}
$$

Since ν is a fuzzy kernel of λ, for all $p, r, k \in Q$, $b \in \mathrm{Supp}(\lambda)$, $x \in X$,

$$
\begin{aligned}
\nu(p * r^{-1}) &\geq \delta(b * k, x, p) \wedge \delta(b, x, r) \wedge \nu(k) \\
&= \delta(b * k, x_0, p) \wedge \delta(b, x_0, r) \wedge \delta(e, x, e) \wedge \nu(k) \\
&= \delta(b, x_0, p) \wedge \delta(k^{-1}, x_0, e) \wedge \delta(b, x_0, r) \wedge \\
&\quad \delta(e, x, e) \wedge \nu(k).
\end{aligned}
\tag{6.5}
$$

From (6.4) and (6.5), it follows that

$$\nu(p * r^{-1}) \geq \delta(q * k, x, p) \wedge \delta(q, x, r) \wedge \nu(k)$$

for all $p, q, r, k \in Q$, $x \in X$. Since by assumption, ν is a fuzzy normal subgroup of Q, ν is a fuzzy kernel of T. ∎

Theorem 6.18.25 *Let $T = (Q, X, \delta)$ be a multiplicative fsa over a finite group with e-input $x_0 \in X$. Then the following statements are equivalent:*

(1) λ is a fuzzy kernel of T.

(2) λ is a fuzzy normal subgroup of Q and $\lambda(q) \geq \delta(p, x_0, q) \wedge \lambda(p)$ for all $p, q \in Q$.

Proof. (1)⇒(2). Since λ is a fuzzy kernel of T,

$$
\begin{aligned}
\lambda(q) &= \lambda(q * e^{-1}) \\
&\geq \delta(e * p, x_0, q) \wedge \delta(e, x_0, e) \wedge \lambda(p) \\
&= \delta(p, x_0, q) \wedge \lambda(p) \\
&\quad (\text{since } \delta(p, x_0, q) \leq \delta(e, x_0, e))
\end{aligned}
$$

for all $p, q \in Q$.

(2)⇒(1). Since T is multiplicative, for all $p, q, r, k \in Q$, $x \in X$,

$$
\begin{aligned}
\delta(q * k, x, p * r) \wedge \delta(q, x, p) \wedge \lambda(k) &= \delta(q * k, x_0, p * r) \wedge \delta(q, x_0, p) \wedge \\
&\quad \delta(e, x, e), \lambda(k) \\
&= (\delta(q, x_0, p) \wedge \delta(k^{-1}, x_0, r^{-1}) \wedge \\
&\quad \delta(e, x, e)) \wedge \lambda(k) \\
&= \delta(q, x_0, p) \wedge \delta(k, x_0, r) \wedge \delta(e, x, e) \\
&\quad \wedge \lambda(k) \quad (\text{since } \delta(k^{-1}, x_0, r^{-1}) = \\
&\quad \delta(e, x_0, e) \wedge \delta(k, x_0, r) \\
&\quad = \delta(k, x_0, r)) \\
&\leq \delta(k, x_0, r) \wedge \lambda(k) \\
&\leq \lambda(r) \quad (\text{by } (6.5)) \\
&\leq \lambda(p * r * p^{-1}) \quad (\text{by } (6.5)).
\end{aligned}
$$

$$(6.6)$$

Now let $p_1, p_2, q, k \in Q$, $x \in X$. Since Q is a group, there exists a unique element $r \in Q$ such that $p_1 = p_2 * r$. Thus

$$
\begin{aligned}
\lambda(p_1 * p_2^{-1}) &= \lambda(p_2 * r * p_2^{-1}) \\
&\geq \delta(q * k, x, p_2 * r) \wedge \delta(q, x, p_2) \wedge \lambda(k) \quad (\text{by } (6.6)) \\
&\geq \delta(q * k, x, p_1) \wedge \delta(q, x, p_2) \wedge \lambda(k).
\end{aligned}
$$

Hence λ is a fuzzy kernel of T. ∎

Theorem 6.18.26 *Let $T = (Q, X, \delta)$ be a multiplicative fsa over a finite group with e-input $x_0 \in X$. There exists a semigroup homomorphism $f : Q \to \mathcal{FP}(Q)$ and a function $g : X \to \mathcal{FP}(Q)$ such that*

$$\delta(q, x, p) = f(q)(p) \wedge g(x)(e)$$

for all $p, q \in Q$, $x \in X$.

Proof. Define $f : Q \to \mathcal{FP}(Q)$ by $f(q) = \delta_q$ for all $q \in Q$, where $\delta_q : Q \to [0,1]$ is defined by $\delta_q(p) = \delta(q, x_0, p)$ for all $p \in Q$. Since T is multiplicative,

$$
\begin{aligned}
\delta_{q_1 * q_2}(p_1 * p_2) &= \delta(q_1 * q_2, x_0, p_1 * p_2) \\
&= \delta(q_1, x_0, p_1) \wedge \delta(q_2^{-1}, x_0, p_2^{-1}) \\
&= \delta(q_1, x_0, p_1) \wedge \delta(q_2, x_0, p_2) \\
&= \delta_{q_1}(p_1) \wedge \delta_{q_2}(p_2)
\end{aligned}
$$

for all $p_1, p_2, q_1, q_2 \in Q$. Hence

$$
\begin{aligned}
(\delta_{q_1} * \delta_{q_2})(p) &= \vee\{\delta_{q_1}(q) \wedge \delta_{q_2}(r) \mid q, r \in Q, \ q * r = p\} \\
&= \delta_{q_1 * q_2}(p)
\end{aligned}
$$

for all $p \in Q$. Thus

$$
f(q_1) * f(q_2) = f(q_1 * q_2). \tag{6.7}
$$

It is well known that $\mathcal{FP}(Q)$ is a semigroup with respect to the product of fuzzy subsets of Q. Hence, it follows from (6.7) that f is a semigroup homomorphism from Q into $\mathcal{FP}(Q)$.

For $x \in X$, define the fuzzy subset λ_x of Q by $\lambda_x(q) = \delta(e, x, q)$ for all $q \in Q$. Define $g : X \to \mathcal{FP}(Q)$ by $g(x) = \lambda_x$ for all $x \in X$. Since T is multiplicative, it follows that $\delta(q, x, p) = f(q)(p) \wedge g(x)(e)$ for all $p, q \in Q, x \in X$. ∎

6.19   Exercises

1. Let $M = (Q, X, \mu)$ be the ffsm, where $Q = \{q_1, q_2\}$, $X = \{a\}$, and μ is defined by $\mu(q_1, a, q_1) = \frac{1}{3}$, $\mu(q_1, a, q_2) = 0$, $\mu(q_2, a, q_1) = 0$, and $\mu(q_2, a, q_2) = \frac{2}{3}$. Determine $E(M)$ and $\overline{E}(M)$.

2. Let $M = (Q, X, \mu)$ be the ffsm, where $Q = \{q_1, q_2\}$, $X = \{a\}$, and μ is defined by $\mu(q_1, a, q_1) = 0$, $\mu(q_1, a, q_2) = \frac{1}{3}$, $\mu(q_2, a, q_1) = 0$, and $\mu(q_2, a, q_2) = \frac{2}{3}$. Determine $E(M)$ and $\overline{E}(M)$.

3. Let $M = (Q, X, \mu)$ be the ffsm, where $Q = \{q_1, q_2\}$, $X = \{a\}$, and μ is defined by $\mu(q_1, a, q_1) = 0$, $\mu(q_1, a, q_2) = \frac{1}{3}$, $\mu(q_2, a, q_1) = \frac{2}{3}$, and $\mu(q_2, a, q_2) = 0$. Determine $E(M)$ and $\overline{E}(M)$.

4. In Example 6.6.4, determine $E(M_1)$, $E(M_2)$, $E(M_1 \times M_2)$, and $E(M_1 \wedge M_2)$ when $c_1 \neq c_2$ and/or $d_1 \neq d_2$.

5. Prove (2), (3), and (4) of Theorem 6.17.1.

6. Let λ and μ be fuzzy subgroups of G such that λ is a fuzzy normal subgroup of μ. Prove that μ/λ is a fuzzy subgroup of $\text{Supp}(\mu)/\text{Supp}(\lambda)$.

7. Prove Theorems 6.18.12 and 6.18.13.

8. (113) Let (Q, X, τ) be a T-generalized state machine. Prove that

$$\tau^+(p, xy, q) = \vee\{\tau^+(p, x, r)\ T\ \tau^+(r, y, q)\ |r \in Q\},$$

for all $p, q \in Q$ and $x, y \in X^+$.

9. (113) Let (Q, X, τ) be a T-generalized state machine. Define $\equiv$ on X^+ by $\forall x, y \in X^+$, $x \equiv y$ if $\tau^+(p, x, q) = \tau^+(p, y, q)\ \forall p, q \in Q$. Prove that $\equiv$ is a congruence relation on X^+.

10. (113) Let $M = (Q, X, \tau)$ be a T-generalized state machine. Let $\equiv$ be a congruence relation defined in Exercise 6. Let $[x] = \{y \in X^+ \mid x \equiv y\}\ \forall x \in X^+$ and $S(M) = \{[x] \mid x \in X^+\}$. Prove that $S(M)$ is a semigroup, where the binary operation on $S(M)$ is defined by $[x][y] = [xy]\ \forall[x], [y] \in S(M)$.

11. (113) Let $M = (Q, X, \tau)$ be a T-generalized state machine, where T is the ordinary product on $[0, 1]$. Show by example that $S(M)$ need not be finite, where $S(M)$ is defined in Exercise 10.

12. (113) Let $M = (Q, X, \tau)$ be a T-generalized state machine. Prove that since Q is finite, $S(M)$ is finite if and only if $\text{Im}(\tau^+)$ is finite.

13. (113) Construct an example of a T-generalized state machine (Q, X, τ) such that $\sum_{q \in Q} \tau^+(p, x, q) > 1$ and $x \in X^+$.

14. (113) A T-generalized state machine (Q, S, ρ) is called a T-**generalized transformation semigroup** if S is a finite semigroup such that (a) $\rho(p, uv, q) = \vee\{\rho(p, u, r)\ T\ \rho(r, v, q) \mid r \in Q\}\ \forall p, q \in Q$ and $\forall u, v \in S$ and (b) $\forall u, v \in S$, $\rho(p, u, q) = \rho(p, v, q)\ \forall p, q \in Q$ implies $u = v$. A t-norm T is said to be T-**generalized transformation semigroup inducible** if $S(M)$ is finite and $\sum_{q \in Q} \tau^+(p, x, q) \leq 1\ \forall p \in Q$ and $x \in X^+$ for every T-generalized state machine (Q, X, τ). Prove that there exists a T-generalized transformation semigroup inducible t-norm T.

Chapter 7

More on Fuzzy Languages

7.1 Fuzzy Regular Languages

In this chapter, we study the concepts of max-min $(FL_\vee(M))$ and min-max $(FL_\wedge(M))$ fuzzy languages recognized by a type of fuzzy automaton M. We show that if λ_1 and λ_2 are $F_\vee$-regular languages, then so are $\lambda_1 \cup \lambda_2$ and $\lambda_1 \cap \lambda_2$. We prove a type of fuzzy pumping lemma. We use this result to give a necessary and sufficient condition for $FL_\vee(M)$ to be nonconstant.

In this section, fuzziness is introduced through the initial and final states.

Definition 7.1.1 *A **partial fuzzy automaton (pfa)** is a 5-tuple* $M = (Q, X, \delta, \iota, \tau)$, *where*

(1) Q is a finite nonempty set of states,

(2) X is a finite nonempty set of input symbols,

(3) $\delta : Q \times X \to Q$ is a function, called the transition function,

*(4) ι is a fuzzy subset of Q, i.e., $\iota : Q \to [0, 1]$, called the **initial fuzzy** state,*

*(5) τ is a fuzzy subset of Q, called the **fuzzy subset of final states**.*

The transition function δ can be extended to $\delta^* : Q \times X^* \to Q$ by

$$
\begin{aligned}
\delta^*(q, \Lambda) &= q \\
\delta^*(q, ua) &= \delta(\delta^*(q, u), a)
\end{aligned}
$$

$\forall\, q \in Q,\ \forall\, u \in X^*,\ \forall\, a \in X$. It can be easily verified that $\delta^*(q, uv) = \delta^*(\delta^*(q, u), v)\ \forall u, v \in X^*$.

Definition 7.1.2 *Let $M = (Q, X, \delta, \iota, \tau)$ be a pfa. The **max-min fuzzy language recognized** by M is the fuzzy subset $FL_\vee(M)$ of X^* defined by*

$$
FL_\vee(M)(u) = \vee_{q \in Q}(\iota(q) \wedge \tau(\delta^*(q, u)))
$$

and the **min-max fuzzy language recognized** by M is the fuzzy subset $FL_\wedge(M)$ of X^* defined by

$$FL_\wedge(M)(u) = \wedge_{q \in Q}(\iota(q) \vee \tau(\delta^*(q,u))).$$

Definition 7.1.3 Let X be a nonempty finite set. A fuzzy subset λ of X^* is called $\mathbf{F_\vee}$-**regular** ($\mathbf{F_\wedge}$-**regular**) if there exists a pfa, $M = (Q, X, \delta, \iota, \tau)$, such that $\lambda = FL_\vee(M)$ ($\lambda = FL_\wedge(M)$).

Theorem 7.1.4 Let L be a regular language on a nonempty finite set X. Then the characteristic function χ_L of L is a $F_\vee$-regular language.

Proof. Since L is a regular language on X, there exists a deterministic partial finite automata (dpfa), $M' = (Q, X, \delta, q_0, F)$, such that the language $L(M')$ recognized by M' is L. Consider the pfa, $M = (Q, X, \delta, \iota, \tau)$, where $\iota : Q \to [0,1]$ is defined by $\iota(q_0) = 1$ and $\iota(q) = 0$ if $q \neq q_0$ and $\tau : Q \to [0,1]$ is defined by $\tau(q) = 1$ if $q \in F$ and $\tau(q) = 0$ if $q \notin F$. Let $u \in X^*$. Then

$$
\begin{aligned}
FL_\vee(M)(u) &= \vee_{q \in Q}(\iota(q) \wedge \tau(\delta^*(q,u))) \\
&= \tau(\delta^*(q_0, u)) \\
&= \begin{cases} 1 \text{ if } \delta^*(q_0, u) \in F \\ 0 \text{ if } \delta^*(q_0, u) \notin F \end{cases} \\
&= \chi_L(u).
\end{aligned}
$$

Thus $FL_\vee(M) = \chi_L$. ∎

Theorem 7.1.5 Let $L \subseteq X^*$. Suppose the characteristic function χ_L of L is a $F_\vee$-regular language. Then L is a regular language on X.

Proof. Since χ_L is a $F_\vee$-regular language, there exists a pfa, $M = (Q, X, \delta, \iota, \tau)$, such that $FL_\vee(M) = \chi_L$. Thus

$$\vee_{q \in Q}(\iota(q) \wedge \tau(\delta^*(q,u))) = \begin{cases} 1 \text{ if } u \in L \\ 0 \text{ if } u \notin L. \end{cases}$$

Now $u \in L$ if and only if there exist $q' \in Q$ such that $\iota(q') = 1$ and $\tau(\delta^*(q', u)) = 1$. Let

$$Q_0 = \{q \in Q | \iota(q) = 1\}$$

and

$$F = \{\delta^*(q, u) \in Q | \tau(\delta^*(q, u)) = 1 \text{ for some } u \in L\}.$$

Then $Q_0 \neq \emptyset$ and $F \neq \emptyset$. Let $M_q = (Q, X, \delta, q, F)$, $q \in Q_0$. Let L_q be the language recognized by M_q. Then $L_q \subseteq L$. Hence $\cup_{q \in Q_0} L_q \subseteq L$. Let $u \in L$. Then $\vee_{q \in Q}(\iota(q) \wedge \tau(\delta^*(q, u))) = 1$. Hence there exists $q \in Q_0$ such that $\iota(q) \wedge \tau(\delta^*(q, u)) = 1$. Then $\tau(\delta^*(q, u)) = 1$ and so $\delta^*(q, u) \in F$. Thus $u \in L_q$. Hence $\cup_{q \in Q_0} L_q = L$. Since each L_q is regular, L is regular. ∎

Theorem 7.1.6 *Let L be a regular language on a nonempty finite set X. Then the characteristic function χ_L of L is a $F_\wedge$-regular language.*

Proof. Since L is a regular language on X, there exists a dpfa, $M' = (Q, X, \delta, q_0, F)$, such that the language $L(M')$ recognized by M' is L. Consider the pfa, $M = (Q, X, \delta, \iota, \tau)$, where $\iota : Q \rightarrow [0,1]$ defined by $\iota(q_0) = 0$ and $\iota(q) = 1$ if $q \neq q_0$ and $\tau : Q \rightarrow [0,1]$ defined by $\tau(q) = 1$ if $q \in F$ and $\tau(q) = 0$ if $q \notin F$. Now

$$FL_\wedge(M)(u) = \wedge_{q \in Q}(\iota(q) \vee \tau(\delta^*(q, u))).$$

Let $u \notin L$. Then $\chi_L(u) = 0$ and $\delta^*(q_0, u) \notin F$. Hence $\iota(q_0) = 0$ and $\tau(\delta^*(q_0, u)) = 0$. Thus $FL_\wedge(M)(u) = 0$. Suppose $u \in L$. Then $\delta^*(q_0, u) \in F$ and hence $\tau(\delta^*(q_0, u)) = 1$. Thus $\iota(q_0) \vee \tau(\delta^*(q_0, u)) = 1$. If $q \neq q_0$, then $\iota(q) = 1$. Thus $\iota(q) \vee \tau(\delta^*(q, u)) = 1$. Hence $FL_\wedge(M)(u) = 1$. Thus $FL_\wedge(M) = \chi_L$. ∎

Theorem 7.1.7 *Let $L \subseteq X^*$. Suppose the characteristic function χ_L of L in X^* is a $F_\wedge$-regular language. Then L is a regular language.*

Proof. Since χ_L is a $F_\wedge$-regular language, there exists a pfa, $M = (Q, X, \delta, \iota, \tau)$, such that $FL_\wedge(M) = \chi_L$. Thus for all $u \in X^*$,

$$\chi_L(u) = \wedge_{q \in Q}(\iota(q) \vee \tau(\delta^*(q, u)))$$
$$= \begin{cases} 1 \text{ if } u \in L \\ 0 \text{ if } u \notin L. \end{cases}$$

Now $u \in L$ if and only if for all $q \in Q$, either $\iota(q) = 1$ or $\tau(\delta^*(q, u)) = 1$. Let

$$Q_0 = \{q \in Q \mid \iota(q) = 1\}$$

and for all $q \in Q \backslash Q_0$,

$$F_q = \cup_{x \in X^*}\{\delta^*(q, x) \mid \tau(\delta^*(q, u)) = 1\}.$$

For all $q \in Q \backslash Q_0$, let $M_q = (Q, X, \delta, q, F_q)$ and let L_q denote the language recognized by M_q. Then $y \in \cap_{q \in Q \backslash Q_0} L_q$ if and only if for all $q \in Q \backslash Q_0$, $y \in L_q$ if and only if for all $q \in Q \backslash Q_0$, $\delta^*(q, y) \in F_q$ if and only if for all $q \in Q \backslash Q_0$, $\tau(\delta^*(q, y)) = 1$ if and only if $y \in L$. Hence $L = \cap_{q \in Q \backslash Q_0} L_q$ and

$$\chi_L(u) = \wedge_{q \in Q \backslash Q_0} \tau(\delta^*(q, u)). \blacksquare$$

Example 7.1.8 *Let $Q = \{q_1, q_2\}$, $X = \{0, 1\}$, $\delta(q_1, 0) = q_1 = \delta(q_2, 0)$, $\delta(q_1, 1) = q_2 = \delta(q_2, 1)$, $\iota(q_1) = 0 = \iota(q_2)$, $\tau(q_1) = 0$, and $\tau(q_2) = 1$. Thus in the proof of Theorem 7.1.7, $Q_0 = \emptyset$ and $F_{q_2} = \{1, 11, 111, \ldots\}$.*

Theorem 7.1.9 *Let λ be a $F_\vee$-regular language on X. Then for all $c \in Im(\lambda)$, λ_c is a regular language on X.*

Proof. Since λ is $F_\vee$-regular, there exists $M = (Q, X, \delta, \iota, \tau)$ such that $FL_\vee(M) = \lambda$. Thus

$$\lambda(u) = \vee_{q \in Q}(\iota(q) \wedge \tau(\delta^*(q, u)))$$

for all $u \in X^*$. Let $c \in \text{Im}(\lambda)$. Let $u \in \lambda_c$. Then $\lambda(u) = \vee_{q \in Q}(\iota(q) \wedge \tau(\delta^*(q, u))) \geq c$. Since Q is finite, there exists $q_u \in Q$, such that $\iota(q_u) \wedge \tau(\delta^*(q_u, u)) \geq c$. Hence for all $u \in \lambda_c$ there exists $q_u \in Q$ such that $\iota(q_u) \wedge \tau(\delta^*(q_u, u)) \geq c$.

For all $u \in \lambda_c$, let $M_{q_u} = (Q, X, \delta, q_u, \tau_c)$ where $\tau_c = \{q \in Q \mid \tau(q) \geq c\}$. Let $L(M_{q_u})$ be the language of M_{q_u}. Let $x \in L(M_{q_u})$. Then $\delta^*(q_u, x) \in \tau_c$ and so $\tau(\delta^*(q_u, x)) \geq c$. Since $\iota(q_u) \geq c$, $\iota(q_u) \wedge \tau(\delta^*(q_u, x)) \geq c$. Hence $\vee_{q \in Q}(\iota(q) \wedge \tau(\delta^*(q, x))) \geq \iota(q_u) \wedge \tau(\delta^*(q_u, x)) \geq c$. Thus $x \in \lambda_c$. Hence $L(M_{q_u}) \subseteq \lambda_c$. Conversely, let $x \in \lambda_c$. Then there exists $q_x \in Q$ such that $\iota(q_x) \wedge \tau(\delta^*(q_x, x)) \geq c$. Thus $\tau(\delta^*(q_x, x)) \geq c$ and so $\delta^*(q_x, x) \in \tau_c$. Hence $x \in L(M_{q_x})$. Thus $\lambda_c \subseteq \cup L(M_{q_x})$. Hence $\lambda_c = \cup L(M_{q_x})$. Since each $L(M_{q_x})$ is regular, λ_c is regular. ∎

Theorem 7.1.10 *Let λ be a finite valued fuzzy subset of X^*. If λ_c is regular for all $c \in \text{Im}(\lambda)$, then λ is a $F_\vee$-regular language on X^*.*

Proof. Let $\text{Im}(\lambda) = \{c_1, c_2, \ldots, c_k\}$ where $c_1 > c_2 > \ldots > c_k$. Then

$$\lambda_{c_1} \subset \lambda_{c_2} \subset \ldots \subset \lambda_{c_k}.$$

Let us denote $\lambda_i = \lambda_{c_i}$, $i = 1, 2, \ldots, k$. Let $M_i = (Q_i, X, \delta_i, q_i, F_i)$ be the dfa for $\lambda_i - \lambda_{i-1}$, $i = 1, 2, \ldots, k$ where $\lambda_0 = \emptyset$. Let τ_i be the characteristic function of $F_i \cup \{d\}$ in $Q_i \cup \{d\}$, where d is a state such that $d \notin Q_i$, $i = 1, 2, \ldots, k$. Let $M = (Q, X, \delta, \iota, \tau)$, where

$$Q = (Q_1 \cup \{d\}) \times (Q_2 \cup \{d\}) \times \ldots \times (Q_k \cup \{d\}),$$

$$\delta((p_1, p_2, \ldots, p_k), a) = (\delta_1(p_1, a), \delta_2(p_2, a), \ldots, \delta_k(p_k, a)),$$

where we set $\delta_i(p_i, a) = d$ if $p_i = d$, $i = 1, 2, \ldots, k$,

$$\tau(p_1, p_2, \ldots, p_k) = \tau_1(p_1) \wedge \tau_2(p_2) \wedge \ldots \wedge \tau_k(p_k)$$

and

$$\iota(p_1, p_2, \ldots, p_k) = \begin{cases} c_i \text{ if } (p_1, \ldots, p_k) = (d, \ldots, d, q_i, d, \ldots, d) \\ 0 \text{ otherwise.} \end{cases}$$

Let $u \in X^*$ and $\lambda(u) = c_i$. Then $u \in \lambda_i - \lambda_{i-1}$. Now

$$\vee_{q \in Q}(\iota(q) \wedge \tau(\delta^*(q, u))) = c_i \wedge 1 = c_i$$

since

$$\iota(q) \wedge \tau(\delta^*(q, u)) = \begin{cases} c_i & \text{if } q = (d, \; \dots, d, q_i, d, \dots, d) \\ 0 & \text{otherwise.} \end{cases}$$

Hence $\lambda(u) = \vee_{q \in Q}(\iota(q) \wedge \tau(\delta^*(q, u)))$. Thus λ is a $F_\vee$-regular language on X^*. ∎

We illustrate Theorems 7.1.9 and 7.1.10 in the next two examples.

Example 7.1.11 *Let $G = (N, T, P, s)$ be the grammar of Example 1.8.4. Then $L(G) = \{b^m ab^n \mid m = 0, 1, \dots; \; n = 1, 2, \dots\}$ (see Example 1.8.6). Define $\lambda : \{a, b\}^* \to [0, 1]$ by $\forall b^m ab^n \in L(G)$, $\lambda(b^m ab^n) = .5$ if $m > 0$, $\lambda(b^m ab^n) = .9$ if $m = 0$, and $\lambda(x) = 0$ for all other $x \in \{a, b\}^*$. Then $\lambda_{.5} = L(G)$ is regular. Now $\lambda_{.9} = \{ab^n \mid n = 1, 2, \dots\}$ is regular since it is generated by the productions $s \to aS$, $S \to bS$, and $S \to b$.*

Example 7.1.12 *Let λ be the fuzzy language of Example 7.1.11. Then $\lambda_{.9} = \{ab^n \mid n = 1, 2, \dots\}$ and $\lambda_{.5} \backslash \lambda_{.9} = \{b^m ab^n \mid m, n = 1, 2, \dots\}$. Let $M_1 = (Q_1, X, \delta_1, s, F)$ and $M_2 = (Q_2, X, \delta_2, s, F)$, where $Q_1 = \{s, S, \emptyset, F\}$, $Q_2 = \{s, S_1, S_2, \emptyset, F\}$, $X = \{a, b\}$, and δ_1 and δ_2 are defined as follows.*

$$\begin{array}{llll}
\delta_1(s, a) &= S & \delta_2(s, a) &= \emptyset \\
\delta_1(s, b) &= \emptyset & \delta_2(s, b) &= S_1 \\
\delta_1(S, a) &= \emptyset & \delta_2(S_1, a) &= S_2 \\
\delta_1(S, b) &= F & \delta_2(S_1, b) &= S_1 \\
\delta_1(\emptyset, a) &= \emptyset & \delta_2(S_2, a) &= \emptyset \\
\delta_1(\emptyset, b) &= \emptyset & \delta_2(S_2, b) &= F \\
\delta_1(F, a) &= \emptyset & \delta_2(\emptyset, a) &= \emptyset \\
\delta_1(F, b) &= F & \delta_2(\emptyset, b) &= \emptyset \\
& & \delta_2(F, a) &= \emptyset \\
& & \delta_2(F, b) &= F.
\end{array}$$

Then M_1 accepts $\lambda_{.9}$ and M_2 accepts $\lambda_{.5} \backslash \lambda_{.9}$. Now $(M_1 \cup \{d\}) \times (M_2 \cup \{d\})$ has 30 elements. Hence we list only a few of the images of δ, τ, and ι.

$$\begin{array}{lll}
\delta((s, d), a) &= (\delta_1(s, a), \delta_2(d, a)) &= (S, d) \\
\delta((S, d), b) &= (\delta_1(S, b), \delta_2(d, b)) &= (F, d) \\
\delta((d, s), a) &= (\delta_1(d, a), \delta_2(s, a)) &= (d, \emptyset) \\
\delta((d, \emptyset), b) &= (\delta_1(d, b), \delta_2(\emptyset, b)) &= (d, \emptyset)
\end{array}$$

$$\begin{array}{lll}
\tau(F, d) &= 1 \\
\tau(d, F) &= 1 \\
\iota(s, d) &= .9 \\
\iota(d, s) &= .5.
\end{array}$$

Now

$$
\begin{aligned}
\vee_{q \in Q}(\iota(q) \wedge \tau(\delta^*(q, ab))) &= (\iota(s, d) \wedge \tau(\delta^*((s, d), ab)) \vee \\
&\quad (\iota(d, s) \wedge \tau(\delta^*((d, s), ab)) \\
&= (.9 \wedge \tau(\delta^*((d, \emptyset), b)) \vee \\
&\quad (.5 \wedge \tau(\delta^*((d, \emptyset), b)) \\
&= (.9 \wedge \tau(F, b)) \vee (.5 \wedge \tau(d, \emptyset)) \\
&= (.9 \wedge 1) \vee (.5 \wedge 0) \\
&= .9.
\end{aligned}
$$

Then

$$
\begin{aligned}
\vee_{q \in Q}(\iota(q) \wedge \tau(\delta^*(q, bab))) &= (\iota(s, d) \wedge \tau(\delta^*((s, d), bab)) \vee \\
&\quad (\iota(d, s) \wedge \tau(\delta^*((d, s), bab)) \\
&= (.9 \wedge \tau(\delta^*((\emptyset, S_1), ab)) \vee \\
&\quad (.5 \wedge \tau(\delta^*((d, S_1), ab)) \\
&= (.9 \wedge \tau(\delta^*((\emptyset, S_2), b)) \vee \\
&\quad \vee (.5 \wedge \tau \delta^*((d, S_2), b)) \\
&= (.9 \wedge \tau(\emptyset, F)) \vee (.5 \wedge \tau(d, F)) \\
&= (.9 \wedge 0) \vee (.5 \wedge 1) \\
&= .5.
\end{aligned}
$$

Theorem 7.1.13 *Let λ_1 and λ_2 be finite valued $F_\vee$-regular languages on X^*. Then $\lambda_1 \cup \lambda_2$ is a $F_\vee$-regular language on X^*.*

Proof. Now $(\lambda_1 \cup \lambda_2)_c = \lambda_{1c} \cup \lambda_{2c}$ for all $c \in [0, 1]$. The result now follows by Theorems 7.1.9 and 7.1.10. ∎

Theorem 7.1.14 *Let λ_1 and λ_2 be $F_\vee$-regular languages on X^*. Then $\lambda_1 \cap \lambda_2$ is a $F_\vee$-regular language on X^*.*

Proof. Since λ_1 and λ_2 are $F_\vee$-regular languages on X^*, there exists $M_1 = (Q_1, X, \delta_1, \iota_1, \tau_1)$ and $M_2 = (Q_2, X, \delta_2, \iota_2, \tau_2)$ such that $FL_\vee(M_1) = \lambda_1$ and $FL_\vee(M_2) = \lambda_2$. Let

$$
M = (Q_1 \times Q_2, X, \delta_1 \times \delta_2, \iota_1 \times \iota_2, \tau = \tau_1 \times \tau_2).
$$

Now

$$
\begin{aligned}
FL_\vee(M)(u) &= \vee_{(p,q) \in Q_1 \times Q_2}((\iota_1 \times \iota_2)(p, q) \wedge \tau((\delta_1 \times \delta_2)^*((p, q), u))) \\
&= \vee_{(p,q) \in Q_1 \times Q_2}(\iota_1(p) \wedge \iota_2(q) \wedge \\
&\quad \tau_1 \times \tau_2((\delta_1^*(p, u), \delta_2^*(q, u))) \\
&= \vee_{(p,q) \in Q_1 \times Q_2}(\iota_1(p) \wedge \iota_2(q) \wedge \tau_1(\delta_1^*(p, u)) \\
&\quad \wedge \tau_2(\delta_2^*(q, u))) \\
&= (\vee_{p \in Q_1}(\iota_1(p) \wedge \tau_1(\delta_1^*(p, u)))) \wedge \\
&\quad (\vee_{q \in Q_2}(\iota_2(q) \wedge \tau_2(\delta_2^*(q, u)))) \\
&= FL_\vee(M_1)(u) \wedge FL_\vee(M_2)(u) \\
&= \lambda_1(u) \wedge \lambda_2(u) \\
&= (\lambda_1 \cap \lambda_2)(u).
\end{aligned}
$$

Hence $FL_\vee(M) = \lambda_1 \cap \lambda_2$. ∎

Theorem 7.1.15 *Let λ be a $F_\vee$-regular language on X^*. Then $\overline{\lambda} = 1 - \lambda$ is a $F_\wedge$-regular language on X^*.*

Proof. Since λ is a $F_\vee$-regular language on X^*, there exists $M = (Q, X, \delta, \iota, \tau)$ such that $FL_\vee(M) = \lambda$. Let $M^c = (Q, X, \delta, \iota^c, \overline{\tau})$ where $\overline{\tau} = 1 - \tau$ and $\overline{\iota} = 1 - \iota$. Now

$$
\begin{aligned}
FL_\wedge(M^c)(u) &= \wedge_{q \in Q}(\overline{\iota}(q) \vee \overline{\tau}(\delta^*(q, u))) \\
&= \wedge_{q \in Q}((1 - \iota(q)) \vee (1 - \tau(\delta^*(q, u)))) \\
&= 1 - \vee_{q \in Q}(\iota(q) \wedge \tau(\delta^*(q, u))) \\
&= 1 - \lambda(u) \\
&= \overline{\lambda}(u)
\end{aligned}
$$

for all $u \in X^*$. Hence $FL_\wedge(M^c) = \overline{\lambda}$. Thus $\overline{\lambda}$ is a $F_\wedge$-regular language on X^*. ∎

Let $M = (Q, X, \delta, \iota, \tau)$ be a pfa. Let $Q = \{q_1, q_2, \ldots, q_n\}$ and $\iota(q_i) = t_i$ for all $1 \leq i \leq n$. For all $p \in Q$ define $p^{-1}\tau : X^* \to [0, 1]$ by

$$
(p^{-1}\tau)(x) = \tau(\delta^*(p, x))
$$

for all $x \in X^*$. For all $p_i \in Q$, $1 \leq i \leq n$ define

$$
(p_1, p_2, \ldots, p_n)^{-1}\tau : X^* \to [0, 1]
$$

by

$$
((p_1, p_2, \ldots, p_n)^{-1}\tau)(x) = \vee_{i=1}^n (c_i \wedge (p_i^{-1}\tau)(x))
$$

for all $x \in X^*$. Let $\mathcal{K} = \{(p_1, p_2, \ldots, p_n)^{-1}\tau \mid p_i \in Q, 1 \leq i \leq n\}$. Since Q is finite, $\mathcal{K}$ is finite.

Let λ be a fuzzy subset of X^*. Let $u \in X^*$. Define

$$
u^{-1}\lambda : X^* \to [0, 1]
$$

by

$$
(u^{-1}\lambda)(v) = \lambda(uv)
$$

for all $v \in X^*$. Let $p \in Q$. Then

$$
(p^{-1}\tau)(v) = \tau(\delta^*(p, v))
$$

for all $v \in X^*$. Suppose that $\lambda = FL_\vee(M)$. Let $x \in X^*$. Then

$$
\begin{aligned}
(u^{-1}\lambda)(x) &= \lambda(ux) \\
&= \vee_{q \in Q}(\iota(q) \wedge \tau(\delta^*(q, ux))) \\
&= (\iota(q_1) \wedge \tau(\delta^*(\delta^*(q_1, u), x))) \vee \ldots \vee (\iota(q_n) \wedge \\
&\quad \tau(\delta^*(\delta^*(q_n, u), x))) \\
&= (c_1 \wedge \tau(\delta^*(\delta^*(q_1, u), x))) \vee \ldots \vee \\
&\quad (c_n \wedge \tau(\delta^*(\delta^*(q_n, u), x))).
\end{aligned}
$$

Let $\delta^*(q_i, u) = p_i$ for all $1 \leq i \leq n$. Thus

$$
\begin{aligned}
(u^{-1}\lambda)(x) &= (c_1 \wedge \tau(\delta^*(p_1, x))) \vee \ldots \vee (c_n \wedge \tau(\delta^*(p_n, x))) \\
&= (c_1 \wedge (p_1^{-1}\tau)(x)) \vee \ldots \vee (c_n \wedge (p_n^{-1}\tau)(x)) \\
&= ((p_1, p_2, \ldots, p_n)^{-1}\tau)(x).
\end{aligned}
$$

Hence $u^{-1}\lambda = (p_1, p_2, \ldots, p_n)^{-1}\tau \in \mathcal{K}$. Thus the set $\mathcal{F} = \{u^{-1}\lambda \mid u \in X^*\}$ contains only a finite number of distinct elements.

Theorem 7.1.16 *Let λ be a fuzzy subset of X^*. The following are equivalent.*

(1) λ is a $F_\vee$-regular language of X^*.

(2) $\mathcal{R} = \{(u, v) \in X^* \times X^* \mid u^{-1}\lambda = v^{-1}\lambda\}$ is a right congruence of finite index.

(3) $\mathcal{P} = \{(u, v) \in X^* \times X^* \mid \lambda(xuy) = \lambda(xvy)$ for all $x, y \in X^*\}$ is a right congruence of finite index.

Proof. (1)$\Rightarrow$(2): Clearly $\mathcal{R}$ is an equivalence relation on X^*. Since $\mathcal{F} = \{u^{-1}\lambda \mid u \in X^*\}$ is a finite set, the index of $\mathcal{R}$ is finite. Let $u, v, x \in X^*$ and $u\mathcal{R}v$. Then $u^{-1}\lambda = v^{-1}\lambda$. Hence $\lambda(uw) = \lambda(vw)$ for all $w \in X^*$. Thus $\lambda(uxy) = \lambda(vxy)$ for all $y \in X^*$. Hence $(ux)^{-1}\lambda = (vx)^{-1}\lambda$. Thus $\mathcal{R}$ is a right congruence.

(2)$\Rightarrow$(1): Let Λ denote the empty word in X^*. Let $M = (Q, X, \delta, \iota, \tau)$ be a pfa, where Q is the set of all distinct $\mathcal{R}$-equivalence classes and $\delta, \iota,$ and τ are defined as follows:

$$
\delta : Q \times X \to Q
$$

$$
\delta([x], a) = [xa]
$$

for all $[x] \in Q, a \in X,$

$$
\iota : Q \to [0, 1]
$$

$$
\iota([x]) = \begin{cases} 1 \text{ if } [x] = [\Lambda] \\ 0 \text{ if } [x] \neq [\Lambda] \end{cases}
$$

for all $[x] \in Q$ and

$$
\tau : Q \to [0, 1]
$$

$$
\tau([x]) = \lambda(x)
$$

for all $[x] \in Q$.

The function δ can be extended to $\delta^* : Q \times X^* \to Q$, where $\delta^*([x], y) = [xy]$ for all $[x] \in Q, y \in X^*$.

Let $[x], [y] \in Q$ and $a \in X$. Suppose $[x] = [y]$. Then $x\mathcal{R}y$. Since $\mathcal{R}$ is a right congruence, $xa\mathcal{R}ya$. Thus $[xa] = [ya]$ and so $\delta([x], a) = \delta([y], a)$. Hence δ is well defined. Now if $[x] = [y]$, then $x^{-1}\lambda = y^{-1}\lambda$. Thus

$$(x^{-1}\lambda)(\Lambda) = (y^{-1}\lambda)(\Lambda)$$

and so $\lambda(x) = \lambda(y)$. This implies that $\tau([x]) = \tau([y])$. Hence τ is well defined. Let $x \in X^*$. Then

$$
\begin{aligned}
FL_\vee(M)(x) &= \vee_{q \in Q}(\iota(q) \wedge \tau(\delta^*(q, x))) \\
&= \tau(\delta^*([\Lambda], x)) \\
&= \tau([x]) \\
&= \lambda(x).
\end{aligned}
$$

Hence $FL_\vee(M) = \lambda$. Thus λ is a $F_\vee$-regular language of X^*.

(1)$\Longleftrightarrow$(3): The proof of this part is similar to (1)$\Longleftrightarrow$(2). ∎

Lemma 7.1.17 (Pumping Lemma) *Let $M = (Q, X, \delta, \iota, \tau)$ be a pfa. Suppose $|Q| = n$. Then for all $x \in X^*$, $|x| \geq n$, there exist $u, v, w \in X^*$ such that $x = uvw$, $|uv| \leq n$, $|v| > 0$ and*

$$FL_\vee(M)(uv^m w) = FL_\vee(M)(x)$$

for all $m \geq 0$.

Proof. Let $x = a_1 a_2 \ldots a_p$, $a_i \in X$, $1 \leq i \leq p$, $p \geq n$. Now

$$FL_\vee(M)(x) = \vee_{q \in Q}(\iota(q) \wedge \tau(\delta^*(q, x))).$$

Let $q \in Q$. Let $\delta(q, a_1) = q_1$, $\delta(q_1, a_2) = q_2$, $\ldots$, $\delta(q_{p-1}, a_p) = q_p$. Since $|Q| = n$, there exist $i \neq j$, $0 \leq i < j \leq n$ such that $q_i = q_j$, where $q_0 = q$. Let $u = a_1 a_2 \ldots a_{i-1}$, $v = a_i a_{i+1} \ldots a_{j-1}$, $w = a_j a_{j+1} \ldots a_p$. Then $x = uvw$, $|uv| \leq n$, $|v| > 0$. Consider $uv^2 w$. Now

$$uv^2 w = a_1 a_2 \ldots a_{i-1} a_i a_{i+1} \ldots a_{j-1} a_i a_{i+1} \ldots a_{j-1} a_j a_{j+1} \ldots a_p.$$

Clearly $\delta^*(q, u) = q_i$, $\delta^*(q_i, v) = q_j = q_i$, and $\delta^*(q_j, w) = q_p$. Now $\delta^*(q_i, v^2) = \delta^*(\delta^*(q_i, v), v) = \delta^*(q_i, v) = q_j = q_i$. By induction it follows that

$$\delta^*(q_i, v^m) = \delta^*(\delta^*(q_i, v^{m-1}), v) = \delta^*(q_i, v) = q_j = q_i$$

for all $m \geq 1$. Hence

$$
\begin{aligned}
\delta^*(q_i, uv^m w) &= \delta^*(\delta^*(q, u), v^m w) \\
&= \delta^*(q_i, v^m w) \\
&= \delta^*(\delta^*(q_i, v^m), w) \\
&= \delta^*(q_j, w) \\
&= q_p \\
&= \delta^*(q_i, uvw)
\end{aligned}
$$

for all $m \geq 0$. Thus

$$
\begin{aligned}
FL_\vee(M)(uv^m w) &= \vee_{q \in Q}(\iota(q) \wedge \tau(\delta^*(q, uv^m w))) \\
&= \vee_{q \in Q}(\iota(q) \wedge \tau(\delta^*(q, uvw))) \\
&= FL_\vee(M)(uvw) \\
&= FL_\vee(M)(x). \blacksquare
\end{aligned}
$$

Theorem 7.1.18 *Let $M = (Q, X, \delta, \iota, \tau)$ be a pfa. Suppose $|Q| = n$. Then $FL_\vee(M)$ is nonconstant if and only if there exist $w_1, w_2 \in X^*$ such that $|w_1| < n$ and $FL_\vee(M)(w_2) < FL_\vee(M)(w_1)$.*

Proof. Suppose that $FL_\vee(M)$ is nonconstant. Then there exist $u, v \in X^*$ such that $FL_\vee(M)(u) \neq FL_\vee(M)(v)$. Let $w \in X^*$ be such that $|w| \geq n$. Now either $FL_\vee(M)(w) \neq FL_\vee(M)(u)$ or $FL_\vee(M)(w) \neq FL_\vee(M)(v)$. Suppose $FL_\vee(M)(w) \neq FL_\vee(M)(u)$.

Case 1: $FL_\vee(M)(w) > FL_\vee(M)(u)$.

Let

$$
S = \{x \in X^* \mid |x| \geq n \text{ and } FL_\vee(M)(x) > FL_\vee(M)(u)\}.
$$

Since $w \in S$, $S \neq \emptyset$. By the well ordering principle there exists $w_0 \in S$ such that $|w_0|$ is smallest. Now $|w_0| \geq n$ and $FL_\vee(M)(w_0) > FL_\vee(M)(u)$. By the pumping lemma, there exist $x, y, z \in X^*$ such that $w_0 = xyz$, $|xy| \leq n$, $|y| > 0$, and

$$
FL_\vee(M)(xy^i z) \geq FL_\vee(M)(w_0)
$$

for all $i \geq 0$. Let $w_1 = xz$. Since $|w_1| < |w_0|$, $w_1 \notin S$. Since $FL_\vee(M)(xz) \geq FL_\vee(M)(w_0) > FL_\vee(M)(u)$ it follows that $|w_1| < n$. Hence w_1 and u are the required words.

Case 2: $FL_\vee(M)(w) < FL_\vee(M)(u)$.

In this case if $|u| < n$, then u and w are the required words. Suppose that $|u| \geq n$. Then proceeding as in Case 1, we can show that there exists words w_1 and w_2 such that $|w_1| < n$ and $FL_\vee(M)(w_2) < FL_\vee(M)(w_1)$.

The converse is immediate. $\blacksquare$

7.2　On Fuzzy Recognizers

In this section, we define and examine the concept of a fuzzy recognizer. If $L(M)$ is the language recognized by an incomplete fuzzy recognizer M, we show that there is a completion M^c of M such that $L(M^c) = L(M)$, Theorem 7.2.14. We also show that if A is a recognizable set of words, then there is a complete accessible fuzzy recognizer M_A such that $L(M_A) = A$, Theorem 7.2.20. Our long-term goal is to determine rational decompositions of recognizable sets. In fact, we wish to determine a decomposition that gives a constructive characterization of a recognizable set. We lay groundwork for this determination by proving Theorems 7.2.28 and 7.2.29.

Let $M = (Q, X, \mu)$ be a fuzzy finite state machine, where Q and X are nonempty sets and μ is a fuzzy subset of $Q \times X \times Q$. Q is called the set of states and X is called the set of input symbols. Let X^* denote the set of all strings of finite length over X. Let $A, B \subseteq X^*$. Then $AB = \{uv \mid u \in A, v \in B\}$. For $x \in X^*$, $xA = \{x\}A$ and $Ax = A\{x\}$.

Definition 7.2.1 *Let* $M = (Q, X, \mu)$ *be a ffsm. Let* $q \in Q, x \in X^*$, *and* $A \subseteq X^*$. *Define* $q * x : Q \to [0, 1]$ *and* $q * A : Q \to [0, 1]$ *by*

$$(q * x)(p) = \mu^*(q, x, p)$$

$\forall p \in Q$ *and*

$$q * A = \cup_{x \in A} q * x.$$

Definition 7.2.2 *Let* $M = (Q, X, \mu)$ *be a ffsm. Let* $\alpha : Q \to [0, 1]$, $x \in X^*$, *and* $A \subseteq X^*$. *Define*
(1) $\alpha * x : Q \to [0, 1]$ *by*

$$(\alpha * x)(p) = \vee\{\alpha(q) \wedge \mu^*(q, x, p) \mid q \in Q\}$$

$\forall p \in Q$,
(2) $\alpha * A : Q \to [0, 1]$

$$\alpha * A = \cup_{x \in A} \alpha * x,$$

(3) $\alpha * x^{-1} : Q \to [0, 1]$ *by*

$$(\alpha * x^{-1})(p) = \vee\{\alpha(q) \wedge \mu^*(p, x, q) \mid q \in Q\}$$

$\forall p \in Q$,
(4) $\alpha * A^{-1} : Q \to [0, 1]$

$$\alpha * A^{-1} = \cup_{x \in A} \alpha * x^{-1}.$$

Lemma 7.2.3 *Let* $M = (Q, X, \mu)$ *be a ffsm. Let* α *be a fuzzy subset of* Q *and* $A \subseteq X^*$. *Then*
(1)

$$(\alpha * A)(p) = \vee\{\vee\{\alpha(q) \wedge \mu^*(q, x, p) \mid q \in Q\} \mid x \in A\},$$

(2)

$$(\alpha * A^{-1})(p) = \vee\{\vee\{\alpha(q) \wedge \mu^*(p, x, q) \mid q \in Q\} \mid x \in A\}$$

$\forall p \in Q$. ∎

Lemma 7.2.4 *Let* $M = (Q, X, \mu)$ *be a ffsm and let* $\alpha : Q \to [0, 1]$. *Let* $S = \{\alpha * x \mid x \in X^*\}$. *Then* S *is a finite set.*

Proof. Let $x \in X^*$. Then $(\alpha * x)(p) = \vee\{\alpha(q) \wedge \mu^*(q, x, p) \mid q \in Q\}$. Since μ is finite valued, it follows that μ^* is finite valued. Also, α is finite valued. It now follows that the number of mappings of the form $\alpha * x : Q \rightarrow [0, 1]$ is finite. Hence S is finite. ∎

Theorem 7.2.5 *Let α be a fuzzy subset of Q, $x, y \in X^*$ and $A, B \subseteq X^*$. Then*

(1) $(\alpha * x) * y = \alpha * (xy)$,

(2) $(\alpha * A) * y = \alpha * (Ay)$,

(3) $(\alpha * A) * B = \alpha * (AB)$,

(4) $(\alpha * x^{-1}) * y^{-1} = \alpha * (yx)^{-1}$,

(5) $(\alpha * A^{-1}) * y^{-1} = \alpha * (yA)^{-1}$,

(6) $(\alpha * A^{-1}) * B^{-1} = \alpha * (BA)^{-1}$.

Proof. (1) Let $p \in Q$. Then

$$
\begin{aligned}
((\alpha * x) * y)(p) &= \vee\{(\alpha * x)(q) \wedge \mu^*(q, y, p) \mid q \in Q\} \\
&= \vee\{\vee(\{\alpha(r) \wedge \mu^*(r, x, q) \mid r \in Q\}) \wedge \mu^*(q, y, p) \\
&\quad \mid q \in Q\} \\
&= \vee\{\alpha(r) \wedge (\vee\{\mu^*(r, x, q) \wedge \mu^*(q, y, p) \mid q \in Q\}) \\
&\quad \mid r \in Q\} \\
&= \vee\{\alpha(r) \wedge \mu^*(r, xy, p) \mid r \in Q\} \\
&= (\alpha * (xy))(p).
\end{aligned}
$$

Hence $(\alpha * x) * y = \alpha * (xy)$.

(2) Let $p \in Q$. Then

$$
\begin{aligned}
((\alpha * A) * y)(p) &= \vee\{(\alpha * A)(q) \wedge \mu^*(q, y, p) \mid q \in Q\} \\
&= \vee\{(\vee\{(\alpha * x)(q) \mid x \in A\}) \wedge \mu^*(q, y, p) \\
&\quad \mid q \in Q\} \\
&= \vee\{\vee\{(\alpha * x)(q) \wedge \mu^*(q, y, p) \mid q \in Q\} \\
&\quad \mid x \in A\} \\
&= \vee\{((\alpha * x) * y)(p) \mid x \in A\} \\
&= \vee\{(\alpha * (xy))(p) \mid x \in A\} \\
&= (\cup_{x \in A} \alpha * (xy))(p) \\
&= (\alpha * (Ay))(p).
\end{aligned}
$$

Thus $(\alpha * A) * y = \alpha * (Ay)$.

(3) $(\alpha * A) * B = \cup_{y \in B}((\alpha * A) * y) = \cup_{y \in B} \alpha * (Ay) = \alpha * (AB)$.

(4) Let $p \in Q$. Then

$$
\begin{aligned}
((\alpha * x^{-1}) * y^{-1})(p) &= \vee\{(\alpha * x^{-1})(q) \wedge \mu^*(p,y,q) \mid q \in Q\} \\
&= \vee\{(\vee\{\alpha(r) \wedge \mu^*(q,x,r) \mid r \in Q\}) \\
&\quad \wedge \mu^*(p,y,q) \mid q \in Q\} \\
&= \vee\{\alpha(r) \wedge (\vee\{\mu^*(q,x,r) \wedge \mu^*(p,y,q) \\
&\quad \mid q \in Q\}) \mid r \in Q\} \\
&= \vee\{\alpha(r) \wedge (\vee\{\mu^*(p,y,q) \wedge \mu^*(q,x,r) \\
&\quad \mid q \in Q\}) \mid r \in Q\} \\
&= \vee\{\alpha(r) \wedge \mu^*(p,yx,r) \mid r \in Q\} \\
&= (\alpha * (yx)^{-1})(p).
\end{aligned}
$$

Hence $(\alpha * x^{-1}) * y^{-1} = \alpha * (yx)^{-1}$.

(5) Let $p \in Q$. Then

$$
\begin{aligned}
((\alpha * A^{-1}) * y^{-1})(p) &= \vee\{(\alpha * A^{-1})(q) \wedge \mu^*(p,y,q) \mid q \in Q\} \\
&= \vee\{(\vee\{(\alpha * x^{-1})(q) \mid x \in A\}) \\
&\quad \wedge \mu^*(p,y,q) \mid q \in Q\} \\
&= \vee\{\vee\{(\alpha * x^{-1})(q) \wedge \mu^*(p,y,q) \mid q \in Q\} \\
&\quad \mid x \in A\} \\
&= \vee\{((\alpha * x^{-1}) * y^{-1})(p) \mid x \in A\} \\
&= \vee\{(\alpha * (yx)^{-1})(p) \mid x \in A\} \\
&= (\cup_{x \in A}\alpha * (yx)^{-1})(p) \\
&= (\alpha * (yA)^{-1})(p).
\end{aligned}
$$

Hence $(\alpha * A^{-1}) * y^{-1} = \alpha * (yA)^{-1}$.

(6) $(\alpha * A^{-1}) * B^{-1} = \cup_{y \in B}((\alpha * A^{-1}) * y^{-1}) = \cup_{y \in B}(\alpha * (yA)^{-1}) = \alpha * (BA)^{-1}$. ∎

Definition 7.2.6 $M = (Q, X, \mu)$ *be a ffsm. Let* $\alpha : Q \to [0,1]$ *and* $\tau : Q \to [0,1]$. *Define*

$$
\alpha^{-1} \circ \tau = \{x \in X^* \mid \alpha(q) \wedge \mu^*(q,x,p) \wedge \tau(p) > 0 \text{ for some } q, p \in Q\}.
$$

Definition 7.2.7 *Let* Q *and* X *be finite subsets. A **fuzzy recognizer** is a five tuple* $M = (Q, X, \mu, \iota, \zeta)$, *where*

(1) Q *is a finite nonempty set of states,*

(2) X *is a finite nonempty set of input symbols,*

(3) $\mu : Q \times X \times Q \to [0,1]$ *is a function, called the **fuzzy transition function**,*

(4) ι *is a fuzzy subset of* Q, *i.e.,* $\iota : Q \to [0,1]$, *called the **initial fuzzy state**, and*

(5) ζ *is a fuzzy subset of* Q, *i.e.,* $\zeta : Q \to [0,1]$, *called the **fuzzy subset of final states**.*

Clearly, if $M = (Q, X, \mu, \iota, \zeta)$ is a fuzzy recognizer, then $N = (Q, X, \mu)$ is a fuzzy finite state machine. We call N the fuzzy finite state machine associated with the fuzzy recognizer M.

Definition 7.2.8 *Let* $M = (Q, X, \mu, \iota, \zeta)$ *be a fuzzy recognizer. Let* $x \in X^*$. *Then* x *is said to be* **recognized** *by* M *if*

$$\vee_{q \in Q}(\iota(q) \wedge (\vee_{p \in Q}\{\mu^*(q, x, p) \wedge \zeta(p)\})) > 0.$$

Lemma 7.2.9 *Let* $M = (Q, X, \mu, \iota, \zeta)$ *be a fuzzy recognizer. Let* $x \in X^*$. *Then* x *is recognized if and only if there exists* $p, q \in Q$ *such that* $\iota(q) \wedge \mu^*(q, x, p) \wedge \zeta(p) > 0.$ ∎

Definition 7.2.10 *Let* $M = (Q, X, \mu, \iota, \zeta)$ *be a fuzzy recognizer. Let*

$$L(M) = \{x \in X^* \mid x \text{ is recognized by } M\}.$$

$L(M)$ *is called the* **language recognized** *by the fuzzy recognizer* M.

Lemma 7.2.11 *Let* $M = (Q, X, \mu, \iota, \zeta)$ *be a fuzzy recognizer. Then*

$$L(M) = \{x \in X^* \mid \iota(q) \wedge \mu^*(q, x, p) \wedge \zeta(p) > 0 \text{ for some } q, p \in Q\}.$$ ∎

Definition 7.2.12 *Let* $M = (Q, X, \mu)$ *be a fuzzy finite state machine.* M *is called* **complete** *if for all* $q \in Q$, $a \in X$, *there exists* $p \in Q$ *such that* $\mu(q, a, p) > 0.$

Definition 7.2.13 *Let* $M = (Q, X, \mu, \iota, \zeta)$ *be a fuzzy recognizer. Then* M *is called* **complete** *if the fuzzy finite state machine associated with* M *is complete.*

Let $M = (Q, X, \mu, \iota, \zeta)$ be a fuzzy recognizer such that M is not complete. Let $Q^c = Q \cup \{t\}$, where t is an element such that $t \notin Q$. For all $q \in Q$, let $0 < m_q \leq 1$. Let $0 < m \leq 1$. Define $\mu^c : \bar{Q} \times X \times \bar{Q} \to [0, 1]$ by for all $p, q \in Q$, $a \in X$, such that $\mu(p, a, q) \neq 0$,

$$\mu^c(p, a, q) = \mu(p, a, q),$$

for all $p \in Q$, $a \in X$

$$\mu^c(p, a, t) = \begin{cases} m_p & \text{if } \vee \{\mu(p, a, q) \mid q \in Q\} = 0 \\ 0 & \text{if } \vee \{\mu(p, a, q) \mid q \in Q\} > 0 \end{cases}$$

and

$$\mu^c(t, a, p) = \begin{cases} m & \text{if } t = p \\ 0 & \text{if } t \neq p. \end{cases}$$

Define $\iota^c : Q^c \to [0, 1]$ by

$$\iota^c(p) = \begin{cases} \iota(p) & \text{if } p \neq t \\ 0 & \text{if } p = t. \end{cases}$$

Define $\zeta^c : Q^c \to [0,1]$ by

$$\zeta^c(p) = \begin{cases} \zeta(p) & \text{if } p \in Q \\ 0 & \text{otherwise.} \end{cases}$$

It is easy to see that the fuzzy recognizer $M^c = (Q^c, X, \mu^c, \bar{\iota}, \zeta^c)$ is complete. Q^c is called a **completion** of M.

Theorem 7.2.14 *Let $M = (Q, X, \mu, \iota, \zeta)$ be an incomplete fuzzy recognizer. Let M^c be a completion of M. Then $L(M) = L(M^c)$.*

Proof. Let $x \in L(M)$. Then $\exists q, p \in Q$ such that $\iota(p) \wedge \mu^*(q, x, p) \wedge \zeta(p) > 0$. This implies that $\iota^c(p) \wedge \bar{\mu}^*(q, x, p) \wedge \zeta(p) > 0$. Hence $x \in L(M^c)$. Thus $L(M) \subseteq L(M^c)$. Now let $x \in L(M^c)$. There exists $q, p \in \bar{Q}$ such that $\iota^c(p) \wedge \bar{\mu}^*(q, x, p) \wedge \bar{\zeta}(p) > 0$. This implies that $\iota^c(p) > 0$. Thus $p \in Q$ and so $\iota(p) = \iota^c(p)$. Suppose $\mu^*(q, x, p) = 0$. Since $\bar{\mu}^*(q, x, p) > 0$ and $\mu^*(q, x, p) = 0$, we must have $p = t$. This is a contradiction since $p \in Q$. Hence $\mu^*(q, x, p) > 0$. Since $p \in Q$, $\zeta(p) = \bar{\zeta}(p) > 0$. Thus $\iota(p) \wedge \mu^*(q, x, p) \wedge \zeta(p) > 0$ and so $x \in L(M)$. Consequently, $L(M) = L(M^c)$. ∎

Definition 7.2.15 *Let $M = (Q, X, \mu, \iota, \zeta)$ be a fuzzy recognizer. Let*

$$S = \{q \in Q \mid \vee \{\vee\{\iota(p) \wedge \mu^*(p, x, q) \mid p \in Q\} \mid x \in X^*\} > 0\}.$$

Definition 7.2.16 *Let $M = (Q, X, \mu, \iota, \zeta)$ be a fuzzy recognizer. Then M is called **accessible** if $S = Q$.*

Theorem 7.2.17 *Let $M = (Q, X, \mu, \iota, \zeta)$ be a fuzzy recognizer. Then M is accessible if and only if $(\iota * X^*)(q) > 0 \;\forall q \in Q$.* ∎

Theorem 7.2.18 *Let $M = (Q, X, \mu, \iota, \zeta)$ be a fuzzy recognizer. Then*
(1) $L(M) = \iota^{-1} \circ \zeta$.
(2) Let $x \in X^*$. Let $A = L(M)$. Then

$$x^{-1}A = (\iota * x)^{-1} \circ \zeta,$$

where $x^{-1}A = \{y \in X^* \mid xy \in A\}$.

Proof. (1) The proof is straightforward.
(2) Let $y \in (\iota * x)^{-1} \circ \zeta$. Then $(\iota * x)(q) \wedge \mu^*(q, y, p) \wedge \zeta(p) > 0$ for some $q, p \in Q$. This implies that $(\iota * x)(q) > 0$ and $\mu^*(q, y, p) \wedge \zeta(p) > 0$. Now $(\iota * x)(q) = \vee_{r \in Q}\{\iota(r) \wedge \mu^*(r, x, q)\} > 0$ and so $\iota(r) \wedge \mu^*(r, x, q) > 0$ for some $r \in Q$. Thus we have $\iota(r) \wedge \mu^*(r, x, q) \wedge \mu^*(q, y, p) \wedge \zeta(p) > 0$ for some $q, p, r \in Q$. This implies that $\iota(r) \wedge \mu^*(r, xy, p) \wedge \zeta(p) > 0$ for some $p, r \in Q$. Thus $xy \in A$ and so $y \in x^{-1}A$. Now let $y \in x^{-1}A$. Then $xy \in A$ and so there exists $p, r \in Q$ such that $\iota(r) \wedge \mu^*(r, xy, p) \wedge \zeta(p) > 0$. This implies that $\mu^*(r, xy, p) > 0$. Now $\mu^*(r, xy, p) = \vee_{q \in Q}\{\mu^*(r, x, q) \wedge \mu^*(q, y, p)\} > 0$. Thus there exists $q \in Q$ such that $\mu^*(r, x, q) \wedge \mu^*(q, y, p) > 0$, i.e., $\mu^*(r, x, q) > 0$ and $\mu^*(q, y, p) > 0$. Now $(\iota * x)(q) = \vee_{s \in Q}\{\iota(s) \wedge \mu^*(s, x, q)\} \geq \iota(r) \wedge \mu^*(r, x, q) > 0$. Hence $(\iota * x)(q) \wedge \mu^*(q, y, p) \wedge \zeta(p) > 0$ for some $q, p \in Q$. Thus $y \in (\iota * x)^{-1} \circ \zeta$. Hence $x^{-1}A = (\iota * x)^{-1} \circ \zeta$. ∎

Definition 7.2.19 *Let $A \subseteq X^*$. Then A is called **recognizable** if $\exists$ a fuzzy recognizer M such that $A = L(M)$.*

Theorem 7.2.20 *Let A be a recognizable subset of X^*. Then there exists a complete accessible fuzzy recognizer M_A such that $L(M_A) = A$.*

Proof. There exists a fuzzy recognizer $M = (Q, X, \mu, \iota, \tau)$ such that $L(M) = A$. Let $Q_A = \{x^{-1}A \mid x \in X^*\}$. Let $0 < m_A,\, n_A,\, c_A \leq 1$. Define $\mu_A : Q_A \times X \times Q_A \to [0,1]$ by

$$\mu_A(x^{-1}A, a, y^{-1}A) = \begin{cases} m_A & \text{if } (xa)^{-1}A = y^{-1}A \\ 0 & \text{otherwise,} \end{cases}$$

$\forall x^{-1}A, y^{-1}A \in Q_A,\, a \in X$. Define $\iota_A : Q_A \to [0,1]$ by

$$\iota_A(x^{-1}A) = \begin{cases} n_A & \text{if } x^{-1}A = A \\ 0 & \text{otherwise,} \end{cases}$$

$\forall x^{-1}A \in Q_A$. Define $\zeta : Q_A \to [0,1]$ by

$$\zeta(x^{-1}A) = \begin{cases} c_A & \text{if } x \in A \\ 0 & \text{otherwise.} \end{cases}$$

Set $M_A = (Q_A, X, \mu_A, \iota_A, \zeta)$. Next we show that M_A is a complete fuzzy recognizer such that $L(M_A) = A$. Let $x^{-1}A, y^{-1}A, u^{-1}A, v^{-1}A \in Q_A, a, b \in X$. Let $(x^{-1}A, a, y^{-1}A) = (u^{-1}A, b, v^{-1}A)$. Then $x^{-1}A = u^{-1}A$, $y^{-1}A = v^{-1}A$, and $a = b$. Now $(xa)^{-1}A = a^{-1}(x^{-1}A) = b^{-1}(u^{-1}A) = (ub)^{-1}A$. Hence $(xa)^{-1}A = y^{-1}A$ if and only if $(ua)^{-1}A = v^{-1}A$. This shows that μ_A is well defined. By Theorem 7.2.18, $x^{-1}A = (\iota * x)^{-1} \circ \tau$ for all $x \in X^*$. By Lemma 7.2.4, $S = \{\iota * x \mid x \in X^*\}$ is finite. It now follows that $D = \{(\iota * x)^{-1} \circ \tau \mid x \in X^*\}$ is finite. Hence Q_A is a finite set. Clearly M_A is a complete accessible recognizer.

Let $x \in A$. Then $\zeta(x^{-1}A) = c_A > 0$. Let $x = a_1 a_2 \ldots a_n,\, a_i \in X$ for all i. Now

$$\begin{aligned} \mu_A^*(A, x, x^{-1}A) \geq\ & \mu_A(A, a_1, a_1^{-1}A) \wedge \mu_A(a_1^{-1}A, a_2, (a_1 a_2)^{-1}A) \\ & \wedge \mu_A((a_1 a_2)^{-1}A, a_3, (a_1 a_2 a_3)^{-1}A) \wedge \ldots \wedge \\ & \mu_A((a_1 a_2 \ldots a_{n-1})^{-1}A, a_n, (a_1 a_2 \ldots a_n)^{-1}A) \\ =\ & m_A \wedge m_A \wedge \ldots \wedge m_A \\ =\ & m_A \\ >\ & 0. \end{aligned}$$

Thus $\iota_A(A) \wedge \mu_A^*(A, x, x^{-1}A) \wedge \zeta(x^{-1}A) > 0$. This implies that $x \in L(M_A)$. Now let $x \in L(M_A)$. There exist $y^{-1}A, z^{-1}A \in Q_A$ such that

$$\iota_A(y^{-1}A) \wedge \mu_A^*(y^{-1}A, x, z^{-1}A) \wedge \zeta(z^{-1}A) > 0.$$

Hence $\iota_A(y^{-1}A) > 0$, $\mu_A^*(y^{-1}A, x, z^{-1}A) > 0$, and $\zeta(z^{-1}A) > 0$. By the definition of ι_A, it follows that $y^{-1}A = A$. Thus

$$\mu_A^*(A, x, z^{-1}A) = \mu_A^*(y^{-1}A, x, z^{-1}A) > 0$$

and so $x^{-1}A = z^{-1}A$ by the definition of μ_A^*. Hence $\zeta(x^{-1}A) = \zeta(z^{-1}A) > 0$ and so $x \in A$. Consequently, $A = L(M_A)$. ∎

Let $M = (Q_A, X, \mu, \iota, \zeta)$. The relation $\sim$ defined on X^* by for all $x, y \in X^*$, $x \sim y$ if and only if $(\mu^*(q, x, p) > 0$ if and only if $\mu^*(q, y, p) > 0$ for all $q, p \in Q)$. Then $\sim$ is a congruence relation by Theorem 6.2.7. Also, if $x \sim y$, then $x \in L(M)$ if and only if $y \in L(M)$. Let $A \subseteq X^*$. Define a relation $\sim_A$ on X^* by for all $x, y \in X^*$, $x \sim_A y$ if and only if $(uxv \in A$ if and only if $uyv \in A$ for all $u, v \in X^*)$. It is easily seen that $\sim_A$ is a congruence relation on X^*. Clearly, for all $x, y \in X^*$, if $x \sim y$, then $x \sim_A y$. However, in general, the converse may not be true, i.e., if $x \sim_A y$, then $x \sim y$.

Theorem 7.2.21 *Let A be a recognizable subset of X^*. Let M_A be the fuzzy recognizer constructed in the proof of Theorem 7.2.20. Define $\sim$ on X^* by for all $x, y \in X^*$, $x \sim y$ if and only if $(\mu_A^*(u^{-1}A, x, v^{-1}A) > 0$ if and only if $\mu_A^*(u^{-1}A, y, v^{-1}A) > 0$ for all $u^{-1}A, v^{-1}A \in Q_A)$. Define $\sim_A$ on X^* by for all $x, y \in X^*$, $x \sim_A y$ if and only if $(uxv \in A$ if and only if $uyv \in A$ for all $u, v \in X^*)$. Then $\forall x, y \in X^*$, $x \sim y$ if and only if $x \sim_A y$.*

Proof. Let $x, y \in X^*$ and $x \sim_A y$. Then $\forall u, v \in X^*$, $uxv \in A$ if and only if $uyv \in A$. Thus $(ux)^{-1}A = (uy)^{-1}A \; \forall u \in X^*$. Hence for all $u^{-1}A, v^{-1}A \in Q_A$, $(ux)^{-1}A = v^{-1}A$ if and only if $(uy)^{-1}A = v^{-1}A$. Thus by the definition of μ_A, $\mu_A^*(u^{-1}A, x, v^{-1}A) > 0$ if and only if $\mu_A^*(u^{-1}A, y, v^{-1}A) > 0$. Hence $x \sim y$. The converse is trivial. ∎

Theorem 7.2.22 *Let $A \subseteq X^*$. The following assertions are equivalent.*

(1) A is recognizable.

(2) $X^* / \sim_A$ is finite, where $X^* / \sim_A$ denotes the set of all congruence classes with respect to $\sim_A$.

(3) A is the union of congruence classes of a congruence relation of X^* of finite index.

Proof. (1)$\Rightarrow$(2): Since A is recognizable, for the fuzzy recognizer M_A, the relations $\sim$ and $\sim_A$ are equivalent. Thus $X^* / \sim_A = X^* / \sim$. Since $X^* / \sim$ is finite by Theorem 6.2.8, $X^* / \sim_A$ is finite.

(2)$\Rightarrow$(3): Since $X^* / \sim_A$ is finite, $\sim_A$ is a congruence relation of finite index. Now $A = \cup\{[x] \mid x \in A\}$, where $[x]$ is the equivalence class with respect to the relation $\sim_A$.

(3)$\Rightarrow$(1): Suppose $\sim$ is a congruence of finite index on X^*. For $x \in X^*$, let $[x]$ denote the equivalence class with respect to $\sim$. Let $A = \cup\{[\alpha_i] \mid \alpha_i \in X^*, i = 1, 2, \ldots, n\}$. Let $Q = X^* / \sim = \{[x] \mid x \in X^*\}$. Since $\sim$ is of finite index, Q is a finite set. Let $0 < m, n, c \leq 1$. Define $\mu : Q \times X \times Q \to [0,1]$ by

$$\mu([x], a, [y]) = \begin{cases} m & \text{if } [xa] = [y] \\ 0 & \text{otherwise,} \end{cases}$$

$\forall [x], [y] \in Q$, $a \in X$. Define $\iota : Q \to [0,1]$ by

$$\iota([x]) = \begin{cases} n & \text{if } [x] = [\lambda] \\ 0 & \text{otherwise,} \end{cases}$$

$\forall [x] \in Q$. Define $\zeta : Q \to [0,1]$ by

$$\zeta([x]) = \begin{cases} c & \text{if } x \in A \\ 0 & \text{otherwise,} \end{cases}$$

$\forall [x] \in Q$. Set $M = (Q, X, \mu, \iota, \zeta)$. Then M is a fuzzy recognizer. Let $x \in A$. Now $\iota([\Lambda]) = n > 0$ and $\zeta([x]) = c > 0$. Let $x = a_1 a_2 \ldots a_n$, $a_i \in X$ for all i. Now $\mu^*([\Lambda], x, [x]) \geq \mu([\Lambda], a_1, [a_1]) \wedge \mu([a_1], a_2, [a_1 a_2]) \wedge \mu([a_1 a_2], a_3, [a_1 a_2 a_3]) \wedge \ldots \wedge \mu([a_1 a_2 \ldots a_{n-1}], a_n, [a_1 a_2 \ldots a_{n-1} a_n]) = m \wedge \ldots \wedge m = m > 0$. Thus $\iota([\Lambda]) \wedge \mu^*([\Lambda], x, [x]) \wedge \zeta([x]) > 0$. This implies that $x \in L(M)$. Now let $x \in L(M)$. There exist $[y], [z] \in Q$ such that $\iota([y]) \wedge \mu^*([y], x, [z]) \wedge \zeta([z]) > 0$. Thus $\iota([y]) > 0$, $\mu^*([y], x, [z]) > 0$, and $\zeta([z]) > 0$. By the definition of ι, it follows that $[y] = [\Lambda]$. Thus $\mu^*([\Lambda], x, [z]) = \mu^*([y], x, [z]) > 0$ and so $[x] = [z]$ by the definition of μ^*. Hence $\zeta([x]) = \zeta([z]) > 0$ and so $x \in A$. Consequently, $A = L(M)$. ∎

Definition 7.2.23 Let $M_1 = (Q_1, X, \mu_1, \iota_1, \zeta_1)$ and $M_2 = (Q_2, X, \mu_2, \iota_2, \zeta_2)$ be fuzzy recognizers. Then $M = M_1 \cup M_2 = (Q_1 \times Q_2, X, \mu_1 \wedge \mu_2, \iota_1 \wedge \iota_2, \zeta_1 \vee \zeta_2)$.

Theorem 7.2.24 Let $M_1 = (Q_1, X, \mu_1, \iota_1, \zeta_1)$ and $M_2 = (Q_2, X, \mu_2, \iota_2, \zeta_2)$ be fuzzy recognizers. Then

$$(\mu_1 \wedge \mu_2)^*((q_1, q_2), x, (p_1, p_2)) = \mu_1^*(q_1, x, p_1) \wedge \mu_2^*(q_2, x, p_2)$$

$\forall ((q_1, q_2), x, (p_1, p_2)) \in (Q_1 \times Q_2) \times X \times (Q_1 \times Q_2)$.

Proof. Let $((q_1, q_2), x, (p_1, p_2)) \in (Q_1 \times Q_2) \times X \times (Q_1 \times Q_2)$ and $|x| = n$. If $n = 0$ or $n = 1$, then the result is true by definition. Assume it is true for $|x| = n$. Let $a \in X$. Now

$$
\begin{aligned}
(\mu_1 \wedge \mu_2)^*((q_1, q_2), xa, (p_1, p_2)) &= \vee\{(\mu_1 \wedge \mu_2)^*((q_1, q_2), x, (r_1, r_2)) \wedge \\
&\quad (\mu_1 \wedge \mu_2)((r_1, r_2), a, (p_1, p_2)) \\
&\quad | (r_1, r_2) \in Q_1 \times Q_2\} \\
&= \vee\{\mu_1^*(q_1, x, r_1) \wedge \mu_2^*(q_2, x, r_2) \wedge \\
&\quad \mu_1(r_1, a, p_1) \wedge \mu_2(r_2, a, p_2) \mid \\
&\quad (r_1, r_2) \in Q_1 \times Q_2\} \\
&= \vee\{\mu_1^*(q_1, x, r_1) \wedge \mu_1(r_1, a, p_1) \\
&\quad | r_1 \in Q_1\} \wedge \{\mu_2^*(q_2, x, r_2) \wedge \\
&\quad \mu_1(r_2, a, p_2) \mid r_2 \in Q_1\} \\
&= \mu_1^*(q_1, xa, p_1) \wedge \mu_2^*(q_2, xa, p_2). \quad \blacksquare
\end{aligned}
$$

Theorem 7.2.25 *Let $M_1 = (Q_1, X, \mu_1, \iota_1, \zeta_1)$ and $M_2 = (Q_2, X, \mu_2, \iota_2, \zeta_2)$ be complete fuzzy recognizers. Then*

$$L(M_1 \cup M_2) = L(M_1) \cup L(M_2).$$

Proof. Let $x \in L(M_1 \cup M_2)$. Then $\exists (q_1, q_2), (p_1, p_2) \in Q_1 \times Q_2$ such that

$$(\iota_1 \wedge \iota_2)(q_1, q_2) \wedge (\mu_1 \wedge \mu_2)^*((q_1, q_2), x, (p_1, p_2)) \wedge (\zeta_1 \vee \zeta_2)(p_1, p_2) > 0.$$

Thus $\exists (q_1, q_2), (p_1, p_2) \in Q_1 \times Q_2$ such that

$$\iota_1(q_1) \wedge \iota_2(q_2) \wedge \mu_1^*(q_1, x, p_1) \wedge \mu_2^*(q_2, x, p_2) \wedge (\zeta_1(p_1) \vee \zeta_2(p_2)) > 0.$$

This implies that $\exists (q_1, q_2), (p_1, p_2) \in Q_1 \times Q_2$ such that

$$(\iota_1(q_1) \wedge \mu_1^*(q_1, x, p_1) \wedge \zeta_1(p_1)) \vee (\iota_2(q_2) \wedge \mu_2^*(q_2, x, p_2) \wedge \zeta_2(p_2)) > 0.$$

Hence $\exists (q_1, q_2), (p_1, p_2) \in Q_1 \times Q_2$ such that either $(\iota_1(q_1) \wedge \mu_1^*(q_1, x, p_1) \wedge \zeta_1(p_1)) > 0$ or $(\iota_2(q_2) \wedge \mu_2^*(q_2, x, p_2) \wedge \zeta_2(p_2)) > 0$. Thus $x \in L(M_1) \cup L(M_2)$. Hence $L(M_1 \cup M_2) \subseteq L(M_1) \cup L(M_2)$. Now let $x \in L(M_1) \cup L(M_2)$. Then $x \in L(M_1)$ or $x \in L(M_2)$. Suppose $x \in L(M_1)$. Then $\exists q_1, p_1 \in Q_1$ such that $\iota_1(q_1) \wedge \mu_1^*(q_1, x, p_1) \wedge \zeta_1(p_1) > 0$. Now $\exists q_2 \in Q_2$ such that $\iota_2(q_2) > 0$. Since M_2 is complete, $\exists p_2 \in Q_2$ such that $\mu_2^*(q_2, x, p_2) > 0$. Thus we have $\iota_1(q_1) \wedge \iota_2(q_2) > 0$, $\mu_1^*(q_1, x, p_1) \wedge \mu_2^*(q_2, x, p_2) > 0$, and $\zeta_1(p_1) \vee \zeta_2(p_2) > 0$. Hence

$$(\iota_1 \wedge \iota_2)(q_1, q_2) \wedge (\mu_1 \wedge \mu_2)^*((q_1, q_2), x, (p_1, p_2)) \wedge (\zeta_1 \vee \zeta_2)(p_1, p_2) > 0$$

and so $x \in L(M_1 \cup M_2)$. Thus $L(M_1) \cup L(M_2) \subseteq L(M_1 \cup M_2)$. Consequently, $L(M_1 \cup M_2) = L(M_1) \cup L(M_2)$. ∎

Definition 7.2.26 *Let $M_1 = (Q_1, X, \mu_1, \iota_1, \zeta_1)$ and $M_2 = (Q_2, X, \mu_2, \iota_2, \zeta_2)$ be fuzzy recognizers. Then $M = M_1 \cap M_2 = (Q_1 \times Q_2, X, \mu_1 \wedge \mu_2, \iota_1 \wedge \iota_2, \zeta_1 \wedge \zeta_2)$.*

Theorem 7.2.27 *Let $M_1 = (Q_1, X, \mu_1, \iota_1, \zeta_1)$ and $M_2 = (Q_2, X, \mu_2, \iota_2, \zeta_2)$ be fuzzy recognizers. Then*

$$L(M_1 \cap M_2) = L(M_1) \cap L(M_2).$$

Proof. Now $x \in L(M_1 \cap M_2)$ if and only if $\exists (q_1, q_2), (p_1, p_2) \in Q_1 \times Q_2$ such that

$$(\iota_1 \wedge \iota_2)(q_1, q_2) \wedge (\mu_1 \wedge \mu_2)^*((q_1, q_2), x, (p_1, p_2)) \wedge (\zeta_1 \wedge \zeta_2)(p_1, p_2) > 0$$

if and only if $\exists (q_1, q_2), (p_1, p_2) \in Q_1 \times Q_2$ such that

$$\iota_1(q_1) \wedge \iota_2(q_2) \wedge \mu_1^*(q_1, x, p_1) \wedge \mu_2^*(q_2, x, p_2) \wedge (\zeta_1(p_1) \wedge \zeta_2(p_2)) > 0$$

if and only if $\exists (q_1, q_2), (p_1, p_2) \in Q_1 \times Q_2$ such that

$$(\iota_1(q_1) \wedge \mu_1^*(q_1, x, p_1) \wedge \zeta_1(p_1)) \wedge (\iota_2(q_2) \wedge \mu_2^*(q_2, x, p_2) \wedge \zeta_2(p_2)) > 0$$

if and only if $x \in L(M_1) \cap L(M_2)$. Hence $L(M_1 \cap M_2) = L(M_1) \cap L(M_2)$. ∎

In the proofs of the next two theorems, we use the fact that if $M = (Q, X, \mu)$ is a ffsm, $q, p \in Q$, and $x = x_1 x_2 \ldots x_n$, where $x_i \in X$, $i = 1, 2, \ldots n$, then

$$\mu^*(q, x, p) = \vee\{\mu(q, x_1, r_1) \wedge \mu(r_1, x_2, r_2) \wedge \ldots \\ \wedge \mu(r_{n-1}, x_n, p) \mid r_i \in Q, \ i = 1, 2, \ldots n - 1\}.$$

The above fact follows from Lemma 6.2.2.

Theorem 7.2.28 *Let $A, B \subseteq X^*$. If A and B are recognizable, then $A \cdot B$ is recognizable.*

Proof. Let $M_1 = (Q_1, X, \mu_1, \iota_1, \zeta_1)$ and $M_2 = (Q_2, X, \mu_2, \iota_2, \zeta_2)$ be recognizers of A and B, respectively. Let

$$I_i = \{q \in Q_i \mid \iota_i(q) > 0\}$$

and

$$\tau_i = \{q \in Q_i \mid \zeta_i(q) > 0\}$$

for $i = 1, 2$. Define $\iota^\triangle : Q_1 \times \mathcal{P}(Q_2) \to [0, 1]$ by $\forall (q, P) \in Q_1 \times \mathcal{P}(Q_2)$,

$$\iota^\triangle(q, P) = \begin{cases} 0 & \text{if } P \neq \emptyset \\ \iota_1(q) & \text{if } P = \emptyset. \end{cases}$$

Define $\zeta^\triangle : Q_1 \times \mathcal{P}(Q_2) \to [0, 1]$ by $\forall (q, P) \in Q_1 \times \mathcal{P}(Q_2)$

$$\zeta^\triangle(q, P) = \vee\{\zeta_2(p) \mid p \in P\}.$$

Define $\mu_1 \Delta \mu_2 : (Q_1 \times \mathcal{P}(Q_2)) \times X \times (Q_1 \times \mathcal{P}(Q_2)) \to [0, 1]$ as follows:

$$\mu_1 \Delta \mu_2((q, P), c, (q', s_2^c(P))) = \begin{cases} \mu_1(q, c, q') \wedge \vee\{\mu_2(p, c, p') \mid \\ \quad p \in P, p' \in s_2^c(P)\} & \text{if } q' \notin \tau_1, \ P \neq \emptyset \\ \\ \mu_1(q, c, q') & \text{if } q' \notin \tau_1, \ P = \emptyset; \end{cases}$$

$$\mu_1 \Delta \mu_2((q, P), c, (q', s_2^c(P) \cup I_2)) = \mu_1(q, c, q') \wedge \vee\{\mu_2(p, c, p') \\ \mid p \in P, p' \in s_2^c(P) \cup I_2\}$$

if $q \in \tau_1$, where $s_2^c(P) = \{p' \in Q_2 \mid \mu_2(p, c, p') > 0, p \in P\}$; and

$$\mu_1 \Delta \mu_2 \text{ is 0 elsewhere.}$$

Let $a \in A$ and $b \in B$. Then $\exists q \in I_1$, $q' \in T_1$, $p \in I_2$, $p' \in T_2$ such that $\mu_1^*(q, a, q') > 0$ and $\mu_2^*(p, b, p') > 0$. Now

$$ab = u_1 \ldots u_{i_1} u_{i_1+1} \ldots u_{i_2} \ldots u_{i_j} u_{i_j+1} \ldots u_{i_k} v$$

where $u_l \in X$, $v \in X^*$, i_j is the smallest such that $i_j > \ldots > i_1 \geq i_0$ and $u_1 \ldots u_{i_j} \in A$ for $j = 1, \ldots, k$, $i_0 = 1$ and there does not exist i_{k+1} such that $u_1 \ldots u_{i_{k+1}} \in A$. Let $\widehat{u_0} = u_1 \ldots u_{i_1}$ and $\widehat{u_j} = u_{i_j+1} \ldots u_{i_{j+1}}$, $j = 1, \ldots, k-1$. For all $x \in X^*$ and $\forall P \in \mathcal{P}(Q_2)$, let

$$s^x(P) = \{p' \in Q_2 \mid \mu_2^*(p, x, p') > 0, p \in P\}.$$

Let

$$
\begin{aligned}
R_{ab} &= s_2^v(s_2^{\widehat{u_{k-1}}}(s_2^{\widehat{u_{k-2}}}(\ldots s_2^{\widehat{u_0}}(I_2)\ldots)))\cup \\
&\quad s_2^v(s_2^{\widehat{u_{k-1}}}(s_2^{\widehat{u_{k-2}}}(\ldots s_2^{\widehat{u_1}}(I_2)\ldots)))\cup \ldots \cup s_2^v(I_2).
\end{aligned}
\tag{7.1}
$$

Let $q'' \in Q_1$ be such that $\mu_1^*(q', b, q'') > 0$. Now $a = u_1 \ldots u_{i_j}$ for some i_j. Let

$$(I_2)_a = s_2^{\widehat{u_{j-1}}}(s_2^{\widehat{u_{j-2}}}(\ldots s_2^{\widehat{u_o}}(I_2)\ldots)).$$

Then

$$
\begin{aligned}
(\mu_1\Delta\mu_2)^*((q,\emptyset), ab, (q'', R_{ab})) &= \vee\{(\mu_1\Delta\mu_2)^*((q,\emptyset), a, (v, R)) \\
&\quad \wedge(\mu_1\Delta\mu_2)^*((v,R), b, (q'', R_{ab})) \mid \\
&\quad (v, R) \in Q_1 \times \mathcal{P}(Q_2)\} \\
&\geq (\mu_1\Delta\mu_2)^*((q,\emptyset), a, (q', (I_2)_a)) \\
&\quad \wedge(\mu_1\Delta\mu_2)^*((q', (I_2)_a), b, (q'', R_{ab})) \\
&> 0
\end{aligned}
$$

by the definition of $(I_2)_a$ and since R_{ab} is a continuation of $(I_2)_a$, i.e., $R_{ab} = ((I_2)_a)_b$.

Conversely, let $x \in L(M_1\Delta M_2)$. Then

$$\iota^\Delta(q, \emptyset) \wedge (\mu_1\Delta\mu_2)^*((q,\emptyset), x, (q'', (I_2)_x)) > 0,$$

where $\mu_1^*(q, x, q'') > 0$ and $(I_2)_x$ is defined as follows: we have that $x = uv$, where $u = \widehat{u_0} \ldots \widehat{u_{k-1}}$ and where the $\widehat{u_j}$ and v are defined as above (k exists since $x \in L(M_1\Delta M_2)$). Then $(I_2)_x$ is the right-hand side of equation (7.1). Since $x \in L(M_1\Delta M_2)$, $(\mu_1\Delta\mu_2)^*((q,\emptyset), x, (q'', (I_2)_x)) > 0$ and so $\exists j$, $\exists p \in I_2$,

$$s_2^{u_{i_j+1}\ldots u_{i_j+1}\widehat{u_{i_j+1}}\ldots\widehat{u_k}v}(p) \in (I_2)_x \cap \tau_2.$$

Thus $u_{i_j+1} \ldots u_{i_j+1}\widehat{u_{i_j+1}} \ldots \widehat{u_k}v \in B$. Since $u_1 \ldots u_{i_j} \in A$, $x \in A \cdot B$. ∎

Theorem 7.2.29 *Let $A \subseteq X^*$ be recognizable. Then A^* is recognizable.*

Proof. Let $M = (Q, X, \mu, \iota, \zeta)$ be a recognizer of A. Define $\mu' : \mathcal{P}(Q) \times X \times \mathcal{P}(Q) \to [0, 1]$ as follows:

$$\mu'(P, c, s^c(P)) = \vee\{\mu(p, c, q) \mid p \in P, q \in s^c(P)\} \text{ if } s^c(P) \cap T = \emptyset$$

$$\mu'(P, c, s^c(P) \cup I) = \vee\{\mu(p, c, q) \mid p \in P, q \in s^c(P) \cup I\} \text{ if } s^c(P) \cap T \neq \emptyset$$

$$\mu'(P, c, P') = 0 \text{ otherwise,}$$

where $c \in X$, $P, P' \in \mathcal{P}(Q)$, $T = \{q \in Q \mid \zeta(q) > 0\}$, $I = \{q \in Q \mid \iota(q) > 0\}$, and $s^c(P) = \{q \in Q \mid \mu(p, c, q) > 0, p \in P\}$. Define $\zeta' : \mathcal{P}(Q) \to [0, 1]$ by $\zeta'(P) = \vee\{\zeta(p) \mid p \in P\}$ and $\iota' : \mathcal{P}(Q) \to [0, 1]$ by $\iota'(I) = \vee\{\iota(q) \mid q \in I\}$, ι' is 0 otherwise, $T' = \{P \in \mathcal{P}(Q) \mid \zeta'(P) > 0\} = \{P \in \mathcal{P}(Q) \mid P \cap T \neq \emptyset\}$. Let $M' = (\mathcal{P}(Q), X, \mu', \iota', \zeta')$. Let $x \in A^*$. Then $x = a_1 a_2 \dots a_n$, where $a_i \in A$ and if $a_i = a_{i_1} \dots a_{i_{n_i}}$ for $a_{i_1}, \dots, a_{i_{n_i}} \in X$, $\not\exists m_i < n_i$ such that $a_{i_1} \dots a_{i_{n_i}} \in A$, $i = 1, 2, \dots, n$. Let $I_x = s^{a_n}(\dots s^{a_2}(s^{a_1}(I) \cup I) \cup \dots) \cup I$. Now $\mu'^*(I, x, I_x) \geq \mu'^*(I, a_1, s^{a_1}(I) \cup I) \wedge \mu'^*(s^{a_1}(I) \cup I, a_2, s^{a_2}(s^{a_1}(I) \cup I) \cup I) \wedge \dots \wedge \mu'^*(s^{a_{n-1}}(\dots s^{a_2}(s^{a_1}(I) \cup I) \cup \dots) \cup I, a_n, I_\alpha) > 0$ since $\exists i \in I$, $\exists p \in s^{a_1}(I) \cap T$ such that $\mu^*(i, a_1, p) > 0$ (because $a_1 \in L(M)$), $\exists i' \in I$, $\exists p' \in s^{a_2}(s^{a_1}(I) \cup I) \cap T$ such that $\mu^*(i', a_2, p') > 0$ (because $a_2 \in L(M)$) and so on. Thus $x \in L(M')$. Hence $A^* \subseteq L(M')$. Let $y \in L(M')$. Then $\exists P \in T'$ such that $\mu'^*(I, y, P) > 0$. Now suppose $\exists k$ such that

$$y = u_1 \dots u_{i_1} u_{i_1+1} \dots u_{i_j} u_{i_j+1} \dots u_{i_k} v,$$

where $u_1 \dots u_{i_1} \in A$ with i_1 smallest and the i_j are the smallest such that $u_{i_{j-1}+1} \dots u_{i_j} \in A$, $j = 2, \dots, k$. Let $\widehat{u_1} = u_1 \dots u_{i_1}$ and $\widehat{u_j} = u_{i_{j-1}+1} \dots u_{i_j}$, $j = 2, \dots, k$. Then $u_1 \dots u_{i_{j-1}} \in A^*$ and $P = (I)_y$, where

$$(I)_\beta = s^v(s^{\widehat{u_k}}(\dots s^{\widehat{u_2}}(s^{\widehat{u_1}}(I) \cup I) \cup \dots I) \cup I) \cup I.$$

There exists j, $\exists q \in I$ such that $\mu^*(q, \widehat{u_j} \dots \widehat{u_k} v, p) > 0$ for some $p \in (I)_y \cap T$. Thus $\mu'^*((I)_{\widehat{u_1} \dots \widehat{u_{j-1}}}, \widehat{u_j} \dots \widehat{u_k} v, (I)_y) > 0$. Hence $\widehat{u_j} \dots \widehat{u_k} v \in A^*$. If no such k exists, then a similar argument shows that $y \in A$. Thus $y \in A^*$. Hence $L(M') \subseteq A^*$. Consequently $L(M') = A^*$. ∎

If $L(M)$ is the language recognized by an incomplete fuzzy recognizer, we showed that there is a completion M^c of M such that $L(M^c) = L(M)$. We also showed that if A is a recognizable set of words, then there is a complete accessible fuzzy recognizer M_A such that $L(M_A) = A$. If A and B are recognizable sets of words, then $A \cdot B$ and A^* are recognizable. These results are significant in that they lay the groundwork for determining methods of decomposing recognizable sets and thus for giving a constructive characterization of recognizable sets. In particular, we hope to show an analog of Kleene's result for fuzzy finite state machines, namely, that the class of recognizable subsets of X^* equals the class of all regular subsets of X^*.

7.3 Minimal Fuzzy Recognizers

In this section, we show that for any fuzzy recognizer M_n there is a deterministic fuzzy recognizer M_d with the same behavior, Theorem 7.3.8. Then we show that there is complete accessible deterministic fuzzy recognizer M_{dA} with the same behavior as M_n and M_d and that is minimal, Theorem 7.3.11. Our long term goal is to develop methods of decomposing a recognizable set of a fuzzy finite state machine. One method would be to follow along the lines of Kleene to give a constructive characterization of a recognizable set. In this section, we lay the foundation for the accomplishment of our goal.

Clearly, if $M = (Q, X, \mu, \iota, \zeta)$ is a fuzzy recognizer, then $N = (Q, X, \mu)$ is a fuzzy finite state machine. We call N the fuzzy finite state machine associated with the fuzzy recognizer M.

Definition 7.3.1 Let $M = (Q, X, \mu, \iota, \zeta)$ be a fuzzy recognizer. Let

$$S = \{q \in Q \mid \vee \{\vee\{\iota(p) \wedge \mu^*(p, x, q) \mid p \in Q\} \mid x \in X^*\} > 0\}.$$

Definition 7.3.2 Let $M = (Q, X, \mu, \iota, \zeta)$ be a fuzzy recognizer. Then M is called **accessible** if $S = Q$.

Definition 7.3.3 Let $M = (Q, X, \mu)$ be a ffsm. Let $\alpha : Q \to [0, 1]$, $x \in X^*$, and $A \subseteq X^*$. Define
(1) $\alpha * x : Q \to [0, 1]$ by

$$(\alpha * x)(p) = \vee\{\alpha(q) \wedge \mu^*(q, x, p) \mid q \in Q\}$$

$\forall p \in Q$,
(2) $\alpha * A : Q \to [0, 1]$

$$\alpha * A = \cup_{x \in A} \alpha * x.$$

Let $M = (Q, X, \mu, \iota, \zeta)$ be a fuzzy recognizer. Then by Theorem 7.2.17, M is accessible if and only if $(\iota * X^*)(q) > 0$ $\forall q \in Q$.

Definition 7.3.4 Let $A \subseteq X^*$. Then A is called **recognizable** if $\exists$ a fuzzy recognizer M such that $A = L(M)$.

Let Q and X be finite nonempty sets and let $\mu : Q \times X \times Q \to [0, 1]$. μ is called a **fuzzy function** of $Q \times X$ into Q if for all $q \in Q$, $a \in X$, if $\mu(q, a, p) > 0$ and $\mu(q, a, p') > 0$ for some $p, p' \in Q$, then $p = p'$.

Theorem 7.3.5 Let $M = (Q, X, \mu, \iota, \tau)$ be a fuzzy recognizer. Then μ is a fuzzy function of $Q \times X$ into Q if and only if μ^* is a fuzzy function of $Q \times X^*$ into Q.

Proof. Suppose μ is a fuzzy function. Let $q \in Q$ and $x \in X^*$. Suppose $\mu^*(q, x, p) > 0$ and $\mu^*(q, x, p') > 0$ for some $p, p' \in Q$. If $x = \Lambda$, then $p = q = p'$. Suppose $x \neq \Lambda$. Let $x = a_1 a_2 \cdots a_n \in X^*$, $a_i \in X$. There exists $q_1, q_2, \ldots, q_{n-1}, q_1', q_2', \ldots, q_{n-1}' \in Q$ such that $\mu(q, a_1, q_1) \wedge \mu(q_1, a_2, q_2) \wedge \ldots \wedge \mu(q_{n-1}, a_n, p) > 0$ and $\mu(q, a_1, q_1') \wedge \mu(q_1', a_2, q_2') \wedge \ldots \wedge \mu(q_{n-1}', a_n, p') > 0$. This implies that $\mu(q, a_1, q_1) > 0$, $\mu(q_i, a_{i+1}, q_{i+1}) > 0$, $i = 1, 2, \ldots, n-2$, $\mu(q_{n-1}, a_n, p) > 0$, $\mu(q, a_1, q_1') > 0$, $\mu(q_i', a_{i+1}, q_{i+1}') > 0$, $i = 1, 2, \ldots, n-2$, $\mu(q_{n-1}', a_n, p') > 0$. Now $\mu(q, a_1, q_1) > 0$ and $\mu(q, a_1, q_1') > 0$. Since μ is a fuzzy function, $q_1 = q_1'$. Suppose $q_j = q_j'$, $j = 1, 2, \ldots, i$, $i < n-2$. Now $\mu(q_i, a_{i+1}, q_{i+1}) > 0$ and $\mu(q_i, a_{i+1}, q_{i+1}') > 0$ implies that $q_{i+1} = q_{i+1}'$ since μ is a fuzzy function. Hence by induction $q_j = q_j'$, $j = 1, 2, \ldots, n-1$. Hence $\mu(q_{n-1}, a_n, p) > 0$ and $\mu(q_{n-1}, a_n, p') > 0$ implies that $p = p'$. Thus μ^* is a fuzzy function. The converse is trivial. ∎

Definition 7.3.6 *A **deterministic fuzzy recognizer** is a fuzzy recognizer $M_d = (Q_d, X, \mu, \iota, \tau)$ such that*

*(1) there exists a unique $s_0 \in Q_d$ such that $\iota(s_0) > 0$; s_0 is called the **initial state**,*

(2) μ is a fuzzy function of $Q \times X$ into Q, and

(3) for all $x \in X^*$, there exists a unique $q_x \in Q_d$ such that $\mu^*(s_0, x, q_x) > 0$.

Let $M_d = (Q_d, X, \mu, \iota, \tau)$ be a deterministic fuzzy recognizer. Let $\zeta_d = \{q \in Q_d \mid \tau(d) > 0\}$. ζ_d is called the set of **final states** of M_d.

Theorem 7.3.7 *Let $M = (Q, X, \mu, \iota, \tau)$ be a fuzzy recognizer. Suppose M is complete and μ is a fuzzy function of $Q \times X$ into Q. Let $s_0 \in Q$. Then the following are equivalent.*

(1) For all $a \in X$, there exists a unique $q_a \in Q$ such that $\mu(s_0, a, q_a) > 0$.

(2) For all $x \in X^*$, there exists a unique $q_x \in Q$ such that $\mu^*(s_0, x, q_x) > 0$.

Proof. $(1) \Rightarrow (2)$: Let $x \in X^*$ and $|x| = n$. If $x = \Lambda$, i.e., $n = 0$, then $\mu^*(s_0, \Lambda, s_0) = 1 > 0$ and if $\mu^*(s_0, \Lambda, p) > 0$, then by the definition of μ^*, $s_0 = p$. Suppose the result is true for all $y \in X^*$ such that $|y| < |x|$, where $|x| = n \geq 1$. Let $x = ya$, where $y \in X^*$, $a \in X$, $|y| = n - 1$. By the induction hypothesis, there exists a unique $q_y \in Q$ such that $\mu^*(s_0, y, q_y) > 0$. Since M is complete, there exists $p \in Q$ such that $\mu(q_y, a, p) > 0$. Thus $\mu^*(s_0, x, p) \geq \mu^*(s_0, y, q_y) \wedge \mu(q_y, a, p) > 0$. Since μ is a fuzzy function of $Q \times X$ into Q, μ^* is a fuzzy function of $Q \times X^*$ into Q by Theorem 7.3.5. Thus if $\mu^*(s_0, x, p) > 0$ and $\mu^*(s_0, x, p') > 0$ for some $p, p' \in Q$, then $p = p'$. It now follows that there exists a unique $q_x \in Q$ such that $\mu^*(s_0, x, q_x) > 0$.

$(2) \Rightarrow (1)$: Immediate. ∎

Theorem 7.3.8 *For each fuzzy recognizer $M_n = (Q, X, \mu, \iota, \zeta)$, one can construct a deterministic fuzzy recognizer $M_d = (Q_d, X, \mu, \iota, \zeta)$ such that $L(M_d) = L(M_n)$.*

Proof. For all $x \in X^*$, set

$$Q_x = \{q' \in Q \mid \exists q \in Q \text{ such that } \iota(q) \wedge \mu^*(q, x, q') > 0\}.$$

Then

$$Q_\Lambda = \{q' \in Q \mid \iota(q') > 0\}.$$

Let $Q_d = \{Q_x \mid x \in X^*\}$. Define $i : Q_d \to [0, 1]$ by $\forall Q_x \in Q_d$,

$$i(Q_x) = \begin{cases} \vee\{\iota(q) \mid q \in Q_\Lambda\} & \text{if } x = \Lambda \\ 0 & \text{if } x \neq \Lambda. \end{cases}$$

Let $\zeta_d = \{Q_x \in Q_d \mid \zeta(q) > 0 \text{ for some } q \in Q_x\}$. Define $\tau : \zeta_d \to [0, 1]$ by $\forall Q_x \in \zeta_d$, $\tau(Q_x) = \vee\{\zeta(q) \mid q \in Q_x\}$. Define $\nu : Q_d \times X \times Q_d \to [0, 1]$ by $\forall(Q_y, a, Q_x) \in Q_d \times X \times Q_d$,

$$\nu(Q_y, a, Q_x) = \begin{cases} \vee\{\mu^*(q, y, q') \wedge \mu(q', a, r) \mid q \in Q_y, \\ \qquad q' \in Q, \ r \in Q_{ya}\} & \text{if } x = ya \\ 0 \quad \text{otherwise}. \end{cases}$$

Let $M_d = (Q_d, X, \nu, i, \tau)$. We now show that $L(M_n) = L(M_d)$.

Now $x \in L(M_n)$ if and only if $\iota(q) \wedge \mu^*(q, x, q') \wedge \zeta(q') > 0$ for some $q, q' \in Q$ if and only if $\zeta(q') > 0$ for some $q' \in Q_x$ if and only if $\tau(Q_x) > 0$. It suffices to show that $\nu^*(Q_\Lambda, x, Q_x) > 0$ for then $x \in L(M_d)$ if and only if $i(Q_\Lambda) \wedge \nu^*(Q_\Lambda, x, Q_x) \wedge \tau(Q_x) > 0$ if and only if $\tau(Q_x) > 0$ (since $i(Q_\Lambda) > 0$ and $\nu^*(Q_\Lambda, x, Q_x) > 0$) if and only if $x \in L(M_n)$. We show $\nu^*(Q_\Lambda, x, Q_x) > 0$ by induction on $|x|$. Suppose $|x| = 0$. Then $x = \Lambda$ and $\nu^*(Q_\Lambda, \Lambda, Q_\Lambda) = 1 > 0$. Suppose $|x| \geq 1$ and the result is true for all $y \in X^*$ such that $|y| < |x|$. Let $x = ya$, where $a \in X$. Then $\nu^*(Q_\Lambda, x, Q_x) = \vee\{\nu^*(Q_\Lambda, y, r) \wedge \nu(r, a, Q_x) \mid r \in Q_d\}$ and $\nu^*(Q_\Lambda, y, Q_y) > 0$ by the induction hypothesis. Hence it suffices to show that $\nu(Q_y, a, Q_x) > 0$, but the latter inequality is true by the definition of ν since $x = ya$. ∎

Let $M_d = (Q_d, X, \mu, \iota, \tau)$ be a complete accessible deterministic fuzzy recognizer and let s_0 denote the initial state of M_d. For $x \in X^*$, we let $q_x \in Q_d$ denote the unique state such that $\mu^*(s_0, x, q_x) > 0$.

Theorem 7.3.9 *Let $M_d = (Q_d, X, \mu, \iota, \tau)$ be a complete accessible deterministic fuzzy recognizer. Let s_0 be the initial state of M_d. For all $q \in Q_d$, let $q^{-1} \circ \tau = \{y \in X^* \mid \mu^*(q, y, p) \wedge \tau(p) > 0 \text{ for some } p \in Q_d\}$. Let $A = L(M_d)$.*

(1) For all $x \in X^*$, $x^{-1}A = q_x^{-1} \circ \tau$.

(2) Let $q \in Q_d$. Then there exists $x \in X^*$ such that $q^{-1} \circ \tau = x^{-1}A$ and $q = q_x$.

(3) Let $q \in Q_d$ be such that $\tau(q) > 0$. Then $q^{-1} \circ \tau = x^{-1}A$ for some $x \in A$.

(4) $A = \Lambda^{-1}A = s_0^{-1} \circ \tau$.

(5) Let $x = a_1 a_2 \cdots a_n \in X^*$, where $a_i \in X$, $i = 1, 2, \ldots, n$. Then

$$\mu^*(s_0, x, q_x) = \mu(s_0, a_1, q_{a_1}) \wedge \mu(q_{a_1}, a_2, q_{a_1 a_2}) \wedge \ldots \wedge \mu(q_{a_1 \ldots a_{n-1}}, a_n, q_x).$$

Proof. (1) Let $y \in x^{-1}A$. Then $xy \in A$ and so $\iota(s_0) \wedge \mu^*(s_0, xy, p) \wedge \tau(p) > 0$ for some $p \in Q_d$. This implies that $\mu^*(s_0, xy, p) > 0$. Thus there exists $q \in Q_d$ such that $\mu^*(s_0, x, q) \wedge \mu^*(q, y, p) > 0$. Hence $\mu^*(s_0, x, q) > 0$. Since M_d is deterministic, $q = q_x$. Hence $\iota(s_0) \wedge \mu^*(s_0, x, q_x) \wedge \mu^*(q_x, y, p) \wedge \tau(p) > 0$. Thus $y \in q_x^{-1} \circ \tau$. Hence $x^{-1}A \subseteq q_x^{-1} \circ \tau$. Now let $y \in q_x^{-1} \circ \tau$. Then $\mu^*(q_x, y, p) \wedge \tau(p) > 0$ for some $p \in Q_d$. This implies that $\iota(s_0) \wedge \mu^*(q_x, y, p) \wedge \tau(p) > 0$. Now $\mu^*(s_0, x, q_x) > 0$. Hence $\iota(s_0) \wedge \mu^*(s_0, x, q_x) \wedge \mu^*(q_x, y, p) \wedge \tau(p) > 0$ for some $p \in Q_d$. Thus $\iota(s_0) \wedge \mu^*(s_0, xy, p) \wedge \tau(p) > 0$ for some $p \in Q_d$. Hence $xy \in A$ or $y \in x^{-1}A$. It now follows that $x^{-1}A = q_x^{-1} \circ \tau$.

(2) Let $q \in Q_d$. Since M_d is accessible, there exists $x \in X^*$ such that $\mu^*(s_0, x, q) > 0$. Since M_d is deterministic, it follows that $q = q_x$. By (1), $x^{-1}A = q_x^{-1} \circ \tau = q^{-1} \circ \tau$.

(3) By (2), $q^{-1} \circ \tau = x^{-1}A$ for some $x \in X^*$. Since $\tau(q) > 0$, $\mu^*(q, \Lambda, q) \wedge \tau(q) > 0$ and so $\Lambda \in q^{-1} \circ \tau$. Thus $\Lambda \in x^{-1}A$ and so $x \in A$.

(4) Let $y \in A$. Then $\iota(s_0) \wedge \mu^*(s_0, y, q) \wedge \tau(q) > 0$ for some $q \in Q_d$. This implies that $\mu^*(s_0, y, q) \wedge \tau(q) > 0$ for some $q \in Q_d$ and so $y \in s_0^{-1} \circ \tau$. Thus $A \subseteq s_0^{-1} \circ \tau$. Now let $y \in s_0^{-1} \circ \tau$. Then $\mu^*(s_0, y, q) \wedge \tau(q) > 0$ for some $q \in Q_d$. Thus $\iota(s_0) \wedge \mu^*(s_0, y, q) \wedge \tau(q) > 0$ for some $q \in Q_d$ and so $y \in A$. It now follows that $A = s_0^{-1} \circ \tau$.

(5) Now $\mu^*(s_0, x, q_x) = \vee\{\mu(s_0, a_1, q_1) \wedge \mu(q_1, a_2, q_2) \ldots \wedge \mu(q_{n-1}, a_n, q_x) \mid q_1, q_2, \ldots, q_{n-1} \in Q_d\}$. Since Q_d is finite, there exists $q_1, q_2, \ldots, q_{n-1} \in Q_d$ such that $\mu^*(s_0, x, q_x) = \mu(s_0, a_1, q_1) \wedge \mu(q_1, a_2, q_2) \ldots \wedge \mu(q_{n-1}, a_n, q_x)$. Since $\mu^*(s_0, x, q_x) > 0$, $\mu(s_0, a_1, q_1) \wedge \mu(q_1, a_2, q_2) \wedge \mu(q_2, a_3, q_3) \wedge \ldots \wedge \mu(q_{n-1}, a_n, q_x) > 0$. This implies that $\mu(s_0, a_1, q_1) > 0$ and so $q_1 = q_{a_1}$ since M_d is deterministic. Now $\mu(s_0, a_1, q_{a_1}) \wedge \mu(q_{a_1}, a_2, q_2) > 0$ implies that $\mu^*(s_0, a_1 a_2, q_2) > 0$ and since M_d is deterministic $q_2 = q_{a_1 a_2}$. We see that an argument by induction will yield $q_i = q_{a_1 \ldots a_i}$ for all $i = 1, 2, \ldots, n-1$. ∎

Theorem 7.3.10 *Let A be a recognizable subset of X^*. Then there exists a complete accessible deterministic fuzzy recognizer M_{dA} such that $L(M_{dA}) = A$.*

Proof. Let $M_d = (Q_d, X, \mu, \iota, \tau)$ be a deterministic fuzzy recognizer such that $A = L(M_d)$. Since M_d is deterministic, there exists a unique $s_0 \in Q_d$ such that $\iota(s_0) > 0$.

Let $Q_{dA} = \{x^{-1}A \mid x \in X^*\}$. Let $m_A = \vee\{\mu(q, a, p) \mid q, p \in Q, a \in X\}$. Then $1 \geq m_A > 0$. Let $0 < t_A \leq 1$. Define $\mu_{dA} : Q_{dA} \times X \times Q_{dA} \to [0, 1]$ by

$$\mu_{dA}(x^{-1}A, a, y^{-1}A) = \begin{cases} m_A & \text{if } (xa)^{-1}A = y^{-1}A \\ 0 & \text{otherwise,} \end{cases}$$

$\forall x^{-1}A, y^{-1}A \in Q_{dA}$, $a \in X$. Define $\iota_{dA} : Q_A \to [0,1]$ by

$$\iota_{dA}(x^{-1}A) = \begin{cases} \iota(s_0) & \text{if } x^{-1}A = A \\ 0 & \text{otherwise,} \end{cases}$$

$\forall x^{-1}A \in Q_A$. Define $\tau_{dA} : Q_A \to [0,1]$ by

$$\tau_{dA}(x^{-1}A) = \begin{cases} c_A & \text{if } x \in A \\ 0 & \text{otherwise.} \end{cases}$$

Set $M_{dA} = (Q_{dA}, X, \mu_{dA}, \iota_{dA}, \tau_{dA})$. Next we show that M_{dA} is a complete accessible fuzzy deterministic recognizer such that $L(M_{dA}) = A$. Clearly $\iota_{dA}(A) = \iota(s_0) > 0$ and $\iota_{dA}(x^{-1}A) = 0$ if $A \neq x^{-1}A$. Thus M_{dA} has a unique initial state.

Let $x^{-1}A, y^{-1}A, u^{-1}A, v^{-1}A \in Q_{dA}$, $a, b \in X$. Let $(x^{-1}A, a, y^{-1}A) = (u^{-1}A, b, v^{-1}A)$. Then $x^{-1}A = u^{-1}A$, $y^{-1}A = v^{-1}A$, and $a = b$. Now $(xa)^{-1}A = a^{-1}(x^{-1}A) = b^{-1}(u^{-1}A) = (ub)^{-1}A$. Hence $(xa)^{-1}A = y^{-1}A$ if and only if $(ua)^{-1}A = v^{-1}A$. This shows that μ_{dA} is well defined.

Let $x^{-1}A, y^{-1}A, z^{-1}A \in Q_{dA}$ and $\mu_{dA}(x^{-1}A, a, y^{-1}A) > 0$ and

$$\mu_{dA}(x^{-1}A, a, z^{-1}A) > 0.$$

Then $y^{-1}A = (xa)^{-1}A = z^{-1}A$. Hence M_{dA} is deterministic.

By Theorem 7.3.9, Q_{dA} is a finite set. Clearly M_{dA} is a complete accessible fuzzy recognizer.

Let $x \in A$. Then $\tau_{dA}(x^{-1}A) = c_A > 0$. Let $x = a_1 a_2 \ldots a_n$, $a_i \in X$ for all i. Now

$$\begin{aligned} \mu_{dA}^*(A, x, x^{-1}A) &\geq \mu_{dA}(A, a_1, a_1^{-1}A) \wedge \mu_{dA}(a_1^{-1}A, a_2, (a_1 a_2)^{-1}A) \wedge \\ &\quad \mu_{dA}((a_1 a_2)^{-1}A, a_2, (a_1 a_2 a_3)^{-1}A) \wedge \ldots \\ &\quad \wedge \mu_{dA}((a_1 a_2 \ldots a_{n-1})^{-1}A, a_n, (a_1 a_2 \ldots a_n)^{-1}A) \\ &= m_A \wedge m_A \wedge \ldots \wedge m_A \\ &= m_A \\ &> 0. \end{aligned}$$

Thus $\iota_{dA}(A) \wedge \mu_{dA}^*(A, x, x^{-1}A) \wedge \tau_{dA}(x^{-1}A) > 0$. This implies that $x \in L(M_{dA})$. Now let $x \in L(M_{dA})$. There exists $z^{-1}A \in Q_{dA}$ such that $\iota_{dA}(A) \wedge \mu_{dA}^*(A, x, z^{-1}A) \wedge \tau_{dA}(z^{-1}A) > 0$. Thus $\iota_{dA}(A) > 0$, $\mu_{dA}^*(A, x, z^{-1}A) > 0$, and $\tau_{dA}(z^{-1}A) > 0$. Now $\mu_{dA}^*(A, x, z^{-1}A) > 0$ implies that $x^{-1}A = z^{-1}A$ by the definition of μ_{dA}^*. Hence $\tau_{dA}(x^{-1}A) = \tau_{dA}(z^{-1}A) > 0$ and so $x \in A$. Consequently, $A = L(M_{dA})$. ∎

Theorem 7.3.11 *Let $A \subseteq X^*$ be recognizable. Suppose $M_d = (Q_d, X, \mu, \iota, \tau)$ is a complete accessible deterministic fuzzy recognizer with behavior A. Let M_{dA} be the complete accessible deterministic fuzzy recognizer as constructed in the proof of Theorem 7.3.10. Then $\exists$ a function $f : Q_d \to Q_{dA}$ such that*

(1) $f(s_0) = A$;

(2) $f^{-1}(F_{dA}) = F_d$, where $F_d = \{q \in Q_d \mid \tau(q) > 0\}$ and $F_{dA} = \{x^{-1}A \in Q_{dA} \mid \tau_{dA}(x^{-1}A) > 0\}$;

(3) $\mu_{dA}(f(q), a, f(p)) \geq \mu(q, a, p)$ for all $p, q \in Q_d$, $a \in X$;

(4) f is surjective.

Proof. Define $f : Q_d \to Q_{dA}$ by $\forall q \in Q_d$, $f(q) = q^{-1} \circ \tau$. By Theorem 7.3.9(3), f maps Q_d into Q_{dA}.

(1) $f(s_0) = s_0^{-1} \circ \tau = A$ by Theorem 7.3.9(4).

(2) Let $q \in F_d$. Then $\tau(q) > 0$. There exists $t \in A$ such that $q^{-1} \circ \tau = t^{-1}A$ by Theorem 7.3.9(3). Now $t \in A = L(M_{dA})$. Hence $\mu_{dA}^*(A, t, y^{-1}A) \wedge \tau_{dA}(y^{-1}A) > 0$ for some $y^{-1}A$. This implies that $\mu_{dA}^*(A, t, y^{-1}A) > 0$ and $\tau_{dA}(y^{-1}A) > 0$. Now $\mu_{dA}^*(A, t, y^{-1}A) > 0$ implies that $t^{-1}A = y^{-1}A$ and $\tau_{dA}(y^{-1}A) > 0$ implies that $y^{-1}A \in F_{dA}$. Hence $f(q) = q^{-1} \circ \tau = t^{-1}A = y^{-1}A \in F_{dA}$, i.e., $q \in f^{-1}(F_{dA})$. Suppose $q \in f^{-1}(F_{dA})$. Then $\tau_{dA}(q^{-1} \circ \tau) = \tau_{dA}(f(q)) > 0$. Now $q^{-1} \circ \tau = t^{-1}A$ for some $t \in X^*$. Thus $\tau_{dA}(t^{-1}A) > 0$ and by the definition of τ_{dA} it follows that $t \in A$. Since $t \in A$, $\Lambda \in t^{-1}A = q^{-1} \circ \tau$. Hence $\mu^*(q, \Lambda, r) \wedge \tau(r) > 0$ for some $r \in Q_d$. This implies that $q = r$. Thus $\tau(q) > 0$ and so $q \in F_d$. Consequently $f^{-1}(F_{dA}) = F_d$.

(3) Let $q, p \in Q_d$. Now $f(q) = q^{-1} \circ \tau$ and $f(q) = p^{-1} \circ \tau$. By Theorem 7.3.9, there exists $x, y \in X^*$ such that $q^{-1} \circ \tau = x^{-1}A$ and $p^{-1} \circ \tau = y^{-1}A$. Let $a \in X$. Suppose $\mu_{dA}(f(q), a, f(p)) = 0$. Then $\mu_{dA}(x^{-1}A, a, y^{-1}A) = 0$ and so $(xa)^{-1}A \neq y^{-1}A$. We claim that $\mu(q, a, p) = 0$.

Suppose $\mu(q, a, p) > 0$. Either there exists $t \in (xa)^{-1}A$ such that $t \notin y^{-1}A$ or there exists $t \in y^{-1}A$ such that $t \notin (xa)^{-1}A$. First suppose that there exists $t \in (xa)^{-1}A$ such that $t \notin y^{-1}A$. Since $xat \in A$, $\mu^*(s_0, xat, r) \wedge \tau(r) > 0$ for some $r \in Q_d$. Thus $\mu^*(s_0, xat, r) > 0$. This implies that there exists $q', p' \in Q_d$ such that $\mu^*(s_0, x, q') \wedge \mu(q', a, p') \wedge \mu^*(p', t, r) > 0$. Since M_d is deterministic, $q' = q$ (note this follows from the proof of Theorem 7.3.9 by the choice of x). Thus $\mu(q, a, p') > 0$ and since M_d is deterministic, it follows that $p = p'$. Hence $\mu^*(p, t, r) > 0$. By the choice of y (as in the proof of Theorem 7.3.9), $\mu^*(s_0, y, p) > 0$. It now follows that $\mu^*(s_0, y, p) \wedge \mu^*(p, t, r) \wedge \tau(r) > 0$ and so $t \in y^{-1}A$, a contradiction. Now suppose there exists $t \in y^{-1}A$ such that $t \notin (xa)^{-1}A$. Then $xat \notin A$. Now $yt \in A$. Thus $\mu^*(s_0, yt, r) \wedge \tau(r) > 0$ for some $r \in Q_d$. Thus there exists $p'' \in Q_d$ such that $\mu^*(s_0, y, p'') \wedge \mu^*(p'', t, r) > 0$. From this it follows that $p = p''$. Thus $\mu^*(p, t, r) > 0$. Hence $\mu^*(s_0, x, q) \wedge \mu(q, a, p) \wedge \mu^*(p, t, r) \wedge \tau(r) > 0$ and so $xat \in A$, a contradiction. Hence $\mu(q, a, p) = 0$. From the definition of μ_{dA}, it now follows that $\mu_{dA}(f(q), a, f(p)) \geq \mu(q, a, p)$.

(4) Let $x^{-1}A \in Q_{dA}$. By Theorem 7.3.9, $x^{-1}A = q_x^{-1} \circ \tau$. Thus f is surjective. ∎

We regard the recognizer M_{dA} as being a **minimal complete recognizer** of the recognizable subset A, where the term "minimal" refers to the properties described in Theorem 7.3.11; in particular, (4) implies that $|Q_{dA}| \leq |Q_d|$.

7.4 Fuzzy Recognizers and Recognizable Sets

In the next five sections, we present the work of [118]. In [118], the definitions of a fuzzy recognizer, Definition 7.4.1, the recognition of a word, Definition 7.6.9, accessibility, Definition 7.7.1, and so on are equivalent to the ones given above, i.e., Definition 7.2.8, Definition 7.2.10, and Definition 7.2.16, respectively. However the approach differs and additional results are obtained. Hence we use notation that is compatible with that of [118].

The idea of a reversal, inverse image, accessible part, coaccessible part, and trim part regarding fuzzy recognizers are introduced and their properties are discussed.

We characterize the words recognized by a fuzzy recognizer and prove fuzzy recognizability of several crisp sets. We prove that if a set of words is recognized by a fuzzy recognizer, then it is also recognized by its accessible and coaccessible parts. We prove similar results for a complete fuzzy recognizer.

We also give a procedure to construct a complete fuzzy recognizer from a given fuzzy recognizer with the property that they recognize the same subset.

Let $M = (Q, X, \mu)$ denote a fuzzy finite state machine and $f : (X')^* \to X^*$ be a monoid homomorphism, where X' is a nonempty set. If $f(X') \subseteq X \cup \{\Lambda\}$, then f is called a **fine homomorphism**.

Definition 7.4.1 *A fuzzy X-**recognizer** of a fuzzy finite state machine $M = (Q, X, \mu)$ is a triple $\mathcal{M} = (M, \iota, \tau)$, where ι and τ are fuzzy subsets of Q. ι is called a fuzzy set of **initial states** and τ is called a fuzzy set of **final states**.*

A fuzzy X-recognizer will be called a recognizer when the set of input symbols X is understood.

Let $M = (Q, X, \mu)$ and $M' = (Q', X', \mu')$ be fuzzy finite state machines. Define

$$\mu \times \mu' : (Q \times Q') \times (X \times X') \times (Q \times Q') \to [0, 1]$$

as follows:

$$\mu \times \mu'((p, p'), (a, a'), (q, q')) = \mu(p, a, q) \wedge \mu'(p', a', q').$$

Recall that the fuzzy finite state machine $M \times M' = (Q \times Q', X \times X', \mu \times \mu')$ is called the **direct product** of M and M'.

Let $M = (Q, X, \mu)$ and $M' = (Q', X', \mu')$ be fuzzy finite state machines. Define

$$\mu \cap \mu' : (Q \times Q') \times X \times (Q \times Q') \to [0, 1]$$

as follows:

$$(\mu \cap \mu')((p, p'), a, (q, q')) = \mu(p, a, q) \wedge \mu'(p', a, q').$$

Recall that the fuzzy finite state machine $M \cap M' = (Q \times Q', X, \mu \cap \mu')$ is called the **restricted direct product** of M and M'.

Definition 7.4.2 Let $M = (Q, X, \mu)$ and $M' = (Q', X, \mu')$ be fuzzy finite state machines such that $Q \cap Q' = \emptyset$. Define

$$\mu \cup \mu' : (Q \cup Q') \times X \times (Q \cup Q') \to [0, 1]$$

as follows:

$$(\mu \cup \mu')(p, a, q) = \begin{cases} \mu(p, a, q) & \text{if } p, q \in Q \\ \mu'(p, a, q) & \text{if } p, q \in Q' \\ 0 & \text{otherwise.} \end{cases}$$

The fuzzy finite state machine $M \vee M' = (Q \cup Q', X, \mu \cup \mu')$ is called the **join** of M and M'.

We recall the definition of a sub-fuzzy finite state machine. Let $M = (Q, X, \mu)$ be a fuzzy finite state machine. A fuzzy finite state machine $M^s = (Q^s, X^s, \mu^s)$ is called a sub-fuzzy finite state machine of M, written $M^s \subseteq M$, if

(i) $Q^s \subseteq Q$, $X^s \subseteq X$, and

(ii) $\mu^s = \mu|_{Q^s \times X^s \times Q^s}$.

Note again that the definition of a sub-fuzzy finite state machine of M differs from that of a submachine of M, Definition 6.7.7, in that $X^s = X$ is not required.

7.5 Operations on (Fuzzy) Subsets

Next, we introduce some notation that will simplify the proofs of some of the theorems.

Let $A \subseteq X^*$, $B \subseteq X^*$, and $a, b \in X^*$. Then

N1. $AB^{-1} = \{x \in X^* \mid xb \in A, \text{ for some } b \in B\}$.

N2. $A^{-1}B = \{x \in X^* \mid ax \in B, \text{ for some } a \in A\}$.

N3. $Ab^{-1} = \{x \in X^* \mid xb \in A\}$.

N4. $a^{-1}B = \{x \in X^* \mid ax \in B\}$.

N5. $AB = \{xy \mid x \in A, y \in B\}$.

N6. $Ab = \{xb \mid x \in A\}$.

N7. $aB = \{ay \mid y \in B\}$.

Clearly, $AB^{-1} = \cup_{b \in B} Ab^{-1}$, $A^{-1}B = \cup_{a \in A} a^{-1}B$, and $AB = \cup_{b \in B} Ab = \cup_{a \in A} aB$.

Let $\delta : Q \to [0, 1]$, $\gamma : Q \to [0, 1]$, $x \in X^*$, $p \in Q$, and $A \subseteq X^*$. Then

N8. $p * x : Q \to [0, 1]$ is defined by $\forall q \in Q$,

$$p * x(q) = \mu^*(p, x, q).$$

N9. $p * A : Q \to [0, 1]$ is defined by $\forall q \in Q$,

$$p * A(q) = \vee\{\mu^*(p, y, q) \mid y \in A\}.$$

N10. $\delta * x : Q \to [0, 1]$ is defined by $\forall q \in Q$,

$$\delta * x(q) = \vee\{\delta(p) \wedge \mu^*(p, x, q) \mid p \in Q\}.$$

N11. $\delta * A : Q \to [0, 1]$ is defined by $\forall q \in Q$,

$$\delta * A(q) = \vee\{\delta(p) \wedge \mu^*(p, x, q) \mid p \in Q, \ x \in A\}.$$

N12. $\delta * x^{-1} : Q \to [0, 1]$ is defined by $\forall q \in Q$,

$$\delta * x^{-1}(q) = \vee\{\delta(p) \wedge \mu^*(q, x, p) \mid p \in Q\}.$$

N13. $\delta * A^{-1} : Q \to [0, 1]$ is defined by $\forall q \in Q$,

$$\delta * A^{-1}(q) = \vee\{\delta(p) \wedge \mu^*(q, x, p) | p \in Q, x \in A\}.$$

Clearly $p * A = \cup_{y \in A} p * y$, $\delta * A = \cup_{y \in A} \delta * y$, and $\delta * A^{-1} = \cup_{y \in A} \delta * y^{-1}$. The following results are easily proven.

Theorem 7.5.1 *Let* $\delta : Q \to [0, 1]$, $x, y \in X^*$, *and* $A, B \subseteq X^*$. *Then*
(1) $(\delta * x) * y = \delta * (xy)$;
(2) $(\delta * A) * y = \delta * (Ay)$;
(3) $(\delta * x) * B = \delta * (xB)$;
(4) $(\delta * A) * B = \delta * (AB)$;
(5) $(\delta * x^{-1}) * y^{-1} = \delta * (yx)^{-1}$;
(6) $(\delta * A^{-1}) * y^{-1} = \delta * (yA)^{-1}$;
(7) $(\delta * x^{-1}) * B^{-1} = \delta * (Bx)^{-1}$;
(8) $(\delta * A^{-1}) * B^{-1} = \delta * (BA)^{-1}$. ∎

Theorem 7.5.2 *Let* $\delta, \delta_1, \delta_2$ *be fuzzy subsets of* Q *and let* $x, y \in X^*$. *Then the following properties hold:*
(1) $\delta * \Lambda = \delta$;
(2) $(\delta_1 \cup \delta_2) * x = (\delta_1 * x) \cup (\delta_2 * x)$;
(3) $(\delta_1 \cap \delta_2) * x = (\delta_1 * x) \cap (\delta_2 * x)$;
(4) $(\delta_1 \cup \delta_2) * x^{-1} = (\delta_1 * x^{-1}) \cup (\delta_2 * x^{-1})$;
(5) $(\delta_1 \cap \delta_2) * x^{-1} = (\delta_1 * x^{-1}) \cap (\delta_2 * x^{-1})$;
(6) $\chi_\emptyset * x = \chi_\emptyset$, *where* $\chi_\emptyset$ *is the characteristic function of* $\emptyset$ *in* Q;
(7) $\chi_Q * x = \cup_{p \in Q} p * x$, *where* χ_Q *is the characteristic function of* Q. ∎

Let $\delta : Q \rightarrow [0,1]$, $\gamma : Q \rightarrow [0,1]$.

N14. Define $\delta \# \gamma : X^* \rightarrow [0,1]$ by $\forall x \in X^*$,

$$\delta \# \gamma(x) = \vee\{\delta(p) \wedge \gamma(q) \wedge \mu^*(p,x,q) \mid p, q \in Q\}.$$

N15. $\delta^{-1} \circ_\mu \gamma = \{x \in X^* \mid \delta * \gamma(x) > 0\}.$
When μ is understood, we sometimes write $\delta^{-1} \circ \gamma$ for $\delta^{-1} \circ_\mu \gamma$.

Theorem 7.5.3 *Let* $\delta : Q \rightarrow [0,1]$ *and* $x \in X^*$. *Then* $\delta * \chi_Q(x) > 0$ *if and only if* $\exists q \in Q$ *such that* $\delta * x(q) > 0$.

Proof. $\delta * \chi_Q(x) > 0 \Leftrightarrow \vee\{\delta(p) \wedge \chi_Q(q) \wedge \mu^*(p,x,q) \mid p,q \in Q\} > 0 \Leftrightarrow \vee\{\delta(p) \wedge \mu^*(p,x,q) \mid p,q \in Q\} > 0 \Leftrightarrow \vee\{\delta(p) \wedge \mu^*(p,x,q) \mid p \in Q\} > 0$ for some $q \in Q \Leftrightarrow \delta * x(q) > 0$. ∎

Theorem 7.5.4 *Let* δ *and* γ *be fuzzy subsets of* Q, $q \in Q$, $A, B \subseteq X^*$, *and* $x, y \in X^*$. *Then the following properties hold:*
(1) $\delta^{-1} \circ (\gamma * A^{-1}) = (\delta^{-1} \circ \gamma)A^{-1}$;
(2) $\delta^{-1} \circ (\gamma * x^{-1}) = (\delta^{-1} \circ \gamma)x^{-1}$;
(3) $(\delta * A)^{-1} \circ \gamma = A^{-1}(\delta^{-1} \circ \gamma)$;
(4) $(\delta * x)^{-1} \circ \gamma = x^{-1}(\delta^{-1} \circ \gamma)$;
(5) $(q * A)^{-1} \circ \gamma = A^{-1}(q^{-1} \circ \gamma)$;
(6) $(q * x)^{-1} \circ \gamma = x^{-1}(q^{-1} \circ \gamma)$.

Proof. We prove (1), (3), and (5). We ask the reader to verify (2), (4), and (6).
(1) $z \in \delta^{-1} \circ (\gamma * A^{-1}) \Leftrightarrow \delta\#(\gamma * A^{-1})(z) > 0 \Leftrightarrow \delta(p_0) \wedge (\gamma * A^{-1})(q_0) \wedge \mu^*(p_0, z, q_0) > 0$ for some $p_0, q_0 \in Q \Leftrightarrow \delta(p_0) \wedge \gamma(t_0) \wedge \mu^*(q_0, x_0, t_0) \wedge \mu^*(p_0, z, q_0) > 0$ for some $p_0, q_0, t_0 \in Q$, $x_0 \in A \Leftrightarrow \delta(p_0) \wedge \gamma(t_0) \wedge [\mu^*(p_0, z, q_0) \wedge \mu^*(q_0, x_0, t_0)] > 0$ for some $p_0, q_0, t_0 \in Q$, $x_0 \in A \Leftrightarrow \delta(p_0) \wedge \gamma(t_0) \wedge \mu^*(p_0, zx_0, t_0) > 0$ for some $p_0, t_0 \in Q$, $x_0 \in A \Leftrightarrow zx_0 \in \delta^{-1} \circ \gamma$ for some $x_0 \in A \Leftrightarrow z \in (\delta^{-1} \circ \gamma)A^{-1}$.
(3) $z \in (\delta * A)^{-1} \circ \gamma \Leftrightarrow ((\delta * A)^{-1} \# \gamma)(z) > 0 \Leftrightarrow (\delta * A)(p_0) \wedge \gamma(q_0) \wedge \mu^*(p_0, z, q_0) > 0$ for some $p_0, q_0 \in Q \Leftrightarrow \delta(t_0) \wedge \mu^*(t_0, x_0, p_0) \wedge \gamma(q_0) \wedge \mu^*(p_0, z, q_0) > 0$ for some $p_0, q_0, t_0 \in Q$, $x_0 \in A \Leftrightarrow \delta(t_0) \wedge \gamma(q_0) \wedge \mu^*(t_0, x_0 z, q_0) > 0$ for some $q_0, t_0 \in Q$, $x_0 \in A \Leftrightarrow x_0 z \in \delta^{-1} \circ \gamma$ for some $x_0 \in A \Leftrightarrow z \in A^{-1}(\delta^{-1} \circ \gamma)$.
(5) $y \in (q * A)^{-1} \circ \gamma \Leftrightarrow (q * A)^{-1} \# \gamma(y) > 0 \Leftrightarrow q * A(p_0) \wedge \gamma(q_0) \wedge \mu^*(p_0, y, q_0) > 0$ for some $p_0, q_0 \in Q$, $y \in A \Leftrightarrow \mu^*(q, x, p_0) \wedge \gamma(q_0) \wedge \mu^*(p_0, y, q_0) > 0$ for some $p_0, q_0 \in Q$, $y \in A \Leftrightarrow \gamma(q_0) \wedge [\mu^*(q, x, p_0) \wedge \mu^*(p_0, y, q_0)] > 0$ for some $p_0, q_0 \in Q$, $y \in A \Leftrightarrow \gamma(q_0) \wedge \mu^*(q, xy, q_0) > 0$ for some $q_0 \in Q$, $y \in A \Leftrightarrow y \in A^{-1}(q^{-1} \circ \gamma)$. ∎
N16. Let $\delta_1 : Q \rightarrow [0,1]$ and $\delta_2 : Q' \rightarrow [0,1]$. Then define $\delta_1 \cap \delta_2 : Q \times Q' \rightarrow [0,1]$ by $\forall(q, q') \in Q \times Q'$,

$$(\delta_1 \cap \delta_2)(q, q') = \delta_1(q) \wedge \delta_2(q').$$

N17. Let $\delta_1 : Q \to [0,1]$, $\delta_2 : Q' \to [0,1]$ be such that $Q \cap Q' = \emptyset$. Then define $\delta_1 \cup \delta_2 : Q \cup Q' \to [0,1]$ by $\forall q \in Q$,

$$(\delta_1 \cup \delta_2)(q) = \begin{cases} \delta_1(q) & \text{if } q \in Q \\ \delta_2(q) & \text{if } q \in Q'. \end{cases}$$

Theorem 7.5.5 *Let* $M = (Q, X, \mu)$ *and* $M' = (Q', X, \mu')$ *be fuzzy finite state machines. Let* δ_1, γ_1 *be fuzzy subsets of* Q *and* δ_2, γ_2 *be fuzzy subsets of* Q'. *Then the following properties hold.*

(1) $(\delta_1 \cup \delta_2)^{-1} \circ_{\mu \cup \mu'} (\gamma_1 \cup \gamma_2) = (\delta_1^{-1} \circ_\mu \gamma_1) \cup (\delta_2^{-1} \circ_{\mu'} \gamma_2);$

(2) $(\delta_1 \cap \delta_2)^{-1} \circ_{\mu \cap \mu'} (\gamma_1 \cap \gamma_2) = (\delta_1^{-1} \circ_\mu \gamma_1) \cap (\delta_2^{-1} \circ_{\mu'} \gamma_2).$

Proof. (1) Let $x \in (\delta_1 \cup \delta_2)^{-1} \circ (\gamma_1 \cup \gamma_2)$. Then $(\delta_1 \cup \delta_2) * (\gamma_1 \cup \gamma_2)(x) > 0$. Thus $(\delta_1 \cup \delta_2)(p) \wedge (\gamma_1 \cup \gamma_2)(q) \wedge (\mu \cup \mu')^*(p, x, q) > 0$ for some $p, q \in Q \cup Q'$.

It follows that both p, q belong to either Q or Q'. For, if $p \in Q$ and $q \in Q'$ and vice versa, then by definition $(\mu \cup \mu')^*(p, x, q) = 0$ and consequently $(\delta_1 \cup \delta_2)(p) \wedge (\gamma_1 \cup \gamma_2)(q) \wedge (\mu \cup \mu')^*(p, x, q) = 0$, a contradiction.

If $p, q \in Q$, then clearly $\delta_1(p) \wedge \gamma_1(q) \wedge \mu^*(p, x, q) = (\delta_1 \cup \delta_2)(p) \wedge (\gamma_1 \cup \gamma_2)(q) \wedge (\mu \cup \mu')^*(p, x, q) > 0$. Therefore, $\delta_1 \# \gamma_1(x) > 0$ and hence $x \in \delta_1^{-1} \circ \gamma_1$. Similarly, if $p, q \in Q'$, then $x \in \delta_2^{-1} \circ \gamma_2$.

On the other hand, let $x \in \delta_1^{-1} \circ \gamma_1$. Then $\delta_1(p) \wedge \gamma_1(q) \wedge \mu^*(p, x, q) > 0$ for some $p, q \in Q$, i.e., $(\delta_1 \cup \delta_2)(p) \wedge (\gamma_1 \cup \gamma_2)(q) \wedge (\mu \cup \mu')^*(p, x, q) > 0$. A similar situation holds if $x \in \delta_2^{-1} \circ \gamma_2$.

(2) $x \in (\delta_1 \cap \delta_2)^{-1} \circ (\gamma_1 \cap \gamma_2) \Leftrightarrow (\delta_1 \cap \delta_2) \# (\gamma_1 \cap \gamma_2)(x) > 0 \Leftrightarrow (\delta_1 \cap \delta_2)(p_1, p_2) \wedge (\gamma_1 \cap \gamma_2)(q_1, q_2) \wedge (\mu \cup \mu')^*((p_1, p_2), x, (q_1, q_2)) > 0$ for some $(p_1, p_2), (q_1, q_2) \in Q \times Q' \Leftrightarrow [\delta_1(p_1) \wedge \gamma_1(q_1) \wedge \mu^*(p_1, x, q_1)] \wedge [\delta_2(p_2) \wedge \gamma_2(q_2) \wedge \mu^*(p_2, x, q_2)] > 0$ for some $p_1, q_1 \in Q$, $p_2, q_2 \in Q' \Leftrightarrow [\delta_1 \# \gamma_1(x)] \wedge [\delta_2 \# \gamma_2(x)] > 0 \Leftrightarrow [\delta_1 \# \gamma_1(x)] > 0$ and $[\delta_2 \# \gamma_2(x)] > 0 \Leftrightarrow x \in \delta_1^{-1} \circ \gamma_1$ and $x \in \delta_2^{-1} \circ \gamma_2 \Leftrightarrow x \in (\delta_1^{-1} \circ \gamma_1) \cap (\delta_2^{-1} \circ \gamma_2)$. ∎

Theorem 7.5.6 *Let* $M = (Q, X, \mu)$ *and* $M' = (Q', X, \mu')$ *be fuzzy finite state machines. Let* δ_1, γ_1 *be fuzzy subsets of* Q *and* δ_2, γ_2 *be fuzzy subsets of* Q'. *Then*

$$(\delta_1 \cap \delta_2)^{-1} \circ_{\mu \times \mu'} (\gamma_1 \cap \gamma_2) = (\delta_1^{-1} \circ_\mu \gamma_1) \times (\delta_2^{-1} \circ_{\mu'} \gamma_2).$$

Proof. $(x, y) \in (\delta_1 \cap \delta_2)^{-1} \circ (\gamma_1 \cap \gamma_2) \Leftrightarrow (\delta_1 \cap \delta_2) * (\gamma_1 \cap \gamma_2)(x, y) > 0 \Leftrightarrow (\delta_1 \cap \delta_2)(p_1, p_2) \wedge (\gamma_1 \cap \gamma_2)(q_1, q_2) \wedge (\mu \times \mu')^*((p_1, p_2), (x, y), (q_1, q_2)) > 0$ for some $(p_1, p_2), (q_1, q_2) \in Q \times Q' \Leftrightarrow [\delta_1(p_1) \wedge \gamma_1(q_1) \wedge \mu^*(p_1, x, q_1)] \wedge [\delta_2(p_2) \wedge \gamma_2(q_2) \wedge \mu'^*(p_2, y, q_2)] > 0$ for some $p_1, q_1 \in Q$, $p_2, q_2 \in Q' \Leftrightarrow [\delta_1 \# \gamma_1(x)] \wedge [\delta_2 \# \gamma_2(y)] > 0 \Leftrightarrow x \in \delta_1^{-1} \circ \gamma_1$ and $y \in \delta_2^{-1} \circ \gamma_2 \Leftrightarrow (x, y) \in (\delta_1^{-1} \circ \gamma_1) \times (\delta_2^{-1} \circ \gamma_2)$. ∎

The proof of the following theorem is similar to the proof of Theorem 7.5.6.

Theorem 7.5.7 *Let* $\delta_1 : Q \to [0,1]$, $\delta_2 : Q' \to [0,1]$ *and* $x \in X^*$. *Then*

(1) $(\delta_1 \cap \delta_2) * x = (\delta_1 * x) \cap (\delta_2 * x);$

(2) $(\delta_1 \cap \delta_2) * x^{-1} = (\delta_1 * x^{-1}) \cap (\delta_2 * x^{-1}).$ ∎

7.6   Construction of Recognizers and Recognizable Sets

Definition 7.6.1 Let $\mathcal{M} = (M, \iota, \tau)$ and $\mathcal{M}' = (M', \iota', \tau')$ be fuzzy X-recognizers of $M = (Q, X, \mu)$ and $M' = (Q', X, \mu')$, respectively. Then the fuzzy X-recognizer $\mathcal{M} \cap \mathcal{M}' = (M \cap M', \iota \cap \iota', \tau \cap \tau')$ is called the **restricted direct product** of M and M'.

Definition 7.6.2 Let $\mathcal{M} = (M, \iota, \tau)$ be a fuzzy X-recognizer of a fuzzy finite state machine $M = (Q, X, \mu)$, and $\mathcal{M}' = (M', \iota', \tau')$ be a fuzzy X'-recognizer of a fuzzy finite state machine $M' = (Q', X', \mu')$. Then the fuzzy $X \times X'$-recognizer $\mathcal{M} \times \mathcal{M}' = (M \times M, \iota \cap \iota', \tau \cap \tau')$ is called the **direct product** of M and M'.

Definition 7.6.3 Let $\mathcal{M} = (M, \iota, \tau)$ and $\mathcal{M}' = (M', \iota', \tau')$ be fuzzy X-recognizers of $M = (Q, X, \mu)$ and $M' = (Q', X, \mu')$, respectively, where $Q \cap Q' = \emptyset$. Then the fuzzy X-recognizer $\mathcal{M} \cup \mathcal{M}' = (M \cup M, \iota \cup \iota', \tau \cup \tau')$ is called the **join** of M and M'.

Definition 7.6.4 The mapping $\rho : X^* \to X^*$ is called a **reversal** of X^* if the following conditions hold:
  (1) $\rho(\Lambda) = \Lambda$,
  (2) $\rho(a) = a$,
  (3) $\rho(xa) = \rho(a)\rho(x)$, and
  (4) $\rho(\rho(x)) = x$,
$\forall a \in X$ and $x \in X^*$.

It follows by induction that $\rho(xy) = \rho(y)\rho(x)$ $\forall x, y \in X^*$.
Let ρ be a reversal of X^*. Define the fuzzy subset

$$\mu^\rho : Q \times X \times Q \to [0, 1]$$

by $\mu^\rho(p, a, q) = \mu(q, \rho(a), p)$ $\forall a \in X$, $\forall p, q \in Q$.

Definition 7.6.5 If $M = (Q, X, \mu)$ is a fuzzy finite state machine, then the fuzzy finite state machine $M^\rho = (Q, X, \mu^\rho)$ is called the **reversal** of M.

Theorem 7.6.6 Let $M^\rho = (Q, X, \mu^\rho)$ be the reversal of $M = (Q, X, \mu)$. Then

$$(\mu^\rho)^*(p, xy, q) = \mu^*(q, \rho(y)\rho(x), p)$$

$\forall x, y \in X^*$. ∎

Let $A \subseteq X^*$ and $\rho : X^* \to X^*$ be the reversal mapping. Let

$$A^\rho = \{\rho(x) \mid x \in A\}.$$

Theorem 7.6.7 *Let δ and γ be fuzzy subsets of Q. Then $(\delta^{-1} \circ \gamma)^\rho = \gamma^{-1} \circ \delta$.*

Proof. $\rho(x) \in (\delta^{-1} \circ \gamma)^\rho \Leftrightarrow x \in \delta^{-1} \circ \gamma \Leftrightarrow \delta \# \gamma(x) > 0 \Leftrightarrow \delta(p) \wedge \gamma(q) \wedge \mu^*(p, x, q) > 0$ for some $p, q \in Q \Leftrightarrow \delta(p) \wedge \gamma(q) \wedge (\mu^\rho)^*(q, \rho(x), p) > 0$ for some $p, q \in Q \Leftrightarrow \gamma(q) \wedge \delta(p) \wedge (\mu^\rho)^*(q, \rho(x), p) > 0$ for some $p, q \in Q \Leftrightarrow \gamma \# \delta(\rho(x)) > 0 \Leftrightarrow \rho(x) \in \gamma^{-1} \circ \delta$. ∎

Definition 7.6.8 *Let $\mathcal{M} = (M, \iota, \tau)$ be a fuzzy X-recognizer. Let $\rho : X^* \to X^*$ be the reversal mapping. Then the fuzzy X-recognizer $\mathcal{M}^\rho = (M^\rho, \tau, \iota)$ of M^ρ is called the **reversal** of M.*

Definition 7.6.9 *Let $\mathcal{M} = (M, \iota, \tau)$ be a fuzzy X-recognizer. Then $x \in X^*$ is said to be **recognized by** $\mathcal{M}$ if $\iota \# \tau(x) > 0$.*

Let $B(\mathcal{M})$ denote the set of all words that are recognized by $\mathcal{M}$. $B(\mathcal{M})$ is called the **behavior** of $\mathcal{M}$.

Theorem 7.6.10 *Let $\mathcal{M}$ be a fuzzy recognizer and $x \in X^*$. Then the following conditions are equivalent.*

(1) $x \in B(\mathcal{M})$;
(2) $x \in \iota^{-1} \circ \tau$;
(3) $\iota \# \tau(x) > 0$;
(4) $\exists p_0 \in Q$ such that $\iota \cap \tau * x^{-1}(p_0) > 0$;
(5) $\exists q_0 \in Q$ such that $\iota * x \cap \tau(q_0) > 0$.

Proof. Clearly (1), (2), and (3) are equivalent.

$(3)\Leftrightarrow(4)$: $\iota \# \tau(x) > 0 \Leftrightarrow \vee\{\iota(p) \wedge \tau(q) \wedge \mu^*(p, x, q) \mid p, q \in Q\} > 0 \Leftrightarrow \exists p_0, q_0 \in Q$ such that $\iota(p_0) \wedge \tau(q_0) \wedge \mu^*(p_0, x, q_0) > 0 \Leftrightarrow \exists p_0, q_0 \in Q$ such that $\iota(p_0) > 0$ and $\tau(q_0) \wedge \mu^*(p_0, x, q_0) > 0 \Leftrightarrow \exists p_0 \in Q$ such that $\iota(p_0) > 0$ and $\vee\{\tau(q) \wedge \mu^*(p_0, x, q) \mid q \in Q\} > 0 \Leftrightarrow \exists p_0 \in Q$ such that $\iota(p_0) > 0$ and $\tau * x^{-1}(p_0) > 0 \Leftrightarrow \exists p_0 \in Q$ such that $\iota \cap (\tau * x^{-1})(p_0) > 0$.

$(3)\Leftrightarrow(5)$: This can be proved by interchanging the roles of p_0 and q_0 and using the definition of $\iota * x$. ∎

Definition 7.6.11 *A subset A of X^* is called X-**recognizable**, if there exists an X-recognizer $\mathcal{M}$ such that $B(\mathcal{M}) = A$.*

Theorem 7.6.12 *Let X be a nonempty finite set. Then the following sets are X-recognizable:*

(1) $\{a\}$, where $a \in X$.
(2) Λ.
(3) A^*, where $A \subseteq X$.
(4) X. ∎

Theorems 7.5.5, 7.6.7, and 7.6.10 immediately lead to the following theorem.

Theorem 7.6.13 *Let $\mathcal{M}$ and $\mathcal{M}'$ be fuzzy recognizers. Then the following assertions hold.*

(1) $B(\mathcal{M} \cup \mathcal{M}') = B(\mathcal{M}) \cup B(\mathcal{M}')$.

(2) $B(\mathcal{M} \cap \mathcal{M}') = B(\mathcal{M}) \cap B(\mathcal{M}')$.

(3) $B(\mathcal{M}^\rho) = B(\mathcal{M})^\rho$.

Proof. (3) We have $B(\mathcal{M})^\rho = (\iota^{-1} \circ \tau)^\rho = \tau^{-1} \circ \iota = B(\mathcal{M}^\rho)$. ∎

The following corollary is immediate from Theorem 7.6.13.

Corollary 7.6.14 *Let A and B be recognizable subsets of X^*. Then the following sets are recognizable.*

(1) $A \cup B$.

(2) $A \cap B$.

(3) A^ρ.

Corollary 7.6.15 *A subset A of X^* is recognizable if and only if A^ρ is recognizable.*

Proof. $(A^\rho)^\rho = A$. ∎

In view of Theorems 7.5.6 and 7.6.10, we obtain the following theorem.

Theorem 7.6.16 *Let $\mathcal{M}$ be a X-recognizer and $\mathcal{M}'$ be a X'-recognizer. Then $B(\mathcal{M} \times \mathcal{M}') = B(\mathcal{M}) \times B(\mathcal{M}')$.* ∎

Corollary 7.6.17 *If A is X-recognizable and B is X'-recognizable, then $A \times B$ is $X \times X'$-recognizable.* ∎

Theorem 7.6.18 *Let A be X-recognizable and B be any subset if X^*. Then the following sets are X-recognizable.*

(1) AB^{-1}.

(2) Ax^{-1}.

(3) $B^{-1}A$.

(4) $x^{-1}A$.

Proof. We prove (1) and (3) only.

Since A is X-recognizable, there exists a fuzzy recognizer $\mathcal{M} = (M, \iota, \tau)$ of a fuzzy finite state machine $M = (Q, X, \mu)$ such that $B(\mathcal{M}) = A$.

(1) Consider a fuzzy recognizer $\mathcal{M}' = (M, \iota, \tau * B^{-1})$. Then by Theorem 7.5.4(1), $\iota^{-1} \circ (\tau * B^{-1}) = (\iota^{-1} \circ \tau)B^{-1} = B(\mathcal{M})B^{-1} = AB^{-1}$. Thus AB^{-1} is recognizable.

(3) Consider the fuzzy recognizer $\mathcal{M}' = (M, \iota, B^{-1} * \tau)$. Then by Theorem 7.5.4(3), $(\iota * B)^{-1} \circ \tau = B^{-1}(\iota^{-1} \circ \tau) = B^{-1}B(\mathcal{M}) = B^{-1}A$. Hence $B^{-1}A$ is recognizable. ∎

Let X and X' be sets and let $f : (X')^* \to X^*$ be a fine homomorphism. If $M = (Q, X, \mu)$ is a fuzzy finite state machine, then define $\mu' : Q \times X' \times Q \to [0,1]$ by

$$\mu'(p, a', q) = \mu(p, f(a'), q)$$

$\forall p, q \in Q, a' \in X'$. This defines a fuzzy finite state machine $M' = (Q, X, \mu')$ and a fuzzy X'-recognizer $\mathcal{M}' = (M', \iota, \tau)$.

Definition 7.6.19 *The fuzzy X'-recognizer $\mathcal{M}'$ defined immediately above is called the **inverse image** of $\mathcal{M}$ and is denoted by $f^{-1}(\mathcal{M})$.*

Theorem 7.6.20 *Let $\delta : Q \to [0,1]$ and $\gamma : Q \to [0,1]$. Then $\delta^{-1} \circ_{\mu'} \gamma = f^{-1}(\delta^{-1} \circ_\mu \gamma)$.*

Proof. $x \in \delta^{-1} \circ_{\mu'} \gamma \Leftrightarrow \delta \# \gamma(x) > 0 \Leftrightarrow \delta(p) \wedge \gamma(q) \wedge \mu'(p, x, q) > 0$ for some $p, q \in Q \Leftrightarrow \delta(p) \wedge \gamma(q) \wedge \mu^*(p, f(x), q) > 0$ for some $p, q \in Q \Leftrightarrow f(x) \in \delta^{-1} \circ_\mu \gamma$. ∎

The following corollaries are easily shown to hold.

Corollary 7.6.21 $B(f^{-1}(\mathcal{M})) = f^{-1}(B(\mathcal{M}))$. ∎

Corollary 7.6.22 *If A is X-recognizable, then $f^{-1}(A)$ is X'-recognizable.* ∎

Let $\mathcal{M} = (M, \iota, \tau)$ be a fuzzy recognizer of a fuzzy finite state machine $M = (Q, X, \mu)$ and let $a \in X$. Consider $Q' = Q \cup \{t'\}$, where $t' \notin Q$. Define $\mu_a : Q' \times X \times Q' \to [0, 1]$ by

$$\mu_a(p, b, q) = \begin{cases} \mu(p, b, q) & \text{if } p, q \in Q, \\ 1 & \text{if } p \in Q, \ \tau(p) > 0, \ b = a \text{ and } q = t', \\ 0 & \text{otherwise.} \end{cases}$$

Clearly, $M_a = (Q', X, \mu_a)$ is a fuzzy finite state machine. Define $\iota_a : Q' \to [0, 1]$ as follows:

$$\iota_a(p) = \begin{cases} \iota(p)) & \text{if } p \in Q, \\ 0 & \text{otherwise.} \end{cases}$$

Let τ_a be the characteristic function of $\{t'\}$. Then $\mathcal{M}_a = (M_a, \iota_a, \tau_a)$ is the fuzzy recognizer of M_a.

Let X be a nonempty finite set, $A \subseteq X^*$, and $u \in X$. Recall that

$$Au = \{xu \mid x \in A\}.$$

Theorem 7.6.23 *Let $\mathcal{M}_a = (M_a, \iota_a, \tau_a)$ be a fuzzy recognizer of M_a. Then $\iota_a^{-1} \circ_{\mu_a} \tau_a = (\iota^{-1} \circ_\mu \tau)a$.*

Proof. Let $x \in \iota_a^{-1} \circ_{\mu_a} \tau_a$. Then $(\iota_a \# \tau_a)(x) > 0$. Thus

$$\iota_a(p_0') \wedge \tau_a(q_0') \wedge \mu_a^*(p_0', x, q_0') > 0$$

for some $p_0', q_0' \in Q'$. Hence $q_0' = t'$ and $p_0' \in Q$.

Now $\mu_a^*(p_0', x, q_0') > 0$ is possible only if either $\tau(p_0') > 0$ and $x = a$ or there exists $y \in X^*$ such that $\mu_a^*(p_0', ya, t') > 0$. We consider these two cases.

Suppose $\tau(p_0') > 0$ and $x = a$.

Now $\iota(p_0') > 0$, $\tau(p_0')$, and $\mu^*(p_0', \Lambda, p_0') > 0$. Hence $\iota(p_0') \wedge \tau(p_0') \wedge \mu^*(p_0', \Lambda, p_0') > 0$, i.e., $\Lambda \in \iota^{-1} \circ_\mu \tau$. Therefore, $x = \Lambda a \in (\iota^{-1} \circ_\mu \tau)a$.

Suppose now that $y \in X^*$ such that $\mu_a^*(p_0', ya, t') > 0$.

Then there exists $t_0' \in Q'$ such that $\mu_a^*(p_0', y, t_0') \wedge \mu_a(t_0', a, q_0') > 0$. This is true only when $\tau(t_0') > 0$, i.e., $t_0' \in Q$. Thus $\iota(p_0') \wedge \tau(t_0') \wedge \mu^*(p_0', y, t_0') > 0$. Therefore, $\iota \# \tau(y) > 0$. Hence $x = ya \in (\iota^{-1} \circ_\mu \tau)a$. Thus $\iota_a^{-1} \circ_\mu \tau_a \subseteq (\iota^{-1} \circ_\mu \tau)a$.

We now show inclusion in the other direction. Let $x = ya \in (\iota^{-1} \circ_\mu \tau)a$. Then $y \in (\iota^{-1} \circ_\mu \tau)$. Therefore, $\iota(p_0) \wedge \tau(q_0) \wedge \mu^*(p_0, y, q_0) > 0$ for some $p_0, q_0 \in Q$. Thus $\iota(p_0) > 0$ and $\tau(q_0) > 0$. Now $\tau(q_0) > 0 \Rightarrow \mu_a(q_0, a, t') = 1$ and $\mu_a^*(p_0, y, q_0) = \mu^*(p_0, y, q_0) > 0$. Hence $\mu_a^*(p_0, ya, t') = \vee\{\mu_a^*(p_0, y, r') \wedge \mu_a(r', a, t') \mid r' \in Q'\} > \mu_a^*(p_0, y, q_0) \wedge \mu_a(q_0, a, t') > 0$. Therefore, $\iota(p_0) \wedge \tau_a(t') \wedge \mu_a^*(p_0, ya, t') > 0$ and hence $\iota_a(p_0) \wedge \tau_a(t') \wedge \mu_a^*(p_0, ya, t') > 0$ since $p_0 \in Q$, i.e., $x = ya \in \iota_a^{-1} \circ_{\mu_a} \tau_a$. Thus $(\iota^{-1} \circ_\mu \tau)a \subseteq \iota_a^{-1} \circ_{\mu_a} \tau_a$.

Hence $\iota_a^{-1} \circ_{\mu_a} \tau_a = (\iota^{-1} \circ_\mu \tau)a$. ∎

Corollary 7.6.24 $B(\mathcal{M}_a) = B(\mathcal{M})a$. ∎

Corollary 7.6.25 *If a subset A of X^* is recognizable and $u \in X$, then Au is recognizable.* ∎

7.7   Accessible and Coaccessible Recognizers

Definition 7.7.1 *Let $\mathcal{M} = (M, \iota, \tau)$ be a fuzzy X-recognizer of a fuzzy finite state machine $M = (Q, X, \mu)$ and $R = \{q \in Q \mid \iota * X^*(q) > 0\}$. If $Q = R$, then $\mathcal{M}$ is called **accessible**.*

Let $\mathcal{M} = (M, \iota, \tau)$ be a fuzzy recognizer of $M = (Q, X, \mu)$. Let $R = \{q \in Q \mid \iota * X^*(q) > 0\}$. Consider $\mu^a = \mu|_{R \times X \times R}$, $\iota^a = \iota|_R$, and $\tau^a = \tau|_R$. Then $M^a = (R, X, \mu^a)$ is a fuzzy finite state machine and $\mathcal{M}^a = (M^a, \iota^a, \tau^a)$ is a fuzzy recognizer of M^a.

Definition 7.7.2 *Let $\mathcal{M} = (M, \iota, \tau)$ be a fuzzy X-recognizer of M. Then $\mathcal{M}^a = (M^a, \iota^a, \tau^a)$ is called the **accessible part** of $\mathcal{M}$.*

Definition 7.7.3 *Let $\mathcal{M} = (M, \iota, \tau)$ be a fuzzy X-recognizer of a fuzzy finite state machine $M = (Q, X, \mu)$. Let $S = \{p \in Q \mid \tau * (X^*)^{-1}(p) > 0\}$. If $Q = S$, then $\mathcal{M}$ is called **coaccessible**.*

Consider $\mu^b = \mu|_{S \times X \times S}$, $\iota^b = \iota|_S$, and $\tau^b = \tau|_S$. Then $M^b = (S, X, \mu^b)$ is a fuzzy finite state machine and $\mathcal{M}^b = (M^b, \iota^b, \tau^b)$ is a fuzzy recognizer of M^b.

Definition 7.7.4 *Let* $\mathcal{M} = (M, \iota, \tau)$ *be a fuzzy X-recognizer of M. Then* $\mathcal{M}^b = (M^b, \iota^b, \tau^b)$ *is called the* **coaccessible** *part of M.*

If $\iota(q) > 0$, then $q \in R$. Thus whenever ι is crisp, so is ι^a and $\iota^a = \iota$ as sets. If $\tau(q) > 0$, then $q \in S$. Hence whenever τ is crisp, so is τ^a and $\tau^a = \tau$ as sets.

Theorem 7.7.5 *Let $\mathcal{M}$ be a fuzzy recognizer. Then the following properties hold:*
  (1) $\iota^{-1} \circ \tau = (\iota^a)^{-1} \circ \tau^a$;
  (2) $\iota^{-1} \circ \tau = (\iota^b)^{-1} \circ \tau^b$.

Proof. (1) Let $x \in \iota^{-1} \circ \tau$. Then $\iota \# \tau(x) > 0$. Therefore,

$$\iota(p_0) \wedge \tau(q_0) \wedge \mu^*(p_0, x, q_0) > 0$$

for some $p_0, q_0 \in Q$. Clearly, $q_0 \in R$ since $\iota(p_0) \wedge \mu^*(p_0, x, q_0) > 0$. Also, $\iota(p_0) > 0$ and $\mu^*(p_0, \Lambda, p_0) = 1 > 0$ and so $p_0 \in R$. Hence

$$\tau^a(q_0) \wedge \iota^a(p_0) \wedge \mu^{a*}(p_0, x, q_0) > 0.$$

But then $\iota^a \# \tau^a(x) > 0$ and thus $x \in (\iota^a)^{-1} \circ \tau^a$. Therefore, $\iota^{-1} \circ \tau \subseteq (\iota^a)^{-1} \circ \tau^a$.

We now show inclusion in the other direction. Let $x \in (\iota^a)^{-1} \circ \tau^a$. Then $\iota^a \# \tau^a(x) > 0$, i.e., $\iota^a(p_0^a) \wedge \tau^a(q_0^a) \wedge \mu^{a*}(p_0^a, x, q_0^a) > 0$ for some $p_0^a, q_0^a \in R$. Therefore, $\iota^a(p_0^a) > 0$, $\tau^a(q_0^a) > 0$, and $\mu^{a*}(p_0^a, x, q_0^a) > 0$. Thus

$$\iota(p_0^a) = \iota^a(p_0^a), \tau(q_0^a) = \tau^a(q_0^a)$$

and

$$\mu^*(p_0^a, x, q_0^a) = \mu^{a*}(p_0^a, x, q_0^a).$$

Therefore,

$$\iota(p_0^a) \wedge \tau(q_0^a) \wedge \mu^*(p_0^a, x, q_0^a) > 0.$$

Hence $\iota \# \tau(x) > 0$ and so $x \in \iota^{-1} \circ \tau$. Thus $(\iota^a)^{-1} \circ \tau^a \subseteq \iota^{-1} \circ \tau$. Consequently, $\iota^{-1} \circ \tau = (\iota^a)^{-1} \circ \tau^a$.
  (2) The proof is similar to (1). ∎

Corollary 7.7.6 *Let $\mathcal{M}$ be a fuzzy recognizer. Then*
  (1) $B(\mathcal{M}) = B(\mathcal{M}^a)$;
  (2) $B(\mathcal{M}) = B(\mathcal{M}^b)$. ∎

Theorem 7.7.7 *Let $\mathcal{M}$ be an X-fuzzy recognizer. Then $\mathcal{M}$ is coaccessible if and only if $\mathcal{M}^p$ is accessible.*

Proof. $\mathcal{M}$ is a coaccessible fuzzy recognizer $\Leftrightarrow Q = \{p \in Q | \tau *$
$(X^*)^{-1}(p) > 0\} \Leftrightarrow \forall p \in Q, \exists x_0 \in X^*$, and $q_0 \in Q$ such that $\tau(q_0) \wedge$
$\mu^*(p_0, x_0, q_0) > 0 \Leftrightarrow \forall p \in Q, \exists x_0^\rho$ namely $\rho(x_0) \in X^*$ and $q_0 \in Q$ such that
$\tau(q_0) \wedge \mu^{\rho*}(p_0, \rho(x_0), q_0) > 0 \Leftrightarrow Q = \{p \in Q \mid \tau * X^*(p) > 0\} \Leftrightarrow \mathcal{M}^\rho$ is an
accessible fuzzy recognizer. ∎

Definition 7.7.8 *A fuzzy X-recognizer $\mathcal{M}$ is called* **trim** *if it is both an
accessible and coaccessible fuzzy X-recognizer.*

Theorem 7.7.9 *Let $\mathcal{M} = (M, \iota, \tau)$ be the fuzzy recognizer of the fuzzy
finite state machine $M = (Q, X, \mu)$ and let $W = \{p \in Q \mid \iota * X^* \cap \tau *
(X^*)^{-1}(p) > 0\}$. Then $\mathcal{M}$ is trim if and only if $Q = W$.* ∎

Consider $\mu^t = \mu|_{W \times X \times W}$, $\iota^t = \iota|_W$, and $\tau^t = \tau|_W$. Then $M^t =
(W, X, \mu^t)$ is a fuzzy finite state machine and $\mathcal{M}^t = (M^t, \iota^t, \tau^t)$ is a fu-
zzy X-recognizer.

Definition 7.7.10 *Let $\mathcal{M}$ be a fuzzy X-recognizer. Then $\mathcal{M}^t = (M^t, \iota^t, \tau^t)$
is called the* **trim part** *of M.*

The following theorem is immediate from Corollary 7.7.6.

Theorem 7.7.11 *If $\mathcal{M} = (M, \iota, \tau)$ is a fuzzy recognizer, then $B(\mathcal{M}) =
B(\mathcal{M}^t)$.* ∎

Corollary 7.7.12 *If $\mathcal{M}$ is trim, then $(\mathcal{M}^a)^b = \mathcal{M}^t = (\mathcal{M}^b)^a$.* ∎

If $\mathcal{M}$ is an accessible (coaccessible, trim) fuzzy X-recognizer, then
$\mathcal{M}^a = \mathcal{M}$ ($\mathcal{M}^b = \mathcal{M}$, $\mathcal{M}^t = \mathcal{M}$, respectively).

7.8　Complete Fuzzy Machines

Recall that a fuzzy finite state machine $M = (Q, X, \mu)$ is called complete if
for all $(p, u) \in Q \times X$, there exists $q \in Q$ such that $\mu(p, u, q) > 0$.

As in Definition 7.2.13, if $M = (Q, X, \mu)$ is a complete fuzzy finite
state machine, then the fuzzy X-recognizer $\mathcal{M} = (M, \iota, \tau)$ of $\mathcal{M}$ is called
complete.

Theorem 7.8.1 *If $\mathcal{M}$ is a complete fuzzy recognizer, then so is $\mathcal{M}^a$.*

Proof. Let $(p^a, u) \in R \times X$. Then $(p^a, u) \in Q \times X$. Since $\mathcal{M}$ is complete,
there exists $q \in Q$ such that $\mu(p^a, u, q) > 0$. Now $p^a \in R \Rightarrow \iota * X^*(p^a) > 0$.
Therefore, $\exists q_0 \in Q$ and $x_0 \in X^*$ such that $\iota(q_0) \wedge \mu^*(q_0, x_0, p^a) > 0$.
This implies that $\iota(q_0) > 0$ and $\mu^*(q_0, x_0, p^a) > 0$. Hence $\mu^*(q_0, x_0, p^a) \wedge
\mu(p^a, u, q) > 0$. Thus $\mu^*(q_0, x_0 u, q) = \vee\{\mu^*(q_0, x_0, t) \wedge \mu(t, u, q) \mid t \in Q\} > 0$.
Hence $\iota(q_0) \wedge \mu^*(q_0, x_0 u, q) > 0$, i.e., $\iota * X^*(q) > 0$ and so $q \in R$. Since
$\mu^a(p^a, u, q) = \mu(p^a, u, q)$, $\mathcal{M}^a$ is complete. ∎

The following definition of completion and the following construction of
a completion differs somewhat from that in Section 7.2.

Definition 7.8.2 *Let* $M = (Q, X, \mu)$ *be a fuzzy finite state machine. A fuzzy finite state machine* $M^c = (Q^c, X^c, \mu^c)$ *is called a* **completion** *of* M, *if the following conditions hold:*

(1) M^c *is a complete fuzzy finite machine, and*

(2) M *is a subfuzzy finite state machine of* M^c.

Let $M = (Q, X, \mu)$ be a fuzzy finite state machine that is incomplete. Consider $M' = (Q', X, \mu')$, where $Q' = Q \cup \{z\}$, $z \notin Q$ and

$$\mu'(p, u, q) = \begin{cases} \mu(p, u, q) & \text{if } p, q \in Q \text{ and } \mu(p, u, q) \neq 0, \\ 1 & \text{if either } \mu(p, u, r) = 0 \ \forall r \in Q \text{ and } q = z \\ & \text{or } p = q = z, \\ 0 & \text{otherwise.} \end{cases}$$

Then $M' = (Q', X, \mu')$ is called the **smallest completion** of M.

Definition 7.8.3 *Let* $\mathcal{M} = (M, \iota, \tau)$ *be a fuzzy X-recognizer of a fuzzy finite state machine* $M = (Q, X, \mu)$ *and* $M^c = (Q^c, X, \mu^c)$ *be a completion of* M. *Then the fuzzy X-recognizer* $\mathcal{M}^c = (M^c, \iota^c, \tau^c)$ *of* M^c *is called the* **completion** *of* $\mathcal{M}$, *where* $\iota^c : Q^c \to [0, 1]$ *and* $\tau^c : Q^c \to [0, 1]$ *are such that*

$$\iota^c(q) = \begin{cases} \iota(q) & \text{if } q \in Q \\ 0 & \text{if } q \notin Q. \end{cases}$$

$$\tau^c(q) = \begin{cases} \tau(q) & \text{if } q \in Q \\ 0 & \text{if } q \notin Q. \end{cases}$$

If M^c *is the smallest completion of* M, *then* $\mathcal{M}^c$ *is called the* **smallest completion** *of* $\mathcal{M}$.

Theorem 7.8.4 *Let* $\mathcal{M} = (M, \iota, \tau)$ *be a fuzzy recognizer and* $\mathcal{M}^c$ *is the smallest completion of* $\mathcal{M}$. *Then* $B(\mathcal{M}) = B(\mathcal{M}^c)$. ∎

Theorem 7.8.5 *If* $\mathcal{M}$ *is an accessible fuzzy recognizer, then so is* $\mathcal{M}^c$.

Proof. Let $\mathcal{M}$ be accessible. Then $Q = \{q \in Q \mid \iota * X^*(q) > 0\}$. Let $q^c \in Q^c$. If $q^c \in Q$, then the desired result holds. Let $q^c = z$. Since M is incomplete, there exists $(p_0, u_0) \in Q \times X$ such that $\mu(p_0, u_0, t) = 0$ for all $t \in Q$. But then $(\mu^c)(p_0, u_0, z) = 1 > 0$. Since $\mathcal{M}$ is accessible and $p_0 \in Q$, there exists $r \in Q$ and $y_0 \in X^*$ such that $\iota(r) \wedge \mu^*(r, y_0, p_0) > 0$. Now $(\mu^c)^*(r, y_0 u_0, z) = \vee\{(\mu^c)^*(r, y_0, s) \wedge (\mu^c)(s, u_0, z) \mid s \in Q\} > (\mu^c)^*(r, y_0, p_0) \wedge (\mu^c)(p_0, u_0, z) = (\mu^c)^*(r, y_0, p_0) = \mu^*(r, y_0, p_0) > 0$. Thus $\iota(r) \wedge (\mu^c)^*(r, y_0 u_0, z) > 0$, i.e., $\iota * X^*(q^c) > 0$. Hence $\mathcal{M}^c$ is accessible. ∎

Let $\mathcal{M} = (M, \iota, \tau)$ be a fuzzy recognizer of a fuzzy finite state machine $M = (Q, X, \mu)$. Recall that $\mathcal{P}(Q)$ is the power set of Q.

Define $\mu^\sim : \mathcal{P}(Q) \times X \times \mathcal{P}(Q) \to [0,1]$ by

$$\mu^\sim(P, u, R) = \begin{cases} 1 & \text{if } P \neq \emptyset \text{ or } R \neq \emptyset \\ \vee\{\mu(p, u, r) \mid p \in P, \ r \in R\} & \text{otherwise,} \end{cases}$$

$\iota^\sim : \mathcal{P}(Q) \to [0,1]$ by

$$\iota^\sim(P) = \begin{cases} \wedge\{\iota(p) \mid p \in P\} & \text{if } P \neq \emptyset \\ 0 & \text{otherwise,} \end{cases}$$

and $\tau^\sim : \mathcal{P}(Q) \to [0,1]$ by

$$\tau^\sim(P) = \begin{cases} \wedge\{\tau(p) \mid p \in P\} & \text{if } P = \emptyset \\ 0 & \text{otherwise.} \end{cases}$$

Then $M^\sim = (\mathcal{P}(Q), X, \mu^\sim)$ is a fuzzy finite state machine and $\mathcal{M}^\sim = (M^\sim, \iota^\sim, \tau^\sim)$ is a complete fuzzy X-recognizer of $M^\sim = (\mathcal{P}(Q), X, \mu^\sim)$.

Theorem 7.8.6 *Let $\mathcal{M} = (M, \iota, \tau)$ be a fuzzy recognizer of a fuzzy finite state machine $M = (Q, X, \mu)$. Then $B(\mathcal{M}) = B(\mathcal{M}^\sim)$.*

Proof. Let $x \in B(\mathcal{M}^\sim)$. Now $x \in B(\mathcal{M}^\sim) \Rightarrow x \in \iota^\sim \circ_{\mu^\sim} \tau^\sim \Rightarrow \iota^\sim \# \tau^\sim(x) > 0 \Rightarrow \iota^\sim(P) \wedge \tau^\sim(R) \wedge \mu^\sim(P, x, R) > 0$ for some $P, R \in \mathcal{P}(Q)$. Clearly, by the definition of $\iota^\sim$ and $\tau^\sim$, both P and R are nonempty. Thus $[\wedge\{\iota(p) \mid p \in P\}] \wedge [\wedge\{\tau(r) \mid r \in R\}] \wedge [\vee\{\mu^*(p, x, r) \mid p \in P, r \in R\}] > 0$ for some $P, R \in \mathcal{P}(Q)$ and so $\iota(p) > 0 \ \forall p \in P$, $\tau(r) > 0 \ \forall r \in R$ and there $\exists t_0 \in P$, $r_0 \in R$ such that $\mu^*(t_0, x, r_0) > 0$ for some $P, R \in \mathcal{P}(Q)$. Hence $\iota(t_0) > 0$, $\tau(r_0) > 0$, and $\mu^*(t_0, x, r_0) > 0$ for some $t_0, r_0 \in Q$. Thus $\iota(t_0) \wedge \tau(r_0) \wedge \mu^*(t_0, x, r_0) > 0$ for some $t_0, r_0 \in Q$. This implies that $\iota \# \tau(x) > 0$ and so $x \in \iota \circ_\mu \tau$. Hence $x \in B(\mathcal{M})$. Thus $B(\mathcal{M}^\sim) \subseteq B(\mathcal{M})$.

Conversely, suppose $x \in B(\mathcal{M})$. Then $x \in \iota \circ_\mu \tau$ and so $\iota * \tau(x) > 0$. This implies that $\iota(p) \wedge \tau(q) \wedge \mu^*(p, x, q) > 0$ for some $p, q \in Q$. Choose $P = \{p\}$ and $R = \{q\}$. It then follows that $B(\mathcal{M}) \subseteq B(\mathcal{M}^\sim)$. ∎

Theorem 7.8.7 *Every fuzzy X-recognizable subset A of X^* is the behavior of a complete fuzzy recognizer.* ∎

7.9 Fuzzy Languages on a Free Monoid

The study of fuzzy grammars, the rules of fuzzy syntaxes, and the recognition ability of fuzzy automata extends the application area of fuzzy set theory. One goal is to reduce the difference between formal languages and natural languages.

In this section, we examine fuzzy regular languages, adjunctive languages, and dense languages. We present their algebraic properties. The results are from [217].

In this and the next section, we let X denote a finite alphabet with at least one element. Recall that $\mathcal{FP}(X)$ denotes the set of all fuzzy subsets of X. We use the superscript T to denote the transpose of a matrix.

Definition 7.9.1 *[160] A finite fuzzy automaton on an alphabet X is a 5-tuple $M = (Q, Y, \{T_u \mid u \in X\}, \sigma_0, \sigma_1)$ such that*
(1) $Q = \{q_1, q_2, \ldots, q_n\}$ is the set of states,
(2) $Y = \{y_1, y_2, \ldots, y_n\} \subseteq Q$ is the set of output symbols,
(3) $\{T_u \mid u \in X\}$ is the set of fuzzy transition matrices,
where $\delta_{q_i q_j} : X \to [0,1]$ and $T_u = [\delta_{q_i q_j}(u)]$, $q_i, q_j \in Q$, $i, j = 1, 2, \ldots, n$,
(4) $\sigma_0 = \begin{bmatrix} i_1 & i_2 & \cdots & i_n \end{bmatrix}$, $i_k \in [0,1]$ for $k = 1, 2, \ldots, n$,
(5) $\sigma_1 = \begin{bmatrix} j_1 & j_2 & \cdots & j_n \end{bmatrix}^T$, $j_k \in [0,1]$ for $k = 1, 2, \ldots, n$.

σ_0 determines the fuzzy subset of initial states and σ_1 determines the fuzzy subset of final subsets.

Let $M = (Q, Y, \{T_u \mid u \in X\}, \sigma_0, \sigma_1)$ be a finite fuzzy automaton. Define $\delta : Q \times X \times Q \to [0,1]$ by

$$\delta(q_i, u, q_j) = \delta_{q_i q_j}(u)$$

for $i, j = 1, 2, \ldots, n$ and $\forall u \in X$. Then δ is a fuzzy transition function.

Let δ^* be defined as usual. Then $\delta^*(q, \Lambda, q') = 1$ if $q = q'$ and 0 otherwise. Also for all $x = u_1 u_2 \cdots u_k \in X^*$, $x \neq \Lambda$,

$$\delta^*(q, x, q') = \vee\{\delta(q, u_1, q_1) \wedge \delta(q_1, u_2, q_2) \wedge \ldots \wedge \delta(q_{k-1}, u_k, q') \mid q_1, q_2, \ldots, q_{k-1} \in Q\}$$

or

$$T_x = T_{u_1} \circ T_{u_2} \circ \cdots \circ T_{u_k},$$

where $\circ$ is the sup-min composition of fuzzy matrices.

Definition 7.9.2 *[160] Any member of $\mathcal{FP}(X^*)$ is called a **fuzzy language** on the free monoid X^*. For all $u \in X$, let χ_u denote the characteristic function of $\{u\}$ in X^*. Then χ_u is called the **basic fuzzy language generated by** u. Let*

$$E = \{\chi_u \mid u \in X\} \cup \{\Lambda\}.$$

Definition 7.9.3 *[160] Let $U \subseteq \mathcal{FP}(X^*)$ be such that the following conditions hold:*
(1) $\forall c \in [0,1]$, $\mu \in U \Rightarrow c \wedge \mu \in U$,
(2) $\mu_1, \mu_2 \in U \Rightarrow \mu_1 \cup \mu_2 \in U$,
(3) $\mu_1, \mu_2 \in U \Rightarrow \mu_1 \circ \mu_2 \in U$, where $(\mu_1 \circ \mu_2)(x) = \vee \{\mu_1(u) \wedge \mu_2(v) \mid uv = x\}$, $\forall x \in X^*$,
(4) $\mu \in U \Rightarrow \mu^\infty \in U$, where μ^∞ is the Kleene's closure of μ, i.e.,

$$\mu^\infty = \mu_0 \cup \mu \cup \mu^2 \cup \cdots,$$

where $\mu_0(x) = 0 \; \forall x \in X^*$.
*Then U is called a **closed family of fuzzy languages on** X^*.*

Let

$$\mathcal{F} = \{U \mid U \text{ is a closed family of fuzzy languages on } X^*\}.$$

Clearly, $\mathcal{F}P(X^*) \in \mathcal{F}$.

Definition 7.9.4 [160] Let

$$\mathbf{FR}(X^*) = \cap_{E \subseteq M \subseteq \mathcal{F}} M.$$

$\mathbf{FR}(X^*)$ is called the **family of fuzzy regular languages** on X^*. If $\mu \in$ $\mathbf{FR}(X^*)$, then μ is called a **fuzzy regular language** on X^*.

Clearly, $\mathbf{FR}(X^*)$ is a subalgebra of $\mathcal{F}P(X^*)$ generated by E with the four operations in Definition 7.9.3.

Definition 7.9.5 [160] Let $M = (Q, Y, T_u, \sigma_0, \sigma_1)$ be a finite fuzzy automaton on X. Define $f_M : X^* \to [0,1]$ by $\forall u \in X^*$,

$$f_M(u) = \sigma_0 \circ T_u \circ \sigma_1.$$

Then f_M is called the fuzzy **language determined** by the fuzzy automaton M.

Example 7.9.6 Let $Q = \{s, S, F\}$, $X = \{a, b\}$, and $\delta : Q \times X \times Q \to [0,1]$ be defined as follows:

$$\begin{aligned}
\delta(s, a, S) &= .9 \\
\delta(s, b, s) &= .5 \\
\delta(S, b, S) &= .9 \\
\delta(S, b, F) &= .9
\end{aligned}$$

and $\delta(q, x, q') = 0$ for any other $(q, x, q') \in Q \times X \times Q$. Let $\sigma_0 = (1, 0, 0)$ and $\sigma_1 = (0, 0, 1)^T$. Then

$$T_a = \begin{array}{c} s \\ S \\ F \end{array} \begin{bmatrix} 0 & .9 & 0 \\ 0 & 0 & 0 \\ 0 & 0 & 0 \end{bmatrix}, \qquad T_b = \begin{array}{c} s \\ S \\ F \end{array} \begin{bmatrix} .5 & 0 & 0 \\ 0 & .9 & .9 \\ 0 & 0 & 0 \end{bmatrix}.$$

Let $M = (Q, Y, \{T_a, T_b\}, \sigma_0, \sigma_1)$, where $Y = Q$. Then

$$f_M(bab) = \sigma_0 \circ T_b \circ T_a \circ T_b \circ \sigma_1 = \begin{bmatrix} 1 & 0 & 0 \end{bmatrix} \begin{bmatrix} 0 & .5 & .5 \\ 0 & 0 & 0 \\ 0 & 0 & 0 \end{bmatrix} \begin{bmatrix} 0 \\ 0 \\ 1 \end{bmatrix} = .5$$

and

$$f_M(ab) = \sigma_0 \circ T_a \circ T_b \circ \sigma_1 = \begin{bmatrix} 1 & 0 & 0 \end{bmatrix} \begin{bmatrix} 0 & .9 & .9 \\ 0 & 0 & 0 \\ 0 & 0 & 0 \end{bmatrix} \begin{bmatrix} 0 \\ 0 \\ 1 \end{bmatrix} = .9.$$

It follows that $f_M(b^m a b^n) = .5$ if $m > 0$ and $f_M(b^m a b^n) = .9$ if $m = 0$. We see that $f_M = \lambda$, where λ is the fuzzy language of Example 7.1.11.

From [160], $\mu \in \mathbf{FR}(X^*)$ if and only if there exists a finite fuzzy automaton M such that $f_M = \mu$.

Definition 7.9.7 *[160] We call P_L a **main congruence** if $\forall x, y \in X^*$, $x \equiv y(P_L)$ if and only if $\forall u, v \in X^*$, $uxv \in L \Leftrightarrow uyv \in L$.*

Example 7.9.8 *Let $X = \{a, b\}$ and $L = \{a^n \mid n = 0, 1, \ldots\}$. Let $x, y \in X^*$. Then $\forall u, v \in X^*$, $uxv \in L$ if and only if $u = a^i$, $x = a^j$, and $v = a^k$ for some $i, j, k \in \mathbb{N} \cup \{0\}$. Thus $x \equiv y(P_L)$ if and only if either $x \in L, y \in L$ or $x \notin L, y \notin L$. Hence the equivalence classes corresponding to $\equiv$ are L and $X^* \backslash L$. Thus the index of P_L is 2.*

Example 7.9.9 *Let $X = \{a, b\}$ and $L = \{b^m ab^n \mid m = 0, 1, \ldots; n = 1, \ldots\}$. Let $x, y \in X^*$. Then $\forall u, v \in X^*$, $uxv \in L$ if and only if either $u = b^i$, $x = b^j ab^k$, $v = b^l$ (not both $k = 0, l = 0$) or $u = b^i ab^j$, $x = b^k$, $v = b^l$ (j, k, l not all 0) or $u = b^i$, $x = b^j$, $v = b^k ab^l$ ($l \neq 0$) for $i, j, k, l \in \mathbb{N} \cup \{0\}$. Thus $x \equiv y(P_L)$ if and only if either $x, y \in L \cup \{b^k a \mid k = 0, 1, \ldots\}$ or $x, y \in \{b^k \mid k = 0, 1, \ldots\}$ or $x, y \in X^* \backslash (L \cup \{b^k a \mid k = 0, 1, \ldots\} \cup \{b^k \mid k = 0, 1, \ldots\})$. Thus the index of P_L is 3.*

Proposition 7.9.10 *[218] An ordinary language $L \subseteq X^*$ is regular if and only if the index of P_L is finite.* ∎

Proposition 7.9.11 *[218] An ordinary language $L \subseteq X^*$ is adjunctive if and only if $\forall x, y \in X^*$, $x \equiv y(P_L) \Rightarrow x = y$.* ∎

Example 7.9.12 *Let $X = \{a, b\}$ and $L = \{a^n b^n \mid n = 0, 1, \ldots\}$. Let $x, y \in X^*$. Then $\forall u, v \in X^*$, $uxv \in L$ if and only if either $u = a^n b^i$, $x = b^j$, $v = b^{n-i-j}$ or $u = a^i$, $x = a^{n-i} b^j$, $v = b^{n-j}$ where $n, i, j \in \mathbb{N} \cup \{0\}$. Thus $x \equiv y(P_L)$ implies $x = y$. Hence L is adjunctive. The index of P_L is not finite and L is not regular.*

Definition 7.9.13 *Let $\mu \in \mathcal{FP}(X^*)$. Then F_μ is called a **fuzzy main congruence** with respect to μ on X^* if*

$$\forall x, y \in X^*, x \equiv y(F_\mu) \Leftrightarrow \forall u, v \in X^*, \mu(uxv) = \mu(uyv).$$

Proposition 7.9.14 *Let $x, y \in X^*$. Then $x \equiv y(F_\mu)$ if and only if $\forall c \in [0, 1]$, $x \equiv y(P_{\mu_c})$, where $\mu_c = \{x \in X^* \mid \mu(x) \geq c\}$.*

Proof. Suppose that $x \equiv y(F_\mu)$. Then $\forall u, v \in X^*$, $\mu(uxv) = \mu(uyv)$. Thus $\forall c \in [0, 1]$, $\mu(uxv) \geq c$ if and only if $\mu(uyv) \geq c$. Hence $\forall u, v \in X^*$, $\forall c \in [0, 1]$, $uxv \in \mu_c$ if and only if $uyv \in \mu_c$. This implies that $x \equiv y(P_{\mu_c})$.

Conversely, suppose that $\forall c \in [0, 1]$, $x \equiv y(P_{\mu_c})$. Then $\forall u, v \in X^*$, $\mu(uxv) \geq c \Leftrightarrow \mu(uyv) \geq c \ \forall c \in [0, 1]$. Thus $x \equiv y(F_\mu)$. ∎

Proposition 7.9.15 *Let $x, y \in X^*$. Then $x \equiv y(F_\mu)$ if and only if $\forall c \in [0, 1]$, $x \equiv y(P_{\mu_{c+}})$, where $\mu_{c+} = \{x \in X^* \mid \mu(x) > c\}$.*

Proof. Suppose that $\forall c \in [0,1]$, $x \equiv y(P_{\mu_c})$. Then $\forall u, v \in X^*$, $\mu(uxv) > c \Leftrightarrow \mu(uyv) > c$. Let $\mu(uxv) = d$. Now for any $\varepsilon > 0$, we have $\mu(uxv) > d - \varepsilon$. Therefore, $\mu(uyv) > d - \varepsilon$ for all $\varepsilon > 0$. Thus $\mu(uyv) \geq d = \mu(uxv)$. Similarly, $\mu(uxv) \geq \mu(uyv)$. Hence, $x \equiv y(F_\mu)$. The converse is easily shown. ∎

Corollary 7.9.16 $F_\mu = \cap_{c \in [0,1]} P_{\mu_c} = \cap_{c \in [0,1]} P_{\mu_{c^+}}$. ∎

7.10 Algebraic Character and Properties of Fuzzy Regular Languages

Proposition 7.10.1 *Let* $\mu \in \mathcal{FP}(X^*)$. *Then* $\mu \in \mathbf{FR}(X^*)$ *if and only if the index of* F_μ *is finite.*

Proof. Suppose that $\mu \in \mathbf{FR}(X^*)$. Then there exists a finite fuzzy automaton $M = (Q, Y, T_u, \sigma_0, \sigma_1)$ such that $f_M = \mu$. Clearly, $f_M = \sigma_0 \circ T \circ \sigma_1$. Define a relation ρ on X^* as follows: $\forall x, y \in X^*$, $x \equiv y(\rho) \Leftrightarrow T_x = T_y$. Clearly, ρ is an equivalence relation. Let $x, y \in X^*$. Then

$$
\begin{aligned}
x \equiv y(\rho) &\Rightarrow T_x = T_y \\
&\Rightarrow \forall u, v \in X^*, T_u \circ T_x \circ T_v = T_u \circ T_y \circ T_v \\
&\Rightarrow \forall u, v \in X^*, T_{uxv} = T_{uyv} \\
&\Rightarrow \forall u, v \in X^*, \sigma_0 \circ T_{uxv} \circ \sigma_1 = \sigma_0 \circ T_{uyv} \circ \sigma_1 \\
&\Rightarrow \forall u, v \in X^*, \mu(uxv) = \mu(uyv) \\
&\Rightarrow x \equiv y(F_\mu).
\end{aligned}
$$

Hence $\rho \subseteq F_\mu$. Since X and Q are finite, $\{T_w \mid w \in X^*\}$ is a finite set. Therefore, since the index of ρ is finite, the index of F_μ is also finite.

Conversely, suppose that the index of F_μ is finite. Let $[\Lambda], [x_1], \ldots, [x_m]$ be the distinct congruence classes of F_μ, where $x_i \in X^*$, $i = 1, 2, \ldots, m$. Let $M = (Q, Y, \{T_u \mid u \in X\}, \sigma_0, \sigma_1)$ be a finite fuzzy automaton, where

$$
Q = Y = \{[\Lambda], [x_1], \ldots, [x_m]\},
$$

the fuzzy transition matrix is defined as $\sigma_{[w]} \circ T_u = \sigma_{[wu]}$, and $\sigma_{[\Lambda]} = [\, 1 \quad 0 \quad \ldots \quad 0 \,]$, $\sigma_{[x_i]} = [\, 0 \quad \ldots \quad 0 \quad 1 \quad 0 \quad \ldots \quad 0 \,]$, $i = 1, 2, \ldots, m$. Set $\sigma_0 = \sigma_{[\Lambda]}$,

$$
\sigma_1 = (\mu([\Lambda]), \mu([x_1]), \ldots, \mu([x_m]))^T.
$$

Since $\forall x \in X^*$, $x \in [x]$, $\mu(x) = \mu([x])$. Also

$$
f_M(x) = \sigma_0 \circ T_x \circ \sigma_1 = \sigma_{[\Lambda]} \circ T_x \circ \sigma_1 = \sigma_{[x]} \circ \sigma_1 = \mu([x]).
$$

Thus $\forall x \in X^*$, $\mu(x) = f_M(x)$. Hence $\mu = f_M$ and so $\mu \in \mathbf{FR}(X^*)$. ∎

Proposition 7.10.2 Let $\mu \in \mathcal{FP}(X^*)$. Then $\mu \in \mathbf{FR}(X^*)$ if and only if $\forall c \in [0,1]$, μ_c is regular, and $|\mathrm{Im}(\mu)| < \infty$.

Proof. Suppose that $\mu \in \mathbf{FR}(X^*)$. Then the index of F_μ is finite. By Corollary 7.9.16, $P_{\mu_c} \supseteq F_\mu$, $\forall c \in [0,1]$. Let $c \in [0,1]$. Hence the index of P_{μ_c} is finite and so μ_c is regular. We may assume that the congruence classes of F_μ are $[x_1], [x_2], \ldots, [x_m]$ since the index of F_μ is finite. Clearly, $\forall i$, $i = 1, 2, \ldots, m$, $\forall u, v \in X^*$, if $u, v \in [x_i]$, then $\mu(u) = \mu(v)$. Therefore, $\mathrm{Im}(\mu) = \{\mu(x_1), \mu(x_2), \ldots, \mu(x_m)\}$ and so $|\mathrm{Im}(\mu)| < \infty$.

Conversely, suppose that $\forall c \in [0,1]$, μ_c is regular, and $|\mathrm{Im}(\mu)| < \infty$. Then the characteristic function χ_{μ_c} of μ_c is a fuzzy regular language. By the resolution theorem of fuzzy sets (see [160]), $\mu = \bigcup_{c \in [0,1]} c \, \chi_{\mu_c}$. Let $\mathrm{Im}(\mu) = \{c_1, c_2, \ldots, c_k\}$. Then $\mu = c_1 \chi_{\mu_{c_1}} \cup c_2 \chi_{\mu_{c_2}} \cup \cdots \cup c_k \chi_{\mu_{c_k}}$. Hence by Definition 7.9.4, $\mu \in \mathbf{FR}(X^*)$. ∎

Proposition 7.10.3 $(\mathbf{FR}(X^*), \cap, \cup, {}^-)$ is a de Morgan algebra, where ${}^-$ denotes complement of fuzzy subsets.

Proof. Since $\mathbf{FR}(X^*) \subset \mathcal{FP}(X^*)$ and $(\mathcal{FP}(X^*), \cup, \cap, {}^-)$ is a de Morgan algebra, it suffices to show that $(\mathbf{FR}(X^*), \cup, \cap, {}^-)$ is a subalgebra of $(\mathcal{FP}(X^*), \cup, \cap, {}^-)$. By Definition 7.9.4, it follows that $\forall \mu, \nu \in \mathbf{FR}(X^*)$, $\mu \cup \nu \in \mathbf{FR}(X^*)$. Also, since $\mu, \nu \in \mathbf{FR}(X^*)$, $\forall c \in [0,1]$, μ_c, ν_c are regular, and $|\mathrm{Im}(\mu)|$ and $|\mathrm{Im}(\nu)|$ are finite. Thus $\forall c \in [0,1]$, $(\mu \cap \nu)_c = \mu_c \cap \nu_c$ is regular, and $|\mathrm{Im}(\mu \cap \nu)| \leq |\mathrm{Im}(\mu)| + |\mathrm{Im}(\nu)|$ is finite. Hence by Proposition 7.10.3, $\mu \cap \nu \in \mathbf{FR}(X^*)$. Let $x, y \in X^*$. Then

$$
\begin{aligned}
x \equiv y(F_\mu) \;&\Leftrightarrow\; \forall u, v \in X^*, \mu(uxv) = \mu(uyv) \\
&\Leftrightarrow\; \forall u, v \in X^*, 1 - \mu(uxv) = 1 - \mu(uyv) \\
&\Leftrightarrow\; \forall u, v \in X^*, \overline{\mu}(uxv) = \overline{\mu}(uyv) \\
&\Leftrightarrow\; x \equiv y(F_{\overline{\mu}}).
\end{aligned}
$$

Thus since the index of F_μ is finite, the index of $F_{\overline{\mu}}$ is finite. Hence, $\overline{\mu} \in \mathbf{FR}(X^*)$. ∎

Proposition 7.10.4 Let $\mu, \nu \in \mathbf{FR}(X^*)$. Then

(1) $\mu \circ \nu \in \mathbf{FR}(X^*)$,

(2) $\nu^{-1}\mu$ (or $\mu\nu^{-1}$) $\in \mathbf{FR}(X^*)$,

where $\nu^{-1}\mu$ (or $\mu\nu^{-1}$) is called the **fuzzy left (right, respectively) quotient** of μ with respect to ν and is defined as follows:

$$(\nu^{-1}\mu)(y) = \vee\{\mu(xy) \wedge \nu(x) \mid x \in \Sigma^*\}, \quad \forall y \in X^*,$$

$$(\mu\nu^{-1})(y) = \vee\{\mu(yx) \wedge \nu(x) \mid x \in \Sigma^*\}, \quad \forall y \in X^*.$$

Proof. (1) Let $c \in [0,1]$. Since $(\mu \circ \nu)_c = \mu_c \nu_c$ and μ and ν are fuzzy regular, μ_c and ν_c are regular. Also since $(\mu \circ \nu)_c = \mu_c \nu_c$ is an adjoin of two regular languages, $(\mu \circ \nu)_c$ is regular. Now the degrees of membership of elements of $\mu \circ \nu$ are obtained from the degrees of membership of elements of μ and ν with the operations max and min. Thus $\text{Im}(\mu \circ \nu) \subseteq \text{Im}(\mu) \cup \text{Im}(\nu)$ and so $\text{Im}(\mu \circ \nu)$ is finite. Hence $\mu \circ \nu \in \mathbf{FR}(X^*)$.

(2) For all $s, t \in X^*$, $s \equiv t(F_\mu)$

$$\Rightarrow \quad \forall u, v \in X^*, \mu(usv) = \mu(utv)$$
$$\Rightarrow \quad \forall u, v, x \in X^*, \mu(xusv) = \mu(xutv)$$
$$\Rightarrow \quad \forall u, v, x \in X^*, \ \mu(xusv) \wedge \nu(x) = \mu(xutv) \wedge \nu(x)$$
$$\Rightarrow \quad \forall u, v \in X^*, \ \vee_{x \in X^*}(\mu(xusv) \wedge \nu(x)) = \ \vee_{x \in X^*}(\mu(xutv) \wedge \nu(x))$$
$$\Rightarrow \quad \forall u, v \in X^*, \ (\nu^{-1}\mu)(usv) = (\nu^{-1}\mu)(utv)$$
$$\Rightarrow \quad s \equiv t(F_{\nu^{-1}\mu}).$$

Thus, $F_\mu \subseteq F_{\nu^{-1}\mu}$. Since $\mu \in \mathbf{FR}(X^*)$, the index of F_μ is finite and so the index of $F_{\nu^{-1}\mu}$ is also finite. Consequently, $\nu^{-1}\mu \in \mathbf{FR}(X^*)$. $\blacksquare$

Note that ν in Proposition 7.10.4(2) may be any fuzzy language.

Proposition 7.10.5 *Let f be a homomorphism from X^* onto X_1^*, where X_1 is a nonempty set. Then*

(1) $\mu \in \mathbf{FR}(X^*) \Rightarrow f(\mu) \in \mathbf{FR}(X_1^*)$;

(2) $\nu \in \mathbf{FR}(X_1^*) \Rightarrow f^{-1}(\nu) \in \mathbf{FR}(X^*)$.

Proof. (1) Let $\mu \in \mathbf{FR}(X^*)$. Let $c \in [0,1]$ and $x_1 \in X_1^*$. Now $x_1 \in (f(\mu))_c$ if and only if $(f(\mu))(x_1) \geq c$ if and only if $\vee\{\mu(x) \mid f(x) = x_1\} \geq c$ if and only if $\exists x', f(x') = x_1$ and $\mu(x') \geq c$ (since $|\text{Im}(\mu)| < \infty$) if and only if $f(x') = x_1$, $x' \in \mu_c$ if and only if $x_1 \in f(\mu_c)$. Hence $(f(\mu))_c = f(\mu_c)$. Since μ_c is regular, $f(\mu_c)$ is regular and so $(f(\mu))_c$ is regular. Since $\text{Im}(\mu)$ is finite, it follows that $\text{Im}(f(\mu)) \subseteq \text{Im}(\mu)$ and so $\text{Im}(f(\mu))$ is finite. Hence $f(\mu) \in \mathbf{FR}(X_1^*)$.

(2) Let $\nu \in \mathbf{FR}(X_1^*)$ and $x, y \in X^*$. Then

$$x \equiv y(F_{f^{-1}(\nu)}) \quad \Leftrightarrow \quad (f^{-1}(\nu))(uxv) = (f^{-1}(\nu))(uyv) \quad \forall u, v \in X^*$$
$$\Leftrightarrow \quad \nu(f(uxv)) = \nu(f(uyv)) \quad \forall u, v \in X^*$$
$$\Leftrightarrow \quad \nu(f(u)f(x)f(v)) = \nu(f(u)f(y)f(v)) \quad \forall u, v \in X^*$$
$$\Leftrightarrow \quad \nu(u_1 f(x)v_1) = \nu(u_1 f(y)v_1) \quad \forall u_1, v_1 \in X_1^*$$
$$\Leftrightarrow \quad f(x) \equiv f(y)(F_\nu).$$

Consequently, the index of $F_{f^{-1}(\nu)}$ in X^* equals the index of F_ν in X_1^*. Since the index of F_ν is finite, the index of $F_{f^{-1}(\nu)}$ is finite and so $f^{-1}(\nu) \in \mathbf{FR}(X^*)$. $\blacksquare$

Definition 7.10.6 *Let $\mu \in \mathcal{FP}(X^*)$ and let $h(\mu) = \vee\{\mu(u) \mid u \in X^*\}$.*

*(1) Then $h(\mu)$ is called the **height** of μ.*

*(2) If there exists $u_0 \in X^*$ such that $h(\mu) = \mu(u_0)$, then u_0 is called a **saddle point** of μ.*

Proposition 7.10.7 Let $\mu \in \mathbf{FR}(X^*)$. Then there exists $n \in \mathbb{N}$ such that μ has a saddle point with length less than or equal to $n - 1$.

Proof. Since $\mu \in \mathbf{FR}(X^*)$, there exists a finite fuzzy automaton $M = (Q, Y, T_u, \sigma_0, \sigma_1)$ such that $\mu = f_M = \sigma_0 \circ T \circ \sigma_1$, where $|Q| = n$, $\sigma_0 = \begin{bmatrix} c_1 & c_2 & \dots & c_n \end{bmatrix}$, and $\sigma_1 = \begin{bmatrix} d_1 & d_2 & \dots & d_n \end{bmatrix}^T$. Clearly,

$$h(\mu) = \vee_{x \in X^*} \mu(x) = \vee \{\vee_{x \in \cup_{k=0}^{n-1} X^k} \mu(x), \ \vee_{x \in X^* \setminus \cup_{k=0}^{n-1} X^k} \mu(x)\},$$

where $X^0 = \{\Lambda\}$. We now show that

$$\vee_{x \in \cup_{k=0}^{n-1} X^k} \mu(x) \geq \vee_{x \in X^* \setminus \cup_{k=0}^{n-1} X^k} \mu(x).$$

Let $v \in X^* \setminus \cup_{k=0}^{n-1} X^k$ be such that $|v| \geq n$. Let $v = v_1 v_2 \cdots v_m$, $m \geq n$, $v_i \in X$, $i = 1, \dots, m$. Then

$$\begin{aligned}
\mu(v) &= \sigma_0 \circ T_v \circ \sigma_1 \\
&= \sigma_0 \circ T_{v_1} \circ T_{v_2} \circ \dots \circ T_{v_m} \circ \sigma_1 \\
&= \vee_{1 \leq i_1, \dots, i_m \leq n} (c_i \wedge r_{i i_1}^{(1)} \wedge r_{i_1 i_2}^{(2)} \wedge \dots \wedge r_{i_{m-1} i_m}^{(m)} \wedge d_{i_m}),
\end{aligned}$$

where $T_{v_k} = (r_{ij}^{(k)})$, $k = 1, 2, \dots, m$. Consider the term

$$d = c_i \wedge r_{i i_1}^{(1)} \wedge r_{i_1 i_2}^{(2)} \wedge \dots \wedge r_{i_{m-1} i_m}^{(m)} \wedge d_{i_m}.$$

The number of different elements in the indices of d is less than or equal to n. Since $m \geq n$, there exists p, q with $p < q$ and $i_p = i_q$. Now

$$d \leq d' = c_i \wedge r_{i i_1}^{(1)} \wedge \dots \wedge r_{i_{p-1} i_p}^{(p)} \wedge r_{i_q i_{q+1}}^{(q+1)} \wedge \dots \wedge r_{i_{m-1} i_m}^{(m)} \wedge d_{i_m}.$$

If the number $m + p - q + 1$ of indices $i, i_1, \dots, i_p, i_{q+1}, \dots, i_m$ is still greater than n, then we repeat the above process and eliminate a part of r until the number of indices is not greater than n. Thus we may assume that $m + p - q + 1 \leq n$. Now

$$\begin{aligned}
d &\leq \vee_{1 \leq i, i_1, \dots, i_p, i_{q+1}, \dots, i_m \leq n} (c_i \wedge r_{i i_1}^{(1)} \wedge \dots \wedge r_{i_{p-1} i_p}^{(p)} \wedge r_{i_q i_{q+1}}^{(q+1)} \wedge \dots \\
&\qquad \wedge r_{i_{m-1} i_m}^{(m)} \wedge d_{i_m}) \\
&= \sigma_0 \circ T_{v_1} \circ \dots \circ T_{v_p} \circ T_{v_{p+1}} \circ \dots \circ T_{v_m} \circ \sigma_1.
\end{aligned}$$

Let $v' = v_1 v_2 \cdots v_p v_{q+1} \cdots v_m$. Clearly, $|v'| = m + p - q \leq n - 1$. Furthermore, $d \leq \sigma_0 \circ T_{v'} \circ \sigma_1 = \mu(v') \leq \vee_{x \in \cup_{k=0}^{n-1} X^k} \mu(x)$. Since d was an arbitrary term in $\mu(v)$, it follows that $\mu(v) \leq \vee_{x \in \cup_{k=0}^{n-1} X^k} \mu(x)$. Thus

$$\vee_{x \in X^* \setminus \cup_{k=0}^{n-1} X^k} \mu(x) \leq \vee_{x \in \cup_{k=0}^{n-1} X^k} \mu(x).$$

This implies that $h(\mu) = \vee_{x \in X^*} \mu(x) = \vee_{x \in \cup_{k=0}^{n-1} X^k} \mu(x)$. Since $\cup_{k=0}^{n-1} X^k$ is a finite set, there exists $u_0 \in X^*$ such that $|u_0| \leq n - 1$ and $h(\mu) = \mu(u_0)$.

∎

Example 7.10.8 *Let λ be the fuzzy language of Example 7.1.11. Then λ is regular, $h(\lambda) = .9$, and ab is a saddle point of λ with length not greater than $n - 1$, where $n = 3$.*

Proposition 7.10.9 *(**Action lemma of the fuzzy pump**) If $\mu \in \mathbf{FR}(X^*)$, then there exists a number n such that for any $u \in X^*$, $u = xyw$ with $|xy| \leq n$, $|y| \geq 1$, and $\forall j \geq 0$, $\mu(xy^j w) \geq \mu(u)$ provided $|u| \geq n$.*

Proof. As the proof of Proposition 7.10.7, let $\mu = f_M$ and $n = |Q|$. Let $u \in X^*$ and let $u = v_1 v_2 \cdots v_m$, $m \geq n$. Now

$$\mu(u) = \vee_{1 \leq i, i_1, \ldots, i_m \leq n}(c_i \wedge r_{i i_1}^{(1)} \wedge \ldots \wedge r_{i_{m-1} i_m}^{(m)} \wedge d_{i_m}).$$

Since m and n are finite, there exists a term d in the above equality such that

$$\mu(u) = d = c_i \wedge r_{i i_1}^{(1)} \wedge \ldots \wedge r_{i_{m-1} i_m}^{(m)} \wedge d_{i_m}.$$

Since the number of indices $i, i_1, \ldots, i_m$ is greater than n, there exist, as in the proof of Proposition 7.10.7, p, q with $p < q$, $i_p = i_q$, and

$$d' = c_i \wedge r_{i i_1}^{(1)} \wedge \ldots \wedge r_{i_{p-1} i_p}^{(p)} \wedge r_{i_q i_{q+1}}^{(q+1)} \wedge \ldots \wedge r_{i_{m-1} i_m}^{(m)} \wedge d_{i_m} \geq \mu(u),$$

where $m + p - q < n$. Furthermore,

$$\vee_{i \leq i, i_1, \ldots, i_p, i_{qm}, \ldots, i_m \leq n} c_i \wedge r_{i i_1}^{(1)} \wedge \ldots \wedge r_{i_{p-1} i_p}^{(p)} \wedge r_{i_q i_{q+1}}^{(q+1)} \wedge \ldots$$
$$\wedge r_{i_{m-1} i_m}^{(m)} \wedge d_{i_m} \geq \mu(u).$$

That is, $\mu(v_1 \cdots v_p v_q \cdots v_m) \geq \mu(u)$. Let $x = v_1 \cdots v_p$, $y = v_{p+1} \cdots v_q$, and $w = v_{q+1} \cdots v_m$. Clearly, $|xy| \leq n$, $|y| \geq 1$, and $\mu(xy^0 w) \geq \mu(u)$.

Also,

$$d_{(i)} = c_i \wedge r_{i i_1}^{(1)} \wedge \ldots \wedge r_{i_{p-1} i_p}^{(p)} \wedge$$
$$\underbrace{r_{i_p i_{p+1}}^{(p+1)} \wedge \ldots \wedge r_{i_{q-1} i_q}^{(q)} \wedge \ldots \wedge r_{i_p i_{p+1}}^{(p+1)} \wedge \ldots \wedge r_{i_{q-1} i_q}^{(q)}}_{\left(r_{i_p i_{p+1}}^{(p+1)}, \ldots, r_{i_{q-1}, i_q}^{(q)} \right) j-\text{times}}$$
$$\wedge r_{i_q i_{q+1}}^{(q+1)} \wedge \ldots \wedge r_{i_{m-1} i_m}^{(m)} \wedge d_{i_m}$$
$$= d$$
$$= \mu(u),$$

where (i) stands for the set of all indices in the above formula. Thus $\mu(v_1 \cdots v_p(v_{p+1} \cdots v_q)^j v_{q+1} \cdots v_m) = \vee\{d_{(i)} \mid 1 \leq (i) \leq n\} \geq \mu(u)$, i.e., $\mu(xy^j w) \geq \mu(u)$, $\forall j \geq 1$. Consequently, $\forall j \geq 0$, $\mu(xy^j w) \geq \mu(u)$. ∎

Corollary 7.10.10 Let $\mu \in \mathbf{FR}(X^*)$. Then $\forall c \in [0,1]$ there exists a positive integer n such that for any $u \in \mu_c$, $u = xyw$ with $|xy| \leq n$, $|y| \geq 1$, and $\forall j \geq 0$, $xy^j w \in \mu_c$ provided $|u| \geq n$. ∎

If A is an ordinary regular language, then its characteristic function $\chi_A \in \mathbf{FR}(X^*)$ and Corollary 7.10.10 becomes the ordinary action lemma of the pump (see [122]). Hence Proposition 7.10.9 is a generalization of the ordinary action lemma of the pump and for this reason is called the action lemma of the fuzzy pump.

We now consider fuzzy adjunctive languages.

Definition 7.10.11 Let $\lambda \in \mathcal{FP}(X^*)$. If $\forall x, y \in X^*$, $x \equiv y(F_\lambda) \Rightarrow x = y$, then λ is called a **fuzzy adjunctive language**.

Example 7.10.12 Let $X = \{a, b\}$ and $L = \{a^n b^n \mid n = 0, 1, \ldots\}$. Define $\lambda : X^* \to [0,1]$ by $\lambda(a^i b^i) = .9$ for $i = 0, 1, \ldots, n$, $\lambda(a^i b^i) = .5$ for $i = n+1, n+2, \ldots$, and $\lambda(x) = 0$ if $x \in X^* \backslash L$. By Example 7.9.12, L is adjunctive. Let $x, y \in X^*$. Then $\forall u, v \in X^*$, $\lambda(uxv) = \lambda(uyv)$ implies $uxv, uyv \in L$ or $uxv, uyv \in X^* \backslash L$. Hence it follows that $x \equiv y(F_\lambda)$ implies $x = y$. Thus λ is a fuzzy adjunctive language.

Proposition 7.10.13 The following statements are equivalent:

(1) $\lambda \in \mathcal{FP}(X^*)$ is a fuzzy adjunctive language.
(2) F_λ is the identity relation.
(3) $\forall x \in X^*$, $[x]_{F_\lambda} = \{x\}$.
(4) $\forall x, y \in X^*$, if $x \neq y$, then there exist $u, v \in X^*$ such that $\lambda(uxv) \neq \lambda(uyv)$. ∎

Proposition 7.10.14 Let $\lambda \in \mathcal{FP}(X^*)$. Then λ is a fuzzy adjunctive language if there exist $c \in [0,1]$ such that λ_c is an ordinary adjunctive language. ∎

Proposition 7.10.15 Let $\lambda \in \mathcal{FP}(X^*)$. Then λ is a fuzzy adjunctive language $\Leftrightarrow \forall x, y \in X^*$, $\{x \equiv y(\bigcap_{c \in (0,1)} P_{\lambda_c}) \Rightarrow x = y\}$. ∎

Lemma 7.10.16 Let f be a homomorphism from X^* onto X_1^*. Then the following properties hold:
(1) $\forall \mu \in \mathcal{FP}(X^*)$, $(f(\mu))_c = f(\mu)_c$, $\forall c \in [0,1]$.
(2) $\forall \nu \in \mathcal{FP}(X^*)$, $(f^{-1}(\nu))_c = f^{-1}(\nu)_c$, $\forall c \in [0,1]$. ∎

Proposition 7.10.17 Let f be a homomorphism from X^* onto X_1^*. Let $\lambda \in \mathcal{FP}(X^*)$ be a fuzzy adjunctive language. Then $f(\lambda) \in \mathcal{FP}(X_1^*)$ is fuzzy adjunctive.

Proof. Since λ is fuzzy adjunctive, $\forall x, y \in X^*, x \equiv y(\bigcap_c P_{\lambda_c}) \Rightarrow x = y$. That is, $\forall c \in (0,1)$, $\forall u, v \in X^*$, $uxv \in \lambda_c \Leftrightarrow uyv \in \lambda_c$ implies that $x = y$. Let $f(x), f(y) \in f(X^*) = X_1^*$ and $f(x) \equiv f(y)(\bigcap_c P_{(f(\lambda)_c)})$. Then $\forall c \in (0,1)$, $\forall u, v \in X^*$, $f(u)f(x)f(v) \in (f(\lambda))_c \Leftrightarrow f(u)f(y)f(v) \in (f(\lambda))_c$. This implies that $f(uxv) \in (f(\lambda))_c \Leftrightarrow f(uyv) \in (f(\lambda))_c$, $\forall c$, $\forall u, v \in X^*$, which in turn implies that $f(uxv) \in f(\lambda_c) \Leftrightarrow f(uyv) \in f(\lambda_c)$, $\forall c$, $\forall u, v \in X^*$. Hence $uxv \in \lambda_c \Leftrightarrow uyv \in \lambda_c$, $\forall c$, $\forall u, v \in X^*$. Thus $x \equiv y(P_{\lambda_c})$, $\forall c$ and so $x \equiv y(\bigcap_c P_{\lambda_c})$. Hence $x = y$ and so $f(x) = f(y)$. Therefore, $f(\lambda)$ is fuzzy adjunctive. ∎

Proposition 7.10.18 *Let* $f : X^* \to X_1^*$ *be an isomorphism. Suppose that* $\nu \in \mathcal{F}P(X_1^*)$ *is a fuzzy adjunctive language. Then* $f^{-1}(\nu) \in \mathcal{F}P(X^*)$ *is fuzzy adjunctive.*

Proof. Let $x, y \in X^*$ and $x \equiv y(\bigcap_c P_{(f^{-1}(\nu))_c})$. Then $uxv \in (f^{-1}(\nu))_c \Leftrightarrow uyv \in (f^{-1}(\nu))_c$ $\forall c$, $\forall u, v \in X^*$ and so $uxv \in f^{-1}(\nu_c) \Leftrightarrow uyv \in f^{-1}(\nu_c)$ $\forall c$, $\forall u, v \in X^*$. Thus $f(uxv) \in \nu_c \Leftrightarrow f(uyv) \in \nu_c$ $\forall c$, $\forall u, v \in X^*$. Hence $f(x) \equiv f(y)(P_{\nu_c})$ $\forall c$ and so $f(x) \equiv f(y)(\bigcap_c P_{\nu_c})$. Thus $f(x) = f(y)$. Since f is one-one, $x = y$. Therefore, $f^{-1}(\nu)$ is fuzzy adjunctive. ∎

Proposition 7.10.19 *Let* $\lambda \in \mathcal{F}P(X^*)$. *If there exists* $\nu \in \mathcal{F}P(X^*)$ *such that* $\nu^{-1}\lambda$ *is adjunctive, then* λ *is fuzzy adjunctive.*

Proof. Since $\nu^{-1}\lambda$ is adjunctive, $\forall s, t \in X^*$, $s \equiv t(F_{\nu^{-1}\lambda}) \Rightarrow s = t$. Suppose that $s \equiv t(F_\lambda)$. Then $\forall u, v \in X^*$, $\lambda(usv) = \lambda(utv)$. Thus $\lambda(xusv) = \lambda(xutv)$ $\forall x, u, v \in X^*$. This implies that $\lambda(xusv) \wedge \nu(x) = \lambda(xutv) \wedge \nu(x)$ $\forall x, u, v \in X^*$ and so

$$\vee\{\lambda(xusv) \wedge \nu(x) \mid x \in X^*\} = \vee\{\lambda(xuyv) \wedge \nu(x) \mid x \in X^*\}$$

$\forall u, v \in X^*$. Thus $(\nu^{-1}\lambda)(usv) = (\nu^{-1}\lambda)(utv)$, $\forall u, v \in X^*$ and so $s \equiv t(F_{\nu^{-1}\lambda})$. Hence $s = t$. Therefore, λ is fuzzy adjunctive. ∎

Proposition 7.10.20 *Let* $\lambda \in \mathcal{F}P(\{z\}^*)$. *Then* λ *is fuzzy adjunctive if and only if* λ *is not fuzzy regular.*

Proof. Suppose λ is fuzzy adjunctive. If λ is fuzzy regular, then the index of F_λ is finite. This contradicts the fact that λ is fuzzy adjunctive and z^* is divided into infinite classes that contain only one element by F_λ.

Conversely, if λ is not fuzzy adjunctive, then there exist i, j with $i \neq j$ and $z^i \equiv z^j(F_\lambda)$. Now $\{z\}^*$ is divided into at most j classes,

$$[\Lambda], [z], [z^2], \ldots, [z^{j-1}].$$

Hence the index of F_λ is finite, i.e., λ is fuzzy regular. ∎

Proposition 7.10.21 Let $\lambda \in F(\{z\}^*)$. If there exists $c \in [0, 1]$ such that
(1) $\forall m \geq 1$, $\exists n \in \mathbb{N}$, $\lambda(z^{m+n}) \geq c$,
(2) $\forall m \geq 1$, $\exists l \in \mathbb{N}$, $\lambda(z^{m+l}) < c$,
(3) $\forall m \geq 1$, $\forall t \in \mathbb{N}$, $\exists s \geq t$, $(\lambda(z^{s+i}) < c$ or $\lambda(z^{s+i}) \geq c$, $\forall i = 1, 2, \dots, m)$,
then λ is a fuzzy adjunctive language.

Proof. It suffices to show that $\forall r \geq 0$, $k \geq 1$, $z^r \not\equiv z^{r+k}(F_\lambda)$. In fact, if $k + 1 = m \geq 1$, then set $s \geq 1$ when $r = 0$ and set $s \geq r$ when $r \geq 1$. From (3), it follows that $\forall i = 1, 2, \dots, m$, $\lambda(z^{s+i}) < c$ or $\lambda(z^{s+1}) \geq c$.

Suppose that $\lambda(z^{s+i}) < c$, $i = 1, 2, \dots, k + 1$. From (1), there exists m_1, the smallest integer such that $\lambda(z^{s+k+1+m_1}) \geq c$. Thus

$$\lambda(z^{s-r+1} z^r z^{m_1}) < c, \qquad \lambda(z^{s-r+1} z^{r+k} z^{m_1}) \geq c,$$

i.e., there exist $z^{s-r+1}, z^{m_1} \in z^*$ such that

$$\lambda(z^{s-r+1} z^r z^{m_1}) \neq \lambda(z^{s-r+1} z^{r+k} z^{m_1}).$$

Hence $z^r \not\equiv z^{r+k}(F_\lambda)$.

Suppose that $\lambda(z^{s+i}) \geq c$, $i = 1, 2, \dots, k + 1$. From (2), there exists m_2, the smallest integer such that $\lambda(z^{s+k+1+m_2}) < c$. Similarly, it follows easily that $z^r \not\equiv z^{r+k}(F_\lambda)$. ∎

Definition 7.10.22 Let $\lambda \in \mathcal{FP}(X^*)$ and $c \in [0, 1]$. Then λ is called a **c-discrete language** if $\forall x, y \in X^*$, $x \neq y$, $\lambda(x) \geq c$ and $\lambda(y) \geq c$ implies that $|x| \neq |y|$.

Example 7.10.23 Let $X = \{a, b\}$ and $L = \{a^n b^n \mid n = 0, 1, \dots\}$. Define the fuzzy subset λ of X^* by $\lambda(x) = 0$ if $x \in X^* \backslash L$ and $\lambda(x) > 0$ if $x \in L$. The λ is c-discrete for $c \in (0, 1]$ since no two distinct elements of L have the same length.

Lemma 7.10.24 Let $|X| \geq 2$ and $\lambda \in \mathcal{FP}(X^*)$. Then λ is a fuzzy adjunctive language if and only if $\forall u, v \in X^*$, $|u| = |v|$ and $u \equiv v(F_\lambda)$ implies that $u = v$. ∎

Proposition 7.10.25 Let $\lambda \in \mathcal{FP}(X^*)$. Suppose that λ is c-discrete and $\forall w \in X^*$, there exist $u, v \in X^*$ such that $\lambda(uwv) \geq c$. Then λ is a fuzzy adjunctive language. Conversely, if λ is a fuzzy adjunctive language, then $\forall w \in X^*$ there exist $u, v \in X^*$ such that $\lambda(uwv) > 0$.

Proof. If $u \equiv v(F_\lambda)$ and $|u| = |v|$, then by the hypothesis there exist $x, y \in X^*$ such that $\lambda(xuy) > c$. Since $u \equiv v(F_\lambda)$, $\lambda(xuy) = \lambda(xvy)$ and so $\lambda(xvy) > c$. Now, since $|xuy| = |xvy|$ and λ is c-discrete, it follows that $xuy = xvy$ and so $u = v$. Hence λ is fuzzy adjunctive.

Conversely, if there exists $w \in X^*$ such that $\forall u, v \in X^*$, $\lambda(uwv) = 0$, then $\forall u, v \in X^*$, $\lambda(uwv) = \lambda(uw^2v) = 0$, i.e., $w \equiv w^2(F_\lambda)$, which contradicts the fact that λ is a fuzzy adjunctive language. ∎

Proposition 7.10.26 *If $\lambda \in \mathcal{F}P(X^*)$ is a fuzzy adjunctive language, then $\forall w \in X^*$, $|\lambda^w| = \infty$, where $\lambda^w = \{x \mid x \in X^*wX^*, \lambda(x) > 0\}$.*

Proof. Since λ is fuzzy adjunctive, $\forall w \in X^*$, $|\lambda^w| \neq 0$. Suppose that $|\lambda^w| < \infty$. Let $u \in \lambda^w$ be such that $|u| = \vee\{|v| \mid v \in \lambda^w\}$. Clearly, $|\lambda^{wu}| = 0$. This contradicts the fact that λ is a fuzzy adjunctive language. ∎

Proposition 7.10.27 *Let $|X| \geq 2$. Then λ is an 0-discrete fuzzy adjunctive language if and only if $\mathrm{Supp}(\lambda)$ is an ordinary discrete adjunctive language.*

Proof. If $\mathrm{Supp}(\lambda)$ is an ordinary discrete adjunctive language, then λ is clearly a 0-discrete fuzzy adjunctive language.

Conversely, suppose that λ is a 0-discrete fuzzy adjunctive language and $\mathrm{Supp}(\lambda)$ is not adjunctive. By Lemma 7.10.24, there exist $x, y \in X^*$ such that $x \neq y$, $|x| = |y|$, and $x \equiv y(P_{\mathrm{Supp}(\lambda)})$. Hence $\forall u, v \in X^*$, $uxv \in \mathrm{Supp}(\lambda) \Leftrightarrow uyv \in \mathrm{Supp}(\lambda)$. Consequently, $\forall u, v \in X^*$, $\lambda(uxv) > 0 \Leftrightarrow \lambda(uyv) > 0$. Since λ is a fuzzy adjunctive language, by Proposition 7.10.25, it follows that $\forall x \in X^*$ there exist $u_0, v_0 \in X^*$ with $\lambda(u_0xv_0) > 0$. Thus $\lambda(u_0yv_0) > 0$. Since $|u_0xv_0| = |u_0yv_0|$ and λ is 0-discrete, $u_0xv_0 = u_0yv_0$ and so $x = y$, a contradiction. Hence $\mathrm{Supp}(\lambda)$ must be an adjunctive language. If $\forall x, y \in X^*$ with $x \neq y$, $x, y \in \mathrm{Supp}(\lambda)$, then clearly $\lambda(x) > 0$ and $\lambda(y) > 0$. Since λ is 0-discrete, $|x| \neq |y|$. Therefore, $\mathrm{Supp}(\lambda)$ is discrete. ∎

We now consider fuzzy dense languages.

Definition 7.10.28 *Let $\mu \in \mathcal{F}P(X^*)$ and $w \in X^*$. Let*

$$\mu_c^w = \{x \mid x \in X^*wX^*, \mu(x) > c\}, \quad c \in [0, 1].$$

*If $\forall w \in X^*$, $|\mu_c^w| \neq 0$, then μ is called a c-**dense language**. In particular, a 0-dense language is called a **fuzzy dense language** and we write $\mu_0^w = \mu^w$.*

Example 7.10.29 *Let $X = \{a\}$. Define the fuzzy subset μ of X^* by $\mu(a^n) = \frac{1}{n+1}$, $n = 0, 1, 2, \ldots$, where $a^0 = \Lambda$. Then μ is a 0-dense language. Now μ is not c-dense for any $c \in (0, 1]$ since there exists n such that $\frac{1}{n+1} < c$ and for $w = a^n$ and $x \in X^*a^nX^*$, $\mu(x) \leq \frac{1}{n+1} < c$.*

Proposition 7.10.30 *Let $\omega \in \mathcal{F}P(X^*)$. Then the following statements are equivalent.*

(1) ω is a fuzzy dense language.

(2) $\forall w \in X^*$, $|\omega^w| = \infty$.

(3) There exists a fuzzy adjunctive language λ with $\omega \supseteq \lambda$.

Proof. $(1) \Rightarrow (2)$ Immediate from Proposition 7.10.26.

$(2) \Rightarrow (3)$ Define an ordering relation on X^* as follows: if $|x| < |y|$, then $x < y$; if $|x| = |y|$, then $x < y$ means that x, y is in the lexicographic ordering of elements of X. Now

$$X^* = \{\Lambda < w_1 < w_2 < \cdots < w_n < \cdots\}.$$

Since $|\omega^w| = \infty$, $\forall w \in X^*$, $|\text{Supp}(\omega)| = \infty$, and the number of elements in $\{[x]_\omega\}$ is greater than 1.

Let $x_1, x_2, \ldots, x_n, \ldots$ be the representative elements from these classes. Define $\lambda \in \mathcal{FP}(X^*)$ as follows:

$$\lambda(x) = \begin{cases} \omega(x) & \text{if } x = x_i \text{ for some } i \\ 0 & \text{if } x \neq x_i \text{ for all } i. \end{cases}$$

Clearly, $\lambda \subseteq \omega$. Now $\forall x, y \in X^*$, $x \neq y$, if $\lambda(x) > 0$, $\lambda(y) > 0$, then $|x| \neq |y|$ and so λ is 0-discrete. Since $|\omega^{w_1}| = \infty$, there exists an element $u_1 w_1 v_1$ with the shortest length in ω^{w_1}. We select $u_1 w_1 v_1$ as the representative element. Then $\lambda(u_1 w_1 v_1) = \omega(u_1 w_1 v_1) > 0$. Similarly, since $|\omega^{w_2}| = \infty$, we can choose $u_2 w_2 v_2 \in \omega^{w_2}$ such that $\lambda(u_2 w_2 v_2) = \omega(u_2 w_2 v_2) > 0$ and $|u_1 w_1 v_1| < |u_2 w_2 v_2|$. Hence since $\forall i \ |\omega^{w_i}| = \infty$, there exists $u_i w_i v_i \in \omega^{w_i}$ such that $\lambda(u_i w_i v_i) = \omega(u_i w_i v_i) > 0$ and $|u_{i-1} w_{i-1} v_{i-1}| < |u_i w_i v_i|$. Thus $|\lambda^{w_i}| \neq 0$, $\forall i$. By Proposition 7.10.25, λ is a fuzzy adjunctive language and $\omega \supseteq \lambda$.

$(3) \Rightarrow (1)$ Now $\omega \supseteq \lambda$ implies that $|\omega^w| > |\lambda^w| \ \forall w \in X^*$. Since λ is fuzzy adjunctive, $|\lambda^w| \neq 0 \ \forall w \in X^*$ by Proposition 7.10.25. Thus $|\omega^w| \neq 0$ $\forall w \in X^*$. Consequently, ω is a fuzzy dense language. ∎

Proposition 7.10.31 *Let* $\mu \in \mathcal{FP}(X^*)$. *Then either* μ *or* $\overline{\mu}$ *must be a fuzzy dense language.*

Proof. If $\forall w \in X^*$, $|\mu^w| = \infty$, then μ is fuzzy dense. Suppose that there exists $w \in X^*$ such that $|\mu^w| < \infty$. Then $|\{x \mid x \in X^* w X^*, \mu(x) > 0\}| < \infty$. Thus $|\{x \mid x \in X^* w u X^*, \mu(x) > 0\}| < \infty \ \forall u \in X^*$ and so $|\{x \mid x \in X^* w u X^*, \mu(x) = 0\}| = \infty \ \forall u \in X^*$. This implies that $|\{x \mid x \in X^* w u X^*, \overline{\mu}(x) = 1 > 0\}| = \infty$. Hence $|(\overline{\mu})^{wu}| = \infty \ \forall u \in X^*$. Since $X^* w u X^* \subseteq X^* u X^*$, $\forall u \in X^*$, $|(\overline{\mu})^{wu}| \leq |(\overline{\mu})^u| = \infty$. Therefore, $\overline{\mu}$ is fuzzy dense. ∎

Proposition 7.10.32 *Let* $\omega = \mu \cup \nu$ *and* $c \in [0, 1]$. *Then* ω *is* c-*dense if and only if either* μ *or* ν *is* c-*dense.*

Proof. Suppose that ω is c-dense and ν is not c-dense. Then there exists $w_0 \in X^*$ such that

$$|\{x \mid x \in X^* w_0 X^*, \nu(x) > c\}| < \infty.$$

This implies that $\forall u \in X^*$, $|\{x \mid x \in X^* w_0 u X^*, \ \nu(x) > c\}| < \infty$. Since ω is c-dense, $\forall w \in X^*$, $|\{x \mid x \in X^* w X^*, \ \omega(x) > c\}| = \infty$, i.e., $|\{x \mid x \in X^* w X^*, \ (\mu \cup \nu)(x) > c\}| = \infty \ \forall w \in X^*$. This is equivalent to $|\{x \mid x \in X^* w X^*, \ \mu(x) \vee \nu(x) > c\}| = \infty \ \forall w \in X^*$. Consequently, $|\{x \mid x \in X^* w_0 u X^*, \ \mu(x) \vee \nu(x) > c\}| = \infty \ \forall u \in X^*$. From this, it follows that $|\{x \mid x \in X^* w_0 u X^*, \ \mu(x) > c\}| = \infty \ \forall u \in X^*$. Hence $\forall u \in X^*$, $|\mu_c^{w_0 u}| = \infty$ and so $|\mu_c^u| = \infty \ \forall u \in X^*$. Thus μ is c-dense.

Conversely, suppose that μ is c-dense. Now $\forall w \in X^*$,

$$|\{x \mid x \in X^* w X^*, \ \mu(x) > 0\}| = \infty$$

and so $|\{x \mid x \in X^* w X^*, \ \mu(x) \vee \nu(x) > c\}| = \infty \ \forall w \in X^*$. This implies that $|\{x \mid x \in X^* w X^*, \ \omega(x) > c\}| = \infty \ \forall w \in X^*$. Hence ω is c-dense. ∎

Proposition 7.10.33 *Let $\omega \in \mathcal{FP}(X^*)$. Then ω is c-dense if and only if $\omega|_{X^* w X^*}$ is c-dense, where $\omega|_{X^* w X^*}$ denotes the restriction of ω to $X^* w X^*$, $w \in X^*$.*

Proof. Suppose that ω is c-dense. Then $\forall w, u \in X^*$, $|\{x \mid x \in X^* w u X^*, \ \omega(x) > c\}| \neq 0$. Thus there exists $x \in X^* w u X^* \subseteq X^* w X^* \cap X^* u X^*$ such that $\omega(x) > c$. Hence $\forall u \in X^*$, there exists $x \in X^* u X^*$ such that $\omega|_{X^* w X^*}(x) > c$. This implies that $\forall u \in X^*$,

$$|\{x \mid x \in X^* u X^*, \omega|_{X^* w X^*}(x) > c\}| \neq 0.$$

Hence $\omega|_{X^* w X^*}$ is c-dense $\forall w \in X^*$.

Conversely, suppose that $\forall w \in X^*$, $\omega|_{X^* w X^*}$ is c-dense. Then $\forall w \in X^*$, $|\{x \mid x \in X^* w X^*, \ \omega|_{X^* w X^*}(x) > c\}| \neq 0$. Thus $\forall w \in X^*$, $|\{x \mid x \in X^* w X^*, \ \omega(x) > c\}| \neq 0$. Hence ω is c-dense. ∎

Proposition 7.10.34 *Let $|X| \geq 2$, $\omega = \mu \cup \nu$, and ω be a fuzzy adjunctive language. Then one of the following statements hold.*
(1) μ or ν is a fuzzy adjunctive language.
(2) μ and ν are fuzzy dense languages.

Proof. Suppose that μ, ν are not fuzzy adjunctive and ν is not fuzzy dense. Then there exists $w \in X^*$ such that

$$|\{x \mid x \in X^* w X^*, \nu(x) > 0\}| < \infty.$$

Since μ is not fuzzy adjunctive, there exist $u, v \in X^*$ such that $u \neq v$, $|u| = |v|$, and $u \equiv v(F_\mu)$. Now $|\nu^w| < \infty$ and so there exists u such that $|u| > \vee\{|z| \mid z \in \nu^w\}$. Since $uw \equiv vw(F_\mu)$, $\forall x, y \in X^*$, $\mu(xuwy) =$

$\mu(xvwy)$, and $|xuwy| = |xvwy| > |u|$. Thus $xuwy, xuvy \notin \nu^w$ and so $\nu(xuwy) = \nu(xvwy) = 0$. Now $\forall x, y \in X^*$,

$$\begin{aligned}
\omega(xuwy) &= \mu(xuwy) \vee \nu(xuwy) \\
&= \mu(xuwy) \\
&= \mu(xvwy) \\
&= \mu(xvwy) \vee \nu(xvwy) \\
&= \omega(xvwy).
\end{aligned}$$

Hence $uw \equiv vw(F_\omega)$. Since ω is a fuzzy adjunctive language, $uw = vw$, which contradicts the fact that $u \neq v$. Thus ν is fuzzy dense. Similarly, it can be shown that μ is fuzzy dense. ∎

Proposition 7.10.35 *Let λ be a fuzzy adjunctive language and $w \in X^*$. Then $\lambda|_{X^*wX^*}$ is a fuzzy adjunctive language.*

Proof. Clearly, $\lambda = \lambda|_{X^*wX^*} \cup \lambda|_{\overline{X^*wX^*}}$. Let $x \in X^*$. Then

$$\begin{aligned}
\lambda|_{\overline{X^*wX^*}}(x) &= \begin{cases} \lambda(x) & x \in \overline{X^*wX^*} \\ 0 & x \notin \overline{X^*wX^*}, \end{cases} \\
&= \begin{cases} \lambda(x) & x \notin X^*wX^* \\ 0 & x \in X^*wX^*. \end{cases}
\end{aligned}$$

Clearly, $(\lambda|_{\overline{X^*wX^*}})^w$ is the empty set and so $\lambda|_{\overline{X^*wX^*}}$ is not fuzzy dense. By Proposition 7.10.34, $\lambda|_{X^*wX^*}$ is a fuzzy adjunctive language. ∎

Proposition 7.10.36 *Let $|X| \geq 2$, λ be a 0-discrete fuzzy adjunction language and $\lambda = \mu \cup \nu$. Then μ or ν is a fuzzy adjunctive language.*

Proof. If μ and ν are not fuzzy adjunctive, then there exist $x_1, x_2, y_1, y_2 \in X^*$ such that $x_1 \neq x_2$, $y_1 \neq y_2$, $|x_1| = |x_2|$, $|y_1| = |y_2|$, and $x_1 \equiv x_2(F_\mu)$, $y_1 \equiv y_2(F_\nu)$. Clearly, $x_1 y_1 \equiv x_2 y_1(F_\mu)$ and $x_1 y_1 \equiv x_1 y_2(F_\nu)$. Since λ is fuzzy adjunctive, there exist $u, v \in X^*$ such that $\lambda(ux_1y_1v) = \mu(ux_1y_1v) \vee \nu(ux_1y_1v) > 0$ by Proposition 7.10.25. If $\mu(ux_1y_1v) \geq \nu(ux_1y_1v)$, then

$$\lambda(ux_1y_1v) = \mu(ux_1y_1v) = \mu(ux_2y_1v) > 0$$

and furthermore $\lambda(ux_2y_1v) = \mu(ux_2y_1v) \vee \nu(ux_2y_1v) > 0$. Since λ is 0-discrete, $|ux_1y_1v| = |ux_2y_1v|$ implies that $ux_1y_1v = ux_2y_1v$, which contradicts that $x_1 \neq x_2$. Similarly, if $\mu(ux_1y_1v) \leq \nu(ux_1y_1v)$, then we can show that $y_1 = y_2$, a contradiction. Hence μ or ν is a fuzzy adjunctive language. ∎

Proposition 7.10.37 *Every 0-discrete fuzzy adjunctive language may be written as the disjoint union of two 0-discrete fuzzy adjunctive languages.*

Proof. Let ω be a 0-discrete fuzzy adjunctive language. Let $L \subseteq X^*$ be an adjunctive language and let

$$B_1 = \text{Supp}(\omega) \cap X^* L X^*,$$

$$B_2 = \text{Supp}(\omega) \cap X^* \overline{L} X^*.$$

Then $B_1 \cup B_2 = \text{Supp}(\omega)$, $B_1 \cap B_2 = \emptyset$, and $\omega = \omega|_{B_1} \cup \omega|_{B_2}$, $\omega|_{B_1} \cap \omega|_{B_2} = \emptyset$. Now since ω is 0-discrete, $\omega|_{B_1}$ and $\omega|_{B_2}$ are also 0-discrete. Since L is an adjunctive language, $\forall w \in X^*$ there exist $u, v \in X^*$ such that $uwv \in L$. Also since ω is fuzzy adjunctive and $uwv \in X^*$, there exist $s, t \in X^*$ with $\omega(suwvt) > 0$. This implies that $x = suwvt \in \text{Supp}(\omega) \cap X^* L X^* = B_1$. Thus $\forall w \in X^*$, $|(\omega|_{B_1})^w| \neq 0$. Hence $\omega|_{B_1}$ is fuzzy dense. By Proposition 7.10.25, $\omega|_{B_1}$ is fuzzy dense. Similarly, $\omega|_{B_2}$ is also fuzzy dense since the adjunctivity of L implies that $\overline{L}$ is adjunctive. ∎

Proposition 7.10.38 *Every fuzzy dense language may be written as the disjoint union of a fuzzy adjunctive language and an 0-discrete fuzzy adjunctive language.*

Proof. Let ω be a fuzzy dense language. First suppose that ω is not fuzzy adjunctive.

Suppose that $|X| > 2$. We consider the case $|X| = 1$ later. Let $X^* = \{u_1, u_2, \ldots, u_n, \ldots\}$. Since ω is fuzzy dense, $\forall u_i$, $|\{x \mid x \in X^* u_i X^*, \omega(x) > 0\}| = \infty$. Let $x_i = s_i u_i t_i \in \omega^{u_i}$, $i = 1, 2, \ldots$ be such that

$$|x_1| < |x_2| < \cdots < |x_n| < \cdots.$$

Let $B = \{x_1, x_2, \ldots, x_n, \ldots\}$. By Definition 7.10.22 and Proposition 7.10.25, $\omega|_B$ is an 0-discrete fuzzy adjunctive language. Next, we show that $\omega|_{\overline{B}}$ is fuzzy adjunctive. Suppose that $\omega|_{\overline{B}}$ is not fuzzy adjunctive.

Since ω is not fuzzy adjunctive, there exist v_1, v_2 such that $v_1 \neq v_2$, $|v_1| = |v_2|$, and $v_1 \equiv v_2(F_\omega)$. Also, since $\omega|_{\overline{B}}$ is not fuzzy adjunctive, there exist w_1, w_2 such that $w_1 \neq w_2$, $|w_1| = |w_2|$, and $w_1 \equiv w_2(F_{\omega|_{\overline{B}}})$. Clearly, $v_1 w_1 \equiv v_1 w_2(F_{\omega|_{\overline{B}}})$. Since $\omega|_B$ is fuzzy adjunctive, $v_1 w_1 \not\equiv v_1 w_2(F_{\omega|_B})$. Thus there exist $s, t \in X^*$ such that $\omega|_B(sv_1 w_1 t) \neq \omega|_B(sv_1 w_2 t)$. Now either $\omega|_B(sv_1 w_1 t) > \omega|_B(sv_1 w_2 t)$ or $\omega|_B(sv_1 w_1 t) < \omega|_B(sv_1 w_2 t)$. To be specific, suppose that $\omega|_B(sv_1 w_1 t) > \omega|_B(sv_1 w_2 t)$. Clearly, $sv_1 w_1 t \in B$. If $\omega|_B(sv_1 w_2 t) > 0$, then $\omega(sv_1 w_1 t) > \omega(sv_1 w_2 t) > 0$, i.e., $v_1 w_1 \equiv v_1 w_2(F_\omega)$. If $\omega|_B(sv_1 w_2 t) = 0$ and $sv_1 w_2 t \in B$, then $\omega(sv_1 w_2 t) = 0$ and $\omega(sv_1 w_1 t) > \omega(sv_1 w_2 t)$, i.e., $v_1 w_1 \not\equiv v_1 w_2(F_\omega)$. If $\omega|_B(sv_1 w_2 t) = 0$ and $sv_1 w_2 t \notin B$, then $sv_1 w_2 t \in \overline{B}$. From $v_1 w_1 \equiv v_1 w_2(F_{\omega|_{\overline{B}}})$ and $sv_1 w_1 t \notin \overline{B}$ we have that $\omega(sv_1 w_2 t) = \omega|_{\overline{B}}(sv_1 w_2 t) = \omega|_{\overline{B}}(sv_1 w_1 t) = 0$. Hence $v_1 w_1 \not\equiv v_1 w_2(F_\omega)$. Therefore, it always holds that $v_1 w_1 \not\equiv v_1 w_2(F_\omega)$, i.e.,

$$\omega(sv_1 w_1 t) > \omega(sv_1 w_2 t).$$

However, $v_1 \equiv v_2(F_\omega)$, $v_1 w_1 \equiv v_2 w_1(F_\omega)$, $v_1 w_2(F_\omega)$, and furthermore, $\omega(s v_2 w_1 t) > \omega(s v_2 w_2 t)$. Since $s v_1 w_1 t \neq s v_2 w_1 t$, $|s v_1 w_1 t| = |s v_2 w_1 t|$, $s v_1 w_1 t \in B$, and by the definition of B, $s v_2 w_1 t \notin B$, i.e., $s v_2 w_1 t \in \overline{B}$. Consequently,

$$\omega(s v_2 w_1 t) = \omega|_{\overline{B}}(s v_2 w_1 t) > \omega(s v_2 w_2 t) \geq \omega|_{\overline{B}}(s v_2 w_2 t),$$

i.e., $w_1 \not\equiv w_2(F_{\omega|_{\overline{B}}})$, a contradiction. Hence $\omega|_{\overline{B}}$ is a fuzzy adjunctive language, $\omega = \omega|_B \cup \omega|_{\overline{B}}$, and $\omega|_B \cap \omega|_{\overline{B}} = \chi_\emptyset$.

Suppose that $|X| = 1$. Since ω is not fuzzy adjunctive, by Definition 7.9.7, ω is a fuzzy regular language. If $\omega|_{\overline{B}}$ is not fuzzy adjunctive, then $\omega|_{\overline{B}}$ is fuzzy regular. Since $\omega = \omega|_{\overline{B}} \cup \omega|_B$, $\omega_c = (\omega|_B)_c \cup (\omega|_{\overline{B}})_c$ $\forall c \in [0,1]$. Clearly, ω_c and $(\omega|_{\overline{B}})_c$ are regular languages $\forall c \in [0,1]$. Since $(\omega|_B)_c \cap (\omega|_{\overline{B}})_c = \emptyset$ $\forall c \in [0,1]$, $(\omega|_B)_c$ is a regular language. Also since ω is fuzzy regular, $|\{\omega(x) | x \in X^*\}| < \infty$ and $|\{(\omega|_B)(x) | x \in X^*\}| < \infty$. By Proposition 7.10.2, $\omega|_B$ is a fuzzy regular language. This contradicts that $\omega|_B$ is a fuzzy adjunctive language. Hence $\omega|_{\overline{B}}$ is also a fuzzy adjunctive language.

Suppose that ω is a fuzzy adjunctive language. Now $\omega = \omega|_B \cup \omega|_{\overline{B}}$, where $\omega|_B$ is a 0-discrete fuzzy adjunctive language. By Proposition 7.10.15, $\omega|_B = \omega|_{B_1} \cup \omega|_{B_2}$ and $\omega|_{B_1} \cap \omega|_{B_2} = \chi_\emptyset$, where $\omega|_{B_1}$ and $\omega|_{B_2}$ are two 0-discrete fuzzy adjunctive languages. Thus

$$\omega = \omega|_{B_1} \cup \omega|_{B_2} \cup \omega|_{\overline{B}}.$$

If $\omega|_{B_2} \cup \omega|_{\overline{B}}$ is a fuzzy adjunctive language, then the result is true. Suppose that $\omega|_{B_2} \cup \omega|_{B^c}$ is not fuzzy adjunctive. Then $\omega|_{B_2}$ is fuzzy adjunctive. By Proposition 7.10.30, $\omega|_{B_2} \cup \omega|_{\overline{B}}$ is a fuzzy dense language. From the preceding arguments, it follows that $\omega|_{\overline{B}}$ is a fuzzy adjunctive language. Since $\omega = \omega|_B \cup \omega|_{\overline{B}}$, the desired result follows. ∎

7.11 Deterministic Acceptors of Regular Fuzzy Languages

The results in this section are from [228]. It is well known that there is a one-to-one relationship between finite automata, regular (type-3) languages, and regular expressions. This relationship demonstrates the use of regular expressions for describing deterministic finite fuzzy automata and regular fuzzy languages. It introduces a normal form for the production of a regular fuzzy grammar when the max-min rule is used.

Specific fuzzy system models based on fuzzy set theory include a description of decision making in a fuzzy environment [14], finite fuzzy automata as learning systems [250], and fuzzy grammars and languages [122, 123].

Fuzzy automata, grammars, and languages lead to a better understanding of nondeterministic algorithms and of pattern recognition tasks using syntactic pattern recognition techniques.

In this section, an algorithm is developed for constructing a deterministic finite automaton that classifies the strings of a language with a regular fuzzy grammar. The derivations of the grammar are governed by the max-min rule [122, 123]. An equivalent unambiguous regular fuzzy grammar with productions in a normal form is developed from an extension of this algorithm.

Definition 7.11.1 *A regular fuzzy grammar is a four-tuple $G = (N, T, S, P)$, where N is a finite set of nonterminals, T is a finite set of terminals, $S \in N$ is the starting symbol, P is a finite set of productions, $N \cap T = \emptyset$, and the elements in P are of the form $A \xrightarrow{k} aB$ or $A \xrightarrow{k} a, A, B \in N$, $a \in T$, $0 < k \leq 1$.*

Definition 7.11.2 *A finite quasi-fuzzy automaton (fqfa) is a six-tuple $M = (Q, X, Y, \delta, w, q_0)$, where Q is a finite set of states, X is a finite input alphabet, Y is a finite output alphabet, $\delta : Q \times X \times [0, 1] \to Q$ is the fuzzy state transition map, $\omega : Q \to Y$ is the output map, and $q_0 \in Q$ is the starting state.*

Definition 7.11.2 differs from the fuzzy machines defined previously in two important ways. First, the interval $[0, 1]$ determines the third component of the ordered triple appearing in δ's domain rather than containing the image of δ. Second the output map is crisp and has Q as its domain rather than $Q \times X$.

A regular fuzzy grammar reduces to a conventional grammar when production weights are all equal to 1. Similarly, a finite quasi-fuzzy automaton reduces to a conventional finite-state Moore machine by restricting the transition weights to the value 1.

A fuzzy subset of T^* is called a **fuzzy language** in the alphabet T.

Given a regular fuzzy grammar G, the membership grade of a string x of T^* in the regular fuzzy language $L(G)$ is the maximum value of any derivation of x, where the value of a specific derivation of x is equal to the minimum weight of the productions used.

From the max-min rule for a fuzzy language, every string of T^* has its highest computable membership grade. It is known [123] that given a regular fuzzy grammar G, a corresponding finite quasi-fuzzy automaton can be constructed that "accepts" the language $L(G)$. The following theorem describes the construction of a deterministic nonfuzzy, finite automaton that computes the membership function of LG). Unless otherwise specified, we use the symbol X to denote both the automaton input alphabet and the language terminals, i.e., $X = T$.

Theorem 7.11.3 *Let G be a regular fuzzy grammar. Then there exists a deterministic Moore sequential machine dfa, with output alphabet $Y \subseteq \{c \mid c$ is a production weight$\} \cup \{0\}$, which computes the membership function $\mu\colon X^* \to [0,1]$ of the language $L(G)$.*

Proof. We give a five-step algorithm for constructing the dfa.

Step 1: Given the regular fuzzy grammar, obtain the corresponding fqfa.

The fqfa is obtained in the same way that a nonfuzzy finite automaton is obtained from a nonfuzzy regular grammar [22], with the exception that a production weight is assigned to the corresponding transition [123].

Step 2: Obtain the set W of possible nonzero membership grades of strings in the language $L(G)$.

W is taken to be the finite set of distinct production weights or, equivalently, the weights of the transitions of the fqfa. (The reasoning is as follows: (a) the max and min operations do not introduce a weight not already assigned to some production and (b) each production weight initially may be the membership grade of some string (or strings) in the language $L(G)$.)

Step 3: For all $c \in W$, obtain the regular expression $F'(c)$ describing those $x \in X^*$ such that $\mu(x) \geq c$.

(It is known that given a regular fuzzy language and a threshold c, $0 \leq c \leq 1$, the nonfuzzy "threshold language"

$$L(c) = \{x \mid x \in X^*,\ \mu(x) \geq c\}$$

is regular.) The regular expression $F'(c)$ can be found in the following method. Examine the transition diagram of the fqfa and retain without weight only those transitions whose weight is equal to or greater than c. This yields a nondeterministic nonfuzzy machine that recognizes the language $L(c)$ and from which the regular expression $F'(c)$ can be obtained directly by standard techniques for nondeterministic transition graphs or by conversion to a deterministic finite automaton and solution of the descriptive equations [22].

Step 4: For all $c \in W$, obtain the regular expression $F(c)$ describing those strings x of X^* such that $\mu(x) = c$. (It is known that if $F'(c_1)$ and $F'(c_2)$ are regular expressions, then so are the Boolean functions of $F'(c_1)$ and $F'(c_2)$. Specifically, if $F'(c_1)$ and $F'(c_2)$ define two threshold languages, then the new regular expression

$$F(c_2) = F'(c_2) \cap \overline{F'(c_1)}$$

defines the regular language consisting of those strings that are in $L(c_2)$ and not in $L(c_1)$.) Consider the finite set W of possible nonzero membership grades in $L(G)$. Let $c_1, c_2 \in W$ be such that $c_1 > c_2$. Then the regular expression

$$F(c_2) = F'(c_2) \cap \overline{F'(c_1)}$$

defines the set of strings

$$
\begin{aligned}
\{x \mid x \in X^*, x \in L(c_2), x \in \overline{L(c_1)}\} &= \{x \mid x \in X^*, \ \mu(x) \geq c_2, \\
&\qquad \mu(x) < c_1\} \\
&= \{x \mid x \in X^*, \ \mu(x) = c_2\}.
\end{aligned}
$$

$F(c_2)$ is an equivalence class of the equivalence relation E on X^* defined by $(x, y) \in E$ if and only if $\mu(x) = \mu(y)$. This procedure, applied to all pairs of adjacent membership grades in W beginning with the lowest value, yields the disjoint regular expressions defining the dfa.

$Step$ 5: Use the regular expressions $F(c)$ to obtain the state transition diagram of the dfa, where $c \in W$. ∎

The procedure for obtaining a state transition diagram by taking derivations of a regular expression is discussed in [22].

Another method of decomposing a fuzzy grammar into nonfuzzy grammars using the concept of level set can be found in [258].

Corollary 7.11.4 *Let G_1 be a regular fuzzy grammar. Then there is an equivalent unambiguous regular fuzzy grammar G_2 in which productions have the form $A \xrightarrow{1} aB$ or $A \xrightarrow{c} a$, where $A, B \in N$, $a \in T$, and $0 < c \leq 1$.*

Proof. Construct the dfa as described in proof of Theorem 7.11.3. Then there is in P a production $A \xrightarrow{1} aB$ (or $A \to aB$, with weight 1 understood) for each transition $\delta(q_A, a) = q_B$. If $\omega(q_B) = c \neq 0$, then there is in P a production $A \xrightarrow{c} a$ for each transition $\delta(q_A, a) = q_B$. The starting symbol S of G_2 corresponds to the starting state q_S of the dfa. Suppose the terminal string $x = ab \ldots de$ causes the dfa started in state q_S to halt in a state with output k. Then there is a derivation

$$S \to aA \to abB \to \cdots \to ab \cdots dD \xrightarrow{c} ab \cdots de$$

in G_2. Conversely, such a derivation in G_2 yields a string that causes the dfa to terminate in a state with output c. G_2 is unambiguous since it is obtained from a deterministic finite automaton. ∎

The following example illustrates Theorem 7.11.3 and Corollary 7.11.4.

If $x, y \in T^*$ in the following example, then we use the notation $x + y$ to denote $\{x, y\}$.

Example 7.11.5 (216) *Consider the regular fuzzy grammar* $G_1 = (N, T, S, P)$, *where* $N = \{S, A, B\}$, $T = \{a, b\}$, *and the productions are as follows:*

$$S \xrightarrow{0.3} aS \qquad S \xrightarrow{0.5} aA \qquad S \xrightarrow{0.7} aB$$
$$S \xrightarrow{0.3} bS \qquad S \xrightarrow{0.2} bA \qquad A \xrightarrow{0.5} b$$
$$B \xrightarrow{0.4} b.$$

Step 1: The corresponding fuzzy machine is shown in Figure 7.1.

Step 2: $W = \{0.7, 0.5, 0.4, 0.3, 0.2\}$.

Step 3:

$$
\begin{aligned}
c &= 0.2, & F'(0.2) &= (a+b)^*(ab+bb) \\
c &= 0.3, & F'(0.3) &= (a+b)^*ab \\
c &= 0.4, & F'(0.4) &= ab \\
c &= 0.5, & F'(0.5) &= ab \\
c &= 0.7, & F'(0.7) &= \emptyset.
\end{aligned}
$$

Step 4:

$$
\begin{aligned}
F(0.2) &= F'(0.2) \cap \overline{F'(0.3)} &= (a+b)^*bb \\
F(0.3) &= (a+b)(a+b)^*ab \\
F(0.4) &= \emptyset \\
F(0.5) &= ab \\
F(0.7) &= \emptyset.
\end{aligned}
$$

Step 5: The deterministic classifier of strings in the regular fuzzy language $L(G)$ *is shown in Figure 7.2. Using the method of Corollary 7.11.4, the productions of the equivalent grammar* G_2 *are as follows:*

$$S \to aA \mid bB$$
$$A \to aC \mid bD \qquad A \xrightarrow{0.5} b$$
$$B \to aC \mid bE \qquad B \xrightarrow{0.2} b$$

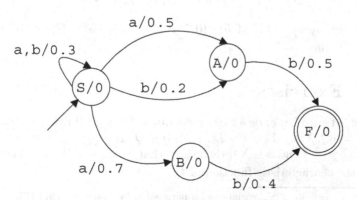

Figure 7.1[1]: fqfa obtained from G_1

$$C \to aC \mid bF \qquad C \xrightarrow{0.3} b$$
$$D \to aC \mid bE \qquad D \xrightarrow{0.2} b$$
$$E \to aC \mid bE \qquad E \xrightarrow{0.2} b$$
$$F \to aC \mid bE \qquad F \xrightarrow{0.2} b.$$

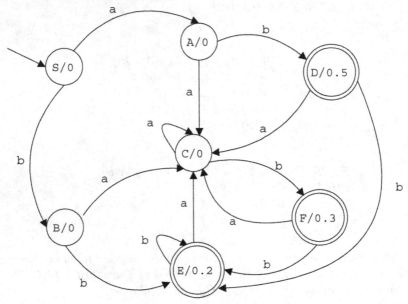

Figure 7.2[2]: dfa obtained from G_1

Example 7.11.6 *[28] Consider the string ab. Using G_1,*

$$S \xrightarrow{0.5} aA \xrightarrow{0.5} ab \text{ and } S \xrightarrow{0.7} aB \xrightarrow{0.4} ab.$$

We have that $\mu(ab) = (0.5 \wedge 0.5) \vee (0.7 \wedge, 0.4) = 0.5$. Using G_2, we have that $S \to aA \xrightarrow{0.5} ab$.

7.12   Exercises

1. Let $\mathcal{M} = (M, \iota, \tau)$ be a fuzzy recognizer of a fuzzy finite state machine $M = (Q, X, \mu)$. Let $A = L(M)$. If $q \in Q$ and $\iota(q) \wedge \mu(p, x, q) > 0$ for some $p \in Q$ and $x \in X^*$, then show that $x^{-1}A = \chi_{\{q\}}^{-1} \circ \tau$, where $\chi_{\{q\}}$ is the characteristic function of $\{q\}$.

---

[1] Figure 7.1 is from [228], reprinted with permission by Copyright 1974 IEEE.
[2] Figure 7.2 is from [228], reprinted with permission by Copyright 1974 IEEE.

2. (121) Let $\mathcal{M} = (M, \iota, \tau)$ be a fuzzy X-recognizer of $M = (Q, X, \mu)$ and let $\mathcal{M}' = (M', \iota', \tau')$ be a fuzzy X'-recognizer of $M' = (Q', X', \mu')$. Let $f : Q \to Q'$ and $g : X \to X'$. Then (f, g) is called a **homomorphism** of $\mathcal{M}$ into $\mathcal{M}'$ if $\forall p, q \in Q$ and $\forall u \in X$, (1) $\mu(p, u, q) \leq \mu'(f(p), g(u), f(q))$, (2) $\iota(q) \leq \iota'(f(p))$, and (3) $\tau(p) \leq \tau'(f(p))$. If (f, g) is a homomorphism of $\mathcal{M}$ into $\mathcal{M}'$ prove that $\forall p, q \in Q$ and $x \in X^*$, $\mu^*(p, x, q) \leq \mu'^*(f(p), g(u_1) \ldots g(u_n), f(q))$, where $x = u_1 \ldots u_n$ and $u_i \in X$, $i = 1, 2, \ldots, n$.

3. (121) Suppose that (f, g) is a homomorphism of $\mathcal{M}$ into $\mathcal{M}'$ and (h, k) is a homomorphism of $\mathcal{M}'$ into $\mathcal{M}''$. Prove that $(h \circ f, k \circ g)$ is a homomorphism of $\mathcal{M}$ into $\mathcal{M}''$.

4. (121) Suppose that $\mathcal{M} = (M, \iota, \tau)$ is a complete accessible fuzzy X-recognizer of the ffsm $M = (Q, X, \mu)$. Let $A = L(M)$. Show that there exists a homomorphism of $\mathcal{M}_A$ into $\mathcal{M}$, where $\mathcal{M}_A$ is the fuzzy X-recognizer of M_A. Show that if $\mathrm{Supp}(\iota) \neq \emptyset$, then $\mathcal{M}_A$ is a homomorphic image of $\mathcal{M}$.

5. (121) Let $\mathcal{M} = (M, \iota, \tau)$ be a complete accessible fuzzy X-recognizer of a ffsm $M = (Q, X, \mu)$. Then $\mathcal{M}$ is called **reduced** if $q_1^{-1} \circ \tau = q_2^{-1} \circ \tau$ implies $q_1 = q_2$. Let $A = L(M)$. Prove that $\mathcal{M}$ is reduced if and only if f is injective, where $(f, g) : \mathcal{M} \to \mathcal{M}_A$ is a homomorphism and g is the identity map. If, in addition, $\mathrm{Supp}(\iota) \neq \emptyset$, prove that $\mathcal{M}$ is reduced if and only if $\mathcal{M} \cong \mathcal{M}_A$.

6. Prove Theorems 7.5.1 and 7.5.2.

7. Prove (2), (4), and (6) of Theorem 7.5.4.

8. Prove Theorem 7.6.6.

9. Prove (2) and (4) of Theorem 7.6.18.

10. Prove (2) of Theorem 7.7.5.

11. Prove Theorem 7.8.4.

12. Prove that $x \equiv y(F_L)$ implies $x = y$ in Example 7.10.12.

13. Compare the fuzzy state transition map $\delta : Q \times X \times [0, 1] \to Q$ in Definition 7.11.2 with a function $\mu : Q \times X \times Q \to [0, 1]$. Show for example that $\delta(q, a, \frac{1}{2}) = q_1$ and $\delta(q, a, \frac{1}{2}) = q_2$ is not possible, while $\mu(q, a, q_1) = \frac{1}{2}$ and $\mu(q, a, q_2) = \frac{1}{2}$ is possible. Show also that $\delta(q, a, \frac{1}{2}) = q_1$ and $\delta(q, a, \frac{1}{4}) = q_1$ is possible, while $\mu(q, a, q_1) = \frac{1}{2}$ and $\mu(q, a, q_1) = \frac{1}{4}$ is not possible.

Chapter 8

Minimization of Fuzzy Automata

8.1 Equivalence, Reduction, and Minimization of Finite Fuzzy Automata

The problem of equivalence, reduction, and minimization is completely solved for deterministic automata. For stochastic automata, the problem is studied in detail in [29, 58, 168]. The case for some types of fuzzy automata is given in Chapters 2 and 3. In this and the following three sections, we present an algebraic approach concerning the minimization of fuzzy automata as developed in [231].

The theoretical foundation [169] of the algorithm of Even [58] and the analogies between some aspects of the theory of rings (resp., modules) and the theory of semirings (resp., semimodules) indicate the way for determining a general solution in the case of fuzzy automata. Using the notion of a noetherian semimodule, we present an algorithm for the equivalence of some kinds of fuzzy automata.

We recall the definitions of semiring and semimodule [27, 56, 59] and some ideas from the theory of fuzzy automata from Chapters 2 and 3 and [262].

Let S be a set and let $*_1$ and $*_2$ be two binary operations on S, i.e., $*_i : S \times S \to S$, $i = 1, 2$. In this chapter, we call the triple $(S, *_1, *_2)$ a (**commutative**) **semiring** if $(S, *_1)$ and $(S, *_2)$ are (commutative) semigroups with identity and $*_2$ is distributive over $*_1$, i.e., $\forall a, b, c \in S$,

$$
\begin{aligned}
a *_2 (b *_1 c) &= (a *_2 b) *_1 (a *_2 c) \\
(b *_1 c) *_2 a &= (b *_2 a) *_1 (c *_2 a).
\end{aligned}
$$

Let E be a set, S be a semiring and let $*_1' : E \times E \to E$ and $*_2' : S \times E \to E$. The triple $(E, *_1', *_2')$ is called a **left semimodule** over S if

for all $a, b \in S$ and $x, y \in E$ the following conditions hold:
(SM.1) $(E, *_1')$ is a commutative semigroup with identity;
(SM.2)

$$a *_2' (x *_1' y) = (a *_2' x) *_1' (a *_2' y)$$
$$(a *_1 b) *_2' x = (a *_2' x) *_1' (b *_2' x)$$

(SM.3)

$$a *_2' (b *_2' x) = (a *_2 b) *_2' x.$$

A function $h : (E, *_1', *_2') \to (E, *_1'', *_2'')$ is called a **morphism of semi-modules** if the following properties hold: $\forall a \in S, \forall x, y \in E$,

$$h(x *_1' y) = h(x) *_1'' h(y) \text{ and } h(a *_2' x) = a *_2'' h(x).$$

We sometimes either use · for the operation $*_2'$ or suppress it all together and we replace $*_1'$ with $+$. We also simply call E a semimodule over S or an S-semimodule.

The notion of a **right semimodule** is defined similarly. If S is a commutative semiring, we call E a semimodule.

Let M be an S-semimodule. A set $X \subseteq M$ is called a **system of generators** for M if X generates M, i.e., if every element of M is expressible in the form $\sum_{i=1}^{n} a_i x_i$, $a_i \in S$, $x_i \in X$, $i = 1, 2, \ldots, n$ for some $n \in \mathbb{N}$. A **quasi-base** is a minimal system of generators for M. If the quasi-base is finite, the dimension of M (denoted dim M) is the cardinality of X.

Let X be a set, not necessarily finite, and let S be a semiring. Let

$$VX = \sum_{x \in X} a_x \cdot x, \ a_x \in S, \ x \in X,$$

where $a_x = 0$ except for a finite number of elements $x \in X$. Then it follows that VX is an S-semimodule, called a **free semimodule**. It also follows that the set X is a minimal system of generators for VX.

Proposition 8.1.1 *Let M be a semimodule over S. Then the following conditions are equivalent.*

(1) Every increasing sequence of sub-S-semimodules of M, i.e., $M_1 \subseteq M_2 \subseteq \ldots \subseteq M_k \subseteq \ldots$, such that $M_i \neq M_{i-1}$, is finite.

(2) For every sub-S-semimodule of M, there exists a finite minimal system of generators.

(3) Every nonempty set G of sub-S-semimodules of M contains a maximal element. ∎

A semimodule that satisfies any of the properties of Proposition 8.1.1 is called **noetherian**.

Proposition 8.1.2 *If X is a nonempty finite set, then the free semimodule VX is noetherian.* ∎

Example 8.1.3 *(1) Let $I = [0, 1]$. Consider the binary operations $*_1 =$ max and $*_2 = $ min on I. Then the triple $(I, *_1, *_2) = ([0, 1], \text{max}, \text{min})$ is a commutative semiring.*

(2) Let L be a distributive lattice. Then the triple $(L, \text{max}, \text{min})$ is a commutative semiring.

(3) Let X be a finite set and VX be the free semimodule generated by X over the semiring from (1). The operations on the free semimodule are as follows:

$$*_1' = + : VX \times VX \to VX, \quad \sum_{x \in X} a_x \cdot x + \sum_{x \in X} b_x \cdot x = \sum_{x \in X} (a_x \vee b_x) \cdot x,$$

$$*_2' = \cdot : [0, 1] \times VX \to VX, \quad c \sum_{x \in X} a_x \cdot x = \sum_{x \in X} (c \wedge a_x) \cdot x.$$

If X is finite, then VX is a noetherian semimodule (see Proposition 8.1.2).

Definition 8.1.4 *A **fuzzy automaton** is a quadruple $A = (Q, X, Y, \mu)$, where Q, X, Y are finite sets and $\mu : Q \times X \times Y \times Q \to [0, 1]$.*

(The definition of a fuzzy automaton in Definition 8.1.4 is the same as that of a max-min sequential-like machine in Definition 2.9.1. We use the differing terminology partially to be consistent with the original papers, but also due to the different approach.)

As usual Q is the set of states, X is the input alphabet, and Y is the output alphabet for the fuzzy automaton A. Let $\mu(q_j, x_i, y_r, q_k) = a_{ij}^{rk} \in [0, 1]$.

If the interval $[0, 1]$ is replaced by the distributive lattice L (see Example 8.1.3(2)), we obtain the more general notion of L-automaton, closely related to fuzzy automaton.

Every fuzzy automaton A defines the free semimodules $V(Q \times X)$ and $V(Y \times Q)$ over the semiring $[0, 1]$. Define the function

$$\widehat{\mu} : V(Q \times X) \to V(Y \times Q)$$

by $\forall q_j \in Q, \forall x_i \in X$,

$$\widehat{\mu}(q_j, x_i) = \sum_{r, k} a_{ij}^{rk}(y_r, q_k).$$

The corresponding array $M_\mu = [a_{ij}^{rk}]$ describes the fuzzy automaton.

Consider the words $x = x_1 \cdot x_2 \ldots \cdot x_p \in X^*$, $y = y_1 \cdot y_2 \ldots \cdot y_p \in Y^*$ and the matrices $M(x_i, y_r) = [m_{jk}(x_i, y_r)]$, where $m_{jk}(x_i, y_r) = a_{ij}^{rk}$. Denote the max-min product of matrices by $\circ$. Then we obtain the expression

$$M(x, y) = M(x_1, y_1) \circ M(x_2, y_2) \circ \ldots \circ M(x_p, y_p).$$

If $P = P(\Lambda, \Lambda)$ is a column matrix with $|Q|$ rows and whose elements are equal to 1, let

$$P(x, y) = M(x, y) \circ P.$$

Let $A = (Q, X, Y, \mu)$ be a fuzzy automaton. We call (Q, ε) the fuzzy set of initial states, i.e., ε is a fuzzy subset of Q. Let $q \in Q$ and $a \in [0, 1]$. Define $\varepsilon_q^1 : Q \to [0, 1]$ and $\varepsilon_q^0 : Q \to [0, 1]$ by $\forall q' \in Q$,

$$\varepsilon_q^1(q') = \begin{cases} 1 & \text{if } q = q' \\ 0 & \text{if } q \neq q' \end{cases} \qquad \varepsilon_q^0(q') = \begin{cases} 0 & \text{if } q = q' \\ a \in [0, 1] & \text{if } q \neq q' \end{cases}$$

with the condition that $\sum_{\overline{q} \in Q} \varepsilon_q^0(\overline{q}) \neq 0$ for ε_q^0. The fuzzy automaton A, denoted in this case (A, ε_q^1) (resp. (A, ε)), is called **initial** (resp. **weakly initial**).

For the fuzzy automaton A, define $S_\varepsilon(x, y) = \varepsilon \circ P(x, y)$. An entry of $S_\varepsilon(x, y)$ indicates the maximal degree of membership for the input word x and the output word y, where (Q, ε) is fixed.

Let $A = (X, Q, Y, \mu)$ and $A' = (X', Q', Y', \mu')$ be fuzzy automata and let ε and ε' be fuzzy subsets of Q and Q' respectively.

Definition 8.1.5 *The initial automata (A, ε) and (A', ε') are said to be* **equivalent,** *written $(A, \varepsilon) \sim (A', \varepsilon')$, if $S_\varepsilon(x, y)_A = S_{\varepsilon'}(x, y)_{A'}$ for all $x \in X^*$ and $y \in Y^*$.*

(1) Let $A = A' = (X, Q, Y, \mu)$. If $(A, \varepsilon) \sim (A', \varepsilon')$, then ε and ε' are said to be **equivalent** *on Q, written $\varepsilon \sim \varepsilon'$.*

(2) If $(A, \varepsilon_q^1) \sim (A', \varepsilon_{q'}^{'1})$, then the states $q \in Q$ and $q' \in Q'$ are said to be **equivalent,** *written $q \sim q'$.*

(3) $A = (X, Q, Y, \mu)$ is said to be **equivalently embedded** *into $A' = (X', Q', Y', \mu')$, written $A \precsim A'$, if for each $q \in Q$, there exists an equivalent state $q' \in Q'$ of A'.*

(4) A is said to be **weakly equivalently embedded** *into A', written $A \underset{\approx}{\precsim} A'$, if for all $\varepsilon : Q \to [0, 1]$ there exists $\varepsilon' : Q' \to [0, 1]$ such that $(A, \varepsilon) \sim (A', \varepsilon')$.*

(5) A and A' are said to be **equivalent,** *written $A \sim A'$, if $A \precsim A'$ and $A' \precsim A$.*

(6) A and A' are called **weakly equivalent,** *written $A \approx A'$, if $A \underset{\approx}{\precsim} A'$ and $A' \underset{\approx}{\precsim} A$.*

Definition 8.1.6 *Let A be a fuzzy automaton.*

(1) A is said to be in **reduced form** *if for all $q, q' \in Q$, $q \sim q'$ implies $q = q'$.*

(2) A' is called a **reduct** *of A if A' is in reduced form and is equivalent to A.*

(3) A is said to be in **minimal form** *if for each $\varepsilon_{q_i}^1$ $(i \leq |Q|)$, there does not exist $\varepsilon_{q_i}^0$ $(i \leq |Q|)$ such that $(A, \varepsilon_{q_i}^1) \sim (A', \varepsilon_{q_i}^0)$.*

(4) A' is called a **minimal** *of A if it is in minimal form and $A \approx A'$.*

Example 8.1.7 *Let $X = \{u\}$ and $Y = \{1\}$. Let $Q_1 = \{q_1, q_2\}$ and $Q_2 = \{s_1, s_2\}$. Define $\mu_1 : Q_1 \times X \times Y \times Q_1 \to [0,1]$ and $\mu_2 : Q_2 \times X \times Y \times Q_2 \to [0,1]$ as in Example 2.10.2. Let $\varepsilon_1 : Q_1 \to [0,1]$ and $\varepsilon_2 : Q_2 \to [0,1]$. Then*

$$M_1(u,1) = \begin{bmatrix} \frac{1}{2} & \frac{1}{3} \\ \frac{2}{3} & 0 \end{bmatrix} \text{ and } M_2(u,1) = \begin{bmatrix} 0 & \frac{1}{2} \\ \frac{2}{3} & 0 \end{bmatrix}. \text{ In fact, } M_1(u^n, 1^n) =$$

$$\begin{bmatrix} \frac{1}{2} & \frac{1}{3} \\ \frac{1}{2} & \frac{1}{3} \end{bmatrix} \text{ for } n = 2, 3, \ldots. \text{ For } n = 2, 4, 6, \ldots, \ M_2(u^n, 1^n) = \begin{bmatrix} \frac{1}{2} & 0 \\ 0 & \frac{1}{2} \end{bmatrix}$$

and for $n = 3, 5, 7, \ldots, \ M_2(u^n, 1^n) = \begin{bmatrix} 0 & \frac{1}{2} \\ \frac{1}{2} & 0 \end{bmatrix}.$ We also have that

$$\begin{aligned}
S_{\varepsilon_1}(u,1)_{A_1} &= \begin{bmatrix} \varepsilon_1(q_1) & \varepsilon_1(q_2) \end{bmatrix} \circ \begin{bmatrix} \frac{1}{2} & \frac{1}{3} \\ \frac{2}{3} & 0 \end{bmatrix} \circ \begin{bmatrix} 1 \\ 1 \end{bmatrix} \\
&= \begin{bmatrix} (\varepsilon_1(q_1) \wedge \frac{1}{2}) \vee (\varepsilon_1(q_2) \wedge \frac{2}{3}) & (\varepsilon_1(q_1) \wedge \frac{2}{3}) \end{bmatrix} \circ \begin{bmatrix} 1 \\ 1 \end{bmatrix} \\
&= (\varepsilon_1(q_1) \wedge \frac{1}{2}) \vee (\varepsilon_1(q_2) \wedge \frac{2}{3}) \vee (\varepsilon_1(q_1) \wedge \frac{1}{3}) \\
&= (\varepsilon_1(q_1) \wedge \frac{1}{2}) \vee (\varepsilon_1(q_2) \wedge \frac{2}{3})
\end{aligned}$$

and

$$\begin{aligned}
S_{\varepsilon_2}(u,1)_{A_2} &= \begin{bmatrix} \varepsilon_2(s_1) & \varepsilon_2(s_2) \end{bmatrix} \circ \begin{bmatrix} 0 & \frac{1}{2} \\ \frac{2}{3} & 0 \end{bmatrix} \circ \begin{bmatrix} 1 \\ 1 \end{bmatrix} \\
&= \begin{bmatrix} (\varepsilon_2(s_2) \wedge \frac{2}{3}) & (\varepsilon_2(s_1) \wedge \frac{1}{2}) \end{bmatrix} \circ \begin{bmatrix} 1 \\ 1 \end{bmatrix} \\
&= (\varepsilon_2(s_2) \wedge \frac{2}{3}) \vee (\varepsilon_2(s_1) \wedge \frac{1}{2}).
\end{aligned}$$

In fact,

$$\begin{aligned}
S_{\varepsilon_1}(u^n, 1^n)_{A_1} &= (\varepsilon_1(q_1) \wedge \frac{1}{2}) \vee (\varepsilon_1(q_2) \wedge \frac{1}{2}) \vee (\varepsilon_1(q_1) \wedge \frac{1}{3}) \vee (\varepsilon_1(q_2) \wedge \frac{1}{3}) \\
&= (\varepsilon_1(q_1) \wedge \frac{1}{2}) \vee (\varepsilon_1(q_2) \wedge \frac{1}{2}),
\end{aligned}$$

for $n = 2, 3, 4, \ldots$. For $n = 2, 4, 6, \ldots,$

$$S_{\varepsilon_2}(u^n, 1^n)_{A_2} = (\varepsilon_2(s_1) \wedge \frac{1}{2}) \vee (\varepsilon_2(s_2) \wedge \frac{1}{2}),$$

and for $n = 3, 5, 7, \ldots,$

$$S_{\varepsilon_2}(u^n, 1^n)_{A_2} = (\varepsilon_2(s_2) \wedge \frac{1}{2}) \vee (\varepsilon_2(s_1) \wedge \frac{1}{2}).$$

We see that if $\varepsilon_1(q_2) \geq \frac{2}{3} \leq \varepsilon_2(s_2)$ and $\varepsilon_1(q_1) \geq \frac{1}{2} \leq \varepsilon_2(s_1)$, then $(A_1, \varepsilon_1) \sim (A_2, \varepsilon_2)$. We also have that for appropriate choices of η_1 and η_2 in Example 2.10.2 and of ε_1 and ε_2 here, $r^{I_1}(u^n, 1^n) = S_{\varepsilon_1}(u^n, 1^n)_{A_1}$ and $r^{I_2}(u^n, 1^n) = S_{\varepsilon_2}(u^n, 1^n)_{A_2} \ \forall n \in \mathbb{N}.$

Example 8.1.8 *Let* $X = \{u\}$ *and* $Y = \{0,1\}$. *Define* $\mu : Q \times X \times Y \times Q \rightarrow [0,1]$ *as in Example 2.10.3, i.e.,* $\mu(q_1, u, 1, q_2) = \frac{1}{2}$ *and* $\mu(q_2, u, 0, q_1) = \frac{2}{3}$ *with* μ *of any other element equal to* 0. *Let* $\varepsilon : Q \rightarrow [0,1]$. *Then* $M(u, 1) =$

$$\begin{bmatrix} 0 & \frac{1}{2} \\ 0 & 0 \end{bmatrix}, \; M(uu, 10) = \begin{bmatrix} \frac{1}{2} & 0 \\ 0 & 0 \end{bmatrix}, \; M(uuu, 101) = \begin{bmatrix} 0 & \frac{1}{2} \\ 0 & 0 \end{bmatrix}, \; M(uuuu,$$

$$1010) = \begin{bmatrix} \frac{1}{2} & 0 \\ 0 & 0 \end{bmatrix}, \dots, \text{ and } M(u, 0) = \begin{bmatrix} 0 & 0 \\ \frac{2}{3} & 0 \end{bmatrix}, \; M(uu, 01) = \begin{bmatrix} 0 & 0 \\ 0 & \frac{1}{2} \end{bmatrix},$$

$$M(uuu, 010) = \begin{bmatrix} 0 & 0 \\ \frac{1}{2} & 0 \end{bmatrix}, \text{ and } M(uuuu, 0101) = \begin{bmatrix} 0 & 0 \\ 0 & \frac{1}{2} \end{bmatrix}, \dots \text{. It follows}$$

that $r^I(x, y) = S_\varepsilon(x, y)_A \; \forall x \in X^*$ *and* $\forall y \in Y^*$. *Now*

$$S_{\varepsilon_{q_1}^1}(u, 1) = \begin{bmatrix} 1 & 0 \end{bmatrix} \circ \begin{bmatrix} 0 & \frac{1}{2} \\ 0 & 0 \end{bmatrix} \circ \begin{bmatrix} 1 \\ 1 \end{bmatrix} = \frac{1}{2}$$

and

$$S_{\varepsilon_{q_2}^1}(u, 1) = \begin{bmatrix} 0 & 1 \end{bmatrix} \circ \begin{bmatrix} 0 & \frac{1}{2} \\ 0 & 0 \end{bmatrix} \circ \begin{bmatrix} 1 \\ 1 \end{bmatrix} = 0.$$

Hence it is not the case that $q_1 \sim q_2$. *Thus* A *is in reduced form.*

8.2 Equivalence of Fuzzy Automata: An Algebraic Approach

Let $A = (Q, X, Y, \mu)$ be a fuzzy automaton. Define a function $t : V(X^* \times Y^*) \rightarrow VQ$ as follows:

$$t(\Lambda, \Lambda) = \sum_{q \in Q} q,$$

$$t(x, y) = \begin{cases} \sum_{q_j \in Q} p_j(x, y) q_j & \text{if } |x| = |y| \\ 0 & \text{if } |x| \neq |y|, \end{cases}$$

where $p_j : X^* \times Y^* \rightarrow [0,1]$. It follows that t is a morphism of semimodules. We denote its corresponding matrix by M_t.

Construct the sequence $E_0 \subseteq E_1 \subseteq \dots \subseteq E$ of subsets of $E = X^* \times Y^*$ as follows:

$$E_0 = \{(\Lambda, \Lambda)\}$$
$$\vdots$$
$$E_i = E_{i-1} \cup \{(x, y) \mid x \in X^*, \; y \in Y^*, \; |x| = |y| = i\}.$$

Proposition 8.2.1 *The following assertions hold:*
  (1) VE_i *is a sub-semimodule of* VE_{i+1} *for* $i = 0, 1, \ldots$.
  (2) tVE_i *is a sub-semimodule of* tVE_{i+1} *for* $i = 0, 1, \ldots$.
  (3) If $tVE_i = tVE_{i+1}$, *then* $tVE_i = tVE_{i+k}$ *for* $k = 0, 1, \ldots$.
  (4) $tVE_k = tVE_{k+1}$ *for some* $k \in \mathbb{N}$.

Proof. (1) Since E_i is a set of generators (quasi-basis) for the semi-module VE_i, $i = 0, 1, \ldots$, and $E_0 \subseteq E_1 \subseteq \ldots \subseteq E$, we have $VE_0 \subseteq VE_1 \subseteq \ldots \subseteq VE$.
(2) The result follows here since t is a morphism.
(3) See [176].
(4) The result follows here since VQ is a semimodule that is noetherian and $VtE \subseteq VQ$. ∎

Theorem 8.2.2 *Let* (A, ε) *and* (A', ε') *be weakly initial fuzzy automata. Then* $(A, \varepsilon) \sim (A', \varepsilon')$ *if and only if* $\varepsilon \circ t = \varepsilon' \circ t'$.

Proof. If $|x| = |y|$ for $(x, y) \in X^* \times Y^*$, then by Definition 8.1.5,

$$S_\varepsilon(x, y)_A = S_{\varepsilon'}(x, y)_{A'}$$

and since $S_\varepsilon(x, y) = \varepsilon \circ (M(x, y) \circ P)$, we have that

$$\varepsilon \circ (M(x, y) \circ P) = \varepsilon' \circ (M'(x, y) \circ P).$$

This expression is equivalent to $\varepsilon \circ t(x, y) = \varepsilon' \circ t'(x, y)$ for each $(x, y) \in X^* \times Y^*$ such that $|x| = |y|$. If $|x| \neq |y|$, then by the definition of t, $t(x, y) = t'(x, y) = 0$, i.e., $\varepsilon \circ t = \varepsilon' \circ t'$.

Conversely, suppose that $\varepsilon \circ t = \varepsilon' \circ t'$. Clearly, $\varepsilon \circ t(x, y) = \varepsilon' \circ t'(x, y)$ for all $(x, y) \in X^* \times Y^*$. Hence $\varepsilon \circ t(x, y) = \varepsilon' \circ t'(x, y)$. However,

$$M_t(x, y) = \begin{cases} M(x, y) \circ P & \text{if } |x| = |y| \\ 0 & \text{if } |x| \neq |y|. \end{cases}$$

It follows that $\varepsilon \circ (M(x, y) \circ P) = \varepsilon' \circ (M'(x, y) \circ P)$, i.e., $S_\varepsilon(x, y)_A = S_{\varepsilon'}(x, y)_{A'}$ for all $(x, y) \in X^* \times Y^*$ such that $|x| = |y|$. ∎
A similar result is given in Chapters 2 and 3.

Corollary 8.2.3 *Let* A *be a fuzzy automaton. Then* $\varepsilon \sim \varepsilon'$ *if and only if* $S_\varepsilon(x, y)_A = S_{\varepsilon'}(x, y)_A$ *for all* $(x, y) \in X^* \times Y^*$ *such that* $|x| = |y| \leq n - 1$.

Proof. If $(A, \varepsilon) \sim (A, \varepsilon')$, then $S_\varepsilon(x, y)_A = S_{\varepsilon'}(x, y)_A$ for $|x| = |y| = 0, 1, \ldots$. Hence $|x| = |y| \leq n$. If $S_\varepsilon(x, y)_A = S_{\varepsilon'}(x, y)_A$ for all $(x, y) \in X^* \times Y^*$ such that $|x| = |y| \leq n - 1$, then by Proposition 8.2.1(4), it follows that $\varepsilon \circ (M(x, y) \circ P) = \varepsilon' \circ (M(x, y) \circ P)$. Thus $(A, \varepsilon) \sim (A, \varepsilon')$. ∎
This is the fuzzy interpretation of the well-known Carlyle theorem [29] for equivalence of stochastic automata.

Corollary 8.2.4 *Let A be a fuzzy automaton. Then the following statements are equivalent:*

(1) $\varepsilon \sim \varepsilon'$.

(2) $\varepsilon \circ M_t = \varepsilon' \circ M_t$. ∎

Corollary 8.2.5 *Let A be a fuzzy automaton. Then the following statements are equivalent:*

(1) $\varepsilon_{q_i}^1 \sim \varepsilon_{q_j}'^1$.

(2) The i-th and j-th rows in the matrix M_t are identical.

Proof. (1)$\Rightarrow$(2) Suppose that $\varepsilon_{q_i}^1 \sim \varepsilon_{q_j}'^1$. By Corollary 8.2.4, $\varepsilon_{q_i}^1 \circ M_t \sim \varepsilon_{q_j}'^1 \circ M_t$. However, by the construction of $\varepsilon_{q_i}^1$, the i-th and j-th rows in M_t are identical.

(2)$\Rightarrow$(1) The proof is straightforward. ∎

Lemma 8.2.6 *If $(A, \varepsilon) \sim (A, \varepsilon')$, then $dim(Im(t)) = dim(Im(t'))$.*

Proof. By Theorem 8.2.2, $\varepsilon \circ t \sim \varepsilon' \circ t' \Leftrightarrow \varepsilon \circ M_t \sim \varepsilon' \circ M_{t'}$. Thus it follows that $dim(Im(t)) = dim(Im(t'))$. ∎

Theorem 8.2.7 *Let A and A' be fuzzy automata. If ε is given, then the problem of finding ε', if it exists, such that $(A, \varepsilon) \sim (A', \varepsilon')$, is algorithmically decidable.* ∎

For a proof of Theorem 8.2.7, see the algorithm in Figure 8.1.

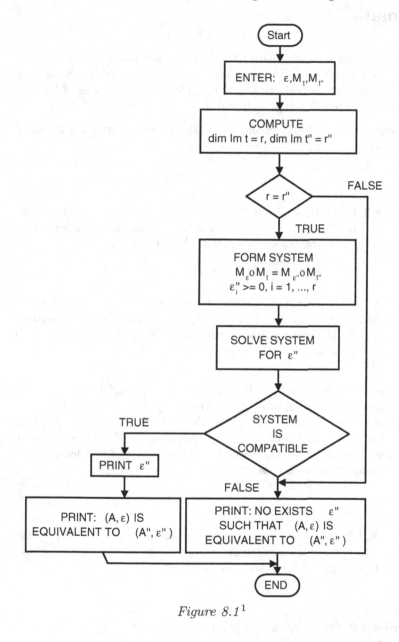

Figure 8.1[1]

It is stated in [231] that the computing program is not easy to realize since useful standard programs are missing.

---

[1] Figure 8.1 is reprinted from [231] with permission by Academic Press.

8.3   Reduction and Minimization of Fuzzy Automata

The reduction and minimization of fuzzy automata are a consequence of the theory of equivalence of fuzzy automata. They are of importance in applications. In this section, we give a completion of the ideas of Santos, which were presented in Chapters 2 and 3.

The next result is closely connected with the problem of reduction of fuzzy automata.

Theorem 8.3.1 *Let M_t be the matrix associated with the fuzzy automaton A. If M_t contains two identical rows, then there exist fuzzy automata A' and A'', with $|Q| - 1$ states each, such that $A \sim A'$ and $A \sim A''$.*

Proof. Suppose that the rows corresponding to the states q_i and q_j are identical in M_t. Let $Q' = Q \backslash \{q_i\}$ and $Q'' = Q \backslash \{q_j\}$. Then the corresponding matrix $M_{t'}$ (resp. $M_{t''}$) for the fuzzy automaton A' (resp. A'') is obtained from M_t by eliminating the i-th (resp. j-th) row. We show that $A \sim A'$ (resp. $A \sim A''$).

The equivalent state to $q \in Q$, $q_i \neq q \neq q_j$ is $q \in Q'$ (resp. $q \in Q''$) and vice versa since $\varepsilon_q^1 \circ M_t = \varepsilon_q'^1 \circ M_{t'}$ (resp. $\varepsilon_q^1 \circ M_t = \varepsilon_q''^1 \circ M_{t''}$). The equivalent state to $q = q_i, q_j \in Q$, respectively, is the state $q_i \in Q'$ (resp. $q_j \in Q''$). The state equivalent to $q_i \in Q'$ (resp. $q_j \in Q''$) is $q_i \in Q$ (resp. $q_j \in Q$). The proof of these conditions is a consequence of the definition of ε_q^1, of the construction of M_t. ∎

Corollary 8.3.2 *For every fuzzy automaton, there exists a reduced fuzzy automaton. All reduced fuzzy automata associated to a given fuzzy automaton have sets of states of the same cardinality.* ∎

Theorem 8.3.3 *For finite fuzzy automata, the relation of equivalence is decidable.* ∎

The block-scheme (Figure 8.2) of the algorithm proving the equivalence of two fuzzy automata A and A' is in fact the proof of Theorem 8.3.3.

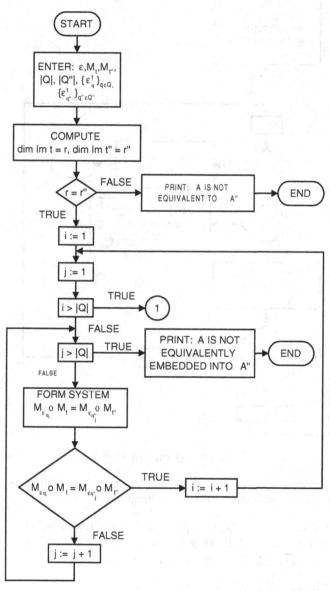

Figure 8.2[2]

---

[2] Figure 8.2 is reprinted from [231] with permission by Academic Press.

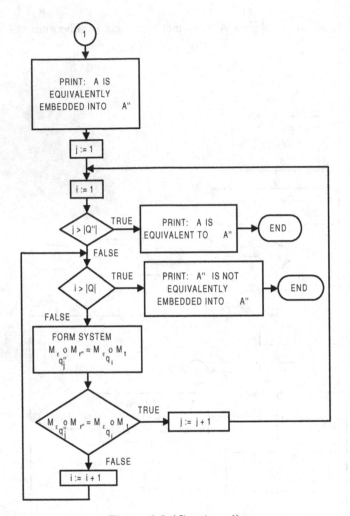

Figure 8.2 (Continued)

The following result pertains to the existence and explicit construction of a minimal fuzzy automaton of a given fuzzy automaton.

Theorem 8.3.4 *Let* $A = (Q, X, Y, \mu)$ *be a fuzzy automaton. If* $\varepsilon^1_{q_m} \sim \varepsilon^0_{q_m}$ *and* $\varepsilon^0_{q_m}$ *contains* $1 \in [0,1]$ *as a component, then there exists a fuzzy automaton* $\overline{A} = (\overline{Q}, X, Y, \overline{\mu})$, *with* $|Q| - 1$ *states such that* $A \approx \overline{A}$.

Proof. Suppose

$$\varepsilon^0_{q_m} = \begin{bmatrix} \varepsilon_1 & \varepsilon_2 & \cdots & \varepsilon_{m-1} & 0 & \varepsilon_{m+1} & \cdots & \varepsilon_n \end{bmatrix}$$

satisfies the condition of the theorem and

$$\varepsilon^1_{q_m} = \begin{bmatrix} 0 & 0 & \cdots & 0 & 1 & 0 & \cdots & 0 \end{bmatrix}$$

is equivalent to $\varepsilon_{q_m}^0$. We construct the fuzzy automaton $\overline{A} = (\overline{Q}, X, Y, \overline{\mu})$ as follows: Let $\overline{Q} = Q \backslash \{q_m\}$ and define $\overline{\mu} : V(\overline{Q} \times X) \to V(Y \times \overline{Q})$ as follows:

$$\overline{\mu}(q_j, x_i) = \sum_{r,k} \overline{a}_{ij}^{rk}(y_r, q_k),$$

where $\overline{a}_{ij}^{rk} = a_{ij}^{rk} \vee (\varepsilon_{q_k}^0 \wedge a_{ij}^{rm})$. We now show that the states $q_j, j \neq m$, with the same indices in A and $\overline{A}$ are equivalent. For the words with length $l = 1$, we have

$$\overline{a}_{ij}^r = \vee_{k \neq m} \{\overline{a}_{ij}^{rk}\} = \vee\{\vee_{k \neq m}\{a_{ij}^{rk}\}, \vee_{k \neq m}(\varepsilon_k^0 \wedge a_{ij}^{rm})\} = \vee_k\{a_{ij}^{rk}\} = a_{ij}^r.$$

In the last equality, recall that $\varepsilon_{q_k}^0$ contains $1 \in [0,1]$ as a component, i.e., $\vee_{k \neq m}(\varepsilon_k^0 \wedge a_{ij}^{rm}) = a_{ij}^{rm}$ since

$$\vee_{k \neq m}(\varepsilon_k^0 \wedge a_{ij}^{rm}) = (\vee_{k \neq m}\{\varepsilon_k^0\}) \wedge a_{ij}^{rm} = a_{ij}^{rm}.$$

Suppose the states $q_j, j \neq m$, are l-equivalent, i.e., for every $x \in X^*, y \in Y^*$ such that $|x| = |y| = l$, the following holds: $\overline{p}_j(x,y)_{\overline{A}} = p_j(x,y)_A$. By the hypothesis, $\vee_{k \neq m}(\varepsilon_k^0 \wedge p_k(x,y)) = p_m(x,y)$. Thus we have

$$\begin{aligned}
\overline{p}_j(x_i x, y_r y) &= \vee_{k \neq m}(\overline{a}_{ij}^{rk} \wedge p_k(x,y)) \\
&= (\vee_{k \neq m}(a_{ij}^{rk} \wedge p_k(x,y))) \vee \\
&\quad (a_{ij}^{rm} \wedge (\vee_{k \neq m}\{(\varepsilon_k^0 \wedge p_k(x,y))\})) \\
&= \vee_k(a_{ij}^{rk} \wedge p_k(x,y)) \\
&= p_j(x_i x, y_r y).
\end{aligned}$$

That is, the states with the same indices for automata A and $\overline{A}$ are $(l+1)$-equivalent and thus equivalent. For each

$$\overline{\varepsilon} = \begin{bmatrix} \varepsilon_1 & \varepsilon_2 & \cdots & \varepsilon_{m-1} & \varepsilon_{m+1} & \cdots & \varepsilon_n \end{bmatrix}$$

for $\overline{A}$, there exists an equivalent

$$\varepsilon = \begin{bmatrix} \varepsilon_1 & \varepsilon_2 & \cdots & \varepsilon_{m-1} & 0 & \varepsilon_{m+1} & \cdots & \varepsilon_n \end{bmatrix}$$

for the automaton A. For a given $\varepsilon = (\varepsilon_1, \varepsilon_2, \ldots, \varepsilon_n)$ for the automaton A, the corresponding equivalent $\overline{\varepsilon} = (\overline{\varepsilon}_1, \overline{\varepsilon}_2, \ldots, \overline{\varepsilon}_n)$ for $\overline{A}$ is defined by the equation $\overline{\varepsilon}_i = \varepsilon_i \vee (\varepsilon_m \wedge \varepsilon_i^0), i \neq m$, where ε_i^0 is the i-th component of the vector $\varepsilon_{q_m}^0$. Thus

$$\begin{aligned}
S_\varepsilon(x,y)_A &= \varepsilon \circ P(x,y) \\
&= \vee_i(\varepsilon_i \wedge p_i(x,y)) \\
&= (\vee_{i \neq m}\{(\varepsilon_i \wedge p_i(x,y))\}) \vee (\varepsilon_m \wedge (\varepsilon_i^0 \wedge p_i(x,y))) \\
&= \vee_{i \neq m}\{(\varepsilon_i \vee (\varepsilon_m \wedge \varepsilon_i^0)) \wedge \overline{p}_i(x,y)\} \\
&= \vee_{i \neq m}\{(\overline{\varepsilon}_i \wedge \overline{p}_i(x,y))\} \\
&= S_{\overline{\varepsilon}}(x,y)_{\overline{A}}.
\end{aligned}$$

∎

8.4   Minimal Fuzzy Finite State Automata

Fuzzy finite automata are used to design complex systems. For example, they are useful for a knowledge-based system designer since a knowledge-based system should solve a problem from fuzzy knowledge and should also provide the user with reasons for arriving at certain conclusions. A design tool is more valuable if there exist guidelines to assist the designer to come up with the best possible design. One of the major criteria for a best design is that it be minimal.

In this section, we show that a fuzzy finite automaton M_1 has an equivalent minimal fuzzy finite automaton M. We also show that M can be chosen so that if M_2 is equivalent to M_1, then M is a homomorphic image of M_2.

In Sections 8.1–8.4, we considered fuzzy automata (Q, X, Y, μ), where $\mu : Q \times X \times Y \times Q \to [0, 1]$. Hence μ served both as the state transition function and the output function. In this section, we consider fuzzy automata such that the state transition function and the output function are distinct. We ask the reader to compare the two approaches in the Exercises.

A **fuzzy finite automaton** (**ffa**) is a five-tuple $M = (Q, X, Y, \mu, \omega)$, where Q, X, and Y are finite nonempty sets, μ is a fuzzy subset of $Q \times X \times Q$, and ω is a fuzzy subset of $Q \times X \times Y$ such that the following conditions hold:

(1) for all $p, q \in Q$ if there exists $a \in X$ such that $\mu(p, a, q) > 0$, then there exists $b \in Y$ such that $\omega(p, a, b) > 0$,

(2) for all $p \in Q$, for all $a \in X$ if there exists $b \in Y$ such that $\omega(p, a, b) > 0$, then there exists $q \in Q$ such that $\mu(p, a, q) > 0$,

(3) for all $p \in Q$, $a \in X$, $\vee\{\mu(p, a, q) \mid q \in Q\} \geq \vee\{\omega(p, a, b) \mid b \in Y\}$.

The elements of Q are called states, and the elements of X and Y are called input and output symbols, respectively. μ is called the fuzzy transition function and ω is called the fuzzy output function.

The approach in this section differs from that of the previous sections in that the fuzzy transition function and fuzzy output functions are distinct in this section.

Definition 8.4.1 *Let $M = (Q, X, Y, \mu, \omega)$ be a ffa.*
(1) Define the fuzzy subset μ^* of $Q \times X^* \times Q$ as follows:

$$\mu^*(p, \Lambda, q) = \begin{cases} 1 & \text{if } p = q \\ 0 & \text{if } p \neq q, \end{cases}$$

$$\mu^*(p, xa, q) = \vee\{\mu(p, a, r) \wedge \mu^*(r, x, q) \mid r \in Q\},$$

for all $p, q \in Q$, $x \in X^*$, and $a \in X$.
(2) Define the fuzzy subsets ω^* of $Q \times X^* \times Y^*$ as follows:

$$\omega^*(p, x, y) = \begin{cases} 1 & \text{if } x = y = \Lambda \\ 0 & \text{if either } x \neq \Lambda \text{ and } y = \Lambda \text{ or } x = \Lambda \text{ and } y \neq \Lambda, \end{cases}$$

$$\omega^*(p, xa, yb) = \vee\{\omega(p, a, b) \wedge \mu(p, a, r) \wedge \ \omega^*(r, x, y)| \ r \in Q\},$$

for all $p \in Q$, $x \in X^*$, $a \in X$, $y \in Y^*$, and $b \in Y$.

Let $M = (Q, X, Y, \mu, \omega)$ be a ffa. Let $p, q \in Q$, $a \in X$, $b \in Y$. Then $\mu^*(p, a, q) = \mu^*(p, \Lambda a, q) = \vee\{\mu(p, a, r) \wedge \mu^*(r, \Lambda, q) \ | \ r \in Q\} = \mu(p, a, q) \wedge \mu^*(q, \Lambda, q)$ (since $\mu^*(r, \Lambda, q) = 0$ if $r \neq q$) $= \mu(p, a, q)$. Also $\omega^*(p, a, b) = \omega^*(p, \Lambda a, \Lambda b) = \vee\{\omega(p, a, b) \wedge \mu(p, a, r) \wedge \omega^*(r, \Lambda, \Lambda)| \ r \in Q\} = \vee\{\omega(p, a, b) \wedge \mu(p, a, r) \ | \ r \in Q\}$ (since $\omega^*(r, \Lambda, \Lambda) = 1$) $= \omega(p, a, b) \wedge (\vee\{\mu(p, a, r) \ | \ r \in Q\}) = \omega(p, a, b)$, where the latter equality holds by condition (3). We have thus proved the following result.

Theorem 8.4.2 Let $M = (Q, X, Y, \mu, \omega)$ be a ffa. Then
(1) $\mu = \mu^*|_{Q \times X \times Q}$,
(2) $\omega = \omega^*|_{Q \times X \times Y}$. ■

Lemma 8.4.3 Let $M = (Q, X, Y, \mu, \omega)$ be a ffa. Then for all $p, q \in Q$, $x, u \in X^*$,

$$\mu^*(p, xu, q) = \vee\{\mu^*(p, x, r) \wedge \mu^*(r, u, q) \ | \ r \in Q\}. \ ■$$

Lemma 8.4.4 Let $M = (Q, X, Y, \mu, \omega)$ be a ffa. For all $p \in Q$, $x \in X^*$, $y \in Y^*$, if $|x| \neq |y|$, then

$$\omega^*(p, x, y) = 0.$$

Proof. Let $p \in Q$, $x \in X^*$, $y \in Y^*$, and $|x| \neq |y|$. Suppose $|x| > |y|$ and let $|y| = n$. We prove the result by induction on n. If $n = 0$, then $y = \Lambda$ and $x \neq \Lambda$. Hence by definition $\omega^*(p, x, y) = 0$. Suppose the result is true for all $u \in X^*$, $v \in Y^*$ such that $|u| > |v|$ and $|v| = n - 1$. Suppose $n \geq 1$. Write $x = ua$, $y = vb$ where $u \in X^*$, $a \in X$, $v \in Y^*$, and $b \in Y$. Then $|u| > |v|$ and $|v| = n - 1$. Now by the induction hypothesis, for all $r \in Q$, $\omega^*(r, u, v) = 0$. Thus $\omega^*(p, x, y) = \omega^*(p, ua, vb) = \vee\{\omega(p, a, b) \wedge \mu(p, a, r) \wedge \omega^*(r, u, v)| \ r \in Q\} = 0$. Hence for all $p \in Q$, $x \in X^*$, $y \in Y^*$, if $|x| > |y|$, then $\omega^*(p, x, y) = 0$. Similarly, by induction, we can show that for all $p \in Q$, $x \in X^*$, $y \in Y^*$, if $|x| < |y|$, then $\omega^*(p, x, y) = 0$. ■

Theorem 8.4.5 Let $M = (Q, X, Y, \mu, \omega)$ be a ffa. Then statements $1(a)$ and $2(a)$ are equivalent as are $1(b)$ and $2(b)$.
(1) *(a)* for all $p, q \in Q$ if there exists $a \in X$ such that $\mu(p, a, q) > 0$, then there exists $b \in Y$ such that $\omega(p, a, b) > 0$;
  (b) for all $p \in Q$, for all $a \in X$ if there exists $b \in Y$ such that $\omega(p, a, b) > 0$, then there exists $q \in Q$ such that $\mu(p, a, q) > 0$.
(2) *(a)* for all $p, q \in Q$ if there exists $x \in X^*$ such that $\mu^*(p, x, q) > 0$, then there exists $y \in Y^*$ such that $\omega^*(p, x, y) > 0$;
  (b) for all $p \in Q$, for all $x \in X^*$ if there exists $y \in Y^*$ such that $\omega^*(p, x, y) > 0$, then there exists $q \in Q$ such that $\mu^*(p, x, q) > 0$.

Proof. (1)$\Rightarrow$(2): (a) Let $p, q \in Q$, and $x \in X^*$ be such that $\mu^*(p, x, q) > 0$. Let $|x| = n$. If $n = 0$, then $x = \Lambda$ and so $p = q$. Also $\Lambda \in Y^*$ and $\omega^*(p, \Lambda, \Lambda) > 0$. Suppose the result is true for all $u \in X^*$ such that $|u| = n - 1$. Let $x = ua$, where $a \in X$ and $u \in X^*$. Then $|u| = n - 1$. Now $\mu^*(p, x, q) = \mu^*(p, ua, q) = \vee\{\mu^*(p, a, r) \wedge \mu^*(r, u, q) \mid r \in Q\} > 0$ implies that $\mu^*(p, a, r) \wedge \mu^*(r, u, q) > 0$ for some $r \in Q$. Hence $\mu(p, a, r) = \mu^*(p, a, r) > 0$ and $\mu^*(r, u, q) > 0$. Hence by 1(a), there exists $b \in Y$ such that $\omega(p, a, b) > 0$. Also by the induction hypothesis, there exists $v \in Y^*$ such that $\omega^*(r, u, v) > 0$. Let $y = vb$. Then $\omega^*(p, x, y) = \omega^*(p, ua, vb) \geq \omega(p, a, b) \wedge \mu(p, a, r) \wedge \omega^*(r, u, v) > 0$. The result now follows by induction.

(b) Let $p \in Q$, $x \in X^*$, and suppose there exists $y \in Y^*$ such that $\omega^*(p, x, y) > 0$. Then by Lemma 8.4.3, $|x| = |y| = n$, say. If $n = 0$, then $x = y = \Lambda$. Also $\mu^*(p, \Lambda, p) > 0$. If $n = 1$, then $x \in X$ and $y \in Y$ and $\omega(p, x, y) = \omega^*(p, x, y) > 0$. Thus by 1(b), there exists $r \in Q$ such that $\mu(p, x, r) > 0$. Hence $\mu^*(p, x, r) = \mu(p, x, r) > 0$. Thus the result is true if $n = 0$ or $n = 1$. Suppose the result is true for all $u \in X^*$, $v \in Y^*$ such that $|u| = |v| = n - 1$. Suppose $n > 1$. Let $x = ua$, $y = vb$, where $a \in X$, $b \in Y$, $u \in X^*$, and $v \in Y^*$. Then $|u| = |v| = n - 1$. Now $\omega^*(p, x, y) = \omega^*(p, ua, vb) > 0$ implies $\omega(p, a, b) > 0$ and there exists $r \in Q$ such that $\mu(p, a, r) > 0$ and $\omega^*(p, u, v) > 0$. By the induction hypothesis, there exists $q \in Q$ such that $\mu^*(r, u, q) > 0$. Hence $\mu^*(p, x, q) = \mu^*(p, ua, q) \geq \mu(p, a, r) \wedge \mu^*(r, u, q) > 0$.

(2)$\Rightarrow$(1): Straightforward. $\blacksquare$

Definition 8.4.6 Let $M_i = (Q_i, X, Y, \mu_i, \omega_i)$ be a ffa, $i = 1, 2$. Let $q_i \in Q_i$, $i = 1, 2$. Then q_1 and q_2 are called **equivalent**, written $q_1 \approx_{Q_1 Q_2} q_2$, if and only if for all $x \in X^*$, $y \in Y^*$ ($\omega_1^*(q_1, x, y) > 0$ if and only if $\omega_2^*(q_2, x, y) > 0$). M_1 and M_2 are said to be equivalent, written $M_1 \approx M_2$, if for all $q_1 \in Q_1$ there exists $q_2 \in Q_2$ such that $q_1 \approx_{Q_1 Q_2} q_2$ and for all $q_2 \in Q_2$ there exists $q_1 \in Q_1$ such that $q_2 \approx_{Q_1 Q_2} q_1$.

If $M_1 = M_2 = M$, say, then we denote the relation $\approx_{Q_1 Q_2}$ by $\approx_{Q_1}$. Clearly, $\approx_{Q_1}$ is an equivalence relation.

Lemma 8.4.7 Let $M = (Q, X, Y, \mu, \omega)$ be a ffa. Let $p, q \in Q$, $x \in X^*$, $y \in Y^*$, $a \in X$, $b \in Y$. Suppose $\mu(q, a, p) > 0$, $\omega^*(p, x, y) > 0$, and $\omega(q, a, b) > 0$. Then $\omega^*(q, xa, yb) > 0$.

Proof. Now $\omega^*(q, xa, yb) \geq \omega(q, a, b) \wedge \mu(q, a, p) \wedge \omega^*(p, x, y) > 0$. $\blacksquare$

Remark 8.4.8 Let $M = (Q, X, Y, \mu, \omega)$ be a ffa. Let $p, q \in Q$, $x \in X^*$, $y \in Y^*$, $a \in X$, $b \in Y$. We assume for the rest of this section that $\omega^*(q, xa, yb) > 0$, $\mu(q, a, p) > 0$, and $\omega(q, a, b) > 0$ implies $\omega^*(p, x, y) > 0$.

Proposition 8.4.9 Let $M_i = (Q_i, X, Y, \mu_i, \omega_i)$ be a ffa $i = 1, 2$. Let $q_i, p_i \in Q_i$, $i = 1, 2$. Suppose $q_1 \approx_{Q_1 Q_2} q_2$ and there exists $a \in X$ such that $\mu_1(q_1, a, p_1) > 0$ and $\mu_2(q_2, a, p_2) > 0$. Then $p_1 \approx_{Q_1 Q_2} p_2$.

Proof. Let $x \in X^*$, $y \in Y^*$. Suppose $\omega_1^*(p_1, x, y) > 0$. By hypothesis, there exists $a \in X$ such that $\mu_1(q_1, a, p_1) > 0$ and $\mu_2(q_2, a, p_2) > 0$. Since M_1 is a ffa, there exists $b \in Y$ such that $\omega_1(q_1, a, b) > 0$. Thus by Lemma 8.4.7, $\omega_1^*(q_1, xa, yb) > 0$. Hence $\omega_2^*(q_2, xa, yb) > 0$ and $\omega_2(q_2, a, b) > 0$ since $q_1 \approx_{Q_1 Q_2} q_2$. Now $\omega_2^*(q_2, xa, yb) > 0$, $\mu_2(q_2, a, p_2) > 0$, and $\omega_2(q_2, a, b) > 0$. Hence by Remark 8.4.8, $\omega_2^*(p_2, x, y) > 0$. Similarly we can show that $\omega_2^*(p_2, x, y) > 0$ implies $\omega_1^*(p_1, x, y) > 0$. Thus $p_1 \approx_{Q_1 Q_2} p_2$. ∎

Remark 8.4.10 Let $M_i = (Q_i, X, Y, \mu_i, \omega_i)$ be ffa, $i = 1, 2, 3$. Let $q_i, p_i \in Q_i$, $i = 1, 2, 3$.

(1) Suppose $q_1 \approx_{Q_1 Q_2} q_2$ and $q_2 \approx_{Q_2 Q_3} q_3$. Then for all $x \in X^*$, $y \in Y^*$, $\omega_1^*(q_1, x, y) > 0$ if and only if $\omega_2^*(q_2, x, y) > 0$ if and only if $\omega_3^*(q_3, x, y) > 0$. Hence $q_1 \approx_{Q_1 Q_3} q_3$.

(2) Let $p_1 \approx_{Q_1} q_1$ and $q_1 \approx_{Q_1 Q_2} q_2$. Then for all $x \in X^*$, $y \in Y^*$, $\omega_1^*(p_1, x, y) > 0$ if and only if $\omega_1^*(q_1, x, y) > 0$ if and only if $\omega_2^*(q_2, x, y) > 0$. Hence $p_1 \approx_{Q_1 Q_2} q_2$.

(3) Let $p_1 \approx_{Q_1 Q_2} q_2$ and $q_1 \approx_{Q_1 Q_2} q_2$. Then for all $x \in X^*$, $y \in Y^*$, $\omega_1^*(p_1, x, y) > 0$ if and only if $\omega_2^*(q_2, x, y) > 0$ if and only if $\omega_1^*(q_1, x, y) > 0$. Hence $p_1 \approx_{Q_1} q_1$.

From Remark 8.4.10, it follows that we can use the same symbol $\approx$ for states to be equivalent, whether states are in the same set or different sets.

Definition 8.4.11 Let $M_i = (Q_i, X, Y, \mu_i, \omega_i)$ be ffa, $i = 1, 2$. Let $f : Q_1 \rightarrow Q_2$ be a function. f is called a **homomorphism** of M_1 into M_2 if

(1) for all $q_1 \in Q_1, a \in X$, $b \in Y$, $\omega_1(q_1, a, b) > 0$ if and only if $\omega_2(f(q_1), a, b) > 0$,

(2) (a) for all $p_1, q_1 \in Q_1$, $a \in X$ if $\mu_1(q_1, a, p_1) > 0$, then

$$\mu_2(f(q_1), a, f(p_1)) > 0,$$

(b) for all $p_1, q_1 \in Q_1$, $a \in X$ if $\mu_2(f(q_1), a, f(p_1)) > 0$, then there exists $r_1 \in Q_1$ such that $\mu_1(q_1, a, r_1) > 0$ and $f(p_1) = f(r_1)$.

M_2 is called a **homomorphic image** of M_1, if there exists a homomorphism $f : M_1 \rightarrow M_2$ such that f is onto.

Definition 8.4.12 Let $M = (Q, X, Y, \mu, \omega)$ be a ffa. M is said to be **minimal** if for all $p, q \in Q$, $p \approx q$ implies $p = q$.

Theorem 8.4.13 Let $M_1 = (Q, X, Y, \mu_1, \omega_1)$ be a ffa. Then there exists a minimal ffa M such that $M_1 \approx M$. Furthermore, M can be chosen so that if M_2 is any ffa such that $M_2 \approx M_1$, then M is a homomorphic image of M_2.

Proof. Set $Q = \{[q] \mid q \in Q_1\}$, where $[q]$ is the equivalence class containing q and is induced by the equivalence relation $\approx_{Q_1}$. Define μ : $Q \times X \times Q \to [0,1]$ and $\omega : Q \times X \times Y \to [0,1]$ by

$$\mu([q], a, [p]) = \vee\{\mu_1(s, a, t) \mid s, t \in Q_1,\ s \approx q,\ t \approx p\},$$

and

$$\omega([q], a, b) = \vee\{\omega_1(s, a, b) \mid s \in Q_1,\ s \approx q\}$$

for all $[q], [p] \in Q$, $a \in X$, and $b \in Y$. Let $([q], a, [p]) = ([q'], a, [p'])$. Then $q \approx q'$ and $p \approx p'$. Hence $\vee\{\mu_1(s, a, t) \mid s, t \in Q_1, s \approx q, t \approx p\} = \vee\{\mu_1(s', a, t') \mid s', t' \in Q_1, s' \approx q', t \approx p'\}$ since $\approx$ is an equivalence relation on Q_1. Thus μ is well defined. Similarly ω is well defined. Let $[q], [p] \in Q$. Suppose that there exists $a \in X$ such that $\mu([q], a, [p]) > 0$. Then there exists $s, t \in Q_1$, $s \approx q$, $t \approx p$ such that $\mu_1(s, a, t) > 0$. Since M_1 is a ffa, there exists $b \in Y$ such that $\omega_1(s, a, b) > 0$. Hence $\omega([q], a, b) \geq \omega_1(s, a, b) > 0$. Now suppose $[q] \in Q$, $a \in X$, and there exists $b \in Y$ such that $\omega([q], a, b) > 0$. This implies $\omega_1(s, a, b) > 0$ for some $s \in Q_1$, $s \approx q$. Since M_1 is a ffa, there exists $t \in Q_1$ such that $\mu_1(s, a, t) > 0$. Now $\mu([q], a, [t]) \geq \mu_1(s, a, t) > 0$.

Let $[q] \in Q$, $a \in X$. Since $\approx$ is an equivalence relation on Q_1, we can write $Q_1 = \cup_{i=1}^{k}[p_i]$, where $Q = \{[p_1], [p_2], \dots, [p_k]\}$ and the $[p_i]$ are distinct. Now $\vee\{\mu([q], a, [p]) \mid [p] \in Q\} = \vee_{i=1}^{k}\mu([q], a, [p_i]) = \vee_{i=1}^{k}(\vee\{\mu_1(s, a, t) \mid s \approx q, t \approx p_i\}$. Also, $\vee\{\omega([q], a, b) \mid b \in Y\} = \vee\{\vee\{\omega_1(s, a, b) \mid s \approx q\} \mid b \in Y\} = \vee\{\vee\{\omega_1(s, a, b) \mid b \in Y\} \mid s \approx q\} \leq \vee\{\vee\{\mu_1(s, a, r) \mid r \in Q_1\} \mid s \approx q\} = \vee(\vee\{\mu_1(s, a, r) \mid s \approx q, r \in \cup_{i=1}^{k}[p_i]\}) = \vee_{i=1}^{k}(\vee\{\mu_1(s, a, r) \mid s \approx q, r \in [p_i]\}) = \vee_{i=1}^{k}(\vee\{\mu_1(s, a, r) \mid s \approx q, r \approx p_i\}) = \vee\{\mu([q], a, [p]) \mid [p] \in Q\}$. Consequently, $M = (Q, X, Y, \mu, \omega)$ is a ffa.

First we show that for all $[q] \in Q$, $x \in X^*$, $y \in Y^*$, $\omega^*([q], x, y) > 0$ if and only if $\omega_1^*(q, x, y) > 0$. Let $[q] \in Q$, $x \in X^*$, $y \in Y^*$, and $\omega_1^*(q, x, y) > 0$. Then $|x| = |y| = n$, say. If $n = 0$, then $x = y = \Lambda$. Thus $\omega^*([q], x, y) = \omega^*([q], \Lambda, \Lambda) = 1 > 0$. If $n = 1$, then $x \in X$ and $y \in Y$. Hence $\omega_1(q, x, y) = \omega_1^*(q, x, y) > 0$. Hence $\omega^*([q], x, y) = \omega([q], x, y) = \vee\{\omega_1(s, x, y) \mid s \in Q_1, s \approx q\} \geq \omega_1(q, x, y) > 0$. Thus if $n = 0$ or $n = 1$, then $\omega_1^*(q, x, y) > 0$ implies $\omega^*([q], x, y) > 0$. Suppose the result is true for all $u \in X^*$, $v \in Y^*$ such that $|u| = |v| = n - 1$. Let $n > 1$. Now $x = ua$, $y = vb$ for some $a \in X$, $b \in Y$, $u \in X^*$, $v \in Y^*$. Then $|u| = |v| = n - 1$. Thus $\omega_1^*(q, x, y) = \omega_1^*(q, ua, vb) = \vee\{\omega_1(q, a, b) \wedge \mu_1(q, a, r) \wedge \omega_1^*(r, u, v) \mid r \in Q_1\} > 0$ implies that there exists $r \in Q$ such that $\omega_1(q, a, b) > 0$, $\mu_1(q, a, r) > 0$, $\omega_1^*(r, u, v) > 0$. By the $n = 1$ case and the induction hypothesis, $\omega([q], a, b) > 0$ and $\omega^*([r], u, v) > 0$. Also $\mu([q], a, [r]) \geq \mu_1(q, a, r) > 0$. Hence $\omega^*([q], x, y) = \omega^*([q], ua, vb) \geq \omega([q], a, b) \wedge \mu([q], a, [r]) \wedge \omega^*([r], u, v) > 0$. The result now follows by induction.

Now suppose $\omega^*([q], x, y) > 0$ for some $[q] \in Q$, $x \in X^*$, $y \in Y^*$. Then $|x| = |y| = n$, say. If $n = 0$, then $x = y = \Lambda$ and so $\omega_1^*(q, x, y) = 1 > 0$. If $n = 1$, then $x \in X$ and $y \in Y$ and so $\omega([q], x, y) = \omega^*([q], x, y) > 0$.

Thus there exists $s \in Q_1$, $s \approx q$ such that $\omega_1(s, x, y) > 0$. Since $s \approx q$ and $\omega_1(s, x, y) > 0$, we must have $\omega_1(q, x, y) > 0$, i.e., $\omega_1^*(q, x, y) > 0$. Thus the result is true for $n = 0$ and $n = 1$. Make the induction hypothesis that the result is true for all $u \in X^*$, $v \in Y^*$ such that $|u| = |v| = n - 1$. Let $n > 1$. Now $x = ua$, $y = vb$ for some $a \in X$, $b \in Y$, $u \in X^*$, $v \in Y^*$. Hence $|u| = |v| = n - 1$. Thus $\omega^*([q], x, y) = \omega^*([q], ua, vb) = \vee\{\omega([q], a, b) \wedge \mu([q], a, [r]) \wedge \omega^*([r], u, v) \mid [r] \in Q\} > 0$. This implies that there exists $[r] \in Q$ such that $\omega([q], a, b) > 0$, $\mu([q], a, [r]) > 0$, $\omega^*([r], u, v) > 0$. Now $\mu([q], a, [r]) > 0$ implies $\mu_1(s, a, t) > 0$ for some $s, t \in Q_1$, $s \approx q$, $t \approx r$. Since M_1 is a ffa, there exists $d \in Y$ such that $\omega_1(s, a, d) > 0$. Thus $\omega_1(q, a, d) > 0$ since $s \approx q$. This implies that there exists $t' \in Q_1$ such that $\mu_1(q, a, t') > 0$ since M_1 is a ffa. Now since $s \approx q$, $\mu_1(s, a, t) > 0$ and $\mu_1(q, a, t') > 0$, we have that $t \approx t'$ by Proposition 8.4.9. Thus $t' \approx r$. Now $\omega([q], a, b) > 0$ implies $\omega_1(q, a, b) > 0$ by the $n = 1$ case and $\omega^*([r], u, v) > 0$ implies $\omega_1^*(r, u, v) > 0$ by the induction hypothesis. Since $r \approx t'$, $\omega_1^*(t', u, v) > 0$. Hence $\omega_1^*(q, x, y) = \omega_1^*(q, ua, vb) \geq \omega_1(q, a, b) \wedge \mu_1(q, a, t') \wedge \omega_1^*(t', u, v) > 0$. This proves our claim.

Let $[q], [p] \in Q$ and $[q] \approx [p]$. Let $x \in X^*$ and $y \in Y^*$. Now $\omega_1^*(q, x, y) > 0$ if and only if $\omega^*([q], x, y) > 0$ if and only if $\omega^*([p], x, y) > 0$ (since $[q] \approx [p]$) if and only if $\omega_1^*(p, x, y) > 0$. Thus $q \approx p$ and so $[q] = [p]$. Hence M is minimal. Since for all $q \in Q_1$, $x \in X^*$, $y \in Y^*$, $\omega_1^*(q, x, y) > 0$ if and only if $\omega^*([q], x, y) > 0$, $q \approx [q]$. Thus $M_1 \approx M$.

Let $M_2 = (Q_2, X, Y, \mu_2, \omega_2)$ be a ffa such that $M_2 \approx M_1$. Define $f : Q_2 \to Q$ as follows: for all $q_2 \in Q_2$ there exists $q_1 \in Q_1$ such that $q_2 \approx q_1$. Define $f(q_2) = [q_1]$. Suppose $q_2, p_2 \in Q_2$ and $q_2 = p_2$. Then there exists $q_1, p_1 \in Q_1$ such that $q_2 \approx q_1$ and $p_2 \approx p_1$. Thus by Remark 8.4.10, $q_1 \approx p_1$ and so $[q_1] = [p_1]$. Hence f is well defined. Now for all $[q_1] \in Q$, $q_1 \approx [q_1]$. There exists $q_2 \in Q_2$ such that $q_2 \approx q_1$. Now $f(q_2) = [q_1]$. Hence f is onto. We now show that f is a homomorphism.

(1) Let $q_2 \in Q_2$, $a \in X, b \in Y$. Suppose $\omega_2(q_2, a, b) > 0$. Since $M_2 \approx M_1$, there exists $q_1 \in Q_1$ such that $q_2 \approx q_1$. Then $\omega_1(q_1, a, b) > 0$. Thus $\omega(f(q_2), a, b) = \omega([q_1], a, b) \geq \omega_1(q_1, a, b) > 0$. Now suppose $\omega(f(q_2), a, b) > 0$. Let $f(q_2) = [q_1]$. Then $q_2 \approx q_1$. Since $\omega([q_1], a, b) = \omega(f(q_2), a, b) > 0$, $\omega_1(s, a, b) > 0$ for some $s \in Q_1$, $s \approx q_1$. Since $s \approx q_1$, $\omega_1(q_1, a, b) > 0$. This implies that $\omega_2(q_2, a, b) > 0$ since $q_2 \approx q_1$. Hence $\omega_2(q_2, a, b) > 0$ if and only if $\omega(f(q_2), a, b) > 0$ for all $q_2 \in Q_2$, $a \in X, b \in Y$.

(2) First suppose $\mu_2(q_2, a, p_2) > 0$ for some $q_2, p_2 \in Q_2$ and $a \in X$. There exists $q_1, p_1 \in Q_1$ such that $q_2 \approx q_1$ and $p_2 \approx p_1$. Thus $f(q_2) = [q_1]$ and $f(p_2) = [p_1]$. Since $\mu_2(q_2, a, p_2) > 0$ and M_2 is a ffa, there exists $b \in Y$ such that $\omega_2(q_2, a, b) > 0$. This implies that $\omega_1(q_1, a, b) > 0$ since $q_2 \approx q_1$. Since M_1 is a ffa, there exists $t_1 \in Q_1$ such that $\mu_1(q_1, a, t_1) > 0$. Since $q_2 \approx q_1, \mu_1(q_1, a, t_1) > 0$ and $\mu_2(q_2, a, p_2) > 0$, we have that $t_1 \approx p_2$ by Proposition 8.4.9. By Remark 8.4.10, $t_1 \approx p_1$. Hence $\mu(f(q_2), a, f(p_2)) = \mu([q_1], a, [p_1]) \geq \mu_1(q_1, a, t_1) > 0$. Now suppose $\mu(f(q_2), a, f(p_2)) > 0$ for some $q_2, p_2 \in Q_2$ and $a \in X$. There exists $q_1, p_1 \in Q_1$ such that $q_2 \approx q_1$ and

$p_2 \approx p_1$, $f(q_2) = [q_1]$ and $f(p_2) = [p_1]$. Thus $\mu([q_1], a, [p_1]) > 0$. This implies $\mu_1(s_1, a, t_1) > 0$ for some $s_1, t_1 \in Q_1$, $s_1 \approx q_1$, and $t_1 \approx p_1$. Since M_1 is a ffa, there exists $b \in Y$ such that $\omega_1(s_1, a, b) > 0$. This implies $\omega_1(q_1, a, b) > 0$ and so $\omega_2(q_2, a, b) > 0$ since $q_2 \approx q_1$. Since M_2 is a ffa, there exists $r_2 \in Q_2$ such that $\mu_2(q_2, a, r_2) > 0$. Finally we show that $f(r_2) = f(p_2)$. Since $s_1 \approx q_1$ and $q_2 \approx q_1$, $s_1 \approx q_2$ by Remark 8.4.10. Hence by Proposition 8.4.9, $t_1 \approx r_2$. Once again by Remark 8.4.10, $p_1 \approx r_2$. Hence $f(r_2) = [p_1] = f(p_2)$. ∎

8.5   Behavior, Reduction, and Minimization of Finite L-Automata

In the remainder of the chapter, we consider [174].

Issues concerning the behavior, equivalence, reduction, and minimization have been completely studied for deterministic, non-deterministic, and stochastic automata [220]. The approach given in Chapter 2 and Sections 8.1–8.3 ([203, 231] also) for fuzzy automata is useful if it is known how to solve systems of linear equations over the bounded chain ([0, 1], max, min). It is established in [203] that if the system is compatible, the solution is among the m-fold variations with repetitions over n elements. Since the number of these variations is m^n, the time complexity function of a search manner algorithm is exponential. Following [170], a polynomial time algorithm for solving systems of linear equations over $\mathbb{L}$ is given. This result is purely algebraic. It allows one to compute the behavior matrix, to study approximate ($\varepsilon-$) equivalence, ε-reduction, ε-minimization, and to prove their algorithmical decidability for each finite L-automaton.

Moreover, the concept of computational complexity and algorithmical decidability as described in [220] and the properties of the chain according to [125] are used here. The terminology for automata theory is as in Sections 8.1–8.3.

8.6   Matrices over a Bounded Chain

We write $\mathbb{L}$ for the bounded chain $\mathbb{L} = (L, \vee, \wedge, 0, 1)$ over the linearly ordered set L with upper and lower bounds 1 and 0, respectively.

Let I, J be index sets and K a finite index set. Let $M_{I \times J}$ denote the matrix $[m_{ij}]$, where $m_{ij} \in L$ for each $(i, j) \in I \times J$. Then $M_{I \times J}$ is referred to as a matrix over $\mathbb{L}$.

The matrix $M_{I \times J} = [m_{ij}]$ is the product of $A_{I \times K} = [a_{ik}]$ and $B_{K \times J} = [b_{kj}]$, if $m_{ij} = \vee_{k \in K}(a_{ik} \wedge b_{kj})$ for each $i \in I$ and each $j \in J$.

Let $f : L \longrightarrow [0, 1]$ be an injective function such that $x \leqslant y \implies f(x) \leqslant f(y)$. The *distance* with respect to f between x and y for $x, y \in L$ is defined as follows:

$$d_f(x, y) = |f(x) - f(y)|.$$

Let $|x - y| = d_f(x, y)$ for $x, y \in L$.

In the remainder of the chapter, we assume f to be fixed.

Let $A_{I \times J} = [a_{ij}]$ and $B_{I \times J} = [b_{ij}]$ be matrices over $\mathbb{L}$ and $\varepsilon \in [0, 1]$ be fixed. We say that the i-th row in A is ε-**close** to the k-th row in B if $|a_{is} - b_{ks}| \leqslant \varepsilon$ for each $s \in J$. We also say that A and B are ε-**close**, written $d(A, B) \leqslant \varepsilon$, if $|a_{ij} - b_{ij}| \leqslant \varepsilon$ for each $i \in I$ and each $j \in J$.

Proposition 8.6.1 *If $A_{n \times 1} = [a_i]$ and $B_{n \times 1} = [b_i]$ are ε-close, then*

$$|\vee\{a_i \mid i \in I\} - \vee\{b_i \mid i \in I\}| \leqslant \varepsilon.$$

Proof. Let $\vee\{a_i \mid i \in I\} = a_k$ and $\vee\{b_i \mid i \in I\} = b_r$. Then $a_k - \varepsilon \leqslant b_k \leqslant b_r \leqslant a_r + \varepsilon \leqslant a_k + \varepsilon$ and so $a_k - \varepsilon \leqslant b_r \leqslant a_k + \varepsilon$. Thus $-\varepsilon \leqslant b_r - a_k \leqslant \varepsilon$ and $|a_k - b_r| \leqslant \varepsilon$. Hence $|\vee\{a_i \mid i \in I\} - \vee\{b_i \mid i \in I\}| = |a_k - b_r| \leqslant \varepsilon$. ∎

The result in Proposition 8.6.1 is valid for the transposed matrices A^T and B^T.

Proposition 8.6.2 *If the products below make sense, then for $A_{I \times J}$, $B_{I \times J}$ the following statements hold:*

(1) $d(A, B) \leqslant \varepsilon \Longrightarrow d(CA, CB) \leqslant \varepsilon$;
(2) $d(A, B) \leqslant \varepsilon \Longrightarrow d(AT, BT) \leqslant \varepsilon$;
(3) $d(A, B) \leqslant \varepsilon \Longrightarrow d(CAT, CBT) \leqslant \varepsilon$.

Proof. Let $A = [a_{ij}]$, $B = [b_{ij}]$, $C_{P \times I} = [c_{ij}]$, $CA = [h_{ij}]$, $CB = [u_{ij}]$. Let h and u denote, respectively, the vectors

$$h = h(i, j) = (h_k(i, j)), k \in I; \quad u = u(i, j) = (u_k(i, j)), k \in I,$$

where $h_k(i, j) = c_{ik} \wedge a_{kj}$ and $u_k(i, j) = c_{ik} \wedge b_{kj}$. Then

$$h_{ij} = \vee_k(c_{ik} \wedge a_{kj}) = \vee_k h_k(i, j), \quad u_{ij} = \vee_k(c_{ik} \wedge b_{kj}) = \vee_k u_k(i, j).$$

From $|a_{ij} - b_{ij}| \leqslant \varepsilon$, it follows that $|h - u| \leqslant \varepsilon$ since

$$|h_k(i, j) - u_k(i, j)| = |c_{ik} \wedge a_{kj} - c_{ik} \wedge b_{kj}| = \begin{cases} |c_{ik} - c_{ik}| = 0 \text{ or} \\ |a_{kj} - b_{kj}| \leqslant \varepsilon \text{ or} \\ |c_{ik} - b_{kj}| \text{ or} \\ |a_{kj} - c_{ik}|. \end{cases}$$

If $|h_k(i, j) - u_k(i, j)| = |c_{ik} - b_{kj}|$, then $c_{ik} \wedge a_{kj} = c_{ik}$ implies $c_{ik} \leqslant a_{kj}$ and $c_{ik} \wedge b_{kj} = b_{kj}$ implies $b_{kj} \leqslant c_{ik}$, i.e., in this case $b_{kj} \leqslant c_{ik} \leqslant a_{kj}$ and $|c_{ik} - b_{kj}| \leqslant |a_{kj} - b_{kj}| \leqslant \varepsilon$. If $|h_k(i, j) - u_k(i, j)| = |a_{kj} - c_{ik}|$ the proof is similar.

By Proposition 8.6.1, $d(h, u) \leqslant \varepsilon$ implies $|\vee_k(h_k) - \vee_k(u_k)| \leqslant \varepsilon$, which is equivalent to $|h(i, j) - u(i, j)| \leqslant \varepsilon$ and $|h_{ij} - u_{ij}| \leqslant \varepsilon$ for each $i \in I$, $j \in J$. The last inequality means $d(CA, CB) \leqslant \varepsilon$.

(2) The proof is similar to that of (1).

(3) The proof follows from (1) and (2). ∎

We now see that the relation ε-closeness of matrices over $\mathbb{L}$ is invariant.

8.7 Systems of Linear Equivalences over a Bounded Chain

We now review the main results in [170] on computing a solution of a system of linear equations over $\mathbb{L}$.

We assume all matrices are over the bounded chain $\mathbb{L}$. Let I and J be finite sets with $|I| = m \in \mathbb{N}$ and $|J| = n \in \mathbb{N}$. By $A \circ X = B$, we mean the system:

$$
\begin{aligned}
(a_{11} \wedge x_1) \vee (a_{12} \wedge x_2) \vee \cdots \vee (a_{1n} \wedge x_n) &= b_1 \\
&\vdots \\
(a_{m1} \wedge x_1) \vee (a_{m2} \wedge x_2) \vee \cdots \vee (a_{mn} \wedge x_n) &= b_m
\end{aligned}
\tag{8.1}
$$

with coefficients $A_{I \times J} = [a_{ij}]$, unknowns $X_{J \times \{1\}} = [x_j]$, and constants $B_{I \times \{1\}} = [b_i]$. We assume that $b_1 \geqslant b_2 \geqslant \cdots \geqslant b_m$.

A matrix $X^0 = [x_j^0]$ is called a (**point**) **solution** of (8.1) if $A \circ X^0 = B$ holds. The system (8.1) is said to be **solvable** if it possesses at least one solution; otherwise it is said to be **unsolvable**.

In order to propose a polynomial time algorithm for solving the system (8.1), we introduce the following symbolic matrix $\mathcal{B} = [b_{ij}]$:

$$
b_{ij} = \begin{cases}
S & \text{if } a_{ij} < b_i, \\
E & \text{if } a_{ij} = b_i, \\
G & \text{if } a_{ij} > b_i.
\end{cases}
$$

We are interested in determining if there exists $x_j \in L$ such that $a_{ij} \wedge x_j \leqslant b_i$. We mark the i-th equation in (8.1) in a marker vector IND if $a_{ij} \wedge x_j = b_i$ holds.

Theorem 8.7.1 *[170] Consider the system $A \circ X = B$. Then the following statements hold.*

(1) If k is the greatest number of the row with a G-type coefficient in the jth column of $\mathcal{B}$, then $x_j = b_k$ implies $a_{ij} \wedge x_j = b_i$ for $i = k$ and for each $i > k$ with $a_{ij} = b_i$ as well as for each $i' < k$ with $a_{i'j} \geqslant b_{i'} = b_i$.

(2) If the j-th column in $\mathcal{B}$ does not contain a G-type coefficient and r is the smallest number of the row with E-type coefficient in the j-th column, then $x_j = b_r$ means $a_{ij} \wedge x_j = b_i$ for each $i \geqslant r$ with $a_{ij} = b_i$.

(3) If the j-th column in $\mathcal{B}$ does not contain either a G-type or an E-type coefficient, then $a_{ij} \wedge x_j < b_i$ for each $x_j \in L$. ∎

The implementation of Theorem 8.7.1 gives the following.

Algorithm 8.7.2 *[170] For computing a solution of the system (8.1):*
1. *Enter the matrices A, B. Form the matrix $\mathcal{B}$. Erase IND.*
2. *$j = 0$.*
3. *$j = j + 1$.*
4. *If $j > n$ go to 8.*
5. *If the j-th column in $\mathcal{B}$ does not contain a G-type coefficient, then go to 6. Otherwise $x_j = b_k$, put marks in IND for $i = k$, for each $i > k$ with $a_{ij} = b_i$ and for each $i < k$ if $a_{ij} \geqslant b_i = b_k$. Go to 3.*
6. *If the j-th column does not contain an E-type coefficient then go to 7. Otherwise $x_j = b_r$, put marks in IND for $i = r$ and for each $i > r$ with $a_{ij} = b_i$. Go to 3.*
7. *$x_j = 1$. Go to 3.*
8. *If there exists at least one unmarked row in IND then the system is unsolvable. If all rows in IND are marked, then the system is solvable and the point solution is determined in steps 5,6, and 7.*

Theorem 8.7.3 *[170] The following problems are algorithmically decidable in polynomial time for the system (8.1):*
(1) whether the system is solvable or not;
(2) computing a point solution if the system is solvable;
(3) obtaining the numbers of the contradictory equations if the system is unsolvable. ∎

An n-tuple $(X_1, \ldots, X_n)$ with $X_i \subset L$ for $i = 1, \ldots, n$ is an interval solution of (8.1) if each $(x_1, \ldots, x_n)$ with $x_i \in X_i$ is a point solution of (8.1) and $(X_1, \ldots, X_n)$ is maximal with respect to this property.

Using Algorithm 8.7.2, the interval solutions of (8.1) can be found. For $j, j' \in J$, define $j \sim j'$ if $b_{ij} = b_{ij'}$ for each $i \in I$. Then $\sim$ is an equivalence relation on J. Let $[j] = \{j' \mid j' \in J \text{ and } j \sim j'\}$.

Algorithm 8.7.4 *[170] For obtaining the interval solutions of the system (8.1), if (8.1) is solvable:*
1. *Compute the point solution by Algorithm 8.7.2.*
2. *Obtain the equivalence classes $[j]$ for J.*
3. *For each $j \in J$:*
(i) if the j-th column in $\mathcal{B}$ does not contain a G-type coefficient, then go to 3(iv);
(ii) if $\|[j]\| = 1$ then $X_j = \{b_k\}$; go to 3 for the next j;
(iii) for each $j' \in [j]$ form $X_{j'}$ as follows:

$$X_{j'} = \begin{cases} \{b_k\} & \text{if } j = j', \\ [0, b_k] & \text{if } j \neq j'; \end{cases}$$

go to 3 for the next j;

(iv) if the j-th column does not contain an E-type coefficient, then go to 3(vii),

(v) if $|[j]| = 1$, then $X_j = [b_r, 1]$; go to 3 for the next j;

(vi) for each $j' \in [j]$ form $X_{j'}$ according to the rule

$$X_{j'} = \begin{cases} [b_r, 1] & \text{if } j = j', \\ L & \text{if } j \neq j'; \end{cases}$$

go to 3 for the next j;

(vii) if the j-th column contains only S-type coefficients, then $X_j = L$ for each $j \in [j]$; go to 3 for the next j.

Theorem 8.7.5 *[170] The time complexity function of each algorithmical realization for finding all interval solutions of (8.1) is exponential.* ■

The column matrix $B_{I \times \{1\}} = [b_i]$ is called a **convex linear combination** of $A_j = [a_{ij}]_{I \times \{1\}}$, $j \in J$, with coefficients $x_j \in L$ if $B = (A_1 \wedge x_1) \vee \cdots \vee (A_n \wedge x_n)$, i.e., $b_i = \vee_{j \in J}(a_{ij} \wedge x_j)$ for each $i \in I$.

The next result follows from Theorem 8.7.3 and Algorithm 8.7.2.

Corollary 8.7.6 *Let A_j, $j \in J$, and B be as above. It is algorithmically decidable whether B is a convex linear combination of A_j, $j \in J$.* ■

Example 8.7.7 *Consider the system $A \circ X = B$, where $A = \begin{bmatrix} .9 & .7 & .1 \\ .7 & 1 & .3 \\ .2 & .5 & .6 \end{bmatrix}$,*

$X = \begin{bmatrix} x_1 \\ x_2 \\ x_3 \end{bmatrix}$, and $B = \begin{bmatrix} .8 \\ .7 \\ .5 \end{bmatrix}$. It can be seen that $\mathcal{B} = \begin{bmatrix} G & S & S \\ E & G & S \\ S & E & G \end{bmatrix}$

and that $\begin{bmatrix} .8 \\ .7 \\ .5 \end{bmatrix}$ is a maximal solution and that $\begin{bmatrix} .8 \\ .5 \\ 0 \end{bmatrix}$ and $\begin{bmatrix} .8 \\ 0 \\ .5 \end{bmatrix}$ are minimal solutions.

8.8 Finite $\mathbb{L}$-Automata-Behavior Matrix

An $\mathbb{L}$-**automaton** is a quintuple $A = (Q, X, Y, M, \mathbb{L})$, where

(1) Q, X, Y are nonempty sets of states, input letters, and output symbols, respectively;

(2) $\mathbb{L} = (L, \vee, \wedge, 0, 1)$ is a bounded chain;

(3) $M = \{M(x, y) = [m_{qq'}(x, y)] \mid x \in X, y \in Y, q, q' \in Q, m_{qq'}(x, y) \in L\}$ is the transition-output matrices (the stepwise behavior) of A.

If Q, X, Y are finite, then A is called finite $\mathbb{L}$-automaton. We denote by $\mathcal{A}_0$ the class of all finite $\mathbb{L}$-automata.

The membership degree $m_{qq'}(x, y)$ determines the stepwise transition-output behavior of A as follows: in step k the automaton is in state q and

receives the input letter x. It outputs letter y in step k and reaches the
state q' in step $k + 1$.

In order to define and consider the complete input-output behavior of
$A = (Q, X, Y, M, \mathbb{L}) \in \mathcal{A}_0$, we introduce the following notation:

As usual, X^* (resp. Y^*) is the free monoid over X (resp. Y) with the
empty word Λ as the identity;

$U = [u_{ij}]$ is the square unit matrix: $u_{ij} = 0$ if $i \neq j$; $u_{ij} = 1$ if $i = j$;

$E = [\ 1 \quad 1 \quad \cdots \quad 1\ ]^T_{Q \times \{1\}}$ is the column matrix with all elements 1;

$(X \times Y)^* = \{(x, y) \mid x \in X^*, y \in Y^*, |x| = |y|\}$.

The matrix $M(x, y)$ defines the transition-output behavior of $A \in \mathcal{A}_0$
for $(x, y) = (x_1 \ldots x_k, y_1 \ldots y_k) \in (X \times Y)^*$ if

$$M(x, y) = \begin{cases} M(x_1, y_1) \circ \cdots \circ M(x_k, y_k) & \text{for } (x, y) \neq (\Lambda, \Lambda), \\ U & \text{if } (x, y) = (\Lambda, \Lambda). \end{cases}$$

The input-output behavior of A for $(x, y) \in (X \times Y)^*$ is defined by the
matrix $T(x, y)$:

$$T(x, y) = \begin{cases} M(x, y) \circ E & \text{if } (x, y) \neq (\Lambda, \Lambda), \\ E & \text{if } (x, y) = (\Lambda, \Lambda). \end{cases}$$

Every element $m_{qq'}(x, y)$ in $M(x, y)$ defines the operation of A under
the input word x beginning at state q, outputting the word y and reaching
the state q'. Every element $t_q(x, y)$ in $T(x, y)$ defines the operation of A
when it receives the word x beginning at state q and outputs the word y.

From the matrices $T(x, y)$, we shall construct the semi-infinite matrix
T of the complete input-output behavior of A and the finite submatrix B
of the behavior of A.

Suppose that a lexicographic order on $(X \times Y)^*$ is given. Let $T(i)$ be a
finite matrix with columns $T(x, y)$ ordered according to the lexicographic
order on $(X \times Y)^*$ and such that $|x| = |y| \leqslant i$. By definition, $T(0) =
T(\Lambda, \Lambda) = E$. Let T be the semi-infinite matrix of the complete input-
output behavior of A with columns $T(x, y)$ indexed according to the above
order. Let $B(i)$ be the finite matrix obtained from $T(i)$ by omitting all
columns that are convex linear combinations of the previous columns. Let
B be the matrix constructed from T by omitting the columns that are
convex linear combinations of the previous columns. The matrix B is
called the **behavior matrix** for the $\mathbb{L}$-automaton A.

For arbitrary matrices C', C'' we write $C' \subseteq C''$ if each column of C' is
a column in C''. If $C' \subseteq C''$ and $C'' \subseteq C'$, then we write $C' \sim C''$.

Clearly, $B(i) \subseteq T(i) \subseteq T(i + 1) \subseteq T$ for each $i = 0, 1, \ldots$.

Let $\mathcal{M} = \{m_{qq'}(u, v) \mid u \in X, v \in Y, q, q' \in Q\}$ and $\mathcal{F} = \{t_q(x, y) \mid
(x, y) \in (X \times Y)^*, q \in Q\}$ be the sets of the distinct entries for the matrices
$M(u, v) \in M$ and T, respectively. For any $A \in \mathcal{A}_0$, we have $\mathcal{F} \subset \mathcal{M} \cup \{1\}$,
$|\mathcal{M}| = r$ and $|\mathcal{F}| = s$ are finite, and $s \leqslant r + 1$. These properties follow
from the chain properties and the algebraic operations for matrices over $\mathbb{L}$.

The following result is proven in Chapter 2 and Sections 8.1–8.3. For $\mathbb{L} = ([0,1], \vee, \wedge, 0, 1)$, the generalization to the next result is easy.

Theorem 8.8.1 *Let* $A \in \mathcal{A}_0$. *Then the following statements hold:*

(1) *If there exists* $i \in \mathbb{N}$ *such that* $T(i) \sim T(i+1)$, *then* $T(i) \sim T(i+l)$ *for all* $l = 0, 1, 2, \ldots$.

(2) *If there exists* $i \in \mathbb{N}$ *such that* $B(i) = B(i+1)$, *then* $B(i) = B(i+l)$ *for all* $l = 0, 1, \ldots$.

(3) *There exists an integer* $k \leqslant s^{|Q|} - 1$ *such that* $T(k) \sim T(k+1)$ *and* $B(k) = B$. ∎

Corollary 8.8.2 *The behavior matrix of any* $A \in \mathcal{A}_0$ *is finite.* ∎

For equivalence, reduction, and minimization of automata, the main question is how to compute B. Thus for any column in T, we must determine whether it is a convex linear combination of the previous columns in order to select B from T. Recalling Algorithm 8.7.2, Theorem 8.7.3, and Theorem 8.8.1(3), we propose the following algorithm for completing the behavior matrix.

Algorithm 8.8.3 *Computation of the behavior matrix:*
  1. *Enter* $M, s, |Q|$.
  2. *Obtain* $k \leqslant s^{|Q|} - 1$ *such that* $T(k) \sim T(k+1)$.
  3. *Use Algorithm 8.7.2 to select* $B(k)$ *from* $T(k)$.
  4. $B = B(k)$.

8.9 ε-Equivalence

Let $A = (Q, X, Y, M, \mathbb{L}) \in \mathcal{A}_0$ be given. Let $\varepsilon \in [0, 1]$. The states q, $q' \in Q$ are called ε-**equivalent**, written $q \varepsilon q'$ if for all $(x, y) \in (X \times Y)^*$, $|t_q(x, y) - t_q(x, y)| \leqslant \varepsilon$, written $q \varepsilon q'$. In particular, for $\varepsilon = 0$ we have $t_q(x, y) = t_{q'}(x, y)$ for all $(x, y) \in (X \times Y)^*$ and the states q, q' are called **equivalent**, written $q \sim q'$.

Theorem 8.9.1 *Let* $A \in \mathcal{A}_0$ *be given. Let* $q, q' \in Q$.

(1) q *and* q' *are* ε-*equivalent if and only if their corresponding rows in* $T(k) \sim T(k+1)$ *for* $k \leqslant s^{|Q|} - 1$ *are* ε-*close;*

(2) q *and* q' *are equivalent if and only if their corresponding rows in* B *are equal.*

Proof. (1) The proof follows from Theorem 8.8.1.

(2) The necessity is clear. Conversely, let $b_{qj} = b_{q'j}$ for each j in $B = [b_{ij}]$. An arbitrary column $t(x, y)$ in T is a convex linear combination of the columns of B, i.e., $t(x, y) = B \circ X$. For $t_q(x, y)$ and $t_{q'}(x, y)$, we have

$$t_q(x, y) = \vee_j (b_{qj} \wedge x_j) = \vee_j (b_{q'j} \wedge x_j) = t_{q'}(x, y). \quad \blacksquare$$

The next algorithm determines whether two states in a given finite $\mathbb{L}$-automaton are ε-equivalent, resp. equivalent.

Algorithm 8.9.2 *1. Enter ε, M, $|Q|$, s.*

2. Construct $T(k)$ for the smallest $k \leqslant s^{|Q|} - 1$ with $T(k) \sim T(k+1)$.

3. If $\varepsilon \neq 0$ check whether the rows corresponding to q and q' are ε-close in $T(k)$ and go to 6.

4. Use Algorithm 8.8.3 to select B from $T(k)$.

5. Check whether the rows corresponding to q and q' are equal in B. Go to 7.

6. Print whether $q \in q'$. Go to 8.

7. Print whether $q \sim q'$.

8. End.

Corollary 8.9.3 *For any $A \in \mathcal{A}_0$, the relation ε-equivalence of states is algorithmically decidable.* ∎

We can define an ε-partition over Q for a given $A \in \mathcal{A}_0$ from the relation ε-equivalence of states. A subset $Q_i \subseteq Q$ is an ε-class with *center* $q_i \in Q_i$ if $q_i \varepsilon q$ holds for each $q \in Q_i$ and $q_i \varepsilon q$ does not hold for each $q \in Q \backslash Q_i$. If every state $q \in Q$ belongs to exactly one ε-class, then the family $\{Q_i \mid i \leqslant |Q|\}$ defines an ε-partition over Q. ε-equivalence of states is not an equivalence relation, but ε-partition as its analogue is a sufficient tool for our purposes.

Algorithm 8.9.4 *Construction of an ε-partition over Q:*

1. Enter Q, ε.

2. $\widehat{Q} = Q$.

3. Denote by $\widehat{q} \in \widehat{Q}$ the element with the smallest index.

4. Form the ε-class $Q_i = \{q_i \mid \widehat{q} \varepsilon q_i, q_i \in \widehat{Q}\}$ using Algorithm 8.9.2.

5. Write Q_i as the i-th ε-class with the center $\widehat{q}$.

6. $\widehat{Q} = \widehat{Q} \backslash Q_i$.

7. If $\widehat{Q} = \emptyset$ go to 9.

8. Go to 3.

9. End.

For $\varepsilon = 0$, we obtain the factor set $Q/\sim$ under the equivalence relation equivalence of states.

Corollary 8.9.5 *For all $A \in \mathcal{A}_0$, the construction of an ε-partition of the set of the states Q is algorithmically decidable.* ∎

8.10 ε-Irreducibility

Let $A = (Q, X, Y, M, \mathbb{L})$ and $A' = (Q', X, Y, M', \mathbb{L})$ be finite $\mathbb{L}$-automata with the same X, Y, L and f. Let $\varepsilon \in [0, 1]$. The state $q \in Q$ is said to be ε-**equivalent** to the state $q' \in Q'$, written $q \varepsilon q'$, if $\left| t_q(x, y) - t'_{q'}(x, y) \right| \leqslant \varepsilon$ for all $(x, y) \in (X \times Y)^*$. A is ε-**equivalent** to A', written $A \varepsilon A'$, if for all $q \in Q$ there exists $q' \in Q'$ such that $q \varepsilon q'$ and vice versa. For $\varepsilon = 0$, A is called **equivalent** to A'.

Theorem 8.10.1 *Let* $A = (Q, X, Y, M, \mathbb{L})$ *and* $A' = (Q', X, Y, M', \mathbb{L})$ *be finite* $\mathbb{L}$*-automata. The states* $q \in Q$ *and* $q' \in Q'$ *are* ε*-equivalent if and only if* $\left| t_q(x, y) - t'_{q'}(x, y) \right| \leqslant \varepsilon$ *for all* $(x, y) \in X^* \times Y^*$ *with* $|x| = |y| \leqslant c^{|Q| + |Q'|} - 1$*, where* $c = |\mathcal{F} \cup \mathcal{F}'|$*.*

   Proof. Consider the automaton $A \oplus A' = (Q \cup Q', X, Y, M \oplus M', \mathbb{L})$, where $Q \cap Q' = \emptyset$ and $M \oplus M' = \{M''(u, v) = [m''_{qq'}(u, v)], u \in X, v \in Y, q, q' \in Q \cup Q'\}$ is defined as follows:

$$m''_{qq'}(u, v) = \begin{cases} m_{qq'}(u, v) & \text{if } q, q' \in Q, \\ m'_{qq'}(u, v) & \text{if } q, q' \in Q', \\ 0 & \text{otherwise.} \end{cases}$$

By Theorem 8.9.1, two states in $A \oplus A'$ are ε-equivalent if and only if the corresponding rows in $T(k) \sim T_{A \oplus A'}$ are ε-close, where $k = c^{|Q| + |Q'|} - 1$, $c = |\mathcal{F} \cup \mathcal{F}'|$ (the number of distinct entries in T and T'). ∎

Corollary 8.10.2 *The relation* ε*-equivalence of finite* $\mathbb{L}$*-automata is algorithmically decidable in the class* $\mathcal{A}_0$*.* ∎

   The proof of Corollary 8.10.2 follows from Theorem 8.10.1 and Algorithm 8.9.2.
   An automaton $A \in \mathcal{A}_0$ is called ε-**irreducible** if for each $q_i, q_j \in Q$, $i \neq j$, $q_i \varepsilon q_j$ implies $q_i = q_j$.
   The next result follows directly from Corollary 8.9.3.

Corollary 8.10.3 *For any* $A \in \mathcal{A}_0$*, it is algorithmically decidable whether* A *is* ε*-irreducible or not.* ∎

   An automaton A_r is an ε-**reduct** of a given automaton A if A_r is ε-irreducible and $A \varepsilon A_r$. For $\varepsilon = 0$, A_r is called a **reduct** of A.

Theorem 8.10.4 *For all automaton* $A \in \mathcal{A}_0$*, there exists an* ε*-reduced automaton* A_r*.*

   Proof. The construction of $A_r = (X, Q_r, Y, M_r, \mathbb{L})$ as an ε-reduct of $A = (X, Q, Y, M, \mathbb{L})$ contains the following steps:
   1. Take $X, Y, \mathbb{L}$ for A_r the same as for A.
   2. Obtain M_r from M as follows: if $M = \{M(u, v) | M(u, v) = [m_{ij}(u, v)], u \in X, v \in Y, i, j \leq |Q|\}$, then M_r contains the square matrices of order $|Q_r|$,

$M_r = \{M'(u,v) \mid u \in X, v \in Y\}$. Here $M'(u,v) = [m'_{q_iq_j}(u,v)]$, where q_i and q_j are the centers of the ε-classes Q_i and Q_j, respectively. Every element $m'_{q_iq_j}(u,v)$ is calculated from the elements of the i-th row of $M(u,v) \in M$, belonging to the ε-class with center $q_j \in Q_j$ according to the rule

$$m'_{q_iq_j}(u,v) = \vee_{q_r \in Q_j}(m_{ir}(u,v)).$$

The automaton A_r is ε-irreducible by construction. Consequently, $A \varepsilon A_r$.

∎

The automaton A_r constructed as in the proof of Theorem 8.10.4 is called a **natural ε-reduct** of $A \in \mathcal{A}_0$.

Corollary 8.10.5 *For every $A \in \mathcal{A}_0$, there are a finite number of natural reducts.* ∎

It follows from Corollary 8.10.5 that all natural ε-reducts can be constructed for any $A \in \mathcal{A}_0$. A method for obtaining all ε-reducts for $A \in \mathcal{A}_0$ is as follows:
1. Compute the behavior matrix B for A using Algorithm 8.8.3.
2. Construct all natural ε-reducts A_r for A (cf. Theorem 8.10.4).
3. Compute the behavior matrix B_r for each A_r.
4. For each natural ε-reduct $A_r = (X, Q_r, Y, M_r, \mathbb{L})$ obtain all ε-reducts $\{A_{r'} = (X, Q_r, Y, M'_r, \mathbb{L})\}$, where $M'_r = \{M'(u,v) \mid u \in X, v \in Y\}$ and each $M'(u,v)$ is a solution of the system $M'(u,v) \circ B_r = M_r(u,v) \circ B_r$ for $M'(u,v)$.

We are interested not only in a point solution of these systems but in each solution. According to Algorithm 8.7.4 and Theorem 8.7.5, there is a continuum of ε-reducts for $A \in \mathcal{A}_0$.

The above results allow for the selection of a suitable reduced automaton having behavior ε-equivalent with the original one. Using fuzzy acceptors as recognizing systems and the appropriate ε-equivalence and ε-reduction, feature extraction can be carried out in order to exhibit properties from primary data and to make reduction of the features characterizing a class of objects.

8.11 Minimization

Let $0_q = [c_i]_{Q \times \{1\}}$ be the column matrix with $c_i = 0$ if $q_i = q$. Let $1_q = [d_i]_{Q \times \{1\}}$ be the column matrix with $d_i = 1$ if $q_i = q$, $d_i = 0$ otherwise.

The automaton $A \in \mathcal{A}_0$ is said to be **state minimal** (resp. ε-**state minimal**) if there does not exist $q \in Q$ such that $1_q \circ T = 0_q \circ T$ (resp. $1_q \circ T \varepsilon 0_q \circ T$) for some column matrix 0_q.

Theorem 8.11.1 *The automaton $A \in A_0$ is state minimal (resp. ε-state minimal) if and only if there does not exist $q \in Q$ such that $1_q \circ B = 0_q \circ B$ (resp. $1_q \circ B \varepsilon 0_q \circ B$) for some column matrix 0_q.*

Proof. Since B is a submatrix of T, the first part of the proof follows easily. Conversely, suppose $1_q \circ B = 0_q \circ B$ holds, i.e., the q-th row in B is a 0_q-convex linear combination of the other rows in B,

$$b_{qj} = \vee_{q' \in Q \setminus \{q\}} (b_{q'j} \wedge c_{q'}) \text{ for each } j.$$

Then we obtain for the state minimization

$$
\begin{aligned}
t_q(u, v) &= \vee_j (b_{qj} \wedge x_j) \\
&= \vee_j ((\vee_{q' \in Q \setminus \{q\}} (b_{q'j} \wedge c_{q'})) \wedge x_j) \\
&= \vee_{q' \in Q \setminus \{q\}} (\vee_j (b_{qj} \wedge x_j) \wedge c_{q'}) \\
&= \vee_{q' \in Q \setminus \{q\}} (t_{q'}(u, v) \wedge c_{q'}).
\end{aligned}
$$

The ε-state minimization follows from Proposition 8.6.2 and Theorem 8.10.1. ∎

Corollary 8.11.2 *For all $A \in \mathcal{A}$, it is algorithmically decidable whether A is state (resp. ε-state) minimal or not.*

Proof. The algorithm for the state minimal part consists of the following steps:

1. Compute the behavior matrix B for A.
2. For all $q \in Q$, solve the system $1_q \circ B = 0_q \circ B$ for 0_q.
3. If there does not exist $q \in Q$ with $1_q \circ B = 0_q \circ B$, then A is state minimal.

For the ε-state minimization, we must change the meaning of the matrix $\mathcal{B}$ in Algorithm 8.7.4 as follows:

$$\mathcal{B} = (b_{ij}), \qquad b_{ij} = \begin{cases} S & \text{if } a_{ij} - \varepsilon < b_i, \\ E & \text{if } a_{ij} - \varepsilon \leqslant b_i \leqslant a_{ij} + \varepsilon \\ G & \text{if } a_{ij} + \varepsilon > b_i, \end{cases}$$

for suitable $\varepsilon \in [0, 1]$. Then as a result of Algorithm 8.7.4 (or Algorithm 8.7.2), we obtain $A \circ X \varepsilon B$. Thus, $1_q \circ B \varepsilon 0_q \circ B$ can be solved for 0_q in steps 2, 3 above. ∎

In [104], ε-equivalence for stochastic automata is examined. The main result is that there is a need to consider ε-equivalence only for a subclass SS_0 of the class of all finite stochastic automata ($A \in SS_0$ if and only if any column in T is a substochastic linear combination of the columns in B). The reason for this restriction is that Proposition 8.6.2 and Theorem 8.9.1 and the related statements are not valid for the stochastic case.

The results show that all results given in Sections 8.10, 8.11, 8.12 above are true again only in SS_0 (for $k = |Q| - 1$).

For $\varepsilon = 0$, we have now obtained the results of Sections 2.6 and Sections 8.1−8.3 as a particular case.

8.12 Exercises

1. Prove Proposition 8.1.1.

2. Prove Proposition 8.1.2.

Let $M = (Q, X, Y, \delta)$ be a fuzzy automaton (Definition 8.1.5) and let $M' = (Q, X, Y, \mu, \omega)$ be a fuzzy finite state automaton (Section 8.5). Call M and M'

(1) **strongly equivalent** if $\forall(q, x, y, q') \in Q \times X \times Y \times Q$, $\delta(q, x, y, q') = \mu(q, x, q') = \omega(q, x, y)$;

(2) **equivalent** if $\forall(q, x, y, q') \in Q \times X \times Y \times Q$, $\delta(q, x, y, q') = \mu(q, x, q') \wedge \omega(q, x, y)$;

(3) **weakly equivalent** if $\forall(q, x, y, q') \in Q \times X \times Y \times Q$, $\delta(q, x, y, q') > 0$ if and only if $\mu(q, x, q') \wedge \omega(q, x, y) > 0$.

3. Prove (3) of Proposition 8.2.1.

4. If M and M' are strongly equivalent, prove that $\forall(q, x, y, q') \in Q \times X^* \times Y^* \times Q$, $\delta^*(q, x, y, q') = \mu^*(q, x, q') = \omega^*(q, x, y)$.

5. If M and M' are equivalent, prove that $\forall(q, x, y, q') \in Q \times X^* \times Y^* \times Q$, $\delta^*(q, x, y, q') = \mu^*(q, x, q') \wedge \omega^*(q, x, y)$.

6. If M and M' are weakly equivalent, prove that $\forall(q, x, y, q') \in Q \times X^* \times Y^* \times Q$, $\delta^*(q, x, y, q') > 0$ if and only if $\mu^*(q, x, q') \wedge \omega^*(q, x, y) > 0$.

7. Compare minimality requirements of M and M' under various equivalence assumptions.

8. Consider the system $A \circ X = B$ of Example 8.7.7. Use the algorithms of Section 8.7 to show that $\begin{bmatrix} .8 \\ .7 \\ .5 \end{bmatrix}$ is a maximal solution and $\begin{bmatrix} .8 \\ .5 \\ 0 \end{bmatrix}$ and $\begin{bmatrix} .8 \\ 0 \\ .5 \end{bmatrix}$ are minimal solutions.

Chapter 9

L-Fuzzy Automata, Grammars, and Languages

9.1 Fuzzy Recognition of Fuzzy Languages

In this chapter, we consider fuzzy automata, grammars, and languages, where the interval $[0, 1]$ is replaced by a lattice. We begin our study by considering the work of [94]. We give an application of the theory of fuzzy recognition to the theory of probabilistic automata and discuss the closure properties of some fuzzy language classes corresponding to machine classes.

In formal language theory, languages are classified by the complexities of machines that recognize them. Typically, the machines are finite automata, push-down automata, linear bounded automata, and Turing machines. It is also of interest to develop the theory of recognition of fuzzy languages by machines and their classification by the complexities of machines that recognize them. The theory should be a reasonable extension of the ordinary language recognition theory. In ordinary language recognition theory, a machine is said to recognize a language L if and only if for every word in L, the machine decides that it is a member of L and for a word not in L, the machine either decides that it is not a member of L or loops forever. That is, a machine may be said to recognize a language if and only if the machine computes the characteristic function of the language. Thus it is natural to define a machine to recognize a fuzzy language if and only if the machine computes its fuzzy membership function. The question arises concerning the meaning for a machine to have a fuzzy membership function. In ordinary language theory, it is defined that for a given input word a machine computes the characteristic function value at 1 if and only if it takes one of special memory-configurations, such as configurations with a final state and configurations with the empty stack for cases where the machine has pushdown stacks. Hence a straightforward extension is to define each fuzzy membership function value to be represented by some memory configuration of the machine. In other words, the memory configuration, which the machine moves into after a sequence of moves for a given word, should be uniquely associated with the membership function value of the word.

Even if the machine obtains a configuration uniquely associated with the membership function value of the word, we must have that the machine computes the membership function value of the word. We can be sure that the machine knows the value if it is to be able to answer exactly any question about the value. Furthermore, it is also felt in [94] that it is essential that the values represented by memory configurations of the machine are lattice elements. Thus, we require that the machine should be able to compare the fuzzy membership function value of a given word, which is stored in its memory, with any lattice element that is given to the machine as a question. Such a lattice element will be called a cutpoint. We assume that a cutpoint is represented by an infinite sequence of symbols in a finite alphabet, following the infinite expansions of decimals. For any cutpoint c, the machine having the memory configuration associated with the fuzzy membership function value $\mu(x)$ of a given input word x should be able to read, as a subsequent input, the infinite sequence corresponding to the cutpoint c sequentially, and after a finite step of deterministic moves, it can determine which of the following four cases is valid, (1) $\mu(x) > c$, (2) $\mu(x) < c$, (3) $\mu(x) = c$, and (4) $\mu(x)$ and c are incomparable.

9.2   Fuzzy Languages

Let X be a finite set of symbols called an alphabet. Let L be a lattice with minimum element 0. An L-**fuzzy language** over X is defined to be a function from X^* into L. At times, we omit the symbol L if its presence is clear. If μ is an L-fuzzy language over X and $x \in X^*$, then $\mu(x)$ represents the membership grade for x to be in the L-fuzzy language.

We consider languages associated with an L-fuzzy language μ and a cutpoint c, namely

$$L_G(\mu, c) = \{x \in X^* \mid \mu(x) > c\},$$

$$L_{GE}(\mu, c) = \{x \in X^* \mid \mu(x) \geqslant c\}.$$

We call such languages **cutpoint languages** for μ and c.

Let Δ be a finite alphabet. Let Δ^∞ be the set of all infinite sequences of symbols in Δ extending infinitely to the right. A one-to-one function r from L into Δ^∞ is called a **representation** of L over Δ if there exists a function $d : r(L) \times r(L) \rightarrow \mathbb{N} \cup \{0\}$ that assigns to any distinct two elements $r(l)$ and $r(m)$ in $r(L)$ a positive integer $d(r(l), r(m))$ such that $d(r(l), r(m)) = d(r(m), r(l))$, $d(r(l), r(l)) = 0$ and conditions (1) and (2) below are satisfied.

(1) Let $l, m \in L$, $l \neq m$. Then for the prefix w_l of $r(l)$ and the prefix w_m of $r(m)$ of length $d(r(l), r(m))$, respectively, and for any $\alpha', \beta' \in \Delta^\infty$, either condition (a) or (b) holds:

(a) Either $w_l \alpha'$ or $w_m \beta'$ is not in $r(L)$.

(b) Both $w_l \alpha'$ and $w_m \beta'$ are in $r(L)$, and

$$r^{-1}(w_l \alpha') > r^{-1}(w_m \beta') \text{ if } l > m$$

$$r^{-1}(w_l \alpha') < r^{-1}(w_m \beta') \text{ if } l < m$$

and $r^{-1}(w_l \alpha')$ and $r^{-1}(w_m \beta')$ are incomparable if l and m are incomparable.

(2) For any l, m, and n in L with $l > m > n$,

$$d(r(l), r(n)) \leqslant d(r(l), r(m)) \wedge d(r(m), r(n)).$$

For $\alpha, \beta \in r(L)$, $d(\alpha, \beta)$ is called the **D-length** of α and β. For $l \in L$, $r(l)$ is called the representation of l with respect to r. A lattice does not always have a representation. However, we consider from now through Section 9.8 only lattices that have a representation over some finite alphabet. Condition (2) means that representations of lattices are restricted to those that have decimal expansions of real numbers. However, there may be many representations for a lattice.

Example 9.2.1 *Let L be a lattice with a finite number of elements. If L has elements $l_1, \ldots, l_k$, let $\Delta = \{l_1, \ldots, l_k\}$. If $r(l_i) = l_i l_i l_i \ldots$ for $1 \leqslant i \leqslant k$, then r is a representation of L over Δ.*

Example 9.2.2 *Let $L_{[0,1]}$ be the set of all real numbers in $[0,1]$ with the usual ordering. A representation r_1 is given as follows: Let $\Delta = \{0, 1, \dot{0}, \dot{1}\}$. For $l \in L_{[0,1]}$, let $e(l)$ be the binary expansion of l not of the form $w111\ldots$ with $w \in \{0,1\}^*0$. For any rational number l in $[0,1]$, set $e(l) = w_0 w_1 w_1 w_1 \ldots$ such that if $e(l) = w_0' w_1' w_1' w_1' \ldots$, then $|w_0| \leqslant |w_0'|$ and $|w_1| \leqslant |w_1'|$. Let $\dot{e}(l) = w_0 \, \dot{w}_1 \, w_1 w_1 \ldots$, where $\dot{w}_1 = \dot{a}_1 \dot{a}_2 \ldots \dot{a}_k$ for $w_1 = a_1 a_2 \ldots a_k$ with $a_i \in \{0,1\}$, $i = 1, 2, \ldots, k$. Then let*

$$r_1(l) = e(l) \text{ if } l \text{ is irrational,}$$

$$r_1(l) = \dot{e}(l) \text{ if } l \text{ is rational.}$$

Example 9.2.3 *Let $L_{[0,1]_Q}$ denote the set of all rational numbers in $[0,1]$ with the usual ordering. Let $\Delta = \{0, 1, \dot{0}, \dot{1}\}$. Define a representation $r_2 : L_{[0,1]_Q} \longrightarrow \Delta^\infty$ as follows: $r_2(0) = \dot{0}\dot{0}\dot{0}, \dots, r_2(1) = \dot{1}\dot{1}\dot{1}, \dots$, and for l such that $e(l) = w000\dots$ with $w \in \{0,1\}^*1$, define $r_2(l) = w\,\dot{0}\dot{0}\dot{0}\dots$ and for other l in $L_{[0,1]_Q}^*$, define $r_2(l) = e(l)$.*

9.3   Fuzzy Recognition by Machines

Machines treated in Sections 9.3–9.8 may be finite automata, pushdown automata, linear bounded automata, and Turing machines. These machines can be represented in the following manner.

A machine has an input terminal that reads input symbols and Λ sequentially, a memory storing and processing device, and an output terminal. Formally, a machine is defined to be an 8-tuple $M = \langle \Phi, \Gamma, \Psi, \theta, \delta, \Omega, \kappa, \gamma_0 \rangle$, where

Φ is a finite set of input symbols;

Γ is a finite set of memory-configuration symbols;

Ψ is a finite set of output symbols;

Ω is a finite set of partial functions $\{\omega_i\}$ from Γ^* into Γ^*;

θ is a partial function from Γ^* into Γ^n, for some $n \geqslant 1$;

(For a memory configuration $\gamma \in \Gamma^*$, $\theta(\gamma)$ designates the instantaneously accessible information of γ by M.);

δ is a partial function from $(\Phi \cup \{\Lambda\}) \times \Gamma^n$ into $\mathcal{P}(\Omega)$;

κ is a partial function from Γ^n to $\Psi \cup \{\Lambda\}$;

γ_0 is an element in Γ^*, called the initial memory configuration.

A memory configuration ξ is said to be derived from a memory configuration γ by $u \in \Phi \cup \{\Lambda\}$, denoted by $\gamma \underset{u}{\Longrightarrow} \xi$, if and only if there exists $\rho = \theta(\gamma)$ and $\omega_i \in \delta(u, \rho)$ such that $\omega_i(\gamma) = \xi$. Let $w \in \Phi^*$. A memory configuration ξ is said to be derived from a memory configuration γ by w, written $\gamma \underset{w}{\Longrightarrow} \xi$, if and only if there exist $u_1, u_2, \dots, u_l$ with u_i in $\Phi \cup \{\Lambda\}$ such that $w = u_1 u_2 \dots u_l$, and there exist $\gamma_0, \gamma_1, \dots, \gamma_l$ with $\gamma_i \in \Gamma^*$ such that $\gamma_0 = \gamma$, $\gamma_l = \xi$, and $\gamma_i \underset{u_{i+1}}{\Longrightarrow} \gamma_{i+1}$ for all $0 \leqslant i \leqslant l - 1$. ($\gamma \underset{\Lambda}{\Longrightarrow} \gamma$ is valid for all $\gamma \in \Gamma^*$.) Given an input word $w \in \Phi^*$ and the initial memory configuration γ_0 first, M reads input symbols or Λ sequentially along w, changes memory configurations step by step possibly in a nondeterministic way and reaches γ such that $\gamma_0 \Longrightarrow \gamma$, and emits the output $\kappa(\theta(\gamma))$.

Clearly, a machine $M = \langle \Phi, \Gamma, \Psi, \theta, \delta, \Omega, \kappa, \gamma_0 \rangle$ can be restricted to a specified family of automata such as Turing machines, pushdown automata, and finite automata for appropriate choices of $\Gamma, \theta, \delta, \Omega, \kappa$, and γ_0.

A machine $M = \langle \Phi, \Gamma, \Psi, \theta, \delta, \Omega, \kappa, \gamma_0 \rangle$ is called a **deterministic machine** if for any memory configuration γ such that $\gamma_0 \underset{x}{\Longrightarrow} \gamma$ for some $x \in \Phi^*$, if $\delta(\Lambda, \theta(\gamma)) \neq \emptyset$, then $\delta(\Lambda, \theta(\gamma))$ contains at most one element and $\delta(u, \theta(\gamma)) = \emptyset$ for any $u \in \Phi$, and if $\delta(\Lambda, \theta(\gamma)) = \emptyset$, then for any $u \in \Phi, \delta(u, \theta(\gamma))$ contains at most one element.

Let L be a lattice with a minimum element 0 and $\mu : X^* \longrightarrow L$ be an L-fuzzy language over an alphabet X. Let r be a representation of L over an alphabet Δ. A machine $M = \langle \Phi, \Gamma, \Psi, \theta, \delta, \Omega, \kappa, \gamma_0 \rangle$ is said to **fuzzy recognize** μ with r if and only if the following conditions hold:

(1) $\Phi = X \cup \Delta \cup \{t\}$, where t is an element not in $X \cup \Delta$.

(2) $\Psi = \{>, <, =, !\}$.

(3) There exists a partial function ν from Γ^* into L such that the conditions (i)−(iv) hold:

(i) Let $S_x = \{\gamma \mid \gamma_0 \underset{xt}{\Longrightarrow} \gamma\}$ for all $x \in X^*$. Then for all $x \in X^*$ such that $S_x \neq \emptyset$, $S_x \subseteq \mathrm{Dom}(\nu)$ and $\nu_x = \vee\{\nu(\gamma) \mid \gamma \in S_x\}$ exists. If $S_x = \emptyset$, ν_x is undefined.

(ii) Let γ be any memory configuration in S_x. Let M_γ denote the machine $\langle \Delta, \Gamma, \Psi, \theta, \delta', \Omega, \kappa, \gamma \rangle$, where δ' is the restriction of δ over $(\Delta \cup \{\Lambda\}) \times \Gamma^n$. Then M_γ is a deterministic machine.

(iii) For any memory configuration $\gamma \in \Gamma^*$, if $\kappa(\theta(\gamma))$ is in Ψ, that is, $\kappa(\theta(\gamma)) \neq \Lambda$, then for any $u \in \Phi \cup \{\Lambda\}$, $\delta(u, \theta(\gamma))$ is empty. Also, $\kappa(\theta(\gamma))$ is in Ψ only if $\gamma_0 \underset{xty}{\Longrightarrow} \gamma$ for some $x \in X^*$ and y in $PRE_\gamma(L)$. (ω_1 is called a **prefix** of a word or an infinite sequence α if $\alpha = \omega_1\beta$ for some β. Let Π be either a set of words or a set of infinite sequences. $PRE\Pi$ is the set of all prefixes of elements in Π.)

(iv) Let $\gamma \in \mathrm{Dom}(\nu)$. For all $l \in L$, there exists a prefix ν of $r(l)$ and γ' in Γ such that $\gamma \underset{\nu}{\Longrightarrow} \gamma'$ and the following properties hold:

$$\kappa(\theta(\gamma')) \text{ is } > \text{ if } \nu(\gamma) > l,$$
$$\kappa(\theta(\gamma')) \text{ is } = \text{ if } \nu(\gamma) = l,$$
$$\kappa(\theta(\gamma')) \text{ is } < \text{ if } \nu(\gamma) < l,$$
$$\kappa(\theta(\gamma')) \text{ is } ! \text{ if } \nu(\gamma) \text{ and } l \text{ are incomparable.}$$

Consider xt in X^*t, where t indicates the end of the input sequence. Then a machine M moves possibly nondeterministically into some memory configuration γ such that $\nu(\gamma)$ is defined. Let S_x denote the set of all such γ's. Consider the maximum value ν_x of $\{\nu(\gamma)|\gamma \in S_x\}$ as the value of x computed by M. If S_x is empty, then the value of x cannot be computed by M. A sequence of moves from the initial memory configuration to a memory configuration in S_x is called a **value computation** for x.

If the machine M completes the value computation for $x \in X^*$, that is, if $S_x \neq \emptyset$, then M is required to have the following ability. Let γ be any memory configuration in S_x. Then M_γ should be able to compare $\nu(\gamma)$ with any element l in L if the representation $r(l)$ of l is presented to M_γ as a question. In other words, when $r(l)$ is given to M_γ, M_γ moves deterministically reading input symbols in $\{\Lambda\} \cup \Delta$ along the infinite sequence $r(l)$ and emits one of $>$, $<$, $=$, and $!$ following the order of $\nu(\gamma)$ and l in L after reading a finite length of a prefix of $r(l)$, and halts (see (iv)).

A sequence of moves of M from $\gamma \in \text{Dom}(\nu)$ to a halting configuration is called an **order-comparing computation** for γ.

Let the fuzzy language $\mu : X^* \longrightarrow L$ be such that

$$\mu(x) = \begin{cases} \nu_x & \text{if } \nu_x \text{ is defined} \\ 0 & \text{otherwise.} \end{cases}$$

Then μ is said to be fuzzy recognized by the machine M with the representation r.

By condition (iii), if M reads a word in Δ^* that is not any prefix of $r(l)$ for any $l \in L$, then M does not emit any of $>$, $<$, $=$, and $!$. In other words, if M is given an illegal question, then M does not answer.

Let T_0, T_1, T_2, and T_3 be the classes of Turing machines, linear bounded automata, pushdown automata, and finite automata, respectively. Let DT_0, DT_1, DT_2, and DT_3 be the classes of deterministic Turing machines, deterministic linear bounded automata, deterministic pushdown automata, and deterministic finite automata, respectively. An L-fuzzy language μ is said to be **fuzzy recognized** by a machine in $T_i(DT_i)$ if μ is fuzzy recognized by a machine in $T_i(DT_i)$ with some representation r, for $i = 0, 2$, and 3. We say that an L-fuzzy language $\mu : X^* \longrightarrow L$ is **fuzzy recognized** by a (deterministic) linear bounded automaton if μ is fuzzy recognized by a (deterministic) Turing machine M with some representation r in the following manner: Let $x \in X^*$ and $l \in L$. If

$$\gamma_0 \underset{y_1}{\Longrightarrow} \gamma$$

for some prefix y_1 of xt, or

$$\gamma_0 \underset{xty}{\Longrightarrow} \gamma$$

for some y in $PRE\gamma(L)$, then $|\gamma| \leqslant t|x|$ for some constant t, where γ_0 is the initial configuration of M. In other words, lengths of memory configurations in M for any $x \in X^*$ are always less than or equal to some constant times $|x|$ throughout the value computation and the order-comparing computation of x with any cutpoint in L.

Example 9.3.1 *Let $X = \{a, b\}$ and $w \in X^*$. Let $n_a(w)$ and $n_b(w)$ denote the numbers of occurrences of a and b in w, respectively. The fuzzy language*

$$\mu_1 : X^* \longrightarrow L_{[0,1]_{\mathbb{Q}}}$$

defined by $\forall w \in X^*$,

$$\mu_1(w) = \frac{1}{2} + \left(\frac{1}{2}\right)^{|n_a(w) - n_b(w)| + 1},$$

is fuzzy recognized by a deterministic pushdown automaton.

A pushdown automaton $M = \langle \Phi, \Gamma, \Psi, \theta, \delta, \Omega, \kappa, \gamma_0 \rangle$ with γ_2 in Example 9.3.1 fuzzy recognizes μ_1, where $\Phi = X \cup \Delta \cup \{t\}$ with $X = \{a, b\}$ and $\Delta = \{0, 1, \dot{0}, \dot{1}\}$, $\Gamma = Q \cup \{z_0, a, b, 1\}$, where $Q = \{q_0, q_1, q_>, q_<, q_=\}$, $\Psi = \{>, <, =\}$, θ is a partial function from Γ^* into Γ^2 such that $\theta(qxu) = (q, u)$ for all $q \in Q$, $x \in \{\Lambda\} \cup z_0\{ab\}^*$ and $u \in \{z_0, a, b, 1, \dot{1}\}$,

$$\Omega = \{\omega_{1a}, \omega_{1b}, \omega_a, \omega_b, \omega_-, \omega_1, \omega_{\dot{1}}, \omega_>, \omega_<, \omega_=\},$$

where $\forall\, x \in \Gamma^*$

$$\gamma_0 = q_0 z_0$$
$$\omega_{1u}(x) = x1u, \text{ for } x \in \Gamma^* \text{ and } u \in \{a, b\}$$
$$\omega_u(x) = xu, \text{ for } x \in \Gamma^* \text{ and } u \in \{a, b\}$$
$$\omega_-(xu) = x, \text{ for } x \in \Gamma^* \text{ and } u \in \{a, b\}$$
$$\omega_1(q_0 1xu) = q_0 1x1, \text{ for } u \in \{a, b\}$$
$$\omega_{\dot{1}}(q_0 z_0) = \omega_{\dot{1}}(q_0 z_0 1) = q_1 z_0 \dot{1}$$
$$\omega_>(q_1 x) = q_> x$$
$$\omega_<(q_1 x) = q_< x$$
$$\omega_=(q_1 x) = q_= x.$$

$\delta : \Phi \times \Gamma^2 \longrightarrow \mathcal{P}(\Omega)$ is defined as follows:

$$
\begin{aligned}
\delta(t, (q_0, z_0)) &= \{\omega_{\dot{1}}\}, \\
\delta(u, (q_0, z_0)) &= \{\omega_{1u}\} && \text{if } u \in \{a, b\}, \\
\delta(u, (q_0, u)) &= \delta(u, (q_0, 1)) = \{\omega_u\} && \text{if } u \in \{a, b\}, \\
\delta(a, (q_0, b)) &= \delta(b, (q_0, a)) = \{\omega_-\} \\
\delta(t, (q_0, u)) &= \{\omega_1\} && \text{if } u \in \{a, b\}, \\
\delta(t, (q_0, 1)) &= \{\omega_{\dot{1}}\} \\
\delta(1, (q_1, 1)) &= \delta(0, (q_1, u)) = \{\omega_-\} && \text{if } u \in \{a, b\}, \\
\delta(0, (q_1, 1)) &= \{\omega_>\}, \\
\delta(1, (q_1, u)) &= \{\omega_<\} && \text{if } u \in \{a, b\}, \\
\delta(\dot{0}, (q_1, z_0)) &= \{\omega_=\}, \\
\delta(u, (q_1, z_0)) &= \{\omega_<\} && \text{if } u \in \{0, 1\}, \\
\delta(u, (q_1, \dot{1})) &= \{\omega_>\} && \text{if } u \in \{0, 1, \dot{0}\}, \\
\delta(\dot{1}, (q_1, \dot{1})) &= \{\omega_=\}.
\end{aligned}
$$

Let the partial function κ from Γ^2 into $\Psi \cup \{\Lambda\}$ be defined as follows:

$$\kappa(q_\eta, u) = \eta \text{ for } \eta \in \Psi \text{ and } u \in \{z_0, a, b, 1, \dot{1}\}.$$

Define $\nu : \Gamma^* \longrightarrow [0, 1]_{\mathbb{Q}}$ by

$$\nu(q_1 z_0 \dot{1}) = 1$$

and

$$\nu(q_1 z_0 1 u^n 1) = \frac{1}{2} + (\frac{1}{2})^{n+1}, \text{ for } u \in \{a, b\} \text{ and } n \geqslant 1.$$

Let $\mu_1' : X^* \longrightarrow L_{[0,1]}$ be such that $\mu_1'(w) = \mu_1(w)$ for $w \in X^*$. Then it follows that μ_1' is fuzzy recognized by a deterministic pushdown automaton with the representation r_1 of a previous example.

9.4   Cutpoint Languages

Let μ be an L-fuzzy language over X, where L is a lattice with minimum element 0. Then $l \in L$ is called an **isolated cutpoint** of μ if one of the following three conditions holds:

(1) There exist l_1 and l_2 in L such that $l_1 < l < l_2$ and $\forall x \in X^*$ such that $\mu(x) \neq l$, either $\mu(x) \leqslant l_1$ or $\mu(x) \geqslant l_2$.

(2) l is a maximum element of L and there exists $l_1 \neq l$ in L such that $\forall x \in X^*$ with $\mu(x) \neq l$, $\mu(x) \leqslant l_1$.

(3) $l = 0$ and there exists $l_2 \neq 0$ in L such that $\forall x \in X^*$ with $\mu(x) \neq 0$, $\mu(x) \geqslant l_2$.

Theorem 9.4.1 *Let L be a lattice with minimum element 0. Let μ : $X^* \longrightarrow L$ be an L-fuzzy language and let l be an isolated cutpoint of μ. Then, for $i = 0, 1, 2, 3$, if μ is fuzzy recognized by a machine in T_i, then $L_{GE}(f, l)$ and $L_G(f, l)$ are recognized by a machine in T_i.*

Proof. Suppose that μ is fuzzy recognized by a machine

$$M = \langle X \cup \Delta \cup \{t\}, \Gamma, \ \Psi, \theta, \delta, \Omega, \kappa, \gamma_0 \rangle$$

in T_i with a representation r over Δ. Since l is an isolated cutpoint of μ, either (1), (2), or (3) holds. We only prove the case when (1) holds. The proofs for other cases are similar. Let l_1 and l_2 in L be such that $l_1 < l < l_2$. Suppose that $\forall x \in X^*$ with $\mu(x) \neq l$, either $\mu(x) \geqslant l_2$ or $\mu(x) \leqslant l_1$. Let d_1 and d_2 be the D-length of $r(l_1)$ and $r(l)$ and that of $r(l)$ and $r(l_2)$, respectively. Let $d_3 = d_1 \vee d_2$ and let $w \in \Delta^*$ be the prefix of $r(l)$ of length d_3. From the definition of fuzzy recognition, the set $L[M, \geqslant]$ is recognized by a machine in T_i, where $L[M, \geqslant]$ is the set of all words of the form xty with $x \in X^*$, and $y \in \Delta^*$ is such that $\gamma_0 \underset{xty}{\Longrightarrow} \gamma$ and $\kappa(\theta(\gamma))$ is $=$ or $>$. We now show that

$$\{x \in X^* | \mu(x) \geqslant l\} = \{x \in X^* | xty \in L[M, \geqslant] \text{ for some } y \in PRE \ w\Delta^*\}.$$

If $\mu(x) \geqslant l$, then there exists $\gamma \in \Gamma^*$ and $y \in PRE\{r(1)\} \subseteq PREw\Delta^*$ such that $\gamma_0 \underset{xty}{\Longrightarrow} \gamma$ and $\kappa(\theta(\gamma))$ is $=$ or $>$. Conversely, suppose that $\gamma_0 \underset{xty}{\Longrightarrow} \gamma$

and $\kappa(\theta(\gamma))$ is $=$ or $>$, where $x \in X^*$, $y \in PREw\Delta^*$, and $\gamma \in \Gamma^*$. If $y \in PREr(l)$, then clearly $\mu(x) \geqslant l$. Otherwise, there exists $l' \in L$ such that $\mu(x) \geqslant l'$ and for some $w' \in \Delta^*$ and $\alpha \in \Delta^\infty$, $r(l') = y\alpha = ww'\alpha$. From the definition of D-length, neither l' and l_1 nor l' and l_2 are incomparable, and also neither $l' < l_1$ nor $l_2 < l'$ holds. Therefore, $l_1 \leqslant l' \leqslant l_2$. This implies that $\mu(x) = l$ or $\mu(x) \geqslant l_2$. Thus $\mu(x) \geqslant l$. Clearly, there exists a gsm-mapping G, [96, p.272], such that

$$L_{GE}(\mu, l) = G(L[M, \geqslant] \cap X^* cPREw\Delta^*).$$

Since classes of recursively enumerable sets, context-free languages, and regular sets are closed under a gsm-mapping operation, respectively, it holds for $i = 0, 2$, and 3 that $L_{GE}(\mu, l)$ is recognized by a machine in T_i.

Suppose that $i = 1$. Then machine M is a Turing machine such that for some constant t, $|\gamma| \leqslant t|x|$ for any $x \in X^*$ and for any memory configuration γ such that $\gamma_0 \underset{xty}{\Longrightarrow} \gamma$ for some $y \in PRE\{r(l) \mid l \in L\}$. A machine M' is a modification of M as follows, M' moves reading xt in the same way as M moves reading xt for $x \in X^*$. After reading xt, M' continues to read Λ and changes sequentially memory configurations in the same way as M reads some $y \in PREw\Delta^*$. In order to move in this way, M' has an autonomous finite state machine as a submachine that generates any $y \in PREw\Delta^*$ nondeterministically. Clearly, $M' \in T_1$ and thus the set $L(M')$ of all words xt ($x \in X^*$) for which M' emits $>$ or $=$ is recognized by a machine in T_1. It follows that $L(M') = \{xt \mid x \in X^*, y \in PREw\Delta^*$ and $xty \in L(M, \geq)\}$. Therefore, $L(M') = L_{GE}(\mu, l)t$. Hence $L_{GE}(\mu, l)$ is recognized by a machine in T_1.

The proof for $L_G(\mu, l)$ is similar. ∎

Corollary 9.4.2 *Let L be a finite lattice and let $\mu : X^* \to L$ be an L-fuzzy language. Then, for $i = 0, 1, 2, 3$, if μ is fuzzy recognized by a machine in T_i, then for any $l \in L$, $L_{GE}(\mu, l)$ and $L_G(\mu, l)$ are recognized by a machine in T_i.*

Proof. Suppose that μ is fuzzy recognized by a machine

$$M = \langle X \cup \Delta \cup \{t\}, \Gamma, \Psi, \theta, \delta, \Omega, \kappa, \gamma_0 \rangle$$

in T_i. Let $L = \{l_1, l_2, \dots, l_s\}$. Then there exists $w_j \in \Delta^*$ such that $r(l_j) \in w_j\Delta^\infty$, but $r(l_j) \notin w_k\Delta^\infty$ for $1 \leq j < k \leq s$. Let $L[M, \geq]$ be the set of all words of the form xty with $x \in X^*$ and $y \in \Delta^*$ such that $\gamma_0 \underset{xty}{\Rightarrow} \gamma$ and $\kappa(\theta(\gamma))$ is $=$ or $>$. Then it follows from the definition of fuzzy recognition that $L[M, \geq]$ is recognized by a machine in T_i. Clearly, it follows that

$$\{x \in X^* \mid \mu(x) \geq l_j\} = \{x \in X^* \mid xty \in L[M, \geq] \text{ for some } y \in PREw_j\Delta^\infty\}$$

for $j = 1, \ldots, s$ such that l_j is not the minimum element of L. Moreover, if l_j is the minimum element of L, then $\{x \in X^* \mid \mu(x) \geq l_j\} = X^*$ is clearly recognized by a machine in T_i. The remainder of the proof is the same as in the proof of Theorem 9.4.1. ∎

Theorem 9.4.3 *Let L be a lattice with minimum element 0 that has a representation r_0 over Δ_0. Let $\mu : X^* \to L$ be an L-fuzzy language such that $\mu(X^*)$ is finite. Then for $i = 0, 1, 2, 3$, if $L_{GE}(\mu, \mu(x))$ is recognized by a machine in T_i $\forall x \in X^*$, then μ is fuzzy recognized by a machine in T_i.*

Proof. Suppose that $\mu(X^*) = \{l_1, l_2, \ldots, l_s\}$. Let M_j be a machine recognizing $L_{GE}(\mu, l_j)$ for $j = 1, \ldots, s$. A machine M that fuzzy recognizes μ with a representation r over Δ is given as follows: Let $\Delta = \Delta_0 \cup \Delta_1 \cup \Delta_2$, where $\Delta_1 = \{l'_1, l'_2, \ldots, l'_s\}$ such that l'_i is a new symbol corresponding uniquely to l_i for $i = 1, \ldots, s$ and $\Delta_2 = \mathcal{P}(\Delta_1) \times \mathcal{P}(\Delta_1)$. Define r as follows:

$$
\begin{aligned}
r(l_j) &= l'_j r_0(l_j) && \text{for } j = 1, \ldots, s \\
r(l) &= (A_l, B_l) r_0(l) && \text{if } l \notin \mu(X^*),
\end{aligned}
$$

where

$$A_l = \{l'_j \mid l_j > l\} \text{ and } B_l = \{l'_j \mid l_j < l\}.$$

Now M_i is a submachine of M, $i = 1, \ldots, s$. For any word xt with $x \in X^*$, M reads first Λ and chooses nondeterministically the initial configuration of any one of the M_j, say M_k, and hereafter M_k moves reading x as an input word. Let Γ_j be the set of all configurations of M corresponding to accepting configurations of M_j, $j = 1, \ldots, s$. From any configuration $\gamma'_r \in \Gamma_j$, $j = 1, \ldots, s$, M moves into the new memory configuration $\widetilde{\gamma}^j_r$ by the input symbol t. No transition into $\widetilde{\gamma}^j_r$ other than the above one is permitted in M. Also, M with the configuration $\widetilde{\gamma}^j_r$, say $M(\widetilde{\gamma}^j_r)$, moves deterministically as follows: For an input symbol l'_k, $M(\widetilde{\gamma}^j_r)$ emits, respectively, $=$, $>$, $<$, and $!$ according to the cases (1) $l_j = l_k$, (2) $l_j > l_k$, (3) $l_j < l_k$, and (4) l_j and l_k are incomparable, and then halts. Reading (A_l, B_l), $M(\widetilde{\gamma}^j_r)$ emits, respectively, $>$, $<$, and $!$ according to the cases (1) $l'_j \in A_l$, (2) $l'_j \in B_l$, (3) $l'_j \notin A_l \cup B_l$, and then halts. Let $\widetilde{\Gamma}_j = \{\widetilde{\gamma}^j_r \mid \gamma^j_r \in \Gamma_j\}$ and let $\nu(\widetilde{\gamma}^j_r) = l_j$ for all $\widetilde{\gamma}^j_r \in \widetilde{\Gamma}_j$ and for all $j = 1, \ldots, s$. If $\mu(x) = l_k$, then $S_x = \cup_{l_j < l_k} \widetilde{\Gamma}_j$. Therefore,

$$\mu(x) = \nu(\widetilde{\gamma}^j_r) = \vee\{\nu(\gamma) \mid \gamma \in S_x\}.$$

Thus μ is fuzzy recognized by M. ∎

The following corollaries follow from Theorems 9.4.1, 9.4.3, and Corollary 9.4.2.

Corollary 9.4.4 *Let L be a totally ordered set with a minimum element and that has some representation or let L be a finite lattice. If μ is an L-fuzzy language over some alphabet X such that $\mu(X^*)$ is finite, then for $i = 0, 1, 2, 3$, μ is fuzzy recognized by a machine in T_i if and only if $\forall x \in X^*$, $\mathbf{L}_{GE}(\mu, \mu(x))$ is recognized by a machine in T_i.* ∎

Corollary 9.4.5 *Let $\mathbf{L}$ be a language over X and let $\chi_{\mathbf{L}} : X^* \to B$ be the characteristic function of $\mathbf{L}$, where B is the Boolean lattice with two elements. For $i = 0, 1, 2, 3$, $\mathbf{L}$ is recognized by a machine in T_i if and only if $\chi_{\mathbf{L}}$ is fuzzy recognized by a machine in T_i.* ∎

Corollary 9.4.5 shows that the representation concept for fuzzy languages introduced in this section is a fairly good extension of the one for ordinary languages.

Example 9.4.6 *Let L be a lattice with a minimum element 0 and a maximum element 1. An L-fuzzy context-free grammar is defined to be a quadruple $G = (N, T, P, S)$, where $N \cup T$ is a finite set of symbols with $N \cap T = \emptyset$, T is the set of terminal symbols, N is the set of nonterminal symbols, $S \in N$, and P is a finite set of production rules of the form*

$$A \xrightarrow{l} x$$

with $A \in N$, $x \in (N \cup T)^*$, and $l \in L$. For any $y, z \in (N \cup T)^*$, we write

$$y \xrightarrow{l} z$$

if there exists $u, v \in (N \cup T)^*$ and $A \xrightarrow{l} z$ is in P such that $y = uAv$ and $z = uxv$. We write

$$y \overset{1}{\underset{*}{\Rightarrow}} y$$

$$y \overset{0}{\underset{*}{\Rightarrow}} z$$

for all $y, z \in (N \cup T)^*$, and

$$y \overset{m}{\underset{*}{\Rightarrow}} z$$

if and only if there exists a sequence of elements $y_0, y_1, \ldots, y_n \in (N \cup T)^*$ such that $y_0 = y$, $y_n = z$, $y_{i-1} \xrightarrow{l_i} y_i$ for $i = 1, \ldots, n$ and $\wedge_{i=1}^n l_i = m$. For any $x \in T^*$, let $l_x = \vee\{l \in L \mid S \overset{l}{\underset{*}{\Rightarrow}} x\}$. The L-fuzzy language μ defined by

$$\mu(x) = l_x, \text{ for all } x \in T^*$$

is said to be generated by G. An L-fuzzy language generated by some L-fuzzy context-free grammar is called an L-fuzzy context-free language. (L-fuzzy phrase structure, L-fuzzy context-sensitive, L-fuzzy regular grammars and languages are similarly defined, respectively.) $L_{[0,1]}$-fuzzy context-free languages were studied in Chapter 4. From Proposition 18 in [261] and Theorem 9.4.3, it follows that any $L_{[0,1]}$-fuzzy context-free language is fuzzy recognized by a pushdown automaton.

Example 9.4.7 *Let $X = \{a, b, c\}$ and let μ_2 be the $L_{[0,1]}$-fuzzy language over X defined by $\forall i, j, k \in \mathbb{N} \cup \{0\}$*

$$\mu_2(a^i b^j c^k) = (\frac{1}{2})^{|i-j|+1} + (\frac{1}{2})^{|j-k|+1}$$

and

$$\mu_2(w) = 0 \ \text{if} \ w \notin a^* b^* c^*.$$

Then 1 is an isolated cutpoint of μ_2, and $\mathbf{L}_{GE}(\mu_2, 1) = \{a^i b^i c^i \mid i \in \mathbb{N} \cup \{0\}\}$ is not recognized by any pushdown automaton. Thus by Theorem 9.4.1, μ_2 is not fuzzy recognized by any pushdown automaton. It follows that μ_2 is fuzzy recognized by some deterministic linear bounded automaton with the representation r_1 of Example 9.2.2.

We now consider regular representations.

The following theorem is clear from the proof of Theorem 9.4.1.

Theorem 9.4.8 *Let L be a lattice with a minimum element. Suppose that the L-fuzzy language μ is fuzzy recognized by a machine in $T_i(DT_i)$ with a representation r and that $r(l)$, $l \in L$, is generated by an autonomous finite automaton sequentially. Then $\mathbf{L}_G(\mu, l)$ and $\mathbf{L}_{GE}(\mu, l)$ are recognized by a machine in $T_i(DT_i)$, where $i = 0, 1, 2, 3$. ∎*

We next introduce the concept of a regular representation. Let L be a lattice with a minimum element. A representation r of L is said to be **regular** if for all $l \in L$, $r(l)$ is generated sequentially by some autonomous finite automaton.

The following corollary follows from Theorem 9.4.8.

Corollary 9.4.9 *Let L be a lattice with a minimum element. For $i = 1, 2, 3$, if an L-fuzzy language μ is fuzzy recognized by a machine in $T_i(DT_i)$ with a regular representation, then for all $l \in L$, $\mathbf{L}_G(\mu, l)$ and $\mathbf{L}_{GE}(\mu, l)$ are recognized by a machine in $T_i(DT_i)$. ∎*

The representation of r_2 of $L_{[0,1]_Q}$ shown in the Example 9.2.3 is regular. Thus the L-fuzzy language μ_1 of Example 9.3.1 and for any $l \in L_{[0,1]_Q}$, $\mathbf{L}_G(\mu, l)$ and $\mathbf{L}_{GE}(\mu, l)$ are context-free languages. Clearly, only lattices of restricted type have regular representations. For example, $L_{[0,1]}$ cannot have a regular representation. However, fuzzy recognition of a fuzzy language with a regular representation is of interest since, by Corollary 9.4.9, it gives a property independent of cutpoints of the fuzzy language with respect to recognition of its cutpoint languages.

9.5   Fuzzy Languages not Fuzzy Recognized by Machines in DT_2

In view of Theorem 9.4.1 and Corollary 9.4.2, L-fuzzy languages not fuzzy recognized by a machine in T_i for $i = 0, 1, 2, 3$ are easily found. However, Theorem 9.4.1 and Corollary 9.4.2 cannot be used to find an L-fuzzy language whose membership function-values distribute densely over L and that is not fuzzy recognized by a machine in T_i for $i = 0, 1, 2$. Hence it would be of interest to find such a language. Although we do not have such an example, we do have the following example.

Example 9.5.1 *Let* $X = \{0, 1\}$ *and let* μ_3 *be an* $L_{[0,1]_\mathbb{Q}}$*-fuzzy language over* X *such that for* $a_i \in X$, $i = 1, 2, \ldots, k$,

$$\mu_3(a_1 a_2 \ldots a_k) = a_1 2^{-1} + a_2 2^{-2} + \ldots + a_k 2^{-k}$$

(binary expansion), and

$$\mu_3(\Lambda) = 0.$$

We now show that μ_3 *is not fuzzy recognized by any deterministic pushdown automaton.*

Suppose that μ_3 *is fuzzy recognized by a deterministic pushdown automaton*

$$M = \langle X \cup \Delta \cup \{t\}, \Gamma, \{<, =, >\}, \theta, \delta, \Omega, \kappa, \gamma_0 \rangle$$

with a representation r *over* Δ. *Let*

$$\mathbf{L}_1 = \{xty \mid \gamma_0 \underset{xty}{\Rightarrow} \gamma \text{ and } \kappa(\theta(\gamma)) = (=)\}.$$

Then $\mathbf{L}_1$ *is a context-free language included in* $X^* t \Delta^*$. *Let* $\mathbf{L}_2 = \mathbf{L}_1 \cap 0^* 1^* t \Delta^*$. *Then* $\mathbf{L}_2$ *is also a context-free language. Since* M *is deterministic, for any* $xty \in \mathbf{L}_1$, *there exists* $\alpha \in \Delta^\infty$ *such that* $r(\mu_3(x)) = y\alpha$. *Due to the pumping lemma of the theory of context-free languages, there exists a constant* k *such that if* $|z| \geq k$ *and* $z \in \mathbf{L}_2$, *then* $z = uvwxy$ *such that* $vx \neq \Lambda$, $|vwx| \leq k$, *and for all* i, $uv^i wx^i y \in \mathbf{L}_2$. *Let* $m \geq k$ *and let* z_p *be an element in* $\mathbf{L}_2$ *of the form* $0^p 1^m t g_p$, $g_p \in \Delta^*$, *for any* $p \in \mathbb{N} \cup \{0\}$. *Then* $z_p = u_p v_p w_p x_p y_p$, *where* $v_p x_p \neq \Lambda$, $|v_p w_p x_p| \leq k$, *and for all* $i \in \mathbb{N} \cup \{0\}$, $u_p v_p^i w_p x_p^i y_p \in \mathbf{L}_2$. *Since* M *is deterministic and halts immediately after it emits* $(=)$, *there does not exist* $x \in X^*$ *and* $y, y' \in \Delta^*$ *with* $y \neq y'$ *such* *that both* $xty, xty' \in \mathbf{L}_2$. *Since* $\mu_3(x) \neq \mu_3(x')$ *for* x *and* x' *in* 0^*1^m *with* $x \neq x'$, *there do not exist* x *and* x' *in* 0^*1^m *with* $x \neq x'$ *and* $y \in \Delta^*$ *such that* $xty, x'ty \in \mathbf{L}_2$. *Thus for all* p, *neither* u_p *nor* y_p *contains* t. *Since* t *cannot occur in either* v_p *or* x_p, w_p *contains* t *for all* p. *Hence* $v_p \neq \Lambda$ *and* $x_p \neq \Lambda$ *for all* $p \geq 0$. *Also* $\forall\, p \geq 0$, *we can write* $v_p = 1^{s_p}$ *for some* $s_p \geq 1$ *and* $w_p = 1^{t_p} t W_p$ *for some* $t_p \geq 0$ *and* $W_p \in \Delta^*$. *Since* $|v_p w_p x_p| \leq k$ *for all* p, *there exist non-negative integers* p *and* q, *and* W *and* $\Lambda \neq x \in \Delta^*$ *such that* $p < q$, $W_p = W_q = W$, $s_p = s_q = s$, *and* $x_p = x_q = x$. *Thus*

$$z_p = 0^p 1^m t W x y_p$$
$$z_q = 0^q 1^m t W x y_q$$

and for all $i \in N \cup \{0\}$

$$0^p 1^{m-s} 1^{s_i} t W x^i y_p \in \mathbf{L}_2$$
$$0^q 1^{m-s} 1^{s_i} t W x^i y_q \in \mathbf{L}_2.$$

There exist $\alpha_0, \alpha_1 \in \Delta^\infty$ *such that*

$$r^{-1}(W y_p \alpha_0) = \mu_3(0^p 1^{m-s}) < \mu_3(0^p 1^m) = r^{-1}(W x y_p \alpha_1).$$

Let d_0 *be the D-length of* $W y_p \alpha_0$ *and* $W x y_p \alpha_1$, *and let* $j > d_0$. *Then there exists* $\alpha_2 \in \Delta^\infty$ *such that*

$$r^{-1}(W x^i y_p \alpha_2) = \mu_3(0^p 1^{m-s} 1^{sj}) > \mu_3(0^p 1^m).$$

Hence the D-length d_1 *of* $W y_p \alpha_0$ *and* $W x^i y_p \alpha_2$ *is less than or equal to* d_0. *Since* $j > d_0$, $j > d_1$. *However, there exists* $\alpha_3 \in \Delta^\infty$ *such that*

$$r^{-1}(W x^j y_q \alpha_3) = \mu_3(0^p 1^{m-s} 1^{sj}) < \mu_3(0^p 1^{m-s}) = r^{-1}(W y_p \alpha_0),$$

which contradicts the definition of a representation. Thus μ_3 *cannot be fuzzy recognized by any deterministic pushdown automaton.*

It is well known in the theory of probabilistic automata [168] that $\mathbf{L}_{GE}(\mu, l)$ and $\mathbf{L}_G(\mu, l)$ are regular sets for any $l \in L_{[0,1]_Q}$. Thus it is not true in general that an L-fuzzy language is fuzzy recognized by a machine in T_3 (even in DT_3) even if all its cutpoint languages are recognized by machines in T_3. Consequently, the converse of Corollary 9.4.9 does not hold.

9.6 Rational Probabilistic Events

In this section, it is shown that any rational probabilistic event is fuzzy recognized by a deterministic linear bounded automaton with a regular representation. A rational probabilistic automaton A with n states over a finite alphabet X is a triple $A = \langle \pi, \{A(u) \mid u \in X\}, \eta \rangle$, where π is a $1 \times n$ matrix $\begin{bmatrix} \pi_1 & \pi_2 & \cdots & \pi_n \end{bmatrix}$ with $\pi_i \in [0,1]_Q$ for $i = 0, \ldots, n$ such that

$\sum_{i=1}^{n} \pi_i = 1$, for all $u \in X$, $A(u)$ is an $n \times n$ stochastic matrix such that all components of $A(u)$ are in $[0,1]_{\mathbb{Q}}$ and η is an $n \times 1$ matrix such that all components of η are in $[0,1]_{\mathbb{Q}}$. An L-fuzzy language $p : X^* \to L_{[0,1]_{\mathbb{Q}}}$ is said to be **realized** or **accepted** by a rational probabilistic automaton A if $p(\Lambda) = \pi\eta$ and for any $m \in \mathbb{N}$ and $u_i \in X$ for $i = 1, \dots, m$

$$p(u_1 u_2 \dots u_m) = \pi A(u_1) A(u_2) \dots A(u_m) \eta.$$

An L-fuzzy language p is called a **rational probabilistic event** if p is realized by some rational probabilistic automaton. The reader is referred to [168] for properties concerning probabilistic automata.

Theorem 9.6.1 *Every rational probabilistic event is fuzzy recognized by a machine in DT_1 with a regular representation.*

Proof. Suppose that $p : X^* \to L_{[0,1]_{\mathbb{Q}}}$ is a rational probabilistic event. Assume that p is realized by a rational probabilistic automaton

$$A = \langle \pi, \{A(u) | u \in X\}, \eta \rangle,$$

with n states. Let all components of π, $A(u)$'s ($u \in X$), and η be represented in the form h/k for $h, k \in \mathbb{N}$ such that k is common to all of them. For any word $x = u_1 u_2 \dots u_m$ of length $m \geq 1$, let $\pi(x) = (\pi_1(x), \pi_2(x), \dots, \pi_n(x))$ be $\pi A(u_1) A(u_2) \dots A(u_m)$ and let $\pi(\Lambda) = \pi$. Then for any word x of length m, $\pi_i(x)$ is represented in the form $h_i(x)/k^{m+1}$ with $0 \leq h_i(x) \leq k^{m+1}$, for $i = 1, \dots, n$, and $p(x)$ is represented in the form $w(x)/k^{m+2}$ with $0 \leq w(x) \leq k^{m+2}$. We denote the binary expansion of a nonnegative integer j by $b(j)$ and the length of $b(j)$ by $|b(j)|$. Then we can construct a Turing machine z such that when z reads an input word $x \in X^*$, z gives the output string

$$T(x) = b(h_1(x)) \# b(h_2(x)) \# \dots \# b(h_n(x)) \# b(w(x)) \# b(k^{|x|+2})$$

on its tape using at most $d|x|$ working spaces, where d is a constant independent of x. Clearly, $T(xu)$ with $u \in X$ can be computed by a Turing machine from $T(x)$ at most a constant multiple of $|b(k^{|xu|+1})|$ spaces and that for any non-negative integer s, $|b(k^s)| \leq d's$ for some constant d'. Thus by induction on the length of an input word, it follows that there exists a Turing machine z that computes $T(x)$ for $x \in X^*$ using at most $d|x|$ spaces.

Subsequent moves of z are defined as follows: If z with a memory configuration $T(x)$ reads t as its next input symbol, then z transforms $T(x)$ into $b(s(x)) \# b(s'(x))$ on its tape, where $s(x)/s'(x) = w(x)/k^{|x|+2}$ and $s(x)$ and $s'(x)$ are relatively prime. (If $w(x) = 0$, z prints 0 and if $w(x) = k^{|x|+2}$, z prints 1 on its tape.) Since division by a binary number not greater than $b(k^{|x|+2})$ can be done with at most a constant multiple of $|b(k^{|x|+2})|$ spaces, z can give $b(s(x)) \# b(s'(x))$ on its tape using $d|x|$ spaces after it reads xt.

We now define a representation $r : L_{[0,1]_Q} \to \Delta'^{\infty}$ as follows, where $\Delta' = \{0,1\} \times \{0,1,\#\}$: Define $\tau_1 : \Delta'^{\infty} \to \{0,1,\#\}^{\infty}$ and $\tau_2 : \Delta'^{\infty} \to \{0,1\}^{\infty}$ by $\tau_i(u_1, u_2) = u_i$ for $(u_1, u_2) \in \Delta'$ and $\tau_i(x) = \tau_i(u_1)\tau_i(u_2)\tau_i(u_3)\ldots$ for $x = u_1u_2u_3\ldots$ with $u_j \in X$, $j \geq 1$, for $i = 1, 2$. For $l \in L_{[0,1]_Q}$ with $0 \neq l \neq 1$, let $l = s/s'$, where s and s' are relatively prime positive integers. Then $r(l)$ is the element in Δ'^{∞} such that $\tau_1(r(l)) = b(s)\#b(s')\#\#\#\ldots$ and $\tau_2(r(l)) = e(l)$, where $e(l)$ was given in Example 9.2.2. Then $r(0) = (0,0)(\#,0)(\#,0)(\#,0)\ldots$ and $r(1) = (1,1)(\#,1)(\#,1)(\#,1)\ldots$. Clearly, r is a representation since the D-length of $r(l)$ and $r(m)$ with $l \neq m$ may be defined as the positive integer k such that $\tau_2(r(l))$ and $\tau_2(r(m))$ differ first at the kth digit. It follows easily that r is a regular representation. If z with $b(s(x))\#b(s'(x))$ on its tape is given a representation $r(l)$ of $l \in L_{[0,1]_Q}$ as an input succeeding xt for $x \in X^*$, z computes the binary expansion of $s(x)/s'(x)$, digit-by-digit, and compares it with $\tau_2(r(l))$. In parallel with this computation, z compares sequentially $b(s(x))\#b(s'(x))$ with $\tau_1(r(l))$. Then either z emits $=$ and halts if $b(s(x))\#b(s'(x)) = b(s)\#b(s')$, where $\tau_1(r(l)) = b(s)\#b(s')\#\#\#\ldots$, or z emits $>$ or $<$ and halts according to the comparison of the first distinguished digit of $\tau_2(r(l))$ and the binary expansion of $s(x)/s'(x)$. Since the digit-by-digit generation of the binary expansion of $s(x)/s'(x)$ can be made using at most a constant multiple of $|b(s'(x))|$ spaces, z can do the above value comparing computation for $x \in X^*$ using at most $d|x|$ spaces. Thus p is fuzzy recognized by z in DT_1 with the regular representation r. ∎

The following result, which was proved in [230], follows directly from Theorem 9.6.1.

Corollary 9.6.2 *Let $p : X^* \to L_{[0,1]_Q}$ be a rational probabilistic event. Then for all $l \in L_{[0,1]_Q}$, $\mathbf{L_G}(\mu, l)$ and $\mathbf{L_{GE}}(\mu, l)$ are recognized by a machine in DT_1.* ∎

9.7   Recursive Fuzzy Languages

The relation between deterministic machines and nondeterministic machines with respect to the fuzzy recognizability of fuzzy languages differs somewhat from that of ordinary languages. We show that in the fuzzy recognition of fuzzy languages, nondeterministic Turing machines are more powerful than deterministic Turing machines. Let $\{t_0, t_1, t_2, \ldots\}$ be an enumeration of deterministic Turing machines. Let L_3 be a lattice with three elements $0, c$, and 1 such that $0 < c < 1$. Let $X = \{u\}$. Let μ_4 and μ_5 be L_3-fuzzy languages over X defined as follows: For $n \in \mathbb{N} \cup \{0\}$,

$$\mu_4(u^n) = \begin{cases} 1 & \text{if } t_n \text{ with the blank tape eventually halts} \\ 0 & \text{otherwise} \end{cases}$$

$$\mu_5(u^n) = \begin{cases} 1 & \text{if } t_n \text{ with the blank tape eventually halts} \\ c & \text{otherwise.} \end{cases}$$

Lemma 9.7.1 *Let μ_4 and μ_5 be defined as above. Then μ_4 is fuzzy recognized by a deterministic Turing machine. μ_5 is fuzzy recognized by a nondeterministic Turing machine, but it is not fuzzy recognized by a deterministic Turing machine.*

Proof. Clearly, there exists a deterministic Turing machine that fuzzy recognizes μ_4. Since $\mathbf{L_{GE}}(\mu_5, 1)$ and $\mathbf{L_{GE}}(\mu_5, c)$ are recursively enumerable languages, it follows from Theorem 9.4.3 that μ_5 is fuzzy recognized by a Turing machine. Suppose that μ_5 is fuzzy recognized by a deterministic Turing machine. Then it follows that the language $\{u^n \mid \mu_5(u^n) = c\}$ is recursively enumerable. Thus the halting problem of Turing machines is solvable, which is a contradiction. Hence μ_5 is not fuzzy recognized by a deterministic Turing machine. ∎

Let $\mathcal{L}(T_i)$ and $\mathcal{L}(DT_i)$ be the sets of L-fuzzy languages fuzzy recognized by a machine in T_i and DT_i, respectively, for $i = 0, 1, 2, 3$.

Theorem 9.7.2 *(1) $\mathcal{L}(T_0) \supset \mathcal{L}(DT_0)$.*
(2) $\mathcal{L}(T_2) \supset \mathcal{L}(DT_2)$.
(3) $\mathcal{L}(T_3) = \mathcal{L}(DT_3)$.

Proof. (1) is immediate from Lemma 9.7.1. (2) follows from Corollary 9.4.4. (3) follows easily. ∎

It is not known whether or not $\mathcal{L}(T_1) \supset \mathcal{L}(DT_1)$.

Considering Lemma 9.7.1, it seems reasonable to define recursive L-fuzzy languages as follows: An L-fuzzy language μ over X is recursive if and only if μ is fuzzy recognized by some machine

$$M = \langle X \cup \Delta \cup \{t\}, \ \Gamma, \ \Psi, \ \theta, \ \delta, \ \Omega, \ \kappa, \ \gamma_0 \rangle$$

in DT_0 with some representation r over Δ with the condition that $S_x \neq \emptyset$ for any $x \in X^*$, where $S_x = \{\gamma \in \Gamma^* \mid \gamma_0 \underset{xt}{\Rightarrow} \gamma\}$.

Clearly, any L-fuzzy language in $\mathcal{L}(T_3)$ is recursive. The proof of the following proposition follows easily.

Proposition 9.7.3 *An L-fuzzy language in $\mathcal{L}(DT_2) \cup \mathcal{L}(T_1)$ is a recursive fuzzy language.* ∎

9.8 Closure Properties

We now consider closure properties of the classes of fuzzy languages corresponding to machine classes T_i's under fuzzy set operations. We show that these properties are the extensions of certain closure properties of ordinary languages such as regular sets, context-free languages, context-sensitive languages, and recursively enumerable sets. We consider $\mathbf{L}_{[0,1]}$-fuzzy languages.

Theorem 9.8.1 *Let μ and ν be $\mathbf{L}_{[0,1]}$-fuzzy languages over X. Let $i \in \{0,1,2,3\}$. If μ and ν are fuzzy recognized by a machine in T_i, then $\mu \cup \nu$ is fuzzy recognized by a machine in T_i.*

Proof. Suppose that μ and ν is fuzzy recognized by a machine M_1 with a representation $r_1 : \mathbf{L}_{[0,1]} \to \Delta_1^\infty$ and a machine M_2 with a representation $r_2 : \mathbf{L}_{[0,1]} \to \Delta_2^\infty$, respectively. Let $\Delta = \Delta_1 \times \Delta_2$. For $j = 1, 2$ define $\tau_j : \Delta^\infty \to \Delta_j^\infty$ as follows: $\tau_j((a_1, a_2)) = a_j$ with $a_j \in \Delta_j$ and for $y = b_1 b_2 b_3 \dots$ with $b_k \in \Delta$, $k \geq 1$, $\tau_j(y) = \tau_j(b_1)\tau_j(b_2)\tau_j(b_3)\dots$. Let $r : \mathbf{L}_{[0,1]} \to \Delta^\infty$ be such that for any $l \in \mathbf{L}_{[0,1]}$, $\tau_1(r(l)) = r_1(l)$ and $\tau_2(r(l)) = r_2(l)$. Then r is a representation of $\mathbf{L}_{[0,1]}$ over Δ since the D-length $d(r(l), r(m))$ of $r(l)$ and $r(m)$ for any distinct $l, m \in \mathbf{L}_{[0,1]}$ can be defined as $d(r(l), r(m)) = d(r_1(l), r_1(m))$. A machine M that fuzzy recognizes $\mu \cup \nu$ with the representation r is given as follows: M has M_1 and M_2 as submachines. For any word xt with $x \in X^*$, M first reads Λ, chooses nondeterministically M_1 or M_2, and simulates the chosen machine hereafter reading xt. If M reads through xt simulating M_j, $j = 1, 2$, and if M is given $r(l)$, $l \in \mathbf{L}_{[0,1]}$, then M moves as M_j does if it is given $\tau_j(r(l)) = r_j(l)$. Clearly, M fuzzy recognizes $\mu \cup \nu$. ∎

Theorem 9.8.2 *Let $i \in \{0, 1, 3\}$. Let μ and ν be $\mathbf{L}_{[0,1]}$-fuzzy languages over X. If μ and ν are fuzzy recognized by a machine in T_i, then $\mu \cap \nu$ is fuzzy recognized by a machine in T_i.*

Proof. We prove the result for $i = 1$. The proofs for the $i = 0$ and 3 cases are similar. Suppose that μ and ν are fuzzy recognized by a machine M_1 in T_1 with a representation $r_1 : \mathbf{L}_{[0,1]} \to \Delta_1^\infty$ and a machine M_2 in T_1 with a representation $r_2 : \mathbf{L}_{[0,1]} \to \Delta_2^\infty$. Let $\Delta = \Delta_1 \times \Delta_2$. Define a representation $r : \mathbf{L}_{[0,1]} \to \Delta^\infty$ as defined in the proof of Theorem 9.8.1. We determine a machine M in T_1 that fuzzy recognizes $\mu \cap \nu$ with the representation r with M having M_1 and M_2 as submachines. When M is given xt with $x \in X^*$ as an input word, submachines M_1 and M_2 of M move in parallel reading xt, and M reaches the configuration γ corresponding to the pair of configurations γ_1 of M_1 and γ_2 of M_2, which they reach after reading xt, respectively. For this computation of M, at most a constant multiple of $|x|$ spaces is necessary. If M with the configuration γ is given $r(l)$ with $l \in \mathbf{L}_{[0,1]}$ as a subsequent input, then M makes M_1 with the configuration γ_1 read $\tau_1(r(l)) = r_1(l)$ and in parallel makes M_2 with the configuration γ_2 read $\tau_2(r(l)) = r_2(l)$. When one of M_1 or M_2 emits an output, i.e., one of $>$, $<$, and $=$, M remembers it in a state and then continues to make another submachine move until it emits an output. If both M_1 and M_2 emit $>$, then M emits $>$. If one of them emits $=$ and another emits $>$ or $=$, then M emits $=$. If either one of them emits $<$, then M emits $<$. M halts immediately after M emits an output. Since M_1 and M_2 are machines in T_1, M needs at most a constant multiple of $|x|$ spaces in order to do the above value-comparing computation for $x \in X^*$. Thus M in T_1 fuzzy recognizes $\mu \cap \nu$. ∎

Since for two context-free languages L_1 and L_2, $L_1 \cap L_2$ is not necessarily a context-free language, it follows from Corollary 9.4.4 that for two L-fuzzy languages μ and ν each of which is fuzzy recognized by a machine in T_2, $\mu \cap \nu$ is not necessarily fuzzy recognized by a machine in T_2. However, a similar proof to that of Theorem 9.8.2 shows that if μ is fuzzy recognized by a machine in T_2 and ν is fuzzy recognized by a machine in T_3, then $\mu \cap \nu$ is fuzzy recognized by a machine in T_2.

9.9   Fuzzy Grammars and Recursively Enumerable Fuzzy Languages

The purpose of this and the next section is to prove that an L-fuzzy language is generated by an L-fuzzy grammar if and only if it is recursively enumerable, that is, its set of L-fuzzy points is recursively enumerable.

As an immediate consequence, we give simple proofs that the union, the intersection, the concatenation of two generated L-fuzzy languages is a generated L-fuzzy language. The results are from [73].

L-fuzzy grammars and generated L-fuzzy languages were introduced in [122] in order to examine ambiguous languages. In the classical case, a language is generated by a grammar if and only if it is recursively enumerable, [96, p. 150]. The question thus arises if such a result holds for generated L-fuzzy languages also. This leads to the question of a suitable definition of recursive enumerability for L-fuzzy subsets.

In [17], an L-fuzzy subset whose set of fuzzy points is recursively enumerable is called recursively enumerable. There are several arguments that justify such a definition. For example, it is in accordance with the concept of computability for L-maps given in [200, 210]. Also, in [72], it was shown that if L is finite, an L-fuzzy subset is recursively enumerable if and only if it is the domain of an L-map computable via a Turing L-machine.

In this and the next section, we prove that if L is finite, then an L-fuzzy language is generated by an L-fuzzy grammar if and only if it is a recursively enumerable L-fuzzy subset. Recursive enumerability is a very manageable concept. Consequently, this also provides a simple tool to investigate the properties of generated L-fuzzy languages. As stated earlier, we give a simple proof that the union, intersection, and the concatenation of two generated L-fuzzy languages is also a generated L-fuzzy language.

This allows the results given in [17, 70, 71] about the relationship between fuzziness and decidability to be extended to L-fuzzy languages.

We call an L-fuzzy subset μ **crisp** if for every $x \in X$, either $\mu(x) = 1$ or $\mu(x) = 0$. We identify the classical subsets of X with the crisp L-fuzzy subsets via characteristic functions. If $c \in L$ and $c \neq 0$, then the c-cut of μ is the set $\mu_c = \{x \mid \mu(x) \geq c\}$. Let $x \in X$ and $c \in L$. Then the L-fuzzy subset x_c of X is defined by $x_c(y) = c$ if $x = y$ and $x_c(y) = 0$ otherwise. An L-fuzzy point x_c is said to **belong** to μ provided that $\mu(x) \geq c$. We denote by $P(X, L)$ and by $P(\mu)$ the set of the L-fuzzy points of X and the set of the L-fuzzy points of μ, respectively.

An **L-fuzzy grammar** is a 4-tuple $G = (N, T, P, S)$ with N and T finite sets, the nonterminal and terminal symbols, respectively, such that $N \cap T = \emptyset$, $S \in N$ (the initial symbol), and P a finite set of L-fuzzy productions, i.e., elements of the form $x \xrightarrow{c} y$ with $c \in L$, $c \neq 0$, and $x, y \in (N \cup T)^*$. We say that an L-fuzzy grammar $G = (N, T, P, S)$ is in **normal form** provided that if $x \xrightarrow{c} y$ is an L-fuzzy production of G, then either $x, y \in N^*$ or $x \in N$ and $y \in T$. In an L-fuzzy production $x \xrightarrow{c} y$, c represents the **membership degree** of the rewriting rule $x \to y$. If $c = 1$, we write $x \to y$ for $x \xrightarrow{c} y$. If wxw' and wyw' are elements of $(N \cup T)^*$ and $x \xrightarrow{c} y$ belongs to P, then we say that wyw' is **directly derivable** from wxw' with degree c in the L-fuzzy grammar G. If w and w' are in $(N \cup T)^*$, a **derivation chain** in G from w to w' is a pair of words $(w_1 \ldots w_p, c_1 \ldots c_{p-1})$ such that $w_1 = w$, $w_p = w'$, and w_{i+1} is directly derivable from w_i with degree c_i. A **derivation** of w is a derivation chain from S to w. The element $c_1 \wedge c_2 \wedge \ldots \wedge c_{p-1}$ is called the **degree** of the derivation. An L-fuzzy language $\mu : T^* \to L$ is **generated** by the L-fuzzy grammar G provided that, for every $w \in T^*$, $\mu(w) = \vee\{c \in L \mid c$ is the degree of a derivation of $w\}$. It follows that an L-fuzzy grammar G utilizes only a finite subset X of elements of L and that the sublattice L' generated by X is finite even in the case that L is infinite. This means that every generated L-fuzzy language is a generated L'-fuzzy language, where L' is finite. As a consequence, in the next section, we assume that the lattice L is finite.

9.10   Recursively Enumerable L-Subsets

An effective codification of T^*, $P(T^*, L)$, and L is possible since T and L are finite. Then concepts such as a partial recursive function from T^* into L and such as recursive enumerability for subsets of T^* and $P(T^*, L)$ are defined, [182]. An L-fuzzy subset μ is said to be **recursively enumerable** if its set of points $P(\mu)$ is a recursively enumerable subset of $P(T^*, L)$ and μ is called **decidable** if it is a recursive function from T^* into L. An L-fuzzy subset μ is called a **projection** of a decidable L-fuzzy relation if there exists a finite set B and a decidable L-fuzzy subset ν of $T^* \times B^*$ such that

$\mu(x) = \vee\{\nu(x,y) \mid y \in B^*\}$. The following proposition is proved in [17].

Proposition 9.10.1 *Let μ be an L-fuzzy subset of T^*. Then the following statements are equivalent.*

(1) μ is recursively enumerable.

(2) Every cut of μ is recursively enumerable.

(3) $\mu(x)$ is a projection of a decidable L-fuzzy relation.

(4) $\mu(x) = \lim \nu(x,n)$ with ν recursive and increasing with respect to n.

Proof. (1)$\Rightarrow$(2): Immediate.

(2)$\Rightarrow$(3): Suppose that $|B| = 1$. Then we can identify B^* with $\mathbb{N}$. It follows that $\mu(x) = \vee\{c \in L\backslash\{0\} \mid x \in \mu_c\}$. Since μ_c is recursively enumerable, a partial recursive function ν_c exists whose domain is μ_c. Let $\nu : T^* \times \mathbb{N} \times L \to L$ be defined as follows:

$$\nu(x,n,c) = \begin{cases} c & \text{if } \nu_c(x) \text{ is convergent in less than } n \text{ steps,} \\ 0 & \text{otherwise.} \end{cases}$$

Clearly, ν is recursive and $\mu(x) = \vee\{\nu(x,n,c) \mid n \in \mathbb{N}, c \in L\}$. By identifying $\mathbb{N} \times L$ with $\mathbb{N}$, and therefore with B^*, via a suitable codification, we obtain (3).

(3)$\Rightarrow$(4): Since we can identify B^* with $\mathbb{N}$, we can assume that $\mu(x) = \vee\{\nu(x,n) \mid n \in \mathbb{N}\}$ with ν recursive. Set $\nu'(x,n) = \nu(x,1) \vee \nu(x,2) \vee \ldots \vee \nu(x,n)$. Then $\mu(x) = \lim \nu'(x,n)$ and ν' is recursive and increasing with respect to n.

(4)$\Rightarrow$(1): It suffices to note that $x_c \in P(\mu)$ if and only if $\mu(x) \geq c$ if and only if there exists $n \in \mathbb{N}$ such that $\nu(x,n) \geq c$. Thus $P(\mu)$ is recursively enumerable. ∎

Proposition 9.10.2 *The set of recursively enumerable L-fuzzy subsets of T^* is a lattice, the minimal lattice containing the recursively enumerable subsets and the L-fuzzy subsets that are constant maps. Namely, an L-fuzzy subset $\mu : T^* \to L$ is recursively enumerable if and only if μ admits a decomposition $\mu = (c_1 \wedge \chi_1) \vee \ldots \vee (c_n \wedge \chi_n)$ with $c_i \in L$ and where χ_i is the characteristic function of a recursive enumerable subset, $i = 1, \ldots, n$.*

Proof. Let μ and ν be recursively enumerable. Then by Proposition 9.10.1(4), $\mu(x) = \lim \mu'(x,n)$ and $\nu(x) = \lim \nu'(x,n)$ with μ' and ν' recursive and increasing with respect to n. Then $\mu(x) \vee \nu(x) = \lim \mu'(x,n) \vee \nu'(x,n)$ and $\mu(x) \wedge \nu(x) = \lim \mu'(x,n) \wedge \nu'(x,n)$. Since $\mu' \vee \nu'$ and $\mu' \wedge \nu'$ are recursive and increasing with respect to n, $\mu \vee \nu$ and $\mu \wedge \nu$ are recursive enumerable This proves that the set of recursively enumerable L-fuzzy subsets forms a lattice.

Let μ be a recursively enumerable L-fuzzy subset. Set

$$\chi_c(x) = \begin{cases} 1 & \text{if } \mu(x) \geq c \\ 0 & \text{otherwise.} \end{cases}$$

Since μ_c is recursively enumerable, χ_c is the characteristic function of a recursively enumerable subset. Clearly, $\mu = \vee \{c \wedge \chi_c \mid c \in L \backslash \{0\}\}$. Thus μ is generated by the constants c and the recursively enumerable crisp L-fuzzy subsets χ_c. ∎

We now prove our main result, i.e., an L-fuzzy language $\mu : T^* \to L$ is generated by an L-fuzzy grammar if and only if it is recursively enumerable. However, first we prove the following lemma.

Lemma 9.10.3 If $\mu : T^* \to L$ and $\nu : T^* \to L$ are L-fuzzy languages generated by normal form L-fuzzy grammars, then $\mu \cup \nu$ is an L-fuzzy language generated by a normal form L-fuzzy grammar.

Proof. Suppose that μ and ν are generated by the normal form L-fuzzy grammar $G' = (N', T, P', S')$ and $G'' = (N'', T, P'', S'')$, respectively. Without loss of generality, we assume that $N' \cap N'' = \emptyset$. Let G be the normal form L-fuzzy grammar $(N' \cup N'' \cup \{S\}, T, P, S)$, where S is a new symbol not in $N' \cup N''$, and P contains $S \to S'$, $S \to S''$, and all production in P' and P''. Let $\lambda : T^* \to L$ be the L-fuzzy language generated by G. Then $\lambda = \mu \cup \nu$. In a sense, the derivations of G consist of the derivation of G' and the derivations of G''. That is, if $(w_1 \ldots w_p, u_1 \ldots u_{p-1})$ is a derivation of G' (of G'') with w_1 equal to S' (to S''), then $(Sw_1 \ldots w_p, 1u_1 \ldots u_{p-1})$ is a derivation of G' (of G''). Furthermore, since N' and N'' are disjoint, every derivation of G can be obtained either from a derivation of G' or from a derivation of G'' as above. Thus it follows that $\lambda = \mu \cup \nu$. ∎

Theorem 9.10.4 Let μ be an L-fuzzy language of T^*. Then the following statements are equivalent.

(1) μ is generated.

(2) μ is recursively enumerable.

(3) μ is generated by a normal form grammar.

Proof. (1)$\Rightarrow$(2): Suppose that μ is generated by the L-fuzzy grammar G. Let $p_1, p_2, \ldots$ be an effective enumeration of the derivation of G. Moreover, let ν be defined by

$$\nu(w, n) = \begin{cases} \text{the degree of } p_n & \text{if } p_n \text{ is a derivation of } w \\ 0 & \text{otherwise.} \end{cases}$$

Then it follows that μ is the projection of ν, and thus, μ is recursively enumerable by Proposition 9.10.1.

(2)$\Rightarrow$(3): Suppose that μ is recursively enumerable. Then by Proposition 9.10.2, $\mu = (c_1 \wedge \chi_1) \vee \ldots \vee (c_n \wedge \chi_n)$, where χ_i is the characteristic function of recursively enumerable subset S_i of T^*, $i = 1, \ldots, n$. Let G_i be a classical grammar in normal form producing S_i and let $\overline{G_i}$ be the normal form L-fuzzy grammar obtained from G_i by substituting each production $x \to y$ with $x \xrightarrow{c} y$. It follows that $\overline{G_i}$ produces the L-fuzzy language $c_i \wedge \chi_i$.

From Lemma 9.10.3, it follows that μ is generated by a normal form L-fuzzy language.

(3)⇒(1): Immediate. ∎

We now consider some closure properties for generated L-fuzzy languages. Recall that the concatenation of two L-fuzzy languages μ and ν is the L-fuzzy language λ defined by $\forall w \in T^*$, $\lambda(w) = \vee\{\mu(x) \wedge \nu(y) \mid w = xy\}$. Moreover, the Kleene closure μ^∞ of μ is such that $\mu^\infty(w) = \vee\{\mu(x_1) \wedge \ldots \wedge \mu(x_q) \mid w = x_1 \ldots x_q, q \in \mathbb{N}\}$.

Corollary 9.10.5 *The class of all generated L-fuzzy languages is a lattice. In particular, it is the minimal lattice containing the generated (classical) language and the L-fuzzy languages that are constant functions. Furthermore, if μ and ν are generated languages, then the concatenation of μ and ν and the Kleene closure of μ are generated.*

Proof. The first part of the corollary follows from Proposition 9.10.2. Suppose that $\mu(x) = \vee_n\mu'(x,n)$ and $\nu(x) = \vee_n\nu'(x,n)$ with μ' and ν' recursive. Then $\lambda(w) = \vee\{\mu'(x,n) \wedge \nu'(x,m) \mid w = xy, n, m \in \mathbb{N}\}$. Since it is possible to codify the set, $\{(x,y,n,m) \mid w = xy, n,m \in \mathbb{N}\}$, λ is a projection of a decidable relation. Thus λ is recursively enumerable and therefore generated.

Likewise, it is possible to express the Kleene closure of μ by the formula
$\mu^\infty(w) = \vee\{\mu'(x_1,n_1) \wedge \ldots \wedge \mu'(x_q,n_q) \mid w = x_1 \ldots x_q, n_1, \ldots, n_q \in \mathbb{N}\}$.
Since it is possible to codify the set $\{(x_1 \ldots x_q, q, n_1, \ldots, n_q) \mid x_1 \ldots x_q = w, q, n_1, \ldots, n_q \in \mathbb{N}\}$, μ^∞ is recursively enumerable and therefore generated.
∎

Theorem 9.10.4 allows for the transfer of results on the relationship among imprecision, decidability, and recursive enumerability given in [17, 70, 71] to L-fuzzy languages. Namely, we assume that L is a finite sublattice of the interval $[0,1]$ containing $\frac{1}{2}$. We call an L-fuzzy language **infinite indeterminate** (**almost-everywhere indeterminate**) provided that the set $\{x \in T^* \mid s(x) = \frac{1}{2}\}$ is infinite (cofinite). If λ and λ' are L-fuzzy languages, we say that λ' is a **sharpened version** of λ or that λ is a **shaded version** of λ' if

$$\lambda(x) > \frac{1}{2} \Rightarrow \lambda'(x) \geq \lambda(x) \text{ and } \lambda(x) < \frac{1}{2} \Rightarrow \lambda'(x) \leq \lambda(x).$$

Hence we conclude the following results from Theorem 9.10.4:

1. A generated infinitely indeterminate L-fuzzy language exists with no decidable sharpened version, [17, Proposition 5.1]. That is, it is not possible to obtain decidability by using the indeterminateness of a generated fuzzy language, in general.

2. An infinitely indeterminate L-fuzzy language exists with no generated shaded versions, [71, Proposition 4.4]. This provides an example of a fuzzy language that is not generated in a strong case.

3. A generated classical language exists whose unique decidable shaded versions are the L-fuzzy languages infinitely indeterminate, [17, Proposition 5.2]. Consequently, it is not possible to shade a generated classical language in order to obtain decidability.

4. A classical language exists whose unique generated shaded versions are the L-fuzzy languages almost everywhere indeterminate, [71, Proposition 5.2]. Therefore, it is not always possible to obtain generated languages by shading a classical language.

9.11   Various Kinds of Automata with Weights

Various types of automata such as fuzzy automata, max-product automata, and integer-valued generalized automata [233, 234] have been introduced as a generalization of well-known deterministic automata, nondeterministic automata, and probabilistic automata. These automata have the common property that they have "weights" associated with the state transitions as well as initial and final distributions. Clearly, probabilistic automata can be considered as automata with "weights." The operations of max and min and product have been used with these automata. By using addition and multiplication as the operations and the probabilities as the weights, probabilistic automata can be defined. Moreover, integer-valued generalized automata have integer weights and addition and multiplication as operations.

We now present the results of [148] in order to present a general formulation of automata with weights by extracting the basic properties common to the existing automata and by incorporating the appropriate algebraic systems with automata systems and by performing the operations of the algebraic systems to the state transition functions and initial and final distribution functions of the pseudoautomata defined later.

We continue the theme of using a lattice L rather than $[0, 1]$. The concept of L-fuzzy relations enables us to define fuzzy automata, l-semigroup automata, lattice automata, dual lattice automata, max-product automata, and so on. Moreover, L-fuzzy relations are important in formulating other kinds of automata with weights such as semiring automata, ring automata, integer-valued generalized automata, and field automata.

Definition 9.11.1 *Let X be a set and L be a lattice. An L-**fuzzy relation** on X is a function μ from $X \times X$ into L, i.e.,*

$$\mu : X \times X \to L. \tag{9.1}$$

We let $\vee$ and $\wedge$ denote the operations of supremum and infimum on L, respectively. In the remainder of the chapter, the structure of the membership space L is assumed to be a complete lattice ordered semigroup and a complete distributive lattice because of the concept of composition of L-fuzzy relations defined below.

If L is a complete lattice that is a semigroup with identity under $*$ and also satisfies the following distributive laws, then it is a **complete lattice ordered semigroup** (l-**semigroup** for short) and is denoted by $L = (L, \vee, *)$. The distributive laws are as follows. For all $x, y, x_i, y_i \in L$, where $i \in I$, an index set,

$$x * (\vee_{i \in I} y_i) = \vee_{i \in I}(x * y_i) \quad \text{and} \quad (\vee_{i \in I} x_i) * y = \vee_{i \in I}(x_i * y).$$

If the semigroup operation $*$ is replaced by $\wedge$ in $L = (L, \vee, *)$, then L becomes a **complete distributive lattice**.

Definition 9.11.2 *Let μ_1 and μ_2 be L-fuzzy relations on X. Then the* **composition** *(or* **product***) of μ_1 and μ_2 on X, denoted by $\mu_1 \mu_2$, is defined as follows:*

(1) If L is a complete lattice ordered semigroup (or l-semigroup) $(L, \vee, *)$, then

$$\mu_1 \mu_2(x, z) = \vee \{\mu_1(x, y) * \mu_2(y, z) \mid y \in X\}, \tag{9.2}$$

where $*$ is the semigroup operation in L.
(2) If L is a complete lattice $(L, \vee, \wedge)$, then

$$\mu_1 \mu_2(x, z) = \vee \{\mu_1(x, y) \wedge \mu_2(y, z) \mid y \in X\}. \tag{9.3}$$

Since the operations $\vee$ and $\wedge$ are dual in a distributive lattice, the different composition of L-fuzzy relations can be defined by replacing $\vee$ by $\wedge$ as follows:

$$\mu_1 \mu_2(x, z) = \wedge \{\mu_1(x, y) \vee \mu_2(y, z) \mid y \in X\}. \tag{9.4}$$

Due to the associativity of composition of L-fuzzy relations, we can write

$$\mu_1 \mu_2 \cdots \mu_n(x_1, x_{n+1}) = \vee \{\mu_1(x_1, x_2) * \mu_2(x_2, x_3) * \ldots * \mu_n(x_n, x_{n+1}) \mid x_2, \ldots, x_n \in X\}, \tag{9.5}$$

where $\mu_1, \mu_2, \ldots, \mu_n$ are L-fuzzy relations. A similar result holds for (9.3)−(9.4).

If there are two elements 0 and 1 in an l-semigroup $L = (L, \vee, *)$ such that for all $x \in L$,

$$\begin{aligned}
x \vee 0 &= x, \\
x * 0 &= 0 * x = 0, \\
x \vee 1 &= 1, \\
x * 1 &= 1 * x = x,
\end{aligned} \tag{9.6}$$

then they are called a **zero** and an **identity** of L, respectively. For example, let L be $([0,1], \vee, \cdot)$, where the operation $\cdot$ is ordinary multiplication. Then L is an l-semigroup with zero 0 and identity 1. Moreover, let the Cartesian product of $[0,1]$ be written as $[0,1]^2$ and the operations $\vee$ and $\cdot$ be defined as

$$\begin{aligned} (a,b) \vee (c,d) &= ((a \vee c), (b \vee d)), \\ (a,b) \cdot (c,d) &= (a \cdot c, b \cdot d), \end{aligned}$$

for each $(a,b), (c,d) \in [0,1]^2$. Then $L = ([0,1]^2, \vee, \cdot)$ is an l-semigroup with zero $(0,0)$ and identity $(1,1)$.

For the l-semigroup L with zero 0 and identity 1, define the identity relation ι by $\forall x, y \in L$,

$$\iota(x,y) = \begin{cases} 1 & \text{if } x = y \\ 0 & \text{if } x \neq y. \end{cases} \tag{9.7}$$

Then we have

$$\iota\mu = \mu\iota = \mu, \tag{9.8}$$

for each L-fuzzy relation μ.

Clearly, every L-fuzzy relation μ over X is representable by a matrix if X is a finite set. Let $X = \{x_1, x_2, \ldots, x_n\}$. Then μ is represented by the $n \times n$ matrix,

$$[\mu(x_i, x_j)], \quad i, j = 1, 2, \ldots, n.$$

We now define what is called a pseudoautomaton in [148] in order to derive various kinds of automata with weights.

Definition 9.11.3 *A **pseudoautomaton** is a 6-tuple*

$$A = (Q, X, W, \mu, \pi, \eta), \tag{9.9}$$

where
  *(1) Q is a finite set of **states**,*
  *(2) X is a finite set of **input symbols**,*
  *(3) W is a **weighting space**,*
  (4) μ is a weighting function such as

$$\mu : Q \times X \times Q \to W, \tag{9.10}$$

*and is called a **state transition function**. The value $\mu(q, a, q')$ of (q, a, q') $\in Q \times X \times Q$ represents the **weight** of transition from state q to state q' when the input symbol is a,*
  *(5) π is an **initial distribution function**, where*

$$\pi : Q \to W, \tag{9.11}$$

  *(6) η is a **final distribution function**, where*

$$\eta : Q \to W. \tag{9.12}$$

We do not consider time and outputs for the sake of simplicity. If we consider time, the state transition function μ, the initial distribution function π, and the final distribution function η are given as

$$\mu : Q \times X \times Q \times T \to W$$

$$\pi : Q \times T \to W$$

$$\eta : Q \times T \to W,$$

where T is a subset of $\mathbb{R}$.

Let Y be a set of output symbols. Then the output function maps $Q \times X \times Y \times T$ into W.

Definition 9.11.4 *A **weighted-automaton** or simply **automaton** A^* is a 6-tuple*

$$A^* = (Q, X, W, \mu^*, \pi, \eta), \tag{9.13}$$

where Q, X, W, π, and η are given in Definition 9.11.3 and

$$\mu^* : Q \times X^* \times Q \to W. \tag{9.14}$$

We now derive various kinds of automata with weights by introducing binary operations on the weighting space W of the pseudoautomata and by giving the extension rules for obtaining μ^* from μ.

L-Semigroup Automata

(1a). Here the weighting space is a complete lattice ordered semigroup (l-semigroup for short) $L = (L, \vee, *)$ with identity 1 and zero 0, where the operation $*$ is the semigroup operation in L. The state transition function μ, the initial distribution function π, and the final distribution function η are given by replacing W in (9.10), (9.11), and (9.12) by L as follows:

$$\mu : Q \times X \times Q \to L \tag{9.15}$$

$$\pi : Q \to L \tag{9.16}$$

$$\eta : Q \to L. \tag{9.17}$$

(1b). Using the concept of composition of L-fuzzy relations (9.2), the state transition function μ^* for input strings in X^* is obtained recursively as follows:

For $\Lambda, x \in X^*$, and $a \in X$,

$$\mu^*(q, \Lambda, q') = \left\{ \begin{array}{ll} 1 & \text{if } q = q' \\ 0 & \text{if } q \neq q', \end{array} \right. \tag{9.18}$$

$$\mu^*(q, xa, q') = \vee \{\mu^*(q, x, q'') * \mu(q'', a, q') \mid q'' \in Q\}, \tag{9.19}$$

where $q, q' \in Q$, and 1 and 0 are identity and zero of L, respectively.

Suppose that the automaton starts from a certain initial state, say, q_0. Then the initial distribution function π is concentrated at q_0, i.e.,

$$\pi(q) = \left\{ \begin{array}{ll} 1 & \text{if } q = q_0 \\ 0 & \text{if } q \neq q_0. \end{array} \right. \tag{9.20}$$

Let $F \subseteq Q$ be a set of final states. Then the final distribution function η is defined as

$$\eta(q) = \left\{ \begin{array}{ll} 1 & \text{if } q \in F \\ 0 & \text{if } q \notin F. \end{array} \right. \tag{9.21}$$

Given the expression (9.19) and the initial distribution π and the final distribution η, the **weight**, denoted by $\omega(x)$, of the string x of the automata is defined by

$$\omega(x) = \vee \{\pi(q) * \mu^*(q, x, q') * \eta(q') \mid q, q' \in Q\}, \tag{9.22}$$

where $x \in X^*$.

Since there exists an order relation $\geq$ in the l-semigroup $L = (L, \vee, *)$, the language $L(A, c)$ accepted by the l-semigroup automaton A with parameter c can be defined by

$$L(A, c) = \{x \in X^* \mid \omega(x) \geq c\}, \tag{9.23}$$

where c is called a threshold (or cutpoint) and is a member of the weighting space L.

Max-Product Automata

(2a) Let the weighting space be $L' = ([0, 1], \vee, \cdot)$ in the l-semigroup automaton of (1a, b), where the operation $\cdot$ represents ordinary multiplication. Then L' is an l-semigroup with identity 1 and zero 0. Moreover, μ, π, and η are obtained by replacing L in (9.15)–(9.17) by $[0, 1]$, i.e.,

$$\mu : Q \times X \times Q \rightarrow [0, 1], \tag{9.24}$$

$$\pi : Q \rightarrow [0, 1], \tag{9.25}$$

$$\eta : Q \to [0, 1]. \tag{9.26}$$

(2b) μ^* and ω are obtained by replacing $*$ by $\cdot$ in (9.18), (9.19), and (9.22).

$$\mu^*(q, \Lambda, q') = \begin{cases} 1 & \text{if } q = q' \\ 0 & \text{if } q \neq q', \end{cases} \tag{9.27}$$

$$\mu^*(q, xa, q') = \vee\{\mu^*(q, x, q'') \cdot \mu(q'', a, q') \mid q'' \in Q\}, \tag{9.28}$$

$$\omega(x) = \vee\{\pi(q) \cdot \mu^*(q, x, q') \cdot \eta(q') \mid q, q' \in Q\}. \tag{9.29}$$

Clearly, a max-product automaton is a special case of the l-semigroup automaton of (1a,b) and is also a special case of the semiring automata of (14a,b), defined later.

Lattice Automata

(3a) A complete distributive lattice $L = (L, \vee, \wedge)$ is the weighting space. Moreover, μ, π, and η are given from (9.10)−(9.12) by

$$\mu : Q \times X \times Q \to L, \tag{9.30}$$

$$\pi : Q \to L, \tag{9.31}$$

$$\eta : Q \to L. \tag{9.32}$$

(3b) By using the concept of composition of L-fuzzy relations (9.3), μ^* and ω are obtained as follows:

$$\mu^*(q, \Lambda, q') = \begin{cases} 1 & \text{if } q = q' \\ 0 & \text{if } q \neq q', \end{cases} \tag{9.33}$$

$$\mu^*(q, xa, q') = \vee\{\mu^*(q, x, q'') \wedge \mu(q'', a, q') \mid q'' \in Q\}, \tag{9.34}$$

$$\omega(x) = \vee\{\pi(q) \wedge \mu^*(q, x, q') \wedge \eta(q') \mid q'' \in Q\}, \tag{9.35}$$

where 1 and 0 are the maximal and minimal elements of the complete distributive lattice L, respectively. Expressions (9.34) and (9.35) are obtained by replacing $*$ by $\wedge$ from (9.19) and (9.22).

A complete distributive lattice is a special case of a complete lattice ordered semigroup. Likewise, a lattice automaton is considered to be a special case of an l-semigroup automaton.

The operations $\vee$ and $\wedge$ are dual in a complete distributive lattice $L = (L, \vee, \wedge)$. Thus the dual automata of lattice automata can be formulated in the following manner.

Dual Lattice Automata

(4a) This is the same as (3a).

(4b) Using the concept of composition of L-fuzzy relations (9.4), μ^* and ω are given as follows:

$$\mu^*(q, \Lambda, q') = \begin{cases} 0 & \text{if } q = q' \\ 1 & \text{if } q \neq q', \end{cases} \tag{9.36}$$

$$\mu^*(q, xa, q') = \wedge\{\mu^*(q, x, q'') \vee \mu(q'', a, q') \mid q'' \in Q\}, \tag{9.37}$$

$$\omega(x) = \wedge\{\pi(q) \wedge \mu^*(q, x, q') \vee \eta(q') \mid q'' \in Q\}. \tag{9.38}$$

Given a certain initial state q_0 and a final state set F, π and η of the lattice automata of (3a,b) are given as follows in the same manner as (9.20) and (9.21).

$$\pi(q) = \begin{cases} 1 & \text{if } q = q_0 \\ 0 & \text{if } q \neq q_0. \end{cases} \tag{9.39}$$

$$\eta(q) = \begin{cases} 1 & \text{if } q \in F \\ 0 & \text{if } q \notin F. \end{cases} \tag{9.40}$$

However, π and η of the dual lattice automata of (4a,b) are defined as follows:

$$\pi(q) = \begin{cases} 0 & \text{if } q = q_0 \\ 1 & \text{if } q \neq q_0. \end{cases} \tag{9.41}$$

$$\eta(q) = \begin{cases} 0 & \text{if } q \in F \\ 1 & \text{if } q \notin F. \end{cases} \tag{9.42}$$

(Pessimistic) Fuzzy Automata [250, 143, 212, 31, 65, 140, 193, 259]

(5a) If the weighting space, $J = ([0, 1], \vee, \wedge)$, is adopted, then J is a complete distributive lattice under the operations $\vee$ (max) and $\wedge$ (min). Furthermore, μ, π, and η are as follows.

$$\mu : Q \times X \times Q \to [0, 1], \tag{9.43}$$

$$\pi : Q \to [0, 1], \tag{9.44}$$

$$\eta : Q \to [0, 1]. \tag{9.45}$$

(5b) μ^* and ω are defined as follows:

$$\mu^*(q, \Lambda, q') = \begin{cases} 1 & \text{if } q = q' \\ 0 & \text{if } q \neq q', \end{cases} \tag{9.46}$$

$$\mu^*(q, xa, q') = \vee\{(\mu^*(q, x, q'') \wedge \mu(q'', a, q')) \mid q'' \in Q\}, \tag{9.47}$$

$$\omega(x) = \vee\{(\pi(q) \wedge \mu^*(q, x, q') \wedge \eta(q')) \mid q, q' \in Q\}. \tag{9.48}$$

Since $J = ([0, 1], \max, \min)$ is a complete distributive lattice, a fuzzy automaton is a special case of a lattice automaton (3a,b). Therefore, μ^* and ω of $(9.46) - (9.48)$ are obtained from $(9.33) - (9.35)$ by replacing $\vee$ with max, $\wedge$ by min.

Optimistic Fuzzy Automata [250, 143, 212, 140]

(6a) This is the same as (5a).
(6b) μ^* and ω are defined as follows:

$$\mu^*(q, \Lambda, q') = \begin{cases} 1 & \text{if } q = q' \\ 0 & \text{if } q \neq q', \end{cases} \tag{9.49}$$

$$\mu^*(q, xa, q') = \wedge\{(\mu^*(q, x, q'') \vee \mu(q'', a, q')) \mid q'' \in Q\}, \tag{9.50}$$

$$\omega(x) = \wedge\{(\pi(q) \vee \mu^*(q, x, q') \vee \eta(q')) \mid q, q' \in Q\}. \tag{9.51}$$

Optimistic fuzzy automata are special cases of dual lattice automata of (4a,b).

Given an initial state q_0 and a final state F, π and η of the fuzzy automata of (5a,b) are obtained from (9.39) and (9.40). However, π and η of an optimistic fuzzy automata are obtained from (9.41) and (9.42) [143].

Mixed Fuzzy Automata [212]

(7a) This is the same as (5a).
(7b) μ^* and ω are given by using the concept of convex combination of fuzzy subsets, i.e.,

$$\mu^*(q, x, q') = a\mu^*_{PF}(q, x, q') + b\mu^*_{OF}(q, a, q'), \tag{9.52}$$

$$\omega(x) = a\omega_{PF}(x) + b\omega_{OF}(x), \tag{9.53}$$

where μ^*_{PF} and μ^*_{OF} are state transition functions defined by the pessimistic fuzzy automata of 9.11 and the optimistic fuzzy automata of (6a,b), respectively. This is the same for ω_{PF} and ω_{OF}. Also a, b are nonnegative real numbers such that $a + b = 1$.

Composite Fuzzy Automata [250, 65]

(8a) This is the same as (5a).

(8b) μ^* is obtained by operating between μ^*_{PF} and μ^*_{OF} with probability p. This is the same for ω_{PF} and ω_{OF}.

Nondeterministic Automata [220]

(9a) $J' = (\{0,1\}, \max, \min\}$ is adopted as the weighting space. Clearly, J' forms a distributive lattice (more precisely, a Boolean lattice) and μ, π, and η are given as follows:

$$\mu : Q \times X \times Q \to \{0,1\}, \tag{9.54}$$

$$\pi : Q \to \{0,1\}, \tag{9.55}$$

$$\eta : Q \to \{0,1\}. \tag{9.56}$$

(9b) This is the same as (5b).

Nondeterministic automata are special cases of the fuzzy automata of (5a,b) (or l-semigroup automata).

Deterministic Automata [220]

(10a) This is the same as (9a) plus the additional constraints that there exists unique $q' \in Q$ such that $\mu(q, a, q') = 1$ for each $q \in Q$ and $a \in X$ and $\mu(q, a, q'') = 0$ for $q'' \neq q$, and there exists unique $q' \in Q$ such that $\pi(q') = 1$ and $\pi(q'') = 0$ for $q'' \neq q'$. As for η, let F be a set of final states. Then

$$\eta(q) = \begin{cases} 1 & \text{if } q \in F \\ 0 & \text{if } q \notin F. \end{cases}$$

(10b) This is the same as (9b).

Clearly, deterministic automata are special cases of the nondeterministic automata of (9a,b) and also of the probabilistic automata of (18a,b) defined later.

Boolean Automata

(11a) Here the weighting space is a complete Boolean lattice $B = (B, \vee, \wedge)$, where the operations $\vee$ and $\wedge$ are supremum and infimum in B. Clearly, a Boolean lattice is a special case of a distributive lattice. Then μ, π, and η are as follows:

$$\mu : Q \times X \times Q \to B, \tag{9.57}$$

$$\pi : Q \to B, \tag{9.58}$$

$$\eta : Q \to B. \tag{9.59}$$

(11b) This is the same as (3b).

Boolean automata and dual Boolean automata, defined next, are special cases of the lattice automata of 9.11 and the dual lattice automata 9.11, respectively.

In [149], a B-fuzzy grammar is defined to be a 7-tuple (N, T, P, S, J, μ, B), where N is the nonterminal alphabet, T is a terminal alphabet, $S \in N$ is an initial symbol, P is a finite set of productions, J is a set of production labels, and $\mu : J \to B$. We summarize some of the results of [149] in the Exercises.

Dual Boolean Automata

(12a) This is the same as (11a).
(12b) This is the same as (4b).

Mixed Boolean Automata

(13a) This is the same as (11a).
(13b) Using the concept of a convex combination of B-fuzzy sets [25], μ^* and ω are defined as follows:

$$\mu^* = (a \wedge \mu_B^*) \vee (\bar{a} \wedge \mu_{DB}^*), \tag{9.60}$$

$$\omega = (a \wedge \omega_B) \vee (\bar{a} \wedge \omega_{DB}), \tag{9.61}$$

where μ_B^* and μ_{DB}^* are state transition functions that are defined in the Boolean automata of (11a,b) and the dual Boolean automata of (12a,b), respectively, and $a, \bar{a} \in B$, where $\bar{a}$ is the complement of a.

Semiring Automata

Recall that a set R with the operations of addition $+$ and multiplication $\times$ is called a **semiring** if the following three conditions are satisfied: (1) $+$ is associative and commutative; (2) $\times$ is associative; (3) $\times$ distributes over $+$, i.e.,

$$a \times (b + c) = a \times b + a \times c,$$

$$(b + c) \times a = b \times a + c \times a,$$

for all $a, b, c \in R$. The semiring R is called a **semiring with identity** 1 and **zero** 0 if 1 is the identity under $\times$ and 0 is identity under $+$ in R.

For example, let $R = ([0, \infty), +, \cdot)$ with ordinary addition $+$ and ordinary product $\cdot$. Then $[0, \infty)$ is a semiring with identity 1 and zero 0. Similarly, the set of natural numbers containing 0 is also a semiring with identity and zero under $+$ and $\cdot$. Also, $R = ([0, 1], \vee, \cdot)$ is a semiring with identity and zero. Note that this R is also an l-semigroup. In general, l-semigroups and complete distributive lattices are special cases of semirings with identity and zero.

(14a) The weighting space is a semiring $R = (R, +, \times)$ with identity 1 and zero 0. Moreover μ, π, and η are given by:

$$\mu : Q \times X \times Q \to R, \tag{9.62}$$

$$\pi : Q \to R, \tag{9.63}$$

$$\eta : Q \to R. \tag{9.64}$$

(14b) μ^* and ω are given as follows:

$$\mu^*(q, \Lambda, q') = \begin{cases} 1 & \text{if } q = q' \\ 0 & \text{if } q \neq q', \end{cases} \tag{9.65}$$

$$\mu^*(q, xa, q') = \sum_{q'' \in Q} \{\mu^*(q, x, q'') \times \mu(q'', a, q')\}, \tag{9.66}$$

$$\omega(x) = \sum_{q, q' \in Q} \{\pi(q) \times \mu^*(q, x, q') \times \eta(q'). \tag{9.67}$$

As a special case of semiring automata, there exist l-semigroup automata of (1a,b), max-product automata of (2a,b), lattice automata of (3a,b), fuzzy automata of (5a,b), nondeterministic automata of (9a,b), Boolean automata of (12a,b), and so on.

Plus-Weighted Automata [188, 235]

(15a) The weighting space is $R = ([0, \infty), +, \cdot)$, where $+$ and $\cdot$ are ordinary addition and multiplication. Clearly, $R = ([0, \infty), +, \cdot)$ is a semiring with identity 1 and zero 0. Then μ, π, and η are defined from $(9.62)-(9.64)$ as follows:

$$\mu : Q \times X \times Q \to [0, \infty), \tag{9.68}$$

$$\pi : Q \to [0, \infty), \tag{9.69}$$

$$\eta : Q \to [0, \infty). \tag{9.70}$$

(15b) μ^* and ω are defined by letting $+$ be ordinary addition and replacing $\times$ by $\cdot$ in (9.65)–(9.67).

$$\mu^*(q, \Lambda, q') = \begin{cases} 1 & \text{if } q = q' \\ 0 & \text{if } q \neq q', \end{cases} \tag{9.71}$$

$$\mu^*(q, xa, q') = \sum_{q'' \in Q} \{\mu^*(q, x, q'') \cdot \mu(q'', a, q')\}, \tag{9.72}$$

$$\omega(x) = \sum_{q, q' \in Q} \{\pi(q) \cdot \mu^*(q, x, q') \cdot \eta(q')\}. \tag{9.73}$$

Weighted automata are special cases of the semiring automata of (14a, b).

Max-Weighted Automata [188]

(16a) The weighting space is $R = ([0, \infty), \max, \cdot)$, with ordinary multiplication $\cdot$. Clearly, R is a semiring with unity and zero. Here, μ, π, and η are given the same way as (9.68)–(9.70).

(16b) μ^* and ω are defined as follows:

$$\mu^*(q, \Lambda, q') = \begin{cases} 1 & \text{if } q = q' \\ 0 & \text{if } q \neq q', \end{cases} \tag{9.74}$$

$$\mu^*(q, xa, q') = \max_{q'' \in Q} \{\mu^*(q, x, q'') \cdot \mu(q'', a, q')\}, \tag{9.75}$$

$$\omega(x) = \max_{q, q' \in Q} \{\pi(q) \cdot \mu^*(q, x, q') \cdot \eta(q')\}. \tag{9.76}$$

Max-weighted automata are special cases of the semiring automata of (14a,b).

Max-product automata of (2a,b) can be obtained from max-weighted automata by replacing $[0, \infty)$ by $[0, 1]$.

Natural Numbered Automata

(17a) The weighting space is $\mathbb{N} \cup \{0\}$ with ordinary addition and multiplication. Moreover, μ, π, and η are given as follows:

$$\mu : Q \times X \times Q \to \mathbb{N} \cup \{0\}, \tag{9.77}$$

$$\pi : Q \to \mathbb{N} \cup \{0\}, \tag{9.78}$$

$$\eta : Q \to \mathbb{N} \cup \{0\}. \tag{9.79}$$

(17b) This is the same as (15b).

Natural numbered automata are special cases of the weighted automata of (15a, b).

Max-natural numbered automata can be easily defined in a manner similar to a max-weighted automata of (16a,b).

Probabilistic Automata [220, 168]

(18a) Here the weighting space is $([0,1], +, \cdot)$. Then μ, π, and η are

$$\mu : Q \times X \times Q \to [0,1], \tag{9.80}$$

$$\pi : Q \to [0,1], \tag{9.81}$$

$$\eta : Q \to [0,1]. \tag{9.82}$$

In addition, the following constraints of μ, π, and η are assumed. For all $q \in Q$ and $a \in X$,

$$\sum_{q' \in Q} \mu(q, a, q') = 1, \qquad \sum_{q' \in Q} \pi(q') = 1. \tag{9.83}$$

For η, let F be a final state set. Then

$$\eta(q) = \begin{cases} 1 & \text{if } q \in F \\ 0 & \text{if } q \notin F. \end{cases} \tag{9.84}$$

(18b) This is the same as (15b).

There exists another definition of η different from (9.84), [233]. That is, in the same way as μ and π of (9.83), we have

$$\sum_{q \in Q} \eta(q) = 1. \tag{9.85}$$

Generalized Probabilistic Automata [233, 164]

(19a) This is the same as (18a) with the assumption that the image of η is not the unit interval $[0,1]$, but the set of real numbers $(-\infty, \infty)$, i.e.,

$$\eta : Q \to (-\infty, \infty). \tag{9.86}$$

(19b) This is the same as (18b).

The language accepted by generalized probabilistic automata is defined by

$$L(A, c) = \{x \in X^* \mid \omega(x) \geq c\}, \tag{9.87}$$

where $c \in (-\infty, \infty)$. As for the probabilistic automata of (18a,b), $c \in [0,1]$.

Rational Probabilistic Automata [234]

(20a) This is the same as (18a) with the assumption that the values $\mu(q, a, q')$ and $\pi(q)$ are rational numbers in $[0, 1]$.

(20b) This is the same as (18b).

Ring Automata

(21a) The weighting space is a ring [125] with identity $R = (R, +, \cdot)$. Here μ, π, and η are

$$\mu : Q \times X \times Q \to R, \tag{9.88}$$

$$\pi : Q \to R, \tag{9.89}$$

$$\eta : Q \to R. \tag{9.90}$$

(21b) This is the same as (14b).

Ring automata are special cases of the semiring automata of (14a,b). The weighted automata of (15a,b) and the max-weighted automata of (16a,b), which are special cases of the semiring automata of (14a,b), are not special cases of ring automata.

Integer-Valued Generalized Automata [233, 234]

(22a) The weighting space is $\mathbb{Z} = (\mathbb{Z}, +, \cdot)$, where $\mathbb{Z}$ is a set of integers and the operations $+$ and $\cdot$ are ordinary addition and product, respectively. Clearly, $\mathbb{Z}$ is a ring with identity. Here μ, π, and η are

$$\mu : Q \times X \times Q \to \mathbb{Z}, \tag{9.91}$$

$$\pi : Q \to \mathbb{Z}, \tag{9.92}$$

$$\eta : Q \to \mathbb{Z}. \tag{9.93}$$

(22b) This is the same as (15b).

Integer-valued generalized automata are a special case of the ring automata of (21a,b).

Field Automata

(23a) Let the weighting space be a field $F = (F, +, \cdot)$, [125]. Then μ, π, and η are

$$\mu : Q \times X \times Q \to F, \tag{9.94}$$

$$\pi : Q \to F, \tag{9.95}$$

$$\eta : Q \to F. \tag{9.96}$$

(23b) This is the same as (21b).

Clearly, field automata are a special case of ring automata of (21a,b). An integer-valued generalized automata ((22a,b)), which is a special case of ring automata, is not a special case of field automata.

(Real-Valued) Generalized Automata [233, 234, 235]

(24a) The weighting space is $F = ((-\infty, \infty), +, \cdot)$, where $(-\infty, \infty)$ is a set of real numbers, and $+$ and $\cdot$ are ordinary addition and multiplication. Here μ, π, and η are

$$\mu : Q \times X \times Q \to (-\infty, \infty), \tag{9.97}$$

$$\pi : Q \to (-\infty, \infty), \tag{9.98}$$

$$\eta : Q \to (-\infty, \infty). \tag{9.99}$$

(24b) This is the same as (15b).

Real-valued generalized automata are a special case of the field automata of (23a,b).

Rational Automata

(25a) The weighting space is $\mathbb{Q} = (\mathbb{Q}, +, \cdot)$, where $\mathbb{Q}$ is a set of rational numbers, and $+$ and $\cdot$ are ordinary addition and multiplication. Here μ, π, and η are

$$\mu : Q \times X \times Q \to \mathbb{Q} \tag{9.100}$$

$$\pi : Q \to \mathbb{Q}, \tag{9.101}$$

$$\eta : Q \to \mathbb{Q}. \tag{9.102}$$

(25b) This is the same as (15b).

Rational automata are a special case of the real-valued generalized automata of (24a,b) and also of field automata of (23a,b).

We have presented various kinds of automata with weights. Some of these automata are lacking of physical images. However, for example, from the fact that the classes of languages defined by rational probabilistic automata of (20a,b) and integer-valued generalized automata of (22a,b) are equal, various problems concerning rational probabilistic automata can be solved by investigating the properties of integer-valued generalized automata [233,234]. Consequently, automata with weights are important in investigating properties of well-known automata such as deterministic automata and probabilistic automata. Furthermore, they are useful models of learning systems, gaming, and pattern recognition as in the case of fuzzy automata [250, 31, 65].

The set of complex numbers forms a field. Hence as a special case of field automata of (23a,b), we can define complex numbered automata. We cannot, however, define a language accepted by these automata in the same way as (9.23) because of the fact that there does not exist an order relation $\geq$ on the set of complex numbers. However, using the concept of a mapping of the set of complex numbers into a certain algebraic system with ordering, say, by transforming the complex number z to the absolute value $|z|$, we can define a language by a complex numbered automata A as $L(A, z) = \{x \in X^* \mid |\omega(x)| \geq |z|\}$. Along these lines, the concept of a valuation of a field can be considered. A valuation of a field F is a function v of $F\backslash\{0\}$ into an additive abelian totally ordered group such that for all $x, y \in F\backslash\{0\}$, $v(xy) = v(x) + v(y)$ and $v(x + y) \geq v(x) \wedge v(y)$. If we do not use this concept of mapping, we would have to restrict our choice of a ring or a field to those with an ordering as weighting spaces, [18, 183].

9.12 Exercises

1. For Theorem 9.4.1, prove the cases, where (2) or (3) holds.

2. Give the proof of Theorem 9.8.2 for $i = 0$ and 3.

3. Show that if a $L_{[0,1]}$-fuzzy language μ is fuzzy recognized by a machine in T_2 and a $L_{[0,1]}$-fuzzy language ν is fuzzy recognized by a machine in T_2, then $\mu \cap \nu$ is fuzzy recognized by a machine in T_2.

4. If μ_1, μ_2, μ_3 are L-fuzzy relations on a set X, prove that $(\mu_1\mu_2)\mu_3 = \mu_1(\mu_2\mu_3)$.

5. (149) Show that type 2 (context-free) B-fuzzy grammars can generate type 1 (context-sensitive) languages although type 2 fuzzy grammars cannot generate type 1 languages [145].

6. (149) Show that the generative power of type 3 (regular) B-fuzzy grammars is equal to that of the ordinary type 3 grammars.

Chapter 10

Applications

10.1 A Formulation of Fuzzy Automata and Its Application as a Model of Learning Systems

In [26], a formulation of a class of stochastic automata on the basis of Mealy's [138] formulation of finite automata has been carried out. In [68], a stochastic automaton as a model of a learning system operating in an unknown environment was employed. The formulation of fuzzy automata is similar to that of the stochastic automata proposed in [26] and the fuzzy class of systems described in [256]. The advantage of using fuzzy set concepts in engineering systems has been discussed in [255, 256]. In [212], a general formulation has been given to cover both fuzzy and stochastic automata. The next five sections are based on [250] and deal with a specific formulation of fuzzy automata and their engineering applications.

10.2 Formulation of Fuzzy Automata

Definition 10.2.1 *A (finite) fuzzy automaton is a quintuple (Q, X, Y, μ, ω), where*

    (1) Q nonempty finite set (the set of internal states),
    (2) X nonempty finite set (the set of input states),
    (3) Y nonempty finite set (the set of output states),
    (4) μ is a fuzzy subset of $Q \times X \times Q$, i.e., $\mu : Q \times X \times Q \to [0, 1]$,
    (5) ω is a fuzzy subset of $Q \times X \times Y$, i.e., $\omega : Q \times X \times Y \to [0, 1]$.

In Definition 10.2.1, μ is called the **fuzzy transition function** and ω the **fuzzy output function**.

Recall that in Section 8.4, a fuzzy finite automaton was defined as in Definition 10.2.1, but with some added conditions.

It is at times convenient to use the notation μ_A for a fuzzy subset of a set S, where A is thought of as a fuzzy set and μ_A gives the grade membership of elements of S in A. At times, A may be merely a description of a fuzzy subset μ of S.

Let $Q = \{q_1, q_2, \ldots, q_n\}$, $X = \{x_1, x_2, \ldots, x_p\}$, and $Y = \{y_1, y_2, \ldots, y_r\}$. Then $\mu_A(q_i, x_j, q_m)$ is the grade of transition (class A) from state q_i to state q_m with input x_j or

$$q(k) = q_i \text{ to } q_m \text{ or } q(k+1) = q_m, \tag{10.1}$$

where the input is x_j or $x(k) = x_j$ and where k denotes the discrete time element. Hence we write

$$\mu_A(q_i, x_j, q_m) = \mu\{q(k) = q_l, \ x(k) = x_j, \ q(k+1) = q_m\}.$$

In order to decide the existence of the transition, a pair of thresholds c, d may be introduced, where $0 < d < c < 1$. This leads to a three level logic such as

(1) $x \in A$, or "true" if $\mu_A(x) \geq c$,

(2) $x \notin A$, or "false" if $\mu_A(x) \leq d$,

(3) x has an indeterminate status relative to A, or "undetermined" if $d < \mu_A(x) < c$,

where A is a fuzzy set defined as the transition between states for a particular input; x denotes the 3-tuple (q_i, x_j, q_m) as defined in (10.1), where $q_l, q_m \in Q$; $x_j \in X$, i.e., $\mu_A(x) = \mu_A(q_l, x_j, q_m)$.

The function may be dependent or independent of k, the number of steps. If μ is independent of k, μ is called a **stationary fuzzy transition function**. As we will see, nonstationary fuzzy transition functions are used to demonstrate learning behavior of a fuzzy automaton.

For now, let μ be independent of k with fuzzy transition matrix T_{x_j} for all $x_j \in X$. The T_{x_j} are of the following form:

$q(k)$	x_j
	$q(k+1)$
	$q_1, q_2, \ldots, q_m, \ldots, q_s$
q_1	$\vdots$
$\vdots$	
q_l	$\cdots \ \mu_A(q_i, x_j, q_m)$
$\vdots$	
q_n	$\vdots$

The entries of T_{x_j} are $\mu_A(q_i, x_j, q_m)$. The fuzzy transition table is as follows:

$q(k)$	$q(k+1)$				$y(k)$			
		$x(k)$				$x(k)$		
	a	b			a	b		
q_1								
q_2	$[A]$	$[B]$		$\cdots$	$[A_0]$	$[B_0]$		$\cdots$
$\vdots$								
q_n								

where $[A]$ denotes the fuzzy matrix whose ith row and jth column is $\mu_A(q_i, a, q_j)$, for all $q_i, q_j \in Q$, and $[A_0]$ denotes the fuzzy matrix formed similarly by $\omega_A(q_i, a, y_j)$ for all $q_i \in Q$, $y_j \in Y$.

The input sequence transition matrix for a particular n-input tape sequence is defined by an n-ary fuzzy relation in the product space $T_1 \times T_2 \times \ldots \times T_n$. The fuzzy transition function is as follows: Let $X_j(k)$ be an input sequence of length j, i.e., $x_1(k)$, $x_2(k+1), \ldots, x_j(k+j-1)$. Then $\mu_A(q_l, X_p(k), q_s) = \mu_A(q_l, x_j, x_0, \ldots, x_t, q_s)$ is the grade of transition (class A) from state q_l or $q(k) = q_l$ to q_s or $q(k+p) = q_s$ when the input sequence is $x(k) = x_j$, $x(k+1) = x_0, \ldots, x(k+p-1) = x_t$. For identical inputs x_s in a sequence of length j, $X_j(k)$ is denoted by $x_s^j(k)$.

The composition of two fuzzy relations μ_A and μ_B on a set S is denoted by $\mu_B \circ \mu_A$ and is the fuzzy relation in S given by either

(1) $\mu_{B \circ A}(x, y) = \vee\{\mu_A(x, v) \wedge \mu_B(v, y) \mid v \in S\}$ or

(2) $\mu_{B \circ A}(x, y) = \wedge\{\mu_A(x, v) \vee \mu_B(v, y) \mid v \in S\}$.

Based on each definition, we have a particular kind of fuzzy automata. When these two kinds of automata operate together, they act as a composite automaton similar to the structure of a zero-sum two-person game. We illustrate this in a later section, where a learning model is proposed. The above concepts have been studied previously. However, we have introduced new notation. Thus we concentrate on developing automata based on both definitions using the new notation.

The composition given in (1) is often called the **pessimistic** case while that of (2) is often called the **optimistic** case. Now

$$\mu_A(q_l, X_2(k), q_r) = \vee\{\mu_A(q_l, x_j, q) \wedge \mu_A(q, x_0, q_r) \mid q \in Q\}.$$

In general,

$$\begin{aligned}
\mu_A^*(q_l, X_j(k), q_m) &= \mu_A^*(q_l, X_{j-1}(k)x_j(k+j-1), q_m) \\
&= \mu_A^*(q_l, x_1(k)x_2(k+1)\ldots x_j(k+j-1), q_m) \\
&= \vee\{\mu_A(q_l, x_1, q_{r_1}) \wedge \mu_A(q_{r_1}, x_2, q_{r_2}) \wedge \ldots \wedge \\
&\quad \mu_A(q_{r_{j-1}}, x_j, q_m) \mid q_{r_i} \in Q,\ i = 1, 2, \ldots, j-1\}.
\end{aligned}$$

Example 10.2.2 *Let T_{x_1} be given by the following table:*

	$q(k+1)$			
$q(k)$	q_1	q_2	q_3	q_4
q_1	u_{11}	u_{12}	u_{13}	u_{14}
q_2	u_{21}	u_{22}	u_{23}	u_{24}
q_3	u_{31}	u_{32}	u_{33}	u_{34}
q_4	u_{41}	u_{42}	u_{43}	u_{44}

Then

$$\mu_A(q_3, x_1, q_2) = u_{32},$$

$$
\begin{aligned}
\mu_A^*(q_3, x_1^2, q_2) &= \vee\{(\mu_A(q_3, x_1, q_j) \wedge \mu_A(q_j, x_1, q_2)) \mid q_j \in Q, \\
&\qquad j = 1, 2, 3, 4\} \\
&= (u_{31} \wedge u_{12}) \vee (u_{32} \wedge u_{22}) \vee (u_{33} \wedge u_{32}) \vee \\
&\qquad (u_{34} \wedge u_{42}),
\end{aligned}
$$

$$
\begin{aligned}
\mu_A^*(q_3, x_1^3, q_2) &= \vee\{(u_{31} \wedge u_{11} \wedge u_{12}), (u_{31} \wedge u_{12} \wedge u_{22}), (u_{31} \wedge u_{13} \wedge u_{32}), \\
&\qquad (u_{31} \wedge u_{14} \wedge u_{42}), (u_{32} \wedge u_{21} \wedge u_{12}), (u_{32} \wedge u_{22} \wedge u_{22}), \\
&\qquad (u_{32} \wedge u_{23} \wedge u_{32}), (u_{32} \wedge u_{24} \wedge u_{42}), (u_{33} \wedge u_{31} \wedge u_{12}), \\
&\qquad (u_{33} \wedge u_{32} \wedge u_{22}), (u_{33} \wedge u_{33} \wedge u_{32}\}, (u_{33} \wedge u_{34} \wedge u_{42}), \\
&\qquad (u_{34} \wedge u_{41} \wedge u_{12}), (u_{34} \wedge u_{42} \wedge u_{22}), (u_{34} \wedge u_{43} \wedge u_{32}), \\
&\qquad (u_{34} \wedge u_{44} \wedge u_{42})\}.
\end{aligned}
$$

Note,

$$
\begin{aligned}
\mu_A^*(q_3, x_1^3, q_2) &= \vee_{q_j}\{\mu_A^*(q_3, x_1^2, q_j) \wedge \mu_A(q_j, x_1, q_2) \mid q_j \in Q, \\
&\qquad j = 1, 2, 3, 4\} \\
&= (\mu_A(q_3, x_1^2, q_x) \wedge u_{12}) \vee (\mu_A(q_3, x_1^2, q_2) \wedge u_{22}) \vee \\
&\qquad (\mu_A(q_3, x_1^2, q_3) \wedge u_{32}) \vee (\mu_A(q_3, x_1^2, q_4) \wedge u_{42}).
\end{aligned}
$$

This last relation serves as an iteration scheme for a particular sequence of input tape. Similar results for the min-max relation (2) hold.

10.3 Special Cases of Fuzzy Automata

We first consider deterministic automata, where the set A is no longer a fuzzy set. A is an ordinary set where μ takes two values, 0 and 1. Hence, the entries for each row of matrices $[A]$, $[B]$, $[C]$, ..., $[A_j]$, $[B_j]$, ... will have only one 1 and the rest 0. Thus the skeleton matrix $[D]$ of a deterministic automaton will be

$$[D] = [A] + [B] + [C] + \ldots + [H],$$

which is a reduction similar to that in the stochastic automata formulated in [26]. For a two-step transition chain, $\mu_A^*(q_l, xy, q_r)$ is determined by finding the maximum of the minimum of pairs of values of four types $(0,0)$, $(0,1)$, $(1,0)$, $(1,1)$. The total number of paths of length 2 from state q_l to state q_r is equal to

$$\bar{d}_{lr} = \sum_{x,y \in X} \mu_A^*(q_l, xy, q_r).$$

For a path of length 3 to exist from q_l to q_r,

$$\bar{d}_{lr} = \sum_{x,y,z \in X} \mu_A^*(q_l, xyz, q_r).$$

It consists of terms such as $(1,1,1)$, $(1,0,1)$, $\dots$, $(0,0,0)$; of course, only $(1,1,1)$ defines the existence of the path.

The following definition of a nondeterministic automaton has been used in [76]. A nondeterministic automaton is a quintuple $A = (Q, X, \delta, S_0, F)$, where

Q is a nonempty finite set of objects (internal states),
X is a nonempty finite set of objects (input states),
δ is a function from $Q \times X$ into $\mathcal{P}(Q)$, and
S_0, F are subsets of Q and S_0 is nonempty.

To fit this model to the fuzzy automata formulation, the set A is no longer a fuzzy set. It is an ordinary set, where μ takes only two values, 0 and 1. The entries for each row of the matrices $[A], [B], [C], \dots$ are either 1 or 0. There is no restriction on the number of 1's in each row of the matrices. If all the entries of a particular row are zero, then the transition is from the state to the empty set $\emptyset$. If a particular row has more than one 1, then the transition function may map the succeeding state into any one of a number of possible states. For example, if T_{x_1} is given as

$$T_{x_1} = \begin{bmatrix} 1 & 1 & 0 \\ 0 & 0 & 0 \\ 1 & 1 & 1 \end{bmatrix},$$

then the state transition is determined as follows:

	$x(k)$
$q(k)$	$q(k+1)$
q_1	$\{q_1, q_2\}$
q_2	$\emptyset$
q_3	$\{q_1, q_2, q_3\}$

It is felt in [250] that the transition function μ is so general that extra constraints may be incorporated. A special restriction is that the row sum of all transition matrices be equal to 1 similar to that of the stochastic automata. That is, $[A], [B], [C], \dots$ have exactly the same structure as that of stochastic matrices. This type of automata is called **normalized fuzzy automata**. There are many properties that can be examined for fuzzy automata similar to those for deterministic automata and stochastic automata [249].

10.4   Fuzzy Automata as Models of Learning Systems

A basic learning system is given in Figure 10.1.

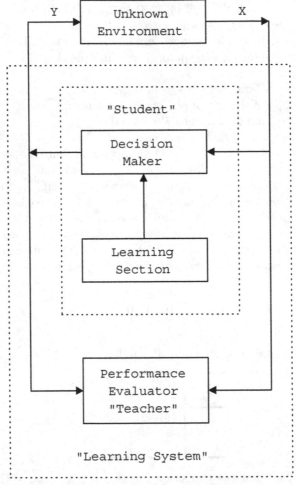

Figure 10.1[1] : Basic Learning Model

[1] Figure 10.1 is from [250], reprinted with permission by Copyright 1969 IEEE.

The proposed model represents a nonsupervised learning system if a proper performance evaluator can be selected [161, 162]. The learning section primarily consists of a composite fuzzy automaton. The performance evaluator serves as an unreliable "teacher" who tries to teach the "student" (the learning section and the decision maker) to make correct decisions.

The decision executed by the decision maker is deterministic. Since on-line operations are required, the decision will be based on the maximum grade of membership. That is, for $\Omega = \{\omega_i \mid i = 1, 2, \dots, r\}$, the set of all pattern classes, it is decided that x is from h_i, or $x \sim \omega_i$ if

$$\mu_{\omega_i}(x) = \vee\{\mu_{\omega_j}(x) \mid \omega_j \in \Omega, \ j = 1, \dots, r\},$$

where $\omega_i \in \Omega$, $i = 1, 2, \dots, r$.

If the decision maker is allowed to stay undecided and defer its decision, especially at the beginning of the first few learning steps, then it will be decided that $x \sim \omega_i$ if

$$\mu_{\omega_i}(x) = \vee\{\mu_{\omega_j}(x) \mid \omega_j \in \Omega, \ j = 1, \dots, r\} \geq c$$

and be undecided if

$$\mu_{\omega_i}(x) = \vee\{\mu_{\omega_j}(x) \mid \omega_j \in \Omega, \ j = 1, \dots, r\} < c,$$

for some predetermined c, $0 < c \leq 1$. Under this condition, the decision maker is allowed to use a pure random strategy for making decisions until it can make a decision with a certain degree of confidence.

For an n-state fuzzy automaton, let $\widehat{\mu}_j(k)$ be the grade of membership for the automaton to be at state q_j in step k. The entries of the fuzzy transition matrix are denoted by

$$\mu_{ij}^l(k) = \mu\{q(k) = q_i, \ x(k) = x_l, \ q(k+1) = q_j\}.$$

Then for input $x(k) = x_l$, we have by Definition 10.2.1 that for $j = 1, 2, \dots, n$,

$$\widehat{\mu}_j(k+1) \quad = \quad \vee\{(\widehat{\mu}_m(k) \wedge \mu_{mj}^l(k)) \mid m = 1, 2, \dots, n; \ m \neq j\}.$$

Using a similar approach as given in [68], the learning behavior is reflected by having nonstationary fuzzy transition matrices with a convergent property. In order to simplify the problem, let

$$\begin{aligned}
\mu_{mj}^l(k) &= \mu_{jj}^l(k-1), \quad \text{for all } m \neq j \\
\mu_{jj}^l(k) &= c_j \mu_{jj}^l(k-1) + (1 - c_j)d_j,
\end{aligned}$$

where $0 < c_j < 1$, $0 < d_j \leq 1$, $j = 1, 2, \dots, n$. The structure of the learning section is illustrated in Figure 10.2.

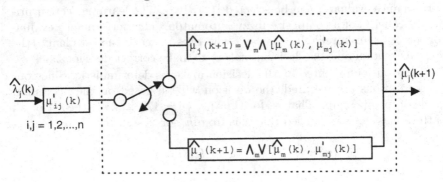

Composite Fuzzy Automaton

Figure 10.2[2]: Learning Section

The composite automaton switched by $p = \frac{1}{2}$ constitutes the major part of the learning section. That is, with $p = \frac{1}{2}$, the composite fuzzy automaton operates between

(1) $\widehat{\mu}_j(k+1) = \vee\{(\widehat{\mu}_m(k) \wedge \mu^l_{mj}(k)) \mid m = 1, 2, \ldots n\}$

and

(2) $\widehat{\mu}_j(k+1) = \wedge\{(\widehat{\mu}_m(k)\vee \mu^l_{mj}(k)) \mid m = 1, 2, \ldots n\}.$

The fuzzy automaton starts with no a priori information $\widehat{\mu}_j(0) = 0$ or 1 for all j, or with a priori information $\widehat{\mu}_j(0) = \widehat{\lambda}_j(0)$ for some $\widehat{\lambda}_j$ and for all j; $0 \le \widehat{\lambda}_j(0) \le 1$. The convergence of the above algorithm can be shown as follows:

$$\widehat{\lambda}_j(k) \to \widehat{\lambda}_j \text{ as } k \to \infty \text{ for some } \widehat{\lambda}_j.$$

Hence $\mu^l_{ij}(k) \to \widehat{\lambda}_j$ as $k \to \infty$. The algorithm is said to converge if $\widehat{\mu}_j(k) \to \widehat{\lambda}_j$ as $k \to \infty$. Thus as $k \to \infty$, we must show that

$$\begin{aligned} \widehat{\lambda}_j &= \vee\{(\widehat{\mu}_m(k) \wedge \widehat{\lambda}_j) \mid m = 1, 2, \ldots, n\} \\ &= \wedge\{(\widehat{\mu}_m(k) \vee \widehat{\lambda}_j) \mid m = 1, 2, \ldots, n\} \end{aligned} \quad (10.2)$$

so that we have the same output $\widehat{\mu}_j(k+1)$. It follows that

$$\wedge\{\widehat{\mu}_m(k) \mid m = 1, 2, \ldots, n\} \le \widehat{\lambda}_j \le \vee\{\widehat{\mu}_m(k) \mid m = 1, 2, \ldots, n\} \quad (10.3)$$

must hold in order for (10.2) to hold. However, $\widehat{\mu}_j(k) \to \widehat{\lambda}_j$ as $k \to \infty$. Therefore, (10.3) holds as $k \to \infty$. Now $\widehat{\mu}_j(k) \to \widehat{\lambda}_j$ can be shown in the following manner. There exists at least one k such that

$$\widehat{\mu}_j(k+1) = \vee\{(\widehat{\mu}_m(k) \wedge \widehat{\lambda}_j) \mid m = 1, 2, \ldots, n\}$$

---

[2]Figure 10.2 is from [250], reprinted with permission by Copyright 1969 IEEE.

and

$$\widehat{\mu}_j(k+2) = \wedge\{(\widehat{\mu}_m(k+1) \vee \widehat{\lambda}_j) \mid m = 1, 2, \ldots, n\}, \quad \text{for large } k.$$

Suppose that $\widehat{\lambda}_j > \vee\{\widehat{\mu}_m(k) \mid m = 1, 2, \ldots, n\}$. For other values of $\widehat{\lambda}_j$, we have $\widehat{\mu}_j(k+1) = \widehat{\lambda}_j$. Then

$$\widehat{\mu}_j(k+1) = \vee\{\widehat{\mu}_m(k) \mid m = 1, 2, \ldots, n\}.$$

However, $\widehat{\mu}_j(k+2) = \wedge\{(\widehat{\mu}_1(k+1) \vee \widehat{\lambda}_j), \ldots, (\vee\{\widehat{\mu}_m(k) \vee \widehat{\lambda}_j \mid m = 1, 2, \ldots, n\}), \ldots, (\widehat{\mu}_n(k+1) \vee \widehat{\lambda}_j)\}$. Thus $\widehat{\mu}_j(k+2) = \widehat{\lambda}_j$. Hence

$$\widehat{\mu}_j(k) \to \widehat{\lambda}_j \text{ as } k \to \infty.$$

10.5 Applications and Simulation Results

The learning model proposed above has been applied to engineering problems. We consider its applications to pattern classification and control systems. Let $Z(k)$ be the instantaneous performance evaluation at the kth step of learning [162]. The performance evaluator $M(Z, k)$ must be such that it is bounded above, i.e., $\exists$ a real number T such that

$$0 < M(Z, k) \leq T < \infty, \quad k = 1, 2, \ldots$$

and it must converge, i.e., $\exists$ a real number M such that

$$\lim_{k \to \infty} M(Z, k) = M > 0.$$

The goal of the system is to maximize or minimize $M(Z, k)$. Hence $M(Z, k)$ is an estimator of $Z(k)$ at the kth step of learning. For example, let

$$M(Z, k) = \frac{1}{k} \sum_{i=1}^{k} Z(i).$$

Then

$$\begin{aligned}
M(Z, k) &= \frac{k-1}{k}\left[\frac{1}{k-1} \sum_{i=1}^{k-1} Z(i)\right] + \frac{Z(k)}{k} \\
&= \frac{k-1}{k} M(Z, k-1) + \frac{Z(k)}{k}, \quad k = 1, 2, \ldots.
\end{aligned}$$

In this example, $M(Z, k)$ is a sample average of $Z(k)$ estimated recursively at each step of learning.

We now give an application to pattern classification.

A pattern recognition system with nonsupervised learning is given in Figure 10.3.

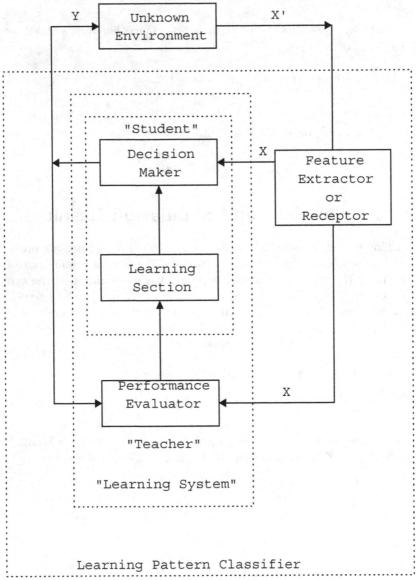

Figure 10.3[3]: Learning Pattern Classifier

The role of the input and output is explained in what follows. The pattern classifier receives a new sample from the unknown environment during each time interval. After the new sample is processed through the receptor,

---

[3]Figure 10.3 is from [250], reprinted with permission by Copyright 1969 IEEE.

the output is fed to both the decision maker for classification and the performance evaluator for performance evaluation. The performance criterion of the system has to be selected so that its maximization or minimization reflects the clustering properties of the pattern classes, i.e., the unknown environment. Because of the natural distribution of the samples, the performance criterion can be incorporated into the system to serve as a teacher of the learning pattern recognizer.

Concerning the problem of pattern classification, the nonsupervised learning model is formulated as follows. It is assumed that the classifier (the decision maker) has available sets of discriminant functions characterized by sets of parameters. The system adapts itself to the best solution with a proper specification of the performance evaluation and without any external supervision. The best solution denotes the set of discriminant functions that gives the minimum misrecognition among the sets of discriminant functions for the given set of training samples. The criterion used in this particular case is based upon the sample averages and the average deviations from the sample averages of the training patterns generated by each set of discriminant functions. The best set of discriminant functions must give the maximum total distance between its sample averages and the minimum total sample deviation (average of the squared deviation from the sample averages).

Let $\Omega = \{\omega_i \mid i = 1, 2, \ldots, r\}$ be the set of pattern classes. Let $X_1, X_2, \ldots$ be the sequence of incoming samples from the unknown environment in $\mathbb{R}^d$. Suppose n sets of discriminant functions are given a priori. The decision maker may assume the structure as shown in Figure 10.4.

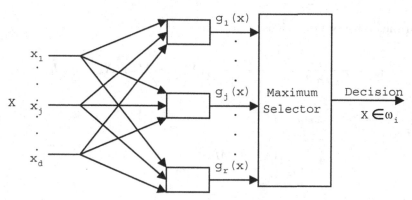

Figure 10.4[4]: *Multiclass Pattern Classifier*

Then

$$K_k^l = |\omega_1[\textstyle\sum_{i=1}^{r} S_{ki}^l] - \omega_2[\sum_{i>j}(M_{ki}^l - M_{kj}^l)'(M_{ki}^l - M_{kj}^l)]|,$$
$$l = 1, 2, \ldots, n,$$

---

[4]Figure 10.4 is from [250], reprinted with permission by Copyright 1969 IEEE.

where K_k^l is the performance evaluation for the lth set of discriminant functions at the kth step of learning, S_{ki}^l is the sample deviation for the ith class of the lth set of discriminant functions at the kth step of learning, M_{ki}^l is the sample average for the ith class of the lth set of discriminant functions at the kth step of learning, and

$$0 \le \omega_1, \omega_2 < \infty.$$

Let the $n_{k,i}$ denote the number of samples belonging to ω_i up to the kth step of learning. The values of the K_k^l, S_{ki}^l, and M_{ki}^l are estimated recursively as follows. For the lth set of discriminant functions, if

$$X_{k+1} \sim \omega_i, \quad i = 1, 2, \dots, r,$$

then

$$M_{k+1,i} = \frac{n_{k,i}}{n_{k,i}+1} M_{k,i} + \frac{X_{k+1}}{n_{k,i}+1},$$

$$S_{k+1,i} = \frac{1}{n_{k,i}+1} \sum_{j=1}^{n_{k,i}+1} (X_j - M_{k+1,i})'(X_j - M_{k+1,i}).$$

From the above two equations, we obtain

$$S_{k+1,i} = \frac{n_{k,i}}{M_{k,i}+1} S_{ki} + \frac{n_{k,i}}{(n_{k,i}+1)^2} (X_{k+1} - M_{ki})'(X_{k+1} - M_{ki}).$$

Also,

$$\begin{aligned} M_{k+1,j} &= M_{kj} \quad \text{for } j \ne i \\ S_{k+1,j} &= S_{kj} \quad \text{for } j \ne i. \end{aligned}$$

The minimum K_k^l, $l = 1, 2, \dots, n$, serves to indicate the best solution among the n sets of discriminant functions.

The patterns used in the computer simulations are the characters $A, B,$ and C. The four features $x_1, x_2, x_3,$ and x_4 are extracted from each character. The extracted features correspond to the number of distinct intersections of the sample character with lines $a_1, a_2, a_3,$ and a_4 as shown in

Figure 10.5.

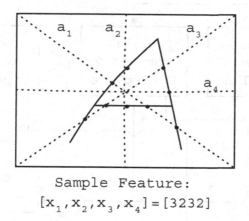

Sample Feature:

$$[x_1, x_2, x_3, x_4] = [3232]$$

Figure 10.5[5]: Description of Feature Extraction

Specific computer simulations are conducted for (1) two equal numbers of training characters from A and B using one hyperplane and (2) three equal numbers of training characters from A, B, and C using three hyperplanes.

The estimation of the n membership functions $\widehat{\lambda}_j(k)$ at the kth step of learning is as follows:

$$\widehat{\lambda}_j(k) = 1 - \frac{K_k^j}{c}, \quad j = 1, 2, \ldots, n; \ k = 1, 2, \ldots;$$

$$c \geq \vee\{K_k^j \mid j = 1, 2, \ldots, n; \ k = 1, 2, \ldots\}$$

with

$$\lim_{k \to \infty} \widehat{\lambda}_j(k) = \widehat{\lambda}_j, \quad 0 \leq \widehat{\lambda}_j < 1.$$

---

[5] Figure 10.5 is from [250], reprinted with permission by Copyright 1969 IEEE.

The details of the computer flow diagram are shown in Figure 10.6.

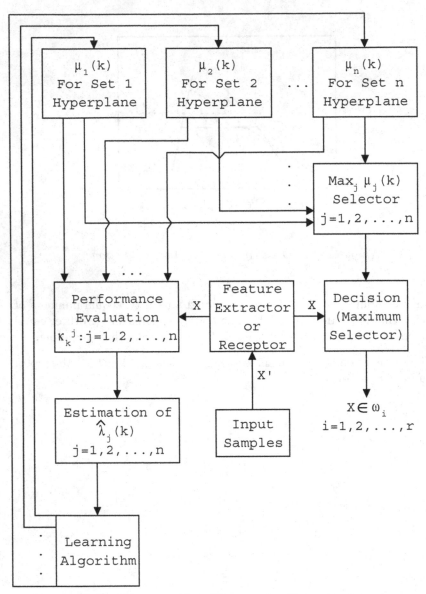

Figure 10.6[6]: Computer Flow Diagram for Character Learning

Ten sets of predetermined hyperplanes are used in both examples with 60 samples from each class. The incoming samples are introduced to the system in a random fashion. The results of the learning curves, $\mu_j(k)$ versus k for each example, are given in [250, Figures 7 and 8, p. 220].

___

[6] Figure 10.6 is from [250], reprinted with permission by Copyright 1969 IEEE.

We now give an application to control systems.

The learning controller presented here is similar to that presented in [161]. We give a brief explanation of it. A time-discrete plant may be described as

$$x(k+1) = \phi_{k+1}[x(k), u(k+1)],$$

where

$$x(k), \ x(k+1) \in \Omega_x = \{x_i \mid i = 1, 2, \ldots, p; \ p < \infty\}$$

and

$$u(k) \in \Omega_u = \{u_i \mid i = 1, 2, \ldots, p; \ p < \infty\}.$$

Here $x(k)$ is the observed response of the plant at the $(k+1)$th instant when the control action $u(k+1)$ is applied. It is assumed that ϕ_k is unknown for $k = 1, 2, \ldots$. The instantaneous performance evaluation of a control action $u(k)$ is given by

$$Z(k+1) = \omega(x(k), u(k+1), x(k+1))$$

with the set $\{Z(k) \mid k = 1, 2, \ldots\}$ bounded above, say

$$0 < Z(k) < T < \infty, \ k = 1, 2, \ldots .$$

The goal of the control is to minimize $\widehat{M}_{k+1}(Z|u(k), x(k), u(k+1))$ and the sample average of Z. The sample average for each control policy is estimated as follows. Let $u(k+1) = u_l$ be applied after observing $u(k) = u_j$ and $x(k) = x_i$. Then

$$\widehat{M}_{k+1}[Z|u_j, x_i, u_l] = \frac{n}{n+1}\widehat{M}_k[Z|u_j, x_i, u_l] + \frac{1}{n+1}Z(k+1)$$

$$\widehat{M}_{k+1}[Z|u_j, x_i, u_h] = \widehat{M}_k[Z|u_j, x_i, u_h], \ h = 1, 2, \ldots, p, \ h \neq l,$$

where $n = n(j, i, l)$ denotes the number of occurrences of $u(k) = u_j$, $x(k) = x_i$, and $u(k+1) = u_l$.

$$\hat{\mu}_{k+1}[u_l|u_j, x_i] = 1 - \frac{\widehat{M}_{k+1}[Z|u_j, x_i, u_l]}{T}. \tag{10.4}$$

By equation (10.4), the system is able to associate the control action with the maximum grade membership with the minimum sample average of Z.

Figure 10.7 illustrates the learning controller.

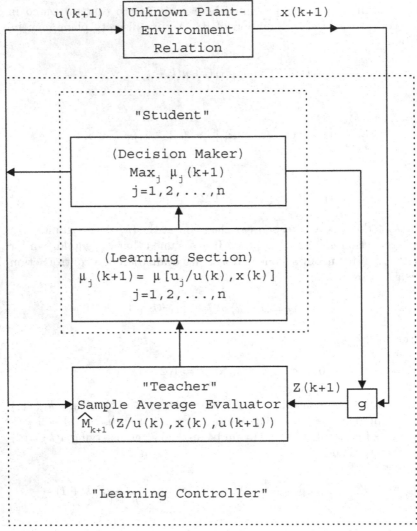

Figure 10.7[7]: *Proposed Learning Controller*

An application of a class of fuzzy automata as a model of learning systems was proposed above. A nonsupervised learning algorithm for a fuzzy automaton was presented, together with its application to pattern recognition and automatic control problems. Computer simulations of character recognition by the authors of [250] showed satisfactory results. The following is a pertinent result.

---

[7] Figure 10.7 is from [250], reprinted with permission by Copyright 1969 IEEE.

Theorem 10.5.1 *A necessary and sufficient condition for*

$$\wedge\{\{u_{ji} \vee u_k \mid i = 1, 2, \ldots, n\} \mid j = 1, 2, \ldots, s\} = \vee\{\{u_{ji} \wedge u_k \mid$$
$$i = 1, 2, \ldots, n\}$$
$$\mid j = 1, 2, \ldots, s\}$$
$$= u_k$$

is that

$$\wedge\{\vee\{u_{ji} \mid i = 1, 2, \ldots, n\} \mid j = 1, 2, \ldots, s\} \leq u_k$$
$$\leq \vee\{\wedge\{u_{ji} \mid i = 1, 2, \ldots,$$
$$n\} \mid j = 1, 2, \ldots, s\}.$$

Proof. We first prove the necessity. We may assume without loss of generality that

(1) $u_{j1} = \vee_i\{u_{ji}\} \geq u_{jn} = \wedge_i\{u_{ji}\}$,
(2) $u_{p1} = \vee_j\{u_{j1}\} \geq u_{qn} = \wedge_j\{u_{jn}\}$,
(3) $u_{mn} = \vee_j\{u_{jn}\}$, $u_{l1} = \wedge_j\{u_{j1}\}$.

Then

$$
\begin{aligned}
F_1 &= \wedge\{\{u_{ji} \vee u_k \mid i = 1, 2, \ldots, n\} \mid j = 1, 2, \ldots, s\} \\
&= (u_{11} \vee u_k) \wedge \ldots \wedge u_{p1} \wedge \ldots \wedge (u_{s1} \vee u_k) \\
F_2 &= \vee\{\{u_{ji} \wedge u_k \mid i = 1, 2, \ldots, n\} \mid j = 1, 2, \ldots, s\} \\
&= (u_{1n} \wedge u_k) \vee \ldots \vee u_{qn} \vee \ldots \vee (u_{sn} \wedge u_k).
\end{aligned}
$$

Assuming the value of u_k at different regions, one can show that if $F_1 = F_2 = u_k$, then $u_{l1} \leq u_k \leq \mu_{mn}$.

The sufficiency is immediate. ∎

10.6 Properties of Fuzzy Automata

We explore some properties similar to those of deterministic automata and probabilistic automata. We consider only ergodic, stationary, periodic, and aperiodic fuzzy transition matrices. The numerical illustrations are given to illustrate a specific property of fuzzy automata and also to furnish computational examples of the concepts discussed above. The computational procedure of Section 10.2 is followed.

After a certain number of iterations of identical inputs, it is possible for the overall fuzzy transition matrix to remain the same. For example, let

$$[T_{i_1}]^{(1)} = \begin{bmatrix} 0.9 & 0.5 & 0.3 \\ 0.2 & 0.4 & 0.95 \\ 0.8 & 0.1 & 0.25 \end{bmatrix}.$$

Then

$$[T_{i_1}]^{(2)} = \begin{bmatrix} 0.9 & 0.5 & 0.5 \\ 0.8 & 0.4 & 0.4 \\ 0.8 & 0.5 & 0.3 \end{bmatrix}$$

$$[T_{i_1}]^{(3)} = \begin{bmatrix} 0.9 & 0.5 & 0.5 \\ 0.8 & 0.5 & 0.4 \\ 0.8 & 0.5 & 0.5 \end{bmatrix}$$

$$[T_{i_1}]^{(4)} = \begin{bmatrix} 0.9 & 0.5 & 0.5 \\ 0.8 & 0.5 & 0.5 \\ 0.8 & 0.5 & 0.5 \end{bmatrix} = [T_{i_1}]^{(5)} = \ldots = [T_{i_1}]^{(n)} = \ldots.$$

A stationary fuzzy transition matrix with the property $[T_{i_1}]^{(n)} \to T$ as $n \to \infty$ is called an **ergodic fuzzy transition matrix**. Certain stationary fuzzy transition matrices have no ergodic property, but have a periodic property. As an illustration, consider the following examples.

Example 10.6.1

$$[T_i]^{(1)} = \begin{bmatrix} 0.3 & 1.0 \\ 0.8 & 0.1 \end{bmatrix}$$

$$[T_i]^{(2)} = \begin{bmatrix} 0.8 & 0.3 \\ 0.3 & 0.8 \end{bmatrix}$$

$$[T_i]^{(3)} = \begin{bmatrix} 0.3 & 0.8 \\ 0.8 & 0.3 \end{bmatrix}$$

$$[T_i]^{(4)} = \begin{bmatrix} 0.8 & 0.3 \\ 0.3 & 0.8 \end{bmatrix}.$$

The matrix keeps oscillating between $\begin{bmatrix} 0.3 & 0.8 \\ 0.8 & 0.3 \end{bmatrix}$ and $\begin{bmatrix} 0.8 & 0.3 \\ 0.3 & 0.8 \end{bmatrix}$. The period of oscillation is 2.

Example 10.6.2

$$[T_i]^{(1)} = \begin{bmatrix} 0.3 & 0.4 & 0.7 \\ 0.9 & 0.1 & 0.6 \\ 0.5 & 0.8 & 0.2 \end{bmatrix}$$

$$[T_i]^{(2)} = \begin{bmatrix} 0.5 & 0.7 & 0.4 \\ 0.5 & 0.6 & 0.7 \\ 0.8 & 0.4 & 0.6 \end{bmatrix}$$

$$[T_i]^{(3)} = \begin{bmatrix} 0.7 & 0.4 & 0.6 \\ 0.6 & 0.7 & 0.6 \\ 0.5 & 0.6 & 0.7 \end{bmatrix}$$

$$[T_i]^{(4)} = \begin{bmatrix} 0.5 & 0.6 & 0.7 \\ 0.7 & 0.6 & 0.6 \\ 0.6 & 0.7 & 0.6 \end{bmatrix}$$

$$[T_i]^{(5)} = \begin{bmatrix} 0.6 & 0.7 & 0.6 \\ 0.6 & 0.6 & 0.7 \\ 0.7 & 0.6 & 0.6 \end{bmatrix}$$

$$[T_i]^{(6)} = \begin{bmatrix} 0.7 & 0.6 & 0.6 \\ 0.6 & 0.7 & 0.6 \\ 0.6 & 0.7 & 0.7 \end{bmatrix}$$

$$[T_i]^{(7)} = \begin{bmatrix} 0.6 & 0.7 & 0.7 \\ 0.7 & 0.6 & 0.6 \\ 0.6 & 0.7 & 0.6 \end{bmatrix}$$

$$[T_i]^{(8)} = \begin{bmatrix} 0.6 & 0.7 & 0.6 \\ 0.6 & 0.7 & 0.7 \\ 0.7 & 0.6 & 0.6 \end{bmatrix}$$

$$[T_i]^{(9)} = [T_i]^{(6)}; \ [T_i]^{(10)} = [T_i]^{(7)}; [T_i]^{(11)} = [T_i]^{(8)}; \ldots .$$

The matrix in Example 10.6.2 has a period equal to 3. From Examples 10.6.1 and 10.6.2, it can be said that ergodicity has a period equal to 1. A fuzzy transition matrix having a period of length 1 is considered as an aperiodic fuzzy transition matrix.

10.7 Fractionally Fuzzy Grammars and Pattern Recognition

A type of fuzzy grammar, called a fractionally fuzzy grammar, is presented in this and the next section. It was first introduced in [44]. These grammars are especially suitable for pattern recognition because they are powerful and are easily parsed.

Formal language theory has been applied to pattern recognition problems in which the patterns contain most of their information in their structure rather than in their numeric values [69, 223, 250, 226, 225]. In order to increase the generative power of grammars and to make grammars more powerful for the purpose that they become more suited to pattern recognition, the concept of a phrase structured grammar can be extended in several ways. One way is to randomize the use of the production rules. This results in stochastic grammars [223, 67, 109] and fuzzy grammars. Languages produced by fuzzy grammars have shown some promise in dealing with pattern recognition problems, where the underlying concept may be probabilistic or fuzzy [250, 226, 225]. A second way of extending the concept of a grammar is to restrict the use of the productions [106, 190, 191]. This results in programmed grammars and controlled grammars. These grammars can generate all recursively enumerable sets with a context-free core grammar. Programmed grammars have the added advantage that they are easily implemented on a computer.

Cursive script recognition experiments [54, 52, 53, 139, 124, 213] in the 1960's and 70's had the major emphasis on recognizing whole words. None used a syntactic approach. A typical method presented in [53] decomposed the words to be recognized in to sequences of strokes that were then combined into letters and into words. In [139], an attempt was made to distinguish words that are similar in appearance such as *fell, feel, foul,* etc. All of these experiments except the ones in [213] input their data on a graphics device and kept the sequence of the points as a part of the data. In [213], pictures of writing were inputted and hence the sequence information was not available.

In the following, we show that the languages generated by the class of type i (Chomsky) fractionally fuzzy grammars properly includes the set of languages generated by type i fuzzy grammars. We also show that the set of languages generated by all type 3 (regular) fractionally fuzzy grammars is not a subset of the set of languages generated by all unrestricted (type 0) fuzzy grammars. We show that context-sensitive fractionally fuzzy grammars are recursive and can be parsed by most methods used for ordinary context-free grammars. We describe a pattern recognition experiment that uses fractionally fuzzy grammars to recognize the script letters i, e, t, and l without the help of the dot on the i or the crossing of the t. We discuss the construction of a fractionally fuzzy grammar based on a training set.

A **fuzzy grammar** FG is a six-tuple $FG = (T, N, S, P, J, \mu)$, where T, N, S, P are, respectively, the terminal alphabet, the nonterminal alphabet, the starting symbol, and the set of production rules as with an ordinary grammar. $J = \{r_i \mid i = 1, 2, \ldots, n\}$ is a set of distinct labels for the productions in P, and μ is a fuzzy membership function, $\mu : J \to [0, 1]$. Let $V = T \cup N$. Suppose that rule r_i is $\alpha \to \beta$. Then we write the application of r_i as follows:

$$\gamma \alpha \delta \xrightarrow[r_i]{\mu(r_i)} \gamma \beta \delta,$$

where $\alpha, \beta, \gamma, \delta \in V^*$. If $\theta_i \in V^*$, $i = 0, 1, 2, \ldots, m$, and

$$S = \theta_0 \xrightarrow[r_1]{\mu(r_1)} \theta_1 \xrightarrow[r_2]{\mu(r_2)} \theta_2 \xrightarrow[r_3]{\mu(r_3)} \theta_3 \ldots \xrightarrow[r_m]{\mu(r_m)} \theta_m = x$$

is a derivation of x in FG, we write

$$S = \theta_0 \xrightarrow[r_1 r_2 r_3 \ldots r_m]{\mu(r_1 r_2 r_3 \ldots r_m)} \theta_m = x,$$

where we write $\mu(r_1 r_2 r_3 \ldots r_m)$ for $\mu(r_1) \wedge \mu(r_2) \wedge \ldots \wedge \mu(r_m)$. The grade of membership of $x \in T^*$ is given by the function $\widehat{\mu} : T^* \to [0, 1]$,

$$\widehat{\mu}(x) = \vee\{\mu(r_1 r_2 r_3 \ldots r_m)\},$$

where the supremum is taken over all derivations of $x \in L(FG)$, the language generated by FG. It follows that ordinary grammars, i.e., those with $\mu(r_i) = 1$ for all $r_i \in J$, are a special case of fuzzy grammars. Fuzzy grammars can be classified according to the form of the production rules. In Chapter 4, it has been shown that for every context-free fuzzy grammar G, there exists two context-free fuzzy grammars G_g and G_c such that $L(G) = L(G_g) = L(G_c)$ and G_g is in Greibach normal form and G_c is in Chomsky normal form.

Example 10.7.1 *Consider the fuzzy grammar $FG = (T, N, S, P, J, \mu)$, where $T = \{a, b\}$, $N = \{S, A, B, C\}$, and P, J, and μ are given as follows:*

$$
\begin{array}{llll}
r_1: & S \to AB & \mu(r_1) &= 1 \\
r_2: & A \to a & \mu(r_2) &= 1 \\
r_3: & B \to b & \mu(r_3) &= 1 \\
r_4: & A \to aAB & \mu(r_4) &= 0.9 \\
r_5: & A \to aB & \mu(r_5) &= 0.5 \\
r_6: & A \to aC & \mu(r_6) &= 0.5 \\
r_7: & C \to a & \mu(r_7) &= 0.5 \\
r_8: & C \to aa & \mu(r_8) &= 0.2 \\
r_9: & A \to B & \mu(r_9) &= 0.2
\end{array}
$$

The language generated by this fuzzy grammar consists of strings of the form $a^n b^m$ with $n, m > 0$. The membership of these strings is given as follows:

$$
\widehat{\mu}(a^n b^m) = \begin{cases}
1 & \text{if } m = n = 1 \\
.9 & \text{if } m = n \neq 1 \\
.5 & \text{if } m = n \pm 1 \\
.2 & \text{if } m = n \pm 2 \\
0 & \text{otherwise.}
\end{cases}
$$

Hence it follows that this grammar generates strings of the form $a^n b^m$, where $|n - m| \leq 2$.

As we have previously seen, a fuzzy grammar can be used to generate ordinary languages by the use of thresholds. One such language with threshold c is the set of strings

$$L(FG, c) = \{x \in L(FG) \mid \widehat{\mu}(x) > c\}.$$

There are two other threshold languages defined in [147], namely, the two-threshold language and the equal-threshold language. They are defined as follows:

$$L(FG, c_1, c_2) = \{x \in L(FG) \mid c_1 < \widehat{\mu}(x) \leq c_2\}$$

and

$$L(FG, =, c) = \{x \in L(FG) \mid \widehat{\mu}(x) = c\}.$$

The language $L(FG, c)$ is most often used to compare the generating power of a fuzzy grammar to that of an ordinary grammar. However, $L(FG, c) = L(G)$, where G is the grammar obtained from the FG by removing all productions whose fuzzy membership is less than or equal to c and then removing the fuzziness from the remaining rules. Hence it is stated in [44] that the use of threshold language seems limited.

10.8   Fractionally Fuzzy Grammars

The patterns in syntactic pattern recognition are strings over the terminal alphabet. These strings must be parsed in order to find the pattern classes to which they most likely belong. Back tracking is required by many parsing algorithms [3]. That is, after applying some rules, it is discovered that the input string cannot be parsed successfully by this sequence of rules. Rather than starting all over again, it is desirable to reverse the action of one or more of the most recently applied rules in order to try another sequence of productions. It is sufficient with ordinary grammars to keep track of the derivation tree as it is generated with each node being labeled with a symbol from $T \cup N$, where T is the set of terminal symbols and N is the set of nonterminal symbols. However, this tree is not sufficient for fuzzy grammars since the fuzzy value at the ith step is the minimum of the value at $(i-1)$th step and the fuzzy membership of the ith rule. If this minimum was the ith rule's membership, there is no way of knowing the fuzzy value at the $(i - 1)$th step. Hence the fuzzy value at each step must also be remembered at each node. Consequently, the memory requirements are greatly increased for many practical problems.

   Another drawback of fuzzy grammars in pattern recognition is that all strings in $L(FG)$ can be classified into a finite number of subsets by their membership in the language. The number of such subsets is limited by the number of productions in the grammar. This is due to the fact that if $x \in L(FG)$ with a membership $\widehat{\mu}(x)$, then there must be a rule in FG with the membership $\widehat{\mu}(x)$ since $\widehat{\mu}(x) = \wedge\{\mu(r_{i_j}) \mid j = 1, 2, \dots, m\}$ for some sequence of rules $r_{i_1} r_{i_2} \dots r_{i_m}$ in P. Thus $L(FG, c)$ for some threshold c is always a language generated by those rules in the grammar with a membership greater than c.

To overcome these drawbacks, we present a method introduced in [44] of computing the membership of a string x that can be derived by the m sequence production rules, $r_1^k r_2^k \ldots r_{l_k}^k$, of lengths l_k, where $k = 1, 2, \ldots, m$. This brings us to the next definition.

Definition 10.8.1 *A **fractionally fuzzy grammar** is a 7-tuple $FFG = (N, T, S, P, J, g, h)$, where N, T, S, P, and J are the nonterminal alphabet, the terminal alphabet, the starting symbol, the set of productions, and a distinct set of labels on the productions as a fuzzy grammar, respectively. The functions g and h map J into $\mathbb{N} \cup \{0\}$ such that $g(r_k) \leq h(r_i) \ \forall r_i \in J$. A string is generated in the same manner as that by a fuzzy grammar. The membership of the derived string is given by $\forall x \in T^*$,*

$$\mu(x) = \vee \left\{ \frac{\sum_{j=1}^{l_k} g(r_j^k)}{\sum_{j=1}^{l_k} h(r_j^k)} \ \middle| \ k = 1, 2, \ldots, m \right\},$$

where $0/0$ is defined as 0.

Since $0/0$ is defined to be 0, it follows that $0 \leq \mu(x) \leq 1 \ \forall x \in T^*$. Clearly, backtracking over a rule r can now be accomplished by simply subtracting $g(r)$ and $h(r)$ from the respective running totals.

We now give an interpretation of Definition 10.8.1 in a heuristic sense. As each rule r is applied, $g(r)$ and $h(r)$ are added to the respective running totals for the numerator and denominator of the fuzzy membership. Clearly, the fuzzy membership of a string could be any rational number in $[0, 1]$. Moreover the number of fuzzy membership levels is not limited by the number of productions. With a view to pattern recognition, it follows that the amount of impact a rule has on the final membership level is proportional to the value of $h(r)$. The membership of the string tends to increase if $g(r)$ is approximately equal to $h(r)$. It tends to decrease if $g(r)$ is much less than $h(r)$ or if $g(r)$ is close to 0. Rules for which $g(r)$ and $h(r)$ are both 0 have no effect on the membership. Hence it is possible to divide the rules into three classes. Those that strongly indicate membership in the class, those that strongly indicate membership in another class, and those that serve little purpose in separating the classes, but that are traits between different classes.

Example 10.8.2 *Consider the fractionally fuzzy grammar that has $T = \{a, b\}$, $N = \{S\}$, and P, J, g, and h given as follows:*

$r_1:$	$S \rightarrow ab$	$g(r_1) = 1$	$h(r_1) = 1$
$r_2:$	$S \rightarrow aSb$	$g(r_2) = 1$	$h(r_2) = 1$
$r_3:$	$S \rightarrow aS$	$g(r_3) = 0$	$h(r_3) = 1$
$r_4:$	$S \rightarrow Sb$	$g(r_4) = 0$	$h(r_4) = 1.$

Consider the string $a^3 b^5$. The following sequences of labels yield three of the ways the string $a^3 b^5$ can be derived: r_2, r_2, r_4, r_4, r_1 and $r_2, r_3, r_4, r_4, r_4,$

r_1 and r_3, r_3, r_4, r_4, r_4, r_4, r_1 yield the string a^3b^5 with associated values $\frac{3}{5}$, $\frac{2}{5}$, and $\frac{1}{5}$, respectively. It follows that $\mu(a^3b^5) = \frac{3}{5} \vee \frac{2}{5} \vee \frac{1}{5} = \frac{3}{5}$.

This grammar generates the fuzzy language $\{a^n b^m \mid n, m \in \mathbb{N}\}$. The membership of the string $a^n b^m$ is given by

$$\mu(a^n b^m) = \frac{n \wedge m}{n \vee m}.$$

This set of strings is the fuzzy set of strings that are "almost" $a^n b^n$. That is, the first pair of rules generates the set of strings $a^n b^n$ for $n > 0$, and the second pair of rules allows for variations in the number of a's and the number of b's. The closer n and m are, the greater the membership of the string. This follows since $g(r) = h(r)$ for the first pair of rules and $g(r) < h(r)$ for the second pair. While the grammar shown in Example 10.7.1 could only measure finite differences between m and n, the grammar here measures membership on percentage difference between m and n. This is similar to the way one would judge whether or not the lengths of two lines were the same.

We now compare the relative generative power of fractionally fuzzy grammars to that of fuzzy grammars.

Theorem 10.8.3 *The set of all languages generated by type i fractionally fuzzy grammars properly contains the set of all languages generated by type i fuzzy grammars, where $i = 0, 1, 2$, and 3.*

Proof. From the preceding remarks, it follows that the number of distinct levels of membership in a language generated by a fuzzy grammar is limited by the number of production rules in the fuzzy grammar and this number is finite. Hence it suffices to show that for every fuzzy grammar of type i, there is a fractionally fuzzy grammar of type i which generates the same language, and that there exists a fractionally fuzzy grammar of type i which generates a language with an infinite number of distinct membership levels.

Let $FG = (N, T, S, P, J, \mu)$ be a fuzzy grammar of type i. We can construct a fractionally fuzzy grammar $FFG = (N', T', S', P', J', g, h)$ as follows: Let c_i be the distinct values of $\mu(r_i)$ for $i = 1, 2, \ldots, m$. There is no loss in generality in assuming that $c_1 > c_2 > \ldots > c_m$. Let g_i and h_i be integers such that $c_i = \frac{g_i}{h_i}$, $i = 1, 2, \ldots, m$. For all $A \in N$, define m distinct symbols A_i, $i = 1, 2, \ldots, m$. These symbols together with the new starting symbol S' make up the set of nonterminals N' for the fractionally fuzzy grammar. Let P' initially be the set of rules

$$r_0^i: \quad S' \to S_i \qquad g(r_0^i) = g_i \qquad h(r_0^i) = h_i,$$

$i = 1, 2, \ldots, m$. For each rule $r_j: \alpha \to \beta$ in P and each c_i, $i = 1, 2, \ldots, m$, if $\mu(r_j) \geq c_i$, add to these rules the rule

$$r_j^i: \quad \alpha_i \to \beta_i \qquad g(r_j^i) = g_i \qquad h(r_j^i) = h_i,$$

where α_i and β_i are the strings α and β with each nonterminal A replaced with the associated new nonterminal symbol A_i. The FFG has, in effect, m subgrammars reachable by the r_0^i rules. The ith subgrammar produces only those strings that would have been generated with a membership of at least c_i in $L(FG)$. In the ith subgrammar, these strings have membership $\frac{g_i}{h_i} = c_i$. Since their membership in $L(FFG)$ is defined as the supremum of all derivations, it follows that each string has equal membership in both languages. That is, $L(FFG) \supseteq L(FG)$.

To show the proper inclusion, consider the following fractionally fuzzy grammar:

$$r_1: \quad S \to a \qquad g(r_1) = 1 \qquad h(r_1) = 1$$
$$r_2: \quad S \to aS \qquad g(r_2) = 0 \qquad h(r_2) = 1.$$

This fuzzy grammar generates strings of the form a^n for $n \in \mathbb{N}$. Since the fuzzy membership of the string a^n is $\frac{1}{n}$, the language clearly has an infinite number of distinct membership levels. Since the grammar is regular, it is also context-free, context-sensitive, and a member of the class of type-0 grammars. ∎

In order for fractionally fuzzy grammars to be of use in pattern recognition, it must be possible to determine whether a given string is a member of the language. That is, in order to apply fractionally fuzzy grammars to pattern recognition, an algorithm is needed that can compute in bounded time the membership of a string in $L(FFG)$. The following lemma leads to Theorem 10.8.7, which proves that context-sensitive fractionally fuzzy grammars are recursive and such an algorithm exists.

Lemma 10.8.4 *Consider a context-sensitive fractionally fuzzy grammar. Suppose a derivation contains the sequence*

$$\ldots \to \theta_i \to \theta_{i+1} \to \ldots \to \theta_{i+k} \to \ldots,$$

where $\theta_i = \theta_{i+k}$. Then either $k \leq n^p$, where $n = |V|$ and $p = |\theta|$ is the length of the string θ_i, or $\theta_{i+j} = \theta_{i+m}$ for $0 \leq j < m < n^p$.

Proof. The result clearly follows because of the noncontracting nature of context-sensitive grammars (i.e., $|\theta_u| \leq |\theta_v|$ for $u \leq v$) and since there are exactly n^p distinct strings over V of length p. ∎

Lemma 10.8.5 *Let FFG be a fractionally fuzzy grammar. Suppose that $x \in L(FFG)$ is derivable by the sequence*

$$S = \theta_0 \to \theta_1 \to \theta_2 \to \ldots \to \theta_n = x.$$

If $\theta_j = \theta_k$ for $j < k$ and $0 \leq j < n$, then the membership of x in FFG is at least

$$\frac{\sum_{m=1}^{j} g(r_{i_m}) + \sum_{m=k+1}^{n} g(r_{i_m})}{\sum_{m=1}^{j} h(r_{i_m}) + \sum_{m=k+1}^{n} h(r_{i_m})} \vee \frac{\sum_{m=j}^{k} g(r_{i_m})}{\sum_{m=j}^{k} h(r_{i_m})}.$$

Proof. The loop in the derivation sequence from θ_j to θ_k can be removed. The first argument of the maximum represents the membership given to x by this shortened derivation. The loop can also be repeated as many times as desired. As the number of times the loop is repeated is increased, the membership of x approaches the second argument. Since there may be other derivations of x, the membership of x in $L(FFG)$ is at least the maximum of these two terms. ∎

Lemma 10.8.5 may be applied repeatedly in a derivation. Hence if a derivation contains a loop nested within a loop, the loops can be considered separately and the membership of the string is the maximum given by the loop-free derivation, the inner loop, and the outer loop.

Lemma 10.8.6 *Let FFG be a context-sensitive fractionally fuzzy grammar with $|V| = n$. Let $R_0^1 = \{S\}$. Let R_0^k be the set of all strings over V of length k that can be directly generated from a string of length less than k. Let R_j^k be the set of all strings of length k that can be directly generated from a string in R_{j-1}^k, $j = 1, 2, \ldots$. Then the set $R^k = R_0^k R_1^k \ldots R_n^k$ contains all strings over V of length k that can be generated by the FFG. Moreover, the derivation needed to generate R^k contains all the simple derivation loops on strings of length k in $L(FFG)$.*

Proof. Suppose that there exists a θ_m such that $\theta_m \notin R^k$, $|\theta_m| = k$, and θ_m is derivable from $\theta_i \in R_0^k$. Let $\theta_i \to \theta_{i+1} \ldots \to \theta_{i+j} = \theta_m$ be the shortest sequence from θ_i to θ_m for FFG. Since $j > n$ by assumption, it follows by Lemma 10.8.4 that this sequence must contain a loop and is thus not the shortest sequence. Hence no such θ_m exists.

Suppose that $\theta_j \to \theta_{j+1} \ldots \to \theta_{j+s} = \theta_j$ is a simple loop in FFG, where $|\theta_j| = k$. Suppose this loop was not detected by the derivations that generate R^k and that this loop can be detected from $\theta_i \in R_0^k$ by the shortest sequence

$$\theta_i \to \theta_{i+1} \ldots \to \theta_{i+r} = \theta_j.$$

By assumption, $r + s > n$ since the loop was not detected by generating R^k. However, the $r + s$ strings $\theta_i, \theta_{i+1}, \ldots, \theta_{j+s-1}$ cannot all be distinct. Since $\theta_i, \theta_{i+1}, \ldots, \theta_{i+r}$ are distinct by assumption and $\theta_j, \ldots, \theta_{j+s-1}$ are also distinct by assumption, we have that $\theta_{i+u} = \theta_{j+v}$ for some $0 \leq u < r$ and $0 \leq v < s$. Thus the derivation sequence

$$\theta_i \to \theta_{i+1} \ldots \to \theta_{i+u} = \theta_{j+v} \to \theta_{j+v+1} \ldots \to \theta_{j+s} = \theta_j \to \ldots \to \theta_{j+v}$$

also detects the simple loop and is shorter than the original sequence. This contradicts the original assumption. Consequently, no such loop can appear in a derivation. ∎

Theorem 10.8.7 *If a fractionally fuzzy grammar FFG is a context-sensitive fractionally fuzzy grammar, then it is recursive.*

Proof. Let $|V| = n$. For any string x, we need only generate $R^1, R^2, \ldots,$ R^k, where $k = |x|$. Let R_i be the set of all strings derivable from the starting symbol in i steps. Since i is finite, R_i can be determined. Since it takes n_j steps to generate R^j from R_0^j, R_m contains R^j if $m \geq 1 + n + n^2 + \ldots + n^j$. Thus in a finite number of steps, all strings of length k can be found and all loops in these derivations can be detected. Hence it can be determined whether or not $x \in L(FFG)$ and its membership if it is. ∎

The following example provides an interesting property of c-fractionally fuzzy grammars.

Example 10.8.8 *For a regular fractionally fuzzy grammar FFG, the language $L(FFG, c)$ is not necessarily a regular language:*

Consider the two fractionally fuzzy grammars $FFG_1 = (T, N, X, P, J,$ $g_1, h_1)$ and $FFG_2 = (T, N, S, P, J, g_2, h_2)$, where $T = \{0, 1\}$, $N = \{S, A\}$, and P is given as follows:

$$
\begin{aligned}
r_1 &: \quad S \to 0S \\
r_2 &: \quad S \to 0 \\
r_3 &: \quad S \to A \\
r_4 &: \quad A \to 1A \\
r_5 &: \quad A \to 1.
\end{aligned}
$$

Let $h_1(r_i) = h_2(r_i) = 1$ for $i = 1, 2, 3, 4, 5$ and let $g_1(r_i) = 1$ for $i = 1, 2, 3$, $g_1(r_i) = 0$ for $i = 4, 5$. Let $g_2(r_i) = 0$ for $i = 1, 2$, and let $g_2(r_i) = 1$ for $i = 3, 4, 5$. Clearly, both the grammars produce strings of the form $0^n 1^m$, where $m, n \geq 0$ and $m + n > 0$. Consider the string $0^2 1^3$. Now

$$S \overset{r_1}{\to} 0S \overset{r_1}{\to} 00S \overset{r_3}{\to} 00A \overset{r_4}{\to} 001A \overset{r_4}{\to} 0011A \overset{r_5}{\to} 00111.$$

Hence $\mu_1(0^2 1^3) = \frac{3}{6} = \frac{2+1}{3+2+1}$ and $\mu_2(0^2 1^3) = \frac{4}{6} = \frac{3+1}{3+2+1}$. It follows that the fuzzy membership of $0^n 1^m$ is given by the following two equations:

$$\mu_1(0^n 1^m) = \frac{n+1}{m+n+1}$$

and

$$\mu_2(0^n 1^m) = \begin{cases} \frac{m+1}{m+n+1} & \text{if } m > 0 \\ 0 & \text{if } m = 0. \end{cases}$$

Assume that the set

$$L(FFG_1, 0.5) = \{0^n 1^m | n \geq m\}$$

and the set

$$L(FFG_2, 0.5) = \{0^n 1^m \mid n \leq m\}$$

are regular. Since regular sets are closed under intersection, the set

$$L(FFG_1, 0.5) \cap L(FFG_2, 0.5) = \{0^n 1^n \mid n > 0\}$$

must be regular. However, it is known that $\{0^n 1^n \mid n > 0\}$ is context-free and not regular. Thus the assumption is false and the desired result holds.

10.9   A Pattern Recognition Experiment

In [44], an experiment was developed to test the usefulness of fractionally fuzzy grammars in pattern recognition. A pattern space consisting of a set of strings was chosen in order to use script writing. The data were input to a computer on a graphics tablet. This data consisted of strings of points in a 2-dimensional space of the tablet. The data were a sample of seven persons' handwritings. Each person was given a list of 400 seven-letter words and was told approximately how large the person should write. The first three persons wrote all 400 words while the last four wrote the first 100 words. The data was digitized in a continuous mode by the computer whenever the pen was down. Each point collected in this manner was compared to the previously stored point to determine if the distance between them was greater than a given threshold. The threshold was chosen to be about 0.04 inch. If the distance was not greater than the threshold, the new point was discarded and a new position of the pen was read. If the threshold was exceeded, this point was added to the data and the process was repeated. This procedure resulted in a record of 250 points in the $X - Y$ plane (with zero fill-in) for each seven-letter word written. Some examples of words input to the computer are shown in Figure 1 of [44, p. 344].

These points were converted into a string of symbols that would comprise the terminal alphabet. This was accomplished by comparing each adjacent pair of points to see the relative direction traveled by the pen at that point. The directions were then classified into one of eight directions. Each direction was separated by 45 degrees with class 0 centered at 0 degrees (the positive X-direction) and the remaining classes being numbered 1 through 7 in a counterclockwise direction. Hence the terminal alphabet consisted of eight octal digits, i.e., $T = \{0, 1, 2, \ldots, 7\}$. Figure 2 of [44, p. 345] shows that quantization of directions introduced some distortion into the data.

The individual letters were separated by an operator using an interactive graphics program. These letters then consisted of strings of octal digits whose lengths varied from 10 to about 70 characters in length. The crossing of t's and the dotting of the i's were deleted since they did not necessarily follow the basic letter without other letters intervening. Hence in order to keep the computer time down, only four letters were used in the test. The machine was asked to separate the i's, e's, t's, and the l's without the dots on the i's and the crossings of the t's. In view of Example 10.8.8, it was also decided to use only regular fractionally fuzzy grammars. The grammars listed in Figure 10.8 were generated by cut and try methods based on the following ideas.

Production Rule	$\frac{g_e}{h_e}$	$\frac{g_i}{h_i}$	$\frac{g_l}{h_l}$	$\frac{g_t}{h_t}$	Comments
$S \to 0S$	0/0	0/0	0/0	0/0	Allows horizontal initial stroke
$S \to 1A$	14/14	10/14	0/18	0/18	Begining of a letter. Initial membership
$S \to 2B$	12/12	8/14	0/18	0/18	Ditto
$A \to 0A$	0/1	0/1	0/1	0/1	Small backtrack in direction (noise)
$A \to 1A$	0/0	0/0	0/0	0/0	Expected in all up strokes No effect
$A \to 2B$	0/0	0/0	0/0	0/0	Ditto
$A \to 2B$	0/0	0/0	0/0	0/0	Ditto
$A \to 3C$	1/1	0/3	4/4	0/3	Top of a letter not pointed
$A \to 4D$	0/1	0/1	4/4	0/1	Ditto
$A \to 5E$	0/16	8/8	0/4	4/4	Top sharply pointed
$A \to 6F$	0/16	8/8	0/4	4/4	Ditto
$B \to 0B$	0/2	0/2	0/2	0/2	Noise up on stroke
$B \to 1B$	0/1	0/1	0/1	0/1	Noise up on stroke
$B \to 2B$	0/0	0/0	0/0	0/0	Expected on up stroke (no effect)
$B \to 3C$	1/1	0/3	4/4	0/3	Rounded top of a letter
$B \to 3J$	0/2	0/2	0/2	0/2	Noise sequence started
$B \to 4D$	0/0	0/0	4/4	0/0	Ditto
$B \to 5E$	0/5	5/5	3/3	5/5	Neutral top of a letter
$B \to 6F$	0/16	8/8	0/4	4/4	Pointed top of a letter
$C \to 3C$	2/2	0/7	4/4	0/7	Rounded top of a letter
$C \to 4D$	5/5	0/5	4/4	0/5	Ditto
$C \to 5E$	1/1	0/3	2/2	0/3	Neutral top of a letter
$C \to 6F$	0/3	3/3	1/1	3/3	Slightly pointed top
$C \to 7G$	0/7	6/6	0/3	6/6	Pointed top
$D \to 4D$	2/2	0/7	5/5	0/7	Very open loop of a letter
$D \to 5E$	1/1	0/3	2/2	0/3	Open loop or a highly slanted letter
$D \to 6F$	0/0	0/0	0/0	0/0	No effect

Production Rule	$\frac{g_c}{h_c}$	$\frac{g_i}{h_i}$	$\frac{g_l}{h_l}$	$\frac{g_t}{h_t}$	Comments
$E \to 5E$	0/1	0/1	2/2	0/1	Possible loop
$E \to 6F$	0/2	0/2	2/2	2/2	Expected part of down stroke
$E \to 7F$	0/1	0/1	0/1	0/1	Noise on down stroke
$F \to 0H$	0/0	0/0	0/0	0/0	End of a letter (tail)
$F \to 0$	0/0	0/0	0/0	0/0	Ditto
$F \to 7G$	0/2	0/2	2/2	2/2	Down stroke
$F \to 7$	0/2	0/2	2/2	2/2	Down stroke. End of a letter
$F \to 6F$	0/1	0/2	2/2	2/2	Down stroke
$F \to 6$	0/2	0/2	2/2	2/2	Down stroke. End of a letter
$F \to 5F$	0/1	0/1	0/1	0/1	Noise on down stroke
$G \to 0H$	0/0	0/0	0/0	0/0	No effect (tail)
$G \to 0$	0/0	0/0	0/0	0/0	End of a tail
$G \to 7G$	0/2	0/2	2/2	2/2	Down stroke
$G \to 7$	0/2	0/2	2/2	2/2	Ditto
$G \to 6G$	0/3	0/3	0/3	0/3	Noise
$G \to 6$	0/2	0/2	0/2	0/2	Noise or end of a letter
$H \to 0H$	0/0	0/0	0/0	0/0	Tail of a letter (no effect)
$H \to 0$	0/0	0/0	0/0	0/0	Ditto
$J \to 2B$	0/0	0/0	0/0	0/0	Noise

Figure 10.8[8] . The List of Production Rules of the Fractionally Fuzzy Grammar for the Experiment.

Since all the letters under consideration started with a near horizontal, left to right stroke (octal direction 0) and continued in a counterclockwise direction (increasing octal direction) until returning to a near horizontal tail, the same set of production rules was used for all classes. The productions used the nonterminal symbols $A, B, \ldots, G$ to represent the highest octal direction so far encountered, A representing 1, B representing 2, and so on. In the ideal case, only higher octal directions and higher nonterminal representations are reachable from any nonterminal symbol. However, to allow for noise in the less curved portions of the letters, the terminal symbol generated was allowed to be one less than the highest so far generated. A change of direction of more than 225 degrees counterclockwise was not allowed since this would never occur in these letters. In order to allow a tail of any length to be affixed to the ideal letter, the nonterminal symbol H was added. The grammar was tested on a training set and was found to accept most of the strings. In order for all strings in the training set to be accepted, minor modifications were made. For example, J was added to the nonterminal alphabet to pick up an unusual noise condition. The fractionally fuzzy membership functions were developed using the following criteria. First, a rule that could not help distinguish one class from another was given the value 0/0 and would then have no effect on the final membership assuming some rule r, for which $h(r) \neq 0$, was also applied.

---

[8] Figure 10.8 is from [44], reprinted with permission from Academic Press.

Second, a rule for which $h(r)$ was small would have little effect on the final membership of any string generated by that rule. Third, any rule for which $h(r)$ was large would have a large effect on the final membership of any string generated by using that rule. Fourth, if rule r was used, the fuzzy membership of the string would be changed in the direction towards the value $\frac{g(r)}{h(r)}$ by that application of rule r. Hence if $\frac{g(r)}{h(r)}$ were close to 1, then the membership of the string would be increased and if $\frac{g(r)}{h(r)}$ were close to 0, the membership of the string would be decreased. Finally, a rule that was used in all strings could be given a membership value that could serve as a starting point from which subtraction could occur by rules with $\frac{g(r)}{h(r)} = 0$ and to which addition could occur by rules with $\frac{g(r)}{h(r)} = 1$. Either the second or the third rule in Figure 10.8 must be used in any valid derivation. Some comments are included in Figure 10.8 to give some insight into why the membership functions for that rule were chosen. For example, the rule $B \rightarrow 6F$ is used when a vertical line changes direction abruptly from up to down. This would indicate a sharply pointed crown and the letters i and t are reinforced while e and l are reduced in membership when this rule is used. After adjustment on the training set to allow a threshold of 0.5 or more to indicate "in the class" and less than 0.5 to indicate "not in the class", the grammars were used on a random sampling of 121 letters from the remainder of the patterns. The strings were parsed in a top down (left to right) manner by a program written in SNOBOL 4 programming language. The results of this test are summarized in Figure 10.9.

Class	E	I	L	T
Method 1 : % error	10	16	28	74
Method 2 : % error	10	4	5	27

Figure 10.9[9] : Results of the Experiment

Two methods of categorizing were tested. The first classified the letter into any class for which the pattern had a fuzzy membership of 0.5 or more. Some letters were not classified while others were classified into more than one class. The method was considered successful if the correct class was included among other classes since a contextual post-processor could be used to find the correct letter. The second method classified the pattern into the class that had the highest fuzzy membership. As expected, the second method had better results, with 90% of the e's, 96% of the i's, 95% of the l's, and 73% of the t's correctly classified. The only distinction between a t and an l is the width of the loop. Since many of the t's were quite wide, they were incorrectly classified as l's. If the presence of one or more t's was detected by the presence of absence of a horizontal line written directly above some portion of the word, most of these incorrect classifications could be corrected by a contextual post-processor such as

---

[9]Figure 10.9 is from [44], reprinted with permission from Academic Press.

described in [54]. The distinction between the e's and l's could have been improved if the data were prescaled to eliminate differences in the average height of the letters generated by the different subjects. Considering the similarities in the four letters tested, the results were considered in [44] to be quite good.

10.10 General Fuzzy Acceptors for Syntactic Pattern Recognition

In [62, 63], the syntactic approach to pattern recognition was examined by using formal deterministic and stochastic languages. In [175], fuzzy regular languages for pattern description in relationship with finite fuzzy acceptors were considered, where max-min was used as the composition of L-fuzzy relations.

In [98], general fuzzy acceptors were considered using sup-$*$-composition for binary operations $*$ preserving the properties concerning ε-equivalence and which has min as the greatest lower bound. We consider [98] in the next two sections.

Let $(L, \vee, \wedge, 0, 1)$ be a complete lattice with upper and lower bound 1 and 0, respectively. Let $f : L \to [0,1]$ be an injective isotonic map, i.e., a one-to-one function such that $u \le v \Rightarrow f(u) \le f(v)$. Here and in the next section, we assume f is fixed.

For $x = (x_1, \dots, x_n)$, $y = (y_1, \dots, y_n) \in L^n$ ($n \in \mathbb{N}$), the distance $\|x - y\|$ with respect to f between x and y is defined as

$$\vee_{i=1}^n |f(x_i) - f(y_i)|.$$

As usual, an L-fuzzy subset of a set S is a function from S into L. Let $L(S) = \{\mu \mid \mu : S \to L\}$. Let I, J, K be arbitrary nonempty sets. Any binary operation $*$ on L can be used for the construction of the composition of L-fuzzy relations [50] as follows. Let $\mu \in L(I \times J)$, $\nu \in L(J \times K)$. Define the sup-$*$-composition $\mu \circ \nu$ by $\forall (i, k) \in I \times K$,

$$\mu \circ \nu(i, k) = \vee\{\mu(i, j) * \nu(j, k) \mid j \in J\}.$$

Let $\sigma \in L(I)$ be an L-fuzzy subset of I. Then an L-fuzzy subset $\sigma \circ \mu$ of J can be defined by

$$\sigma \circ \mu(j) = \vee\{\sigma(i) * \mu(i, j) \mid i \in I\}, \ \forall j \in J.$$

If $\rho \in L(J)$, $\mu \circ \rho$ is defined similarly. Moreover for $\sigma' \in L(I)$, $\sigma \circ \sigma'$ can be defined to be a scalar.

In order to consider the complete transition behavior of a general acceptor and how to compute it, the following definition is needed, [19, 91]. A triple $(L, *, \le)$, denoted by $\mathcal{L}$, is called a **commutative cl-semigroup** if $(L, \vee, \wedge, 0, 1)$ is a complete lattice with upper and lower bounds 1 and 0, respectively, and $(L, *)$ is a commutative semigroup such that the following property holds:

$$a * (\vee A) = \vee\{a * b \mid b \in A\}, \ \forall A \subseteq L.$$

The real unit interval equipped with an averaging operator $M = *$, [141], gives an example of a commutative cl-semigroup as does a complete Heyting algebra with $\wedge = *$.

Definition 10.10.1 *A **fuzzy acceptor** A over $\mathcal{L}$ is a six-tuple $A = (Q, X, \iota, \tau, \mathcal{M}, \mathcal{L})$, where*
  (1) Q is a nonempty set of states;
  (2) X is a nonempty set of input symbols;
  (3) $\iota : Q \to L$ is the initial state distribution;
  (4) $\tau : Q \to L$ is the final state evaluation;
  (5) $\mathcal{M} = \{M(x) \mid Q \times Q \to L \mid x \in X\}$ is the set of transition functions;
  (6) $\mathcal{L}$ is a commutative cl-semigroup.

If $* = \wedge$ and Q and X are finite sets, then this is the same as the definition in [175].

We consider the complete transition behavior of an acceptor and how to compute it. We now extend the stepwise transition behavior of A. Let X be the set of input symbols. Let $X^1 = X$ and $X^j = X \times X \times \ldots \times X$ (j times), $j \in \mathbb{N}$. Let $S[X]$ denote the disjoint union of the sets $X^1, X^2, \ldots$. Then the elements of $S[X]$ are sequences $(x_{i_1}, x_{i_2}, \ldots, x_{i_m})$, $x_{i_j} \in X$, $m = 1, 2, \ldots$. Define a multiplication on $S[X]$ by juxtaposition, i.e.,

$$(x_{i_1}, x_{i_2}, \ldots, x_{i_m})(x_{j_1}, x_{j_2}, \ldots, x_{j_n}) = (x_{i_1}, x_{i_2}, \ldots, x_{i_m}, x_{j_1}, x_{j_2}, \ldots, x_{j_n}).$$

Then $S[X]$ is a free semigroup with respect to this operation. Let $\mathcal{F}[X]$ denote $S[X] \cup \{\Lambda\}$. Then $\mathcal{F}[X]$ is a free monoid generated by X with identity Λ.

Every element of $\mathcal{F}[X]$ is called a **word** on X. Every word $x \in S[X]$ can be represented by $x = x_{i_1} x_{i_2} \ldots x_{i_k}$, where $x_{i_1}, x_{i_2}, \ldots, x_{i_k} \in X$. Then x is said to have length k.

Let $A = (Q, X, \iota, \tau, \mathcal{M}, \mathcal{L})$ be a fuzzy acceptor. For any input word $x \in S[X]$, the transition function $M(x)$ is computed by the expression

$$M(x) = M(x_{i_1}) \circ M(x_{i_2}) \circ \ldots \circ M(x_{i_k}),$$

where $x = x_{i_1} x_{i_2} \ldots x_{i_k} \in S[X]$ and where $M(\Lambda)$ may be regarded as the identity. The expression represents the complete behavior of A in k consecutive steps. Every element $M(x)_{qq'}$ is the grade of membership if the input word is $x \in X_k$ with the beginning state $q \in Q$ at instant t and the last state $q' \in Q$ at instant $t + k$. The set of all transition behavior fuzzy relations describes the complete behavior of A and we use the notation

$\mathcal{M}^*(A) = \{M(x) \mid x \in \mathcal{F}[X]\}$. If $x \in \mathcal{F}[X]$, $\iota \circ M(x) \circ \tau$ is the analytic extension of the stepwise behavior of A.

Let $L = Q = [a, b]$ be an interval. Then $\forall x \in M$, the transition function $M(x)$ is a fuzzy subset of $[a, b] \times [a, b]$. The initial state distributions and the final state evaluations are fuzzy subsets of $[a, b]$. Hence fuzzy acceptors may be extended to continuous types. In particular, if a state set Q is a subset of $[a, b]$ such that $|Q| = n$ for some $n \in \mathbb{N}$, then the transition functions can be represented as $n \times n$ matrices. The initial state distributions and the final state evaluations also are vectors of dimension n.

10.11 ε-Equivalence by Inputs

Definition 10.11.1 *Let* $\mu, \mu' : I \times J \to L$, *and let* $\varepsilon \in [0, 1]$ *be fixed. Then* μ *and* μ' *are called* ε-**close**, *denoted by* $\mu \varepsilon \mu'$, *if* $\|\mu(i, j) - \mu'(i, j)\| \leq \varepsilon$, $\forall(i, k) \in I \times J$.

The following lemma and definition are needed to study the properties of ε-closeness.

Lemma 10.11.2 (95) *Let* f *and* g *be bounded, real-valued functions on a set* X. *Then*

$$\left| \vee_{x \in X} f(x) - \vee_{x \in X} g(x) \right| \leq \vee_{x \in X} \left| f(x) - g(x) \right|,$$

and

$$\left| \wedge_{x \in X} f(x) - \wedge_{x \in X} g(x) \right| \leq \vee_{x \in X} \left| f(x) - g(x) \right|.$$

Definition 10.11.3 *A binary operation* $*$ *on* L *is called* **contractive** *if*

$$\left| x * y - x' * y' \right| \leq \left| (x, y) - (x', y') \right|,$$

$\forall x, x', y, y' \in L$.

Fuzzy acceptors with contractive operations $*$ are called **contractive fuzzy acceptors**. It follows that min and max are simple contractive operations that are associative. To construct contractive fuzzy acceptors $\mathcal{L}$, $(L, *)$ must be a commutative semigroup.

In [141], averaging operators between min and max are summarized, and pictorial representations are given. Averaging operators are defined as follows:

An **averaging operator** is a function $M : [0, 1] \times [0, 1] \to [0, 1]$ such that the following conditions hold $\forall x, y \in [0, 1]$:

(1) $x \wedge y \leq M(x, y) \leq x \vee y$, $M \notin \{\wedge, \vee\}$;
(2) $M(x, y) = M(y, x)$ (commutative);
(3) M is increasing and continuous;

(4) $M(0,0) = 0$, $M(1,1) = 1$.

It is known that there exist no associative strictly increasing averaging operators [47]. An associative averaging operator that is not strictly increasing is, for fixed $p \in [0,1]$,

$$M(x,y) = med(x,y,p) = \begin{cases} y & \text{if } x \le y \le p \\ p & \text{if } x \le p \le y \\ x & \text{if } p \le x \le y. \end{cases} \tag{10.5}$$

This median operator is contractive.

It follows that min and max are the least and greatest contractive operations of associative averaging operators, respectively.

The concept of fuzzy subsets has been incorporated into the syntactic approach at two levels. The pattern primitives are themselves considered to be labels of fuzzy sets, e.g., such subpatterns as "almost circulars arcs", "gentle", "fair", "sharp" curves are considered. Also, the structural relations among the subpatterns may be fuzzy, so that the formal grammar is fuzzified by the weighted production rules and the grades of memberships of a string are obtained by max-min composition of the grades of the productions used in the derivation. When the pattern primitives are extracted from an image with low quality or a deformed pattern, the min operation in the max-min composition of the grades of production is sensitive to distortion of primitives. In such a case, the above median operator, which is well known to be useful for noise suppression, preserves the grade of membership of primitives better than min operators. Consequently, sup-med-composition rather than max-min may work well if the parameter p in (10.5) is decided appropriately.

Proposition 10.11.4 *Let $*$ be a contractive operation on L. Let μ, μ', ν, ν' be L-fuzzy relations. If compositions below make sense, then the following properties hold:*

(1) $\mu \varepsilon \mu' \Rightarrow \nu \circ \mu \circ \nu' \varepsilon \nu \circ \mu' \circ \nu'$ (invariance),

(2) $\mu \varepsilon \mu', \nu \varepsilon \nu' \Rightarrow \mu \circ \nu \varepsilon \mu' \circ \nu'$ (coordination).

Proof. Since property (1) can be easily obtained from (2), we only prove (2). Let the L-fuzzy relations μ and μ' on $I \times J$ be ε-close, and let the L-fuzzy relations ν and ν' on $J \times K$ be ε-close. Then for $\forall (i,k) \in I \times K$,

$$
\begin{aligned}
\|(\mu \circ \nu)(i,k) - (\mu' \circ \nu')(i,k)\| &= \| \vee_{j \in J} \mu(i,j) * \nu(j,k) - \\
& \quad \vee_{j \in J} \mu'(i,j) * \nu'(j,k) \| \\
&\le \vee_{j \in J} \| \mu(i,j) * \nu(j,k) - \\
& \quad \mu'(i,j) * \nu'(j,k) \| \\
&\le \vee_{j \in J} \| \mu(i,j) - \mu'(i,j) \| \vee \\
& \quad \| \nu(j,k) - \nu'(j,k) \| \\
&\le \varepsilon,
\end{aligned}
$$

where the first inequality comes from Lemma 10.11.2 and the second inequality comes from the assumption that $*$ is contractive. Thus $\mu \circ \nu \varepsilon \mu' \circ \nu'$. ∎

To obtain an ε-partition on X, we define ε-equivalence, keeping in mind the definition for ε-closeness.

Definition 10.11.5 Let $A = (Q, X, \iota, \tau, \mathcal{M}, \mathcal{L})$ be a fuzzy acceptor and $\varepsilon \in [0, 1]$. Then $x, y \in \mathcal{F}[X]$ are said to be ε-**equivalent,** written $x\varepsilon y$, if the following conditions hold:
  (1) $x, y \in X^k$ for some $n \in \mathbb{N}$,
  (2) $M(x)\varepsilon M(y)$.

The following corollary follows from Proposition 10.11.4(1).

Corollary 10.11.6 Let $A = (Q, X, \iota, \tau, \mathcal{M}, \mathcal{L})$ be a contractive fuzzy acceptor and let $x, y \in \mathcal{F}[X]$. If $x\varepsilon y$, then $\iota \circ M(x) \circ \tau \varepsilon \iota \circ M(y) \circ \tau$. ∎

Proposition 10.11.7 Let $A = (Q, X, \iota, \tau, \mathcal{M}, \mathcal{L})$ be a contractive fuzzy acceptor and $\varepsilon \in [0, 1]$. Then the following properties hold:
  (1) if $x\varepsilon y$ for $x, y \in \mathcal{F}[X]$, then $(uxv)\varepsilon(uyv)$, $\forall u, v \in X^k$,
  (2) if $x\varepsilon x'$ and $y\varepsilon y'$ for $x, x', y, y' \in \mathcal{F}[X]$, then $xy\varepsilon x'y'$.

Proof. (1) Let $x\varepsilon y$. Then $M(x)\varepsilon M(y)$. By Proposition 10.11.4(1),

$$M(uxv) = M(u) \circ M(x) \circ M(v)\varepsilon M(u) \circ M(y) \circ M(v) = M(uyv).$$

Thus $(uxv)\varepsilon(uyv)$.
  (2) The proof uses Proposition 10.11.4(2) and is similar to that of (1). ∎

Note that the relation ε-equivalence is reflexive and symmetric, but not transitive. However, an ε-partition on X is needed for syntactic pattern recognition. Thus we need the following definition.

Definition 10.11.8 Let $A = (Q, X, \iota, \tau, \mathcal{M}, \mathcal{L})$ be a fuzzy acceptor and $\varepsilon \in [0, 1]$ be fixed. For every $x \in X$, an ε-**class** $[x]$ with the center x is defined to be $[x] = \{x' \in X \mid x\varepsilon x'\}$. A family of ε-classes is called an ε-**partition** of X if it consists of disjoint subsets that cover X.

By choosing a suitable operation $*$, we can compute an ε-partition on X through the following Algorithm from Section 8.9 and [174].

Algorithm 10.11.9 1. Define a contractive binary operation $*$.
  2. For every $x \in X$, compute $\iota \circ M(x) \circ \tau$.
  3. Choose a center with the greatest value $\iota \circ M(x) \circ \tau$.
  4. Construct the ε-class $[x]$.
  5. Replace X with $X \backslash [x]$. If $X \neq \emptyset$, go to step 3.
  6. Print all ε-classes.

From this algorithm, we compute an ε-partition of X. The choice of the center x for an ε-class $[x]$ depends on the task under examination and on the user's need.

To compute only an ε-partition of X, it suffices to choose an operation $*$ that is contractive for each argument, i.e., $\|x * y - x' * y\| \leq \|x - x'\|$ and $\|x * y - x * y'\| \leq \|y - y'\|$ for all $x, x', y, y' \in L$. This yields (1) of Proposition 10.11.4. For example, let $*$ be Yager's t-norm with $r \geq 1$:

$$*(x, y) = 1 - (1 \wedge \sqrt[r]{(1 - x)^r + (1 - y)^r}).$$

Then $\|\partial * / \partial x\| \leq 1$ and $\|\partial * / \partial y\| \leq 1$. This implies that it is contractive in each argument. Hence by using Yager's t-norm, we can obtain an ε-partition on X.

Proposition 10.11.10 *Let* $A = (Q, X, \iota, \tau, \mathcal{M}, \mathcal{L})$ *be a contractive fuzzy acceptor. Let* $[v] = \{v' \in X^k \mid v \varepsilon v'\}$ *and* $[x] = \{x' \in X \mid x \varepsilon x'\}$ *be ε-classes with center* $v \in X^k$ *and* $x \in X$, *respectively. Then* vx *and* xv *are centers of ε-classes on* X^{k+1}.

Proof. The proof follows from Proposition 10.11.4(2). ∎

Let X^k / ε denote the quotient set on X^k, $k = 1, 2, \ldots$. As a natural extension of Algorithm 10.11.9 by Proposition 10.11.7, we have the following algorithm by Proposition 10.11.10. It forms an ε-partition on X^k for each $k \in \mathbb{N}$.

Algorithm 10.11.11 *1. Input* $k \in \mathbb{N}$.
2. For $i = 1$, *compute* $X / \varepsilon = X^1 / \varepsilon$ *by using Algorithm* 10.11.9.
3. If $i > k$, *go to step 5.*
4. For $i = i + 1$, *compute* X^{i+1} / ε *by using Proposition* 10.11.10. *Go to step 3.*
5. Print all ε-classes of X^k / ε.

10.12 Fuzzy-State Automata: Their Stability and Fault Tolerance

Special branches of systems theory have achieved a high level of development and sophistication. Attempts to unify these branches have been made, [105, 23, 251, 178]. In the next five sections, we present the work of [36], which introduces geometrical and stability concepts from dynamical system theory into the theory of abstract automata, but without giving up the two main features accounting for the performance of automata or sequential machines, i.e., discrete state spaces and sequential operation at consecutive discrete values of time. The concept of tolerance space, introduced in [178], is used. This concept is carried over to the state space of abstract automata.

We first present a brief account of the elementary properties of tolerances and tolerance spaces in Section 10.13. We then introduce in Section 10.14 a special class of finite automata with given tolerance, the class of fuzzy-state automata. Their state transitions exhibit stability properties. We examine the class of fuzzy-state automata with respect to its order structure. In Section 10.15, stability of fuzzy-state automata is studied. The appearance of approximate fixed point and attractors in the state space of fuzzy-state automata and almost periodicity of their state transitions are studied. The application of the previously developed framework is dealt with in Section 10.16. For this, the field of fault-tolerant computing is chosen, i.e., the ability of certain real automata to execute, within a given tolerance, specific programs regardless of failures in their performance or hardware.

The goal of [36] was to understand how biological systems work, in particular, how they control malfunctions. With this in mind this capability was simulated on an abstract level. With respect to fault tolerance, von Neumann noted the differences between biological and artificial systems. He states that in a biological system, not every error has to be caught, explained, and corrected; see [239, p. 71].

10.13   Relational Description of Automata

The concept of a tolerance space is due to the authors of [178], [265], and [179], although these authors used different names for tolerance spaces. In [5] and [43], tolerance spaces were used in connection with automata and control theory. Some reasons for studying tolerance spaces are the nature of our perception of space and time, [178], and state space properties of man-made systems like digital computers. Strong similarities between tolerance spaces and topological spaces exist. There are substantial and very important differences, however.

A relation on X is called a **tolerance relation** if it is reflexive and symmetric. A **tolerance space** (X, τ) (or simply X) is a set X with a tolerance relation τ on X. If $(x, x') \in \tau$, we say x is within tolerance of x' and write $x\tau x'$; $iX = X \times X$ is called the **big tolerance** and $\delta X = \{(x, x) \mid x \in X\}$ is called the **little tolerance** on X.

Example 10.13.1 *(1)* $(\mathbb{N}^0, \simeq)$ *is a tolerance space, where* $\mathbb{N}^0 = \mathbb{N} \cup \{0\}$ *and* $\simeq$ *is the relation* $\{(m, n) \mid |m - n| \leq 1, \ m, n \in \mathbb{N}^0\}$.
(2) $|n| = (\{0, 1, \ldots, n\}, i)$ *is called the standard* n-***simplex.***

Let $\rho, \sigma \subseteq X \times Y$, $S \subseteq X$, and τ and ς be tolerances on X and Y, respectively. Then we introduce the following notation:
(1) $\rho^{-1} = \{(y, x) \mid (x, y) \in \rho\}$.
(2) $S \cdot \rho = \{y \mid \exists x \in S, \ x\rho y\}$ and $x\rho = \{x\} \cdot \rho$.

(3) $\rho \cdot \sigma = \{(x,y) \mid \exists x'$ such that $x\rho x', x'\sigma y\}; (\rho,\rho)\tau = \rho^{-1} \cdot \tau \cdot \rho$, the image of τ under ρ acting on $Q \times Q$. Similarly, $\rho_* \tau = \delta Y \cup (\rho,\rho)\tau$ and $\rho^* \varsigma = \rho \cdot \varsigma \cdot \rho^{-1}$.

(4) $\rho^0 = \delta X; \rho^1 = \rho; \rho^n = \rho^{n-1} \cdot \rho$.

(5) $\rho \backslash \sigma = \{(x,y) \in \rho \mid (x,y) \notin \sigma\}; \bar{\rho} = Q \times Q \backslash \rho; \overset{o}{\rho} = \rho^0 \cup \rho \cup \rho^{-1}$.

We use δ (ι) throughout for the little (big) tolerance and omit indicating the underlying tolerance space if it is understood.

Definition 10.13.2 *Let* τ *and* σ *be tolerances on* X *and* Y, *respectively. A relation* $\rho \subseteq X \times Y$, $\rho \neq \emptyset$, *is called a* **fuzrelation** *(or fuzzy) if* $\rho_* \tau \subseteq \sigma$. *A complete and univalued fuzrelation is called a* **fuzmap**. *If* $f : (X,\tau) \to Y$ *is a set theoretic map,* $f_* \tau$ *is called the* **coinduced** *tolerance on* Y. *If* $\iota : Y \to X$ *is an inclusion, the induced tolerance* $\iota^* \tau$ *on* Y *is called the* **subspace tolerance** *of* Y *and* $(Y, \iota^* \tau)$ *(or loosely* (Y,τ) *or* Y) *is a subspace of* X.

It follows that $f_* \tau$ is the least tolerance on Y such that $f : (X,\tau) \to (Y, f^* \tau)$ is a fuzmap and it is the unique tolerance on Y that has the universal property that for all tolerances σ on Z and for all set-theoretic maps $g : Y \to (Z,\sigma)$, $f \cdot g : (X,\tau) \to (Z,\sigma)$ is a fuzmap if and only if $g : (Y, f_* \tau) \to (Z,\sigma)$ is a fuzmap, [179]. Similarly, if $g : Y \to (X,\tau)$ is a set-theoretic map, the induced tolerance on Y, $g^* \tau$, is the biggest tolerance such that $g : (Y, g^* \tau) \to (X,\tau)$ is a fuzmap. It has a universal property dual to that of the coinduced tolerance. Tolerance spaces and fuzmaps form a category denoted by Fuz, [179].

Example 10.13.3 *Let* $X = Y = \mathbb{N}^0$ *and* $\tau = \,\simeq\, = \{(m,n) \mid |m-n| \le 1, m,n \in \mathbb{N}^0\}$. *Let* $\rho = \{(m, m+1) \mid m \in \mathbb{N}^0\}$. *Then* $(\rho,\rho)\tau = \rho^{-1} \cdot \tau \cdot \rho = \{(1,1),(1,2)\} \cup \{(m,n) \mid |m-n| \le 1, m \in \mathbb{N}\backslash\{1\}, n \in \mathbb{N}\}$. *(Note that* $(0,n) \notin \rho^{-1} \,\forall n \in \mathbb{N}$.) *Clearly,* $\rho_* \tau = \{(0,0)\} \cup (\rho,\rho)\tau$ *and* $\rho_* \tau \subseteq \tau$. *It follows that* $\rho^* \tau = \rho \cdot \tau \cdot \rho^{-1} = \{(0,0),(0,1)\} \cup \{(m,n) \mid |m-n| \le 1, m \in \mathbb{N}, n \in \mathbb{N}^0\}$. *Let* $f : \mathbb{N}^0 \to \mathbb{Z}$ *be such that* $f(m) = m \,\forall m \in \mathbb{N}^0$. *Then* $f_* \tau = \delta \mathbb{Z} \cup f^{-1} \cdot \tau \cdot f = \delta \mathbb{Z} \cup \tau$. *Let* $\iota : \mathbb{N} \to \mathbb{N}^0$ *be such that* $\iota(m) = m$ $\forall m \in \mathbb{N}$. *Then* $\iota^* \tau = \iota \cdot \tau \cdot \iota^{-1} = \{(m,n) \mid |m-n| \le 1, m \in \mathbb{N}, n \in \mathbb{N}^0\} = \tau \backslash \{(0,0),(0,1)\}$.

Definition 10.13.4 *Let* (X_1, τ_1) *and* (X_2, τ_2) *be tolerance spaces. The* **union** *of* (X_1, τ_1) *and* (X_2, τ_2) *is defined to be the pair* $(X_1 \cup X_2, \tau = (\iota_{1*}\tau_1) \cup (\iota_{2*}\tau_2))$, *where* ι_j *is the injection* $\iota_j : X_j \to X_1 \cup X_2$, $j = 1,2$. *Their* **product** *is defined to be the pair* $(X_1 \times X_2, \tau_1\tau_2 = \cap(pr_j \,^*\tau_j))$, *where* pr_j *is the projection* $pr_j : X_1 \times X_2 \to X_j$, $j = 1,2$.

The tolerance τ in Definition 10.13.4 is the least tolerance such that all injections ι_j are fuzmaps and $\tau_1\tau_2$ is the biggest tolerance such that all projections pr_j are fuzmaps, [179].

Definition 10.13.5 *Let* $x \in X$. *The* **tolerance neighborhood** *of* x *is defined to be the subspace* $N(x) = (x\tau, \iota^* \tau)$, *where* $\iota : x\tau \to X$ *is an injection. For* $A \subseteq X$, *let* $N(A) = \cup_{x \in A} N(x)$.

If ρ is a fuzrelation and $x\rho \neq \emptyset$, then $N(x)\rho \subseteq N(x\rho)$.

Let (X, τ) and (Y, σ) be tolerance spaces and let $\rho \subseteq X \times Y$ be a fuzrelation. Then $\rho \neq \emptyset$ and $\rho_* \tau \subseteq \sigma$. Let $x \in X$. Then $N(x\rho) = x\rho\sigma = \{y \in Y \mid \exists y' \in Y, (x, y') \in \rho \text{ and } (y', y) \in \sigma\}$ and $N(x)\rho = \{y \in Y \mid \exists x' \in \iota^* \tau, x'\rho y\}$. Let $y \in N(x)\rho$. Then there exists $x' \in \iota^* \tau$ such that $(x', y) \in \rho$. Thus $(y, x') \in \rho^{-1}$. Now $x\rho \neq \emptyset$ so $(x, y') \in \rho$ for some $y' \in Y$. Since $(x, x') \in \tau$, $(x', x) \in \tau$. From $(y, x') \in \rho^{-1}$, $(x', y) \in \tau$, and $(x, y') \in \rho$, we conclude that $(y, y') \in \rho^{-1}\tau\rho \subseteq \sigma$. Hence $(y', y) \in \sigma$. Thus $y \in N(x)\rho$. Therefore, $N(x)\rho \subseteq N(x\rho)$.

Example 10.13.6 *Let* $X = \{x, y, z\}$ *and* $\tau = \{(x, x), (y, y), (z, z), (x, y), (y, x), (y, z), (z, y)\}$. *Then* (X, τ) *is a tolerance space. Now* $x\tau = \{x, y\}$. *Let* $\iota : \{x, y\} \to X$ *be such that* $\iota(x) = x$ *and* $\iota(y) = y$. *Then* $\iota^*\tau = \iota\tau\iota^{-1} = \{(x, x), (x, y), (y, y), (y, x)\}$.

Definition 10.13.7 *Let* $g, f : (X, \tau) \to (Y, \sigma)$ *be fuzmaps. The* f *and* g *are called* **homotopic**, *written* $g \simeq f$, *if there exists* $m \in \mathbb{N}^0$ *and a sequence* $\{F_i \mid i = 0, 1, \dots, m\}$ *of fuzmaps* $F_i : (X, \tau) \to (Y, \sigma)$ *such that the following conditions hold:*

(1) $f = F_0$, $g = F_m$,
(2) $(F_i, F_{i+1})\tau \subseteq \sigma$, $i = 0, 1, \dots, m - 1$.

A fuzmap $f : (X, \tau) \to (Y, \sigma)$ is said to be **null-homotopic** if it is homotopic to a constant map and two tolerance spaces X and Y are homotopy equivalent if there exist fuzmaps $f : X \to Y$ and $g : Y \to X$ such that $f \cdot g \simeq \delta X$ and $g \cdot f \simeq \delta Y$.

There is a natural metric on a tolerance space (X, τ), namely, the hopmetric $d : X \times X \to \mathbb{N}^0 \cup \{\infty\}$, where $d(x, x') = \wedge\{m \mid \exists \text{ a fuzmap } \omega : (\mathbb{N}^0, \simeq) \to (X, \tau) \text{ such that } \omega(0) = x, \omega(m) = x'\}$ and $d(x, x') = \infty$ if there is no such ω.

Definition 10.13.8 *Let* A, B *be subspaces of* (X, τ). *The* **interior** *of* A, *written* $\text{int}A$, *is defined to be the set,* $\text{int}A = \{x \in A \mid d(x, X \backslash A) > 1\}$, *where* $d(x, A) = \wedge_{x' \in A}\{d(x, x')\}$. *The* **boundary** *of* A, *written as* ∂A, *is defined to be the set* $\partial A = N(A) \backslash \text{int}A$. *The tolerance space* X *is called* τ**-connected** *if for all* $(x, x') \in X \times X$, $d(x, x') \neq \infty$ *and is called* **contractible** *if it is homotopy equivalent to a point, i.e., to a tolerance space* (Y, δ) *with* $|Y| = 1$. *The component* $C(x')$ *of* x' *in* X *is the maximal* τ-*connected subspace of* X *containing* x'.

Disregarding transition probabilities, we represent the state transitions of an automaton and its outputs by its next-state relation $\Delta \subseteq Q \times X \times Q$ and its output relation $\omega \subseteq Q \times Y$. Q is the nonempty set of states of the automaton, X its input, and Y its output alphabet. The quintuple

$$A_\omega = (Q, X, Y, \Delta, \omega)$$

is called an (abstract) **automaton with output**. The triple

$$A = (Q, X, \Delta)$$

is called an (abstract) **automaton**. Moreover, A is assumed to operate sequentially on a discrete time scale.

Definition 10.13.9 *An automaton A is called **deterministic** if $\forall (q, x) \in Q \times X$, (q, x, r), $(q', x', r') \in Q \times X \times Q$, $q = q'$ and $x = x'$ implies $r = r'$. It is called **complete** if $\forall (q, x) \in Q \times X$ there exists $r \in Q$ such that $(q, x, r) \in \Delta$ and **finite** if $|Q|$ is finite.*

In general, A is a nondeterministic automaton. Let $x \in X$. Let $\delta_x \subseteq Q \times Q$ be such that $(q, q') \in \delta_x$ if and only if $(q, x, q') \in \Delta$. For computational convenience, the relation Δ is often given as a family D of binary state-transition relations: $D = \{ \delta_x \mid x \in X \}$.

Definition 10.13.10 *Let $x = x_1 x_2 \ldots x_n \in X^*$, where $x_i \in X$, $i = 1, 2, \ldots n$. Then the (state) transition relation of A_x under the action of input word x is defined as $\delta_\Lambda = \delta$, $\delta_x = \delta_{x_1} \cdot \delta_{x_2} \cdot \ldots \cdot \delta_{x_n}$, its output relation as $\omega_x = \delta_x \cdot \omega$, and $X^l = \{ x \in X^* \mid |x| = l \}$, $|x|$ the length of x.*

A conventional function-type description of an automaton appears in many cases not to be adequate when dealing with complex systems. For example, transition modification and memory errors may turn a deterministic automaton into a nondeterministic one. Hence a relational description of automaton has been chosen here.

Definition 10.13.11 *An automaton $\widehat{A} = (\widehat{Q}, \widehat{X}, \widehat{Y}, \widehat{\Delta}, \widehat{\omega})$ is called a **subautomaton** of an automaton A if $\widehat{X} = X$, $\widehat{Y} = Y$, $\widehat{Q} \subseteq Q$, $\widehat{\Delta} = \Delta \cap (\widehat{Q} \times X \times \widehat{Q})$, and $\widehat{\omega} = \omega \cap (\widehat{Q} \times Y)$.*

Every subset of Q determines a unique (in general incomplete) subautomaton of A. For example, a special subautomaton of A is given by the successor set of state $q \in Q$, where we recall that this set is defined by $R(q) = R(\{q\}) = \{ q' \mid \exists x \in X^* \text{ such that } q' \in q\delta_x \}$.

Recall that a cover on Q with the substitution property is a family $S = \{ Q_i \mid i \in I \}$ (I an index set) of nonempty distinct subsets Q_i of Q such that $\cup_{i \in I} Q_i = Q$ and for all $x \in X$ and for all $Q_i \in S$, there exist $Q_j \in S$ such that $Q_i \cdot \delta_x \subseteq Q_j$.

Definition 10.13.12 *Let $A = (Q, X, \Delta)$ be a complete finite automaton and S a cover on Q with the substitution property. The state-transition relations of the automaton $A/S = (X, S, \Delta^S)$ are defined as follows: for all $x \in X$ and $Q_i, Q_j \in S$, $Q_i \delta_x^S Q_j$ if and only if $Q_i \cdot \delta_x \subseteq Q_j$.*

Definition 10.13.13 *Let $A = (Q, X, \Delta)$ be an automaton and τ a toler-ance on Q. Then (A, τ) is called an **automaton with tolerance**. The tolerance space $(\mathbb{N}^0, \simeq)$ is considered as the underlying time set of (A, τ).*

Let σ be a tolerance on the output alphabet Y of A_ω, expressing in-distinguishability of outputs and let ω be a set-theoretic function. Then $(A, \omega^* \sigma)$ is an automaton with natural tolerance.

As another example, let M be a digital computer whose states are given by the contents of its registers. Two states of M may be defined as being within tolerance if and only if they differ in the contents of a limited number of registers [5].

10.14   Fuzzy-State Automata

In this section, we introduce a class of automata with tolerance whose state transitions exhibit stability properties.

Definition 10.14.1 *An automaton with tolerance (A, τ), $A = (Q, X, \Delta)$, with fuzzy state-transition relations is called a **fuzzy-state automaton**.*

We prove in the following lemma that the successors of states within tolerance of a fuzzy-state automaton all stay within tolerance under the action of any input word. We make use of this property when we analyze the effects of temporary and permanent state errors on the performance of fuzzy-state automata.

Lemma 10.14.2 *Let (A, τ), $A = (Q, X, \Delta)$, be an automaton with toler-ance.*

(1) Then A is a fuzzy-state automaton if and only if $\delta_{x*}\tau \subseteq \tau$ for all $x \in X$.

(2) Suppose that A is deterministic and complete. Then (A, τ) is a fuzzy-state automaton if and only if $\tau \subseteq \delta_x^* \tau$ for all $x \in X$.

(3) Both formulas for τ hold for all $x \in X$ if and only if they hold for all $x \in X^*$.

Proof. (1) The proof here is immediate.

(2) Since $\tau \subseteq \delta_x^* \tau$ and $\delta_x^{-1} \cdot \delta_x \subseteq \delta$, we have that $\delta_x^{-1} \cdot \tau \cdot \delta_x \subseteq \delta_x^{-1} \cdot \delta_x \cdot \tau \cdot \delta_x^{-1} \cdot \delta_x \subseteq \tau$. If A is a complete fuzzy-state automaton, then $\tau \subseteq \delta_x \cdot \delta_x^{-1} \cdot \tau \cdot \delta_x \cdot \delta_x^{-1} \subseteq \delta_x \cdot \tau \cdot \delta_x^{-1}$ for all $x \in X$.

(3) The desired result follows from the fact that

$$(\delta_x \cdot \delta_y)^* \tau = \delta_x^* (\delta_y^* \tau)$$

and $(\delta_x \cdot \delta_y)_* \tau \subseteq \delta_{x*}(\delta_{y*}\tau)$ for all $x, y \in X$. ∎

Lemma 10.14.2(1) can be simply restated as follows: $A = (Q, X, \Delta)$ is a fuzzy-state automaton if and only if $\forall x \in X$, $(q, x, q') \in \Delta$, $(q, q'') \in \tau$, $(q'', x, q''') \in \Delta$ implies $(q', q''') \in \tau$.

Lemma 10.14.2 implies that the transition relations of a fuzzy-state automaton are fuzrelations for all $x \in X^*$ and thus are metric decreasing, i.e., $\forall x \in X^*$, $q, \overline{q} \in Q$, $d(q, \overline{q}) \geq d(q^*, \overline{q}^*)$, where $q^* \in q\delta_x$ and $\overline{q}^* \in \overline{q}\delta_x$. This clearly represents a stability property. Let $A = (Q, X, \Delta)$ be a finite automaton. Let $F(A)$ be the set of all fuzzy-state automata with A as first coordinate. Let $F(A, \tau) = \{(A, \sigma) \in F(A) \mid \sigma \subseteq \tau, \sigma$ a tolerance on $Q\}$. Then $F(A)$ forms a complete, distributive lattice with respect to the ordering of automata with tolerance given by $(A, \sigma) \leq (A^*, \tau)$ if and only if A is a subautomaton of A^* and $\sigma \subseteq \tau$. This follows from the next result.

Theorem 10.14.3 *Let $A = (Q, X, \Delta)$ be an automaton. Then the following assertions hold.*

(1) The partially ordered set F_A of symmetric and reflexive binary relations ρ on Q with $\delta_{x*}\rho \subseteq \rho$ for all $x \in X$ (ordered by set inclusion) forms a complete, distributive lattice, which is a sublattice of the lattice of all tolerances on Q.

(2) F_A is closed under the relative product if A is complete. F_A is then a lattice ordered monoid if A is complete and deterministic.

Proof. (1) The proof follows from [253].

(2) If A is complete, $\delta \subseteq \delta_x \cdot \delta_x^{-1}$, $\delta_x^{-1} \cdot \rho_1 \cdot \rho_2 \cdot \delta_x \subseteq \delta_x^{-1} \cdot \rho_1 \cdot \delta_x \cdot \delta_x^{-1} \cdot \rho_2 \cdot \delta_x \subseteq \rho_1 \cdot \rho_2$, $\rho_1, \rho_2 \in F_A$. If A is deterministic, $\delta_x^{-1} \cdot \delta_x \subseteq \delta$, $\delta_x^{-1} \cdot \delta \cdot \delta_x \subseteq \delta$, and $\delta_x \cdot \delta = \delta \cdot \delta_x = \delta_x$ $\forall x \in X$. ∎

It follows from Theorem 10.14.3 that $F(A, \tau) \neq \emptyset$ has a unique maximal element. This element is denoted by (A, τ^*). It plays an important role later. The unit of $F(A)$ is (A, ι). If A is deterministic, $F(A)$ has zero (A, δ) and so $F(A, \tau) \neq \emptyset$.

Example 10.14.4 *We consider coarsening of a complete, finite fuzzy-state automaton (A, τ). The tolerance τ defines a (unique) cover S_τ on Q with the substitution property as follows: Let $Q_i \subseteq Q$. Then $Q_i \in S_\tau$ if and only if (a) $q\tau q^*$ for all $q, q^* \in Q_i$ and (b) if $q\tau q^*$ for all $q^* \in Q_i$, then $q \in Q_i$.*

The coarsening of (A, τ), namely $\widehat{A} = (A/S_\tau, \widehat{\tau})$, is a fuzzy-state automaton, where $\widehat{\tau}$ is defined below.

In order to construct tolerance $\widehat{\tau}$, define $\rho \subseteq S_\tau \times Q$ by $Q_i\rho q$ if and only if $q \in Q_i$. It follows that $(\widehat{\delta}_x, \delta_x)\rho \subseteq \rho$ for all $x \in X$. Define $\widehat{\tau}$ as $\widehat{\tau} = \rho^*\tau$. Clearly, $\widehat{\tau}$ is reflexive and symmetric and $\widehat{\delta}_{x*}\widehat{\tau} \subseteq \widehat{\tau}$ $\forall x \in X$ since $(\widehat{\delta}_x, \delta_x)\widehat{\tau} \subseteq \widehat{\delta}_x^{-1} \cdot \rho \cdot \delta_x \cdot \delta_x^{-1} \cdot \tau \cdot \delta_x \cdot \delta_x^{-1} \cdot \rho^{-1} \cdot \widehat{\delta}_x \subseteq \rho \cdot \tau \cdot \rho^{-1}$.

The automaton with tolerance appearing in the next result is a generalization of the concept of a tolerance automaton, introduced first in [5].

Example 10.14.5 *Let (A, τ) be an automaton with tolerance such that $pr_{13}\Delta \subseteq \tau$. Then (A, τ) is called a **tolerance automaton**.*

We now prove the following statement: If (A, τ) is a finite tolerance automaton, then there exists $n \in \mathbb{N}$ with $1 \leq n \leq |Q| - 1$, such that (A, τ^n) is a fuzzy-state tolerance automaton:

(a) Since $\tau = \delta \cup \tau$, it follows by induction that $\tau^m \subseteq \tau^{m+1}$, $m = 1, 2, \ldots$. Therefore, there exists $n \in \mathbb{N} \cup \{0\}$ such that $\tau^n = \tau^{n+j}$, $j = 0, 1, 2, \ldots$, since $Q \times Q$ is finite. Moreover, since τ can be represented by a $|Q| \times |Q|$ Boolean matrix having only ones as diagonal elements, τ has a characteristic exponent $n \leq |Q| - 1$.

(b) Let $q\tau^m q'$, $q^* \in pr_2(q\Delta)$, and $q'^* \in pr_2(\bar{q}\Delta)$. Then $q^*\tau q$, $q'^*\tau q'$, and so $q^*\tau^{2+m}q'^*$, i.e., $\delta_{x*}\tau^m \subseteq \tau^{2+m} \ \forall x \in X$. Let n be as in part (a). Then $\delta_{x*}\tau^n \subseteq \tau^{2+n} = \tau^n$ for all $x \in X$. Thus (A, τ^n) is a fuzzy-state automaton with $pr_{13}\Delta \subseteq \tau \subseteq \tau^n$.

Any state of a tolerance automaton is in tolerance with its predecessors and any input word x of a tolerance automaton defines a fuzmap

$$\omega_x : (\mathbb{N} \cup \{0\}, \simeq) \to (Q, \tau),$$

i.e., a path in Q. The "phase space velocity" $c = \frac{d(q_t, q_{t'})}{d(t, t')}$ of A is less than or equal to 1. In this sense, a tolerance automaton has inertia that gives rise to stable behavior. The example closing Section 10.13 is also an example of a tolerance automaton.

Example 10.14.6 *The automaton A^* given in Table 10.1 determines 64 automata with tolerance. This is easy to see from the following reasoning: Since there are 4 states, a tolerance can be represented by a 4×4-matrix whose entries consist of 0's and 1's. Since a tolerance is reflexive and symmetric only 6 positions of the matrix determine a tolerance. Thus there are $64 = 2^6$ possible tolerances.*

Table 10.1: Next State Relation of A^*

	Δ^*	q_0	q_1	q_2	q_3	states of A^*
inputs	x_1	q_0	q_0	q_1	q_1	*next states*
of A^*	x_2	q_2	q_2	q_0	q_0	*of A^**

$F(A^*)$ is shown in Figure 10.10. (A^*, τ_{19}) is the only nontrivial tolerance

automaton of $F(A^*)$.

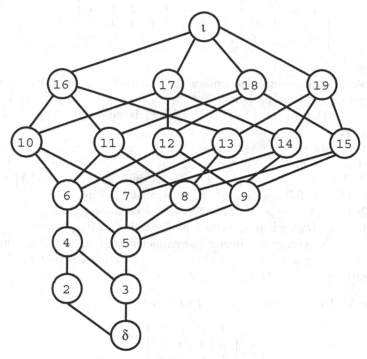

Figure 10.10^{10} : Lattice $F(A^*)$

The following Boolean matrices (with respect to the ordering of the states given by (q_0, q_1, q_2, q_3)) represent the basic tolerances of $F(A^*)$.

$$\tau_2 \qquad \begin{bmatrix} 1 & 0 & 0 & 0 \\ 0 & 1 & 0 & 0 \\ 0 & 0 & 1 & 1 \\ 0 & 0 & 1 & 1 \end{bmatrix} \qquad\qquad \tau_3 \qquad \begin{bmatrix} 1 & 1 & 0 & 0 \\ 1 & 1 & 0 & 0 \\ 0 & 0 & 1 & 0 \\ 0 & 0 & 0 & 1 \end{bmatrix}$$

$$\tau_5 \qquad \begin{bmatrix} 1 & 1 & 1 & 0 \\ 1 & 1 & 0 & 0 \\ 1 & 0 & 1 & 0 \\ 0 & 0 & 0 & 1 \end{bmatrix} \qquad\qquad \tau_7 \qquad \begin{bmatrix} 1 & 1 & 1 & 1 \\ 1 & 1 & 0 & 0 \\ 1 & 0 & 1 & 0 \\ 1 & 0 & 0 & 1 \end{bmatrix}$$

---

[10] Figure 10.10 is from [36], reprinted with permission by Kluwer Academic/Plenum Publishers.

$$
\tau_8 \quad
\begin{bmatrix}
1 & 1 & 1 & 0 \\
1 & 1 & 0 & 1 \\
1 & 0 & 1 & 0 \\
0 & 1 & 0 & 1
\end{bmatrix}
\qquad
\tau_9 \quad
\begin{bmatrix}
1 & 1 & 1 & 0 \\
1 & 1 & 1 & 0 \\
1 & 1 & 1 & 0 \\
0 & 0 & 0 & 1
\end{bmatrix}
$$

We now determine (A, τ^*).

The following procedure determines the maximal symmetric binary relation ρ_τ on Q, with the property that $(\delta_x, \delta_x)\rho_\tau \subseteq \rho_\tau \subseteq \tau \ \forall x \in X$. Thus $(A, \rho_\tau) = (A, \tau^*)$ (unit of $F(A, \tau)$) if $F(A, \tau)$ is nonempty, e.g., if A is deterministic.

Define the relation $\rho(k)$ as follows:

Step 1: $\rho(1) := \tau$.

Step 2: For all $k \geq 1$, $q\rho(k+1)p$ if and only if $\forall x \in X \cup \{\Lambda\}$ and $\forall (q', p') \in Q \times Q$ it follows from $q\delta_x q'$, $p\delta_x p'$ that $q'\rho(k)p'$.

Step 3: If $\rho(k+1) = \rho(k)$ go to Step 4, else go to Step 2.

Step 4: $\rho_\tau = \rho(k_0)$, where k_0 is the smallest index k with $\rho(k) = \rho(k+1)$.

Clearly, ρ_τ is symmetric since τ is symmetric and ρ_τ is the maximal relation with the properties stated above since $\rho \subseteq \tau$ implies that $\rho_\sigma \subseteq \rho_\tau$. Consequently, $\sigma \subseteq \rho_\tau$ if $(\delta_x, \delta_x)\sigma \subseteq \sigma \ \forall x \in X$.

Example 10.14.7 *Consider Figure 10.10. Let*

$$
\tau =
\begin{bmatrix}
1 & 0 & 1 & 0 \\
0 & 1 & 0 & 1 \\
1 & 0 & 1 & 1 \\
0 & 1 & 1 & 1
\end{bmatrix}.
$$

Then $q_0\delta_{x_1}q_0$ and $q_2\delta_{x_1}q_1$, but $(q_0, q_1) \notin \rho(1)$. Thus $(q_0, q_2) \notin \rho(2)$. Moreover, $q_1\delta_{x_1}q_0$ and $q_3\delta_{x_1}q_1$. Since $(q_0, q_1) \notin \rho(1)$, $(q_0, q_3) \notin \rho(2)$. Now $(q_2, q_3) \in \rho(2)$ since $q_2\delta_{x_1}q_1$, $q_3\delta_{x_1}q_1$, $q_2\delta_{x_2}q_0$, and $q_3\delta_{x_2}q_0$. It follows that $(A, \tau^*) \notin F(A^*)$ and $\rho(2) = \rho(3) = \tau_2$. Hence $(A^*, \tau^*) = (A^*, \tau_2)$.

Consider an automaton with tolerance (A, τ). Tolerance τ^* determines the maximal set of pairs of automata states that are in tolerance τ and whose successors under any input sequence are also in tolerance τ. Furthermore, if two states are not related by τ^*, then none of their predecessor pairs contains states within tolerance τ^* since $\overline{\tau}^* = \iota \backslash \tau^*$ is reflexive, symmetric, and $\delta_x \ ^*\overline{\tau}^* \subseteq \overline{\tau}^* \ \forall x \in X$, [253].

10.15   Stable and Almost Stable Behavior of Fuzzy-State Automata

In this section, we consider questions about the stable behavior of finite fuzzy state automata that have counterparts in the theory of topological dynamical systems. With this in mind, we state the following definition.

Definition 10.15.1 *Let* (A, τ), $A = (Q, X, \Delta)$, *be an automaton with tolerance,* S *a subspace of* Q, *and* $V \subseteq X^*$. *The subspace* S *is said to be* V-*stable if its neighborhood* $N(S)$ *is* V-*invariant, i.e., if* $N(S) \cdot \delta_v \subseteq N(S)$ $\forall v \in V$. S *is said to be* **almost** V-**stable** *if there is a natural number* l *such that*

$$N(S) \cap \cup_{i=0}^{l} q \delta_v^{n+i} \neq \emptyset$$

holds for all $n \in \mathbb{N} \cup \{0\}$, $q \in S$, *and* $v \in V$. *If* S *is* X^*-*stable and accessible from every state of* $T \subseteq Q$, *then* S *is called an* **attractor set** *of* T.

Clearly a set S of automata states is almost V-stable if its neighborhood is V-invariant, i.e., if S is V-stable. Stable and almost stable sets of automata states characterize the recurrent state motions of automata with tolerance. The property of attractor sets can be illustrated in the following manner. Let automaton (A, τ) be driven by a stationary random input source that generates words such that the probability of any input $x \in X$ following an arbitrary word is greater than k, where $k > 0$. Let A be at time $t_0 = 0$, say, in T, and $p(S, t)$ be the probability of its state at time t belonging to the neighborhood of an attractor set S of T, if it exists. Then $p(S, t \geq |R(T)|) \geq k^{|R(T)|-1}$ and $p(S, t \geq |R(T)|) = 1$ for an autonomous automaton (hence the name "attractor set"). Clearly, Q is an attractor set of any $T \subseteq Q$. The set of reset states of an identity-reset automaton with tolerance δ is an attractor set of its state set and every state of automaton (A^*, τ_7) is almost X-stable with $l_a = 0$, $l_b = l_c = l_d = 1$ (see Figure 10.10).

In the following, we assume that the automaton $A = (Q, X, \Delta)$ is complete and deterministic. The following theorem is an adaptation of Poston's "approximate fixed point theorem" to finite automata.

Theorem 10.15.2 *Suppose that the state space of a finite, fuzzy-state automaton is contractible. Then* $\forall x \in X^*$, *it contains an* x-*invariant and* x-*stable subspace,* $P_x \neq \emptyset$, *whose elements are mutually within tolerance* τ *(approximated fixed points).* ∎

Let automaton A be autonomous and Q be the union of contractible, $\widehat{\delta}_x$-connected subspaces, where $x \in X$. Let τ be the natural tolerance on Q with respect to an output relation ω. Then P_x (which is maximal with respect to the properties of Theorem 10.15.2) is an attractor set of Q. Now $p(P_x, t) = 1$ for t sufficiently large. Because of unobservable (unmeasurable) differences of outputs, automaton A seems to be caught in a final state. However, this may not be the case if $|P_x| > 1$. Another example of an attractor set can be determined in the following manner. By Lemma 10.15.3, for any $q \in Q$, $\partial R(q)$ is an attractor set of $R(q)$ if $\partial R(q) \neq \emptyset$ and if $\text{int} R(q)$ is empty, or else strongly connected, and $Q \backslash R(q)$ is X-invariant. Since $R(q)$ is X-invariant, ∂R is then X^*-stable and either $q \in \partial R(q)$ and so $R(q) \subseteq \partial R(q)$, or else $q \in \text{int} R(q)$. However, this implies that $\partial R(q)$ is reachable from any state of

$R(q)$. Clearly, both conditions hold if all $R(q)$ with $\partial R(q) \neq \emptyset$ are strongly connected, e.g., if (A, τ) is a fuzzy-state permutation automaton.

Lemma 10.15.3 *Let (A, τ) be a fuzzy-state automaton and S be a subspace of its state space. Then the boundary ∂S of S is X^*-stable if S and $Q \backslash S$ are X-invariant.*

Proof. If $|x| = 0$ or $\partial S = \emptyset$, then ∂S is x-invariant since $\delta_A = \delta$. Suppose ∂S is x-invariant for all x with $|x| \leq n$ and $\partial S \neq \emptyset$.

(1) If $\overline{q} \in \partial S \backslash S$, then there is a state $q \in S$ such that $(q, \overline{q}\delta_x) \in \tau$ for $x \in X^n$. Thus $(q\delta_x, \overline{q}\delta_{xa}) \in \tau$, $q\delta_x \in S$, and $\overline{q}\delta_{xa} \notin S$, i.e. , $\overline{q}\delta_{\overline{x}} \in \partial S$ for all $\overline{x} \in X^{n+1}$.

(2) If $q \in \partial S \cap S$, then there is a state $\overline{q} \notin S$ such that $(\overline{q}, q\delta_x) \in \tau$ for $x \in X^n$. It follows that $(\overline{q}\delta_a, q\delta_{xa}) \in \tau$ for all $a \in X$, i.e., $d(\overline{q}\delta_a, q\delta_{xa}) \leq 1$. Suppose that $q\delta_{xa} \in \text{int}S$. Then $\overline{q}\delta_a \in S$ and $R(\overline{q}) \cap S \neq \emptyset$. This is a contradiction since $\overline{q}\delta_a \in S$. Hence $q\delta_{xa} \in S \backslash \text{int}S$. Since $(A, \tau) \in F(A)$, ∂S is X^*-invariant and X^*-stable. ∎

Lemma 10.15.4 *Let A be a complete, deterministic, connected, and autonomous automaton with tolerance τ. Then (A, τ) is almost periodic if and only if every state of A is in tolerance with a periodic state of A.*

Proof. Let $q \in Q$. Let $X = \{x\}$ since A is autonomous. The orbit of q under x, i.e., $O = \{q\delta_x^i\}_{i=0}^{\infty}$, is the union of two sets O_t and O_p ($p = |Q_p|$), where O_p is nonempty and permuted by δ_x and $O_t \cdot \delta_x^{\overline{s}} \subseteq O_p$ for $\overline{s} \geq t = |O_t|$.

(1) Let $q\tau\overline{q}$, $\overline{q}$ be periodic. Then $\overline{q} \in O_p$. There exist natural numbers r and $s < p$ such that $q\delta_x^r = \overline{q}\,\delta_x^s$ since A is connected and deterministic. Hence $q\delta_x^{r+p-s} \in N(q)$ and

$$\cup_{i=0}^{r+p-s-1} q\delta_x^{n+i} \cap N(q) \neq \emptyset, \ n = 0, 1, 2, \ldots .$$

That is, if $\overline{q}$ has period p, then q has an almost period not greater than $p + r - s - 1$.

(2) Suppose that $q \in Q$ is almost periodic. Then it follows that

$$N(q) \cap \cup_{i=0}^{l} q\delta_x^{t+i} \neq \emptyset.$$

Thus $N(q) \cap O_p \neq \emptyset$. Hence $N(O_p) = Q$ if A is almost periodic. Furthermore, $N(O_p) \cdot \delta_x \subseteq N(O_p \cdot \delta_x) = N(O_p)$ if $(A, \tau) \in F(A)$ for A arbitrary. ∎

An automaton (A, τ) is said to be **almost periodic** if every state of A is almost X-stable. The connection between almost periodic and permutation automata can be seen from the next result.

Theorem 10.15.5 *The state space of a deterministic, almost periodic fuzzy-state automaton is the union of a finite number of neighborhoods of closed stable orbits (cycles).*

Proof. The proof follows from part (2) of the proof of Lemma 10.15.4. ∎

The greatest common divisor of the lengths of all proper cycles of a deterministic automaton A is called the **period** of A, [75]. Let d divide the period of A. Let π be a path from state q to state q' and $|\pi|$ denote its length. The distance from q to q', written $|q, q'|_d$, is defined as $|\pi|$ (modulo d) if q and q' are connected and is defined to be ∞ otherwise. It follows that this distance is unique. Assume that the period of A is greater than 1. Now tolerances τ_d^s on Q may be defined as follows. For all $q, q' \in Q$, $q\tau_d^s q'$ if and only if $q = q'$ or $|q, q'|_d = s$ or $d - s$. Automata (A, τ_d^s) are almost periodic. Moreover, (A, τ_d^1) is a tolerance automaton and (A, τ_2^0) is a fuzzy-state automaton if the period of A is even and A is complete.

Example 10.15.6 *Let the automaton $\widehat{A}$ be as specified in Table 10.2.*

Table 10.2: Next State Relation of $\widehat{A}$

$\widehat{\Delta}$	q_0	q_1	q_2	q_3	q_4
x_1	q_1	q_2	q_1	q_0, q_2	q_3
x_2	q_1	q_4	q_1	q_2	q_3

The lengths of the proper cycles are 2 and 4. Thus $d = 2$.

Now $|q_0, q_3|_2 = 3 (\mathrm{mod}\, 2) = 1$ and $|q_0, q_4|_2 = 2 (\mathrm{mod}\, 2) = 0$. For $s = 0$, $s \neq |q_0, q_3|_2 \neq d - s$. For $s = 1$, $|q_0, q_3|_2 = s$. Hence $q_0 \tau_2^1 q_3$. For $s = 0$, $|q_0, q_4|_2 = s$. Thus $q_0 \tau_2^0 q_4$. For $s = 1$, $|q_0, q_4|_2 \neq s$ or $d - s$. It follows that the tolerances τ_2^1 and τ_2^0 are represented by the following Boolean matrices.

$$
\tau_2^1 \qquad\qquad\qquad \tau_2^0
$$

$$
\begin{bmatrix} 1 & 1 & 0 & 1 & 0 \\ 1 & 1 & 1 & 0 & 1 \\ 0 & 1 & 1 & 1 & 0 \\ 1 & 0 & 1 & 1 & 1 \\ 0 & 1 & 0 & 1 & 1 \end{bmatrix}
\qquad
\begin{bmatrix} 1 & 0 & 1 & 0 & 1 \\ 0 & 1 & 0 & 1 & 0 \\ 1 & 0 & 1 & 0 & 1 \\ 0 & 1 & 0 & 1 & 0 \\ 1 & 0 & 1 & 0 & 1 \end{bmatrix}
$$

An effective procedure for evaluating the period of any finite automaton is given in [75].

10.16 Fault Tolerance of Fuzzy-State Automata

An actual machine occasionally makes errors computing its next state. Consequently, it is unreliable to some extent We now consider this unreliability. It is possible to emphasize the essential aspects of more concrete situations within our abstract framework. A machine exhibits stability in some sense if it overcomes the influence of its errors, i.e., if after some time these influences become "tolerable." We will show that fuzzy-state automata behave stably in this sense with respect to certain faulty state transitions.

Let $A = (Q, X, \Delta)$ be a (not necessarily deterministic) finite automaton, τ a tolerance on Q, and $Q^p = pr_3\Delta$.

Definition 10.16.1 *A binary relation ϕ on Q with $pr_1\phi \supseteq Q^p$ together with tolerance $\iota^*\tau\tau$, where $\iota : \phi \subseteq Q \times Q$, is called a **state relation** of (A, τ). A state relation ϕ is called **compatible** if $\phi \subseteq \tau$, **inessential** if $\phi \subseteq \rho_\tau$ (cf. Section 10.14), and **consistent** if it is fuzzy. We say that ϕ changes q if $q(\phi \backslash \delta) \neq \emptyset$. ϕ is called (ρ, l)-**bounded** (by (A, τ)) if for all $x \in X^*$ with $|x| \geq p$,*

$$\delta_{x*}\phi \subseteq \tau^l \tag{10.6}$$

and (10.6) does not hold for $(p-1, l)$ or for $(p, l-1)$.

If ϕ is (p, l)-bounded, then $\phi \cup \phi^{-1}$ is also (p, l)-bounded. In general, if the bounds of two state relations ϕ and ϕ^+ are known, it is usually not difficult to derive bounds for relations such as $\phi \cup \phi^+$, $\phi \cap \phi^+$, and $\phi \cdot \phi^+$. For example, $\phi \cup \phi^+$ is $(\max(p, p^+), \max(l, l^+))$-bounded. We present below an algorithm for determining the bound of a state relation.

State transition errors may be described by a state relation. We then visualize an element $(q_i, q_j) \in \phi$ as follows. Automaton A goes with some nonzero probability into state q_j when it should go into state q_i, due to a permanent or temporary modification in the state transitions of A. Clearly, error $e \in \phi \cap \delta$ is an improper error.

Definition 10.16.2 *A state relation $(\phi, \iota^*\tau\tau)$ is called a **permanent transition modification** (**t-modification**) if it modifies automaton A, i.e., if automaton $A^\phi = (Q, X, \Delta^\phi := \Delta \cdot \phi)$ replaces automaton A. Automaton A^ϕ is called the **modification** of A due to ϕ and A the reference automaton. The state relation ϕ is called a (**temporary**) **error** of A if A is not modified by ϕ.*

Most input errors can be interpreted as state transition errors, i.e., memory errors, [84]. The is also true for errors in the combinational output logic of sequential circuits. Consequently, we concentrate on memory errors. Within the relational framework, we are less interested in the physical causes of errors, but rather in the qualitative aspects of the role errors play in the performance of modifiable systems. In what follows, we assume that the meaning of tolerance τ is that a single error $e \in \phi$ is tolerable in some appropriate sense if $e \in \tau$. Examples can be found in [37].

Lemma 10.14.2 states that compatible errors of a fuzzy-state automaton (A, τ) are inessential and remain inessential under the action of any input word. The maximal compatible state relation of (A, τ) is $TQ = (\tau, \tau\tau)$, and $(\rho_\tau, \tau\tau)$ is its maximal inessential state relation if $pr_1\rho_\tau \supseteq Q^p$. Since the tangent bundle of the tolerance space (Q, τ) is the composite map $t : TQ \subseteq Q \times Q \to^{pr_1} Q$, [179], and the tangent space T_qQ to Q at state

q is the tolerance space $T_q Q = (tq, \tau\tau)$, the compatible changes of state q determine the tangent space at q in a sense. In [37], such "geometric" properties of errors and t-modifications are studied. There modification tolerance of automata is considered, i.e., with masked and correctable t-modification of an automaton. However, in the following, relational (set-theoretic) properties of temporary errors are considered.

We now consider bounded state errors.

Fault tolerance is an important design parameter. The goal is to design systems that stay operational despite failures and, in fact, can repair themselves in response to their own failure. Intuitively, automaton (A, τ) overcomes the influence of error ϕ if after some time it takes a state that is, and from then on stays, in tolerance with the correct state.

Definition 10.16.3 *An automaton (A, τ) is said to τ-**correct** (to correct) error ϕ if there exists $p \in \mathbb{N}$ such that ϕ is $(p, 1)$-bounded $((p, 0)$-bounded) by (A, τ).*

Correction (self-synchronization) by deterministic finite automata has been studied in [85], [84], [42], [224], and elsewhere. Algorithms have been given for determining which temporary state errors are corrected by a deterministic automaton within a certain amount of time (assuming tacitly that these errors do not alter state transitions.) Correction of input errors has been studied in [252] and [84]. There is a strong connection between the capability of correcting an input error and that of correcting the temporary state errors caused by this input error.

Example 10.16.4 *Consider the fuzzy-state automaton (A^*, τ_5) given in Figure 10.10. The errors ϕ_1 and ϕ_2 are given by*

$$
\phi_1 = \begin{bmatrix} 0 & 1 & 0 & 0 \\ 1 & 0 & 0 & 0 \\ 0 & 0 & 1 & 0 \\ 0 & 0 & 0 & 0 \end{bmatrix}, \quad
\phi_2 = \begin{bmatrix} 0 & 0 & 1 & 1 \\ 0 & 0 & 0 & 1 \\ 1 & 0 & 0 & 0 \\ 1 & 1 & 0 & 0 \end{bmatrix}.
$$

The error ϕ_1 is $(0, 1)$- and $(1, 0)$-bounded since ϕ_1 is compatible and $(A^*, \tau_5) \in F(A)$. The error ϕ_2 is $(1, 1)$-bounded.

Bounds of an error and error-correcting input words can be determined by the following error graph procedure as long as the next state relation of an automaton is not too large.

Error graph. (1) The vertices are given as (l/ab), $a, b \in Q$, if $a\tau^l b$ and as $(-/ab)$ if $(a, b) \notin \tau^l$, $l = 0, 1, 2, \dots$, $((0/qq) \equiv (q))$.

(2) An oriented i-edge points from vertex (l/ab) to vertex (l'/cd) if and only if the (unordered) state pair $\{a, b\}$ goes into state pair $\{c, d\}$ under

input x_i. The error graph of (A^*, τ_5) is given in Figure 10.11.

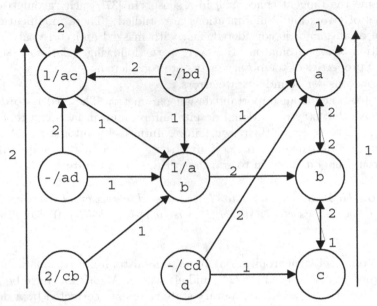

Figure 10.11[11] : *Error Graph of* (A^*, τ_5)

Test A. The following algorithm gives a test by which one can use to decide whether or not a given temporary error ϕ of automaton (A, τ) is (p, l)-bounded and, hence, τ-connected by (A, τ).

Step 1: $p := 0$; $l_p := |Q|$.

Step 2: $\phi_p := \cup_{|w|=p} (\delta_w, \delta_w) \phi$; $\phi_p(1) := \phi_p$; $\phi_p(k+1) := \phi_p(k) \cup \{(q, q^+) \mid \exists x \in X, (\overline{q}, \overline{q}^+) \in \phi_p(k) \text{ such that } \overline{q} \delta_x q, \overline{q}^+ \delta_x q^+)$.

Step 3: $\phi_p^+ := \cup_{k \geq 1} \phi_p(k)$.

Step 4: If there exist $l^i \in \mathbb{N}$ with $1 \leq l^i < l_p$ such that $\phi_p^+ \subseteq \tau^{l^i}$, set $l_p = \min\{l^i\}$ and ϕ is (p, l_p)-bounded, else $l_p := l_p - 1$.

Step 5: $p := p + 1$ if $p \leq |Q|^2 - 1$ go to Step 2, else stop and ϕ is unbounded.

It follows that ϕ_p^+ is the minimal binary relation on Q such that (i) $\phi_p \subseteq \phi_p^+$; (ii) $(\delta_x, \delta_x) \phi_p^+ \subseteq \phi_p^+$ for all $x \in X^*$, [253]. Thus for all $z = xv$ with $|x| = p$, $(\delta_z, \delta_z) \phi \subseteq \phi_p^+$. In order to determine if automaton A recovers from error ϕ after a finite amount of time, it must be determined if there is some p, $0 \leq p \leq |Q|^2 - 1$, such that ϕ is $(p, 1)$-bounded. If ϕ is τ-corrected by (A, τ) for A complete, then it is also τ-connected by any deterministic automaton $(\widehat{A}, \sigma) \geq (A, \tau)$. The precedessors of a single error that is not

---

[11] Figure 10.11 is from [36], reprinted with permission by Kluwer Academic/Plenum Publishers

τ-corrected by a fuzzy-state automaton (A, τ) are also not τ-connected by automaton (A, τ) (see the comments at the end of Section 10.14).

The capability of an automaton to recover from errors is a strong restriction on its state transitions and tolerance. Somewhat weaker notions are given in the next definition.

Definition 10.16.5 *Error ϕ is l-corrected (eventually l-corrected) by automaton (A, τ) if for all $e \in \phi$ (and all $z \in X^*$) there is a word x such that $(\delta_x, \delta_x)e \cap \tau^l \neq \emptyset$ $((\delta_{zx}, \delta_{zx})e \cap \tau^l \neq \emptyset)$.*

Example 10.16.6 *Consider the automaton (A^*, τ_5). The error i is eventually 0-corrected by A^* since $x = x_1 x_1$ is a reset word of A^* (cf. Figure 10.11).*

Test B: In order to decide whether a temporary error ϕ is eventually l-corrected by a fuzzy-state automaton (A, τ), the maximal eventually l-corrected set of state pairs ψ of A is constructed by the following procedure:

Step 1: Construct relation $\psi_m = \{(q, \overline{q}) \mid \exists x \in X^* \text{ such that } (\delta_x, \delta_x)(q, \overline{q}) \cap \tau^l \neq \emptyset\}$. If $(A, \tau) \in F(A)$, then $\psi_1 \subseteq \psi_2 \subseteq \ldots$ and $\psi_j = \psi_{j+1}$ for some $j \in \mathbb{N}$; $\psi_o^+ := \psi_j$.

Step 2: Construct relation ψ_m^+ such that $\psi_m^+ \subseteq \psi_{m-1}^+$ and $\delta_x^{-1} \cdot \psi_m^+ \cdot \delta_x \subseteq \psi_{m-1}^+$ for all $x \in X$. Since $\psi_0^+ \supseteq \psi_1^+ \supseteq \ldots$, $\psi_t^+ = \psi_{t+1}^+$ for some $t \in \mathbb{N} \cup \{0\}$; $\psi := \psi_t^+$. Then error ϕ is eventually l-corrected if and only if $\phi \subseteq \psi$.

The proof of the next result can be obtained by modifying the proof of a corresponding result in [252].

Theorem 10.16.7 *Let A be a finite deterministic automaton. An error ϕ of A is eventually l-corrected if and only if A is l-synchronized with respect to ϕ. That is, A driven by a random source and started either in state q or else in a state $\overline{q} \in q\phi$ is brought, in the long run, into states within tolerance τ^l. However, if A is a fuzzy-state automaton, its correcting input word can be chosen independently of $e \in \phi$, and thus ϕ is (p, l)-bounded for some $l \in \mathbb{N}$.* ∎

We now present a more general result.

Theorem 10.16.8 *All temporary errors of a complete fuzzy-state automaton (A, τ), $A = (Q, X, \Delta)$, are l-correctable, for some $l \leq |Q| - 1$, if and only if there is a (generalized) reset $x \in X^*$ for automaton A such that $Q^x \subseteq C(\widehat{q})$, $\widehat{q} \in Q$, where $Q^x := Q \cdot \delta_x$.*

Proof. It follows that $Q^x \subseteq C(q)$ if and only if there is some $r \in \mathbb{N} \cup \{0\}$ such that $Q^x \subseteq q\tau^r$. Thus since $Q^x \subseteq C(q)$, $q \in Q$, we have that for any error ϕ,

$$(\delta_x, \delta_x)\phi \subseteq \delta_x^{-1} \cdot \iota \cdot \delta_x \subseteq \tau^r q \times q\tau^r \subseteq \tau^{2r} \subseteq \tau^{|Q|-1}.$$

Suppose that any error of A is l-correctable. Let $|Q| = n$ and $Q_m = \{q_i\}_{i=0}^{m}$, $1 \leq m \leq n$. There exists $x \in X^*$ and $q_i^x \in q_i \delta_x$, $i = 0, 1$, such that $q_0^x \tau^l q_1^x$, $q_i \delta_x \subseteq q_i^x \tau \subseteq q_0^x \tau^{l+1}$ since δ_x is fuzzy. Thus $Q_1^x \subseteq q_0^x \tau^{l+1}$. Suppose now that there is some $x \in X^*$ such that $Q_m^x \subseteq \widehat{q}_0 \tau^{l+1}$, $1 \leq m < n$, where $\widehat{q}_i \in q_i \delta_x$. Let $(\widehat{q}_0, \widehat{q}_{m+1})$ be l-corrected by $\widehat{x} \in X^*$. There exists $\widehat{\widehat{q}}_i \in \widehat{q}_i \delta_{\widehat{x}}$, $i = 0, m+1$, such that $\widehat{\widehat{q}}_0 \tau^l \widehat{\widehat{q}}_{m+1}$. Again for $q_i \in Q_m$, $q_i \delta_{x\widehat{x}} \subseteq (\widehat{q}_0 \tau^{l+1}) \cdot \delta_{\widehat{x}} \subseteq \widehat{\widehat{q}}_0 \tau^{l+1}$ and $q_{m+1} \delta_{x\widehat{x}} \subseteq \widehat{\widehat{q}}_{m+1} \tau$ and $Q_{m+1}^{x\widehat{x}} \subseteq \widehat{\widehat{q}}_0 \tau^{l+1}$. This implies that there exists $x \in X^*$, $\widehat{q} \in Q$ such that $Q^x \subseteq \widehat{q} \tau^{l+1}$. $\blacksquare$

If Q^x is τ-connected or if Q is contractible, then there exists $q \in Q$ such that $Q^x \subseteq C(q)$. Hence all errors of A are l-correctable for some $l \leq |Q| - 1$. For topological spaces, Q is contractible if and only if $\delta = \delta_A$ is null-homotopic. The following result gives a converse.

Corollary 10.16.9 *If all temporary errors of an automaton A are l-correctable, $l < 2$, then $(Q^x, i^* \tau)$ is t-connected and (Q, τ) is contractible if, in addition, $\delta_x \simeq \delta_A$ (since then $\delta_x \simeq c$, c a constant map, and $Q^x = \widehat{q} \tau^{l+1}$).* $\blacksquare$

Moreover, if all errors of an automaton A are l-correctable, then every error is also eventually l-corrected and so A is l-synchronized with respect to every error.

Modification tolerance of abstract automata and the problem of finding meaningful tolerance between automata states are studied in [37]. These tolerances should be based on the minimal amount of time needed in order to correct state errors. The approach presented here may also be extended to failure tolerance with respect to spatial propagation of errors in iterative networks.

10.17   Clinical Monitoring with Fuzzy Automata

A framework for an intelligent bedside monitor is presented in this section. The material is from [222]. The purpose of the monitor is to derive an abstraction of the current status of a patient by performing fuzzy state transitions on pre-processed input continuously supplied by clinical instrumentation. An implementation called DiaMon-1 has been used for off-line evaluation of data of patients suffering from adult respiratory distress syndrome.

Intensive care monitors display more and more information as new devices for on-line sampling of physiological data become available. Hence the clinical staff is faced with the problem of monitoring the monitor. It is difficult for humans to perceive and interpret a large number of time-varying parameters, [38, 86, 238]. If parameters interact in such a way that only certain groupings provide hints for critical conditions, then the situation becomes more complicated. When the meaning of a value depends on the patient history, another complicating factor occurs. Context-specific alarms for which no absolute thresholds can be established provide a good example of this [38, 214].

In [222], a formal framework was presented for the design of monitors that have the following properties:

(1) abstraction from objectively observed (quantitative) parameters to (qualitative) stages of a disease,

(2) early indication of improvement in or deterioration of the patient's state by providing smooth transitions between stages, and

(3) consideration of previous events, i.e., history-based interpretation of data.

A state monitor traces the patient's change of state with time, i.e., that records the progress of his illness. A state is considered to be an abstraction of the patient's status that accounts for a specific stage of a disease. Possible paths from one state to the next are provided by the transitions. The transitions depend on input, i.e., events that need to occur or conditions that need to be satisfied for a transition to take place. The input is obtained by processing objectively and preferably automatically acquired data.

A state monitor is an abstract model of the medical knowledge in a specific area. Medical decision making is based on knowledge that must consider uncertainty. Thus judgment of the current state of a patient is often a matter of degree [2]. Hence transition from one stage of a disease to another is usually smooth rather than abrupt. The state monitor presented here is based on the concept of a fuzzy automaton rather than on the concept of a conventional one in order to allow for smooth transitions.

In the following definition, we use the same terminology to define a different notion of a fuzzy automaton. This definition is confined to this section and hence no confusion should arise.

Definition 10.17.1 *A **fuzzy automaton** is a quadruple* $\widetilde{A} = (Q, \widetilde{q}_0, X, \delta)$, *where* Q *is a finite set of states,* $\widetilde{q}_0$ *is a fuzzy subset of* Q *called the **fuzzy initial state**,* X *is a finite set of input symbols, and* $\delta : Q \times X \to Q$ *is a transition function. Let* $\widetilde{i}_t$ *be a fuzzy subset of* X, $t = 0, 1, 2, \ldots$. *Define* $\widetilde{q}_{t+1} : Q \to [0, 1]$ *by for each* $q \in Q$,

$$\widetilde{q}_{t+1}(q) = \begin{cases} \vee\{\widetilde{q}_t(q') \wedge \widetilde{i}_t(x) \mid \delta(q', x) = q, \ q' \in Q, \ x \in X\}, & \text{if } \delta^{-1}(q) \neq \emptyset \\ 0 & \text{if } \delta^{-1}(q) = \emptyset. \end{cases}$$

Define $\widetilde{\delta} : \widetilde{Q} \times \mathcal{FP}(X) \to \mathcal{FP}(Q)$ *by*

$$\widetilde{\delta}(\widetilde{q}_t, \widetilde{i}) = \widetilde{q}_{t+1}, \tag{10.7}$$

where $\widetilde{Q} = \{\widetilde{q}_t \mid t = 0, 1, \ldots\}$.

A sequence of fuzzy states is denoted by $\{\widetilde{q}_t\}_{t=0}^{\infty}$ and is said to be **increasing** if $\widetilde{q}_{t+1} \supseteq \widetilde{q}_t$ for all t.

Since monitoring is a continuous process that is terminated on the exhaustion of input data rather than on arrival at a certain state, a set of final states is not defined here. The definition of a fuzzy automaton that appears here differs from the usual ones; see [49,51,212,250] for example. The transition function is crisp rather than fuzzy. Uncertainty expressed in a fuzzy input alone results in a partial transition from one state to another. A general statement about how strongly two states are related is not possible. Except for the fuzzy initial state, the fuzzification of the automaton is completely determined by the extension principle.

A fuzzy automaton exhibits properties quite different from those of a crisp one. The current state is a fuzzy subset of the set of states. Consequently, the automaton can perform different (partial) transitions simultaneously and therefore track parallel paths. Another different property is that crisp automata report an error on input not accounted for at the current state while a fuzzy automaton reacts on low or zero membership grades in the fuzzy input with continuously decreasing membership grades in its current state as obtained by (10.7). This phenomenon causes a decrease in certainty that seems consistent with all repeated applications of fuzzy set operations.

Example 10.17.2 *Let* $Q = \{q_1, q_2, q_3, q_4, q_5\}$ *and* $X = \{a, b, c\}$. *Define* $\delta : Q \times X \to Q$ *as follows:*

$$\begin{aligned}
\delta(q_1, a) &= q_3, & \delta(q_3, c) &= q_4, \\
\delta(q_1, b) &= q_4, & \delta(q_4, a) &= q_5, \\
\delta(q_1, c) &= q_2, & \delta(q_4, b) &= q_4, \\
\delta(q_2, a) &= q_2, & \delta(q_4, c) &= q_4, \\
\delta(q_2, b) &= q_4, & \delta(q_5, a) &= q_5, \\
\delta(q_2, c) &= q_2, & \delta(q_5, b) &= q_5, \\
\delta(q_3, a) &= q_3, & \delta(q_5, c) &= q_5, \\
\delta(q_3, b) &= q_5.
\end{aligned}$$

Define $\widetilde{q_0} : Q \to [0, 1]$ *as follows:* $\widetilde{q_0}(q_1) = \frac{3}{4}$, $\widetilde{q_0}(q_2) = \frac{1}{4}$, *and* $\widetilde{q_0}(q_i) = 0$ *for* $i = 3, 4, 5$. *Define* $\widetilde{i_0} : X \to [0, 1]$ *by* $\widetilde{i_0}(a) = \frac{1}{2}$, $\widetilde{i_0}(b) = \frac{1}{4}$, *and* $\widetilde{i_0}(c) = \frac{1}{4}$. *Let* $\widetilde{i_t} = \widetilde{i_0}$ *for* $t = 1, 2, \ldots$. *Then*

$$\begin{aligned}
\widetilde{q_1}(q_1) &= 0, \\
\widetilde{q_1}(q_2) &= (\widetilde{q_0}(q_1) \wedge \widetilde{i_0}(c)) \vee (\widetilde{q_0}(q_2) \wedge \widetilde{i_0}(a)) \vee (\widetilde{q_0}(q_2) \wedge \widetilde{i_0}(c)) \\
&= \frac{1}{4}, \\
\widetilde{q_1}(q_3) &= (\widetilde{q_0}(q_1) \wedge \widetilde{i_0}(a)) \\
&= \frac{1}{2}, \\
\widetilde{q_1}(q_4) &= (\widetilde{q_0}(q_1) \wedge \widetilde{i_0}(b)) \vee (\widetilde{q_0}(q_2) \wedge \widetilde{i_0}(b)) \\
&= \frac{1}{4} \\
\widetilde{q_1}(q_5) &= 0.
\end{aligned}$$

In a similar manner, we obtain

$$\widetilde{q_2}(q_1) = 0,$$
$$\widetilde{q_2}(q_2) = \tfrac{1}{4},$$
$$\widetilde{q_2}(q_3) = \tfrac{1}{2},$$
$$\widetilde{q_2}(q_4) = \tfrac{1}{4},$$
$$\widetilde{q_2}(q_5) = \tfrac{1}{4}.$$

In fact, $\widetilde{q_t}(q_i) = \widetilde{q_2}(q_i)$ for $i = 1, 2, 3, 4, 5$, and $t = 2, 3, \ldots$.

Automatically acquired data are generally precise and hence no source of fuzzy input is required by the automaton defined above. Also, if every single parameter value acquired represented an input on its own, (1) the number of possible input symbols needed to be accounted for would be too large for the automaton to handle and (2) the automaton would continuously change its state in order to react to a certain input; otherwise the input would remain unconsidered and hence become lost. Therefore, a fuzzy automaton by itself is not very appropriate to perform monitoring. The data are therefore pre-processed by a function that abstracts from single input parameters by generating fuzzy events that are passed on to the automaton.

Definition 10.17.3 *Let $\widetilde{A} = (Q, \widetilde{q_0}, X, \delta)$ be a fuzzy automaton, R_1 through R_n be the parameter ranges, where n is the number of parameters observed, $P = R_1 \times \ldots \times R_n$ be the parameter value space, and $f : P \to \mathcal{FP}(X)$ be a function that maps parameter tuples to fuzzy subsets of the input alphabet of $\widetilde{A}$. Then $\widetilde{M} = (\widetilde{A}, P, f)$ is called a **state monitor**.*

Note that $\mathcal{FP}(X)$ specifies the interface between preprocessing of data through f and interpretation of input through $\widetilde{A}$. Thus f can therefore be replaced by any computable method that yields a suitable fuzzy subset.

From the definition of δ, it follows that other than the state membership values of $\widetilde{q_0}$, those of $\widetilde{q_t}$ can only be introduced through fuzzy input.

The following lemma follows from (10.7).

Lemma 10.17.4 *If a fuzzy automaton $\widetilde{A}$ is repeatedly fed with constant fuzzy input $\widetilde{i}$, then the set of fuzzy states it transitions between is finite.* ∎

Define the **height** of $\widetilde{q_t}$, written $\mathrm{hgt}(\widetilde{q_t})$, to be $\vee\{\widetilde{q_t}(q) \mid q \in Q\}$ for $t = 0, 1, 2, \ldots$. The sequence $\{\mathrm{hgt}(\widetilde{q_t})\}_{t=0}^{\infty}$ is a decreasing function of t. This reflects a loss of certainty in the automaton. Even if the input does not change, $\mathrm{hgt}(\widetilde{q_t})$ can decrease rapidly. Hence in practice, when the monitor is provided input in rapid succession, once the current state is the empty set, it can never recover. As a matter of fact, if the automaton does not contain any feedback loops, i.e., does not provide circular transitions, it will arrive at the empty state after at most as many steps as there are states. Rather than leaving the responsibility for providing appropriate feedback loops to the designer of the automaton to overcome this undesirable property, we next define a property that overcomes this inadequacy.

Definition 10.17.5 *A fuzzy automation is said to provide a **peak hold** if*

$$\forall q', q \in Q \text{ and } \forall x \in X, \quad \delta(q', x) = q \text{ implies } \delta(q, x) = q. \tag{10.8}$$

The condition in Definition 10.17.5 says that there is a transition for every state on every input that leads to the state. The condition implies that no state can be entered and left on the same input, else δ would not be single-valued. The peak hold guarantees that the maximum evidence for a state provided by its predecessors is memorized and held as long as input of ongoing transitions can support it. However, the peak hold may also be sustained by an input other than the one that initially led to that state since a state does not remember its predecessor. Thus the grade of membership can unintentionally remain high.

Peak hold has a positive side effect in that the automaton cannot oscillate on constant input [250], a property that would clearly not be acceptable in the clinical setting since stable input should be reflected in stable output. This can be seen from the next result.

The fuzzy automaton in Example 10.17.2 provides a peak hold.

Theorem 10.17.6 *The fuzzy state of a fuzzy automaton with peak hold always becomes stable after a finite number of repetitions of the same input.*

Proof. We show that for $t \geq 1$, there is a positive integer r such that

$$\widetilde{q}_t \subseteq \widetilde{q}_{t+1} \subseteq \dots \subseteq \widetilde{q}_{t+r} = \widetilde{q}_{t+r+1} = \dots. \tag{10.9}$$

This is accomplished by showing that
(1) $\{\widetilde{q}_t\}_{i=0}^{\infty}$ is increasing, i.e., $\widetilde{q}_t \subseteq \widetilde{q}_{t+1} \subseteq \dots$ and
(2) $\exists r$ such that $\delta(\widetilde{q}_{t+r}, \widetilde{i}) = \widetilde{q}_{t+r}$. For all subsequent states, (10.9) follows from the fact that δ is a function.

For every fuzzy state $\widetilde{q}_t$ with $t > 0$ and every input $\widetilde{i}$, it follows by (10.7) that for every state q, there is a transition that determines its membership value, i.e., $\forall q \in Q, \delta^{-1}(q) \neq \emptyset$. $\forall q \in Q, \exists q'$ and $\widetilde{i}$ such that

$$\delta(q', x) = q \text{ and } \widetilde{q}_t(q) = \widetilde{q}_{t-1}(q') \wedge \widetilde{i}(x).$$

Thus, $\widetilde{q}_t(q) \leq \widetilde{i}(x)$ and so $\widetilde{q}_t(q) \wedge \widetilde{i}(x) = \widetilde{q}_t(q)$. Repeated input of $\widetilde{i}$ and $\delta(q, x) = q$ then implies

$$\widetilde{q}_{t+1}(q) = \widetilde{q}_t(q) \vee (\vee\{\widetilde{q}_t(q') \wedge \widetilde{i}(x) \mid \delta(q', x) = q\}),$$

$$\widetilde{q}_{t+1}(q) \geq \widetilde{q}_t(q),$$

and so $\widetilde{q}_{t+1} \supseteq \widetilde{q}_t$.

By Lemma 10.17.4, there is no infinite sequence of fuzzy states $\{\widetilde{q}_t\}_{t=0}^{\infty}$ such that

$$\widetilde{q}_{t+1} = \widetilde{\delta}(\widetilde{q}_t, \widetilde{i}) \text{ and } \widetilde{q}_{t+1} \supset \widetilde{q}_t.$$

Consequently, there exists an r such that $\widetilde{q}_{t+r+1} \subseteq \widetilde{q}_{t+r}$ for $t = 0, 1, 2, \ldots$. Since $\{\widetilde{q}_t\}_{t=0}^{\infty}$ is increasing, $\widetilde{q}_{t+r+1}$ must equal $\widetilde{q}_{t+r}$. ∎

The proof of Theorem 10.17.6 shows that $\{\widetilde{q}_t\}_{t=0}^{\infty}$ does not converge to the empty state. It can take several steps for A to become stable since the fuzzy input can cause a propagation of higher membership grades along a sequence of transitions. The fact that nonfuzzy deterministic automata with peak hold are stable after one step provides another example of how fuzzification yields more general results.

The height of the current state is still decreasing even with the peak hold property since high grades of membership cannot be regained once they are lost. A situation where the grade of membership of one state decays while its successor's rises is a source of loss.

This behavior does not model the natural decision process correctly because once a decision has been made, it is usually pursued rather uncritically until there is sufficient evidence for another decision to be made.

By introducing the idea of a threshold, the state monitor can be modified to adopt this kind of behavior. A state is called **active** when its grade of membership in the current state exceeds a certain threshold c. An active state is defined to remain active until there is a transition that induces activity of one of its successors, i.e.,

$$\widetilde{q}_{t+1}(q) = \begin{cases} \widetilde{q}_t(q) & \text{if } \widetilde{q}_t(q) \geq c \text{ and } \nexists q', x \text{ such that} \\ & \delta(q, x) = q' \text{ and } \widetilde{i}_t(x) \geq c \\ (10.7) & \text{otherwise.} \end{cases} \tag{10.10}$$

Thus once a state has reached a certain grade of membership, it keeps it until a transition can pass it on to one of its successors. Thus the height of the current state is always greater than c. Therefore, a certain level of uncertainty is thus always being maintained.

For $c = 1$, (10.10) implies that there is always at least one state q such that $\widetilde{q}_t(q) = 1$. This accounts for the fact that the patient is considered to be at least in one state at a time, even if no successor with more evident support could yet be determined.

We note that (10.10) has only slight effect on the proof of Theorem 10.17.6. In (1) of the proof of Theorem 10.17.6, $\{\widetilde{q}_t\}_{t=0}^{\infty}$ is still increasing since the peak hold also works for $\widetilde{q}_t(q) \geq c$ and $\widetilde{q}_t(q)$ can only drop below c once, namely on the first input. Also, (2) of the proof of Theorem 10.17.6 still holds because Lemma 10.17.4 is not affected.

Moreover, (10.10) cannot prevent the automation from oscillation without peak hold even though the height is kept above c.

Further discussion of clinical monitoring can be found in [222]. We note that another approach using techniques from fuzzy set theory can be found in [214]. Nonfuzzy approaches can be found in [87, 238].

10.18   Fuzzy Systems

In the final section, we consider fuzzy systems for two reasons. First, they resemble fuzzy automata and, second, they have interesting applications to information retrieval [160].

System theory provides a framework for describing general relationships of the empirical world. We are mainly interested in the concept of reachability, observability, stability, and realization.

Let Q be the state space, X the input space, and Y the output space. A deterministic dynamic system (time invariant) is a complex $S = \{Q, Y, \delta, \beta\}$, where $\delta : Q \times X \to Q$ is the dynamics, $q_{t+1} = \delta(q_t, u_t)$, and $\beta : Q \to Y$ is the output map.

A nondeterministic system is the complex $S = \{Q, X, Y, \delta, \beta\}$ with the dynamics $\delta : Q \times X \to \mathcal{P}(Q)$, $q_{t+1} \in \delta(q_t, u_t)$, and the output map $\beta : Q \to \mathcal{P}(Y)$.

This definition can be generalized by considering not only the states, but also the inputs and outputs as being subsets.

Definition 10.18.1 *An **abstract system** is a complex*

$$S_a = \{Q, X, Y, \delta, \beta\}$$

such that $\delta : \mathcal{P}(Q) \times \mathcal{P}(X) \to \mathcal{P}(Q)$, $\beta : \mathcal{P}(Q) \to \mathcal{P}(Y)$.

We next present the application given in [160] of fuzzy systems to information retrieval systems.

Example 10.18.2 *We consider an Information Retrieval system defined in terms of its response to a request for information. An IR system reports on the existence and location of information items relating to a request and does not change the knowledge of the user on the subject of the request. If the search criteria are based on the contents of an information item, then it becomes necessary to use content identification such as a set of descriptors attached to each item. In such cases, it is customary to assign descriptors, normally chosen from a controlled list of allowable terms. That is, a request is defined to be recovered from the store. An IR system compares the specification of required items with the descriptions of the stored items, and retrieves, or lists, all the items that correspond in some defined way to that specification. Consequently, the IR system can be characterized by having as inputs a subset of a set D of information items and a subset of a set R of requests. If the system is presented with new documents, the system must process them to obtain descriptions. The same sort of thing must be done* *with the requests. The next step is the comparison. The ultimate response of the system to a request is a partial ordering on the set of information items.*

Let $P_f(S)$ be the set of all finite subsets of a set S.

Let ξ, η be functions $\xi : D \to Q$, $\eta : R \to Q$, where Q is the set of descriptions. Let $A = P_f(D)$, $B = P_f(R)$, $C = P_f(Q)$, and $\mathcal{R}$ be the set of partial orderings on $P_f(D)$, i.e., for $\rho \in \mathcal{R}$, $M, N \in P_f(D)$, $M\rho N$ if and only if N is more relevant to a given request than M.

We use the following variables to describe the state of the IR system:

q_1 is the set of incoming documents waiting to be processed;

q_2 is the document file;

q_3 is the document description file;

q_4 is the set of incoming requests waiting to be processed;

q_5 is the request currently being processed;

q_6 is the request that was just processed;

q_7 is a partial ordering on $P_f(D)$ (i.e., in $\mathcal{R}$) induced by the request q_6.

The input variables are

$u_1 = $ the documents coming to the system;

$u_2 = $ the requests coming to the system;

and the output variables are

$y_1 = q_6$ is the request that was just processed;

y_2 is a subset of the document file that is maximal with respect to q_7.

Using the above notation, the input space is $\mathcal{X} = A \times B$, the output space is $\mathcal{Y} = R \times A$, and the state space is

$$\mathcal{H} = A \times A \times C \times B \times R \times R \times \mathcal{R}.$$

Let the initial state be given as follows:

$q_1(0) = \{d_i \in D \mid i = 1, 2, \dots, m\}$

$q_2(0) = \{e_1, e_2, \dots, e_n\}$

$q_3(0) = \{w_1, w_2, \dots, w_n\}$ with $w_i = \xi(e_i)$, $i = 1, 2, \dots, n$

$q_4(0) = \{r_1, \dots, r_l\}$, $r_i \in R$

$q_5(0) = r'$

$q_6(0) = r''$

$q_7(0) \in \mathcal{R}$.

The state equation of the IR system can be written in the form: $q(k + 1) = \delta(q(k), u(k))$, $k = 0, 1, 2, \dots$, where $\delta : \mathcal{H} \times \mathcal{X} \to \mathcal{H}$ is the dynamics and the output is $y(k) = \beta(q(k))$, where $\beta : \mathcal{H} \to \mathcal{Y}$ is the output map.

The state equations can be written as follows:

$$q_1(k + 1) = \begin{cases} (q_1(k) \cup u_1(k)) \backslash \{d_{k+1}\}, & \text{if } q_1(k) \neq \emptyset \\ u_1(k), & \text{if } q_1(k) = \emptyset \end{cases}$$

$$q_2(k + 1) = q_2(k) \cup u_1(k)$$

$$q_3(k+1) = \begin{cases} q_3(k) \cup \{\xi(d_{k+1})\}, & \text{if } q_1(k) \neq \emptyset \\ q_3(k), & \text{if } q_1(k) = \emptyset \end{cases}$$

$$q_4(k+1) = \begin{cases} [q_4(k) \cup u_2(k)]\setminus\{r_{k+1}\}, & \text{if } q_4(k) \neq \emptyset \\ u_2(k), & \text{if } q_4(k) = \emptyset \end{cases}$$

$$q_5(k+1) = \begin{cases} r_{k+1}, & \text{if } q_4(k) \neq \emptyset \\ \emptyset & \text{if } q_4(k) = \emptyset \end{cases}$$

$$q_6(k+1) = q_5(k)$$

$$q_7(k+1) = \overline{r}(\eta(q_5(k))),$$

where $\overline{r}$ is the ordering in A induced by $\eta(q_5(k))$.

The output equations are as follows:

$$y_1(k) = q_6(k)$$

and

$$y_2(k) \subseteq q_2(k).$$

The notation d_{k+1} in the first equation means that between k and $k+1$, d_{k+1} is processed, $k = 0, 1, \ldots$. The same thing holds for r_{k+1}.

It follows that $q_7(k+1)$ is an ordering in A and $y_2(k)$ is the subset of documents that give a "best response" at the request $q_6(k)$.

It follows that the IR system is a complex,

$$\mathcal{S}_{IR} = \{\mathcal{H}, \mathcal{X}, \mathcal{Y}, \delta, \beta\},$$

with δ and β as above.

The dynamics δ can be extended to $\mathcal{H} \times \mathcal{X}^*$ considering sequences of pairs ("documents," "requests").

If U and V are two sets, there is an injection

$$i : (U \times V)^* \to U^* \times V^*$$

such that

$$i((u_1, v_1)(u_2, v_2) \ldots (u_n, v_n)) = (u_1 u_2 \ldots .u_n, v_1 v_2 \ldots .v_n).$$

It follows that an element of $\mathcal{X}^*$ can be considered as a pair of the form ("sequence of documents," "sequence of requests"), where the sequences have the same length.

The system $\mathcal{S}_{IR}$ is **reachable**, from the state $q(0) \in \mathcal{H}$ if each state in $\mathcal{H}$ can be reached with a suitable input sequence. This means that the input space (i.e., documents and requests) is rich enough to cover each state (i.e., document file, requests, orderings, etc.).

The system $\mathcal{S}_{IR}$ is **observable** if, knowing the system's response (documents that are relevant to the request), for each sequence of inputs (requests and documents), the state from which the system is started can be uniquely determined.

We are thus led to the concept of a fuzzy system.

Definition 10.18.3 *A **fuzzy system** is a complex*

$$\mathcal{S}_f = \{Q, X, Y, \delta, \beta\}$$

with $\delta : \mathcal{FP}(Q) \times \mathcal{FP}(X) \rightarrow \mathcal{FP}(Q)$, *the **fuzzy dynamics**, and* $\beta : \mathcal{FP}(Q) \rightarrow \mathcal{FP}(Y)$, *the **fuzzy output map**.*

The state equation can be written as follows:

$$q_{t+1} = \delta(q_t, u_t), u_t \in \mathcal{FP}(X), q_t, q_{t+1} \in \mathcal{FP}(Q),$$

where q_t, q_{t+1} are, respectively, the fuzzy states at time $t, t + 1$ and u_t the fuzzy input at time t.

The output equation becomes

$$y_t = \beta(q_t), q_t \in \mathcal{FP}(Q), y_t \in \mathcal{FP}(Y),$$

where y_t is the fuzzy output at time t.

Let $(\mathcal{FP}(X))^*$ denote the free monoid generated by $\mathcal{FP}(X)$, i.e., the set of sequences of fuzzy inputs. Then δ can be extended to $\mathcal{FP}(Q) \times (\mathcal{FP}(X))^*$ by defining

$$\delta : \mathcal{FP}(Q) \times (\mathcal{FP}(X))^* \rightarrow \mathcal{FP}(Q)$$

as follows:

(1) $\delta(\sigma, \Lambda) = \sigma, \forall \sigma \in \mathcal{FP}(Q)$,

(2) $\delta(\sigma, \lambda^*\mu) = \delta(\delta(\sigma, \lambda^*), \mu), \forall \sigma \in \mathcal{FP}(Q), \lambda^* \in (\mathcal{FP}(X))^*, \mu \in \mathcal{FP}(X)$.

If the initial state $\sigma_0 \in \mathcal{FP}(Q)$ is fixed, we can define the function:

$$\delta_{\sigma_0} : (\mathcal{FP}(X))^* \rightarrow \mathcal{FP}(Q)$$

by $\forall \lambda^* \in (\mathcal{FP}(X))^*$,

$$\delta_{\sigma_0}(\lambda^*) = \delta(\sigma_0, \lambda^*).$$

Hence $\delta(\sigma_0, \lambda^*)$ is computed by starting the system in state σ_0, feeding in the input sequence λ^*, and determining the final state.

With these definitions, we can express two basic concepts of systems theory, namely, reachability and observability.

Definition 10.18.4 *The fuzzy system* $\mathcal{S}_f$ *is called* **reachable** *from the state* σ_0 *if* δ_{σ_0} *is onto, i.e.,*

$$\forall \sigma \in \mathcal{F}P(Q), \ \exists \lambda^* \in (\mathcal{F}P(X))^* \ such \ that \ \delta_{\sigma_0}(\lambda^*) = \sigma.$$

The image of δ_{σ_0}, *Im* $\delta_{\sigma_0} \subseteq \mathcal{F}P(Q)$, *is called the* **reachability set** *of the system* $\mathcal{S}_f$ *from* σ_0.

Composing δ_{σ_0} with β we obtain the fuzzy response function (or behavior) of the system $\mathcal{S}_f$, i.e.,

$$f_{\sigma_0} = \beta \circ \delta_{\sigma_0}.$$

Thus

$$f_{\sigma_0} : (\mathcal{F}P(X))^* \to \mathcal{F}P(Y)$$

and so

$$f_{\sigma_0}(\lambda^*) = \beta(\delta(\sigma_0, \lambda^*)) \ \forall \lambda^* \in (\mathcal{F}P(X))^*.$$

Definition 10.18.5 *A fuzzy system* $\mathcal{S}_f$ *is called* **observable** *if the assignment* $\sigma \mapsto f_\sigma$ *is one-to-one.*

The interested reader is urged to see [160] for further details and interesting ideas including the concepts of stability and realization.

10.19   Exercises

1. Let $a, a_{ij} \in \mathbb{R}$ for $i = 1, 2, \ldots, m$ and $j = 1, 2, \ldots, n$. Prove that $(\vee\{\wedge\{a_{ji} \mid i = 1, 2, \ldots, m\} \mid j = 1, 2, \ldots, n\}) \wedge \{a\} = \vee\{\wedge\{a_{ji} \wedge a \mid i = 1, 2, \ldots, m\} \mid j = 1, 2, \ldots, n\}$ and that $(\wedge\{\vee\{a_{ji} \mid i = 1, 2, \ldots, m\} \mid j = 1, 2, \ldots, n\}) \vee \{a\} = \wedge\{\vee\{a_{ji} \wedge a \mid i = 1, 2, \ldots, m\} \mid j = 1, 2, \ldots, n\}$.

2. Prove that the matrix in Example 10.6.2 has period equal to 3.

3. Prove that $\mu(a^n b^m) = \frac{n \wedge m}{n \vee m}$ in Example 10.8.2.

4. Prove Lemma 10.11.2.

5. Let M be the median operator of (10.5). Show that M is associative, not strictly increasing, and contractive.

6. Prove (2) of Proposition 10.11.7.

7. Show that $(\mathbb{N}^0, \simeq)$, where $\mathbb{N}^0 = \mathbb{N} \cup \{0\}$ and $(\{0, 1, \ldots, n\}, i)$, of Example 10.13.1 are tolerance spaces.

8. For f in Definition 10.13.2, prove that $f_*\tau$ is the least tolerance on Y such that $f : (X,\tau) \to (Y, f^*\tau)$ is a fuzmap. Prove also that it is the unique tolerance on Y such that $\forall$ tolerances σ on Z and $\forall$ set-theoretic maps $g : Y \to (Z,\sigma)$, $f \cdot g : (X,\tau) \to (Z,\sigma)$ if and only if $g : (Y, f_*\tau) \to (Z,\sigma)$ is a fuzmap.

9. If $g : Y \to (X,\tau)$ is a set-theoretic map, prove that $g^*\tau$ is the biggest tolerance on Y such that $g : (Y, g^*\tau) \to (X,\tau)$ is a fuzmap.

10. In Definition 10.13.4, prove that τ is the least tolerance such that all injections ι_j are fuzmaps, and $\tau_1\tau_2$ is the largest tolerance such that all projections pr_j are fuzmaps.

11. Prove (1) of Theorem 10.14.3.

12. In Example 10.14.6, show that τ_7 and τ_9 yield $\tau_{14} = \begin{bmatrix} 1 & 1 & 1 & 1 \\ 1 & 1 & 1 & 0 \\ 1 & 1 & 1 & 0 \\ 1 & 0 & 0 & 1 \end{bmatrix}$

and τ_8 and τ_9 yield $\tau_{15} = \begin{bmatrix} 1 & 1 & 1 & 0 \\ 1 & 1 & 1 & 1 \\ 1 & 1 & 1 & 0 \\ 0 & 1 & 0 & 1 \end{bmatrix}$. Then determine τ_{19} from τ_{14} and τ_{15}.

13. In Example 10.15.6, show that $\delta Q \cup \delta_x^{-1}\tau_2^0\delta_x \subseteq \tau_2^0$ for $x \in \{x_1, x_2\}$.

14. Prove that the condition in Definition 10.17.5 implies that no state can be entered and left on the same input, else δ would not be single-valued.

References

1. Ábrahám, S., Some questions of phrase structure grammars, *Computational Linguistics*, 4(1965) 61-70.

2. Adlassing, K.P., Fuzzy set theory in medical diagnosis, *IEEE Transaction on Systems, Man, and Cybernetics, SMC* 16(1986) 260-265.

3. Aho, A., and Ulman, J., *Theory of Parsing, Translation and Compiling*, Vol. 1, Prentice-Hall, Englewood, NJ, 1972.

4. Arbib, M., Realization of stochastic systems, in IEEE Conference Record of the Seventh Annual Symposium on Switching and Automata Theory, Vol. 7, 262-266. Institute of Electrical and Electronic Engineers, New York, 1966.

5. Arbib, M.A., Tolerance automata, *Kybernetika*, 3(1967) 223-233.

6. Arbib, M.A., *Algebraic Theory of Machines, Languages, and Semigroups*, Academic Press, New York, 1968.

7. Arbib, M.A. and Manes, E.G., Fuzzy morphism in automata theory, *Lecture Notes in Computer Science*, 25(1975) 80-86.

8. Arden, D.N., Delayed logic and finite state machines, Theory of Comp. Machine Design, Univ. of Michigan, Ann Arbor, (1960) 1-35.

9. Auterbert, I.M., Pushdown-automata and families of languages generating cylinders, *Lecture Notes in Computer Science*, 53(1977) 231-239.

10. Bacon, G. C., Minimal-state stochastic finite systems, *IEEE Trans. Computers*, CT-11 (1964), 107-108.

11. Baer, R.M. and Spainer, E.H., Referenced automata and metaregular families, *J. Comput. System Sci.*, 3(1969) 423-446.

12. Bavel, Z., *Introduction to the Theory of Automata*, Reston Publishing Co., Inc., Reston, Virginia, 1983.

13. Bellman, R.E., Kalaba, R., and Zadeh, L.A., Abstract pattern classification, *J. Math Appl.*, 13(1966) 1-7.

14. Bellman, R.E. and Zadeh, L.A., Decision-making in a fuzzy environment, *Management Sci.*, 17-B(1970) 141-164.

15. Berstel, J., Some recent results on recognizable formal power series, *Lecture Notes in Computer Science*, Vol. 56, 39-48.

16. Bezdek, J. and Harris, J., Fuzzy partitions and relations − an axiomatic basis for clustering, *Fuzzy Sets and Systems*, 1(1978) 111-126.

17. Biacino, L. and Gerla, G., Recursively enumerable L-sets, *Z. Math. Logik Grundlag Math.*, 33(1987) 107-113.

18. Birkhoff, G., *Lattice Theory*, AMS Colloquium Publications 25, 2nd edition, Providence, RI, 1948.

19. Blanchard, N., (Fuzzy) fixed point property in L-underdeterminate sets, *Fuzzy Sets and Systems*, 30(1989) 11-26.

20. Boassan, L., Classification of the context-free languages, *Lecture Notes in Computer Science*, 53(1977) 34-43.

21. Booth, T.L., Statistical properties of random digital sequences, in IEEE Conference Record of the Seventh Annual Symposium on Switching and Automata Theory, Vol. 7, 251-261. Institute of Electrical and Electronics Engineers, New York, 1966.

22. Booth, T.L., *Sequential Machines and Automata Theory*, Wiley, New York, 1968.

23. Brauer, W., Zu den Grundlagen einer Theorie Topologischer Sequentieller Systeme und Automaten, *Gesellschaft für Mathematik und Datenverarbeitung, Bonn*, 31(1970).

24. Brown, J.G., Fuzzy sets on Boolean lattices, *Rept. No., 1957, Ballastic Research Laboratories, Aberdeen, MD*, 1969.

25. Brown, J.G., A note on fuzzy sets, *Inform. and Control*, 18(1971) 32-39.

26. Bruce, G.D. and Fu, K.S., A model for finite-state probabilistic systems, *Proc. 1st Ann. Allerton Conf. Circuit and Systems Theory*, 1963.

27. Brunner, J., Zur Theorie der R-fuzzy Automaten, *Wiss. Z. Tech. Univ. Dresden*, 27(1978) 693-695.

28. Brunner, J. and Wechler, W., Zur Theorie der R-fuzzy Automaten I, *Wiss. Z. Tech. Univ. Dresden*, 26(1977) 647-652.

29. Carlyle, J.W., Reduced forms for stochastic sequential machines, *J. Math. Anal. Appl.*, 7(1963) 167-175.

30. Carlyle, J.W., Stochastic finite-state system theory, in *System Theory* (L.A. Zadeh and E. Polak, Eds.), McGraw-Hill, New York, 1969, 387-423.

31. Chang, S.K., On the execution of fuzzy programs using finite state machines, *IEEE Trans. Computers*, 21(1972) 241-253.

32. Cheng, S.C. and Mordeson, J.N., Applications of fuzzy algebra in automata theory and coding theory, *Proceedings Fifth IEEE Conference on Fuzzy Systems, Vol.* 1, (1996) 125-129.

33. Cho, S.J., Kim, J.G., and Kim, S.T., On T-generalized subsystems of T-generalized state machines, *Far East J. Math. Sci.*, 5(1997), 131-151.

34. Cho, S.J., Kim, J.G., and Kim, H.D., Sums and joins of fuzzy transformation semigroups, preprint.

35. Cho, S.J., Kim, J.G., and Lee, W.S., Decompositions of T-generalized transformation semigroups, *Fuzzy Sets and Systems*, to appear.

36. Cin, M.D., Fuzzy-state automata: their stability and fault tolerance, *International Journal of Computing and Information Sciences*, 4(1975) 63-80.

37. Cin, M. D., Modification tolerance of fuzzy-state automata, *Informational Journal of Computer and Information Sciences*, 4(1975) 81-93.

38. Coiera, E., Intelligent monitoring and control of dynamic physiological systems, *Artificial Intelligence in Medicine*, 5(1993) 1-8.

39. Comer, S. D., Polygroups derived from cogroups, *J. Alg.*, 89(1984), 397-405.

40. Conway, J.E., *Regular Algebra and Finite Machines*, Chapman & Hall, London and Colchester, 1971.

41. Das, P., On some properties of fuzzy semiautomation over a finite group, *Inform. Sci.*, 101(1997) 71-84.

42. Dauber, P.S., An analysis of errors in finite automata, *Inf. Control*, 8(1965) 295-303.

43. De Backer, W. and Verbeek, L., Study of analog, digital and hybrid computers using automata theory, *ICC Bulletin*, 5(1966) 215-244.

44. DePalma, G.F. and Yau, S.S., Fractionally fuzzy grammars with application to pattern recognition, in *Fuzzy Sets and Their Application to Cognitive and Decision Processes*, L.A. Zadeh, K.S. Fu, K. Tanaka, and M. Shimura, eds., Academic Press, New York (1975) 329-351.

45. Dharmadhikari, S.W., Functions of finite Markov chains, *Ann. Math. Statist.*, 34(1963) 1022-1032.

46. Dirmitrescu, D., Hierarchical pattern classification, *Fuzzy Sets and Systems*, 28(1988) 145-162.

47. Dombi, J., Basic concepts for theory of evaluation: The aggregative operator, *European J. Oper. Res.*, 10(1982) 282-294.

48. Dörfler, W., The cartesian composition of automata, *Math. Systems Theory*, 11(1978) 239-257.

49. Dougherty, E.R. and Giardina, C.R., *Mathematical Methods for Artificial Intelligence and Autonomous Systems*, Prentice-Hall, Englewood Cliffs, NJ, 1988.

50. Drewnaik, J., Fuzzy relation equations and inequalities, *Fuzzy Sets and Systems*, 14(1984) 237-247.

51. Dubois, D. and Prade, H., *Fuzzy Sets and Systems: Theory and Applications*, Academic Press, New York, 1980.

52. Earnest, L.D., Machine recognition of cursive writing, *Information Processing (IFIP)*, C.M. Popllewell, ed., North-Holland, Amsterdam, 1962.

53. Eden, M., Handwriting and pattern recognition, *IRE Trans. on Information Theory*, IT 8(1962) 160-166.

54. Ehrich, R., A contextual post-processor for cursive script recognition, *Proc. 1st International Joint Conf. on Pattern Recognition, October 30,* (1973) 169-171.

55. Ehrig, H., Kiermeier, K.D., Kreowaski, H.J., and Kühnel, W., *Universal Theory of Automata*, Teubner Verlag, Stuttgart, 1974.

56. Eilenberg, S., *Automata, Languages and Machines*, Vol. A, B, Academic Press, New York, 1974, 1976.

57. Ellis, C.A., *Probabilistic Languages and Automata*, Ph.D. Dissertation, Univ. of Illinois, Urbana, IL, 1969.

58. Even, S., Comments on the minimization of stochastic machines, *IEEE Trans. Electron. Comp.*, 14(1965) 634-637.

59. Faith, C., *Algebra: Rings, Modules, Categories*, Springer-Verlag, New York/Heidelberg/Berlin, 1973.

60. Fong, Y. and Clay, J.R., Computer programs for investigating syntactic near rings of finite group semiautomata, *Academia Sinica*, 16(1988) 295-304.

61. Friš, I., Grammars with partial ordering of the rules, *Inform. and Control*, 12(1968) 415-425.

62. Fu, K.S., *Syntactic Methods in Pattern Recognition*, Academic Press, New York, 1974.

63. Fu, K.S., *Syntactic Pattern Recognition and Applications*, Prentice-Hall, Englewood Cliffs, NJ, 1982.

64. Fu, K. S. and Huang, T., Stochastic grammars and languages, *J. Comput. Inform. Sci.*, 1 (1972) 135-170.

65. Fu, K.S. and Li, T.J., Formulation of learning automata and automata games, *Inform. Sci.*, 1(1969) 237-256.

66. Fu, K.S. and Li, T.J., On stochastic automata and languages, *3rd Conference of Information and System Sciences, Princeton University*, (1969) 338-377.

67. Fu, K.S., and Li, T.J., On stochastic automata and languages, *Inform. Sci.*, 1(1969) 403-419.

68. Fu, K.S. and McLaren, R.W., An application of stochastic automata to the synthesis of learning systems, School of Elec. Eng., Purdue University, Tech. Rept. TR-EE65-17, 1965.

69. Fu, K.S. and Swain, P.H., On syntactic pattern recognition, *Software Engineering*, Julius Tou, ed., Academic Press, 1971, 155-182.

70. Gerla, G., Sharpness relation and decidable fuzzy sets, *IEEE Trans. Automat. Contr.*, AC-27(1982) 1113.

71. Gerla, G., Decidability, partial decidability and sharpness relation for L-subsets, *Studia Logica*, 46(1987) 227-238.

72. Gerla, G., Turing L-machines and recursive computability for L-maps, *Studia Logica*, 48(1989) 179-192.

73. Gerla, G., Fuzzy grammars and recursively enumerable fuzzy languages, *Inform. Sci.*, 60(1992) 137-143.

74. Gill, A., *Introduction to the Theory of Finite State Machines*, McGraw-Hill, New York, 1962.

75. Gill, A. and Flexer, J.R., Periodic decomposition of sequential machines, *J. Assoc. Comput. Mach.*, 14(1967) 666-676.

76. Ginsburg, S., *An Introduction to Mathematical Machine Theory,* Addison-Wesley, Reading, MA, 1962.

77. Ginsburg, S., *The Mathematical Theory of Context-free Languages,* McGraw-Hill, New York, 1966.

78. Ginsburg, S., *Algebraic and Automata-Theoretic Properties of Formal Languages,* North-Holland, Amsterdam and Oxford, American Elsevier, New York, 1975.

79. Ginsburg, S. and Greibach, S., Abstract family of languages, in Studies in Abstract Families of Languages, American Mathematical Society, Memoir No. 87(1969), 1-32.

80. Ginsburg, S. and Spanier, E.H., Control sets on grammars, *Math. Systems Theory,* 2(1968) 159-177.

81. Goguen, J., *L*-fuzzy sets, *J. Math. Anal. Appl.*, 18(1967) 145-174.

82. Harkleroad, L., Fuzzy recursion, RET's and isola, *Z. Math. Logik Grundlog Math.*, 30(1984) 425-436.

83. Harrison, M.A., *Introduction to Switching and Automata Theory,* McGraw-Hill, New York, 1965.

84. Harrison, M.A., On the error of correcting capacity of finite automata, *Inf. Control,* 8(1965) 430-450.

85. Hartmanis, J. and Stearns, R.E., A study of feedback and errors in sequential machines, *IEEE Trans. Electron. Comput.*, EC-12(1963) 223-232.

86. Hayashi, Naito, and Wakami, A proposal of a learning type fuzzy aggregation operator that uses the method of steepest descent, *Japan Fuzzy Society Journal,* 5(1993) 1131-1141.

87. Hayes-Roth, B., Washington, R., Hewett, R., Hewett, M., and Seiver, A., Intelligent monitoring and control, *Proc. IJCAI,* 89(1989) 243-249.

88. Hirai, H., Asai, K., and Katajima, Fuzzy automaton and its application to learning control systems, *Mem. of the Fac. of End., Osaka City Univ.*, 10(1968) 67-73.

89. Hirokawa, S. and Miyano, S., A note on the regularity of fuzzy languages, *Mem. Fac. Sci., Kyushu Univ. Ser.* A 32, No. 1(1978) 61-66.

90. Hofer, G., Near rings and group automata, Ph.D. thesis, J. Kepler University, Linz, 1987.

91. Höhle, U., Quotients with respect to similarity relations, *Fuzzy Sets and Systems*, 27(1988) 31-44.

92. Holcombe, W.M.L., *Algebraic Automata Theory*, Cambridge University Press, New York, 1982.

93. Honda, N. and Nasu, M., Recognition of fuzzy languages, in *Fuzzy Sets and Their Application to Cognitive and Decision Processes*, L.A. Zadeh, K.S. Fu, K. Tanaka, and M. Shimura, eds., Academic Press, NJ, (1975) 279-299.

94. Honda, N., Nasu, M., and Hirose, S., *F*-recognition of fuzzy languages, in *Fuzzy Automata and Decision Processes*, M.M. Gupta, G.N. Saridis, and B.R. Gaines, eds., North-Holland, Amsterdam (1977) 149-168.

95. Hong, D.H. and Hwang, S.Y., A note on the value similarity of fuzzy system variables, *Fuzzy Sets and Systems*, 66(1984) 383-386.

96. Hopcroft, J.E. and Ullman, J.D., *Formal Languages and Their Relation to Automata*, Addison-Wesley, Reading, MA, 1969.

97. Huang, T. and Fu., K.S., On stochastic context-free languages, *Inform. Sci.*, 3(1971) 67-73.

98. Hwang, S.Y., Lee, H.S., and Lee, J.J., General fuzzy acceptors for syntactic pattern recognition, *Fuzzy Sets and Systems*, 80(1996) 397-401.

99. Inagaki, Y. and Fukumara, T., On the description of fuzzy meaning of context-free languages, in *Fuzzy Sets and Their Applications to Cognitive and Decision Processes*, L.A. Zadeh, K.S. Fu, K. Tanaka, and M. Shimura, eds., Academic Press, New York, 1975, 301-328.

100. Inagaki, Y., Fukumara, T., and Matuura, H., Some aspects of linear automata, *Inform. Cont.* 20(1972) 439-479.

101. Ito, H., Inagaki, Y., and Fukumura, T. Hierarchical studies of dendrolanguages with application to characterizing derivation trees of phrase structure grammars, *Memoirs of the Faculty of Engineering*, Nagoya University, 25 1(1973) 1-46.

102. Jantosciak, J., Homomorphisms, equivalences and reductions in hypergroups, *Rivista Di Matematica Pura Ed Applicata* 9(1991) 23-47.

103. Johnsonbaugh, R., *Discrete Mathematics*, 5th edition, Prentice-Hall, New York, 2001.

104. Jürgensen, H. and Peeva, K.G., Approximation of stochastic automata, in *Proceedings, Automata, Languages and Systems,* Sofia, 1981.

105. Kalman, R.E., Falb, P.L., and Arbib, M.A., *Topics in Mathematical System Theory,* McGraw-Hill, New York, 1969.

106. Kandel, A., Codes over languages, *IEEE Trans. on Systems, Man and Cybernetics,* SMC-4(1974) 135-138.

107. Kandel, A. and Lee, S. C., *Fuzzy Switching and Automata: Theory and Applications,* Crane Russak, 1980.

108. Kandel, A., *Fuzzy Mathematical Techniques with Applications,* Addison-Wesley, Reading, MA, 1986.

109. Kanst, R. Finite state probabilistic languages, *Inform. Sci.,* 1(1969) 403-419.

110. Kaufmann, A., *Introduction to the Theory of Fuzzy Subsets,* Vol. 1, Academic Press, Orlando, FL, 1973.

111. Kherts, M.M., Entropy of languages generated by automated or context-free grammars with a single-valued deduction, *Nauchno-Tekhnicheskaya Informatsya,* Ser. 2, No. 1 (1968).

112. Kim, H.H., Mizumoto, M., Toyoda J., and Tanaka, K., Lattice grammars, *Trans. Inst. Elect. Commun. Engrs. Japan,* 57-D, No. 5 (1974) 253-260 (in Japanese).

113. Kim, Y.H., Kim, J.G., and Cho, S.J., Products of T-generalized state machines and T-generalized transformation semigroups, *Fuzzy Sets and Systems,* 93(1998) 87-97.

114. Klir, G.J. and Folger, T.A., *Fuzzy Sets, Uncertainty and Information,* Prentice Hall, Englewood Cliffs, NJ, 1988.

115. Klir, G.J. and Yuan, B., *Fuzzy Sets and Fuzzy Logic, Theory and Applications,* Prentice-Hall, Upper Saddle River, NJ, 1995.

116. Knuth, D.E., Semantics of context-free languages, *Math. Sys. Theory,* 2(1968) 127-145.

117. Kumbhojkar, H.V. and Chaudhri, S.R., On proper fuzzification of fuzzy finite state machines, *J. Fuzzy Math.,* to appear.

118. Kumbhojkar, H.V. and Chaudhri, S.R., Fuzzy recognizers and recognizable sets, *Fuzzy Sets and Systems,* to appear.

119. Kumbhojkar, H.V. and Chaudhri, S.R., On covering of products of fuzzy finite state machines, *Fuzzy Sets and Systems,* to appear.

120. Kumbhojkar, H.V. and Chaudhri, S.R., Decomposition of fuzzy finite state machines, preprint.

121. Kumbhojkar, H.V. and Chaudhri, S.R., Homomorphisms of fuzzy recognizers, preprint.

122. Lee, E.T. and Zadeh, L.A., Note on fuzzy languages, *Inform. Sci.*, 1 (1969) 421-434.

123. Lee, E.T. and Zadeh, L.A., Fuzzy languages and their acceptance by automata, in *4th Princeton Conf. Information Science and Systems,* 1970, Princeton University Press, 399.

124. Lindgren, N., Machine recognition of human language, Part III-Cursive script recognition, *IEEE Spectrum,* 2(1965) 104-116.

125. MacLane, S. and Brikhoff, G., *Algebra,* MacMillan, New York, 1979.

126. Malik, D.S. and Mordeson, J.N., On fuzzy recognizers, *Kybernets,* 28(1999) 47-60.

127. Malik, D.S. and Mordeson, J.N., Minimal fuzzy recognizer, *J. Fuzzy Math.,* 7(1999) 381-389.

128. Malik, D.S. and Mordeson, J.N., *Fuzzy Discrete Structures,* Studies in Fuzziness and Soft Computing, Physica-Verlag 58, 2000.

129. Malik, D.S., Mordeson, J.N., and Nair, P.S., Fuzzy normal subgroups in fuzzy subgroups, *J. Korean Math. Soc.,* 29(1992) 1-8.

130. Malik, D. S., Mordeson, J. N. and Sen, M. K., Semigroups of fuzzy finite state machines, *Advances in Fuzzy Theory and Technology,* ed. P.P. Wang, Vol. II, Bookswright Press, 1994, 87-98.

131. Malik, D. S., Mordeson, J. N. and Sen, M. K., Submachines of fuzzy finite state machines, *J. Fuzzy Math,* 2(1994) 781-792.

132. Malik, D. S., Mordeson, J. N. and Sen, M. K., On subsystems of a fuzzy finite state machine, *Fuzzy Sets and Systems,* 68(1994) 83-92.

133. Malik, D. S., Mordeson, J. N. and Sen, M. K., The cartesian composition of fuzzy finite state machines, *Kybernets,* 24(1995) 98-110.

134. Malik, D. S., Mordeson, J. N. and Sen, M. K., On regular fuzzy languages, *Inform. Sci.,* 88(1996) 263-273.

135. Malik, D. S., Mordeson, J. N. and Sen, M. K., Products of fuzzy finite state machines, *Fuzzy Sets and Systems,* 92(1997) 95-102.

136. Matthews, G.H., A note on a symmetry in phase structure grammars, *Inform. and Control,* 7(1964) 360-365.

137. Maurer, H., *Theoretische Grudlagen der Programmiersprachen*, Bobliographisches Institut, Mannheim, Wein and Zurich, 1969.

138. Mealy, G.H., Methods for synthesizing sequential circuits, *Bell Sys. Tech. J. 34(1955)*.

139. Mermelstein, P. and Eden. M., Experiments on computer recognition of connected handwritten words, *Inform. and Control*, 7(1964) 255-270.

140. Mizumoto, M., Fuzzy automata and fuzzy grammars, Ph.D. thesis, Osaka University, 1971.

141. Mizumoto, M., Pictorial representations of fuzzy connectives, part I: Cases of *t*-norms, *t*-conorms and averaging operators, *Fuzzy Sets and Systems*, 31(1989) 217-242.

142. Mizumoto, M. and Tanaka, K., Fuzzy-fuzzy automata, *Kybernets* 5(1976) 107-112.

143. Mizumoto, M., Toyoda, J., and Tanaka, K., Some considerations on fuzzy automata, *J. Comput. System Sci.*, 3(1969) 409-422.

144. Mizumoto, M., Toyoda, J., and Tanaka, K., Fuzzy languages, *Systems-Computers-Controls*, 1(1970) 333-340.

145. Mizumoto, M., Toyoda, J., and Tanaka, K., General formulation of formal grammars, *Inform. Sci.* 4(1972) 87-100.

146. Mizumoto, M., Toyoda J., and Tanaka, K., Examples of formal grammars with weights, *Inform. Proc. Letters*, 2(1973) 74-78.

147. Mizumoto, M., Toyoda J., and Tanaka, K., *N*-fold fuzzy grammars, *Inform. Sci.*, 5(1973) 25-43.

148. Mizumota, M., Toyoda, J., and Tanaka, K., Various kinds of automata with weights, *Journal of Computer and System Sciences*, 10(1975) 219-236.

149. Mizumoto, M., Toyoda, J., and Tanaka, K., *B*-fuzzy grammars, *Int. J. Comput. Math. Sect. A.* 4(1975) 343-368.

150. Mordeson, J. N. and Nair, P.S., Retrievability and connectedness in fuzzy finite state machines, *Proceedings Fifth IEEE Conference on Fuzzy Systems, Vol.* 3, (1996) 1586-1590.

151. Mordeson, J. N. and Nair, P.S., Connectedness in systems theory, *Proceedings Fifth IEEE Conference on Fuzzy Systems, Vol.* 3, (1996) 2045-2048.

152. Mordeson, J. N. and Nair, P.S., Successor and source of (fuzzy) finite state machines, *Inform. Sci.*, 95(1996) 113-124.

153. Mordeson, J. N. and Nair, P.S., Fuzzy Mealy machines, *Kybernets*, 25(1996) 18-33.

154. Mordeson, J. N. and Nair, P.S., Fuzzy Mealy machines: homomorphisms, admissible relations and minimal machines, *Int. J. of Uncertainty, Fuzziness, and Knowledge-Based Systems*, 4(1996) 27-43.

155. Mordeson, J. N. and Nair, P.S., Primaries of infinite automata, Proceedings NAFIPS 97, Can Isik and Valerie Cross, eds., (1997) 279-282.

156. Mordeson, J. N. and Nair, P.S., Cartesian composition of transition spaces, *J. Fuzzy Math.*, 6(1998) 511-521.

157. Mukherjee, N.P. and Bhattacharya, P., Fuzzy normal subgroups and fuzzy cosets, *Inform. Sci.*, 34(1984) 225-239.

158. Nasu, M. and Honda, N., Fuzzy events realized by finite probabilistic automata, *Inform. and Control*, 12(1968) 284-303.

159. Nasu, M. and Honda, N., Mappings induced by PGSM-mappings and sine recursively unsolvable problems of finite probabilistic automata, *Inform. Cont.*, 15(1969) 250-273.

160. Negoita, C.V. and Ralescu, D.A., *Application of Fuzzy Sets to Systems Analysis*, Halsted Press, New York, 1975.

161. Nikolic, Z.J. and Fu, K.S., An algorithm for learning without external supervision and its application to learning control systems, *IEEE Trans. Automatic Control*, AC-11(1966) 414-422.

162. Nikolic, Z.J. and Fu, K.S., A mathematical model of learning in an unknown random environment, *Proc. 1966 NEC*, 22(1966) 607-612.

163. Ott, G., Theory and applications of stochastic sequential machines, Sperry Rand Res. Center, Rept. N. SRRC-RR-66-39, 1966.

164. Page, C.V., Equivalence between probabilistic and deterministic sequential machines, *Inform. and Control*, 9(1966) 469-520.

165. Paz, A., Some aspects of probabilistic automata, *Inform. and Control*, 9(1966) 26-60.

166. Paz. A., Fuzzy star functions, probabilistic automata, and their approximation by nonprobabilistic automata, *Journal of Computing and System Sciences*, 1(1967) 371-390.

167. Paz, A., Homomorphisms between stochastic sequential machines and related problems, *Math Systems Theory*, 2 (1968) 223-245.

168. Paz, A., *Introduction to Probabilistic Automata*, Academic Press, New York/London, 1971.

169. Peeva, K., *Categories of Stochastic Automata*, Ph.D. dissertation, Center of Applied Mathematics, Sofia, 1977.

170. Peeva, K.G., Systems of linear equations over some ordered structures, Math. Gesellschaft DDR, *Proc. Theory of Semigroups*, Greifswald, 1984, 101-105.

171. Peeva, K., On fuzzy automata and fuzzy grammars, *MTA SZT AKI, Közleményck*, Budapest, 33(1985) 55-67.

172. Peeva, K., On approximate reductions and minimization of automata, *EECS'87*, Predela, Bulgaria (1987) 27-28.

173. Peeva, K., Categories of stochastic and fuzzy acceptors for pattern recognition, J. Kacprzyk and A. Steaszak, eds., *Proceedings IFSA-EC and EUR-WG*, Warsaw (1988) 286-296.

174. Peeva, K., Behavior, reduction and minimization of finite L-automata, *Fuzzy Sets and Systems*, 28(1988) 171-181.

175. Peeva, K., Fuzzy acceptors for syntactic pattern recognition, *Internat. J. Approximate Reasoning*, 5(1991) 291-306.

176. Peeva, K., Equivalence, reduction, and minimization of finite automata over semirings, *Theoretical Computer Science*, 88(1991) 269-285.

177. Perles, M., Rabin, M.O., and Shamir, E., The theroy of definite automata, *I.R.E. Trans. Comput.*, EC-12(1963) 233-243.

178. Poincaré, H., *The Value of Science* (1913; reprinted by Dover, New York, 1958), p.37.

179. Poston, T., *Fuzzy Geometry*, Ph.D. thesis, University of Warwick, 1971.

180. Rabin, M.O., Probabilistic automata, *Inform. and Control*, 6(1963) 230-245.

181. Rabin, M.O. and Scott, D., Finite automata and their decision problems, *IBM J. Res. Develop.*, 3(1959) 114-125.

182. Rogers, H., *Theory of Recursive Functions and Effective Computability*, McGraw-Hill, New York, 1967.

183. Rosenfeld, A., *An Introduction to Algebraic Structures*, Holden-Day, San Francisco, 1968.

184. Rosenfeld, A., Fuzzy groups, *J. Math. Anal. Appl.*, 35(1971) 512-517.

185. Rosenkrantz, D.J., Programmed grammars and classes of formal languages, *J. Assoc. Comput. Mach.*, 16(1969) 107-131.

186. Salomaa, A., *Ann. Univ. Turkuensis,* Ser. A1 69, 1964.

187. Salomaa, A., On finite automata with a time-variant structure, *Inform. and Control*, 13(1968) 85-98.

188. Salomaa, A., Probabilistic and weighted grammars, *Inform. and Control*, 15(1969) 529-544.

189. Salomaa, A., *Theory of Automata*, Pergamon Press, Oxford, 1969.

190. Salomaa, A., On grammars with restricted use of productions, *Ann. Acad. Sci. Fennicae,* Vol. A.1 454(1969) 3-32.

191. Salomaa, A., On some families of formal languages, *Ann. Acad. Sci. Fennicae,* Vol. A.1 479(1970) 3-18.

192. Salomaa, A., *Formal Languages*, Academic Press, New York, San Francisco and London, 1973.

193. Santos, E.S., Maximin automata, *Inform. Control,* 13(1968) 363-377.

194. Santos, E.S., Maximin, minimax, and composite sequential machines, *J. Math. Anal. Appl.*, 24 (1968), 246-259.

195. Santos, E.S., Maximin sequential-like machines and chains, *Math. Systems Theory*, 3(1969) 300-309.

196. Santos, E.S., Fuzzy sequential functions, *J. Cybernetics*, 3(1973) 15-31.

197. Santos, E.S., On sets of tapes acceptable by stochastic sequential machines, unknown.

198. Santos, E.S., Probabilistic Turing machines and computability, *Proc. Amer. Math. Soc.*, 22(1969) 704-710.

199. Santos, E.S., Maximin sequential chains, *J. Math. Anal. Appl.*, 26(1969) 28-38.

200. Santos, E.S., Fuzzy algorithms, *Inform. and Control,* 17(1970) 326-339.

201. Santos, E.S., Computability by probabilistic Turing machines, *Trans. Amer. Math. Soc.,* 159(1971) 165-184.

202. Santos, E.S., Max-product machines, *J. Math. Anal. Appl.*, 37(1972) 677-686.

203. Santos, E.S., On reductions of maximin machines, *J. Math. Anal. Appl.*, 40(1972) 60-78.

204. Santos, E.S., Probabilistic grammars and automata, *Inform. and Control*, 21(1972) 27-47.

205. Santos, E.S., Regular probabilistic languages, *Inform. and Control*, 23(1973) 58-70.

206. Santos, E.S., Context-free fuzzy languages, *Inform. and Control*, 26 (1974) 1-11.

207. Santos, E.S., Realizations of fuzzy languages by probabilistic, max-product and maximin automata, *Inform. Sci.*, 8(1975) 39-53.

208. Santos, E.S., Max-product grammars and languages, *Inform. Sci.*, 9(1975) 1-23

209. Santos, E.S., Fuzzy automata and languages, *Inform. Sci.* 10 (1976) 193-197.

210. Santos, E.S., Fuzzy and probabilistic programs, *Inform. Sci.* 10 (1976) 331-335.

211. Santos, E.S., Regular fuzzy expressions, in *Fuzzy Automata and Decision Processes*, M.M. Gupta, G.N. Saridis, and B.R. Gaines, eds., North-Holland, Amsterdam, 1977, 169-175.

212. Santos, E.S. and Wee, W.G., General formulation of sequential machines, *Inform. and Control*, 12(1968) 5-10.

213. Sayre, K.M., Machine recognition of handwritten words: A project report, *Pattern Recognition*, 5(1973) 213-228.

214. Schecke, T., Rau, G., Popp, H.-J, Kasmacher, H., Klaff, G., and Zimmerman, H.-J, A knowledge-based approach to intelligent alarms in anesthesia, *IEEE Engineering in Medicine and Biology*, 10(1991) 38-43.

215. Schützenberger, P.M., On the definition of a family of automata, *Inform. and Control*, 4(1961) 245-270.

216. Scott, D., Some definitional suggestion for automata theory, *J. Comput. Systems Sci.*, 1(1967) 187-212.

217. Shen, J. Fuzzy language on free monoid, *Inform. Sci.*, 88 (1996) 149-168.

218. Shyr, H.J., *Free Monoids and Languages*. Taipei, Taiwan, 1979.

219. Stanat, D., A homomorphism theorem for weighted context-free grammars, *J. Comput. System Sci.*, 6(1972) 217-232.

220. Starke, P.H., *Abstract Automata*, North-Holland, Amsterdam, 1972.

221. Starke, P.H. and Thiele, H., On asynchronous stochastic automata, *Inform. and Control*, 17(1970) 265-292.

222. Steimann, F. and Adlassnig, K., Clinical monitoring with fuzzy automata, *Fuzzy Sets and Systems,* 61(1994) 37-42.

223. Swain, P.H. and Fu, K.S., Stochastic programmed grammars for syntactic pattern recognition, *Pattern Recognition*, 4(1972) 83-100.

224. Takaoka, T., in *Proc. 3rd. Hawaii Int. Conf. Syst. Sci.*, (1970) 76-79.

225. Tamura, S. and Tanaka, K., Learning of fuzzy formal languages, *IEEE Trans. on Systems, Man and Cybernetics*, SMC-3(1973) 98-102.

226. Thomason, M.G., Finite fuzzy automata, regular fuzzy languages, and pattern recognition, *Pattern Recognition,* 5(1973) 383-390.

227. Thomason, M.G., Fuzzy syntax-directed translations, *J. Cybern.* 4, No.1 (1974) 87-94.

228. Thomason, M.G. and Marinos, P.N., Deterministic acceptors of regular fuzzy languages, *IEEE Trans. Sys., Man, and Cyber.*, (1974) 228-230.

229. Thum, M. and Kandel, A., On the complexity of growth of the number of distinct fuzzy switching functions, *Fuzzy Sets and Systems,* 13(1984) 125-137.

230. Tokura, N., Fujii, M., and Kasami, T., Some considerations on linear automata, *Records Natl. Convention, IECE, Japan*, S8-2, (1968) (in Japanese).

231. Topencharov, V.V. and Peeva, K.G., Equivalence, reduction and minimization of finite fuzzy automata., *J. Math. Anal. Appl.*, 84(1981) 270-281.

232. Turakainen, P., On stochastic languages, *Inform. and Control*, 12(1968) 304-313.

233. Turkainen, P. On probabilistic automata and their generalizations, *Ann, Acad. Sci. Fennicae, A.I.*, 429(1968) 1-53.

234. Turakainen, P., On languages representable in rational probabilistic automata, *Ann. Acad. Sci. Fenn.*, (1969) Ser. AI, 439.

235. Turakainen, P., Generalized automata and stochastic languages, *Proc. Amer. Math. Soc.*, 21(1969) 303-309.

236. Turakainen, P., On time-variant probabilistic automata with monitors, *Ann. Univ. Turku* Ser. AI, (1969), 130.

237. Turakainen, P., Some closure properties of the family of stochastic languages, *Inform. Cont.*, 14(1971) 319-357.

238. Uckun, S. Model-based diagnosis in intensive case monitoring: The YAQ approach, *Artificial Intelligence in Medicine*, 5(1993) 31-48.

239. von Neumann, J., *Theory of Selfreproducing Automata*, in A.W. Burks, ed., U. of Illinois Press, Urbana, IL, 1966.

240. Wechler, W., *R*-fuzzy automata with a time-variant structure, in Mathematical Foundations of Computer Science, A. Blikle, ed., *Lecture Notes in Computer Science*, 28(1974) 73-76.

241. Wechler, W., Zur Verallgemeinarung des Theorems von Kleene-Schützenberger auf zeitvariable Automaten. *Elektron, Informationsvera beit. Kybernetic,* 11(1975) 439-445.

242. Wechler, W., Automaten über Inputkategorien, *Elektron, Informationsverabeit. Kybernetic,* 11(1975) 681-685.

243. Wechler, W., *R*-fuzzy grammars, in Mathematical Foundations of Computer Science, A. Blikle, ed., *Lecture Notes in Computer Science*, 32(1975) 450-456.

244. Wechler, W., A hierarchy of *n*-relational languages, TU-Preprint 07-24-76, Technische Univ. Dresden, 1976.

245. Wechler, W., Families of *R*-fuzzy languages, *Lecture Notes in Computer Science*, 56(1977) 117-186.

246. Wechler, W., The concept of fuzziness in the theory of automata, in *Modern Trends in Cybernetics and Systems*, J. Rose and C. Biliciu, eds., Springer-Verlag, Berlin, Heidelberg, New York, 1977.

247. Wechler, W., *The Concept of Fuzziness in Automata and Language Theory*, Akademic-Verlag, Berlin, 1978.

248. Wechler, W. and Dimitrov, *R*-fuzzy automata, in I*nformation Processing*, 74, North-Holland, Amsterdam, 1974, 657-660.

249. Wee, W.G., On generalizations of adaptive algorithm and application of the fuzzy sets concept to pattern classification, Ph.D. dissertation, Purdue University, Lafayette, IN, 1967.

250. Wee, W.G., and Fu, K.S., A formulation of fuzzy automata and its application as a model of learning systems, *IEEE Trans. on Systems Science and Cybernetics,* SSC-5(1969) 215-223.

251. Windeknechet, T.G., *General Dynamical Processes, A Mathematical Introduction,* Academic Press, New York, 1971.

252. Winograd, S., Input error limiting automata, *J. Assoc. Comput. Mach.,* 11(1964) 338-351.

253. Yeh, R.T., On the relational homomorphisms of automata, *Inf. Control,* 13(1968) 140-155.

254. Yuan, B., Homomorphisms and isomorphisms of fuzzy subalgebras, *J. Shanghai Teachers Univ.,* 2(1987) 1-9.

255. Zadeh, L.A., Fuzzy sets, *Inform. and Control,* 8(1965) 338-353.

256. Zadeh, L. A., Fuzzy sets and systems, Symposium on System Theory, Polytechnic Institute of Brooklyn, New York, 29 - 37 (1965).

257. Zadeh, L.A., Toward a theory of fuzzy systems, *ERL Rept., No.*69-2, Electronics Research Laboratories, University of California, Berkeley, 1969.

258. Zadeh, L.A., Fuzzy languages and their relation to human and machine intelligence, Electronics Research Laboratories, University of California, Berkeley, Rep. ERL-M302, August 1971.

259. Zadeh, L.A., Toward a theory of fuzzy systems, in *Aspects of Network and System Theory,* R.E. Kalman and N. Declaris, eds., Holt, Rinehart & Winston, New York, 1971.

260. Zadeh, L.A., Quantitative fuzzy semantics, *Inform. Sci.,* 3(1971) 159-176.

261. Zadeh, L.A., Fuzzy languages and their relation to human and machine intelligence, M. Marinos, ed., *Man and Computer, Proc. Int. Conf. Bordeaux, Paris,* 1970, Karger, Basel, 1972, 130-165.

262. Zadeh, L.A., The concept of a linguistic variable and its application to approximate reasoning, Memo. ERL-M 411, Univ. of California, Berkeley, California, 1973.

263. Zadeh, L.A., Fuzzy logic and approximate reasoning, *Synthese,* 30(1975) 407-428.

264. Zariski, O. and Samuel, P., *Commutative Algebra,* Vol. I, D Van Nostrand Reinhold, 1958.

265. Zeeman, E.C., in *Topology of 3-manifolds*, K. Fort, ed., Prentice-Hall, Englewood Cliffs, NJ, 1961, 240-256.

266. Zimmermann, H.-J, *Fuzzy Set Theory and its Applications*, Kluwer Academic Publishers, Boston, 1991.

INDEX

1-combination, 225
2-combination, 225
3-combination, 225

a-transducer, 142
abstract system, 522
accepted, 29, 38, 175, 177, 437
accepting state, 15
accessible, 339, 347, 362
accessible initial distribution, 49
accessible part, 362
accessible state, 48
action, 247
active state, 521
admissible, 60
admissible partition, 297
admissible relation, 245
almost periodic, 510
almost V-stable, 509
almost-everywhere indeterminate,
    445
AMA, 110
    finitary, 110
    normalized, 110
    synchronous, 110
ambiguous grammar, 23
APA, 209
    bounded, 210
    synchronous, 210
applicable, 38
associative, 9
asynchronous max-product automa-
    ton, 110
asynchronous probabilistic automa-
    ton, 209
attractor set, 509

automaton, 46, 449, 503
automaton with output, 503
automaton with tolerance, 504
averaging operator, 496

basic fuzzy language generated by,
    367
basis, 67, 94, 271
behavior, 359
behavior matrix, 415
belong, 442
big tolerance, 500
bijective function, 6
binary operation, 9
blank, 178
boundary, 502
bounded, 205, 512
bounded APA, 210
bounded probabilistic grammar, 205
bounded probabilistic pushdown
    automaton, 218
bounded PTM, 214

c-cut, 8
c-dense language, 378
c-discrete language, 377
Cartesian composition, 289
Cartesian cross-product, 2
cascade product, 259, 296
CFFL, 139
chain, 6
characteristic function, 8
Chomsky normal form, 20, 141
closed family of fuzzy languages,
    367
CMG, 137

coaccessible, 362

coaccessible part, 363

coinduced, 501

column recomposition, 230

combination

    convex, 225

    max-min, 225

    max-product, 225

commutative, 293

commutative cl-semigroup, 494

commutative semiring, 391

compatible, 512

complete, 242, 338, 364, 503

complete distributive lattice, 447

complete lattice ordered semigroup, 447

completion, 339, 365

compositewise equivalent, 74, 97

compositewise irreducible, 75, 98

compositewise minimal, 80, 99

compositewise ND-simple, 88

compositewise simple, 83

composition, 5, 447

composition of functions, 7

composition of relations, 5

concatenation, 127

concentrated, 73, 96

configuration, 37

congruence relation, 10

    left, 10

    right, 10

connected, 52, 273, 287

connected component, 288

connected with threshold c, 52

consistent, 512

context-free fuzzy language, 139

context-free grammar, 22, 130, 208

context-free language, 23

context-free max-product grammar, 106, 137

context-sensitive grammar, 22, 130, 208

context-sensitive language, 23

context-sensitive max-product grammar, 106

contractible, 502

contractive, 496

contractive fuzzy acceptors, 496

control word, 192

convex combination, 225

convex linear combination, 414

convex max-min combination, 59

convex max-min set, 61

convex max-min span, 60

convex MP-combination, 92

covering, 253

crisp, 442

cutpoint languages, 424

cyclic, 281

D-length, 425

decidable, 442

defined, 175, 177

definite, 56

definite with threshold c, 56

degree, 442

depth, 150

derivable, 21, 128

derivation, 21, 442

derivation chain, 128, 442

derive, 128

deterministic, 503

deterministic fuzzy recognizer, 348

deterministic machine, 426

deterministic pseudoautomaton, 233

deterministic sequential-like machine, 86

diagraph, 13

direct product, 50, 353, 358

direct sum, 50, 305

direct union, 271

directly derivable, 21, 128, 442

discrete, 46

discrete fuzzy production, 107

distributionwise equivalent, 74, 97

distributionwise equivalent with threshold c, 55

distributionwise irreducible, 75, 98

distributionwise irreducible with threshold c, 55

distributionwise minimal, 55
distributive lattice, 6
DSLM, 86

ϵ-approximate, 179, 186
ε-class, 498
ε-close, 411, 496
ϵ-cover, 180
ϵ-cover induced by, 187
ε-equivalent, 416, 418, 498
e-input, 320
ε-irreducible, 418
ε-partition, 498
ε-state minimal, 419
element
    image of, 6
    preimage of, 6
empty set, 1
epimorphism, 10, 304
equivalence
    class, 3
    relation, 3
equivalence relation, 5
equivalent, 24, 29, 73, 97, 129,
    177, 394, 406, 416, 418,
    421
    compositewise, 74, 97
    distributionwise, 74, 97
    statewise, 74, 97
equivalent acceptors, 177
equivalent F-CFDS, 150
equivalent finite-state automata,
    19
equivalent grammars, 24
equivalent initial automata, 394
equivalent states, 394
equivalently embedded, 394
ergodic fuzzy transition matrix, 480
error, 512
eventually l-corrected, 515
extended transition function, 68,
    96

F-CFDL, 149
F-CFDS, 148

F-CFG, 154
fa, 175
fa equivalent to pa, 177
fac, 175
faithful, 247, 252
faithful fuzzy transformation semi-
    group represented, 248
family of fuzzy regular languages,
    368
ffa, 404
ffsm, 237
FG, 482
final configuration, 38
final distribution function, 448
final state, 159
final states, 348, 353
fine homomorphism, 353
$F_\wedge$-regular, 326
finitary, 61, 103
finitary AMA, 110
finitary context-free fuzzy language,
    139
finitary fuzzy language, 139
finitary max-product grammar, 103
finite, 503
finite acceptor, 175
finite automaton, 175
finite fuzzy automaton, 367, 463
finite index, 183
finite quasi-fuzzy automaton, 384
finite set, 1
finite tree domain, 147
finite-state automaton, 15
finite-state machine, 13
formal language, 20
fqfa, 384
fractionally fuzzy grammar, 485
free, 271
free monoid generated by a set, 12
free semigroup, 10
free semigroup generated by a set,
    12
free semimodule, 392
fsa, 313
fsac, 175

fsf, 175
$F_\vee$-regular, 326
fts, 247
full direct product, 254
function, 6
    bijective, 6
    composition of, 7
    injective, 6
    one-one, 6
    onto, 6
    single-valued, 6
    surjective, 6
    well defined, 6
fundamental, 62, 66, 83
fuzmap, 501
fuzrelation, 501
fuzzy acceptor, 495
fuzzy adjunctive language, 375
fuzzy automaton, 393, 463, 517
fuzzy context-free dendrolanguage,
    149
fuzzy context-free dendrolanguage
    generating system, 148
fuzzy context-free grammar, 154
fuzzy dense language, 378
fuzzy direct transition function, 159
fuzzy dynamics, 525
fuzzy finite automaton, 404
fuzzy finite state machine, 237
fuzzy function, 347
fuzzy grammar, 482
fuzzy initial state, 517
fuzzy kernel, 314, 318
fuzzy language, 139, 367, 384
    finitary, 139
fuzzy left quotient, 371
fuzzy main congruence, 369
fuzzy normal subgroup, 312, 313
fuzzy output function, 463
fuzzy output map, 525
fuzzy power set, 8
fuzzy production, 103
    strict, 103
fuzzy production rules, 154
fuzzy recognize, 427

fuzzy recognized, 428
fuzzy recognizer, 337
fuzzy regular language, 368
fuzzy rewriting rules, 148
fuzzy right quotient, 371
fuzzy semiautomaton, 313
fuzzy set of fuzzy derivation trees,
    154
fuzzy set of trees recognized by
    a fuzzy tree automaton,
    159
fuzzy set T of trees, 148
fuzzy simple tree transducer, 163
fuzzy star acceptor, 175
fuzzy star function, 175
fuzzy subgroup, 312
fuzzy subsemiautomaton, 314
fuzzy subset, 8
    c-cut, 8
    complement of, 8
    intersection of, 8
    support of, 8
    union of, 8
fuzzy subset of final states, 325,
    337
fuzzy syntactic categories, 128
fuzzy system, 525
fuzzy transformation semigroup,
    247
fuzzy transformation semigroup as-
    sociated, 249
fuzzy transition function, 337, 463
fuzzy translation rule, 163
fuzzy tree automaton, 158
fuzzy tree translation, 162
    domain of, 162
    range of, 162
fuzzy-state automaton, 504

general direct product, 254
generalized state machine, 242
generate, 125, 276
generated, 10, 270, 442
generating set, 276
generator, 270, 281

generators, 10
genetrix, 205
grammar, 20
    context-free, 22, 130, 208
    context-sensitive, 22, 130, 208
    equivalent, 24
    phrase-structure, 20
    regular, 22, 130, 208
    type 0, 20
    type 1, 22, 130
    type 2, 22, 130
    type 3, 22, 130
    weakly regular, 208
greatest lower bound, 6
Greibach normal form, 20, 141

height, 372, 519
homomorphic image, 407
homomorphism, 10, 122, 242, 243,
    250, 313, 389, 407
homomorphism with threshold c,
    54
homotopic, 502

i.d., 96
identity, 9, 448
identity map, 7
immediate successor, 268
IMPSM, 96
IMSLM, 73
increasing, 517
increasing sequence of fuzzy states,
    517
index, 10
index set, 2
indistinguishable with threshold c,
    53
induced by partition, 5
inessential, 512
infinite indeterminate, 445
infinite set, 1
initial, 394
initial configuration, 38
initial distribution, 96
initial distribution function, 448

initial fuzzy state, 325, 337
initial nonterminal node symbols,
    149
initial nonterminal symbols, 154
initial state, 13, 348
initial states, 353
initialized max-min sequential-like
    machine, 73
initialized max-product sequential-
    like machine, 96
injective function, 6
input string, 14
input symbols, 13, 448
input tape, 95
instantaneous description, 213
interior, 502
internal direct union, 271
intersection, 8, 127
intersection of sets, 2
inverse image, 361
inverse relation, 5
irreducible, 303
    compositewise, 75, 98
    distributionwise, 75, 98
    statewise, 75, 98
isolated cutpoint, 430
isomorphic, 10, 83, 100
isomorphism, 10, 243, 251, 304
isomorphism with threshold c, 54

join, 354, 358

k-admissible set, 225
k-behavior, 51
k-prefix, 175
k-suffix, 175
Kleene closure, 128

$\mathbb{L}$-automaton, 414
l-corrected, 515
L-fuzzy grammar, 442
L-fuzzy language, 424
L-fuzzy relation, 446
l-semigroup, 447
language, 20
    context-free, 23

context-sensitive, 23
regular, 23
language determined by the fuzzy
    automaton, 368
language generated by, 21
language generated by G with con-
    trol language C, 193
language recognized, 38
language recognized by, 338
lattice, 6
leaf node set, 147
leaf symbols, 148, 158, 163
least upper bound, 6
left congruence relation, 10
left semimodule, 391
leftmost, 206
leftmost derivation, 23
level set, 8
linear bounded automaton, 179
linear order, 5
little tolerance, 500
lower bound, 6

M-equivalent, 73, 97
MA, 114
main congruence, 369
mapping, 6
mathematical system, 9
max-min algebra, 59
max-min combination, 59, 225
max-min fuzzy language recognized,
    325
max-min pre-tables, 49
max-min sequential-like machine,
    67
max-min span, 60
max-min tables, 49
max-product algebra, 91
max-product automaton, 114
max-product combination, 225
max-product grammar, 103
    context-free, 106
    context-sensitive, 106
    regular, 106
    type 0, 106

weak regular, 106
max-product sequential-like ma-
    chine, 95
maximal admissible partition, 303
maximal connected component, 288
membership degree, 442
membership grade of a string, 384
min-max fuzzy language recognized,
    326
minimal, 394
    compositewise, 80, 99
    statewise, 80, 99
minimal complete recognizer, 352
minimal ffa, 407
minimal form, 394
minimal generating set, 277
mirror-image language, 41
modification, 512
monoid, 9
monomorphism, 10, 304
morphism of semimodules, 392
MP-combination, 92
MP-set, 93
MP-span, 92
MPSM, 95
MSLM, 67
μ-orthogonal, 300
multiplicative, 320

n-simplex, 500
natural ε-reduct, 419
ND-simple
    compositewise, 88
    statewise, 88
node symbols, 148, 163
noetherian semimodule, 392
nondeterministic finite-state automa-
    ton, 27
nondeterministic sequential-like ma-
    chine, 86
nonprobabilistic, 179
nonterminal node symbols, 148,
    163
nonterminal symbols, 20, 154
normal, 152

normal form, 442
normalized, 103
normalized AMA, 110
normalized fuzzy automata, 468
normalized max-product grammar,
        102
np, 179
NSLM, 86
null set, 1
null-homotopic, 502

observable, 525, 526
one-one function, 6
onto function, 6
optimistic, 465
order-comparing computation, 428
output function, 13
output string, 14
output symbols, 13
output tape, 95

p definite, 56
p definite with threshold c, 56
pa, 176
pac, 177
partial fuzzy automaton, 325
partial order, 5
partially ordered set, 6
partially ranked alphabet, 147
partition
        of a set, 4
path representing, 18
path representing a string, 29
PDA, 37
peak hold, 520
perfect, 293
period, 511
periodically time-variant context-
        free languages, 203
permanent transition modification,
        512
pessimistic, 465
pfa, 325
phrase-structure grammar, 20, 116
polysemigroup, 253

polytransformation semigroup, 252
poset, 6
power set, 2
PPA, 218
prefix, 427
primary submachine, 274
probabilistic acceptor, 177
probabilistic automaton, 176
probabilistic grammar, 205
probabilistic production, 205
probabilistic pseudoautomaton, 223
probabilistic pushdown automaton,
        218
probabilistic Turing machine, 213
product, 312, 447, 501
production, 20
projection, 442
proper submachine, 270
proper subset, 1
pseudoautomaton, 45, 223, 448
        probabilistic, 223
pseudoterms, 148
PTM, 213
ptvcf, 203
pushdown acceptor, 37
pushdown stack, 37

q-related, 272
q-twins, 272
quasi-base, 392
quasi-definite, 181, 189
quasi-retrievable, 272
quotient fuzzy finite state machine,
        298
quotient subgroup, 320

Rabin (probabilistic) pseudoautoma-
        ton, 228
Rabin pseudoautomaton, 228
random control set, 207
random domain, 209
random function, 205
random language, 207
random language generated, 207
random range, 209

rank, 275
rational probabilistic event, 437
rational stochastic language, 198
rational weighted grammar, 194
rational weighted probabiliistic grammar, 194
reachability set, 526
reachable, 525, 526
read, 38
realized, 437
recognizable, 340, 347
recognized, 338
recognized by, 359
recomposition
    column, 230
    row, 230
recursive, 129
recursively enumerable, 442
reduced, 139, 389
reduced form, 394
reduct, 394, 418
regular, 51, 434
regular fuzzy grammar, 384
regular grammar, 22, 130, 208
regular language, 23
regular max-product grammar, 106
regular set, 175
relation, 2
    antisymmetric, 5
    binary, 2
    composition of, 5
    domain of, 3
    equivalence, 3, 5
    image of, 3
    inverse, 5
    reflexive, 3
    symmetric, 3
    transitive, 3
relative complement of a set, 2
replacement function, 104, 206
representation, 424
response function, 48, 73, 97
restricted direct product, 254, 354, 358
restricted max-min automata, 47

retrievable, 272
reversal, 358, 359
right congruence relation, 10
right semimodule, 392
(ρ, l)-bounded, 512
row recomposition, 230

saddle point, 372
scan, 38
sd, 73
semigroup, 9
semiring, 391, 455
semiring with identity, 455
semiring with zero, 455
separable, 259, 261
separated, 273
sequential transducer, 122
set, 1
    empty, 1
    finite, 1
    infinite, 1
    intersection of, 2
    null, 1
    partition of, 4
    power, 2
    proper subset of, 1
    relative complement of, 2
    subset of, 1
    union of, 2
set difference, 2
set of generators, 61, 93
set of productions, 20
set of vertices, 61, 93, 125
shaded version, 445
sharpened version, 445
simple, 100, 285
    compositewise, 83
    statewise, 83
singly generated, 270
smallest completion, 365
solution, 412
solvable, 412
stack, 37
standard form, 291
starting symbol, 21

state distribution, 73
state independent, 294
state minimal, 419
state monitor, 519
state relation, 512
state symbols, 158
state transition function, 13, 448
states, 448
statewise D-simple, 88
statewise equivalent, 74, 97
statewise equivalent with threshold c, 55
statewise irreducible, 75, 98
statewise irreducible with threshold c, 55
statewise minimal, 55, 80, 99
statewise ND-simple, 88
statewise simple, 83
stationary, 46
stationary fuzzy transition function, 464
stochastic language, 197
stochastic matrix, 175
stochastic under maximal interpretation, 198
strict CMG, 138
strict fuzzy production, 103
strict max-product automaton, 114
strict max-product grammar, 103
strictly accessible initial distribution, 49
strictly connected, 52
string, 14, 95
string accepted by, 29
strong covering, 305
strong homomorphism, 242, 243, 251, 314
strong homomorphism with threshold c, 54
strong isomorphism, 243, 251
strong subsystem, 283
strongly connected, 269
strongly equivalent, 421
subautomaton, 503
subfuzzy finite state machine, 305

submachine, 269
submachine generated by, 270
submonoid, 10
subsemigroup, 10
subspace tolerance, 501
substitution, 122, 142
subsystem, 277
successor, 268
sum, 305
super cyclic, 282
surjective function, 6
synchronous AMA, 110
synchronous APA, 210
synchronous PTM, 214
system of generators, 392

τ-connected, 502
τ-correct, 513
T-generalized state machine, 242
T-generalized transformation semigroup, 323
T-generalized transformation semigroup inducible, 323
t-modification, 512
t-norm, 95
temporary error, 512
terminal node symbols, 158
terminal symbols, 20, 154
tolerance automaton, 505
tolerance neighborhood, 501
tolerance relation, 500
tolerance space, 500
top stack symbol, 37
total, 209
total bounded probabilistic pushdown automaton, 218
total order, 5
total probabilistic production, 205
total probabilistic pushdown automaton, 218
total random function, 205
totally connected, 52
totally connected with threshold c, 52
totally ordered set, 6

transformation semigroup, 247
tree, 147
trim, 364
trim part, 364
Turing automaton, 178
Turing machine
    probabilistic, 213
type 0 grammar, 20, 130
type 0 max-product grammar, 106
type 1 grammar, 22, 130
type 2 grammar, 22, 130
type 3 grammar, 22, 130

union, 8, 127, 271, 501
union of sets, 2
unsolvable, 412
upper bound, 6

V-stable, 509
value computation, 427

weak covering, 263, 310
weak homomorphism, 313
weak isomorphism, 313
weak regular max-product gram-
        mar, 106
weakly equivalent, 394, 421
weakly equivalently embedded, 394
weakly initial, 394
weakly p definite, 56
weakly p definite with threshold
        c, 56
weakly regular grammar, 208
weight, 448, 450
weighted grammar of type i, 119,
        194
weighted-automaton, 449
weighting space, 448
word, 14, 495
word recognized by, 338
wreath product, 261
write, 38

X-recognizable, 359
X-recognizer, 353

zero, 448

Printed in the United States
by Baker & Taylor Publisher Services

Printed in the United States
by Baker & Taylor Publisher Services